FOOD SCIENCE AND TECHNOLOGY

NATURAL PRODUCTS AND THEIR ACTIVE COMPOUNDS ON DISEASE PREVENTION

Food Science and Technology

Additional books in this series can be found on Nova's website under the Series tab.

Additional e-books in this series can be found on Nova's website under the e-book tab.

Public Health in the 21st Century

Additional books in this series can be found on Nova's website under the Series tab.

Additional e-books in this series can be found on Nova's website under the e-book tab.

FOOD SCIENCE AND TECHNOLOGY

NATURAL PRODUCTS AND THEIR ACTIVE COMPOUNDS ON DISEASE PREVENTION

M. MOHAMED ESSA
A. MANICKAVASAGAN
AND
E. SUKUMAR
EDITORS

Nova Science Publishers, Inc.
New York

For permission to use material from this book please contact us:
Telephone 631-231-7269; Fax 631-231-8175
Web Site: http://www.novapublishers.com

Library of Congress Cataloging-in-Publication Data

Natural products and their active compounds on disease prevention / editors, M. Mohamed Essa, A. Manickavasagan, E. Sukumar.
p. ; cm.
Includes bibliographical references and index.
ISBN 978-1-62100-153-9 (hardcover)
I. Essa, M. Mohamed. II. Manickavasagan, A. III. Sukumar, E.
[DNLM: 1. Biological Products--pharmacology. 2. Biological Products--therapeutic use. 3. Functional Food. 4. Phytotherapy. QW 800]
615.3'21--dc23

2011032385

Published by Nova Science Publishers, Inc. † New York

CONTENTS

FOREWORD

Gilles J. Guillemin
Head of the Neuroinflammation group, University of New South Wales,
Department of Pharmacology, Sydney, Australia

Over the centuries, several ancient civilizations such as China, Greece and India, to name a few, have accumulated evidence indicating that foods can have significant therapeutical properties either as a treatment and/orasprevention fora large range ofdiseases. Hippocrates, who is considered as one of the fathers of the Western medicine,came up with following the theory "*Let food be thy medicine and medicine be thy food*". Over the last decades, there has been literally an explosion of interest forthese functional foods and variousphyto active compounds known for their medicinal properties.For example, in the United States, nearly sixty percent of the population takes at least one type of natural enhancing health compounds.

These active molecules derived from plants, known as phytochemicals or nutraceuticals,have also increasingly gained the interest of scientists and clinicians searching for new and less toxic (less side effects) compounds for the treatment of major diseases.

They have been looking at phytochemical moleculesable to provide long-term medicinal benefits at low doses.These nutraceuticals include micronutrients such as antioxidants, vitamins E and C, resveratrol and other polyphenols, curcumin, green tea extract, co-enzyme Q10, lycopene, and a large number of other phytochemical-rich foods. This phenomenon is currently leading to the generation of a new era for human health.

This book compilesa number of significant scientific studies or clinical reports based on the therapeutic and/or preventive effects of food such as fruits, nuts, seeds, herbs and algae and highlighting their medicinal and nutritive mechanisms of action.

FOREWORD

Professor Mustaq A Memon
Washington State University,
Pullman, WA, USA

With the increasing awareness of the potential health benefits of natural products, the book provides a comprehensive overview of the many products and their active compounds. The main question asked bythe user of natural products is the validity of these compounds and the science behind it. The book chapters are written by scientists and practitioners of the natural products. Many of the research findings mentioned in the chapters are an outcome of the studies funded by government agencies/or private foundations, resulting in un-biased study conclusions. The authors'affiliation and their work location from various countries give a true global flavor to the book. Other unique feature of the book is the science-based studies incorporating basic laboratory-based as well as applied and clinical research.

Many of the topics include products of daily use including tae, coffee.Other chapters are focused on the benefits of natural products in prevention and cure of diseases like cancer and Parkinson's disease.

The book is suitable for users of the natural products, students, trainees, researchers and others involved in health related professions.

PREFACE

It is well documented that most natural products are enriched with bio-active components that have protective action. There is currently a growing body of evidence that supplementing the human diet with natural products is of major benefit for human health and well-being. Nowadays, the use of complementary/alternative medicine, functional food and especially the consumption of natural products have been increasing rapidly worldwide, mostly because of the supposedly less frequent side effects. Both in conventional and traditional medicines, natural products continue to provide valuable therapeutic agents. The issues regarding the efficacy and safety of currently available modern medicine agents have prompted the search for safer and more effective alternatives. This book focuses on implications of traditional and precessed foods for health and disease prevention.

Chapter 1 - Plant products play a significant role in human diet as they maintain human healthand improve the quality of life. The health-promoting effects of plant products form part of the rich history of ancient civilisations. The non-nutritive but potentially bioactive secondary metabolites in fruits, vegetables, herbs, spices, teas and wines reduce oxidative stress, inflammation, risk of cardiovascular diseases and type 2 diabetes, age-related cognitive decline, risk of major neurodegenerative disorders, including Alzheimer's diseaseand cancer. This reduced risk is associated with the most commonly consumed dietary phyto-chemicals.Metabolic syndrome has been classically defined as the clustering of inter-related risk factors for cardiovascular disease and type2 diabetes including hyperglycaemia, insulin resistance, hypertension, hypertriglyceridaemia, decreased HDL-cholesterol concentration and obesity. The prevalence of metabolic syndrome is reaching pandemic proportions worldwide withdiets rich in saturated fat, cholesterol and refined carbohydrates implicated in the increased risk of cardiovascular disease and metabolic syndrome. From a dietary perspective, a plant-based diet could therefore serve as an alternative intervention for the prevention or treatment of metabolic syndrome. We have previously reviewed the possible therapeutic responses of commonly consumed Indian spices in metabolic syndrome. This review evaluates the effects of further frequently-used medicinal and dietary plants from the Indian sub-continent on the risk factors of metabolic syndrome using data from controlled human, animal and *in vitro* studies. The major bioactive constituents of these plants, mostly identified as herbs, are also discussed.

Chapter 2 - Plant based products are widely used for various ailments, either as a food supplement, prescription drug or over the counter prescription. Scientific facts on these products are warranted, which would help us to develop novel, safe and effective formulation for humans, for wider global acceptance. Though many formulations on herbal products are available, scientific rationale and periodical evaluation of these formulations are lacking.

Arogh, a polyherbal formulation, is a product manufactured by Rumi Herbals, Chennai, which was scientifically validated. Arogh is a herbal tea formulation made out of *Nelumbo nucifera, Hibiscus rosa-sinensis, Rosa alba, Terminalia chebula, Hemidesmus indicus, Glycyrrhiza glabra*, *Zingiber officinale*, *Quercus infectoria* and *Eclipta alba.* Scientific validation revealed that the formulation was effective against hypercholesteremia, anxiety, hyperglycemia, Obesity, Stress, Myocardial infraction and Oxidative stress. The scientific validation of Arogh, revealed as a cardiotonic and a remedy to cater cardiac and associated ailments.

Chapter 3 - Diabetes mellitus was unknown in the prehistoric times, as long as man was a hunter and food gatherer. It was recognized only with the advent of civilizations in Egypt, Rome, Greece and India. Early medical writers reported weight loss, excessive urination and sweet taste of urine as the primary symptoms. Roman Artaeus (70 AD) noted polydipsia and polyuria and named the condition ‘diabetes’, meaning to ‘flow through’. Centuries later, Thomas Willis (1965), a London physician described the sweet taste of urine and introduced the term ‘mellitus’ meaning ‘honey-like’. Diabetes is one of the most common non-communicable diseases and challenging health problems of the 21st century. It is a physiological state when sugar in the blood cannot be utilised by the cells in the body. Hence not only blood sugar, but also the metabolism of carbohydrates, fat and protein increase. Insulin is the hormone essential for the utilization of glucose and is produced by the beta cells of the islets of Langerhans of the pancreas. In diabetes it is either not produced in adequate quantities or does not act properly so that glucose keeps accumulating in the blood but the cells are starved of it.

Chapter 4 - Cardiovascular diseases are rising world-wide and World Health organization (WHO) expects that by year 2020, cardiovascular diseases will cause about 25 million deaths globally. Atherosclerosis is the major underlying cause of the congestive heart failure, myocardial ischemia or peripheral vascular disease. In 2005, one in every five American death was due to the coronary heart disease (CHD). Atherosclerosis is characterized by hardening of the arteries and buildup of plaques in the inner lining of arteries. Plaques are formed by the deposition of fatty substances, cholesterol, cellular debris and calcium from the blood components. It is a slow and complex disease involving many cells such as endothelial cells, vascular smooth muscle cells and inflammatory immune cells. The plaques grow in time and cause obstruction to the blood flow and usually the symptoms are not evident until the later stages where there is complete obstruction to blood flow leading to negligible oxygen supply and nutrients to the myocardial or endothelial tissue.

Chapter 5 - Cancer refers to over 100 different types, but the central dogma is that certain body cells multiply in an uncontrolled manner. Normal cells grow at a predetermined rate, perform specific functions and ‘pass away” by what is called apoptosis and is replaced by other normal cells. Cancer in general, results when a single cell develops mechanisms by which it can multiply rapidly out of control by normal checks and balances, and replaces normal cells along with loss of biological functions.

Chapter 6 - Omega-3 fatty acids, eicosapentaenoic acid (EPA) and docosahexaenoic acid (DHA), are one of the essential fatty acids. As these are necessary for human health but the human body can’t synthesize it. It should be supplied through our diets. Omega-3 fatty acids (FAs) found mainly in fish, such as salmon, tuna, and halibut and seafood including algae and krill, some plants, and nut oils. It has been drawn the attention in the scientific community due to their capability to affect several processes in the body such as neurological, cardiovascular, and immune functions, and cancer. The American Heart Association recommends eating fish (particularly fatty fish such as mackerel, lake trout, herring, sardines, albacore tuna, and salmon) at least 2 times a week. The 2010 Dietary Guidelines for Americans recommends 8 oz. of a variety of seafood per week providing an average daily consumption of 250 mg EPA+DHA. Omega-3 fatty acids are highly concentrated in the brain and appear to be important for cognitive (brain memory and performance) and behavioral function. In fact, those infants are at risk for developing vision and nerve problems that do not get enough omega-3 fatty acids from their mothers during pregnancy. Important deficiency symptoms of omega-3 FAs include fatigue, poor memory, dry skin, heart problems, mood swings or depression, and poor circulation. Further, it is necessary to maintain the balance between of omega-6 /omega-3 ratio in our diets. As omega-6 FAs increase the inflammatory responses in the body.

Chapter 7 - Parkinson's disease (PD) is the second most common neurodegenerative disorder, after Alzheimer's disease, affecting 1% of the population by the age of 65 years and 4–5% of the population by the age of 85 years. PD is found to have either familial or non-familial etiology. The interactions between external toxins (which arise from environmental, dietary and lifestyle factors), internal toxins arising from normal metabolism and the genetic (nuclear genes) and epigenetic (mitochondria, membranes, and proteins) components of neurons occur continuously and could initiate degeneration in DA neurons. The motor disabilities characterizing PD are primarily due to the loss of dopaminergic neurons in the substantia nigra and depletion of dopamine in corpus striatum.

Chapter 8 - *Introduction:* The main objective of any anti-cancer therapy is to provide maximal cancer cell killing while at the same time minimizing the damage to normal cells and tissues. Herbs and dietary phytochemicals have been increasingly recognized in prevention and treatment of human diseases including cancer. There exist enormous prospect for the evaluation of phytochemicals to increase radioprotection of normal cells during cancer radiotherapy. Our work have focused on the mechanism of variety of dietary phytochemicals, namely, ferulic acid, curcumin, sesamol, lycopene and quercetin etc., on normal cells with view to design effective protocols in practical radioprotection. Recent reports have shown that phytochemicals induces radiosensitization of tumor cells by generating reactive oxygen species (ROS). Further, phytochemicals modulate G2/M checkpoint, thus leaving the cancer cells less time to repair the radiation-induced DNA damage and eventually leads them to apoptotic processes. Further it is emphasized that modulation of the apoptotic and signaling pathways may help to achieve efficient destruction of cancer cells, this may provide a new approach in developing effective treatment for cancer. Phytochemicals can therefore be considered as adjunct in radiation therapy. *Conclusion:* This chapter critically explains the role phytochemicals as radioprotectors in normal cells and its suitability as radiosensitizers in cancer cells.

Chapter 9 - The diseases confronting human race are many and a number of them, which are invasive, can be grouped into three major categories - degenerative, parasitic and microbial. Historically, herbal preparations have been used to manage invasive human diseases. At present, the majority of chemical substances commercially available for treating invasive diseases are dietary supplements, natural products, and synthetic drugs modeled after natural products. This chapter provides an outlook for the role and mode of action of natural products and selected herbs used to treat invasive diseases linked to oxidative stress, cancer, malaria, onchocerciasis, leishmaniasis, and microbes. Considering the therapeutic spectrum and potency of natural products currently in use for the treatment of invasive diseases, a combination of dietary supplements, herbs, natural products, synthetic drugs and life style will continue to play important roles in the treatment of invasive diseases.

Chapter 10 - Cigarette/tobacco smoking is associated with an increased incidence and severity of lung inflammation, and impaired lung function. It is a major etiological factor for chronic obstructive pulmonary disease (COPD), a leading cause of mortality and morbidity worldwide. Corticosteroids are considered as the most effective anti-inflammatory agent available in clinical use at present. However, smokers and patients with COPD develop insensitivity to corticosteroid treatment, which leads to considerable management problems. Therefore, there is a great need for the novel anti-inflammatory agent without any secondary side effects. The use of phytochemicals to control cigarette/tobacco smoke-induced lung inflammation may have a potential benefit and is reviewed in this chapter. Plant-derived chemical compounds, such as curcumin, resveratrol, epigallocatechin-3-gallate, ursolic acid, zerumbone, parthenolide, helenalin, baicalin, caffeic acid phenethyl ester, garcinol, vitamin C, vitamin E, β-carotene, sulforaphane and apocynin showed anti-inflammatory effect against cigarette/tobacco smoke exposure. Depending on the molecular target, the phytochemicals were classified into NF-κB inhibitors (which reduce NF-κB-mediated pro-inflammatory gene transcription), histone acetyltransferase inhibitors or histone deacetylase activators (which reduce transcription factor-DNA binding and thereby reduce pro-inflammatory gene transcription), and antioxidants (which scavenge reactive oxygen species-induced pro-inflammatory gene transcription and tissue damage). Since a variety of pathways are involved in the pathogenesis of lung inflammation and COPD, therapeutic administration of a phytochemical with multiple molecular targets or mixture of multiple phytochemicals will be effective in management of lung inflammation and COPD.

Chapter 11 - Natural products have been the most productive source of leads for the development of drugs. Natural products provide an unlimited opportunity for new drug leads because of the unparalleled chemical diversity. Many pharmaceutical agents have been discovered by screening natural products from plants, animals, marine organisms and microorganisms. These biologically derived antibacterial compounds could have reasonable value in controlling antibiotic resistant pathogens also. The prevalence of antibacterial natural products drugs may be due to the evolution of secondary metabolites as biologically active chemicals that conferred selective advantages to the producing organisms. This chapter provides an overview of natural products, their sources, their role in increasing antibiotic susceptibility of drug resistant bacteria and the synergism between natural products and antibiotics.

Chapter 12 - Gastric cancer is the fourth most common cancer and the second leading cause of cancer death worldwide. Therefore it is necessary to determine an effective therapy for gastric cancer. Natural products are very important sources for the development of novel gastric cancer therapeutics. Plant polyphenols and many flavonoids have several beneficial actions on human gastric cancer. However, the actual molecular interactions of polyphenols with biological systems remain mostly speculative. This chapter deals with the potential cellular and molecular mechanisms of some selected polyphenols and its actions on gastric cancer cells. Those mechanisms include regulation of signal transduction pathways, transcription factors and related activities; modulation of cell-cycle regulation or induction of apoptosis, affecting cell differentiation, proliferation, metastasis, immune response, antioxidant, suppression of angiogenesis and chemical metabolism. A better understanding about the nature and biological consequences of polyphenol interactions with gastric cancer cell components will certainly contribute to develop nutritional and pharmacological strategies oriented to prevent the onset and/or the consequences of gastric cancer.

Chapter 13 - This chapter discusses the role of bioactive compounds from *Tribulus terrestris* L. (Zygophyllaceae) (TT) in disease prevention and treatment. *Tribulus terrestris* is gaining popularity in the media because it is projected as libido enhancer, treats male sexual dysfunction, gives stamina and confidence to perform. A large number of research papers have been published on this plant and there is a need to critically review, analyze and summarize the literature to draw a line between myth and reality. Since ancient times *Tribulus terrestris* has been used in the Indian and Chinese traditional medicine to treat hypertension, premature ejaculation, erectile dysfunction, vitiligo, and kidney and eye problems. It has anti-urolithiatic, diuretic, antiacetylcholine, aphrodisiac properties and can stimulate spermatogenesis and libido. The phytochemistry of the extract reveals the presence of alkaloids, steroidal saponins, furostanol saponins, flavonoid glycosides which impart the medicinal properties to this plant. The research papers published until 2010 on *Tribulus terrestris* are thoroughly reviewed and summerized under the following captions: Botany, Phytochemisty, the effects on hypertension, hormones, oxalate metabolism, endocrine sensitive organs, androgen receptors, diuretic and contractile effects, its use as an aphrodisiac, protective agent in diabetes, use as a nutritional supplement, its anti cancer activity, anti microbial activity, analgesic, anthelmintic properties and its side effects. In each section emphasis is laid on identifying the bioactive compound, its specific use and the mode of action.

Chapter 14 - Catechins belong to the flavan-3-ol class of flavonoids, the most abundant polyphenolic compounds found in green tea (*Camellia sinensis*) that have been shown to bioactively affect the pathogenesis of several diseases. Catechins have drawn a lot of attention due to the variety of properties they possess. Among the most studied properties are induction of apoptosis, antioxidative, anti-inflammatory, antiviral, antimicrobial, antiobesity and anti-diabetic properties. Several *in vitro* and *in vivo* studies have attributed the molecular mechanisms by which polypenolic tea catechins (PTCs) exhibit these properties with their potential to increase the expression and phosphorylation of certain proteins, cytokines, upregulation of endogenous free radical scavengers and regulation of signal transduction pathways. This chapter will give a narrative review of the various health promotion, disease prevention properties and molecular mechanisms of action of polyphenolic tea catechins present in a selected database.

Chapter 15 - Plant based medicinal research is well recognized world over as a viable healthcare component. An overwhelming body of evidence has collected in recent years to show the immense potential of the medicinal plants used in various traditional systems. Fenugreek, an annual herb of the *Leguminosae* family has been quoted in Indian, Arabic and Chinese medicine as a treatment for diabetes and as a general tonic to improve metabolism and health. This plant has received attention as an antidiabetic agent and has undergone extensive research in clinical and animal models of diabetes which have clearly documented its blood glucose lowering property. However the plant, especially the seeds, has many benefits beyond that. The present chapter aims to compile the data on wide range functional benefits of this plant demonstrated through the research activities using modern scientific approaches and innovative tools.

Chapter 16 - Adult stem cells hold great promise for the treatment of a spectrum of disorders including chronic, degenerative and malignant diseases. A growing body of evidence indicates that natural compounds with potent stem cell stimulatory mechanisms are invaluable for therapeutic strategies as an alternative or adjuvant to tissue transplantation therapy. Although many herbal stem cell stimulators have been identified, comprehensive knowledge on the development of natural therapeutics is fairly limited. Moreover, in spite of recent developments, underlying cellular and molecular mechanisms by which certain natural products enhance the self-renewal and proliferation of tissue stem cells are still poorly understood. Unraveling the mechanisms of herbal stem cell stimulators provide promising therapeutic approaches for spectrum of disorders. This chapter illustrates emerging trends in the identification and characterization of medicines derived from natural components in regenerative medicine and stem cell-based therapies. Our main intention in this chapter is to highlight the enormous potential of herbal stem cell stimulators that remain unexplored despite their substantial medicinal values.

Chapter 17 - Papaya (*Carica papaya* L.) is a deliciously sweet tropical fruit with musky undertones and a distinctive pleasant aroma. It was first cultivated in Mexico several centuries ago but is currently being cultivated in most of the tropical countries. Everything in papaya plant such as roots, leaves, peel, latex, flower, fruit and seeds have their nutritional and medicinal significance. Papaya can be used as a food, a cooking aid, and in medicine. Papaya is considered as a low calorie nutrient dense fruit. The fresh fruit is commonly used as a carminative, stomachic, diuretic and antiseptic in many parts of the world. The nutrients and phytochemicals contained in papaya help in digestion, reduce inflammation, support the functioning of cardiovascular, immune and digestive systems and may also help in prevention of colon, lung and prostate cancers. Overall, the papaya can act as a detoxifier, activator of metabolism, rejuvenating the body and in the maintenance of body's homeostasis because it is rich in antioxidants, B vitamins, folate and pantothenic acid, and potassium and magnesium as well as fiber. Because of its high vitamin A and carotenoids contents, it can help in preventing the cataract and age-related macular degeneration. Papaya pastes can be used externally as a treatment for skin wounds and burns. This paper discusses the nutritional and medicinal value of papaya (*Carica papaya* L.) and its relationship to human health.

Chapter 18 - Effect of *Hibiscus sabdariffa* (is an edible medicinal plant, indigenous to India, China and Thailand and is used in Ayurveda and traditional medicine), leaf extract (HSEt) on the levels blood ammonia, and serum lipid profiles (cholesterol, triglycerides, phosphor lipids, free fatty acids) were studied for its protective effect during ammonium chloride induced hyperammonemia in Wistar rats. Ammonium chloride (AC) treated rats showed a significant increase in the levels of circulatory ammonia and lipid profiles. These changes were significantly decreased in HSEt and AC treated rats. Our results indicate that HSEt offers protection by influencing the levels of ammonia and lipid profiles in experimental hyperammonemia and this could be due to its (i) ability to detoxify excess ammonia, urea and creatinine, (ii) free radical scavenging property both in vitro and in vivo by means of reducing lipid peroxidation and the presence of natural antioxidants. Hence, it may be concluded that the hypolipidaemic and antihyperlipidaemic effects produced by the HSEt may be due to the presence of flavonoids and other polyphenolic compounds. But the exact underlying mechanism is remains to be elucidated.

Chapter 19 - Walnuts *(Juglansregia* L.) belong to the family Juglandaceae and a good source of fat, protein, vitamins and minerals along with phenolics which act as antioxidants. It is ranked second after blackberries for their antioxidant activities. Walnut has high levels of melatonin (sleep hormone and antioxidant) and vitamin E. It is also a rich source of L-arginine, phospholipids, proteins, tocopherols, polysterols, squalene and unsaturated fatty acids. Whole walnut tree including the seeds, leaves, husks, nuts and kernels are rich in active phytochemicals and natural antioxidants, which are proved to ameliorate/prevent many diseases including cardiovascular disease, diabetes, obesity, cancer, neuorological diseases etc.Walnut is unique among nuts as it is highest source of alpha-linolenic acid (ALA - omega-3fatty acid).Hence, it is gaining more importance in medicinal and pharmaceutical industries.In the last few decades, there has been tremendous interest in the use of walnuts as evidenced by the voluminous work along with consumer's health awareness. This chapter will describe the medicinal propertiesand importance of walnuts.

Chapter 20 - The high prevalence of coffee drinking and coronary heart diseases (CHD) in many developed countries has led to studies on coffee drinking as an etiological factor for CHD. The first major epidemiological study suggests a coffee-coronary disease link was the work of Paul *et al.* Two large case control studies were conducted in the early 1970s, suggested a positive association between coffee drinking and myocardial infarction. In the ensuing decade, additional reports were essentially negative, resulting in abatement of concern about the role of coffee in CHD. Interest was reawakened in the 1980s by reports of an association between coffee drinking and higher serum cholesterol level; additional studies attributed the relation to high low-density lipoprotein cholesterol.

Chapter 21 - Herbal medicines are regulated in many countries and accepted to be integrated in healthcare system. Herbal medicine is an affordable health care resource for many countries.Among the World Health Organization (WHO) efforts for promoting the use of alternative medicines is the creation of awareness about safe and effective alternative medicine therapies among the public and consumers. For cardiovascular diseases, herbal treatments have been used in patients with atherosclerosis, (which occurs when fatty deposits clog and harden arteries), coronary heart disease, (caused by the reduced blood supply to the heart muscle),stroke, (caused by inadequate blood flow to the brain leading to the death of brain cells), hypertension, (occurs when blood pressure is higher than the normal range), cardiac arrhythmias, (which are irregular or abnormal heartbeats).

Chapter 22 - Bioactive compounds from marine organisms have attracted attention of scientists for only a few decades. Nowadays, more than 10,000 new bioactive molecules that exhibited anti-microbial, anti-viral, anti-fungal, anti-cancer, anti-inflammation, anti-fouling and other properties have been isolated from the marine organisms. Only a few of these compounds have been transformed into drugs that have appeared on the market or undergone clinical trials. Some of the marine derived compounds, like omega-3 fatty acids, are important nutraceuticals that provide health and medical benefits including treatment of diseases. In this chapter we reviewed some of the marine derived pharmaceuticals, highlighted challenges of marine drug discovery and outlined important future directions.

Chapter 23 - The fruit of *Punica granatum L. Punicaccae* has been widely used since ancient times. In the past decade; scientists have been researching on pomegranate fruit by analyzing its nutrition, chemistry, pharmaceuticals, medical properties, or even cosmetic properties. These studies are the results of increasing awareness of the health benefits of functional fruits. Pomegranate is a nutritious fruit containing carbohydrates, minerals, vitamins and most importantly, antioxidants. Evidence from literature showed medicinal effects of different parts of pomegranate to prevent varied chronic or common illnesses. In this chapter, a focus on the potential of pomegranate in medicine, food industry, and pharmaceuticals is presented, with emphasis on its nutrition and active components, and possible toxicity of the pomegranate skin. Pomegranate showed protective effects, such as breast cancer, menopausal syndrome, thyroid dysfunctions, injured cells, diabetes, inflamemations, influenza, prostate cancer, brain damage, Alzheimer's disease, hyperlipidemia, hypertension, artery stenosis, dental problems, male infertility, erectile dysfunction, obesity and other health issues.

Chapter 24 - The global burden of non-communicable diseases (NCDs) has been an increasing public health concern. Non-communicable diseases (NCDs) account for 60% of the global mortality. Of the 35 million deaths in attributable to NCDs annually, about 80% are in low- and middle-income countries (LMIC). From 2006 to 2015, deaths due to NCDs are expected to increase by 17%. Dietary approaches hold promise as effective and preventive interventions for NCDs. Dietary factors represent the most potent determinants of metabolic health and have been shown to mitigate specific physiological mechanisms in various disease conditions. Recent epidemiological and experimental studies suggest that healthy dietary pattern, including increased consumption of natural products, whole grains, fruits can favorably influence the risk of NCDs. Increase in dietary fiber (DF) intake has been recommended for a healthy life. Cereals and cereal products, particularly from whole grains forms staple diet in most countries. Moreover, in addition to being a source of carbohydrates whole grains especially wheat, rice, and oats, provides protein and essential fatty acids and possesses unique and beneficial combinations of many micronutrients, polyphenolics and DF. Among the whole grains, Oats had gained a unique position, because of its diverse health benefits to the humans. This chapter mainly deals with the health benefits of oats in relation to the prevention of NCDs.

Chapter 25 - *Mangifera indica L.* (Mango) belonging to family Anacardiaceae. It is an indigenous to Indian subcontinent and an important fruit crop cultivated in tropical and subtropical regions. Its each part like pulp, peel, seed, leaves, flowers and the bark are important due to their medicinal uses. Different part of mango contains many biotic compounds like polyphenolics which can control many degenerative diseases due to their antioxidant activities. Hence, it is gaining more importance in medicinal and pharmaceutical industries. There has been tremendous interest in this plant as evidenced by the voluminous work in last few decades. This chapter will cover the medicinal uses of mango.

Chapter 26 - Some endemic species of medicinal and culinary herbs are of particular interest due to presence of phytochemicals with significant antioxidant capacities and health benefits. Phytochemical rich plant materials are increasingly of interest in the food and medical industry as they are helpful in oxidative retardation of lipids as well as due to their preservative action against microorganisms. Many medicinal plants are rich with large amounts of antioxidants other than vitamin C, vitamin E, and carotenoids. Basils come with loads of health benefits as it is a rich source of key nutrients like Vitamin A, Vitamin C, calcium, phosphorus, beta carotene. Basil leaves are helpful in sharpening memory. Basil is also useful in treatment of fever, common cold, stress, purifying blood, reducing blood glucose, risk of heart attacks and cholesterol level, mouth ulcer and arthritis. Anti-inflammatory properties of basil are also well known.

Chapter 27 - All over the world, health authorities and government agencies have been insisting to include more fruits and vegetables (eight to ten servings) in daily diet for the promotion and maintenance of good health.Fruits and vegetables may help in reducing the risk of high blood pressure, stroke and other cardiovascular diseases,type 2 diabetes, some cancers,developing kidney stones, bone losses and many other diseases. Bioactive compounds and nutrients (such aspotassium, dietary fiber, folate, vitamin A, vitamin E, vitamin C and so on)in fruits and vegetables are responsible for the desired health benefits. In recent years, fruits and vegetables have been subjected to various treatments to develop new products, increase shelf life, blend with other products (such as dairy products) and for other reasons. During processing, the micro elements present in fruits and vegetables which are responsible for health benefits are also receiving various treatments and encounter losses and conversion into other forms. In addition to industrial processing, the storage and cooking conditions also make significant changes in these micro bioactive compounds. To gain the expected health benefits from the consumption of fruits and vegetables, the bioactive compounds should be preserved till they reach the consumer's table. Many studies are being conducted around the world about the process optimization and minimal processing approaches to prevent or minimize the losses of bioactive compounds from fruits and vegetables during processing, storage and cooking conditions. This chapter describes some of the new techniques which are becoming popular in food industries,and their role in the retention of bioactive compounds.

Chapter 28 - All over the world spices have been used in food for a long time to enhance flavor and taste. Most of the spices have potential chemo preventive properties. They have been widely studied for their medicinal values such as influence in lipid metabolism, fat absorption, hypotriglyceridemic and hypocholesterolemic activity, cholesterol turnover to bile acid, anti-lithogenic activity, anti-diabetic activity, antioxidant potential, anti-inflammatory activity, anti-mutagenic and anti-carcinogenic activity, digestive stimulant action, influence in platelet aggregation, protection of erythrocyte integrity, metabolic disposition of active principles and so on. Species, cultivar, agro-climatic field condition, postharvest processing, storage and cooking conditions greatly influence the availability of bioactive compounds in spices. New initiatives are being taken place in processing of spices to minimize the losses of bioactive compounds.

In: Natural Products and Their Active Compounds ... ISBN: 978-1-62100-153-9
Editors: M. Essa, A. Manickavasagan and E. Sukumar

Chapter 1

TRADITIONAL INDIAN MEDICINES FOR METABOLIC SYNDROME

***Vishal Diwan*[1]*, Hemant Poudyal* [1] *and Lindsay Brown**[*2]

[1] School of Biomedical Sciences,
The University of Queensland,
Brisbane, QLD Australia
[2]Department of Biological and Physical Sciences,
University of Southern Queensland,
Toowoomba, QLD Australia

ABSTRACT

Plant products play a significant role in human diet as they maintain human health and improve the quality of life. The health-promoting effects of plant products form part of the rich history of ancient civilisations. The non-nutritive but potentially bioactive secondary metabolites in fruits, vegetables, herbs, spices, teas and wines reduce oxidative stress, inflammation, risk of cardiovascular diseases and type 2 diabetes, age-related cognitive decline, risk of major neurodegenerative disorders, including Alzheimer's disease and cancer. This reduced risk is associated with the most commonly consumed dietary phytochemicals. Metabolic syndrome has been classically defined as the clustering of interrelated risk factors for cardiovascular disease and type2 diabetes including hyperglycemia, insulin resistance, hypertension, hypertriglyceridaemia, decreased HDL-cholesterol concentration and obesity. The prevalence of metabolic syndrome is reaching pandemic proportions worldwide with diets rich in saturated fat, cholesterol and refined carbohydrates implicated in the increased risk of cardiovascular disease and metabolic syndrome. From a dietary perspective, a plant-based diet could therefore serve as an alternative intervention for the prevention or treatment of metabolic syndrome. We have previously reviewed the possible therapeutic responses of commonly consumed Indian spices in metabolic syndrome. This review evaluates the effects of further frequently-used medicinal and dietary plants from the Indian sub-continent on the

[*]Address for correspondence: Professor Lindsay Brown, Department of Biological and Physical Sciences, University of Southern Queensland, Toowoomba 4350, QLD, Australia; Tel: +61 7 4731 1319; Fax: +61 7 4631 1530; Email: Lindsay.Brown@usq.edu.au

risk factors of metabolic syndrome using data from controlled human, animal and *in vitro* studies. The major bioactive constituents of these plants, mostly identified as herbs, are also discussed.

Keywords: Indian traditional medicine, Metabolic Syndrome, hypertension, obesity, diabetes, inflammation, antioxidants

INTRODUCTION

Plants are a necessary component of the human diet. In addition, plant products have played a significant role in attempts to maintain human health and improve the quality of human life throughout human history. The health-promoting effects of plant products have been ascribed, in part, to the non-nutritive but potentially bioactive secondary metabolites in fruits, vegetables, herbs, spices, teas and wines [1]. A plant-based diet rich in fruits, vegetables and legumes reduces the risk of cardiovascular diseases and cancer [2, 3]. This reduced risk is associated with the most commonly consumed dietary phytochemicals including flavonoids, phenolic acids, phytoestrogens, carotenoids, organosulphur compounds, plant sterols, dietary fibers, isothiocyanates and monoterpenes[4].

Increased dietary intake of antioxidant phytochemicals reduces the risk of type 2 diabetes, although no reduction in the risk of type 2 diabetes has been associated with increased consumption of fruits or vegetables [5, 6]. Diets rich in plant-derived antioxidants and anti-inflammatory compounds, for example fruits, nuts and vegetables, such as the Mediterranean diet, reduce both age-related cognitive decline and the risk of major neurodegenerative disorders, including Alzheimer's disease [7, 8].

Metabolic syndrome has been defined as the clustering of interrelated risk factors for cardiovascular disease and type-2 diabetes [9]. Major risk factors include hyperglycaemia, insulin resistance, hypertension, hypertriglyceridaemia, decreased HDL-cholesterol concentration and obesity [9]. The prevalence of metabolic syndrome is reaching pandemic proportions worldwide [10].

According to the National Cholesterol Education Program (NCEP) criteria, the prevalence of metabolic syndrome was 24.0% in males and 23.4% in females among US adults [11]. In a Chinese cohort, the prevalence of metabolic syndrome associated with hypertension was 32.9% in men and 53.1% by NCEP definition [12]. In an urban Indian population, 31.6% of the sample population was diagnosed with metabolic syndrome by NCEP definition, with the prevalence again being higher in women (39.9%) than in men (22.9%) [13].

Lipid mediators of inflammation play an important role in obesity as a chronic low-grade inflammatory state [14]. Increased expression of TNF-α occurred in the adipose tissue of insulin resistant, genetically obese rodents [15] with a direct relationship between TNF-α and insulin resistance. NF-κB regulated target gene transcription of pro-inflammatory mediators such as iNOS, COX-2, IL-6, IL-12 including TNF-α [16, 17]. PPAR-α activation plays an important clinical role in the control of the cellular redox balance and inflammatory responses such as iNOS and COX-2 by decreasing NF-κB transcriptional activity. PPAR-α regulates the transcription of genes involved in lipid, cholesterol, lipoprotein and glucose energy metabolism, as well as insulin sensitivity [18].

Diets rich in saturated fat, cholesterol and refined carbohydrates are implicated in the increased risk of cardiovascular disease and metabolic syndrome [19]. Evidence from the past three decades indicates that virtually all diseases, including the metabolic syndrome, have multifactorial dietary elements that underlie their aetiology, along with other environmental variables and genetic susceptibility [20]. From a dietary perspective, a plant-based diet could therefore serve as an alternative intervention for the prevention or treatment of metabolic syndrome.

Additionally, within the evidence-based framework necessary to substantiate health claims related to all therapy, including diet, identification of unique bioactive compounds would expedite novel lead compound discovery for drug development [21].

Many currently available drugs have been derived from plants including digoxin from *Digitalis lanata*, salicin (the source of aspirin) from *Salix alba*, reserpine from *Rauwolfia serpentina*, atropine from *Atropa belladonna*, quinine from *Cinchona officinalis*, codeine from *Papaversomniferum*, vincristine from *Catharanthus roseus* and taxol from *Taxus brevifolia* [22], although these plants have not been part of the human diet.

Although herbs and spices represent only a small proportion in the diets of the 21st century, a wide range and significant amounts of herbs and spices are used in traditional diets, for example from Europe, Egypt, China and India [21].

Current evidence suggests that increased consumption of culinary herbs and spices could modify disease state or delay the onset of many diseases such as cancer, bacterial and viral infections, cardiovascular diseases, dyslipidaemia, diabetes, cognitive and neurodegenerative disorders and osteoarthritis [3, 21, 23].

We have previously reviewed the possible therapeutic responses of commonly consumed Indian spices in metabolic syndrome [23]. Consequently, this chapter will evaluate the effects of further frequently used medicinal and dietary plants from the Indian sub-continent on the risk factors of metabolic syndrome using data from controlled human, animal and *invitro* studies. The major bioactive constituents of these plants, mostly identified as herbs, are also discussed. The sources of these plants, their active ingredients and proposed uses have been collected in Table 1.

INDIAN GOOSEBERRY

Latin Name: *Embolic officinalis*

Indian gooseberry could decrease the symptoms of metabolic syndrome through lipid-lowering, antidiabetic and anti-inflammatory properties. Extracts decreased blood glucose concentrations in rat models of type 1 diabetes (alloxan-induced, [24]) and type 2 diabetes (high fructose-fed rats, [25]).

PPAR-α expression increased and the lipid profile was improved in aged rats [26], Sprague-Dawley rats [27], fructose-fed rats [25] and in rabbits fed a high cholesterol diet [28]. In fructose-fed rats, blood pressure was reduced and NF-κB activation and TNF-α concentrations were decreased [25].

Table 1. Comprehensive information about sources of traditional medicines

Plant	Common Names	Habitat	Major active constituents	Uses
Indian gooseberry	Amla (Hindi), Amlaki(Sanskrit)	Tropical and subtropical parts of China, India, Indonesia and the Malay Peninsula[99]	Phyllembelic acid, phyllemblin, rutin, curcuminoides, emblicol, rutin, vitamin C [99] (Figure 1)	Anti-atherogenic[100], Anti-tussive[101], Immunomodulatory[102] Hypolipidaemic [28] Anti inflammatory[29]Hepatoprotective[103]
Winter Cherry	Aswagandha (Hindi), Varahakarni (Sanskrit)	Drier parts of India, Baluchistan, Pakistan, Afghanistan, Sri Lanka, Congo, South Africa, Egypt, Morocco and Jordan [104]	Withanine, somniferine, somnine, somniferinine, withananine, glycowithanoloids (sitoindoside IX and sitoindoside X), acyl sterylglucosides (sitoindoside VII and sitoindoside VIII) [104](Figure 2)	Antidiabetic[31] Antihyperlipidemic[33] Antioxidant [32] Cardioprotective [105][32]
Bael fruit	Bael (Hindi), Bilva (Sanskrit)	India[42]	Cineole, citral, citronellal, d-limonene, eugenol, lupeol,rutin, umbelliferone [43](Figure 3)	Anti-inflammatory [35] Reduce blood urea and liver glucogen[41] Antihyperlipidaemic [42] Hypoglycaemic [36]
Black Plum	Jamun (Hindi), Jambul (Sanskrit)	India from the sub-Himalayan tract to extremesouth, Thailand and Philippines [46]	Gallic acid, methylgallate, kaempferol, myricetin, ellagic acidmyricetrin, quercetrin [106](Figure 4)	Hypoglycaemic and antidiabetic [46-49, 107]] Antihyperlipidaemic[47] Antioxidant[51]
Bitter Melon	Karela (Hindi), Karavella (Sanskrit)	Brazil, China, Colombia, Cuba, Ghana, Haiti, India, Mexico,Malaya, New Zealand, Nicaragua, Panama and Peru[108]	Momorcharins, cucurbitins,gentisic acid, goyaglycosides and goyasaponins[108], caffeic acid and ferulic acid [109], fisetin and isorhamnetin[110](Figure 5)	Hypoglycaemic and anti-diabetic [52],[54-56] Decrease adipose tissue and visceral fat [52] Antioxidant and free radical scavenger [64]

Table 1. (Continued)

Plant	Common Names	Habitat	Major active constituents	Uses
Holy Basil	Tulsi(Hindi), Vranda, Vishnu priya (Sanskrit)	Throughout India, up to an altitude of 1,800 m in the Himalayas[73]	Eugenol [75, 111], ursolic acid, apigenin, luteolin[112, 113](Figure 6)	Antidiabetic [69] Hypotensive [73] Hypoglycaemic [68] Antihyperlipidaemic [69] Antioxidant [67] Anti-inflammatory [71] Vasorelaxant[75]
Gymnema	Gur-mar (Hindi) Madhunaashini (Sanskrit)	Deccan peninsula of western India,tropical Africa, Vietnam, Malaysia, Sri Lanka and is widely availablein Japan, Germany and the USA [79]	Gymnemagenin, gymnemicacid-III, -IV, -V, -VIII, and –IX[78, 81] (Figure 7)	Antidiabetic[76] Hypoglycaemic[77] Increase insulin release [80]
Fenugreek	Methi (Hindi) Methika (Sanskrit)	India, Asian,African and European countries [89]	Quercetin, luteolin, kaempferol,tricin, gallic acid[91](Figure 8)	Hepatoprotective[91] Antioxidant [85] Hypoglycaemic[86],[87] Improves antioxidant enzyme status [90] Improves lipid profile [88] Improves glucose handling and insulin resistance [94] Anti-inflammatory [95]

An extract of the fruits reduced inflammation in indomethacin-induced enterocolitis in rats [29] while a leaf extract reduced inflammation in hind paw oedema in rats [30].

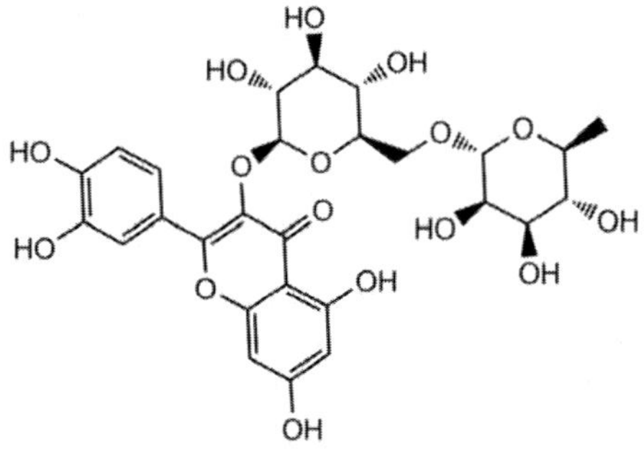

Rutin

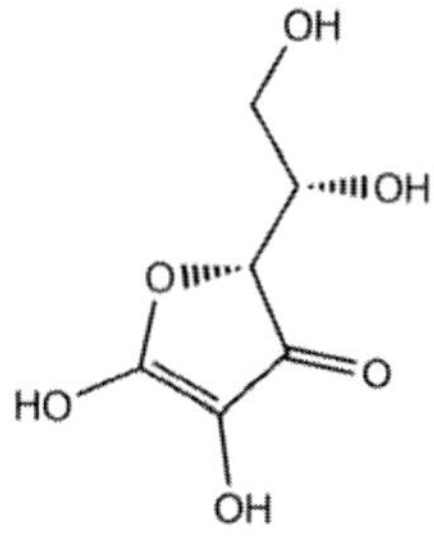

Vitamin C

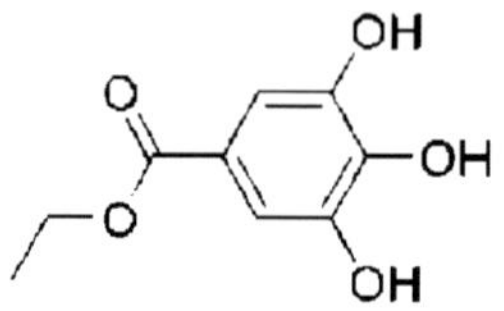

Phyllemblin

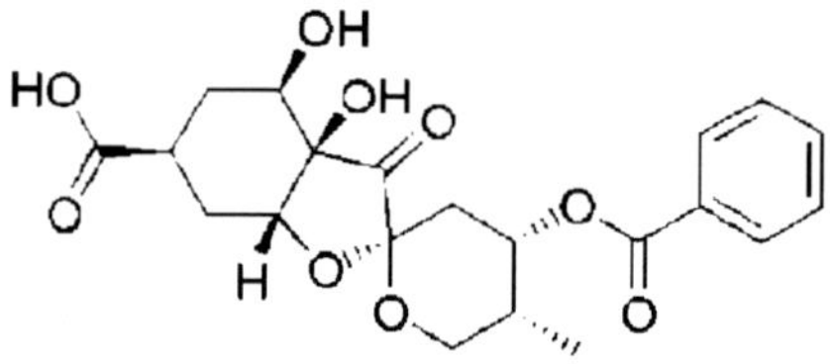

Phyllembellic acid

Figure 1. Major bioactive chemical constituents of Indian gooseberry.

WINTER CHERRY

Latin Name: *Withania somnifera*

Winter cherry is widely used in India as an anti-diabetic [31], antioxidant [32] and cardio-protective agent [32]and has been listed as an official drug in the Indian Pharmacopoeia. Aqueous root extract reduced blood glucose and glycosylated haemoglobin, and improved oral glucose tolerance and insulin sensitivity in streptozotocin-induced type 1 diabetes in rats [31]. In hyper cholesterolaemic rats, dietary root powder decreased plasma total lipids, cholesterol and triglycerides and increased plasma HDL cholesterol concentrations, HMG-CoA reductase activity and bile acid content in the liver [33].*Withania-somnifera* roots improved mean arterial blood pressure, heart rate and left ventricular function in rats with isoprenaline-induced myocardial injury demonstrating cardio protective effects[32]. Antioxidant status (malondialdehyde, glutathione, glutathione peroxidase, [32])was improved suggesting that free radical scavenging properties were effective against cellular injury mediated by reactive oxygen species [34]. These studies showing that winter cherry extracts alleviate some of the symptoms of the metabolic syndrome suggest that there may be some benefit in humans, but there are no clinical trials to prove this.

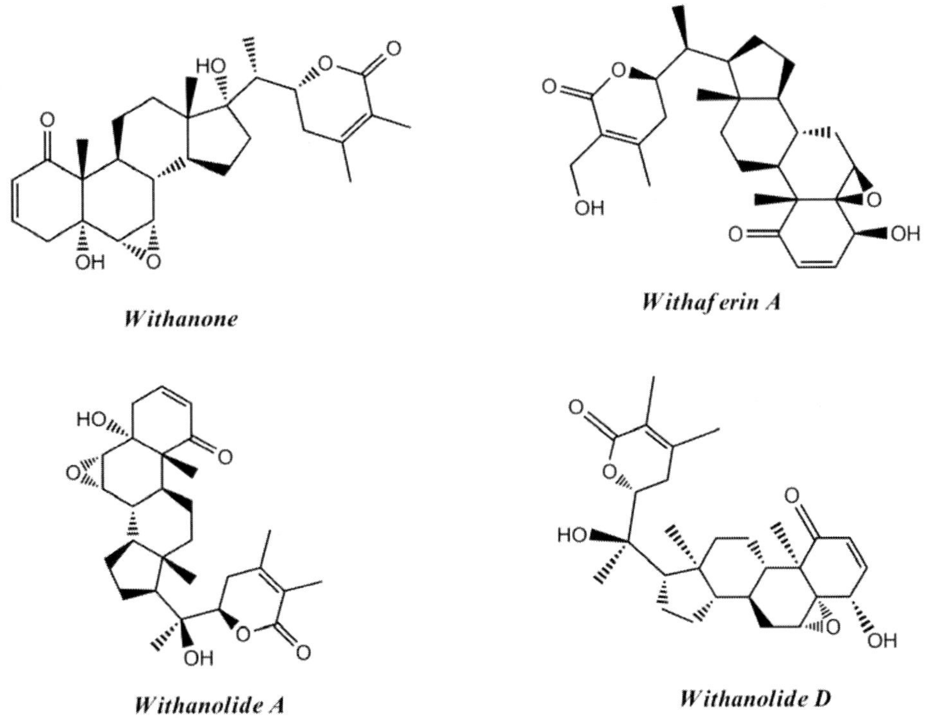

Withanone

Withaferin A

Withanolide A

Withanolide D

Figure 2. Some important phytochemicals from winter cherry.

BAEL

Latin Name: *Aegle marmelos*

The anti-inflammatory, anti-oxidant, hypoglycaemic and hypolipidaemic responses reported in animal models suggest that extracts of different parts of *Aegle marmelos* could produce useful responses in many symptoms of the human metabolic syndrome. Extracts of leaves produced anti-inflammatory effects in carrageenan-induced paw oedema and cotton-pellet granuloma model in rats [35]. Extracts also decreased blood glucose concentrations in glucose-induced hyperglycaemic Wistar rats[36], Sprague-Dawley rats [37] and streptozotocin-induced type-1 diabetic Wistar rats[38].Further, responses to 5-hydroxytryptamine (5-HT) through 5-HT_2A receptors in streptozotocin-diabetic rats were up-regulated [39] as sero-tonergic pathways directly affect glucose homeostasis through regulation of autonomic efferents and peripheral tissues [40]. Fruit extracts reduced plasma thiobarbituric acid reactive substances, hydroperoxides, ceruloplasmin and α-tocopherol and increased plasma reduced glutathione and vitamin C [38]. Leaf extract normalised the liver glycogen, blood urea and serum cholesterol concentrations in type 1 diabetic rats (alloxan-induced) [41]. Further, the seed extract reduced total cholesterol, low density lipoproteins and triglycerides and increased high density lipoproteins [42]. Further characterisation of the responses to extracts from *Aegle marmelos* in appropriate rat models of the metabolic syndrome, followed by trials in humans, may increase the role of this plant in treating the metabolic syndrome.

Aegeline 2, an alkaloidalamide from the leaves [43], lowered blood glucose in sucrose-challenged streptozotocin-induced diabetic rats and decreased plasma triglyceride concentrations, total cholesterol and free fatty acids accompanied by an increase in high density lipoproteins in dyslipidaemic hamsters [44]. Umbelliferone, another important constituent, returned lipid peroxidation markers, nonenzymic and enzymic antioxidants to near normalcy in streptozotocin-diabetic rats [45].

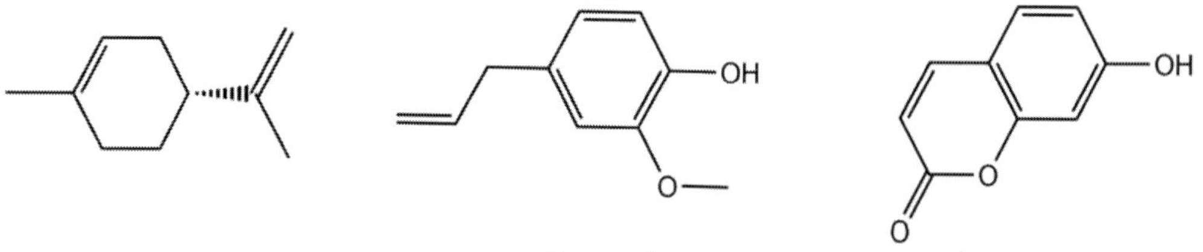

Limonene *Eugenol* *Umbelliferone*

Figure 3. Major chemical constituents of Bael fruit.

BLACK PLUM

Latin Name: *Syzygium cumini*

Plants such as the black plum, a widely used anti-diabetic herb [46], show anti-oxidant responses that may broaden the therapeutic responses to include the metabolic syndrome. Black plum extracts induced insulin release from the pancreas [47], possibly mediating its glucose-lowering effects in streptozotoc in-induced type 1 diabetic mice [47], alloxan-induced diabetic rabbits [46, 48] and type II diabetes [49]. Moreover, extracts reduced plasma lipid concentrations [46, 48] by increasing PPAR-γ activity[49] which directly affects glucose and lipid metabolism [18, 50] in streptozotoc in-induced diabetic mice [47] and rats[49]. Moreover, these extracts also account for antioxidant effects in streptozotocin-induced diabetic rats. The elevated concentrations of vitamin E, lipid peroxides and decreased concentrations of vitamin C and reduced glutathione were reversed by this treatment [51].

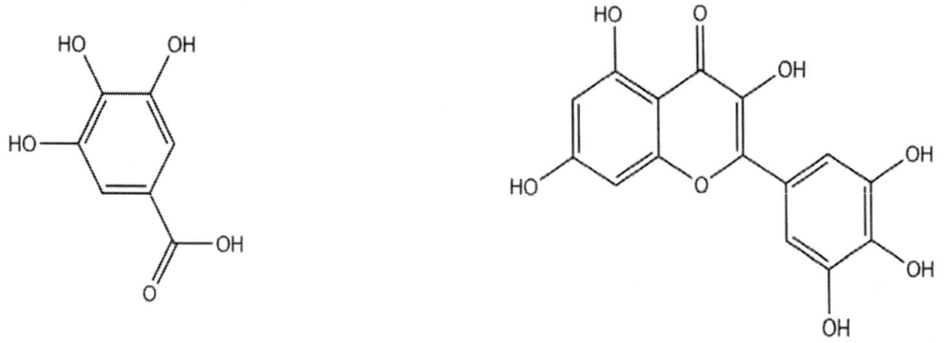

Gallic acid *Myricetin*

Figure 4. Principle chemical constituents of black plum.

Bitter Melon

Latin Name: *Momordica charantia*

Bitter melon is used as a vegetable as well as a traditional medicine. Bitter melon reduced high fat diet-induced hyperglycaemia, hyperleptinaemia, glycated haemoglobin (HbA1c) and free fatty acids in mice [52], fasting blood glucose in alloxan-induced diabetic rats [53] and mice [54], hyperglycaemia and hyperinsulinaemia in fructose-rich diet-fed rats [55], possibly by increasing pancreatic β-cell number in streptozotocin-induced diabetic rats[56]. Studies suggested insulin-like actions and promotion of insulin release [57], extra-pancreatic effects [58] which included increased GLUT4 transporter protein in muscles [59] and GLUT4 translocation to the cell membrane[60], increased glucose utilisation in the liver and muscle [58], inhibition of glucose-6-phosphatase and fructose-1,6-bisphosphatasein liver and stimulation of erythrocytic and hepatic glucose-6-phosphate dehydrogenase activities [61]. Clinical studies also showed glucose-lowering effects in type 2 diabetic patients [57]. In contrast, another clinical study showed no effect on HbA_{1c} and fasting blood glucose concentrations [62].

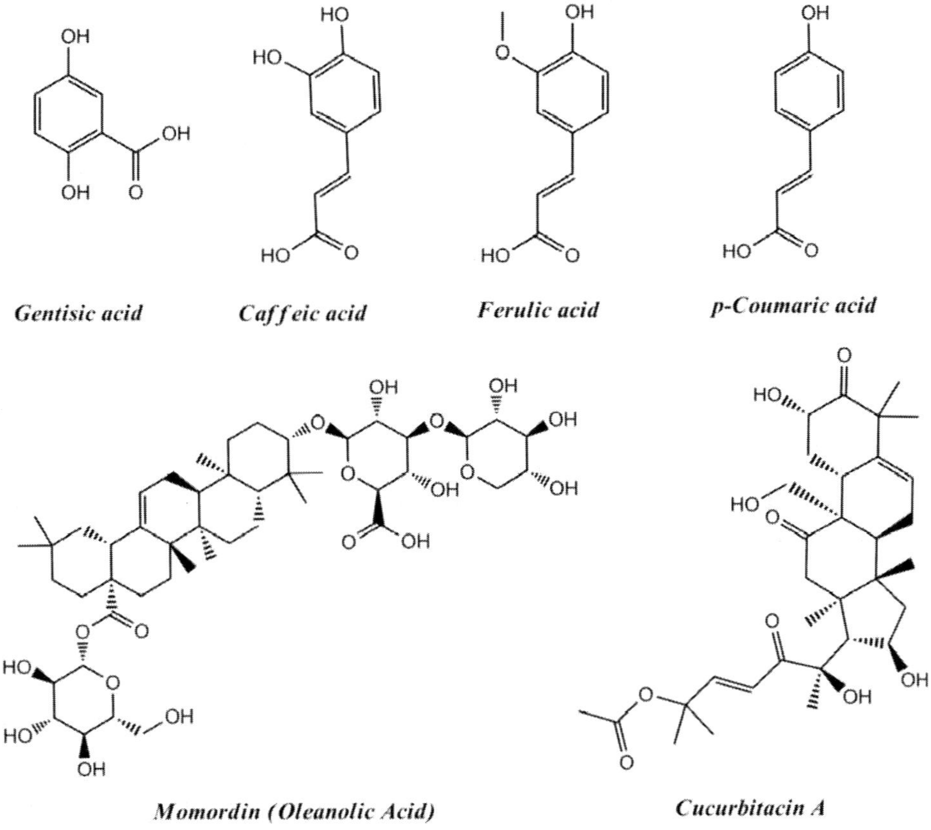

Gentisic acid *Caffeic acid* *Ferulic acid* *p-Coumaric acid*

Momordin (Oleanolic Acid) *Cucurbitacin A*

Figure 5. Principle chemical constituents of Bitter Melon.

Further, biochemical studies indicated that bitter melon-mediated effects were regulated by signaling pathways in pancreatic β-cells, adipocytes and muscles as an extract of bitter melon activated PPARs-α and-γ [63].These studies give a clear mechanism for the use of bitter melon to treat diabetes and hyperglycaemia in metabolic syndrome. Bitter melon decrease depididymal white adipose tissue and visceral fat[52], and adipose leptin and resistin mRNA levels [52].Extracts showed potent antioxidant and free radical scavenging activities [64].Hence, there is experimental evidence, but no clinical trials, that bitter melon could be useful in patients with the metabolic syndrome.

HOLY BASIL

Latin Name: *Ocimum sanctum*

Holy basil is an Indian medicinal plant that has potential for the treatment of the broad range of signs of the metabolic syndrome. An extract of leaves prevented lipid peroxidation in rats (ischaemia and hypo perfusion-induced cerebral injury[65]), rabbits (stress-induced by anaemic hypoxia[66]), improved antioxidant status in rats (chronic restraint[67]), and reduced fasting blood glucose concentrations and improved oral glucose tolerance in alloxan-induced diabetic rats [68]. Moreover, leaf powder in the diet reduced plasma concentrations of glucose, uronic acid, total amino acids, total cholesterol, triglycerides and total lipids in alloxan-induced diabetic rats [69]. Fresh leaves lowered plasma total cholesterol, triglyceride, phospholipid and low density lipoprotein-cholesterol concentrations and increased high density lipoprotein-cholesterol concentrations in normal albino rabbits[70].

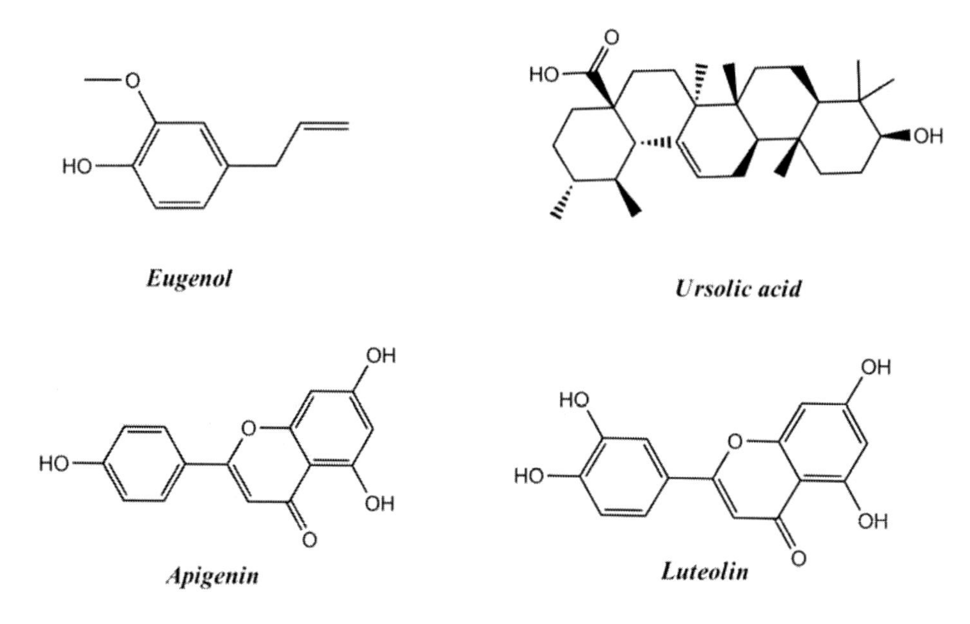

Figure 6. Major chemical constituents of Holy Basil.

Leaf extracts have shown anti-inflammatory responses in rats (carrageenan-induced pedal edema and croton oil-induced granuloma [71]).Holy basil inhibited the cyclooxygenase and lipoxygenase pathways of arachidonic acid metabolism, important enzymes in the production of proinflammatory mediators [72]. The fixed oil obtained from *Ocimum sanctum* lowered blood pressure in anaesthetised dogs, which seems to be due to its peripheral vasodilatory action, and increased the blood-clotting time comparable to aspirin [73].

The oil obtained from holy basil is predominantly eugenol (about 65%) [74]. As a 3% emulsion with liquid parrafin, this oil improved serum lipid profile (cholesterol, low and high density lipoprotein-cholesterol and triglycerides),decreased serum lactate dehydrogenase and creatine kinase concentrations and improved antioxidant status (thiobarbituric acid reactive substances, glutathione peroxidase and superoxide dismutase) in rats fed with a high cholesterol diet [74]. Moreover, eugenol relaxed blood vessels [75]. Hence, these studies with eugenol support further studies into the therapeutic potential of holy basil for cardio protective, hypolipidaemic and hypotensive actions in patients with the metabolic syndrome.

GYMNEMA

Latin Name: *Gymnema sylvestre*

Gymnema primarily produces antidiabetic responses [76]. Various leaf extracts lowered blood glucose concentrations in normal rats after a loading dose of glucose [77] and in type 1 diabetic rats [78]. Dihydroxygymnemic triacetate, an active principle, reduced blood glucose, increased muscle and liver glycogen, improved lipid profile, improved liver function enzymes and increased plasma insulin concentrations in type 1 diabetic rats [78, 79]. The GS4 fraction of dried leaf extract increased insulin release from β-cells of pancreas *in vitro*[80]. Further, gymnemic acid IV, a compound derived from leaves, reduced blood glucose concentrations in streptozotocin-diabetic mice [81]. Leaf extract improved insulin secretion by increasing the number of secretory cells in islets of alloxan-induced [79] and streptozotocin-induced [82] diabetic rats. The ability of leaf extracts to increase faecal excretion of cholesterol and bile acids in normal rats [83] could be useful in obese patients. Gymnema did not reduce blood pressure in sugar-induced hypertension in Spontaneously Hypertensive Rats [84]. Hence, gymnema may be useful as a partial treatment of raised blood glucose and lipid concentrations in the metabolic syndrome, without responses on other key signs such as hypertension.

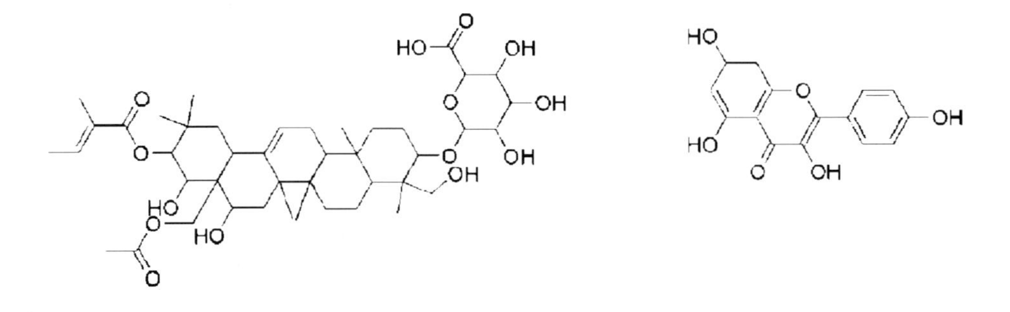

Figure 7. Important bioactive compounds of Gymnema.

FENUGREEK

Latin Name: *Trigonella Foenum-Graecum*

Fenugreek is not only a flavouring agent but is used as a folk medicine. Seed extracts improved the antioxidant status in plasma and liver of alcohol-fed rats [85]. The extract of leaves [86] and seeds [87] reduced blood glucose concentrations in streptozotocin and alloxan-induced diabetic rats, respectively. Further, leaves [86], seeds [88] and germinated seeds [89] reduced oxidative stress in type 1 diabetic rats [86] and high cholesterol-fed rats[88] and improved the antioxidant enzyme status [90]. The seeds improved the lipid profile in high cholesterol-fed rats [88]. Aqueous extracts of germinated seeds reduced lipid peroxidation, oxidative stress and improved serum markers of liver function in cypermethrin-induced hepatotoxicity [91]. 4-hydroxyisoleucine, one of the prime bioactive contents, also improved liver function and reduced blood glucose concentrations in both type 1 and type 2 diabetic rats [92]. The seed extract improved the glucose uptake *in vitro* in cell culture as it increased the GLUT-4 activity mediated through phosphatidylinositol 3-kinase and protein kinase C-dependent insulin pathways [93], thereby lowering blood glucose concentrations. Diosgenin, another important bioactive compound from fenugreek, and fenugreek improved adipocyte differentiation, reduced obesity-related inflammation in *in vitro* assays and in adipocytes of KK-Ay mice and improved glucose handling and insulin resistance in obese diabetic mice [94]. TNF-α mediated NF-κB activation was attenuated by diosgenin, which could be responsible for the anti-inflammatory action of fenugreek [95]. All the above properties of fenugreek could be exploited to reduce some of the risks and signs of metabolic syndrome, with the notable exception of hypertension.

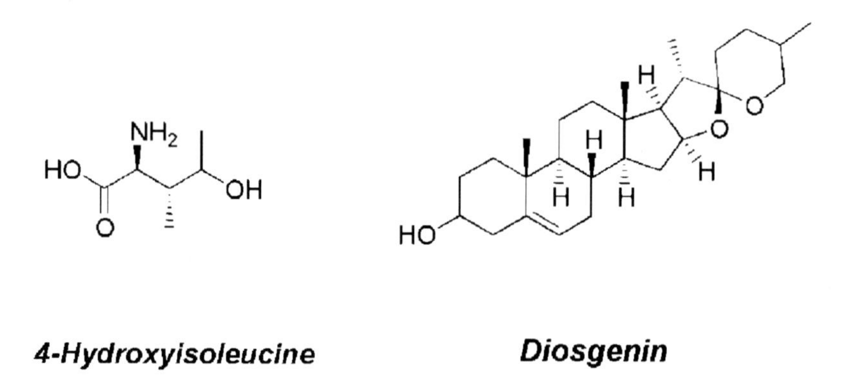

Figure 8. Important bioactive compounds of Fenugreek.

FUTURE OF NATURAL PRODUCTS

Natural products often lead to the discovery of useful drugs. As an example, almost 78% of commercial antibacterials are derived from natural sources [96]. The constant need for newer therapeutic agents has also led to the development of sensitive and robust screening assays [96] and quality control techniques [97]. Additionally, with advances in DNA technology, it is now possible to use plants as a bioreactor to upscale the production of selected secondary metabolites with therapeutic value [96].

Purified compounds derived from natural products for use as drugs undergo stringent testing, quality control and commercialisation conditions. This is rarely the case for the use of the source organism as a dietary supplement or complementary medicine. This is probably due to poor regulation of commercialisation in the complementary and alternative medicine industry [98]. The ultimate end-point for all potential therapeutic natural products or their derivatives is commercialisation, clearly backed by strong evidence from animal and human trials. Commercialisation of natural products is rarely discussed mostly because the primary interests in this field are confined within academic and research institutions. Commercialisation of natural products poses some important and debatable questions. These include a well-defined mechanism of action, pharmacokinetics, dosage, safety and toxicity, standardisation of the product to ensure consistency, purification of the potent active ingredients from natural sources or in some case preparations by chemical synthesis, or more often by partial synthesis from naturally occurring precursors. These procedures require verifiable good laboratory and manufacturing processes. These issues then support the setting up of an effective global regulatory body to implement a stringent set of conditions to ensure the safety of the products.

Conclusion

Indian traditional medicines provide a rich source of potential therapeutic agents as holistic treatments of the metabolic syndrome. Extended pre-clinical studies are required to define the likely targets, together with likely mechanisms of action, dosage and pharmacokinetics. Clinical trials are then required to determine the potential in humans, based on scientific evidence. Traditional medicines will remain in widespread use as therapeutic agents without clinical evidence, but rather based on anecdotal reports. Solid clinical evidence for effectiveness without toxicity of products from Indian herbs would allow intervention with these medicines in the current epidemic of metabolic syndrome throughout the world, not restricted to India and nearby countries.

References

[1] Ostertag, LM; O'Kennedy, N; Kroon, PA; Duthie, GG and de Roos, B. Impact of dietary polyphenols on human platelet function--a critical review of controlled dietary intervention studies, *Mol. Nutr. Food Res,* 2010, *54 (1)*, 60-81.

[2] Bazzano, LA; He, J; Ogden, LG; Loria, CM; Vupputuri, S; Myers, L and Whelton, PK. Fruit and vegetable intake and risk of cardiovascular disease in US adults: the first National Health and Nutrition Examination Survey Epidemiologic Follow-up Study, *Am. J. Clin. Nutr.*2002, *76 (1)*, 93-99.

[3] Huang, WY; Cai, YZ and Zhang, Y. Natural phenolic compounds from medicinal herbs and dietary plants: potential use for cancer prevention, *Nutr. Cancer.*2010, *62 (1)*, 1-20.

[4] Kris-Etherton, PM; Hecker, KD; Bonanome, A; Coval, SM; Binkoski, AE; Hilpert, KF; Griel, AE and Etherton, TD. Bioactive compounds in foods: their role in the prevention of cardiovascular disease and cancer, *Am. J. Med.*2002, *113 Suppl 9B*, 71S-88S.
[5] Hamer, M and Chida, Y. Intake of fruit, vegetables, and antioxidants and risk of type 2 diabetes: systematic review and meta-analysis, *J. Hypertens.*2007, *25 (12)*, 2361-2369.
[6] Dembinska-Kiec, A; Mykkanen, O; Kiec-Wilk, B and Mykkanen, H. Antioxidant phytochemicals against type 2 diabetes, *Br. J. Nutr.*2008, *99 E Suppl 1*, ES109-117.
[7] Sofi, F; Macchi, C; Abbate, R; Gensini, GF and Casini, A. Effectiveness of the Mediterranean diet: can it help delay or prevent Alzheimer's disease?, *J. Alzheimers Dis. 20 (3)*, 795-801.
[8] Joseph, J; Cole, G; Head, E and Ingram, D. Nutrition, brain aging, and neurodegeneration, *J. Neurosci.*2009, *29 (41)*, 12795-12801.
[9] Alberti, KG; Eckel, RH; Grundy, SM; Zimmet, PZ; Cleeman, JI; Donato, KA; Fruchart, JC; James, WP; Loria, CM and Smith, SC, Jr. Harmonizing the metabolic syndrome: a joint interim statement of the International Diabetes Federation Task Force on Epidemiology and Prevention; National Heart, Lung, and Blood Institute; American Heart Association; World Heart Federation; International Atherosclerosis Society; and International Association for the Study of Obesity, *Circulation.*2009, *120 (16)*, 1640-1645.
[10] Cameron, AJ; Shaw, JE and Zimmet, PZ. The metabolic syndrome: prevalence in worldwide populations, *Endocrinol. Metab. Clin. North Am.*2004, *33 (2)*, 351-375..
[11] Ford, ES; Giles, WH and Dietz, WH. Prevalence of the metabolic syndrome among US adults: findings from the third National Health and Nutrition Examination Survey, *JAMA.*2002, *287 (3)*, 356-359.
[12] Li, WJ; Xue, H; Sun, K; Song, XD; Wang, YB; Zhen, YS; Han, YF and Hui, RT. Cardiovascular risk and prevalence of metabolic syndrome by differing criteria, *Chin. Med. J. (Engl),* 2008, *121 (16)*, 1532-1536.
[13] Gupta, R; Deedwania, PC; Gupta, A; Rastogi, S; Panwar, RB and Kothari, K. Prevalence of metabolic syndrome in an Indian urban population, *Int. J. Cardiol.*2004, *97 (2)*, 257-261.
[14] Iyer, A; Fairlie, DP; Prins, JB; Hammock, BD and Brown, L. Inflammatory lipid mediators in adipocyte function and obesity, *Nat. Rev. Endocrinol.*2010, *6 (2)*, 71-82.
[15] Hotamisligil, GS; Budavari, A; Murray, D and Spiegelman, BM. Reduced tyrosine kinase activity of the insulin receptor in obesity-diabetes. Central role of tumor necrosis factor-alpha, *J. Clin. Invest.*1994, *94 (4)*, 1543-1549.
[16] Baldwin, AS, Jr. Series introduction: the transcription factor NF-kappaB and human disease, *J. Clin. Invest.*2001, *107 (1)*, 3-6.
[17] Li, Q and Verma, IM. NF-kappaB regulation in the immune system, *Nat. Rev. Immunol.*2002, *2 (10)*, 725-734.
[18] Berger, JP; Akiyama, TE and Meinke, PT. PPARs: therapeutic targets for metabolic disease, *Trends Pharmacol. Sci.*2005, *26 (5)*, 244-251.
[19] Siri-Tarino, PW; Sun, Q; Hu, FB and Krauss, RM. Saturated fat, carbohydrate, and cardiovascular disease, *Am. J. Clin. Nutr.*2010, *91 (3)*, 502-509.
[20] Cordain, L; Eaton, SB; Sebastian, A; Mann, N; Lindeberg, S; Watkins, BA; O'Keefe, JH and Brand-Miller, J. Origins and evolution of the Western diet: health implications for the 21st century, *Am. J. Clin. Nutr.*2005, *81 (2)*, 341-354.

[21] Tapsell, LC; Hemphill, I; Cobiac, L; Patch, CS; Sullivan, DR; Fenech, M; Roodenrys, S; Keogh, JB; Clifton, PM; Williams, PG; Fazio, VA and Inge, KE. Health benefits of herbs and spices: the past, the present, the future, *Med. J. Aust.*2006, *185 (4 Suppl)*, S4-24.

[22] Winslow, LC and Kroll, DJ. Herbs as medicines, *Arch. Intern. Med.*1998, *158 (20)*, 2192-2199.

[23] Iyer, A; Panchal, S; Poudyal, H and Brown, L. Potential health benefits of Indian spices in the symptoms of the metabolic syndrome: areview, *Indian J. Biochem. Biophys.*2009, *46 (6)*, 467-481.

[24] Sabu, MC and Kuttan, R. Anti-diabetic activity of medicinal plants and its relationship with their antioxidant property, *J. Ethnopharmacol.*2002, *81 (2)*, 155-160.

[25] Kim, HY; Okubo, T; Juneja, LR and Yokozawa, T. The protective role of amla (*Emblica officinalis* Gaertn.) against fructose-induced metabolic syndrome in a rat model, *Br. J. Nutr.*2010, *103 (4)*, 502-512.

[26] Yokozawa, T; Kim, HY; Kim, HJ; Okubo, T; Chu, DC and Juneja, LR. Amla (*Emblica officinalis* Gaertn.) prevents dyslipidaemia and oxidative stress in the ageing process, *Br. J. Nutr.*2007, *97 (6)*, 1187-1195.

[27] Anila, L and Vijayalakshmi, NR. Flavonoids from *Emblica officinalis* and *Mangifera indica*-effectiveness for dyslipidemia, *J. Ethnopharmacol.*2002, *79 (1)*, 81-87.

[28] Mathur, R; Sharma, A; Dixit, VP and Varma, M. Hypolipidaemic effect of fruit juice of Emblica officinalis in cholesterol-fed rabbits, *J. Ethnopharmacol.*1996, *50 (2)*, 61-68.

[29] Deshmukh, CD; Pawar, AT and Bantal, V. Effect of *Emblica officinalis* methanolic fruit extract on indomethacin induced enterocolitis in rats, *Res. J. Med. Plant,* 2010, *4*, 141-148.

[30] Asmawi, MZ; Kankaanranta, H; Moilanen, E and Vapaatalo, H. Anti-inflammatory activities of *Emblica officinalis* Gaertn leaf extracts, *J. Pharm. Pharmacol.*1993, *45 (6)*, 581-584.

[31] Anwer, T; Sharma, M; Pillai, KK and Iqbal, M. Effect of *Withania somnifera* on insulin sensitivity in non-insulin-dependent diabetes mellitus rats, *Basic Clin. Pharmacol. Toxicol.*2008, *102 (6)*, 498-503.

[32] Mohanty, I; Arya, DS; Dinda, A; Talwar, KK; Joshi, S and Gupta, SK. Mechanisms of cardioprotective effect of *Withania somnifera* in experimentally induced myocardial infarction, *Basic Clin. Pharmacol. Toxicol.*2004, *94 (4)*, 184-190.

[33] Visavadiya, NP and Narasimhacharya, AV. Hypocholesteremic and antioxidant effects of *Withania somnifera* (Dunal) in hypercholesteremic rats, *Phytomedicine.* 2007, *14 (2-3)*, 136-142.

[34] Panda, S and Kar, A. Evidence for free radical scavenging activity of Ashwagandha root powder in mice, *Indian J. Physiol. Pharmacol.*1997, *41 (4)*, 424-426.

[35] Arul, V; Miyazaki, S and Dhananjayan, R. Studies on the anti-inflammatory, antipyretic and analgesic properties of the leaves of *Aegle marmelos* Corr, *J. Ethnopharmacol.*2005, *96 (1-2)*, 159-163.

[36] Sachdewa, A; Raina, D; Srivastava, AK and Khemani, LD. Effect of *Aegle marmelos* and *Hibiscus rosa sinensis* leaf extract on glucose tolerance in glucose induced hyperglycemic rats (Charles foster), *J. Environ. Biol.*2001, *22 (1)*, 53-57.

[37] Karunanayake, EH; Welihinda, J; Sirimanne, SR and Sinnadorai, G. Oral hypoglycaemic activity of some medicinal plants of Sri Lanka, *J. Ethnopharmacol.*1984, *11 (2)*, 223-231.

[38] Kamalakkannan, N and Prince, PS. Hypoglycaemic effect of water extracts of *Aegle marmelos* fruits in streptozotocin diabetic rats, *J. Ethnopharmacol.*2003, *87 (2-3)*, 207-210.

[39] Abraham, PM; Paul, J and Paulose, CS. Down regulation of cerebellar serotonergic receptors in streptozotocin induced diabetic rats: Effect of pyridoxine and *Aegle marmelos*e, *Brain Res. Bull.*2010, *82(1-2),* 87-94.

[40] Lam, DD and Heisler, LK. Serotonin and energy balance: molecular mechanisms and implications for type 2 diabetes, *Expert Rev. Mol. Med.*2007, *9 (5)*, 1-24.

[41] Ponnachan, PT; Paulose, CS and Panikkar, KR. Effect of leaf extract of *Aegle marmelose* in diabetic rats, *Indian J. Exp. Biol.*1993, *31 (4)*, 345-347.

[42] Kesari, AN; Gupta, RK; Singh, SK; Diwakar, S and Watal, G. Hypoglycemic and antihyperglycemic activity of *Aegle marmelos* seed extract in normal and diabetic rats, *J. Ethnopharmacol.*2006, *107 (3)*, 374-379.

[43] Karawya, MS; Mirhom, YW and Shehata, IA. Sterols triterpenes, coumarins and alkaloids of *Aegle marmelos* correa, cultivated in Egypt.*Egyptian J. of Pharm. Sci.* 1980, *21 (3-4)*, 239-248.

[44] Narender, T; Shweta, S; Tiwari, P; Papi Reddy, K; Khaliq, T; Prathipati, P; Puri, A; Srivastava, AK; Chander, R; Agarwal, SC and Raj, K. Antihyperglycemic and antidyslipidemic agent from *Aegle marmelos*, *Bioorg. Med. Chem. Lett.*2007, *17 (6)*, 1808-1811.

[45] Ramesh, B and Pugalendi, KV. Antioxidant role of umbelliferone in STZ-diabetic rats, *Life Sci.*2006, *79 (3)*, 306-310.

[46] Sharma, SB; Nasir, A; Prabhu, KM and Murthy, PS. Antihyperglycemic effect of the fruit-pulp of *Eugenia jambolana* in experimental diabetes mellitus, *J. Ethnopharmacol.*2006, *104 (3)*, 367-373.

[47] Sharma, B; Viswanath, G; Salunke, R and Roy, P. Effects of flavonoid-rich extract from seeds of *Eugenia jambolana* (L.) on carbohydrate and lipid metabolism in diabetic mice, *Food Chemistry.*2008, *110 (3)*, 697-705.

[48] Ravi, K; Rajasekaran, S and Subramanian, S. Antihyperlipidemic effect of *Eugenia jambolana* seed kernel on streptozotocin-induced diabetes in rats, *Food Chem. Toxicol.*2005, *43 (9)*, 1433-1439.

[49] Sharma, B; Balomajumder, C and Roy, P. Hypoglycemic and hypolipidemic effects of flavonoid rich extract from *Eugenia jambolana* seeds on streptozotocin induced diabetic rats, *Food Chem. Toxicol.*2008, *46 (7)*, 2376-2383.

[50] Blanquart, C; Barbier, O; Fruchart, JC; Staels, B and Glineur, C. Peroxisome proliferator-activated receptors: regulation of transcriptional activities and roles in inflammation, *J. Steroid Biochem. Mol. Biol.*2003, *85 (2-5)*, 267-273.

[51] Ravi, K; Ramachandran, B and Subramanian, S. Effect of *Eugenia jambolana* seed kernel on antioxidant defense system in streptozotocin-induced diabetes in rats, *Life Sci.*2004, *75 (22)*, 2717-2731.

[52] Shih, CC; Lin, CH and Lin, WL. Effects of *Momordica charantia* on insulin resistance and visceral obesity in mice on high-fat diet, *Diabetes Research and Clinical Practice.*2008, *81 (2)*, 134-143.

[53] Virdi, J; Sivakami, S; Shahani, S; Suthar, AC; Banavalikar, MM and Biyani, MK. Antihyperglycemic effects of three extracts from *Momordica charantia*, *Journal of Ethnopharmacology*.2003, *88 (1)*, 107-111.
[54] Han, C; Hui, Q and Wang, Y. Hypoglycaemic activity of saponin fraction extracted from *Momordica charantia* in PEG/salt aqueous two-phase systems, *Nat. Prod. Res*.2008, *22 (13)*, 1112-1119.
[55] Kubola, J and Siriamornpun, S. Phenolic contents and antioxidant activities of bitter gourd (*Momordica charantia* L.) leaf, stem and fruit fraction extracts in vitro, *Food Chemistry*.2008, *110 (4)*, 881-890.
[56] Ahmed, I; Adeghate, E; Sharma, AK; Pallot, DJ and Singh, J. Effects of *Momordica charantia* fruit juice on islet morphology in the pancreas of the streptozotocin-diabetic rat, *Diabetes Research and Clinical Practice*.1998, *40 (3)*, 145-151.
[57] Welihinda, J; Karunanayake, EH; Sheriff, MH and Jayasinghe, KS. Effect of *Momordica charantia* on the glucose tolerance in maturity onset diabetes, *J. Ethnopharmacol*.1986, *17 (3)*, 277-282.
[58] Sarkar, S; Pranava, M and Marita, R. Demonstration of the hypoglycemic action of *Momordica charantia* in a validated animal model of diabetes, *Pharmacol. Res*.1996, *33 (1)*, 1-4.
[59] Miura, T; Itoh, C; Iwamoto, N; Kato, M; Kawai, M; Park, SR and Suzuki, I. Hypoglycemic activity of the fruit of the *Momordica charantia* in type 2 diabetic mice, *J. Nutr. Sci. Vitaminol. (Tokyo)*, 2001, *47 (5)*, 340-344.
[60] Tan, MJ; Ye, JM; Turner, N; Hohnen-Behrens, C; Ke, CQ; Tang, CP; Chen, T; Weiss, HC; Gesing, ER; Rowland, A; James, DE and Ye, Y. Antidiabetic activities of triterpenoids isolated from bitter melon associated with activation of the AMPK pathway, *Chem. Biol*.2008, *15 (3)*, 263-273.
[61] Shibib, BA; Khan, LA and Rahman, R. Hypoglycaemic activity of *Coccinia indica* and *Momordica charantia* in diabetic rats: depression of the hepatic gluconeogenic enzymes glucose-6-phosphatase and fructose-1,6-bisphosphatase and elevation of both liver and red-cell shunt enzyme glucose-6-phosphate dehydrogenase, *Biochem. J*.1993, *292 (Pt 1)*, 267-270.
[62] Dans, AM; Villarruz, MV; Jimeno, CA; Javelosa, MA; Chua, J; Bautista, R and Velez, GG. The effect of *Momordica charantia* capsule preparation on glycemic control in type 2 diabetes mellitus needs further studies, *J. Clin. Epidemiol*.2007, *60 (6)*, 554-559.
[63] Chuang, CY; Hsu, C; Chao, CY; Wein, YS; Kuo, YH and Huang, CJ. Fractionation and identification of 9c, 11t, 13t-conjugated linolenic acid as an activator of PPARalpha in bitter gourd (*Momordica charantia* L.), *J. Biomed. Sci*.2006, *13 (6)*, 763-772.
[64] Wu, SJ and Ng, LT. Antioxidant and free radical scavenging activities of wild bitter melon (*Momordica charantia* Linn. var. *abbreviata* Ser.) in Taiwan, *Lwt-Food Science and Technology*.2008, *41 (2)*, 323-330.
[65] Yanpallewar, SU; Rai, S; Kumar, M and Acharya, SB. Evaluation of antioxidant and neuroprotective effect of *Ocimum sanctum* on transient cerebral ischemia and long-term cerebral hypoperfusion, *Pharmacology Biochemistry and Behavior*.2004, *79 (1)*, 155-164.
[66] Sethi, J; Sood, S; Seth, S and Talwar, A. Protective effect of Tulsi (*Ocimum Sanctum*) on lipid peroxidation in stress induced by anemic hypoxia in rabbits, *Indian J. Physiol. Pharmacol*.2003, *47 (1)*, 115-119.

[67] Sood, S; Narang, D; Thomas, MK; Gupta, YK and Maulik, SK. Effect of *Ocimum sanctum* Linn. on cardiac changes in rats subjected to chronic restraint stress, *J. Ethnopharmacol.*2006, *108 (3)*, 423-427.

[68] Vats, V; Grover, JK and Rathi, SS. Evaluation of anti-hyperglycemic and hypoglycemic effect of *Trigonella foenum-graecum* Linn, *Ocimum sanctum* Linn and *Pterocarpus marsupium* Linn in normal and alloxanized diabetic rats, *J. Ethnopharmacol.*2002, *79 (1)*, 95-100.

[69] Rai, V; Iyer, U and Mani, UV. Effect of Tulasi (*Ocimum sanctum*) leaf powder supplementation on blood sugar levels, serum lipids and tissue lipids in diabetic rats, *Plant Foods Hum. Nutr.*1997, *50 (1)*, 9-16.

[70] Sarkar, A; Lavania, SC; Pandey, DN and Pant, MC. Changes in the blood lipid profile after administration of *Ocimum sanctum* (Tulsi) leaves in the normal albino rabbits, *Indian J. Physiol. Pharmacol.*1994, *38 (4)*, 311-312.

[71] Godhwani, S; Godhwani, JL and Vyas, DS. *Ocimum sanctum*: an experimental study evaluating its anti-inflammatory, analgesic and antipyretic activity in animals, *J. Ethnopharmacol.*1987, *21 (2)*, 153-163.

[72] Singh, S; Majumdar, DK and Rehan, HM. Evaluation of anti-inflammatory potential of fixed oil of *Ocimum sanctum* (Holybasil) and its possible mechanism of action, *J. Ethnopharmacol.*1996, *54 (1)*, 19-26.

[73] Singh, S; Rehan, HM and Majumdar, DK. Effect of *Ocimum sanctum* fixed oil on blood pressure, blood clotting time and pentobarbitone-induced sleeping time, *J. Ethnopharmacol.*2001, *78 (2-3)*, 139-143.

[74] Suanarunsawat, T; Devakul Na Ayutthaya, W; Songsak, T; Thirawarapan, S and Poungshompoo, S. Antioxidant activity and lipid-lowering effect of essential oils extracted from *Ocimum sanctum* L. leaves in rats fed with a high cholesterol diet, *J. Clin. Biochem. Nutr.*2010, *46 (1)*, 52-59.

[75] Nishijima, H; Uchida, R; Kameyama, K; Kawakami, N; Ohkubo, T and Kitamura, K. Mechanisms mediating the vasorelaxing action of eugenol, a pungent oil, on rabbit arterial tissue, *Japanese Journal of Pharmacology.*1999, *79 (3)*, 327-334.

[76] Leach, MJ. *Gymnema sylvestre* for diabetes mellitus: a systematic review, *J. Altern. Complement. Med.*2007, *13 (9)*, 977-983.

[77] Yadav, M; Lavania, A; Tomar, R; Prasad, GB; Jain, S and Yadav, H. Complementary and comparative study on hypoglycemic and antihyperglycemic activity of various extracts of *Eugenia jambolana* seed, *Momordica charantia* fruits, *Gymnema sylvestre*, and *Trigonella foenum graecum* seeds in rats, *Appl. Biochem. Biotechnol.*2010, *160 (8)*, 2388-2400.

[78] Daisy, P; Eliza, J and Mohamed Farook, KA. A novel dihydroxy gymnemic triacetate isolated from *Gymnema sylvestre* possessing normoglycemic and hypolipidemic activity on STZ-induced diabetic rats, *J. Ethnopharmacol.*2009, *126 (2)*, 339-344.

[79] Ahmed, AB; Rao, AS and Rao, MV. *In vitro* callus and *in vivo* leaf extract of *Gymnema sylvestre* stimulate beta-cells regeneration and anti-diabetic activity in Wistar rats, *Phytomedicine*. 2010, *17(13)*, 1033-1039.

[80] Persaud, SJ; Al-Majed, H; Raman, A and Jones, PM. *Gymnema sylvestre* stimulates insulin release in vitro by increased membrane permeability, *J. Endocrinol.*1999, *163 (2)*, 207-212.

[81] Sugihara, Y; Nojima, H; Matsuda, H; Murakami, T; Yoshikawa, M and Kimura, I. Antihyperglycemic effects of gymnemic acid IV, a compound derived from *Gymnema sylvestre* leaves in streptozotocin-diabetic mice, *J. Asian Nat. Prod. Res.*2000, *2 (4)*, 321-327.

[82] Shanmugasundaram, ER; Gopinath, KL; Radha Shanmugasundaram, K and Rajendran, VM. Possible regeneration of the islets of Langerhans in streptozotocin-diabetic rats given *Gymnema sylvestre* leaf extracts, *J. Ethnopharmacol.*1990, *30 (3)*, 265-279.

[83] Nakamura, Y; Tsumura, Y; Tonogai, Y and Shibata, T. Fecal steroid excretion is increased in rats by oral administration of gymnemic acids contained in *Gymnema sylvestre* leaves, *J. Nutr.*1999, *129 (6)*, 1214-1222.

[84] Preuss, HG; Jarrell, ST; Scheckenbach, R; Lieberman, S and Anderson, RA. Comparative effects of chromium, vanadium and *Gymnema sylvestre* on sugar-induced blood pressure elevations in SHR, *J. Am. Coll. Nutr.*1998, *17 (2)*, 116-123.

[85] Kaviarasan, S; Sundarapandiyan, R and Anuradha, CV. Protective action of fenugreek (*Trigonella foenum graecum*) seed polyphenols against alcohol-induced protein and lipid damage in rat liver, *Cell Biol. Toxicol.*2008, *24 (5)*, 391-400.

[86] Annida, B and Stanely Mainzen Prince, P. Supplementation of fenugreek leaves reduces oxidative stress in streptozotocin-induced diabetic rats, *J. Med. Food.*2005, *8 (3)*, 382-385.

[87] Mowla, A; Alauddin, M; Rahman, MA and Ahmed, K. Antihyperglycemic effect of *Trigonella foenum-graecum* (fenugreek) seed extract in alloxan-induced diabetic rats and its use in diabetes mellitus: a brief qualitative phytochemical and acute toxicity test on the extract, *Afr. J. Tradit. Complement Altern. Med.*2009, *6 (3)*, 255-261.

[88] Belguith-Hadriche, O; Bouaziz, M; Jamoussi, K; El Feki, A; Sayadi, S and Makni-Ayedi, F. Lipid-lowering and antioxidant effects of an ethyl acetate extract of fenugreek seeds in high-cholesterol-fed rats, *J. Agric Food Chem. 58 (4)*, 2116-2122.

[89] Dixit, P; Ghaskadbi, S; Mohan, H and Devasagayam, TP. Antioxidant properties of germinated fenugreek seeds, *Phytother. Res.*2005, *19 (11)*, 977-983.

[90] Genet, S; Kale, RK and Baquer, NZ. Alterations in antioxidant enzymes and oxidative damage in experimental diabetic rat tissues: effect of vanadate and fenugreek (*Trigonellafoenum graecum*), *Mol. Cell Biochem.*2002, *236 (1-2)*, 7-12.

[91] Sushma, N and Devasena, T. Aqueous extract of *Trigonella foenum graecum* (fenugreek) prevents cypermethrin-induced hepatotoxicity and nephrotoxicity, *Hum. Exp. Toxicol. 29 (4)*, 311-319.

[92] Haeri, MR; Izaddoost, M; Ardekani, MR; Nobar, MR and White, KN. The effect of fenugreek 4-hydroxyisoleucine on liver function biomarkers and glucose in diabetic and fructose-fed rats, *Phytother. Res.*2009, *23 (1)*, 61-64.

[93] Vijayakumar, MV; Singh, S; Chhipa, RR and Bhat, MK. The hypoglycaemic activity of fenugreek seed extract is mediated through the stimulation of an insulin signalling pathway, *Br. J. Pharmacol.*2005, *146 (1)*, 41-48.

[94] Uemura, T; Hirai, S; Mizoguchi, N; Goto, T; Lee, JY; Taketani, K; Nakano, Y; Shono, J; Hoshino, S; Tsuge, N; Narukami, T; Takahashi, N and Kawada, T. Diosgenin present in fenugreek improves glucose metabolism by promoting adipocyte differentiation and inhibiting inflammation in adipose tissues, *Mol. Nutr. Food Res.* 2010,*54(11)*, 1596-1608.

[95] Shishodia, S and Aggarwal, BB. Diosgenin inhibits osteoclastogenesis, invasion, and proliferation through the downregulation of Akt, I kappa B kinase activation and NF-kappa B-regulated gene expression, *Oncogene.*2006, *25 (10)*, 1463-1473.
[96] Gullo, VP; McAlpine, J; Lam, KS; Baker, D and Petersen, F. Drug discovery from natural products, *J. Ind. Microbiol. Biotechnol.*2006, *33 (7)*, 523-531.
[97] Yap, KY; Chan, SY; Weng Chan, Y and Sing Lim, C. Overview on the analytical tools for quality control of natural product-based supplements: a case study of ginseng, *Assay Drug Dev. Technol.*2005, *3 (6)*, 683-699.
[98] Iyer, A; Panchal, S; Poudyal, H and Brown, L. Potential health benefits of Indian spices in the symptoms of the metabolic syndrome: a review, *Indian J. Biochem. Biophys.*2009, *46 (6)*, 467-481.
[99] Poltanov, EA; Shikov, AN; Dorman, HJ; Pozharitskaya, ON; Makarov, VG; Tikhonov, VP and Hiltunen, R. Chemical and antioxidant evaluation of Indian gooseberry (*Emblica officinalis* Gaertn., syn. *Phyllanthus emblica* L.) supplements, *Phytother. Res.*2009, *23 (9)*, 1309-1315.
[100] Duan, W; Yu, Y and Zhang, L. Antiatherogenic effects of *Phyllanthus emblica* associated with corilagin and its analogue, *Yakugaku Zasshi.*2005, *125 (7)*, 587-591.
[101] Nosal'ova, G; Mokry, J and Hassan, KM. Antitussive activity of the fruit extract of *Emblica officinalis* Gaertn. (Euphorbiaceae), *Phytomedicine,* 2003, *10 (6-7)*, 583-589.
[102] Sai Ram, M; Neetu, D; Yogesh, B; Anju, B; Dipti, P; Pauline, T; Sharma, SK; Sarada, SK; Ilavazhagan, G; Kumar, D and Selvamurthy, W. Cyto-protective and immunomodulating properties of Amla (*Emblica officinalis*) on lymphocytes: an in-vitro study, *J. Ethnopharmacol.*2002, *81 (1)*, 5-10.
[103] Tasduq, SA; Kaisar, P; Gupta, DK; Kapahi, BK; Maheshwari, HS; Jyotsna, S and Johri, RK. Protective effect of a 50% hydroalcoholic fruit extract of *Emblica officinalis* against anti-tuberculosis drugs induced liver toxicity, *Phytother. Res.*2005, *19 (3)*, 193-197.
[104] Kulkarni, SK and Dhir, A. Withania somnifera: an Indian ginseng, Prog. Neuropsychopharmacol. Biol. Psychiatry.2008, 32 (5), 1093-1105.
[105] Gupta, SK; Mohanty, I; Talwar, KK; Dinda, A; Joshi, S; Bansal, P; Saxena, A and Arya, DS. Cardioprotection from ischemia and reperfusion injury by *Withania somnifera*: a hemodynamic, biochemical and histopathological assessment, *Mol. Cell Biochem.*2004, *260 (1-2)*, 39-47.
[106] Mahmoud, II; Marzouk, MS; Moharram, FA; El-Gindi, MR and Hassan, AM. Acylated flavonol glycosides from *Eugenia jambolana* leaves, *Phytochemistry.*2001, *58 (8)*, 1239-1244.
[107] Middleton, E, Jr.; Kandaswami, C and Theoharides, TC. The effects of plant flavonoids on mammalian cells: implications for inflammation, heart disease, and cancer, *Pharmacol. Rev.*2000, *52 (4)*, 673-751.
[108] Grover, JK and Yadav, SP. Pharmacological actions and potential uses of *Momordica charantia*: a review, *J. Ethnopharmacol.*2004, *93 (1)*, 123-132.
[109] Raj, SK; Khan, MS; Singh, R; Kumari, N and Prakash, D. Occurrence of yellow mosaic geminiviral disease on bitter gourd (*Momordica charantia*) and its impact on phytochemical contents, *Int. J. Food Sci. Nutr.*2005, *56 (3)*, 185-192.
[110] Lako, J; Trenerry, VC; Wahlqvist, M; Wattanapenpaiboon, N; Sotheeswaran, S and Premier, R. Phytochemical flavonols, carotenoids and the antioxidant properties of a

wide selection of Fijian fruit, vegetables and other readily available foods, *Food Chemistry.*2007, *101 (4)*, 1727-1741.

[111] Prakash, P and Gupta, N. Therapeutic uses of *Ocimum sanctum* Linn (Tulsi) with a note on eugenol and its pharmacological actions: a short review, *Indian J. Physiol. Pharmacol.*2005, *49 (2)*, 125-131.

[112] Norr, H and Wagner, H. New constituents from *Ocimum-sanctum*, *Planta Medica,* 1992, *58 (6)*, 574-574.

[113] Gupta, P; Yadav, DK; Siripurapu, KB; Palit, G and Maurya, R. Constituents of *Ocimumsanctum* with antistress activity, *J. Nat. Prod.*2007, *70 (9)*, 1410-1416.

In: Natural Products and Their Active Compounds ... ISBN: 978-1-62100-153-9
Editors: M. Essa, A. Manickavasagan, and E. Sukumar © 2012 Nova Science Publishers, Inc.

Chapter 2

EVIDENCE BASED THERAPY ON "AROGH", A HERBAL TEA FORMULATION

Anoop Austin*[*] *and P. Thirugnanasambantham
Rumi Herbals R&D Centre, Ohri Salai, Mugappair East,
Chennai, Indua

ABSTRACT

Plant based products are widely used for various ailments, either as a food supplement, prescription drug or over the counter prescription. Scientific facts on these products are warranted, which would help us to develop novel, safe and effective formulation for humans, for wider global acceptance. Though many formulations on herbal products are available, scientific rationale and periodical evaluation of these formulations are lacking. Arogh, a polyherbal formulation, is a product manufactured by Rumi Herbals, Chennai, which was scientifically validated. Arogh is a herbal tea formulation made out of *Nelumbo nucifera, Hibiscus rosa-sinensis, Rosa alba, Terminalia chebula, Hemidesmus indicus, Glycyrrhiza glabra*, *Zingiber officinale*, *Quercus infectoria* and *Eclipta alba.* Scientific validation revealed that the formulation was effective against hypercholesteremia, anxiety, hyperglycemia, Obesity, Stress, Myocardial infraction and Oxidative stress. The scientific validation of Arogh, revealed as a cardiotonic and a remedy to cater cardiac and associated ailments.

INTRODUCTION

Nature has provided us with an excellent storehouse of remedies, to cure all the ailments of mankind (Marderosian and Beutler, 2000). In ancient days, almost all the medicines used were from natural sources, particularly from plants. Plants continue to be an important source of new drugs even today (Chevallier, 1996). The importance of botanical, chemical and pharmacological evaluation of plant-derived agents used in the treatment of human ailments

[*] Anoop Austin: Rumi Herbals R & D centre, 40/41, Spartan Avenue, Mugappair East, Chennai – 600 037; India; Email: anoopaustin@gmail.com

has been increasingly recognized in the last decades (Evans, 2005). Herbal remedies are widely used for the treatment and prevention of various diseases and often contain highly active multitude of chemical compounds (Lucas, 1977). Modern research is now focusing greater attention on the generation of scientific validation of herbal drugs, based upon their folklore claim (Murray, 1995). In this modern era, a large Indian population still relies on the traditional system of medicine, which is mostly plant based (Satyavati, 1988).

Further, there is little doubt that Traditional Medicines have been utilized since antiquity in the health care. However, with the advent of the pharmaceutical industry early in this century, the popularity of traditional/herbal medicine declined, in spite of the fact that twenty five percent of all prescription drugs still contain ingredients isolated from plants. The resources now do exist, which can help and assist for greater understanding of the ways in which herbs can facilitate health and restore balance in disease (Murray and Pizzorno, 1991). The global herbal cornucopia represents an eclectic collection of the most authentic early medicines that even today continued to prevent and cure diseases. A major portion of the global population in developing countries still relies on botanical drugs to meet its health needs. The attention paid by health authorities to the use of herbal medicines has increased considerably, both because they are often the only medicine available in less developed areas and because they are becoming a popular alternative treatment in more developed areas. Thus herbal medicines have been given a valuable status and readily available products for primary health care, and WHO has endorsed their safe and effective use (Anonymous, (1993).

It is one of the peculiarities of herbal drugs that their indications have for the most part been determined empirically. The reason is easily understood because most herbal drugs have been used for a long time to alleviate or cure illnesses and more especially disorders. Their introduction in therapeutics happened at a time when "Pharmacodynamics" and "Pharmacokinetics" were unknown concepts, when there was no "Medicine Act" to require proof of the quality, efficacy, and safety of herbal medicines. Today, when introducing a new medicine, extensive investigations are required in the interest of safety. The requirements for the proof of activity of a drug appears to be superfluous; but nevertheless, as a representative of a scientifically oriented pharmaceutical science, one strives to pluck herbal drugs out of their present level of pure empiricism and by elucidating their active principles give their application a more secure basis (Wichtl, 1994).

Herbal drug/s constitutes mainly traditional medicines that primarily use medicinal plant preparations for therapy. Indian subcontinent is a vast repository of medicinal plants that are used in traditional medical treatments (Chopra *et al.,* 1956), which also forms a rich source of knowledge (Cox and Balick, 1994). The various indigenous systems such as Siddha, Ayurveda, Unani and Allopathy use several plant species to treat different ailments (Rabe and Staden, 1997).

In India around 20,000 medicinal plant species have been recorded recently (Dev, 1997), but more than 500 traditional communities use about 800 plant species for curing different diseases (Kamboj, 2000). Currently 80 % of the world population depends on plant-derived medicine for the first line of primary health care for human alleviation because it has no side effects (Farnsworth and Bingel, 1997).

Plants are important sources of medicines and presently about 25% of pharmaceutical prescriptions in the United States contain at least one plant-derived ingredient. In the last century, roughly 121 pharmaceutical products were formulated based on the traditional knowledge obtained from various sources (Vinoth Kumar *et. al.,* 2010).

Medicinal plants play a vital role in the development of new drugs. During 1950-1970 around 100 plants based new drugs were introduced in the International drug market including Deserpidine, Reseinnamine, Reserpine, Vinblastine and Vincristine which are derived from higher plants. From 1971 to 1990, new drugs such as Ectoposide, Guggulsterone, Teniposide, Nabilone, Plaunotol, Z-guggulsterone, Lectinan, Artemisinin and Ginkgolides appeared all over the world. 2 % of drugs were introduced from 1991 to 1995 including Paciltaxel, Toptecan, Gomishin, Irinotecan etc. (Cox, 2009). Plant based drugs provide outstanding contribution to modern therapeutics; for example: serpentine isolated from the root of Indian plant *Rauwolfia serpentina* in 1953, was a revolutionary event in the treatment of hypertension and lowering of blood pressure. Vinblastine isolated from the *Catharanthus rosesus* (Farnsworth and Blowster, 1967) is used for the treatment of Hodgkins, Chorio-carcinoma, Non-Hodgkins Lymphomas, Leukemia in children, Testicular and neck cancer. Vincristine is recommended for acute lymphocytic leukemia in childhood advanced stages of Hodgkins, Lymophosarcoma, small cell lung, cervical and breast cancer (Farnsworth and Bingel, 1997).

Phophyllotoxin is a constituent of *Phodophyllum emodi* currently used against testicular, small cell lung cancer and lymphomas. Indian indigenous tree of *Mappia foetida* are mostly used in Japan for the treatment of cervical cancer. Plant derived drugs are used to cure mental illness, skin diseases, tuberculosis, diabetes, jaundice, hypertension and cancer (Perumal Samy and Ignacimuthu, 1998).

Antidote activity against *Daboia russellii* and *Naja kaouthia* venom was neutralized by lupeol acetate isolated from *Hemidesmus indicus* R.Br (Alam et al., 1994; Chatterjee, et al., 2006). *Strychnus nux-vomica* seed extract also inhibited viper venom induced lipid peroxidation (Chatterjee *et al.,* 2004). Medicinal plants play an important role in the development of potent therapeutic agents. Plant derived drugs came into use in the modern medicine through the uses of plant material as indigenous cure in folklore or traditional systems of medicine (Adailkan and Gauthaman, 2001; Heinrich, 2000; Pfister et al., 2002).

With this background, the formulated polyherbal tea preparation Arogh was selected and subjected for various scientific validation, pertaining to drug/formulation standardization, safety and efficacy studied with respect to biological studies with respect to stress, anxiety, cholesterol, ischemic heart disease, diabetes, obesity, hypertension etc. Scientific validation of this formulation revealed many interesting findings through scientific studies, which helped the product for better segmentation and novel findings more appropriately.

INGREDIENTS

This formulation is a polyherbal formulation made out of Medicinal Plants, widely practiced by Indian Systems of Medicine, This is a coarse powder, formulated using the dried petals of *Nelumbo nucifera, Hibiscus rosa-sinensis and Rosa alba,* fruit rind of *Terminalia chebula,* roots of *Hemidesmus indicus* and *Glycyrrhiza glabra*, rhizomes of *Zingiber officinale*, galls from *Quercus infectoria* and the whole plant of *Eclipta alba.*

The formulation has to be prepared like a hot drink, by taking 2.5 gm of the coarse powder boiled with 100 ml of water and filtered. To the filtrate milk and sweetener has to be added and consumed as an herbal hot drink.

Standardization

Standardization is an integral part of drug development (Kokate, 1994). The formulation appear to be a coarse powder, characterized by rough coarse bits with roasted Sarsaparilla and Citrus flavour, having sour taste. Under day light it appears as light rosy brown, under UV light pinkish brown, which with 50 % HCl under UV as light fluorescent green pink specs and dark fluorescent green colour with 10 % NaOH under UV. The ash content varies from 8.6 – 11.5 %, acid insoluble matter from 2 – 4 %, water solubility from 15.2 – 20 %, loss on drying at 105° C from 4.7 – 9.2 %, pH at 5 % w/v solution from 4.2 – 4.5, Iron content from 1.1 – 1.4 % and content of calcium from 0.4 – 0.7 %. The specifications laid down are characterized towards in-house standards and can be adopted as a basis for the formulation, which can help in sustained efficacy of the formulation.

Heavy Metals

Heavy metals and its safety issues are being dealt in a very serious manner, worldwide. To meet the emergency enforced by the declaration of developed nations and their fear proclaimed by them on our system of medicine. The content of heavy metals is an important area in plant products, as the plant can be contaminated by pesticides, land and water pollutants, which in turn would get stored in the plant products and produce cumulative adverse toxic events. Heavy metals were estimated with Arogh, using Atomic Absorption Spectroscopy which revealed that the formulations were found to be free from Lead, Arsenic, Mercury and Silica, which is a safety index in any herbal formulation, advised by WHO (Anonymous, 1993).

Pesticidal Residue

Medicinal plant materials are liable to contain pesticide residues which accumulate from agricultural practices, such as spraying, treatment of soils during cultivation, and administration of fumigants during storage. It is therefore recommended that every country producing medicinal plant materials (naturally grown or cultivated) should have at least one control laboratory capable of performing the determination of pesticides in accordance with the procedure outlined below.

Since many medicinal preparations of plant origin are taken over long periods of time, limits for pesticide residues should be established following the recommendations of the Food and Agriculture Organization of the United Nations (FAO) and the World Health Organization (WHO) which have already been established for food and animal feed (Wichtl, 1994).

These recommendations include the analytical methodology for the assessment of specific pesticide residues. Presence of pesticides (organochlorine) was screened by thin layer chromatography, for Arogh and was found to be negative. This further confirms the safety of the formulation for human consumption.

MICROBIAL LOAD

Medicinal plant materials normally carry a great number of bacteria and moulds, often originating in soil. While a large range of bacteria and fungi form the naturally occurring microflora of herbs, aerobic spore-forming bacteria frequently predominate. Current practices of harvesting, handling and production may cause additional contamination and microbial growth. The determination of *Escherichia coli* and moulds may indicate the quality of production and harvesting practices. In addition, the presence of aflatoxins in plant material can be hazardous to health if absorbed even in very small amounts. They should therefore be determined after using a suitable clean-up procedure. The formulation was free from *E. coli, Salmonella* and the total bacterial count was less than 1000 CFU/gm. The yeast and moulds was found to be less than 100 CFU/gm. Aflotoxins were also found to be negative. The findings revealed the safety of the formulation with respect to its microbial load.

FINGER PRINTING

Standardization of traditional medicine has become mandatory in the present National and International scientific scenario, as they have to stand competing with stringent regulatory methods and also clinically. HPTLC is one of the versatile chromatographic methods presently available for the rapid analysis of herbal drugs due to several reasons. Firstly, the time required for the demonstration of the most of the characteristic constituents of a drug is very quick and short. Secondly, in addition to qualitative detection, HPTLC also provides semi-quantitative information on the major active constituents of a drug, thus enabling an assessment of drug quality. Thirdly, the fingerprint obtained is suitable for monitoring the identity and purity of drugs and for detecting adulteration and substitution. The distribution of phyto-constituents depends on various factors such as soil, time of collection period of storage, etc. So, it is necessary to standardize the formulation for sustained effects. HPTLC serves as a convenient tool for finding out the distribution pattern of phyto constituents, which are unique for plant based products.

The chromatogram of the formulation was developed from methanol using hot percolation method using soxhlet apparatus. Silica gel 60 F 254 used as stationary phase and Methanol, Chloroform, Water and acetic acid in the ratio 2:7:0.5:0.5 was used as mobile Phase. The chromatogram revealed 3 peaks with Rf of 0.04, 0.08 and 0.11. The peak corresponding to Rf 0.04 had maximum area of 54.46 %, whereas Rf 0.08 had an area of 41.76 % and the remaining area of 3.77 % corresponding to Rf 0.11 respectively.

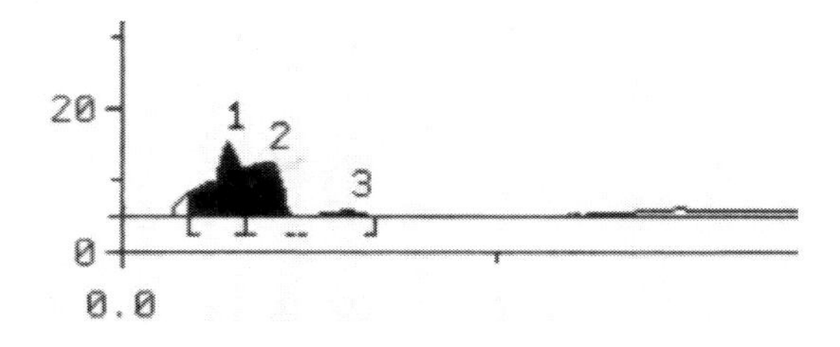

Chromatogram.

STABILITY STUDY

Inadequate storage and distribution of pharmaceutical products can lead to their physical deterioration and chemical decomposition, resulting in reduced activity and, occasionally, in the formation of toxic degradation products.

Degradation is particularly likely to occur under tropical conditions of high ambient temperature and humidity; and it is not widely recognized that, because of the potential for chemical interaction between the active ingredients and excipients, drug dosage forms can be more vulnerable to degradation than pure drug substances.

The stability of a specific product is dependent, in a large measure, on its formulation, and its expiry date should be determined on the basis of stability studies Guidelines for stability testing of pharmaceutical products containing well established drug substances in conventional dosage forms were adopted by the WHO Expert Committee on Specifications for Pharmaceutical Preparations.

Recognizing that stability testing represents the evaluation of a pharmaceutical formulation in its final container, the Expert Committee emphasized that the same fundamental approach should be used for all products irrespective of whether the active ingredient was an established drug substance.

Where sufficient information was already available on the chemical stability of the active ingredient, however, this could be taken into account. The availability of these guidelines was considered to be of special importance since they include advice on the stability testing of products for use in the more extreme climatic conditions found in many developing countries.

Accelerated stability studies were carried out on long- established essential drug substances under standardized conditions (e.g. 30 days' exposure to air at a temperature of 40 to 45° C and a relative humidity of 80 %).

The substances should additionally exposed to a temperature of 70° C under the same humidity conditions for a further 3-5 days. Negative results provided conclusive proof of the stability of the substance even under highly adverse conditions. All tests were carried out with light excluded since it is easy to protect substances from light during storage.

Accelerated stability studies carried out at an ambient temperature of 45° C and 80 ± 2 % relative humidity for a period of 30 days, revealed intactness of the formulation. Colour was unaltered and it was light brown colour coarse powder material with characteristic smell and tastes bitter.

With various reagents it developed different colours. Under UV it developed light pinkish brown colour and with 50 % HCl under UV light fluorescent green pink specs and with 10 % NaOH under UV Dark fluorescent green colour. Loss on drying revealed a gradual increase due to drying nature ie from 10.14 – 10.42 %, where as total ash, acid insoluble matter and water solubility were unaltered. Bulk density there was a reduction from 0.18 to 0.17 gm/ml after 30 days.

The content of tannins was reduced from 176 μg to 164 μg after 30 days. The study revealed intactness of the formulation and can be fixed an expiry date of 1 year for which real time stability will be more appropriate.

Pharmacological Studies

Cholesterol Lowering Effect

Cholesterol and atherosclerosis are interlinked (Muldoon *et al.* 1990) and later established that hyperlipidemia as an independent risk factor for coronary artery disease. But now hyperlipidaemia is being proved an as independent risk factor for ischemic stroke. Additional evidences from the prospective studies (Gotto, 1985) have shown the relationship between plasma cholesterol levels and risk of stroke. Reduction in the plasma cholesterol (Rossouw *et al.,* 1990) is accompanied by significant decrease in the incidence of coronary artery disease and stroke. Though an array of hypolipidemics is available they have documented ill effects. This herbal tea formulation is used in the management of obesity and hypertension, which made to conduct this study. Ganesh *et al.,* (2006) studied the effect of Arogh, in Cholesterol fed Wistar albino rats at a dose of 750 and 1500 mg and compared its activity with Gemfibrisol at a dose of 100 mg. The study revealed a remarkable reversal of body weight in comparison with control group, and was comparable with Gemfibrisol. Arogh at 750 mg has 9.7 % of increase, and 1500 mg had 10 % of weight gain, whereas Gemfibrisol had 11 % of weight gain within 60 days. Apparent decrease in cholesterol level with low dose and significant decrease with 1500 mg dose. Elevation of serum triglycerides were effectively prevented both by Arogh and gemfibrisol. The increase in LDK and VLDL cholesterol due to cholesterol was also reverted by Arogh and HDL cholesterol level was not much altered.

The benefits of hypolipidemic therapy may be effected through alteration in one or more parameters of lipid profile. A reduction in total cholesterol, triglycerides, LDL and VLDL or an increase in HDL may be affected by hypolipedemic agents. In this context, Arogh treatment lowered all the parameters of lipid profile except modifying HDL levels. The activity might be exerted pertaining to its hypolidpidemic activity by multi-dimensional manner either by interfering with intestinal absorbsion of cholesterol, altering lipid metabolism, in liver and other tissues or by preventing oxidative damage of LDL and other lipoptoteins.

Another study was conducted by Asokan *et al.* (2010) with Arogh at a dose of 500 mg, 1 gm and 2 gm which was compared with Atorvastin at a dose of 5 mg revealed that the formulation significantly reduced the high cholesterol diet induced changes *viz.* TC by 28%, TGL by 19%, LDL by 41%, VLDL significantly. Significant increase in HDL was evident. The histopathological changes induced by high cholesterol fed diet were almost reversed to near normal in poly herbal formulation treated animals. High cholesterol diet elicited 24 % increase in the body weight. Arogh decreased the body weight with high cholesterol induced diet (ranging from 7 % to 19 %) and Atorvastin reduced by 6 %. In lipid profile, the total cholesterol increased by 17 %. A dose dependant response was observed. Similar to total cholesterol, TGL (27 %) and LDL (56 %) levels were increased in high cholesterol diet fed animals, which were antagonized by Arogh (TGL 19 %, LDL 41 %) in a dose related manner. High Density Lipoprotein (HDL) was reverted to a maximum of 21 % but, the effect was not dose related. With respect to Very Low Density Lipoprotein (VLDL), high cholesterol diet produced an increase, (25%) which was attenuated by atorvastin (5%). In contrast, a dose related reduction in VLDL by poly herbal formulation (17 %) was noticed. In this study also the observations with Arogh treatment were comparable with atorvastin, and was able to reverse increased body weight, lipid profile, total histo-pathological changes in the aorta which were produced by high cholesterol rich diet.

Reconfirming earlier studies, histopathological studies carried out by Ganesh *et al.,* (2006) on the heart tissuee also revealed thickening of vessels with inflammatory infiltrate and hyalinization of fibres and perivascular inflammatory infiltrates and atheromatorus changes, whcih were developed due to the administration of cholesterol rich diet were reversed by Arogh at 750 and 1500 mg dosages. The blood vessels developing lipid laden intimal smooth muscle cells, accumulation of macrophages, elastic fibres and intracellular and extracellular lipid deposits due to cholesterol rich diet were reverted by Arogh. Cholesterol rich diet in liver developed prominent fatty changes of hepatocytes, focal fatty changes with nuclear condensation of hepatocytes were also reverted by Arogh, whereas, Kidney section, revealed maintenance of its normal architecture with Arogh treatment.

Yet another study by Asokan *et al.* (2010) in different dose pattern were also in line with our above findings. Prominent fatty changes of hepatocytes with kuffer cell hyperplasia and diffuse fatty degeneration of the liver in high cholesterol diet fed animals were reverted by Arogh. In aorta, high cholesterol diet induced diffuse sub minimal fatty changes with proliferation of tunica media. These changes were minimal in Arogh treated group. With 2 gm of Arogh treated animals, a normal histopathology of aorta was noticed and apparently in no significant change in myocardial cells.

Vaibhav Patel, *et al.,* (2008) carried out another study by injecting Triton for inducing hypercholesteremia. Serum parameters including total cholesterol, HDL cholesterol, LDL cholesterol, VLDL cholesterol, triglycerides, total lipids, total protein, HDL ratio, LDL ratio, C:P ratio, L:P ratio and atherogenic index were estimated. Liver was estimated for total cholesterol, triglycerides, total lipids and total protein. Change in body weight was also measured. Treatment with Arogh special at three dose levels 100 mg, 300 mg and 1 gm/kg p.o. was given for 28 days prior to triton injection. Atorvastin at 5 mg/kg was given as a standard drug. Significant increase was found in serum and liver parameters with triton injection whereas drug treatment was found to prevent it significantly. The body weigh was not decreased in this study. Total cholesterol was not much altered with 100 mg dose, whereas 300 and 1000 mg dose significantly reduced and were comparable with standard. HDL was increased in a dose dependant manner. LDL was unaltered whereas VLDL was reduced. The raise in triglyceride was prevented by Arogh. Total lipids were prevented. Cholesterol protein ration was unaltered in 100 mg dose, but it was prevented at 300 and 1000 mg dosages, whereas, on the contrary lipid protein ratio, it was effective in all the three doses tested. The liver lipids were also prevented by the drug. Arogh significantly prevents the raise of serum and liver lipid levels. The effect on LDL was not marked, whereas it was effective against total cholesterol, triglycerides, total lipids and VLDL. Further Arogh might act at one stage of cholesterol turn over in the body, which might be at absorption of cholesterol or action on the oxidation of LDL, or increased secretion of cholesterol from body.

Clinical study carried out by Anoop Austin *et al.* (2006) in human subjects also revealed, similar observations. Twenty patients were scrutinized and followed fairly uniform dietary patterns and complied with routine interviews and biochemical follow-ups. Routine follow-up resulted in a good overall dietary compliance and accounted for the reversal of elevated cholesterol profile. Treatment revealed a reduction of cholesterol from 216.25 to 191.30 mg/dl., Triglycerides levels reduced from 215 to 185.65 mg/dl. And on the contrary HDL was increased. Similarly, LDL reduced from 134.30 to 114.45 mg/dl. VLDL levels reduced from

40.34 to 37.60 mg/dl. The effect of the formulation can be attributed to the synergistic effect (Anoop Austin and Jegadeesan, 2003) of the ingredients present in the formulation in reducing the elevated LDL, triglycerides and elevating HDL.

Anxiety and Stress Reliever

Anxiety is an unpleasant emotional state characterized by apprehension and nervousness, which is a normal behavior. Whenever it becomes severe or chronic, it becomes pathological and can precipitate or aggravate cardiovascular and psychiatric disorders. The etiology of most anxiety disorders although not fully understood, it has come into sharper focus in recent days. Balaraman *et al.* (2007) studied the anxiolytic activity of Arogh at a dose of 30, 100, 300 and 500 mg using various behavioural paradigms, such as Elevated plus maze (EPM), Light/dark apparatus (LDA), Open field apparatus (OFA) and Hole board apparatus (HBA). Diazepam (1 mg/kg) was used as a standard anxiolytic drug. The effect of Arogh at a dose of 100 and 300 mg was further evaluated using serotonin, dopamine and noradrenaline mediated behavioural studies using lithium induced head twitches in rats, haloperidol induced catalepsy in mice and clonidine induced hypothermia in rats respectively. The studies illustrated many interesting features for better acceptability of the formulation. Elevated plus maze revealed significant increase in the time spent in open arms and the number of entries in open arms. Light/dark apparatus studies, significantly increased the time spent in lit zone. Open field apparatus evaluated significant increase in the number of assisted rearing and the number of squares traversed. Hole board apparatus study, significantly increased the number of head poking. On the other hand significant decrease in the number of head twitches was noticed in lithium induced head twitches. Haloperidol induced catalepsy study demonstrated decrease in the duration of catalepsy significantly at 60 min. On the contrary clonidine-induced hypothermia, Arogh even at the dose of 300 mg/kg did not modify the effect of hypothermia. It clearly indicates that Arogh, possess anxiolytic activity, by virtue of its diminished serotonergic transmission and decreased duration of catalepsy indicating potentiation of dopaminergic transmission and modified 5-HT and DA mediated behaviour.

Anoop Austin *et al.,* (2004) carried out swim induced immobility in mice for anti-anxiety effects. Arogh (0.5 -100mg/kg) administered 30 min prior to swim exercise significantly reduced the immobility time, which was also dose related (270 ± 10.5 to 143.8 ± 4.5 sec). A near 50 % inhibition was recorded with 100 g/kg dose. Using 10 g/kg, time of administration was altered from 30 – 300 min and the response was found to be duration dependant. When administered through oral route a comparable effect (44 % inhibition) to that observed after i.p. route was noted after 240 min. One of the major contributing factor to the genesis of hypertension is stress / anxiety. With the model employed in this study, immobility model, Arogh, exhibited a dose, duration and route related anti anxiety response.

A controlled clinical trial by Anoop Austin *et al.* (2011) was carried out in 10 volunteers attending the out patient department of Rohini Holistic Health Centre, Chennai, between the age groups of 20 to 60. They were given 3 gm of Arogh Plus, twice daily, after food along with their regular diet and life style. Daily intake of the formulation for a period of 30 days were found to decrease the symptoms observed due to stress, which was reduced and their work performance was found to be increased. The drug was found to revert back the decreased serum cortisol level to normal after the treatment, which in turn, increased catecholamine neuro transmitters, such as serotonin and dopamine. The study further reinforced, our previous studies and on the contrary, it was found that, the drug was not able to reduce the strees levels on those suffering from metabolic disorders. The performace score was well elucidated and their activity were improved after the study period. The clinical study revealed the formulation to relieve stress due to physiological and mental stress.

Hypertension Therapy

A clinical study was carried out by Babitha *et. al.,* (2003) with ten volunteers with hypertension. Arogh was given as a decoction made out of 2.5 gm twice daily, for a period of 60 days. Anthropometric and Biochemical parameters were examined during the initial day, 30 days and 60th day. The treatment was advocated in a holistic way incorporating Yoga, Diet restriction and regular meditation. The volunteers were found to get reduction in the elevated Systolic and Diastolic pressures. The mental performance were found to be encouraging and were able to carry out their work more than before they were. The study developed the product as a regular use herbal tea to take care of mild to moderate degree of hypertension, whereas the desired effect was not notice for those Systolic pressure were more than 200 mmHg. Their performance score was improved and the status of well being was also well appreciated. The study conformed that the drug might act by way of its anxiolytic activity. The draw back of the study was that the sample was minimum and detailed studies were not incorporated to understand the formulation more appropriately.

Myocardial Protection

Myocardial infraction is a clinical syndrome arising from sudden and persistent curtailment of myocardial blood supply resulting in the necrosis of the myocardium (Szárszoi *et. al.*, 2001), which is followed by numerous pathophysiological and biochemical changes such as lipid peroxidation, hyperglycemia and hyper lipidemia (Al Makdessi *et al.,* 1996). Experimentally, isoproterenol, a β-adrenergic agonist and synthetic catecholamine is capable of producing gross, microscopic myocardial necrosis and depletion of tissue enzymes in the heart (Ceriana, 1992). Isoproterenol treated myocardial infraction serves as a well established model to study the beneficial effects of drugs in cardiac function and mimics clinical conditions of myocardial infarction due to ischemia in humans (Harada *et. al.,* 1993).

A study carried out with isoproterenol induced myocardial damage in wistar rats revealed that Arogh pretreatment maintained the level of lipid peroxide and the activities of diagnostic marker enzymes (Dwivedi and Agarwal, 1994; Dwivedi and Jauhari 1997) like creatine kinase, lactate dehydrogenase and transaminases in serum and heart tissues to near normal levels (Figures 1 and 2). Increase in lipid peroxides due to isoproterenol at a dose of 20 mg/100 gm s.c. was reverted by Arogh effectively at a dose of 150 mg /100 gm p.o.. It further prevented, decrease in the level of those marker enzymes by isoproterenol treatment and they were maintained to near normal values (Suchalatha and Shyamala Devi, 2004) evaluated from heart tissue enzymes.

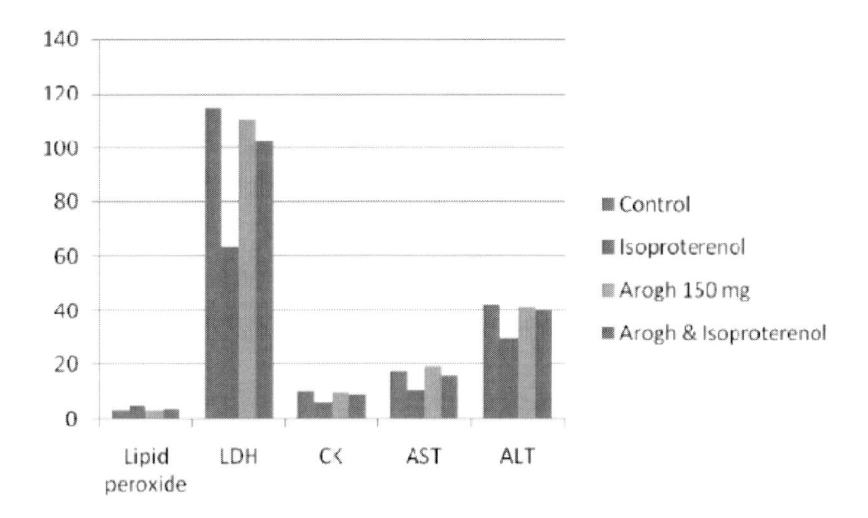

Figure 1. Effect of Arogh on Isoproterenol induced damage in heart enzymes.

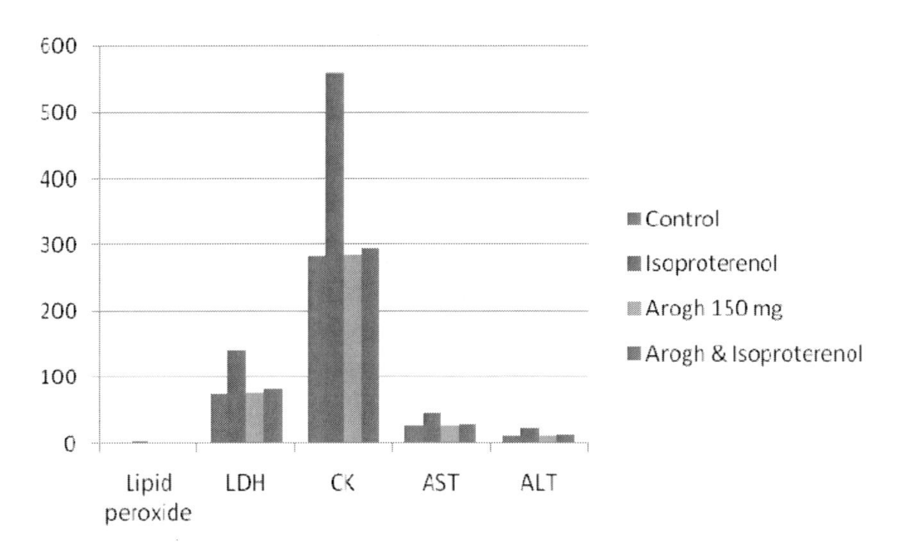

Figure 2. Effect of Arogh on Isoproterenol induced damage in serum enzymes.

The levels of lipid peroxidation, LDH, CK, AST and ALT in heart enzymes were found to be 3.98 ± 0.25, 102.71 ± 5.7, 8.82 ± 0.32, 16.0 ± 0.60 and 40.06 ± 1.8 with Arogh treatment, whereas it was 5.12 ± 0.15, 63.15 ± 1.86, 5.72 ± 0.2, 10.3 ± 0.48 and 29.73 ± 1.12 respectively with isoproterenol treated group, which was statistically significant ($p<0.001$).

Interestingly on the contrary, Arogh pretreatment *per se* did not alter the activity of the serum marker enzymes such as LDH, CK, AST, ALT or the level of lipid peroxides (Figure 2), whereas isoproterenol administered developed a significant elevation in the level of lipid peroxides and activity of marker enzymes. In Arogh pretreated isoproterenol administered rats, the alterations in the level of lipid peroxides was minimized and the activity of marker enzymes were retained at near normal levels (Suchalatha and Shyamala Devi, 2004). The levels of lipid peroxidation, LDH, CK, AST and ALT in heart enzymes were found to be 230 ± 0.08, 83.27 ± 2.86, 293.96 ± 0.8, 29.23 ± 0.80 and 13.70 ± 0.45 with Arogh treatment, whereas it was 4.36 ± 0.13, 140.25 ± 5.2, 559 ± 2.34, 47.08 ± 1.82 and 24.22 ± 0.76 respectively with isoproterenol treated group, which was statistically significant ($p<0.001$).

The observations pertaining to myocardial injury were further confirmed by electrophoretic pattern of LDH isoenzymes, which are considered to be a controlling step for lactate metabolism (Ballo and Messer, 1968) by the heart (Joseph *et al.,* 1995). In acute myocardial infarction, an estimate of the infarct size may be formed by measuring the rate of appearance and disappearance of LDH1 in the blood (Dawson *et al.,* 1964). Pretreatment with Arogh at a dose of 150 mg/100 gm p.o. decreased the intensity of elevation of LDH isoenzymes. The Bands developed were less prominent suggestive of minimal damage to the myocardium which confirms the cardio-protective effect of Arogh, whereas no change in LDH isoenzyme pattern of Arogh alone pretreated animals. Study revealed that Arogh offered protection to myocardium by preventing lipid peroxidation of membrane bound poly-unsaturated fatty acids, thus ensuring myocardial membrane structural integrity and function (Pöpping *et al.,* 1995; Schüssler *et al.,* 1995;.Tripathi *et al.,* 1984).

The Oxidative stress in experimental myocardial infarction was carried out by Suchalatha *et al.* (2004). Arogh at a dose of 150 mg possessed inhibitory activity on lipid peroxidation. Further it also maintained the levels of superoxide dismutase and catalase when treated with Arogh (Figures 3 and 4). Since these enzymes are structurally and functionally impaired free radicals (Nasa *et al.,* 1993), which results in myocardial damage (Leuchtgens, 1993). It also enhanced the activity of glutathione peroxidase and glutathione-s-transferase, which were reverted by the damage caused by isoproterenol. The levels of vitamin E, vitamin C, ceruloplasmin and glutathione were increased due to the drug treatment, which can scavenge superoxide radicals and prevent free radical formation and lipid peroxidation. The synergistic effect of Arogh pretreatment, significantly suppressed the alterations induced by isoproterenol alone in rats. Cytological studies on the heart tissues carried out by Suchalatha and Shyamala Devi (2004) revealed minimal damage, with mild swelling of muscle cells and focal cardiac muscle fibres of the cardiac tissues and near normal architecture of the myocardium, which further confirmed its safety.

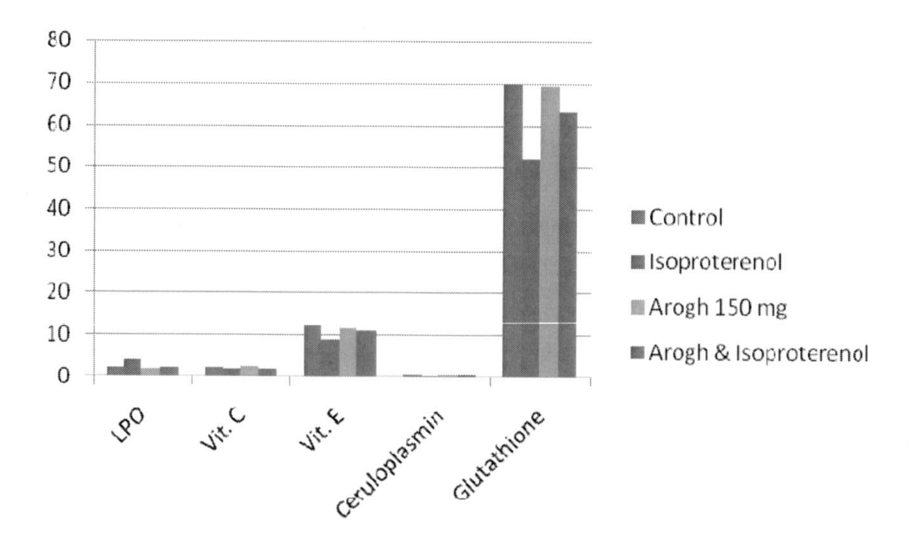

Figure 3. Effect of Arogh on the serum levels of peroxide and antioxidants.

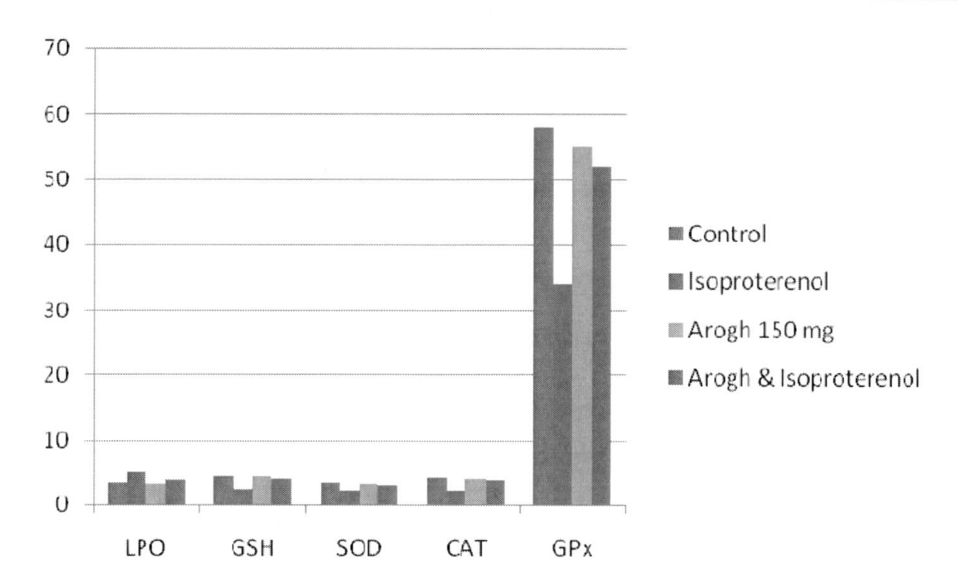

Figure 4. Effect of Arogh on peroxide and antioxidants in heart tissue.

Interestingly, the observations for Arogh, on myocardium was further confirmed by the perfusion studies carried out by Vinoth Kumar *et al.,* 2010) on the *H. indicus* and *H. rosa-sinensis* by Langendorff-perfused rat heart model at a dose of 90, 180, and 360 μg/mL (Pasdois *et al.,* 2007) where, Functional recovery (left ventricular developed pressure (LVDP), and rate of development of pressure), reperfusion arrhythmias, and infarct size by TTC staining (Bae and Zhang, 2005; Hobbs *et al.,* 2004) were observed. A transient increase in LVDP (32%–75%) occurred at all concentrations of *H. indicus*, while coronary flow (CF) was significantly increased after *H. indicus* 180 and 360 μg/mL. Only a moderate increase in LVDP (21% and 55%) and a tendency to increase CF was observed at HRS 180 and 360 μg/mL. *H. indicus* and *H. rosa-sinensis* at 180 and 360 μg/mL significantly improved postischemic recovery of LVDP and reduced the numbers of ectopic beats and duration of ventricular tachycardia. The size of infarction was significantly decreased by *H. indicus* at 360 μg/mL, while *H. rosa-sinensis* significantly reduced the infarct size at all concentrations in a dose-dependent manner. The significantly better recovery of LVEDP and attenuation of post-ischemic diastolic dysfunction at all three doses of *H. indicus* infers that *H. indicus* could improve myocardial relaxation and may reduce the edema caused by ischemica and reperfusion injury (Schaff *et al.,* 1981). In contrast, *H. rosa-sinensis* did not have any significant activity towards the relaxation of the cardiac muscle at any dose. *H. indicus* had a dose-dependent effect on the recovery of LVDP and +dP/dt. *H. rosa-sinensis* had a similar effect on the recovery of LVDP and +dP/dt at HI 180 as in HRS 360 and no protection was observed at *H. rosa-sinensis* 90, suggesting that *H. rosa-sinensis* 180 is the minimum dose required to increase the recovery of contractile function.

H. indicus could protect the heart from arrhythmias at all doses in a dose-dependent manner manifested by a reduced number of PVB (extra heart beat caused by abnormal electrical activity). In addition, a significantly lower total duration of episodes of VT (rapid heart rhythm) was observed at higher doses. Interestingly, *H. rosa-sinensis* at all doses had a significant protection against arrhythmias. The size of infarction (death of a macroscopic area of cardiac tissue) was significantly reduced by *H. indicus* at the highest dose, while *H. rosa-*

sinensis significantly lowered the infarct size at all the doses. In comparison, after 25-min, *H. indicus* exerted a higher protective activity against functional deterioration and a moderate protection against arrhythmias and infarct size, while HRS had a moderate effect on functional recovery and a stronger protection in terms of antiarrhythmic effect and infarct size limitation. *H. indicus* might cause vasodilation, positive inotropic effect, and cardio-protection, while *H. rosa-sinensis* might cause these effects at higher concentrations (Vinoth Kumar *et al.,* 2010). Suppression of arrhythmias which results in smaller size of infarction can be achieved by the protection against contractile dysfunction.

The antioxidant effects of *H. indicus* [Mary *et al* 2003; Rao, *et al.,* 2005] may be associated with tannins, one of the main constituents [Hong *et al.,* 1995]. Likewise, saponins have also been shown to have beneficial effects on cardiovascular diseases [Matsuura, 2001]. Flavonoids produce vasodilation by regulating endothelial nitric oxide (NO) production [Schmitt and Dirsch, 2009] and interaction with ion channels [Akhlaghi and Bandy, 2009]. Moreover, flavonoids are known to protect the I/R-induced myocardial injury by their multifaceted properties, such as antioxidant, antiinflammatory, vasodilatory, and antiplatelet aggregation [Akhlaghi and Bandy, 2009]. Therefore, it is conceivable that the cardio-protective effect can be related to the combined effects of saponins, tannins, and flavonoids. *H. rosa-sinensis* has been shown to enhance the endogenous antioxidant activity and protect the heart from isoproterenol-induced injury [Gauthaman, *et al.,* 2006]. These confirm the effect of Arogh can be attributed towards its effected on myocardium, which can help in myocardial infraction and associated ailments.

ARTICULAR PROTECTION

Injury, fracture, degenerative disorders and autoimmune disorders are the common ailments which results in pain and inflammation and affect our activities (Roy *et al.*, 1982.). Though many pain relievers are available, they are having documented side effects like ulceration, chronic renal failure, liver ailments, blood diathesis etc. (Guyton and Hall, 2000). Many plant derived products are advocated for this condition (Niké *et al.* 2003). The interesting findings observed is that apart from analgesic and anti-inflammatory activities, they were also found to be cytoprotective and free from other side effects observed with synthetic drugs (Kohli *et al.,* 2005). Arogh also demonstrated similar observations. A recent study carried out by Mohan *et al.* (2009) revealed analgesic activity at a dose of 30, 100, 300, and 500 mg/kg, p.o. in swiss albino mice using acetic acid induced writhing method, tail immersion method and hot plate methods (Turner, 1971).

Inhibition of analgesia was observed in acetic acid induced writing method and it demonstrated Arogh, at a dose of 500 mg/kg, possessed maximum inhibition of 49.35 %, which was comparable to Aspirin (20 mg/kg). Lower doses of 30 mg developed 11.53 % of inhibition, 100 mg with 25 % of inhibition and 300 gm with 33.33 % of inhibition.

In tail immersion method, Arogh increased latency to flick tail. The highest nociception inhibition of stimulus was exhibited at 500 mg/kg. Arogh was able to demonstrate maximum latency to tail flick which was maximum after 30 minutes, after which there was a gradual decline in the time taken for tail flick. All tested dosages of Arogh revealed the same observations and pentozocaine at a dose of 17.5 μg i.p developed at 15 minutes and reduced gradually.

Analgesia induced by hot plate method, revealed that Arogh elevated mean basal reaction time. The highest nociception inhibition of stimulus exhibited at 500 mg/kg at 30 min. Similar observations were observed like tail immersion method and confirmed that maximum nociception was attained at 30 minutes.

Anti-inflammatory activities were studied at a dose of 30, 100, 300 and 500 mg/kg, p.o. in Wistar albino rats using carrageenan induced hind paw edema and formalin induced rat paw edema methods. Arogh significantly inhibited carrageenan induced hind paw edema and formalin induced rat paw edema. In carrageenan induced rat paw edema maximum inhibition of paw edema was observed with 500 mg/kg dosage at 4 hours. The inhibition was 43.11 %, where as aspirin at a dose of 20 μg was able to produce 49.54 % of inhibition. Interestingly Arogh exhibited its inhibition significantly from 100 mg dose onwards. Similar observations were notice with formalin induced rat paw edema also confirming its anti-inflammatory activity. The drug response was dose related and maximum inhibition of 42.96 was observed at 500 mg dose which was higher that the standard drug aspirin, which produced only 35.5 % of inhibition. The effects observed in the formulation are attributed to its synergetic actions of the ingredients in the formulation (Wheeler-Aceto and Cowan, 1991). Arogh exhibited an array of analgesic and anti-inflammatory in a dose dependent manner in various animal models of pain and inflammation. The activity of the formulation would probably mediated via inhibition of prostaglandin synthesis (Mohan *et al.,* 2009) as well as central inhibitory mechanism and may have a potential benefit for the management of pain and inflammation (Vinegar *et al.,* 1969).

Carrageenan induced rat paw edema has been a popular inflammatory model to investigate nonsteroidal anti-inflammatory effect of compounds (Vogel, 1969). The first phase is due to release of histamine and serotonin (5-HT) (0-2 h), plateau phase is maintained by kinin like substance (3 h) and second accelerating phase of swelling is attributed to PG release (4 h.). Interestingly Arogh was able to reduce the edema induced by carrageenan in all three phases. Formalin induced edema also shows a biphasic response and originate mainly from neurogenic inflammation followed by participation of kinins and leukocytes with their pro-inflammatory factors including PGs[23], According to Yuh-Fung *et.al.* (1995), acute inflammation induced by formalin results from cellular damages, which provides the production of endogenous mediators.

DIABETES

Yet another interesting finding was carried out by Balaraman *et al.* (2008) inorder to evaluate the antihypertensive and antihyperglycemic activity of Arogh in experimental conditions. Streptozotocin (60 mg/kg i.p.)-nicotinamide (120 mg/kg i.p.) were used to induce Type II Diabetes mellitus. Streptozotocin was given i.p. 15 mints followed by nicotinamide administration. Blood sugar level taken at 3rd and 7th day, rats with blood sugar level >120 mg/dl known to be diabetic and those with blood sugar level >200 mg/dl were included for this study. Various parameters studied includes fasting blood glucose level, serum lipid profile, glycated heamoglobin, liver hexokinase, glucose-6-phosphate dehydrogenase, oxidative stress markers in liver and kidney, blood pressure, body weight, endothelial damage histologically, histopathology of pancreas in diabetic, diabetic treated and normal rats. Arogh was administered at two dose levels ½ and 1 gm/kg. p.o. for a period of one month. Significant decrease in fasting blood glucose level, blood pressure, HbA1c levels observed in diabetic treated rats as compared to diabetic rats. The total cholesterol and serum triglycerides levels, glucose-6-phosphat dehydrogenase, and oxidative stress were also significantly reduced and the HDL levels, liver hexokinase, glycogen levels were significantly increased. The possible mechanism would be by its hypoglycemic activity, potentiating the insulin effect of plasma either increasing the pancreatic secretion of insulin form the existing beta cells or by releasing from the bound form. The significant decrease in serum lipids, can be attributed to improvement in insulin levels by Arogh. The increase in glycogen level can be due to improvement in insulin secretion and glycemic control. The increase in glycogen content is due to reactivation of glycogen synthase mechanism, which helps in glycogenesis process. Significant increase in hexokinase activity and reduction of G6PD activity helps to decrease oxidative stress damages. Regeneration of β cells were observed in the pancreas. Arogh possess it antidiabetic and antihypertensive effects by sequential metabolic correlation between glycolysis and decrease glyconeogenesis apart from it antioxidant property. The effect of Arogh in hypertension is further conformed by earlier report made by Babitha *et. al.* (2003).

Insulin resistance is a characteristic feature of type 2 diabetes mellitus, but other manifestations include hypertension, obesity, hypercoagulable state and dyslipidemia (Balaraman *et al.* 2008). The dyslipedemia associated with insulin-resistant status is characterized by hypertriglyceridemia, an increase in hepatic VLDL secretion, and a decrease in peripheral triglyceride clearance. Arogh (2.5 gm made into decoction) was found to be protecting insulin resistance which is getting generated with fructose. Arogh decreased the serum glucose level upto 84.99± 2.35and decreased serum insulin. Fasting Insulin Resistant Index was also decreasing well with the drug. Total cholesterol, triglycerides, LDL, VLDL reduced significantly and HDL was not elevated substantially. Liver and Muscle glycogen stores were found to be improving with the treatment. Liver oxidative stress parameters like SOD, catalase, reduced glutathione, were improving. Lipid peroxidation was found to be decreasing with treatment. Blood pressure was normalized. The study clearly elucidates that Arogh possess antidiabetic and antioxidant activities. The improvement in insulin resistance might be due to its antioxidant property. Further it may also increase expression of glut-5 or insulin receptor mRNA, which is evident by increase in stores of liver and muscle glycogen. The improvement in blood pressure is through increment in sensitivity of insulin to vascular tissue.

CONCLUSION

Periodical studies carried out by various authors demonstrated Arogh as a positive candidate as an hypocholesteremic, anxiolytic, antihypertensive, cardio-protective from myocardial damage, antioxidant, analgesic and anti-inflammatory and as a anti diabetic. Since the study reveal to be a panacea for many common complaints, it can be much more concentrated in cardiac related ailments, as if most of the activities can be directly or indirectly attributed. Though, the effect on cholesterol is promising, its effect on HDL is not clear. Since cardiac ailments are prone after mid age, normally and can develop other problems like hypertension, diabetes, cholesterol, joint pain and stress, the formulation can be a suitable and can be found useful. Further, a dose related study will throw more light on this product for better understanding and to evaluate its probable mode of action.

REFERENCES

Adailkan, P.G. and Gauthaman, K. (2001) History of herbal medicines with an insight on the pharmacological properties of *Tribulus terrestris*. *The Aging Male.* 4 : 163–169.

Akhlaghi, M. and Bandy, B., (2009). Mechanisms of flavonoid protection against myocardial ischemia-reperfusion injury. *Journal of Molecular and Cellular Cardiology.* 46 (3): 309–317.

Al Makdessi, S., Sweidan, H., Mullner, S. and Jacob, R. (1996). Myocardial protection by pretreatment with Crataegus oxycantha: An assessment by means of the release of lactate dehydrogenase by the ischemic and reperfused Langendorff heart. *Rzneimittelforschung.* 46: 25-27.

Alam, M.I., Auddy, B. and Gomes, A. (1994) Isolation and partial characterization of viper venom inhibiting factor from the root extract of the Indian medicinal plant sarsaparilla (*Hemidesmus indicus* R.Br.). *Toxicon.* 32 : 1551–1557.

Anonymous (1993). *WHO Research Guidelines for Evaluating the Safety and Efficacy of Herbal Medicines.* Published by World Health Organization, Regional Office for the Western Pacific, Manila.

Anoop Austin and Jegadeesan, M. (2003). Biochemical studied on the ulcerogenic potential of *Hemidesmus indicus* var. *indicus* R.Br., *Journal of Ethnopharmacology.* 84 : 149-156.

Anoop Austin, Senthilvel, G., Thirugnanasambantham, P. and Mayisvren, E. (2006) Clinical efficacy of a Polyherbal Instant Formulation (Arogh) in the management of Hyperlipidaemia. *The Cardiology.* 2 (2) : 36-38.

Anoop Austin, Thirugnanasambantham, P. Mayisvren, E. and Viswanathan, S. (2004). Anti-anxiety effect of *Arogh*, a polyherbal preparation, in mice, Recent Advances in Pharmacology, Organised by Department of Cardiology, All India Institute of Medical Sciences (AIIMS), New Delhi.

Anoop Austin, Elsie, C.S. and Thirugnanasambantham, P. (2011). Evaluating the clinical efficacy of a polyherbal formulation Arogh Plus on Stress – A randomised clinical Study, *Journal of Stress Physiology and Biochemistry,* 7 (1) : 66-78.

Anye Nike, B.C., Njamen, D., Wandji, J., Zacharias, T., Dongmo, F.A., Nguelefack, T.B., Albert, D.W. and Kamanyi, A. (2003). Anti-inflammatory and analgesic effect of Drypemolundein A, a Sesquiterpene Lactone from *Drypetes molunduana. Pharmaceutical Biology.* 41: 26-30

Asokan, B.R, Jaikumar, S, Ramaswamy, S, Thirugnanasambantham, P. and Nirmala, P. (2010). Anti-hyperlipidemic activity of a polyherbal formulation in experimental models. *Pharmacologyonline.* 1: 433-442.

Babitha, K., Anoop Austin, Thirugnanasambantham, B., Mayisvren, M. and Senthilvel, G. (2004) Hypertension in Holistic approach, Seminar on Holistic Healing XVII, Holistic Health Care for the 21st Century, at Bangalore.

Bae S. and Zhang, L. (2005) Gender differences in cardio-protection against ischemia/reperfusion injury in adult rat hearts: focus on AKT and Protein Kinase C signaling. *Journal of Pharmacology and Experimental Therapeutics.* 315 (3) : 1125–1135.

Balaraman, R., Anoop Austin and Thirugnanasambantham, P. (2008) Antihyperglyceridemic, antihyperglycemic and antihypertensive effect of Arogh, a polyherbal formulaion in Fructose induced insulin resistant rats, *Indian Pharmacological Congress.* Jaipur.

Balaraman, R., Mohan, M, Aurangabadkar, V.M., Jadhav, G.B., Anoop Austin and Thirugnanasambatham P. (2007) Effect of a polyherbal formulation on anxiety and behavior mediated via monoamine neurotransmitters, *Oriental Pharmacy and Experimental Medicine.* 7 (4) : 409 – 417.

Balaraman, R.., Anoop Austin and Thirugnanasambantham, P. (2008) Effect of Arogh special, a polyherbal formulation on streptozotocin-nicotinamide induced type II Diabetic hypertensive rats, *Indian Pharmacological Congress.* Jaipur.

Ballo, J.M. and Messer, J.R., (1968) Lactate dehydrogenase isoenzymes in human hearts having decreased oxygen supply. *Biochemistry and Biophysics Research Communication.* 33 : 487-491.

Ceriana, P. (1992) Effect of myocardial ischemia reperfusion on granulocyte elastase release. *Anaesthesia Intensive Care.* 20 : 187-190.

Chatterjee, I., Chakravarty, A.K. and Gomes, A. (2004) Antisnake venom activity of ethanolic seed extract of *Stychnos nux-vomica* Linn. I*ndian Journal of Experimental Biology.* 42 : 468–475.

Chatterjee, I., Chakravarty, A.K. and Gomesa, A. (2006) *Daboia russellii* and *Naja kaouthia* venom neutralization by Lupeol acetate isolated from the root extract of Indian sarsaparilla *Hemidesmus indicus* R.Br. *Journal of Ethnopharmacology.* 106 : 38-43.

Chevallier, A. (1996) *Encyclopedia of Medicinal Plants*, DK Publishing, New York, pp.141, 273.

Chopra, R.N., Nayar, S.L. and Chopra, I.C. (1956) *Glossary of Indian medicinal plants, Vol. I.* Council of Scientific and Industrial Research, New Delhi, pp. 197.

Cox, P, and Balick, M. (1994) *The ethnobotanical approach to drug discovery.* Scientific American, pp. 82-87.

Cox, P.A. (2009) *Ethnopharmacology and the search for new drugs.* (Eds. Chadwick, D.J. and J. Marsh). In Bioactive Compounds from Plants. Ciba Foundation Symposium 154, John Wiley and Sons, Chichester, pp. 40-55.

Dawson, D.M., Goodfriend, T.L. and Kaplan, N.O. (1964) Lactate dehydrogenases functions of the two types, *Science.* 143 : 929-933.

Dev, S. (1997) Ethnotherapeutic and modern drug development: The potential of Ayurveda. *Current Science.* 73: 909-928.

Dwivedi, S. and Agarwal, M.P. (1994) Antianginal and cardio protective effects of *Terminalia arjuna*, an indigenous drug, in coronary artery disease, *Journal of Association of Physicians India.* 42: 287-289.

Dwivedi, S. and Jauhari, R. (1997) Beneficial effects of *Terminalia arjuna* in coronary artery disease. *Indian Heart Journal.* 49 : 507-510.

Evans, W.C. (2005). *Plants in medicine: the origins of Pharmacognosy*. In: Trease and Evans pharmacognosy, 15th edition, published by Saunders, An Imprint of Elsevier, US, pp 3-4.

Farnsworth, N.R. and Bingel, A.S. (1997) *Problems and prospects of discovery new drugs from higher plants by pharmacological screening.* In: H.Wagner and P.Wolff (eds.), New Natural products and plant drugs with pharmacological, biological and therapeutical activity, Springer Verlag, Berlin pp. 1-22.

Farnsworth, N.R., Blowster, R.N., Darmratoski, D., Meer, W.A. and Cammarato, L.V. (1967) Studies on *Catharanthus* alkaloids, Evaluation by means of TLC and Ceric Ammonium sulphate spray reagent, *Lloydia.* 27: 302-314.

Ganesh, R., Narayanan, N., Thirugnanasambantham, P., Viswanathan, S. and Parvathavarthini, S. (2006). Effect of Arogh on Hyperlipidemia, *International Journal of Tropical Medicine.* 1 (1) : 18 – 22.

Gauthaman, K.K., Saleem, M.T.S., Thanislas P.T., Prabhu, V.V., Krishnamoorthy, K.K., Devaraj, N.S. and Jayaprakash S Somasundaram, J.S. (2006) Cardioprotective effect of the *Hibiscus rosa sinensis* flowers in an oxidative stress model of myocardial ischemic reperfusion injury in rat. *BMC Complementary and Alternative Medicine.* 6 : 32.

Gotto, A.M. (1985) Some reflections on arteriosclerosis: Past, present, and future. *Circulation*. 72: 8-17.

Guyton and Hall, (2000). *Textbook of Medical Physiology*, 10th Edn. Philadelphia pp. 397-398.

Harada, K., Futaka, Y., Miwa, A., Kaneta, S., Fukushima, H. and Ogawa, N. (1993). Effect of KRN 2391, a novel vasodilator, on various experimental angina models in rats. *Japan Journal of Pharmacology.* 63 : 35.

Heinrich, M. (2000) Plant resources of south-east Asia medicinal and poisonous plants, *Phytochemistry.* 53 : 619–620.

Hobbs, A., Foster, P., Prescott, C., Scotland, R. and Ahluwalia, A. (2004) Natriuretic peptide receptor-C regulates coronary blood flow and prevents myocardial ischemia/reperfusion injury: novel cardioprotective role for endothelium-derived C-type natriuretic peptide, *Circulation*. 110 (10) : 1231–1235.

Joseph, G., Zhao, Y. and Klaus, W. (1995) Pharmacologic action profile of *Crataegus* extract in comparison to epinephrine, amirinone, milrinone and digoxin in the isolated perfused guinea pig heart. *Arzneimittelforschung.* 45: 1261-1265.

Kamboj, V.P. (2000) Herbal medicine. *Current Science* 78 : 35-39.

Kim, N.J. and Hong, N.D. (1996) Studies on the processing of crude drugs on the constituents and biological activities of glycyrhiza. *Korean Journal of Pharmacognosy.* 27 (3) : 196-206.

Kohli, K., Ansari, J., Ali, J. and Reaman, Z. (2005) *Curcumin*: A natural anti-inflammatory agent. *Indian Journal of Pharmacology.* 37 (3): 141-147.

Kokate, C. K. (1994) *Practical Pharmacognosy*, 3rd Edn., Vallabh Prakashan, New Delhi, pp. 107-109.

Leuchtgens, H. (1993). *Crataegus special extract WS 1442 in NYHA II heart failure. A placebo controlled randomized double-blind study*. Fortschreter. Medicina, 111: 352-354.

Lucas, R. (1977). *Nature's medicines*. Wilshire Book Company, California, USA, pp.13-16.

Marderosian, A. and Beutler, J.A., (2000) *The review of natural products,* 3rd edition published by Facts and Comparisons, St. Louis, Missouri, USA, p.357.

Mary, N.K. Achuthan, C.R. Babu, B.H. and Padikkala, J. (2003) *In vitro* antioxidant and antithrombotic activity of *Hemidesmus indicus* (L) R.Br. *Journal of Ethnopharmacology.* 87 (2-3) : 187–191.

Matsuura, H. (2001) Saponins in garlic as modifiers of the risk of cardiovascular disease. *Journal of Nutrition.* 131 (3) : 1000–1005.

Mohan, M., Gulecha, V.S., Aurangabadkar, V.M., Balaraman, R., Anoop Austin and Thirugnanasampathan, P. (2009). Analgesic and anti-inflammatory activity of a polyherbal formulation (Arogh). *Oriental Pharmacy and Experimental Medicine.* 9 (3) : 232-237.

Muldoon, M.F., Manuck, S.B. and Matthews, K.A. (1990) Lowering cholesterol concentrations and mortality: a quantitative review of primary prevention trials. *British Medical Journal.* 301: 309-314.

Murray, M.T. (1995) *The Healing Power of Herbs* (2nd Edition), Prima Publishing Murray, M.T. and Pizzorno, J.E. (1991). Atherosclerosis. In: Encyclopedia of Natural Medicine published by Prima Publishing, CA, pp.156- 170.

Nasa, Y., Hashizume, H., Hoque, A.N. and Abiko, Y. (1993) Protective effect of *Crataegus* extract on the cardiac mechanical dysfunction in isolated perfused working rat heart. *Arzneimittelforschung.* 43: 945-949.

Pasdois, P., Quinlan, C.L. Rissa, A., Tariosse, L., Vinassa, B., Costa, A.D., Pierre, S.V., Dos Santos, P. and Garlid, K.D. (2007) Ouabain protects rat hearts against ischemia-reperfusion injury via pathway involving src kinase, mitoKATP, and ROS. *American Journal of Physiology.* 292 (3) : 1470–1478.

Perumal Samy, R. and Ignacimuthu, S. (1998) Screening of 34 Indian medicinal plants for antibacterial properties. *Journal of Ethnopharmacology.* 62 : 173-182.

Pfister, J.A., Ralphs, M.H., Gardner, D.R., Stegeleier, B.L., Manners, G.D. and Panter, K.E. (2002) Management of three toxic Delphinium species based on alkaloid concentrations. *Biochemical Systematics and Ecology.* 30 : 129–138.

Pöpping, S., Rose, H. and Ionescu, I. (1995) Effect of a hawthorn extract on contraction and energy turnover of isolated rat cardiomyocytes. *Arzneimittelforschung.* 45: 1157-1161.

Rabe, T. and Staden, J.V. (1997) Antibacterial activity of South African plants used for medicinal purposes. *Journal of Ethnopharmacology.* 56: 81-87.

Rabe, T. and Staden, J.V. (1997). Antibacterial activity of South African plants used for medicinal purposes. *Journal of Ethnopharmacology.* 56 : 81-87.

Rao, G.M.M., Venkateswararao, C.H., Rawat, A.K.S., Pushpangadan, P. and Shirwaikar, A. (2005) Antioxidant and antihepatotoxic activities of *Hemidesmus indicus.* R. Br. *Acta Pharmaceutica Turcica.* 47 (2) : 107–113.

Rossouw, J.E., Lewis, B. and Rifkind, B.M. (1990) The value of lowering cholesterol after myocardial infarction, *New England Journal of Medicine.* 323: 1112-1119.

Roy, A., Gupta, J.K. and Lahiri, S.C. (1982) Further studies on anti-inflammatory activity of two potent Indane-1-acetic acids, *Indian Journal of Physiology and Pharmacology.* 26: 206-214.

Saluja, M.S., Sangameswaran, B., Sharma, A., Manocha, N. and Husain, A. (2010) Analgesic and anti-inflammatory activity of a marketed poly herbal formulation (PHF), *International Journal of Pharma Professional's Research.* 1 (1) : 19-23.

Satyavati, G.V. (1988) Gum Guggul (*Commiphora mukul*) – The success of an ancient insight leading to a modern discovery. *Indian Journal of Medicine.* 87 : 327-335.

Schaff, H.V., Gott, V.L., Goldman, R.A., Frederiksen, J.W. and Flaherty, J.T. (1981) Mechanism of elevated left ventricular end-diastolic pressure after ischemic arrest and reperfusion. *The American Journal of Physiology*. 240 (2) : 300–307.

Schmitt, C. A. and Dirsch, V. M. (2009) Modulation of endothelial nitric oxide by plant-derived products. *Nitric Oxide*. 21 (2) : 77–91.

Schüssler, M., Hölzl, J., Rump, A.F. and Fricke, U. (1995) Functional and antiischemic effects of monoacetylvitexinrhamnoside in different *in vitro* models. *General Pharmacology*. 26: 1565-1570.

Suchalatha, S. and Shyamala Devi, C.S. (2004) Effect of Arogh - a polyherbal formulation on the marker enzymes in isoproterenol induced myocardial injury. *Indian Journal of Clinical Biochemistry*. 19 (2) : 184-189.

Suchalatha, S., Thirugnanasambandam, P., Maheswaran, E. and Shyamala Devi, C.S., (2004). Role of Arogh, a polyherbal formulation to mitigate oxidative stress in experimental myocardial infarction. *Indian Journal of Experimental Biology*. 42 : 224-226.

Szárszoi, O., Asemu G., Vaněček, J., Ošt'ádal, B. and Kolář, F. (2001) Effects of melatonin on ischemia and reperfusion injury of the rat heart. *Cardiovascular Drugs and Therapy*. 15 (3) : 251–257.

Tripathi, S.N., Upadhyaya, B.N. and Gupka, V.K. (1984) Beneficial effect of *Inula racemosa* (Pushkarmoola) in angina pectoris: A preliminary report. *Indian Journal of Physiology and Pharmacology*. 28 : 73-75.

Turner, R.A. (1971) *Screening Methods in Pharmacology*, Academic Press: New York, pp. 100-113.

Vaibhav Patel, Balaraman, R., Anoop Austin, Thirugnanasambantham, P. and Balasubramanian, M. (2008) Effect of Arogh special, a polyherbal formulation on triton WR 1339 induced hypercholesterolemia in rats, Paper Presented in *Indian Pharmacological Congress*. held at Jaipur.

Vinegar, R., Schreiber, W. and Hugo, R.J. (1969) Biphasic development of carrageenan edema in rats. *Journal of Pharmacology and Experimental Therapy*. 166: 96-103.

Vinoth Kumar, M.K., Balaraman, R., Pancza, D. and Ravingerova, T. (2010) Hemidesmus indicus *and* Hibiscus rosa-sinensis *affect Ischemia reperfusion injury in isolated rat hearts.* Evidence-Based Complementary and Alternative Medicine, *1 : 1-8.*

Vogel, H.G. (2002) *Drug discovery and evaluation, pharmacological Assay*, 2nd Edn., New York, Springer, p. 670.

Wheeler-Aceto, H. and Cowan, A. (1991) Neurogenic and tissue mediated components of formalin induced edema agents actions. *Fitoterapia*. 34 : 264.

Wichtl, M. (1994) *Herbal Drugs and Phyto-pharmaceuticals. A handbook for practice on a scientific basis,* (Editor: Norman Grainger Bisset), Medpharm, Stuttgart, Germany, pp.5-7.

Yuh-fung, C., Haei-Yann, T. and Tian-Shung, W. (1995) Anti-inflammatory and analgesic activities form roots of *Angelica pubescens*. *Planta Medica*. 61: 2-8.

In: Natural Products and Their Active Compounds ... ISBN: 978-1-62100-153-9
Editors: M. Essa, A. Manickavasagan, and E. Sukumar © 2012 Nova Science Publishers, Inc.

Chapter 3

COMBATING DIABETES MELLITUS – NATURE'S WAY

S. Premakumari, M. Amirthaveni and M. S. Subapriya
Department of Food Science and Nutrition,
Avinashilingam University for Women,
Coimbatore, India

INTRODUCTION

Diabetes mellitus was unknown in the prehistoric times, as long as man was a hunter and food gatherer. It was recognized only with the advent of civilizations in Egypt, Rome, Greece and India. Early medical writers reported weight loss, excessive urination and sweet taste of urine as the primary symptoms. Roman Artaeus (70 AD)[1] noted polydipsia and polyuria and named the condition 'diabetes', meaning to 'flow through'. Centuries later, Thomas Willis (1965)[2], a London physician described the sweet taste of urine and introduced the term 'mellitus' meaning 'honey-like'. Diabetes is one of the most common non-communicable diseases and challenging health problems of the 21st century. It is a physiological state when sugar in the blood cannot be utilised by the cells in the body. Hence not only blood sugar, but also the metabolism of carbohydrates, fat and protein increase. Insulin is the hormone essential for the utilization of glucose and is produced by the beta cells of the islets of Langerhans of the pancreas. In diabetes it is either not produced in adequate quantities or does not act properly so that glucose keeps accumulating in the blood but the cells are starved of it.

DEFINITION

Diabetes mellitus is defined as the silent killer disease of the 21st century. It is an epidemic of the millennium and affects nearly 150 million adults, worldwide. It is a disease with excessive sugar in the urine, thirst and emaciation. The term 'diabetes' refers to the large volume of urine that is passed and the term 'mellitus' describes the sweet taste of urine in this condition. Diabetes mellitus is a chronic metabolic condition characterized by major derogatory changes in the metabolism of glucose and abnormalities in the metabolism of fat, protein and other nutrients. Hyperglycemia, glucosuria, ketosis, acidosis, diabetic coma (unconsciousness), polyuria, weight loss in spite of polyphagia (condition of increased appetite) and polydipsia (condition of increased thirst) are the abnormal characteristics of diabetes. But the principal abnormalities are an increased liberation of glucose in circulation from the liver and a reduced entrance of glucose in peripheral tissues due to deficiency of intracellular glucose and excess of extracellular glucose. Thus, diabetes mellitus is a disorder of metabolism characterized by high blood sugar level and excretion of sugar in urine. However, pathological changes of small blood vessels of eyes, kidneys and other tissues and degeneration of peripheral nerves develop with time. This condition is usually inherited and results from an absence or relative deficiency of insulin.

Prevalence of Diabetes Mellitus

The worldwide prevalence of Diabetes mellitus was estimated to be 2.6 per cent (171mn) in 2000. It is projected to reach 4.4 per cent (36mn) in 2030 (Wild *et al,* 2004)[3]. The global number of people with diabetes mellitus is currently 220mn and is expected to reach 324mn by 2025 (WHO, 2005) [4].

The countries with the largest number of adult diabetics are India, China, US, Pakistan, Japan, Indonesia, Mexico, Egypt, Brazil and Italy. The first three countries account for nearly half of the world's population with diabetes and those in the Western Pacific account for another 44mn diabetics (WHO, 2007)[5].

India leads the world with the largest number of diabetics earning the dubious distinction of being termed the 'diabetes capital of the world". The number of diabetics in India is currently around 40.9 million, of whom, 90 per cent are Type 2 diabetics and is expected to rise to 60.9 million by 2025 and 79.4mn by 2030 unless urgent preventive steps are taken (International Diabetes Federation, 2006)[6]. Indians are generally more susceptible to diabetes than other ethnic groups. The prevalence of diabetes is found to increase with improvement in socio-economic status.

Classification

Diabetes mellitus is a group of diseases characterized by high blood glucose concentration resulting from defects in insulin secretion, action or both. Insulin is a polypeptide secreted by the beta cells of the Islets of Langerhans of the pancreas. The average daily secretion of insulin is about 2mg (50 units) and its half life is about 5 minutes. The World Health Organisation (2007)[7] recommends the most recent aetiological classification of diabetes mellitus, which is as follows:

Type I diabetes mellitus or juvenile diabetes, previously termed insulin-dependent diabetes mellitus (IDDM) is characterized by beta cell destruction through genetic predisposition, which is caused by an auto immune process leading to absolute insulin deficiency and is hence called immune mediated diabetes. Markers of immune destruction are present in this type of diabetes. It accounts for only 2-5 per cent of all cases of diabetes and usually develops during childhood or adolescence. The patients are either normal or undernourished. They are usually lean with excessive thirst, frequent urination and significant weight loss, dehydration, electrolyte disturbance and ketoacidosis. The rate of beta cell destruction is variable, proceeding rapidly in infants and children and slowly in adults.

The capacity of a healthy pancreas to secrete insulin is far in excess of what is needed normally. Therefore clinical onset of diabetes may be preceded by and extensive, asymptomatic period of months to years during which beta cells undergo rapid destruction. Type I diabetes has two forms-Immune Mediated Diabetes mellitus which results from autoimmune destruction of the beta cells of the pancreas and Idiopathic Type I diabetes mellitus, which has no etiology.

Type II diabetes is a progressive disorder present long before it is diagnosed. Hyperglycemia develops gradually, not often severe enough in the early stages for the patient to notice any of the classic symptoms of the disease. These individuals are at increased risk of developing macro vascular and micro vascular complications. The disease relates to disorders of nutrient storage, predominantly associated with insulin resistance with relative insulin deficiency. The pancreas normally continues to secrete insulin until very late in the disease process. Insulin resistance is related strongly to over weight and physical activity as well as genetic predisposition. Blood glucose becomes elevated even in the presence of high circulating insulin.

Predisposing Factors

The risk factors may be genetic or environmental and include family history of diabetes, older age, obesity especially the intra abdominal type, physical inactivity, prior history of gestational diabetes, impaired glucose homeostasis, race or ethnicity, total adiposity and longer duration of adiposity. However, type II diabetes can occur in non obese and many obese may never develop type II diabetes. Hence it is suggested that obesity combined with genetic predisposition may be necessary for type II diabetes to occur.

Another possibility is that a genetic disposition leads to both obesity and insulin resistance. Type II diabetes results from a combination of insulin resistance and beta cell failure but the extent to which each of these factors contribute to the development of the disease is unclear. Yet another contributing factor is explained by the 'thrift gene theory' or the 'Asian phenotype' as it is called worldwide. The so called "Asian phenotype" refers to certain clinical and biochemical abnormalities in India which include increased insulin resistance, adiposity, i.e., highest waist circumference, despite lower body mass index. This phenotype makes Asians more prone to diabetes and coronary artery disease (NCEP,2001)[8].

Objectives of Management

If hyperglycemia is the hall mark of diabetes, then acquiring and maintaining euglycemia is the fundamental objective and diet, exercise and education are the three cornerstones of management in order to prevent the long term complications of the disease. According to Bamji *et al(2006)*[9], Diabetes management should therefore encompass the following:

1) reduce and maintain near normal blood and urine sugar
2) maintain ideal body weight
3) relieve typical physiological symptoms
4) maintain normal range of serum lipids
5) provide nutrient adequacy
6) avoid acute complications
7) prevent long term micro and macro vascular complications and
8) improve the overall quality of life.

Effective dietary intervention is necessary to achieve adequate metabolic control. However, the diabetic diet need not deviate from the normal diet. It is necessary that the normal diet is consumed in less quantity. The normal Indian diet offers greater scope for better dietary management with the diverse and nutrient dense foods of plant origin that it liberally includes, unlike the 'refined-cereal-dominated' typical western diet. Indian dietaries are diverse, with versatile combinations of macro and micro nutrients including fibre. In addition, the climatic conditions prevalent in India not only facilitate larger intakes of fluids, but also provide wider scope for physical exercise compared to temperate countries, which is vital to the health of diabetics.

INDIGENOUS FOODS AND HERBAL PRODUCTS IN MANAGEMENT

Cereal and Pulse Products

Cereals and Millets are suggested as staples in diabetic diets owing to their high fibre content. Studies with wheat *(Triticum aestivum)* and millets such as Thenai *(Setaria italica)* and Varagu *(Paspalum scorbiculatum)* have been carried out to assess their hypoglycemic potentials.

Wheat Products (*Triticum Aestivum*)

Wheat and its products such as, germ, bran and grass (Plates 1, 2 and 3) are indicated to possess hypoglycemic potential. They have considerable amounts of fibre and other components which have a lowering effect on blood sugar. To prove these claims, a study was conducted by Premakumari and Haripriya (2008)[10].

Table 1 presents the nutritive value of wheat germ, bran and grass as reported by the study. In this study, six groups (A to F) of 15 diabetics each were fed with wheat germ, grass and bran, alone and in combination for a period of six months. Group A received 60g of wheat germ, group B, 20g of wheat bran, group C, 100g of wheat bran, group D, germ and bran (60g+20g), group E, germ and grass (60g+100g) and group F, bran and grass (20g+100g) as supplements.

Plate 1. Wheat germ.

Plate 2. Wheat bran.

Plate 3. Wheat grass.

Table 1. Nutritive Value of Wheat Products (Per 100g)

Nutrients	Wheat germ	Wheat bran	Wheat grass
Carbohydrates (g)	51.8	64.5	3.1
Dietary fibre (g)	13.2	42.8	17.3
Protein (g)	23.1	15.5	20.5
Total fat (g)	9.7	4.3	0.0
Saturated fat (g)	1.7	0.6	0.0
Monosaturated fat (mg)	1.4	0.6	0.0
Polyunsaturated fat (mg)	6.0	2.2	0.0
Vitamin C (mg)	9.2	9.5	14.1
Thiamine (mg)	1.9	0.43	0.08
Riboflavin (mg)	0.5	0.11	0.13
Niacin (mg)	6.8	4.3	0.11
Vitamin B6 (mg)	1.3	1.2	0.2
Folic acid (mcg)	281	100	86
Vitamin B12 (mcg)	2.3	103	99
Calcium (mg)	39	73	242
Iron (mg)	6.3	10.6	0.61
Magnesium (mg)	239	611	24
Phosphorus (mg)	842	1013	1210
Zinc (mg)	12.3	7.3	0.33
Copper (mg)	0.8	1.0	0.2
Manganese (mg)	13.3	11.5	10.2
Selenium (mcg)	79.2	77.6	52.3

Thirty grams of wheat germ in 100 ml of toned milk was consumed twice a day (at midmorning and at bedtime). Ten grams of wheat bran mixed in 50g of wheat flour was prepared as chappathi and taken twice a day (breakfast and dinner). Seventh day wheat grass was crushed using a stone grinder so as to avoid the mechanical disintegration of the chlorophyll molecule in wheat grass and 100ml of the same was distributed to the subjects for consumption at midmorning.

Symptoms of polyuria, nocturia polydipsia polyphagia and constipation were drastically reduced. The changes in biochemical picture after the supplementation are given in Table 2.

The investigators observed a reduction in fasting blood glucose (FBS) by 22 to 28mg/dl, post prandial blood glucose (PPS) by 39.8 to 52.87mg/dl and glycosylated haemoglobin (HbA_{1C}) by 1.99 to 3.19 per cent (at $p<0.01$) in all the supplemented groups.

Serum superoxide dismutase, glutathione, glutathione peroxidase, vitamin C, zinc and selenium increased in the subjects while, serum malondialdhyde and copper decreased significantly ($p<0.01$). In general, groups who received combinations of wheat products (D, E and F), particularly group E (wheat germ and grass) showed greater reductions in blood sugar.

Table 2. Changes in Mean Serum Fasting and Post Prandial Glucose (Mg/Dl) and Glycosylated Haemoglobin (%) Levels

Groups	Fasting blood glucose	Post prandial blood glucose	Glycosylated haemoglobin
A Initial	123.2	164.67	8.42
Final	101.1**	123.93**	6.35**
B Initial	123.6	164.6	8.37
Final	100.8**	124.8**	6.41**
C Initial	123.47	163.4	8.4
Final	101.4**	115.4**	6.41**
D Initial	123.8	163.67	8.39
Final	100.53**	115.47**	5.49**
E Initial	124.2	165.94	8.45
Final	95.87**	113.07**	5.26**
F Initial	123.47	165.67	8.41
Final	11.01**	114.20**	5.34**

**Significant at one per cent level.

Thinai (Setaria Italica) and Varagu (Paspalum Scorbiculatum)

Millets are nutritious and healthy food grains to be included in diet of all sections of society and form the dietary components of vegetarian diets, especially in the developing countries. They are important food crops grown in India, tropical Africa and China and are often considered to be a 'poor man's cereals'. Thenai and varagu are grown in Southern parts of India. They are mildly sweet with nut like flavour and contain a myriad of beneficial nutrients.

They contain nearly 50-65 per cent starchy components, 15 per cent protein, high amounts of carbohydrates, fibre and B complex vitamins, including niacin, thiamine, riboflavin and essential aminoacids like methionine and lecithin and vitamin E. They are particularly high in minerals like iron, magnesium, phosphorus and potassium. The grains are rich in phytochemicals including phytic acid and phytates. They are cheaper sources of proteins, but their carbohydrates minerals and vitamins are nutritionally superior to rice and wheat. The active components are coumarin, setarin, flavonoids- Beta sitosterol and kaempferol. Both the millets have been reported to be highly beneficial to the diabetic subjects.

Green Gram (Vigna Radiata)

Whole grams are valued for their fibre rich seed coats. Green gram is specially recommended for fever, diarrheoa, sprue, bleeding disorders and general weakness, enlargement of abdomen, giddiness, beriberi, burning sensation, gout, ulcers, skin diseases, psychological diseases and diabetes. It is particularly rich in enzymes, phosphoglucomutase, arabinolunase, galactokinase, saponins-I, II and III. It is recommended that the diabetic subjects included boiled whole grams in their daily diet to check the rise in blood sugar level.

Thinai, varagu and green gram have been studied in combination with herbal products and their hypoglycemic effect is presented under the heading herbal products in this chapter.

Soybean (Glycine Max)

Soybean is a pulse and an oilseed. It is an important food crop in China, Japan and Korea. It is called the 'Golden Cow' of China and the Cinderella crop of India and contains proteins (vegetable origin) of superior quality and is hence recommended for diabetics, especially vegetarians. For others, it could be an inexpensive substitute for expensive meat products. It is estimated that one hectare of land used for grazing will produce enough meat to satisfy one man's protein needs for only 190 days, planted, it will provide enough protein for 2,167 days, but if planted with soybeans, it will yield enough protein for 5,496 days.

Soybean contains 43.2g of protein, 19.5g of fat, 4.6g of minerals, 3.7g of fibre, 20.9g of carbohydrates, 432kcal of energy, 240mg of calcium, 690mg of phosphorus and 10.4mg of iron per 100g. Nazni and Amirthaveni (2007)[11] studied the effect of soybean supplementation on 20 diabetics. For this purpose, the amount of soybean given per diabetic was 25g per day, incorporated into rice flour as rice noodles, a common south Indian recipe. One serving of rice noodles was prepared with 45g of rice flour and 25g of soya flour (mixed with a few drops of lemon juice) and distributed to the diabetics each day for three months.

The impact of soybean supplementation on selected biochemical parameters is given in Table 3. After supplementation, Fasting Blood Glucose decreased by 27.88mg/dl, PPS by 30.03mg/dl, and HbA_{1C} by 4.69 per cent. Serum total cholesterol decreased by 19mg/dl, LDL cholesterol by 13.77mg/dl VLDL cholesterol by 4.77mg/dl and serum triglycerides by 23.85mg/dl while HDL cholesterol increased by 8.48mg/dl. These differences were statistically significant at one per cent level.

Table 3. Effect of Soy Bean Supplementation on Blood Parameters

Parameters	Desirable levels	Before	After
Blood glucose FBS (mg/dl)	80-115#	163.16	135.28**
PPS (mg/dl)	120-160#	264.74	234.71**
HbA_1C (%)	<8#	10.38	5.69**
Serum lipids (mg/dl) Total Cholesterol	150-200^	244.01	225.01**
HDL Cholesterol	30-60^	37.01	45.49**
LDL Cholesterol	66-178^	164.49	150.72**
VLDL Cholesterol	6-30^	33.22	28.45**
Triglycerides	30-17^0	166.11	142.26**

#Bamji *et al* (2006)[9]; ^- NCEP (2001)[8]; **- Significant at one per cent level.

Leafy Products

India is known for its rich flora most of which have been used in the diet from ancient times. These leafy vegetables are rich in vitamins and minerals and are commonly used in the traditional Indian cuisine. Many of them have special medicinal significance.

Drumstick Leaves (*Moringa Oleifera*)

Drumstick leaves is the cheapest and most versatile crop common in every rural backyard or urban slum in India. According to Trees for Life (2007)[12], India's ancient tradition of

Ayurveda claims that *Moringa* leaves can prevent 300 diseases. Scientific research shows that *Moringa* leaves are in fact a power house of nutrients. The nutrient equation of *Moringa* leaves in comparison with other common foods is depicted below (Plate 4):

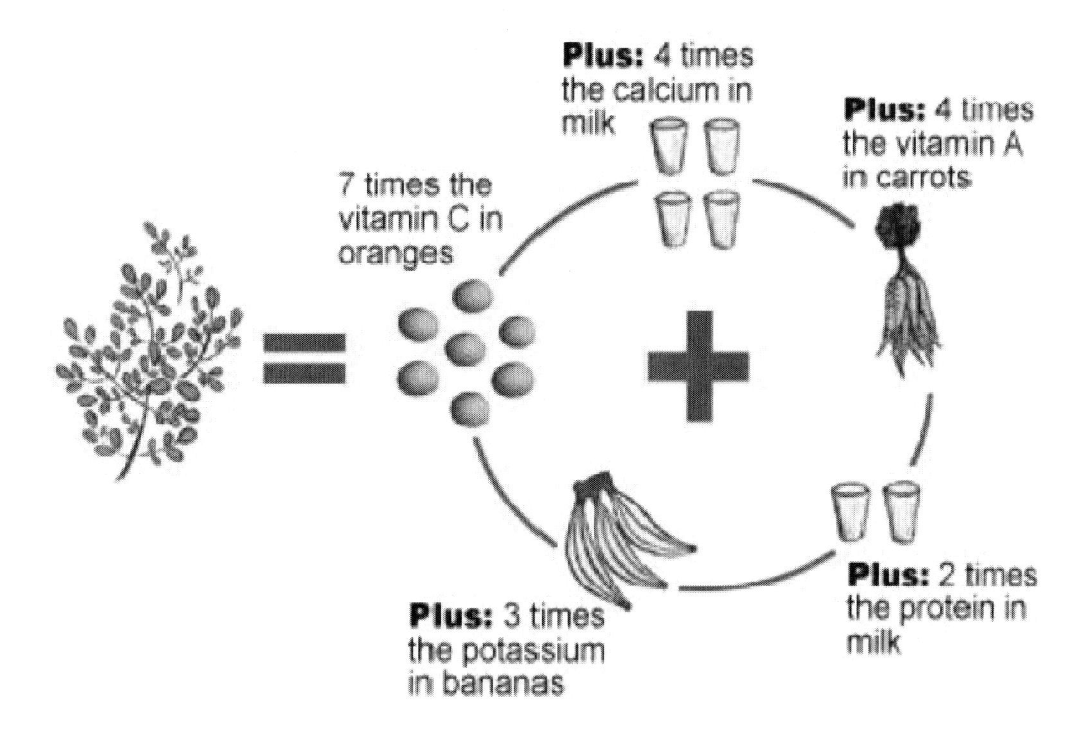

Plate 4. Nutrient equation of *Moringa* leaves [12].

Fenugreek (Trigonella Foenum-Graecum)

Fenugreek or methi is the dried ripe fruit of an annual herb, native of south-eastern Europe and west Asia. India is a major producer and exporter of fenugreek. Fenugreek seeds contain proteins, starch, sugars, mucilage, mineral matter, volatile oil, vitamins, enzymes, saponins and hypoglycemic phytochemicals. Whole fenugreek seeds are about 50 per cent fibre with 20 per cent of mucilage because of which they possess hypoglycemic properties.

A 1:1 hypoglycemic mix was prepared with fenugreek and drumstick leaves powder by Subapriya and Sasikala (2005)[13]. The nutritive value of the hypoglycemic mix in comparison with drumstick leaves powder is given in Table 4.

The energy, fibre, carbodhydrate and calcium content of the hypoglycemic food mix increased well above the levels found in *Moringa* leaf powder. The effect of supplementation of the hypoglycemic mix (10g/day/person) was assessed on 20 diabetics, divided into two groups A and B given *Moringa* leaf powder and hypoglycemic food mix respectively for a period of three months. Distribution of patients in different weight categories before and after supplementation is given in Table 5.

Table 4. Nutrient Content of Hypoglycaemic Mix (Per 100g)

Details	Energy (Kcal)	Protein (g)	Fat (g)	Fibre (g)	Carbo hydrate (g)	Cal cium (g)	Iron (mg)	Thia mine (mg)	Ribo flavin (mg)
Moringa leaves powder	92	24.28	1.70	2.55	12.50	1.52	2.55	0.12	0.05
Hypoglycemic Food Mix	212.5	25.24	3.75	5.15	28.30	80.76	3.85	0.23	0.21

Table 5. Distribution of Patients in Different Weight Categories (N=10 per Group)

Groups/Category	Group A		Group B	
	Before	After	Before	After
Over weight	50	20	10	-
Ideal weight	10	50	30	50
Under weight	40	30	60	50

After supplementation more number of patients moved into the ideal weight category in both groups. There was a decrease in the number of persons either under or over weight. The effect of supplementation on physiological parameters is shown in Table 6.

Table 6. Effect of Supplementation on Physiological Symptoms of Diabetic Subjects (N=10 per Group)

Symptoms	Group A		Group B	
	Before	After	Before	After
Polydipsia	4	-	3	-
Polyuria	3	-	2	-
Polyphagia	5	-	6	-
Constipation	1	-	-	-
Insomnia	4	-	2	-
Shivering	1	-	1	-
Giddiness	2	-	2	-
Burning sensation in extremities	-	-	1	-
Burning sensation during micturition	-	-	1	-
Frequent micturition	2	-	3	-
Multiple symptoms	10	-	6	-

It was encouraging to note that all the physiological symptoms of diabetes disappeared in both the groups after the supplementation period. The effect of supplementation on biochemical parameters is presented in Table 7. Blood glucose, both fasting and post prandial, serum total cholesterol, LDL and VLDL cholesterol, HbA_{1C} and triglycerides decreased significantly ($p<0.01$) while HDL cholesterol increased significantly ($p<0.01$) after supplementation.

Table 7. Effect of Supplementation of Moringa Leaf Powder (A) and Hypoglycemic Mix (B) on Biochemical Parameters (Mg/Dl)

Parameters/Groups	Before	After
Blood glucose A - Fasting	111.8	104.1**
Post prandial	206.7	170.8**
B- Fasting	107.0	99.7**
Post prandial	214.5	154.0**
Serum Total cholesterol- A	265.4	254.4*
B	254.6	239.3**
LDL Cholesterol – A	150.5	148.4**
B	149.8	134.6**
HDL Cholesterol – A	44.38	45.94**
B	48.7	55.9**
VLDL Cholesterol – A	30.3	25.1**
B	29.9	22.6**
Triglycerides – A	172.2	161.3**
B	170.07	153.8**

** Significant at one per cent level.

Bay Leaves (Laurus Nobilis)

The bay tree is native to the Mediterranean region and Asia Minor. Bay leaf or Laurel leaf is the dried leaves of an evergreen shrub or more rarely a tree attaining a height of 15-20m. The aroma of the crushed leaves is delicate and fragrant and taste is aromatic and bitter. The size of the leaves ranges from 2.5 to 7.5cm in length and 1.6 to 2.5cm in breadth. The shape is elliptical and tapers to a point at the base and tip of the leaves.

Bay leaves are robust, strongly aromatic with a woody astringent flavour and a pleasant slightly minty aroma. They are used to flavour all kinds of meat and vegetable dishes, soups and sauces. Bay leaves contain active phenolic phytochemicals with more than three fold potentiated insulin activity which probably helps to improve glucose metabolism. The chromium concentration of spices ranges from 4-1818ng/g, which however, does not correlate with their insulin potentiating activity.

Five hundred milligrams of bay leaves powder was incorporated into 30g of wheat flour, prepared into chappatis (pancakes made out of unleavened wheat dough and distributed to 25 diabetics daily for a period of three months. The effect of supplementation was assessed on blood glucose and serum lipid levels of the diabetics by Nazni and Amirthaveni (2007)[11]. Table 8 presents the effect of supplementation on various biochemical parameters. FBS, PPS, HbA_1C, total cholesterol, triglycerides, LDL and VLDL cholesterol decreased while HDL cholesterol increased significantly ($p<0.01$) after supplementation.

Table 8. Effect of Supplementation of Bay Leaf Powder on Blood Parameters (N=25)

Parameters	Before supplementation	After supplementation
FBS	164.35	146.53**
PBS	251.93	232.58**
HbA_{1C}	10.43	7.61**
Total Cholesterol	244.69	231.62**
HDL Cholesterol	38.33	46.08**
LDL Cholesterol	163.9	149.94**
VLDL Cholesterol	32.92	29.26**
Triglycerides	164.6	146.32**

** Significant at one per cent level.

Stevia Leaves (Stevia Rebaudina)

Stevia is the most health restoring plant on earth. It is a tropical shrub from Brazil, a large number of species of which are employed medicinally. It is widely used for the treatment of diabetes mellitus and has been investigated thoroughly in terms of botany, phytochemistry and pharmacology and proved to be safe for human consumption.

It is the sweetener of the future with zero calories, numerous nutrients and antioxidant properties. The natural sweet compounds in the stevia leaves are called diterpene glycosides or steviol glycosides. The normal proportions of these glycosides are stevioside (10%), rebaudioside A (14%), rebaudioside C (2%) and dulcoside (1%).

Among the various glycosides, stevioside is used as an anti hyperglycemic agent. It has a potential action on skeletal muscle, the major site of glucose disposal by which it reduces the high blood glucose level. Stevia regulates blood sugar by its high natural chromium content. It helps in increasing the number of islet cells on the pancreas for better insulin/blood sugar coordination.

Therapeutic doses of 1.25g of stevia leaves which contain 125mg of the active principle, the stevioside are recommended for hypoglycemic effect. Hence 1.25g of stevia leaf powder was given per day per individual in the form of tea (100ml) daily for a period of three months in a study by Nazni and Amirthaveni (2007)[11]. Supplementation brought about significant reduction in FBS, PBS, HbA_1C and improve lipid profile of subjects.

Insulin Plant Costus Pictus D. Don)

Insulin plant munched daily (2-3 leaves i.e., 6-9g) is strongly believed to exert hypoglycemic effect. Studies have indicated that these leaves are non toxic. A study was undertaken to unravel the hypoglycemic potential of insulin plant leaves by Premakumari and Nair (2006)[14] on alloxan induced diabetic albino rats and type II diabetics. The blood of the animals and human subjects were nearly euglycemic (112.9 and 116.97mg/dl respectively), eucholesterolemic (192.91 and 192.20mg/dl) and eutriglyceridemic (179.88 and161.9 respectively) when on insulin plant leaves.

Other Vegetables

A wide variety of gourd vegetables widely popular in Indian dietaries are acclaimed to possess antidiabetic potentials. Gourd vegetables belong to the family cucurbitaceae. They are the largest group of summer vegetable crops with considerable economic value. They are rich in iron and amino acids and are wonderful diuretics of great value in diabetes.

Bitter gourd *(Momordica charantia)* acclaimed to exert a three fold effect on blood sugar -it acts as a mediator between the body's cells and insulin, blocks the glucose formation in the blood stream and breaks down the barrier that prevents cells from using their own insulin.

Ivy gourd *(Coccinia indica)* is popular in Asian and in Thai Ayurvedic herbalism. It inhibits the activity of enzyme glucose-6-phosphatase which helps to balance blood glucose levels. Pectin from ivy gourd has significant hypoglycemic activity. Bottle gourd *(Lagenaria vulgaris)* is valuable against urinary disorders.

Since only limited studies are available to support the antidiabetic property of bitter gourd and ivy gourd and studies are lacking about the hypo activity of other gourd vegetables, studies were carried out using 75g each of five gourd vegetables namely, bitter gourd *(Momordica charantia)*, ivy gourd *(Coccinia indica)*, snake gourd *(Trichosanthes anguina)*, bottle gourd *(Lagenaria vulgaris)* and ridge gourd *(Luffa acutangula)* by Radhapriya and Lakshmi (2009)[15] on 36 adults with NIDDM. Maximum reduction in FBS was recorded for Ivy gourd (61.71mg/dl) while maximum reduction in PPS was observed in the bitter gourd group (66.10mg/dl). Ivy gourd was most effective in controlling triglyceride and VLDL cholesterol levels while bottle gourd had the highest effect on total cholesterol and HDL and LDL cholesterol levels.

White Pumpkin (*Benincasa Hispida*) with Curry Leaves (*Murraya Koenigii)*

White pumpkin is native to tropical America and is grown throughout India. Pumpkins are rich repositories of Beta carotene, calcium, potassium and fibre. In traditional (Indian and Chinese) medicine, pumpkins are said to be cooling, diuretic and good for stomach aspects. They are believed to regulate blood sugar levels and stimulate the pancreas. Three compounds, beta sitosterol, alpha amyrin and quercetin are reported to exert hypoglycemic effect in alloxan induced diabetic rats. The alcoholic extract at 200mg/kg significantly reduces the blood glucose levels equipotent to the standard drug, tolbutamide.

Curry leaves are used almost everyday in all dishes in the Indian cuisine. Feeding of diabetic rats containing various doses of curry leaves showed varying hypoglycemic and anti hyperglycemic effect. Plasma insulin was significantly high which suggests that the hypoglycemic effect may be mediated through stimulating insulin synthesis and / or secretion the beta cells of the pancreatic islets of Langerhans.

In a study by Priya and Amirthaveni (2009)[16], ash gourd and curry leaves were washed and cleaned. The outer hard skin was removed and cut into small pieces and fresh curry leaves were cut into very small pieces. Fresh salad was made from 100g of ash gourd and one gram of curry leaves (10 leaves) and curd prepared from five grams of skimmed milk powder. Pepper and salt were added to taste. The fresh salad was distributed to 20 diabetics as mid morning supplement for a period of three months.

Curry leaves in combination with white pumpkin in the form of a salad supplementation resulted in significant reduction in blood glucose levels (FPG and PPG) and improved lipid profile of the subjects. Also it was noted that initially 14 subjects were in Obesity Grade I and the number reduced to 4 at the end of the study.

Artichoke (Cynara Scolymus)

Artichoke is a member of the thistle family although its leaves are bigger, broader and softer rather than prickly. Artichoke plants are beautiful, like giant ferns-six feet in diameter and 3-4 feet high. The heart of the flower bud is prized for its delicate flavour and its versatility. The flower petals and fleshy flower buttons are eaten as a vegetable throughout the world which has led to its commercial cultivation in many parts of South and North America and Europe. Artichoke was used as a food and medicine by the ancient Egyptians.

Artichoke is nutrient dense and a medium sized flower will provide 25kcal of energy and trace amounts of minerals such as magnesium, chromium, manganese, potassium, phosphorus, iron and calcium. It contains 12 per cent fibre, 10 per cent vitamin C and 10 per cent folate with no fat or cholesterol. The active chemical components are cyanarin, flavonoids, sesquiterpene lactones, polyphenols and caffeoyl quinic acids. Silymarin is a powerful antioxidant that may help the liver to regenerate healthy tissue.

In a study by Nazni and Amirthaveni (2007)[11], six grams of artichoke powder was given per day for an individual incorporated into 20g of wheat flour and prepared as sugar free biscuits. Four biscuits were prepared and distributed daily as mid morning and evening snacks, two at a time for three months to 30 NIDDM subjects. Over a period of three months the investigators observed a significant improvement in the blood parameters in terms of blood glucose and lipid profile.

Seeds and Spices

Cinnamon (Cinnamonum Zeylanicum)

Cinnamon (Plate 5) is one of the oldest spices known. It was so highly treasured that it was considered more precious than gold.

Cinnamon also received much attention in China which is reflected in its mention in one of the earliest books on Chinese botanical medicine, dated around 2,700 B.C. Currently, it is grown in Srilanka, India and other countries. Cinnamon has a long history both as a spice and as a medicine. It is the bark of the cinnamon tree which is available in its dried tubular form. It is commonly used as a beverage, flavouring and medicinal agent. Cinnamaldehyde or cinnamic aldehyde has been well researched for its effect on blood platelets. It helps to prevent unwanted clumping of blood platelets.

Plate 5. Cinnamon.

This health protective act is accomplished by inhibiting the release of arachidonic acid from platelet membranes. The most active compound of cinnamon is Methyl Hydroxy Chalcone Polymer (MHCP) which increases the conversion of glucose to energy, 20 times and inhibits the formation of dangerous free radicals during metabolic activities. Two grams of encapsulated cinnamon powder was provided for three months to 15 NIDDM subjects in a study by Balasasirekha R and Lakshmi UK (2008)[17]. Fasting and post prandial blood sugar decreased by 47.07 and 56.0mg/dl respectively, while triglycerides, total cholesterol, LDL and VLDL cholesterol decreased by 36.2, 53.73, 53.62 and 7.24mg/dl respectively. HDL cholesterol increased by 7.13mg/dl.

Cumin Seeds (Cuminum Cyminum)

Cumin seeds are extensively used in Ayurvedic medicine for treatment of dyspepsia, diarrhoea and jaundice. It is known to enhance the activity of pancreatic lipase and amylase. It is inferred that this positive influence on the activity of enzymes may have a supplementary role in the overall digestive stimulant action, besides enhancing the enzymes in pancreatic tissue.

The main constituents of cumin seeds are 2.5-4.0 per cent essential oil and 25-35 per cent cumin aldehyde, which is the lead compound, proved to be a new agent for anti diabetic therapeutics. Five grams of roasted cumin seed powder was added to 45g of wheat flour made into masala chapathis and administered for dinner for three months to a group of 25 NIDDM subjects, 45 to 50 years old by Nazni and Amirthaveni (2007) [11].

Table 9. Effect of Supplementation of Cumin Seeds on Blood Parameters

Parameters (mg/dl)	Before supplementation	After supplementation
FBS	160.93	140.14**
PBS	259.61	236.72**
HbA_{1C}	9.56	7.0**
Total Cholesterol	246.35	223.98**
HDL Cholesterol	36.09	40.19**
LDL Cholesterol	162.45	151.05**
VLDL Cholesterol	35.60	32.67**
Triglyceride	175.29	163.4**

** Significant at 1 per cent level.

Table 9 presents the effect of supplementation on various biochemical parameters at the end of the study period. There was a positive impact due to administration of cumin seeds. Hence it could be used as a very effective household remedy to control diabetes mellitus.

Onion (Allium Cepa)

Onions are extensively cultivated in India. The underground bulbs, which constitute the crop, vary in size, colour, firmness and strength of flavour. Onion is valuable and is used either as a salad or spice or for cooking with other vegetables. Its characteristic pungency is due to volatile sulphur compounds.

One hundred grams of onion contributes 50kcal of energy, 1.2 g of protein, 0.1g of fat, 11.1g of carbohydrates, 0.6g of fibre, 46.9mg of calcium, 50mg of phosphorus, 0.6mg of iron, 0.08mg of thiamine, 0.01mg of riboflavin, 0.4mg of niacin and 11.0mg of vitamin C (Gopalan *et al*, 2004)[18]. The principal active ingredients in onions are Allyl Propyl Disulphide (ADPS) although other constituents such as flavonoids also play a role in reducing blood glucose level gradually.

In a study by Nazni and Amirthaveni (2007)[11] 50g of onion was used to extract fresh juice which was diluted in 200 ml of diluted butter milk and administered to the diabetics daily for a period of three months. Apart from a significant reduction in the mean fasting and post prandial glucose levels, mean glycosylated haemoglobin level reduced from the initial level of 9.9 per cent to 6.4 per cent which was significant at one per cent level. The mean HDL cholesterol level increased from 38.22mg/dl to 48.61mg/dl while the triglyceride, LDL cholesterol and VLDL cholesterol decreased to an appreciable extent.

Fruit Products

Amla (Emblica Officinalis) along with Turmeric (Curcuma Longa)

Amla fruit also known as Indian gooseberry is one of the richest sources of bioflavonoids and vitamin C which are resistant to storage and heat damage due to cooking. It is highly nutritious with thrice the protein and twice the ascorbic acid concentration as in apples. It is traditionally used as a laxative, eyewash, appetizer, stimulant and an antidote to anorexia, diarrheoa, anaemia, indigestion and jaundice.

Amla is the richest source of vitamin C which is an antioxidant that blocks some of the damage caused by free radicals. It contains 600mg of vitamin C per 100 grams, which is 20 times more than the vitamin C present in oranges. Vitamin C is helpful for people with diabetes because, it combats the free radicals which cause atherosclerosis and decreased level of antioxidants including vitamin A. Insulin helps the body to take up vitamin C that is needed for proper body function. Thus extra vitamin C supplements may be helpful in diabetes mellitus.

Turmeric (Plate 6) is a condiment acclaimed to be rich in various powerful antioxidants free of side effects and toxicity. It has been used in a multitude of ailments.

To authenticate the claims scientifically, Premakumari and Coworkers (1993)[19] distributed weekly doses of turmeric and amla *(Emblica officinale)* to a group of 36 NIDDM subjects for three months. The diabetics were divided into six groups of six subjects each. Members include groups A and B consumed amla juice, C and D consumed turmeric made into paste with water and E and F consumed a paste of turmeric with amla juice. They consumed these components each day before breakfast and after dinner (15g of the herb per day divided into two parts). Turmeric and amla either alone or in combination were effective in reducing blood sugar (both fasting and post prandial).The best effect was observed in group given 15g of amla followed by 15g of amla and turmeric in combination. The reduction was to the tune of 11.0 to 21.1mg/dl.

Plate 6. Turmeric.

Jumbolin Seeds (Syzygium Cumini)

The jumbolin is native in India, Burma, Srilanka and the Andaman islands. There are different types of fruits which vary in colour and size. The ripe fruit is widely eaten in India. The leaf juice is used in the treatment of dysentery, either alone or in combination with the juice of mango leaves. The leaves may be helpful in the treatment of skin diseases. Jumbolin of good size and quality having a sweet or sub acid flavour and minimum astringency are eaten raw and may be made into tarts, sauces, jam, sherbet, syrup and squash. The fruit has received far more recognition in folk medicine and in pharmaceutical trade than in any other field.

Jumbolin seeds are found to contain an alkaloid, jambosine and an glycoside, jambolin or antimellin, which halts the diastatic conversion of starch into sugar. One serving of two dosas (pan cakes) were prepared with 50g of rice batter with five grams of jambolin seed powder and administered to NIDDM subjects daily for a period of three months by Nazni and Amirthaveni (2007) [11]. The levels of FBS, PBS, HbA_1C decreased significantly at the end of three months study. Also the blood lipid parameters changed favorably at the end of the study.

NUTS AND OILSEEDS

Flax Seeds (Linum Usitatissimum)

Flax seed is an ancient blue flowering crop, called linseed. It is the hard, tiny seed of the flax plant and has been widely used for thousands of years as a source of food. It is praised as the neutraceutical food of the 21st century. Flax seed is rich in fatty acid and fibre content and contains nearly 28 per cent fibre, six per cent carbohydrates, 21 per cent protein, 42 per cent fat, and 30 per cent PUFA. Of this, 24 per cent is omega 3 fatty acids and six per cent is omega 6 fatty acids. It is rich in phenolic compounds, lignin and oil which may help lower cholesterol and prevent atherosclerosis. It is the richest plant source of lignan precursors. Flax seed contains Secoisolariciresinal Diglucoside (SDG), a potent antioxidant.

Table 10. Effect of Supplementation of Flax Seed on Blood Parameters of NIDDM Subjects

Parameters (mg/dl)	Before supplementation	After supplementation
FBS	165.08	157.28**
PBS	264.44	254.10**
HbA_{1C}	9.17	7.89**
Total Cholesterol	245.19	240.71**
HDL Cholesterol	37.44	42.35**
LDL Cholesterol	163.86	142.48**
VLDL Cholesterol	33.06	29.72**
Triglycerides	165.32	147.48**

** Significant at one per cent level.

Five grams of flax seed powder was incorporated into wheat flour and prepared as wheat bread and given to 25 NIDDM subjects daily for a period of three months by Nazni and Amirthaveni (2007)[11]. Table 10 presents the effect of supplementation of flax seed on various biochemical parameters of the subjects studied. There was a very impressive change in the blood parameters of NIDDM subjects.

Chindil Kodi (Tinospora Cordiflora)

Chindil kodi is a climbing shrub with heart shaped leaves and is believed to enhance longevity, promote intelligence and prevent taste. It has a sweet, bitter and acid taste. According to Indian legend, it is known as 'amrita' or the heavenly elixir to protect the celestial people and keep the angels eternally young. The leaves are rich in protein, calcium and phosphorus. It is indicated to be hypoglycemic, antibacterial, antimicrobial, antipyretic, anti-inflammatory, antiarthritic, antiallergic, hepatoprotective, analgesic, immunosuppressive, immunostimulant, antileishmanial, antioxidant, antiendotoxic, hypotensive, diuretic, antiulcer, bronchodialator, and antirheumatic. In a study by Kalaivani and Amirthaveni (2008) [20], leaves, stem and roots of Tinospora cordiflora were collected, shade dried and powdered. Two grams of the powder was administered after breakfast, each day, to 20 diabetics for a period of three months. Body measurements and blood glucose levels was used as criteria to study the impact. The distribution of diabetics according to BMI before and after the study period is given in Table 11.

Table 11. Distribution of Diabetics before and after Study

Criteria	Before	After
BMI*	Number	Number
18.5-24.9 25.0-30.0 >30.0	9 8 3	10 8 2
WHR Low risk	8	12
Moderate risk	8	6
High risk	4	2

* Classification (WHO, 2005)[4]

A gradual reduction in the body weight of the subjects who were over weight or obese initially was observed in this study. Number of subjects with high risk Waist Hip Ratio (WHR) entered into moderate risk category and many in the moderate risk category entered into low risk category. The fasting and post prandial blood glucose levels of the diabetics initially as well as monthly over a period of 90 days are given in Table 12. The results showed remarkable improvement in the biochemical parameters tested.

Table 12. Fasting and Post Prandial Blood Glucose Level (Mg/Dl) before and after *Tinospora Cordiflora* Supplementation

Blood glucose level	Normal # value	Initial	Final			Mean difference	t value
			30 days	60 days	90 days		
Fasting	80-110	118.0	180.3	172.8	166.8	21.2	14.5**
Post prandial	120-160	257.9	248.0	240.8	235.4	22.5	10.2**

Bamji (2006)[8]; ** Significant at one per cent level.

BEVERAGES

Green Tea (Camelia Sinensis)

The medicinal effects of tea have a long and rich history, the first references dating back to 5000 years. It has been consumed literally for thousands of years. Its long safety record makes it an attractive target for drug discovery. In traditional Indian medicine, green tea has been recorded as a mild excitant, stimulant, diuretic and astringent and the leaf infusion was used as a remedy for fungal infections. In traditional Chinese medicine, it has been used as an astringent, cardiotonic, central nervous system stimulant and diuretic. It has been used in the treatment of flatulence, regulation of body temperature, promotion of digestion and improvement of mental processes. Green tea contains organic acids-1.5 per cent, caffeine-3.5 per cent, carbohydrates – 25 per cent; lipids-2 per cent, polyphenols- 37 per cent, protein-15 per cent, lignin-6.5 per cent, ash-5.0 per cent and chlorophyll-0.5 per cent.

A study was conducted to assess the hypoglycemic potential of green tea by Subapriya and Archana (2006)[21]. Ten grams of green tea was supplemented per person per day to 20 mild diabetics. This was divided into two doses of five grams each taken in the morning and evening. To brew one teaspoon of green tea, the leaves were combined with one cup of boiling water and steeped for three to five minutes.

The effect of supplementation on the parameters studied is given in Table 13.

Table 13. Effect of Supplementation of Greentea on Body Parameters of Mild Diabetics

Parameters	Male (N=10)		Female (N=10)	
	Before	After	Before	After
Weight (kg)	72.16	71.08**	63.0	62.12**
BMI	26.58	25.9NS	26.65	26.2NS
Systolic pressure	137.5	123.0**	135.5	124.0**
Diastolic pressure	85.5	81.0**	85.5	80.0**
Fasting Blood Glucose (mg/dl)	171.4	153.82**	173.34	154.61**
Total cholesterol (mg/dl)	201.26	190.8**	201.85	188.36**
HDL Cholesterol (mg/dl)	54.88	65.45**	67.16	78.80**
LDL Cholesterol (mg/dl)	107.43	86.85*	93.15	72.48**
VLDL Cholesterol (mg/dl)	38.89	38.48 ^{NS}	42.57	38.17^{NS}
Serum Triglycerides (mg/dl)	194.63	192.4^{NS}	202.91	184.86^{NS}
Serum calcium (mg/dl)	9.29	10.63**	9.29	10.70**
Total protein (g/dl)	6.35	7.63**	6.31	7.31**

**Significant at one per cent level; NS Not Significant.

While body weight, systolic and diastolic pressures, FBS, total and LDL cholesterol decreased significantly, HDL cholesterol, serum calcium and serum total protein levels increased significantly ($p<0.01$). However statistically insignificant changes were observed in BMI, VLDL cholesterol and serum triglyceride levels.

Herbal Products

Nowadays people understand the importance of herbs and natural foods. Many products commonly used today are of herbal origin and nearly 80 per cent of the world population use herbal products for health care.

Aavaram flower (Cassia auriculata) (Plate 7) commonly called Tanner's cassia is widely used in Indian folk medicine for treatment of diabetes mellitus.

Kadazhalinjil (Salacia reticulata), (Plate 8) ponkoranthi is used in Ayurveda since ages for treating diabetes. It has the property to obstruct and restrain an enzyme which compounds glucose in intestinal wall.

Kotakaranthai (Sphaeranthus indicus) (Plate 9) is an alternative and general nerve tonic. It is known to help in sexual debility, general debility, nervous exhaustion, loss of memory, loss of muscular energy.

Korai kizhangu (Cyprus rotundus) (Plate 10) is used for dyspepsia and as a galactogogue. It contains saponins, alkaloids, proteins, starch, tannins, mucilage and diosgenin as the active constituents. Hence it is compared with a modern drug, metociophramide, which is used in dyspepsia to reduce gastric emptying time and to increase milk secretion in lactating women.

Vilvai leaves (Aegle marmelos) (Plate 11) possesses insulin like action. They are antibilous, anti parasitic, antipyretic, aphrodiasic, aromatic, astringent, digestive stimulant, febrifuge, haemostatic and laxative.

Plate 7. Aavaram flower.

In a study by Radha and Amirthaveni (2007)[22], the above herbs were tried out along with cleaned Thinai and coarsely powdered green gram in different forms of south Indian recipes- porridge, pongal and pan cake. These were tested for their glycemic index and pongal preparation with added herbal powder mixes was selected for the supplementation study since it was locally popular as a breakfast item and had low glycemic index.

Plate 8. Kadazhalinjil.

Plate 9. Kotakaranthai.

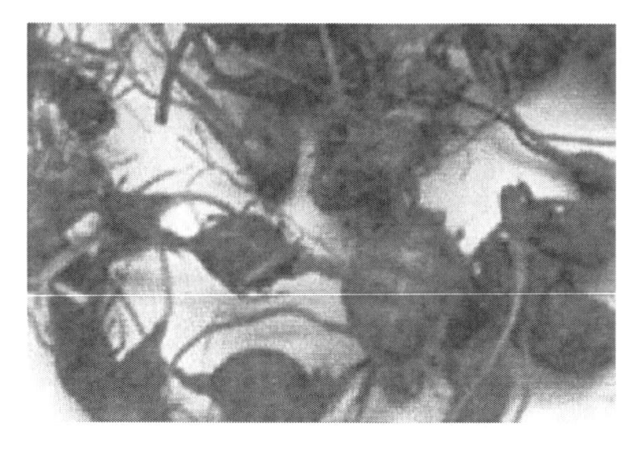

Plate 10. Korai kizhangu.

Plate 11. Vilvai leaves.

Comparison of the effect of supplementation of Thinai, green gram and various herbal powders on blood parameters is given in Table 14. Thinai green gram avaram flower combination was found to have the highest hypoglycemic potential followed by thinai green gram vilvai leaves combination. The other herbal products - millet green gram combinations were also effective in reducing the blood sugar.

Table 14. Comparison of Different Herbal Supplementations on Blood Glucose Level

Fasting Blood Glucose		Post prandial Blood Glucose		Glycosylated Haemoglobin	
Groups	Mean difference (mg/dl)	Groups	Mean difference (mg/dl)	Groups	Mean difference (mg/dl)
TG-AF	-20.67	TG-AF	-26.04	HC-AF	-2.2
TG-VL	-17.33	TG-VL	-21.4	TG-AF	-2.07
HC-KK	-17.2	HC-AF	-18.97	TG-KK	-1.98
HC-AF	-16.07	HC-KK	-17.27	HC-VL	-1.65
TG-KL	-15.7	TG-KL	-16.17	TG-VL	-1.35
TG-KK	-14.8	HC-VL	-15.3	TG_KL	-1.3
HC-VL	-14	TG-KKZ	-14.67	HC-KK	-1.25
TG-KKZ	-9.26	TG	-12.96	TG-KKZ	-1.14
HC-KL	-8.26	HC-KK	-11.9	HC-KL	-1.1
HC-KKZ	-6.2	HC-KL	-9.1	HC-KKZ	-0.68
TG	-5.67	HC-KKZ	-8.1	TG	-0.67

TG-Thinai, Green Gram, HC-Herbal Capsules AF – Aavaram Flower, KL-Kadazhalinjil bark, KK-Kotta Karranthai Plant, KKZ- Korai Kizhangu, VL – Vilvai Leaves.

Ekanayakam (*Salacia reticulata*) is a predominantly found herb native in India. It belongs to the family Hippocrataceae many of which have been extensively used in the treatment of diabetes in the Ayurvedic system of Indian traditional medicines. Premakumari and Coworkers (2007)[23] studied the hypoglycemic effect of Ekanayakam on albino rats and human diabetics. Salacia reticulate did not produce any significant toxic effect in rats over a period of 90 days. Human and rat studies revealed significant hypoglycemic effect in terms of both FBS ad PPS.

Conclusion

It is thus evident that the safest and most sustainable strategies to combat diabetes mellitus are vested in 'God's pharmacy', i.e., nature. Natural foods are known to contain several phyto chemicals and antioxidants which exert beneficial synergism in the human body.

They are also considered safe since they do not exhibit harmful side effects and thus have an edge over allopathic mixtures. The above mentioned plant foods have proven antidiabetic potentials. Many other foods of plant origin not dealt with in the studies cited above may also possess good hypoglycemic potentials.

Research on such foods is the urgent need of the hour. Such research would be a boon to control diabetes in a country like India which is reeling under the multiple burdens of poverty, illiteracy, population explosion, rapid urbanization and burgeoning non communicable diseases, especially diabetes mellitus.

References

[1] Arteaus (70 AD) Maurice E.Shils, James A. Olson, Moshe Shike. *Modern Nutrition in Health and Disease.* VIII Edition. , Lea and Febiger A Waverly Company, USA. PP. 1259-1261.

[2] Thomas Willis (1965), Maurice E.Shils, James A. Olson, Moshe Shike. *Modern Nutrition in Health and Disease.* VIII Edition. , Lea and Febiger A Waverly Company, USA. PP. 1259-1261.

[3] Wild S, Roglic G, Green A, Sicree R and King H (2004) Global prevalence of diabetes-Estimates for the Year and Projections for 2030, *Diabetes Care,* Vol. 27, No. 10, pp 2568 and 2569.

[4] World Health Organisation (2005) *Diabetes and CVD.* 2004-2005, PP 602-605.

[5] World Health Organisation (2007) *International Cardiovascular Disease Statistics.* pp 7 and 8.

[6] International Diabetes Federation (2006) Diagnosis and classification of diabetes, *Diabetes Care.* 30, No.1, pp. 42-47.

[7] World Health Organisation (2009) Raghuram TC Textbook of Human Nutrition, 3rd Edition, Oxford and IBH Publishing Company Pvt. Ltd , New Delhi, India. pp. 357-359.

[8] National Cholesterol Education Programme (2001), NIH News Release, 15th May 2001.

[9] Bamji M.S, Pralhad Rao N and Vinodini Reddy (2006), Text Book of Human Nutrition, Oxford And IBH Publishing Company Pvt Ltd, New Delhi, pp.334,338.

[10] Premakumari S and Haripriya S (2008-2010) Effect of Supplementation of Wheat Germ, Wheat Bran and Wheat Grass to Subjects with Specific Health Issues. *Report of the UGC Minor Research Project,* pp 11, 15, 19, 34, 35, 40.

[11] Nazni P and Amirthaveni M (2007) Effect of Supplementation of Selected Nutraceuticals on Type II Diabetics. A Thesis submitted to Avinashilingam University for Women, Coimbatore in Partial fulfillment of the requirement for the Degree of Doctor of Philosophy in Food Science and Nutrition, pp 39-53, 57.

[12] Trees for Life (http://www.treesforlife.org/documents/moringa/presentation/ Moringa%20Presentation%20(General)%20screen.pdf) (2007)

[13] Sasikala S and Subapriya M.S (2005) Effect of supplementation of a hypoglycemic food mix on Type II diabetes mellitus patients. A Thesis submitted to Avinashilingam University for Women, Coimbatore in Partial fulfillment of the requirement for the Degree of Master of Science in Food Science and Nutrition, pp. 18, 20-22, 26, 50, 51,53.

[14] Premakumari S and Divya G. Nair (2006) Hypoglycemic potential of insulin plant (*Costus pictus D. Don)* in Alloxan induced diabetic rats and type II diabetics. *Indian Journal of Nutrition and Dietetics,* 44, 168.

[15] Radhapriya S and Lakshmi UK (2009) Effect of supplementation of gourd vegetables on diabetic patients. A Thesis submitted to Avinashilingam University for Women, Coimbatore in Partial fulfillment of the requirement for the Degree of Master of Philosophy in Food Science and Nutrition. Pp. 28,41.

[16] Priya V and Amirthaveni M (2009) Hypoglycemic and Hypolipidemic Effect of Ash Gourd *(Benincasa hispida*) and Curry leaves (*Murraya koenigii*). A Thesis submitted to Avinashilingam University for Women, Coimbatore in Partial fulfillment of the requirement for the Degree of Master of Philosophy in Food Science and Nutrition pp. 26, 29, 43, 87, 89 and 91.

[17] Balasasirekha R and Lakshmi UK (2008) Effect of selected spices on hyperlipidimic and diabetic adults. A Thesis submitted to Avinashilingam University for Women, Coimbatore in Partial fulfillment of the requirement for the Degree of Doctor of Philosophy in Food Science and Nutrition. Pp. 37,39, 45, 112, 115, 118, 121, 124, 127, 129 and 130.

[18] Gopalan C, Rama Sastri and Balasubramanian S.C Revised and updated by Narasinga Rao, Deosthale YGand Pant KC (2007) Nutritive Value of Indian Foods, NIN, ICMR, India. Pp. 50.

[19] Premakumari S, Padmavathy C and Ponne S (1993) Hypoglycemic potential of Amla (*Emblica officinalis*) and Turmeric (*Curcuma Longa). Indian Journal of Nutrition and Dietetics,* 22, 154.

[20] Kalaivani S and Amirthaveni M (2008) Effect of supplementation of chindil kodi (*Tinospora cordiflora)* on type II diabetes mellitus. A Thesis submitted to Avinashilingam University for Women, Coimbatore in Partial fulfillment of the requirement for the Degree of Master of Science in Food Science and Nutrition. Pp. 11, 38, 71, 74, 76.

[21] Archana P.V. and Subapriya M. S (2006) Health profile of the diabetics in the Nilgiris. A Thesis submitted to Avinashilingam University for Women, Coimbatore in Partial fulfillment of the requirement for the Degree of Master of Science in Food Science and Nutrition, pp. 6, 31, 67, 68.

[22] Radha R and Amirthaveni M (2007) Effect of supplementation of herbs on NIDDM patients. A Thesis submitted to Avinashilingam University for Women, Coimbatore in Partial fulfillment of the requirement for the Degree of Doctor of Philosophy in Food Science and Nutrition. Pp. 28, 54, 57.

[23] Premakumari S, Kowsalya S, Sailaavanya S and Mujumdar V V (2009*). Indian Journal of Nutrition and Dietetics,* 46, 1.

In: Natural Products and Their Active Compounds ... ISBN: 978-1-62100-153-9
Editors: M. Essa, A. Manickavasagan, and E. Sukumar © 2012 Nova Science Publishers, Inc.

Chapter 4

DEVELOPMENT OF ISCHEMIC DISEASES AND STEM CELL THERAPY

Reeva Aggarwal, Jingwei Lu, Vincent J. Pompili and Hiranmoy Das*

Cardiovascular Stem Cell Research Laboratory,
The Dorothy M. Davis Heart and Lung Research Institute,
The Ohio State University Medical Center, Columbus, OH, US

INTRODUCTION

Cardiovascular diseases are rising world-wide and World Health organization (WHO) expects that by year 2020, cardiovascular diseases will cause about 25 million deaths globally (Chockalingam A, 1999). Atherosclerosis is the major underlying cause of the congestive heart failure, myocardial ischemia or peripheral vascular disease. In 2005, one in every five American death was due to the coronary heart disease (CHD) (Lloyd-Jones et al., 2009). Atherosclerosis is characterized by hardening of the arteries and buildup of plaques in the inner lining of arteries. Plaques are formed by the deposition of fatty substances, cholesterol, cellular debris and calcium from the blood components. It is a slow and complex disease involving many cells such as endothelial cells (Luscher and Barton, 1997), vascular smooth muscle cells (Libby, 2000) and inflammatory immune cells (Galkina and Ley, 2007). The plaques grow in time and cause obstruction to the blood flow and usually the symptoms are not evident until the later stages where there is complete obstruction to blood flow leading to negligible oxygen supply and nutrients to the myocardial or endothelial tissue.

Plaques protrude into arterial lumen when grow bigger in size, rupture and cause thrombosis leading to hypertension, acute myocardial infarction or stroke (Ross, 1999). Medications such as anticoagulants, vasodilator drugs, chemokine-based therapies are not

* Address correspondence to: Hiranmoy Das, PhD; Associate Professor of Internal Medicine; Director, Cardiovascular Stem Cell Research Laboratories; The Dorothy M. Davis Heart and Lung Research Institute; The Ohio State University Medical Center; 460 W. 12th Avenue, BRT 382, Columbus, Ohio 43210; (614) 688-8711 office, (614) 293-5614 fax; hiranmoy.das@osumc.edu

very reliable treatments as cellular damage has already occurred. It is believed that regenerative capacity of the myocardial tissue is inadequate to revert any damage to its cellular components.

Thus, currently regenerative strategies involving stem cell, which can differentiate into endothelial progenitor cells (EPCs) and vascular smooth muscle cells (VSMCs) are being investigated. This article mainly focuses on the biology of atherosclerosis, dietary component of the atherosclerosis, its progression causing mycocardial ischemia and treatment modalities involving/ focusing the stem cells.

Factors Responsible for Atherosclerosis

Factors that cause the development of atherosclerosis are the elevated levels of oxidized low density lipoproteins (ox-LDL), very low density lipoprotein lipids (VLDL), low levels of high density lipoprotein (HDL), hypertension, dietary fat, lack of physical activity, genetic disposition or age. HDL, LDL and VLDL transport endogenous lipids/cholesterol and chylomicrons transport the dietary lipids to adipose tissue and muscles. Increased risk of cardiovascular disease is related to increased levels of LDL cholesterol.

The atherosclerotic lesion develop in the endothelium of the large arteries, which causes recruitment of macrophages and subsequent uptake of LDL-derived cholesterol contributing to fatty streak formation (Navab et al., 1996). The LDL is oxidized by the several enzymatic and non-enzymatic processes in the matrix of artery wall which is thought to elicit immunological responses (Steinberg et al., 1989). These particles are called oxidized LDL (ox-LDL) and are recognized by receptors present on SMC, EC and macrophages (Freeman, 1997).

These activities further leads to chemotactic recruitment of monocytes to the intimal layer of the blood vessel and causes their adhesion to the vascular smooth muscle cells through specific adhesion molecules (discussed in detail later). These monocytes are differentiated into macrophages that engulf ox-LDL and ultimately become foam cells. Thus, Ox-LDL plays a multifunctional role in modulation of inflammation to the endothelial cells, which are implicated to promote thrombogenesis. The thrombogenic events cause stenosis or blockage of the major blood vessels supplying heart or peripheral organs.

Dietary Fat

Atherosclerosis is directly linked to the diet and diet is considered to be one of the major underlying causes of cardiovascular diseases. Foods rich in cholesterol/fat, sugar sweetened products, daily positive calorie balance and sedentary lifestyle pose major risk to the individual due to the development of atherosclerosis. Processed foods or fried food are responsible for increased level of oxidized cholesterol in the blood, which also increases the risk of cardiovascular disease and can cause formation of plaques that ultimately may block the arteries (Lichtenstein et al., 1998). Foods such as eggs, dairy products and meats being processed with the heat treatment, induce lipid oxidation in foods. Diet in the Western countries has high concentrations of oxidized lipid products, which are potentially atherogenic when present in circulation (Yagi et al., 1986).

Dietary fat is categorized as triglycerides (fats and oils), sterols (cholesterol) and phospholipids. The majority of the dietary fat is triglycerides, which are made up of three fatty acids esterified to a glycerol molecule. Triglycerides get hydrolysed to diglycerides, monoglycerides, free fatty acids and glycerol molecules. These hydrolysed molecules are absorbed into the intestine and are reassembled into a triglyceride molecule and are trans-ported to the liver or peripheral tissues via incorporation into chylomicron particles.

Fatty acids (FA) can be classified based on the length of the carbon chain or according to the presence or absence of the double bonds. The former is essential for absorption in body and later is important in regards to health related risk factors. FA having carbon atoms joined together by double or triple bonds are called unsaturated FA. For example, monounsaturated FA (FA with one double or triple bond) are found in nuts such as almonds, pecans or peanuts and in olive oil. Polyunsaturated FA (FA with more than one double or triple bond in single molecule of FA) is found in food such as fish, plant products, walnuts, sunflower seeds and safflower oil. Polyunsaturated FA (PUFA) is known to lower the cholesterol levels in the blood and beneficial to health.

Positions of the double bonds in PUFA are associated with the cardiovascular risk. If, the first double bond is at the third carbon, then FA is called omega-3 FA and if it is at sixth carbon, then it is called omega-6 FA. Each of these has different physiological functions. It was shown that omega-6 FA compete with omega-3 FA for common metabolic enzymes thereby increasing the production of prothrombic, inflammatory leukotrienes and prostaglandins (Simopoulos, 1999). The sources of omega-3 and 6 FA are most of the vegetable oils, egg yolk and fish. It is speculated that omega-6 FA may counteract the beneficial effects of omega-3 FA and thus raises concerns about the dietary intake of the ratios of these FA (Kris-Etherton et al., 2000).

Number of studies and tests on humans and mice showed reduction in development of atherosclerosis related events in vasculature when treated with omega-3 FAs (Kromann and Green, 1980; Kromhout et al., 1985). Cholesterol and PUFA form oxidation products when exposed to high temperatures, air, light and oxidizing agents and when consumed give rise to high levels of oxidized LDL leading to atherosclerosis.

The diet rich in saturated fatty acids (SFA), e.g., palmitate raises the LDL levels compared to the HDL (high density lipids) considered as good cholesterol. It was thought to be thrombogenic and could be replaced with other fatty acids (Kelly et al., 2001).

ENDOTHELIAL CELLS DYSFUNCTION

In normal condition, endothelial cells inhibits oxidation of LDL, smooth muscle proliferation and migration and several pro-atherogenic related processes (Shimokawa, 1999). Atherosclerosis is thought to originate from injury of the endothelial cell layer that lines the lumen of the arteries (Ross, 1993). These injuries arise from conditions such as hyperlipidemia, hypertension, smoking and disease condition like diabetes. The hemo-dynamics of the arterial blood flow was implicated to induce atherogenic effects on the endothelial cell layer. Reduced oscillatory shear stress is exerted on endothelial cells in athero-susceptible regions compare to high, unidirectional shear stress in the athero-protected blood vessels (Barakat et al., 1997).

Research showed that high amount of cholesterol in blood caused activation of the endothelial cells, leading to infiltration of LDL into the intima. Within the intima LDL is being oxidized, and is, converted to ox-LDL. The types of lipids, which are modified by oxidation, are phospholipids such as 1-palmitoyl-2-arachidooyl-sn-glycero-3-phophoryl-choline, epoxy-isoprostane phosphorylcholine and some others (Watson et al., 1999).

Release of these phospholipids caused upregulation of the adhesion molecules such as vascular cell adhesion molecule (VCAM), intercellular adhesion molecule-1 (ICAM-1), P- and E-selectins and monocyte chemotactic protein (MCP-1) (Leitinger, 2003). MCP-1 caused the recruitment of the monocytes and T cells to the vessel wall. The circulating monocytes get attached to the endothelial layer, infiltrate the endothelial layer to reach the intima. These activated monocytes differentiate into macrophages and uptake ox-LDL through their receptors called scavenger receptors (SR-A, CD36) (Janeway and Medzhitov, 2002; Peiser et al., 2002).

These filled macrophages are then become foam cells, which form the fatty streaks of the atherosclerotic lesions. Inflammatory processes mediated by the T cells and macrophages further cause the development of the lesion that stimulates the vascular smooth muscle cells.

Vascular Smooth Muscle Cell Dysfunction

Vascular smooth muscle cells (VSMCs) provide flexibility, elasticity and maintain integrity of the blood vessels. They secrete extra cellular matrix proteins, which comprises of collagen, elastin and proteoglycans (Raines and Ross, 1993). VSMC is the principal cells in the later stages of the atherosclerotic lesion (Stary et al., 1992). The earlier stages of atherosclerotic lesion are dominated by the immune cells. Cytokines such as matrix mellatoproteinases (MMP), interferon- γ (IFN- γ) secreted by the immune cells in the lesion cause VSMC growth and migration to the intimal layer of the blood vessel. VSMCs produce and secrete fibrous cap covering the lesion and is called "fibrofatty" lesion (Stary et al., 1992).

There are distinct changes of metabolic state in the smooth muscle cell phenotype and metabolic state in normal and atherosclerotic conditions. The smooth muscle cells of the in the normal uninjured artery are filled with myofilaments, arranged in concentric layers around the blood vessel and has poorly developed synthetic machinery (rough endoplasmic reticulum (RER) and golgi apparatus). However, in the atherosclerotic state, VSMCs have abundant RER and golgi apparatus but have lost almost all the myofilaments (Thyberg et al., 1990).

There is increased endocytosis of low density lipoprotein (LDL) achieved by apolipoprotein (apo) B/E receptors (Campbell et al., 1983). Apolipoprotein receptor caused binding and degradation of very low-density lipoprotein (VLDL) and thus increased progression towards atherosclerosis.

ACTIVATION OF IMMUNE CELLS

The reasons for the occurrence of coronary thrombosis or peripheral/ myocardial ischemia were shown to be plaque rupture. Plaque rupture occurs when the fibrous cap of the atheroma become thin due to the lytic activities of MMPs, which were predominantly secreted by VSMCs (van der Wal et al., 1994).The rupture exposes the platelet-adhesive molecules, tissue factors and phospholipids and activated immune cells to the blood stream. This is potentially dangerous as it leads to coagulation events, stenosis, heart attack and death in certain cases (Falk and Fernandez-Ortiz, 1995). As mentioned earlier, activation of endothelial layer mediates upregulation of the adhesion molecules and chemokines, which helps to recruit inflammatory immune cells. Further, unidirectional, high shear stress was reported to mediate expression of Kruppel-like factor (KLF-2), a transcription factor that inhibited the expression of immunogenic molecules but was down-regulated by the cytokines (Das et al., 2006; Dekker et al., 2002).

Chemokines released from activated intima causes differentiation of monocytes to macrophage. This step is accompanied by up-regulation of the pattern-recognition receptors such as toll-like receptors and scavenger receptors (Janeway and Medzhitov, 2002). Absence of toll-like receptors (TLRs) in genetically modified mouse inhibited atherosclerosis and suggested that plaque inflammation may directly depend on this pathway (Bjorkbacka et al., 2004). Also, other immune cells at the sites of the plaque rupture such as T cells, mast cells produce cytotoxic radicals, vasoactive substances or proteases such as cysteine proteases (Liu et al., 2004) or MMPs (mentioned above). These proteases destabilize the lesion by attacking the collagen of the cap.

The atherosclerotic lesion is enriched with CD4+, CD8+ and Natural Killer (NK) T cells. CD4+ T cells are reactive to the ox-LDL, heat shock proteins (Stemme et al., 1995) and CD8+ T cells augments the progression of atherosclerosis by causing apoptosis/ necrosis of the arterial cells, especially in genetically modified mice (Ludewig et al., 2000). Cytokines present in atherosclerotic lesions cause activated T cells to differentiate to Th1 effector cells and to produce interferon-γ (Frostegard et al., 1999). Interferon-γ has multiple effects on other immune cells such as it increases the efficiency of antigen presentation, activates macrophages and up-regulates synthesis of tumor necrosis factor-α (TNF-α) and interleukin-1 (Szabo et al., 2003) and interleukin-6 (IL-1 and IL-6).

Thus, systemically IL-6 mediate the induction of the liver related acute-phase protein called c-reactive protein (CRP), which can be diagnosed clinically (Hansson, 2001). Adipose tissue is also induced to cause the production of the related inflammatory cytokines indicative of atherosclerotic conditions.

INHIBITION OF ATHEROSCLEROSIS

Diet-Mediated Inhibition of Atherosclerosis

The relationship between dietary fats and cardiovascular heart disease is well known and termed as "lipid hypothesis". It is based on the principle that low-density lipoprotein (LDL) cholesterol is a major risk for the cardio-related diseases (Ramsden et al., 2009). It was believed that this risk factor could be modified through diet as it plays crucial role in progression of inflammation, endothelial activation ((De Caterina et al., 2006), smooth muscle cell proliferation, plaque rupture, adipogenesis or oxidative stress (Moreno et al., 2008).

Diet rich in linoleic acid (e.g., olive oil) prevent cardiovascular diseases as they are the structural components of the plasma membrane and also are precursors of the regulatory metabolic compounds (Viles and Gottenbos, 1989). Study conducted on healthy men who consumed diet containing 20% monounsaturated fatty acids (MUFA) for 28 days showed significant reduction in macrophage mediated uptake of ox-LDL, when compared to those individuals who consumed saturated fatty acid diet (Moreno et al., 2008). Monounsaturated fatty acid (MUFA) was shown to improve insulin senstivity and reduced thrombogenesis (Galgani et al., 2008) (Figure 1).

It was also shown that reducing total fat intake may not lower the level of serum cholesterol unless saturated fat intake is reduced. (Barr et al., 1992). Long-chain omega-3 poly-unsaturated fatty (PUFA) acids such as eicosapentaenoic and docosahexaenoic acid improved endothelial function, stabilized plaques in late stages of the atherosclerosis (Aarsetoy et al., 2006) and reduced platelet aggregation (Dyerberg and Bang, 1978).

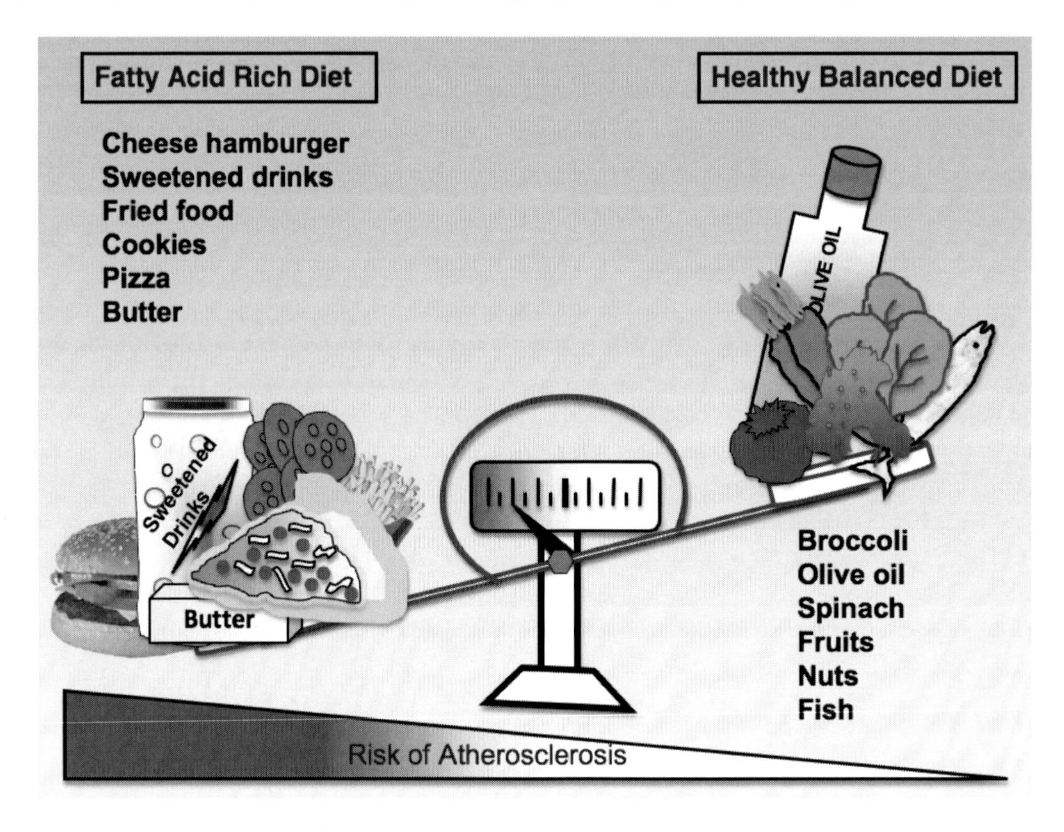

Figure 1. Diet rich in saturated fatty acids increases the risk of atherosclerosis. High consumption of foods rich in saturated fatty acids, such as cheese, sugar, oils, processed foods and meat that raises serum lipid levels causing increased chances of fatty acid deposition in the arteries (left panel), hence, pro-atherogenic. However, diets, which consist of unsaturated fatty acids were shown to have anti-atherogenic properties and prevents the development of cardiovascular diseases. Foods obtained from

plants such as green leafy vegetables (spinach), broccoli, carrots, fruits, olive oil, fish, walnuts etc (right panel).

Survival of VSMCs was critical as the loss of the VSMCs destabilized the plaques, which increased the risk for thrombosis (Kockx and Herman, 1998). Recently, fish oil intake was linked with reduced apoptosis of the smooth muscle cells in the region of atherosclerosis (Perales et al., 2010). At molecular level, the ratio of anti-apoptotic gene to pro-apoptotic genes increased significantly in the smooth muscle cells when fed with fish-oil enriched diet compared to the cholesterol diet (Perales et al., 2010)

Dietary antioxidants such as sulforaphanes found naturally in the vegetables such as brocolli, cabbage, brussell sprouts and collards were shown to activate the expression of transcription factor, Nrf2, and thus suppressed expression of adhesion molecules in the endothelial cells *in-vitro* (Chen et al., 2006) and *in-vivo* (Zakkar et al., 2009). This reduced expression of adhesion molecules was due to the inhibition of phosphorylation/ activation of p38 MAPKinase (Chen et al., 2006).However, since the above study was conducted with isolated compounds, further *in vivo* studies need to be conducted to confirm whether Nrf-2 can be stimulated by the vegetables containing sulforaphanes.

Studies have also found that consumption of magnesium rich diets consisting of more vegetables, nuts, unrefined grains and fruits to have cardio-protective properties (Maier et al., 2004). Magnesium was shown to control vasomotor tone, blood pressure and certain cardiac functions such as cardiac excitability (Champagne, 2008).

Drugs for Atherosclerosis

There are multiple drugs for the treatment of atherosclerosis commercially available in the market. The most commonly used drugs are lipid-lowering drugs and are classified into two groups, statins and fibrates. Statins lowered the cholesterol by inhibiting the enzyme [hydroxy-methylglutaryl-coenzyme A reductase (HMG-CoA reductase)] that is synthesized in the liver (Stepien et al., 2009).These statins are most commonly prescribed drugs in USA (Pletcher et al., 2009). Different types of statins are available such as Lipitor, Crestol or Pravacol. These drugs primarily cause the up-regulation of the LDL receptors (LDLR), which lowers the plasma lipids. Patients who are lacking LDLR expressions genetically, there is no beneficial effect of this drug, that is the limitation.

Fibrates stimulate β-oxidation of fatty acids leading to decrease in triglyceride levels and VLDL production (Barbier et al., 2002) . It also causes stimulation of the HDL mediated reverse cholesterol transport (Berthou et al., 1996). Fibrates are specific agonists of peroxisome proliferator activated receptors (PPARs). Activation of PPAR causes metabolism of the fatty acids in tissues such as liver, heart, kidney and muscles and this causes lowering of the plasma triglycerides levels (Auboeuf et al., 1997). Another lipid modifying drug was shown to boost HDL levels is, CSL-111. It was made with reconstituted human plasma HDL apolipoprotein A1 and reportedly used for reduction of the plaques in the arteries.

Drugs such as acyl-CoA cholesterol acyltransferase (ACAT) inhibitors, cholesteryl ester transporter protein (CETP) inhibitors or torcetrapib are other cholesterol lowering drugs present in the market. The drugs such as ACAT inhibitors were shown to have off-target effects and caused adrenal gland toxicity and others caused hypertension in some percentage of the patients.

Antibody based drugs are also being evaluated for their effects on the lipid metabolism. LDLR was shown to be regulated/ degraded by an enzyme called pro-protein convertase subtilisin/ kexin type 9 (PCSK9). The action of PCSK9 caused degradation of LDLR and hence lowered the levels of LDLR, which caused abnormal increase in the lipid levels. Study performed in non-human primates, provided evidence that infusion of anti-PCSK9 monoclonal antibody blocked the binding of PCSK9-LDLR interaction, increasing the LDLRs and thus decreasing LDL by 80% (Cohen et al., 2006) (Table 1).

Table 1. Drugs are in use for the treatment of atherosclerosis

Drug	Mode of Action
Statins	Inhibit cholesterol biosynthetic pathways Upregulate LDL receptor Inhibit SMC migration Inhibits pro-inflammatory cytokines Inhibits activation of NF-κB.
Fibrates	Stimulate β-oxidation of fatty acids Cause reverse transport of cholesterol Lower plasma triglyceride levels
CSL-111	Boosts good cholesterol
Cholesteryl ester transporter protein (CETP)	Lipid lowering drugs
Antibody-based drugs e.g., anti- Proprotein convertase subtilisin/ kexin type 9 (anti-PCSK9)	Increase LDLR and decreases LDL
Immuno-suppressive drugs	Inhibit activation of T cells, Inhibit proliferation of smooth muscle cells Lock the progression of intimal lesions
Niacin	Raises good cholesterol (HDL)
Ezetimibe	Inhibits intestinal cholesterol absorption
Torcetrapib	Inhibits cholesterol ester transfer protein

Anti-Inflammatory Inhibition of Atherosclerosis

High-density lipoprotein is often called as good cholesterol because of its potential of anti-atherogenic properties. HDL particles are heterogeneous in shape, size and density. The larger spherical HDL particles contain hydrophobic core of cholesteryl ester and triglyceride.

The smaller HDL contain particles contain apolipoprotein-1 (apoA1) in a lipid monolayer and a free cholesterol. HDL prevents oxidation of LDL and inhibits the expression of the adhesion molecules of the endothelial cells (Clay et al., 2001).

HDL unloads the excessive cholesterol from the peripheral tissues (such as lipid laden macrophages) and transports it to the liver for its catabolism (Bruce et al., 1998). This process is called reverse cholesterol transport and prevents accumulation of cholesterol to arterial wall (Assmann et al., 1996). In the presence of apoA1 LDL becomes resistant to oxidation by lipoxygenase, an enzyme that causes oxidation of LDL. Also, the antioxidant enzymes present on HDL, such as paraoxonase (PON) and acetyl-hydrolase, platelet activation factor were shown to exert inhibitory effects on oxidation of LDL (Navab et al., 1996; Navab et al., 2009).

The Th1 pathway discussed earlier in this chapter can be regulated or inhibited pharmacologically in animals and reduced the progression of atherosclerosis (Laurat E et al., 2001). Immunosuppresive drugs were shown to inhibit the activation of T cells, inhibit the proliferation of smooth muscle cells and lock the progression of intimal lesions (Jonasson et al., 1988)(Gallo et al., 1999).The commonly used imunosuppressive drugs are cyclosporine and sirolimus. These drugs are also used to coat the stents to prevent restnosis after arterial or cardiac surgical procedures (Marx and Marks, 2001).

Physical activity such as exercise was shown to have anti-atherogenic effects. Direct effects of functional improvement in endothelial cells have been reported. Exercise caused increased production of endothelial nitric oxide synthase (eNOS) and reduced turnover of the nitric oxide (NO) (Ambrecht R et al., 2008). NO causes vasodilation of the arteries for longer time periods and thus increased laminar sheer stress (Niebauer and Cooke, 1996) and enhanced blood flow. Anti-inflammatory effects of exercise mediated almost 20 to 30 % reduction in the CRP proteins (Kasapis and Thompson, 2005), endothelial adhesive molecules, TNF-α and other inflammatory cytokines (Wilund, 2007). Exercise was shown to cause mobilization of the endothelial progenitor cells (EPCs) also (Laufs et al., 2004). The role of these stem cells is discussed in the section below.

DEVELOPMENT OF ISCHEMIA FROM ATHEROSCLEROSIS

Atherosclerosis is one of the primary causes of cardiovascular and cerebrovascular ischemia. Briefly, atherosclerotic plaques generate stenosis that causes limited or no supply of oxygen and nutrients to the tissues, developing ischemic conditions, which may lead to cell death and impairment of the normal function of the tissues. For example, in case of myocardial tissue, ischemia may be resulted to heart attacks, arrythmias, heart bypass surgery, angioplasty, heart transplant or even death. In case of peripheral tissues, ischemia of the peripheral vasculature leads to peripheral arterial disease, affecting blood supply to the limbs as a result oxygen and nutrient supply is ceased and in extreme cases amputation is the only solution as necrosis of the limb starts to occur.

Stem Cell-Mediated Therapy for Ischemia

The loss of cardiomyocytes and other surrounding cells in the heart tissue leads to permanent damage of the patient's heart. Conventional therapies hardly address this issue. Cell transplantation has been proposed to address the fundamental problem of cell loss (Segers and Lee, 2008). Administration of progenitor cells from bone marrow was reported to significantly improve the recovery of left ventricular contractile function (Assmus et al., 2006). Mechanisms, such as angiogenesis, cell fusion and paracrine effect were found to be important in functional improvement of the heart after stem cell transplantation. However, limitations of low number of biologically functional stem cells and their effective differentiation *in vivo* after stem cell transplantation needs to be overcome.

On an average, over one billion cells undergo apoptosis or necrosis in the event of ischemia (Saver, 2006). However, it was estimated that less than 20% of transplanted cells might eventually survive and contribute to the functional recovery (Saporta et al., 1999). For clinical application in human, it could be conceived that hundreds of millions of cells are needed for single injection (Al-Radi et al., 2003). This indeed needs a development of an *ex-vivo* expansion technique that can promote the expansion of HSCs.

Ex-Vivo Expansion of Hematopoietic Stem Cells

During the past decade, significant efforts were made for *ex-vivo* expansion of hematopoietic stem cells and many mechanisms controlling HSCs fate are still under investigation. Two requirements need to be fulfilled in order to achieve an optimal number of *ex-vivo* expanded cells, which are suitable for transplantation: (i) a large scale expansion without compromising the pluripotency and long-term repopulation capacity (self-renewal), (ii) the expansion should provide biologically functional, safe and transplantable HSCs. This requires the expansion should consist of stem cells only, without any contamination of feeder cells, serum proteins or microbial agents.

Exposure of human tissues to xenogenic products raises the risk of contamination of infectious agents. Studies showed that human hESCs cultured with animal-derived serum replacements on mouse feeder layers could take up non-human sialic acid Neu5Gc from the culture medium. Transplantation using these ESCs might be compromised, as most adults have circulating antibodies against Neu5Gc (Martin et al., 2005).

Previously, it was shown that murine HSCs could be amplified *in vitro* in stromal-based long-term culture (Fraser et al., 1990) . Later, Miller and Evaes' reported a three-fold net increase of long-term lymphomyeloid repopulating cells following culture in serum-free media with interleukin-11, Ftl3-ligand and steel factor (SF) without impairing *in vivo* regenerative potential. This showed the possibility to expand HSCs *in vitro* without using feeder cells (Miller and Eaves, 1997). The major problem with *in vitro* expansion of HSCs is inefficient long-term culture period and low expansion efficiency. Even though, short-term expansion of HSCs was shown to be possible, long-term expansion usually resulted in differentiation and reduced engrafting potential. For example, SCID repopulating cells (SRCs) could be maintained in serum-free condition and showed a 2 to 4-fold increase after 4 days of culture. However, all the SRCs were lost after 9 days of culture (Bhatia et al., 1997).

The poor expansion was thought to be due to asymmetrical cell division and inadequate long term SRC survivability (Glimm and Eaves, 1999).

Currently, different combinations of growth factors are being used in various experiments, and other factors such as soluble form of hedgehog factor, sonic hedgehog may also help in HSCs expansion. However, the optimal choice of these factors is not yet established. Great promise of HSC *ex-vivo* expansion was shown by the usage of biomaterials in stem cell *ex-vivo* expansion following the discovery of 3D tantalum-coated porous biomaterial (TCPB). TCPB was able to support long-term culture of hematopoietic progenitor cells (HPC) and expanded HPCs up to 1.5-fold in numbers after 1 week, and 6.7-fold increase in colony-forming ability after 6 weeks without any cytokines (Bagley et al., 1999). These results indicated that biomaterials might enhance the long-term survivability of HSCs without the aid of cytokines. Various materials without modification were tested for their ability to support HPCs expansion in serum-free medium (LaIuppa et al., 1997). In order to enhance the *ex-vivo* expansion efficiency, investigators tried to modify the biomaterials and improve their design so as to mimic the bone marrow niche.Current research focuses on two major aspects of biomaterial design: the topology of the biomaterial and the interaction between HPCs and nano-segments. Using poly ethyl sulfone (PES) nanofiber matrices, we showed over 200-fold expansion of CD34+ cells purified from human umbilical cord blood within10 days of expansion in serum free expansion media.

We observed 20-fold increase in the expression level of CXCR4 in nanofiber-expanded stem cells, which significantly enhanced its migration and homing ability. *In vivo* tests showed that nanofiber expanded CD34+ cells enhanced therapeutic effect compared to freshly isolated CD133+/CD34+ cells in rat hind limb ischemia model (Aggarwal et al., 2010) and rat infarct models (Das et al., 2009a; Das et al., 2009b; Lu et al., 2010) (Figure 2).

GENETIC MODIFICATION OF HEMATOPOIETIC STEM CELLS

Delivery of genetic materials to the target cells before transplantation has been shown to be able to significantly increase the migration ability, survival ability, differentiation ability, engraftment ability of target cells (Dilber, 1998). Both viral and non-viral methods have been developed to introduce foreign DNA into the nucleus of stem cells. Viral methods have been extensively studied to introduce foreign DNA into nucleus of hematopoietic stem cells. Various types of lentiviruses were shown to transfect non-dividing cells. This gives lentiviruses, a big advantage in clinical application of HSCs as they were used to transfect quiescent HSCs also (Guenechea et al., 2000). Other vectors developed from spumaviruses and adenoviral vectors were also proposed due to their less apparent side effects on host cells (Linial, 1999; Van Tendeloo et al., 2001). Non-viral methods including lipofection, electroporation, and nucleofection have been proposed. However, those methods are hindered by lower efficiency than viral methods ((Rolland, 1998); (Phillips and Tang, 2008)).

It was shown that transplantation of cord blood derived CD34+ cells could enhance neovascularization and use of pro-angiogenic factors such as vascular endothelial cell growth factor (VEGF), platelet derived growth factor (PDGF) could enhance the angiogenic processes (Krupinski et al., 1993; Schlaeger et al., 1997; Zhang et al., 2000) We utilized a bicistronic vector, which contained both VEGF and PDGF genes with internal ribosomal entry site under CMV promoter. Nanofiber expanded CD34+ cells were transfected using nucleofection. Transfection efficiency with pmaxGFP vector was more than 90% and cell viability was ~70%. However, when used VEGF and PDGF containing (VIP) vector transfection efficiency was ~60%, and cell viability was also ~60%, probably due to the large vector size (Das et al., 2009a; Das et al., 2009b) (Figure 2).

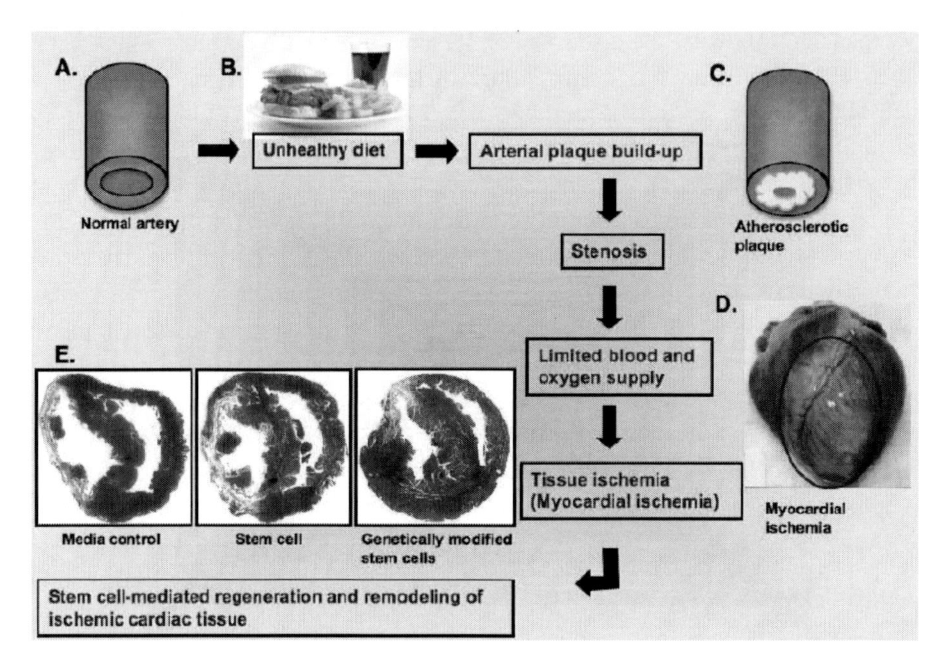

Figure 2. Development of atherosclerosis to ischemia and its therapy using stem cells. Consumption of unhealthy food rich in lipids and saturated fatty acids leads to arterial plaque formation. Rupture of this plaque causes thrombosis leading to blockage of arterial blood flow. This generates ischemic blood vessels and the tissue (e.g., myocardial tissue. Stem cell-mediated therapy can potentially revive degenerated cells of the tissue and thus could restore the functionality of the infracted myocardium via remodeling. A. Schematic represents healthy artery with normal arterial diameter. B. Figure depicts a meal consisting of fried food, sweetened beverage and burger filled with meat and very few vegetables. C. The lumen of artery has reduced due to the fatty deposition on the arterial wall (yellow). D. Representative picture of typical ischemic pig heart, which was induced by ligation of left anterior descending artery. This resembles with the actual human myocardial ischemic heart, where left ventricular tissue become fibrotic and thin, most of the cells are either necrotic or dead, as a result functionality is compromised. E. Represents cross-sectional area of heart tissue from 3 different experimental groups. Myocardial Ischemia was generated in the three groups of rats. The control group received media only after the ischemia. The stem cell group received nanofiber-expanded stem cell for ischemic therapy. The other group received genetically modified stem cell with the pro-angiogenic factors as a therapy. The fibrotic tissue (blue) decreased in the stem cell therapy groups compared to the control. However, in genetically-modified stem cells therapy group, the improvement was significantly higher, as a result functionality of heart was also better.

The possible higher efficiency in our experiment compared with other reports could partially come from the proliferation status of the nano-fiber expanded cells. Indeed, we showed that CD34+ cells transfected with VEGF and PDGF could better promote neovascularization compared with fresh isolated or nanofiber expanded HSCs (Das et al., 2009a; Das et al., 2009b).

Future Directions

It is evident that there is a strong correlation between the dietary food consumption and development of atherosclerosis over a period of time. It was suggested that in earlier stages, atherosclerosis develops as a silent disease and presents itself as major health problem in the later stages. Thus, it would be beneficial to develop techniques that could detect the atherosclerosis at earlier stages of progression. One of the earliest events is the activation of the endothelium that causes the up-regulation of the adhesion molecules triggering the inflammatory cascade. This cascade further causes the migration of the monocytes and SMCs, as a result plaque buildup takes place that causes downstream apoptosis of the endothelial cells and SMCs. If, we could detect these events of activation or apoptosis of the arterial cells, then we could possibly halt the progression of atherosclerosis thus the diseased state could be reversed by changing the diet habit and applying effective drugs. However, once, it reaches to the ischemic level, application of stem cell therapy seems to be effective for improving functionality of the ischemic tissues via neovascularization. Combination of genes of angiogenic potential along with stem cells enhances the beneficial effects of reverting ischemia in shorter time period. Future optimization of number of stem cells for each individual, choosing best route of application for stem cells and efficacy of allogenic stem cell transplantation will provide a maximum benefit of stem cell therapy.

Acknowledgments

This work was supported in part by National Institutes of Health grants, K01 AR054114 (NIAMS), SBIR R44 HL092706-01 (NHLBI), Third Frontier Projects, Ohio Technology, BRCP Grant and The Ohio State University start-up fund for stem cell research. The funders had no role in study design, data collection and analysis, decision to publish or preparation of the manuscript.

References

Aarsetoy, H., Brugger-Andersen, T., Hetland, O., Grundt, H., and Nilsen, D.W. (2006). Long term influence of regular intake of high dose n-3 fatty acids on CD40-ligand, pregnancy-associated plasma protein A and matrix metalloproteinase-9 following acute myocardial infarction. *Thromb. Haemost. 95*, 329-336.

Aggarwal, R., Pompili, V.J., and Das, H. (2010). Genetic Modification of ex-vivo expanded stem cells and potential for clinical application. Frontiers in Bioscience. *Invited Review; In Press.*

Al-Radi, O.O., Rao, V., Li, R.K., Yau, T., and Weisel, R.D. (2003). Cardiac cell transplantation: closer to bedside. *Ann. Thorac. Surg. 75*, S674-677.

Assmann, G., Schulte, H., von Eckardstein, A., and Huang, Y. (1996). High-density lipoprotein cholesterol as a predictor of coronary heart disease risk. The PROCAM experience and pathophysiological implications for reverse cholesterol transport. *Atherosclerosis. 124 Suppl*, S11-20.

Assmus, B., Honold, J., Schachinger, V., Britten, M.B., Fischer-Rasokat, U., Lehmann, R., Teupe, C., Pistorius, K., Martin, H., Abolmaali, N.D., *et al.* (2006). Transcoronary transplantation of progenitor cells after myocardial infarction. *N. Engl. J. Med. 355*, 1222-1232.

Auboeuf, D., Rieusset, J., Fajas, L., Vallier, P., Frering, V., Riou, J.P., Staels, B., Auwerx, J., Laville, M., and Vidal, H. (1997). Tissue distribution and quantification of the expression of mRNAs of peroxisome proliferator-activated receptors and liver X receptor-alpha in humans: no alteration in adipose tissue of obese and NIDDM patients. *Diabetes. 46*, 1319-1327.

Bagley, J., Rosenzweig, M., Marks, D.F., and Pykett, M.J. (1999). Extended culture of multipotent hematopoietic progenitors without cytokine augmentation in a novel three-dimensional device. *Exp. Hematol. 27*, 496-504.

Barakat, A.I., Karino, T., and Colton, C.K. (1997). Microcinematographic studies of flow patterns in the excised rabbit aorta and its major branches. *Biorheology. 34*, 195-221.

Barbier, O., Torra, I.P., Duguay, Y., Blanquart, C., Fruchart, J.C., Glineur, C., and Staels, B. (2002). Pleiotropic actions of peroxisome proliferator-activated receptors in lipid metabolism and atherosclerosis. *Arterioscler. Thromb. Vasc. Biol. 22*, 717-726.

Barr, S.L., Ramakrishnan, R., Johnson, C., Holleran, S., Dell, R.B., and Ginsberg, H.N. (1992). Reducing total dietary fat without reducing saturated fatty acids does not significantly lower total plasma cholesterol concentrations in normal males. *Am. J. Clin. Nutr. 55*, 675-681.

Berthou, L., Duverger, N., Emmanuel, F., Langouet, S., Auwerx, J., Guillouzo, A., Fruchart, J.C., Rubin, E., Denefle, P., Staels, B., *et al.* (1996). Opposite regulation of human versus mouse apolipoprotein A-I by fibrates in human apolipoprotein A-I transgenic mice. *J. Clin. Invest. 97*, 2408-2416.

Bhatia, M., Bonnet, D., Kapp, U., Wang, J.C., Murdoch, B., and Dick, J.E. (1997). Quantitative analysis reveals expansion of human hematopoietic repopulating cells after short-term ex vivo culture. *J. Exp. Med. 186*, 619-624.

Bjorkbacka, H., Kunjathoor, V.V., Moore, K.J., Koehn, S., Ordija, C.M., Lee, M.A., Means, T., Halmen, K., Luster, A.D., Golenbock, D.T., *et al.* (2004). Reduced atherosclerosis in MyD88-null mice links elevated serum cholesterol levels to activation of innate immunity signaling pathways. *Nat. Med. 10*, 416-421.

Bruce, C., Chouinard, R.A., Jr., and Tall, A.R. (1998). Plasma lipid transfer proteins, high-density lipoproteins, and reverse cholesterol transport. *Annu. Rev. Nutr. 18*, 297-330.

Campbell, J.H., Popadynec, L., Nestel, P.J., and Campbell, G.R. (1983). Lipid accumulation in arterial smooth muscle cells. Influence of phenotype. *Atherosclerosis. 47*, 279-295.

Champagne, C.M. (2008). Magnesium in hypertension, cardiovascular disease, metabolic syndrome, and other conditions: a review. *Nutr. Clin. Pract. 23*, 142-151.

Chen, X.L., Dodd, G., Thomas, S., Zhang, X., Wasserman, M.A., Rovin, B.H., and Kunsch, C. (2006). Activation of Nrf2/ARE pathway protects endothelial cells from oxidant injury and inhibits inflammatory gene expression. *Am. J. Physiol. Heart Circ. Physiol. 290*, H1862-1870.

Chockalingam A, B.-V.I. (1999). Impending global pandemic of cardiovascular diseases. Prous Science.

Clay, M.A., Pyle, D.H., Rye, K.A., Vadas, M.A., Gamble, J.R., and Barter, P.J. (2001). Time sequence of the inhibition of endothelial adhesion molecule expression by reconstituted high density lipoproteins. *Atherosclerosis. 157*, 23-29.

Cohen, J.C., Boerwinkle, E., Mosley, T.H., Jr., and Hobbs, H.H. (2006). Sequence variations in PCSK9, low LDL, and protection against coronary heart disease. *N. Engl. J. Med. 354*, 1264-1272.

Das, H., Abdulhameed, N., Joseph, M., Sakthivel, R., Mao, H.Q., and Pompili, V.J. (2009a). Ex vivo nanofiber expansion and genetic modification of human cord blood-derived progenitor/stem cells enhances vasculogenesis. *Cell Transplant. 18*, 305-318.

Das, H., George, J.C., Joseph, M., Das, M., Abdulhameed, N., Blitz, A., Khan, M., Sakthivel, R., Mao, H.Q., Hoit, B.D., *et al.* (2009b). Stem cell therapy with overexpressed VEGF and PDGF genes improves cardiac function in a rat infarct model. *PLoS One. 4*, e7325.

Das, H., Kumar, A., Lin, Z., Patino, W.D., Hwang, P.M., Feinberg, M.W., Majumder, P.K., and Jain, M.K. (2006). Kruppel-like factor 2 (KLF2) regulates proinflammatory activation of monocytes. *Proc. Natl. Acad. Sci. U.S.A. 103*, 6653-6658.

De Caterina, R., Zampolli, A., Del Turco, S., Madonna, R., and Massaro, M. (2006). Nutritional mechanisms that influence cardiovascular disease. *Am. J. Clin. Nutr. 83*, 421S-426S.

Dekker, R.J., van Soest, S., Fontijn, R.D., Salamanca, S., de Groot, P.G., VanBavel, E., Pannekoek, H., and Horrevoets, A.J. (2002). Prolonged fluid shear stress induces a distinct set of endothelial cell genes, most specifically lung Kruppel-like factor (KLF2). *Blood. 100*, 1689-1698.

Dilber, M.S. (1998). Gene transfer into hematopoietic cells: progress, problems and prospects. *Turk. J. Pediatr. 40*, 307-336.

Dyerberg, J., and Bang, H.O. (1978). Dietary fat and thrombosis. *Lancet. 1*, 152.

Falk, E., and Fernandez-Ortiz, A. (1995). Role of thrombosis in atherosclerosis and its complications. *Am. J. Cardiol. 75*, 3B-11B.

Fraser, C.C., Eaves, C.J., Szilvassy, S.J., and Humphries, R.K. (1990). Expansion in vitro of retrovirally marked totipotent hematopoietic stem cells. *Blood. 76*, 1071-1076.

Freeman, M.W. (1997). Scavenger receptors in atherosclerosis. *Curr. Opin. Hematol. 4*, 41-47.

Frostegard, J., Ulfgren, A.K., Nyberg, P., Hedin, U., Swedenborg, J., Andersson, U., and Hansson, G.K. (1999). Cytokine expression in advanced human atherosclerotic plaques: dominance of pro-inflammatory (Th1) and macrophage-stimulating cytokines. *Atherosclerosis. 145*, 33-43.

Galgani, J.E., Uauy, R.D., Aguirre, C.A., and Diaz, E.O. (2008). Effect of the dietary fat quality on insulin sensitivity. *Br. J. Nutr. 100*, 471-479.

Galkina, E., and Ley, K. (2007). Leukocyte influx in atherosclerosis. *Curr. Drug Targets. 8*, 1239-1248.

Gallo, R., Padurean, A., Jayaraman, T., Marx, S., Roque, M., Adelman, S., Chesebro, J., Fallon, J., Fuster, V., Marks, A., *et al.* (1999). Inhibition of intimal thickening after balloon angioplasty in porcine coronary arteries by targeting regulators of the cell cycle. *Circulation. 99*, 2164-2170.

Glimm, H., and Eaves, C.J. (1999). Direct evidence for multiple self-renewal divisions of human in vivo repopulating hematopoietic cells in short-term culture. *Blood. 94*, 2161-2168.

Guenechea, G., Gan, O.I., Inamitsu, T., Dorrell, C., Pereira, D.S., Kelly, M., Naldini, L., and Dick, J.E. (2000). Transduction of human CD34+ CD38- bone marrow and cord blood-derived SCID-repopulating cells with third-generation lentiviral vectors. *Mol. Ther. 1*, 566-573.

Hansson, G.K. (2001). Immune mechanisms in atherosclerosis. *Arterioscler. Thromb. Vasc. Biol. 21*, 1876-1890.

Janeway, C.A., Jr., and Medzhitov, R. (2002). Innate immune recognition. *Annu. Rev. Immunol. 20*, 197-216.

Jonasson, L., Holm, J., and Hansson, G.K. (1988). Cyclosporin A inhibits smooth muscle proliferation in the vascular response to injury. *Proc. Natl. Acad. Sci. US. 85*, 2303-2306.

Kasapis, C., and Thompson, P.D. (2005). The effects of physical activity on serum C-reactive protein and inflammatory markers: a systematic review. *J. Am. Coll. Cardiol. 45*, 1563-1569.

Kelly, F.D., Sinclair, A.J., Mann, N.J., Turner, A.H., Abedin, L., and Li, D. (2001). A stearic acid-rich diet improves thrombogenic and atherogenic risk factor profiles in healthy males. *Eur. J. Clin. Nutr. 55*, 88-96.

Kockx, M.M., and Herman, A.G. (1998). Apoptosis in atherogenesis: implications for plaque destabilization. *Eur. Heart J. 19 Suppl G*, G23-28.

Kris-Etherton, P.M., Taylor, D.S., Yu-Poth, S., Huth, P., Moriarty, K., Fishell, V., Hargrove, R.L., Zhao, G., and Etherton, T.D. (2000). Polyunsaturated fatty acids in the food chain in the United States. *Am. J. Clin. Nutr. 71*, 179S-188S.

Kromann, N., and Green, A. (1980). Epidemiological studies in the Upernavik district, Greenland. Incidence of some chronic diseases 1950-1974. *Acta Med. Scand. 208*, 401-406.

Kromhout, D., Bosschieter, E.B., and de Lezenne Coulander, C. (1985). The inverse relation between fish consumption and 20-year mortality from coronary heart disease. *N. Engl. J. Med. 312*, 1205-1209.

Krupinski, J., Kaluza, J., Kumar, P., Wang, M., and Kumar, S. (1993). Prognostic value of blood vessel density in ischaemic stroke. *Lancet. 342*, 742.

Laluppa, J.A., McAdams, T.A., Papoutsakis, E.T., and Miller, W.M. (1997). Culture materials affect ex vivo expansion of hematopoietic progenitor cells. *J. Biomed. Mater Res. 36*, 347-359.

Laufs, U., Werner, N., Link, A., Endres, M., Wassmann, S., Jurgens, K., Miche, E., Bohm, M., and Nickenig, G. (2004). Physical training increases endothelial progenitor cells, inhibits neointima formation, and enhances angiogenesis. *Circulation. 109*, 220-226.

Leitinger, N. (2003). Oxidized phospholipids as modulators of inflammation in atherosclerosis. *Curr. Opin. Lipidol. 14*, 421-430.

Libby, P. (2000). Changing concepts of atherogenesis. J Intern Med *247*, 349-358.

Lichtenstein, A.H., Kennedy, E., Barrier, P., Danford, D., Ernst, N.D., Grundy, S.M., Leveille, G.A., Van Horn, L., Williams, C.L., and Booth, S.L. (1998). Dietary fat consumption and health. *Nutr. Rev. 56*, S3-19; discussion S19-28.

Linial, M.L. (1999). Foamy viruses are unconventional retroviruses. *J. Virol. 73*, 1747-1755.

Liu, J., Sukhova, G.K., Sun, J.S., Xu, W.H., Libby, P., and Shi, G.P. (2004). Lysosomal cysteine proteases in atherosclerosis. *Arterioscler. Thromb. Vasc. Biol. 24*, 1359-1366.

Lloyd-Jones, D., Adams, R., Carnethon, M., De Simone, G., Ferguson, T.B., Flegal, K., Ford, E., Furie, K., Go, A., Greenlund, K., *et al.* (2009). Heart disease and stroke statistics--

2009 update: a report from the American Heart Association Statistics Committee and Stroke Statistics Subcommittee. *Circulation. 119*, e21-181.

Lu, J., Pompili, V., and Das, H. (2010). Hematopoietic stem cells: ex-vivo expansion and therapeutic potential for myocardial ischemia (invited Review). *Stem Cells and Cloning: Advances and Applications. 3, 57-68.*

Ludewig, B., Freigang, S., Jaggi, M., Kurrer, M.O., Pei, Y.C., Vlk, L., Odermatt, B., Zinkernagel, R.M., and Hengartner, H. (2000). Linking immune-mediated arterial inflammation and cholesterol-induced atherosclerosis in a transgenic mouse model. *Proc. Natl. Acad. Sci. U.S.A. 97*, 12752-12757.

Luscher, T.F., and Barton, M. (1997). Biology of the endothelium. *Clin. Cardiol. 20*, II-3-10.

Maier, J.A., Bernardini, D., Rayssiguier, Y., and Mazur, A. (2004). High concentrations of magnesium modulate vascular endothelial cell behaviour in vitro. *Biochim. Biophys. Acta. 1689*, 6-12.

Martin, M.J., Muotri, A., Gage, F., and Varki, A. (2005). Human embryonic stem cells express an immunogenic nonhuman sialic acid. *Nat. Med. 11*, 228-232.

Marx, S.O., and Marks, A.R. (2001). Bench to bedside: the development of rapamycin and its application to stent restenosis. *Circulation. 104*, 852-855.

Miller, C.L., and Eaves, C.J. (1997). Expansion in vitro of adult murine hematopoietic stem cells with transplantable lympho-myeloid reconstituting ability. *Proc. Natl. Acad. Sci. U. S. A. 94,* 13648-13653.

Moreno, J.A., Lopez-Miranda, J., Perez-Martinez, P., Marin, C., Moreno, R., Gomez, P., Paniagua, J.A., and Perez-Jimenez, F. (2008). A monounsaturated fatty acid-rich diet reduces macrophage uptake of plasma oxidised low-density lipoprotein in healthy young men. *Br. J. Nutr. 100*, 569-575.

Navab, M., Berliner, J.A., Watson, A.D., Hama, S.Y., Territo, M.C., Lusis, A.J., Shih, D.M., Van Lenten, B.J., Frank, J.S., Demer, L.L., *et al.* (1996). The Yin and Yang of oxidation in the development of the fatty streak. A review based on the 1994 George Lyman Duff Memorial Lecture. *Arterioscler. Thromb. Vasc. Biol. 16*, 831-842.

Navab, M., Reddy, S.T., Van Lenten, B.J., Anantharamaiah, G.M., and Fogelman, A.M. (2009). The role of dysfunctional HDL in atherosclerosis. *J. Lipid Res. 50 Suppl*, S145-149.

Niebauer, J., and Cooke, J.P. (1996). Cardiovascular effects of exercise: role of endothelial shear stress. *J. Am. Coll. Cardiol. 28*, 1652-1660.

Peiser, L., Mukhopadhyay, S., and Gordon, S. (2002). Scavenger receptors in innate immunity. *Curr. Opin. Immunol. 14*, 123-128.

Perales, S., Alejandre, M.J., Palomino-Morales, R., Torres, C., and Linares, A. (2010). Influence of cholesterol and fish oil dietary intake on nitric oxide-induced apoptosis in vascular smooth muscle cells. *Nitric Oxide. 22*, 205-212.

Phillips, M.I., and Tang, Y.L. (2008). Genetic modification of stem cells for transplantation. *Adv. Drug Deliv. Rev. 60*, 160-172.

Pletcher, M.J., Lazar, L., Bibbins-Domingo, K., Moran, A., Rodondi, N., Coxson, P., Lightwood, J., Williams, L., and Goldman, L. (2009). Comparing impact and cost-effectiveness of primary prevention strategies for lipid-lowering. *Ann. Intern. Med. 150*, 243-254.

Raines, E.W., and Ross, R. (1993). Smooth muscle cells and the pathogenesis of the lesions of atherosclerosis. *Br. Heart J. 69*, S30-37.

Ramsden, C.E., Faurot, K.R., Carrera-Bastos, P., Cordain, L., De Lorgeril, M., and Sperling, L.S. (2009). Dietary fat quality and coronary heart disease prevention: a unified theory based on evolutionary, historical, global, and modern perspectives. *Curr. Treat Options Cardiovasc. Med. 11*, 289-301.

Rolland, A.P. (1998). From genes to gene medicines: recent advances in nonviral gene delivery. *Crit. Rev. Ther. Drug Carrier Syst. 15*, 143-198.

Ross, R. (1993). The pathogenesis of atherosclerosis: a perspective for the 1990s. *Nature. 362*, 801-809.

Ross, R. (1999). Atherosclerosis--an inflammatory disease. *N. Engl. J. Med. 340*, 115-126.

Saporta, S., Borlongan, C.V., and Sanberg, P.R. (1999). Neural transplantation of human neuroteratocarcinoma (hNT) neurons into ischemic rats. A quantitative dose-response analysis of cell survival and behavioral recovery. *Neuroscience. 91*, 519-525.

Saver, J.L. (2006). Time is brain--quantified. *Stroke. 37*, 263-266.

Schlaeger, T.M., Bartunkova, S., Lawitts, J.A., Teichmann, G., Risau, W., Deutsch, U., and Sato, T.N. (1997). Uniform vascular-endothelial-cell-specific gene expression in both embryonic and adult transgenic mice. *Proc. Natl. Acad. Sci. U.S.A. 94*, 3058-3063.

Segers, V.F., and Lee, R.T. (2008). Stem-cell therapy for cardiac disease. *Nature. 451*, 937-942.

Shimokawa, H. (1999). Primary endothelial dysfunction: atherosclerosis. *J. Mol. Cell Cardiol. 31*, 23-37.

Simopoulos, A.P. (1999). Essential fatty acids in health and chronic disease. *Am. J. Clin. Nutr. 70*, 560S-569S.

Stary, H.C., Blankenhorn, D.H., Chandler, A.B., Glagov, S., Insull, W., Jr., Richardson, M., Rosenfeld, M.E., Schaffer, S.A., Schwartz, C.J., Wagner, W.D., *et al.* (1992). A definition of the intima of human arteries and of its atherosclerosis-prone regions. A report from the Committee on Vascular Lesions of the Council on Arteriosclerosis, American Heart Association. *Circulation. 85*, 391-405.

Steinberg, D., Parthasarathy, S., Carew, T.E., Khoo, J.C., and Witztum, J.L. (1989). Beyond cholesterol. Modifications of low-density lipoprotein that increase its atherogenicity. *N. Engl. J. Med. 320*, 915-924.

Stemme, S., Faber, B., Holm, J., Wiklund, O., Witztum, J.L., and Hansson, G.K. (1995). T lymphocytes from human atherosclerotic plaques recognize oxidized low density lipoprotein. *Proc. Natl. Acad. Sci. US. 92*, 3893-3897.

Stepien, M., Banach, M., Mikhailidis, D.P., Gluba, A., Kjeldsen, S.E., and Rysz, J. (2009). Role and significance of statins in the treatment of hypertensive patients. *Curr. Med. Res. Opin. 25*, 1995-2005.

Szabo, S.J., Sullivan, B.M., Peng, S.L., and Glimcher, L.H. (2003). Molecular mechanisms regulating Th1 immune responses. *Annu. Rev. Immunol. 21*, 713-758.

Thyberg, J., Hedin, U., Sjolund, M., Palmberg, L., and Bottger, B.A. (1990). Regulation of differentiated properties and proliferation of arterial smooth muscle cells. *Arteriosclerosis. 10*, 966-990.

van der Wal, A.C., Becker, A.E., van der Loos, C.M., and Das, P.K. (1994). Site of intimal rupture or erosion of thrombosed coronary atherosclerotic plaques is characterized by an inflammatory process irrespective of the dominant plaque morphology. *Circulation. 89*, 36-44.

Van Tendeloo, V.F., Van Broeckhoven, C., and Berneman, Z.N. (2001). Gene therapy: principles and applications to hematopoietic cells. *Leukemia. 15*, 523-544.

Viles, R., and Gottenbos, J. (1989). Nutritional characteristics and food uses of vegetable oils, in Oil Crops of the World (New York, McGraw Hill).

Watson, A.D., Subbanagounder, G., Welsbie, D.S., Faull, K.F., Navab, M., Jung, M.E., Fogelman, A.M., and Berliner, J.A. (1999). Structural identification of a novel pro-inflammatory epoxyisoprostane phospholipid in mildly oxidized low density lipoprotein. *J. Biol. Chem. 274*, 24787-24798.

Wilund, K.R. (2007). Is the anti-inflammatory effect of regular exercise responsible for reduced cardiovascular disease? *Clin. Sci.* (Lond) *112*, 543-555.

Yagi, K., Kiuchi, K., Saito, Y., Miike, A., Kayahara, N., Tatano, T., and Ohishi, N. (1986). Use of a new methylene blue derivative for determination of lipid peroxides in foods. *Biochem. Int. 12*, 367-371.

Zakkar, M., Van der Heiden, K., Luong le, A., Chaudhury, H., Cuhlmann, S., Hamdulay, S.S., Krams, R., Edirisinghe, I., Rahman, I., Carlsen, H., *et al.* (2009). Activation of Nrf2 in endothelial cells protects arteries from exhibiting a proinflammatory state. *Arterioscler. Thromb. Vasc. Biol. 29*, 1851-1857.

Zhang, Z.G., Zhang, L., Jiang, Q., Zhang, R., Davies, K., Powers, C., Bruggen, N., and Chopp, M. (2000). VEGF enhances angiogenesis and promotes blood-brain barrier leakage in the ischemic brain. *J. Clin. Invest. 106*, 829-838.

In: Natural Products and Their Active Compounds ... ISBN: 978-1-62100-153-9
Editors: M. Essa, A. Manickavasagan, and E. Sukumar © 2012 Nova Science Publishers, Inc.

Chapter 5

PREVENTION AND TREATMENT OF BREAST CANCER BY LIGHT AND FOOD

***P. D. Gupta*[*1] *and K. Pushkala*[2]**
[1]Manipal University, Manipal, India.
[2]SDNB Vaishnav College for Women,
Chromepet, Chennai, India

Keywords: Night light, Various diets, Shift workers, Blind women model, Circadian rhythms

INTRODUCTION

Cancer refers to over 100 different types, but the central dogma is that certain body cells multiply in an uncontrolled manner. Normal cells grow at a predetermined rate, perform specific functions and 'pass away" by what is called apoptosis and is replaced by other normal cells. Cancer in general, results when a single cell develops mechanisms by which it can multiply rapidly out of control by normal checks and balances, and replaces normal cells along with loss of biological functions.

Breast cancer is regulated by sex steroids hormones (Pushkala and Gupta, 2001), which in turn is regulated by pineal hormone, melatonin. Melatonin is produced by dark reaction in the body during sleep, and peaks at 2 a.m. (Pushkala and Gupta, 2011). Breast cancer is the leading cause of death in women all over the world. According to the report of World Health Organization (WHO) 519, 000 deaths occur every year worldwide (WHO, 2009). About one out of every eight women will get breast cancer during their lifetime. Early diagnosis of cancer aids for its management. Some women under 50 do get breast cancer, but it is rare. Prolonged exposure of the breast tissue to estradiol was associated with cancer predisposition in menopausal women to the disease (Gupta and Pushkala, 2006; Pushkala and Gupta, 2009; Gupta *et al.,* 2010).

[*] Corresponding author: E-mail: pdg2000@hotmail.com

Many women have an emotional reaction to breast cancer, made all the more understandable because they have family or friends whose lives were up-ended by breast cancer.

SYNTHETIC LIGHT VS NATURAL SUNLIGHT

Long before life began on the earth, Natural sunshine was there. As a source of heat and energy, sunlight powers a majority of the planet's biological activities. When humans came to existence, their activities were started with the sunrise and ceased with the sunset. However, humans, the grand manipulators, have not been content to cede control of their activity with the Sun cycle. People have spent eons developing ever better means to artificially extend the day. Thanks to a widespread electrification and color-corrected, high-watt light bulbs, synthetic sunlight can now bombard city dwellers around-the-clock. Now cities do not sleep.

Availability of light at night has changed our "Life Style". Exposure to light at night changed our hormonal system, body's rhythm, working capacities, eating and sleeping timings, diminishes the effectiveness of the body's immune system etc. According to Reiter (Reiter *et al.,* 2007), functionally "light is a drug" and that by abusing it, we risk imperiling our health." Long back, Albert Einstein also said that blue light is more dangerous than the yellow incandescent light.

Light entering the eye allows our brain to sense the shape, size, color, and motion of objects around us. It also summons, albeit imperceptibly, a cadre of other biological sentinels. These go on to trumpet light's presence to distant tissues organs and cells lacking the means to detect illumination directly.

Lewy and his colleagues (1980) showed that melatonin production could be shut down in man by waking and exposing them to 2,500 lux of white light at 2 a.m., when synthesis of the hormone was at its peak. (For perspective, 100 lux may be found in a comfortably dim living room, whereas sunlight at high noon on a cloudless day can blast the eyes with 100,000 lux.). However, later it was found that just 50 lux could do the same trick if it is green light.

RISK FACTORS FOR BREAST CANCER

Causative factors for breast cancer are genetic and environmental. The effect of each individual risk factor in developing the disease may not be very significant but the cumulative effect of individual factors may increase the rate up to 99%. So far, about 19 genes have been identified to involve in developing breast cancer. However, only few have been studied in detail.

Breast cancer is one of the most common forms of the disease among other cancerous diseases and probably the one that has been most studied, however, causes of the breast cancer are still largely unknown. Recent research on breast cancer has emphasized on a few natural as well as manmade factors that are believed to have a huge effect in the formation of breast cancer cells (Barnett, *et al.,* 2008 and Gupta *et al.,* 2010) In this chapter we will discuss mainly two factors, *i.e.* Light, especially Light At Night (LAN) and food as important factors for the development of this dreadful disease.

INCIDENCES OF BREAST CANCERS

Since 1940, breast cancer incidence rates have been steadily rising in the world. There is growing evidence for the possible effects of exposure to LAN on cancer risk due to the increased use of modern electric lighting (Costa, 1996; Brown *et al.*, 2009; Schernhammer *et al.*, 2006). Epidemiological observations indicate that breast cancer risk is lower in women who are visually impaired, as compared to the sighted population and that the risk may be inversely correlated with degree of visual impairment (Pushkala and Gupta, 2009; Pukkala *et al.*, 2003; Pukkala *et al.*, 2005).

NIGHT LIGHT

Several studies have suggested night shift workers - for example, in hospitals, graveyards and airports may have high risk of developing breast cancer. This observation has further substantiated by our recent evidence that blind subjects are showing 100 folds less prevalence of breast cancer (Pushkala and Gupta, 2009). Some researchers speculated that the effect may be due to the changes in levels of melatonin, a hormone which is produced by dark reaction is affected by the body's exposure to light (Pukkala *et al.*, 2003).

Melatonin is involved significantly in the metabolic activities such as regulation of the circadian rhythm, sleep, hormonal expression of darkness, seasonal reproduction, retinal physiology, antioxidant, free-radical scavenging, cardiovascular regulation, lipid and glucose metabolism, immune activity and control of cancer. It is also a new member of an expanding group of regulatory factors that control cell proliferation and apoptosis and is the only known chronobiotic hormonal regulator of neoplastic cell growth. At physiological concentrations, melatonin suppresses cell growth and multiplication, and inhibits cancer cell proliferation *in vitro* through specific cell-cycle effects. At pharmacological concentrations, melatonin suppresses cancer cell growth and multiplication. At these levels melatonin acts as a differentiating agent in some cancer cells and lowers their invasive and metastatic status by altering adhesion molecules and maintaining gap junction intercellular communication. In other cancer cell types, melatonin alone or with other agents, induces programmed cell death (Blast *et al.*, 2009).

During night time, melatonin is secreted by the pineal gland. Normally, melatonin production is 5–20 times higher at night in the dark than during the day (Stevens, 2009). Melatonin reaches the hormone producing glands through blood, the ovaries and the pituitary gland and stops their production of hormones. The hormones that result from these glands, estrogen, progesterone, follicle stimulating hormone (F.S.H), and the lutenizing hormone (L.H.) can cause rapid breast tissue cell growth. Cellular growth and turnover can lead to errors in their genetic code, leading to cancer. Continual growth increases the chances for error, and so, greater chances of occurrence of cancer. Normal elevation of melatonin at night will be suppressed by exposure to artificial light of sufficient intensity. Even weak LAN impedes the pineal gland from producing melatonin. Without melatonin, glands which produce hormones are out of control, and leads to greater chance for breast cancer development.

From the schematic representation as in Figure 1 the following interpretations can be made:

- Diminished functions of the pineal gland lower its melatonin production
- Lesser melatonin production allows for ovarian growth and greater hormonal production
- Greater hormonal production leads to earlier sexual maturation, early puberty and greater breast cancer development.

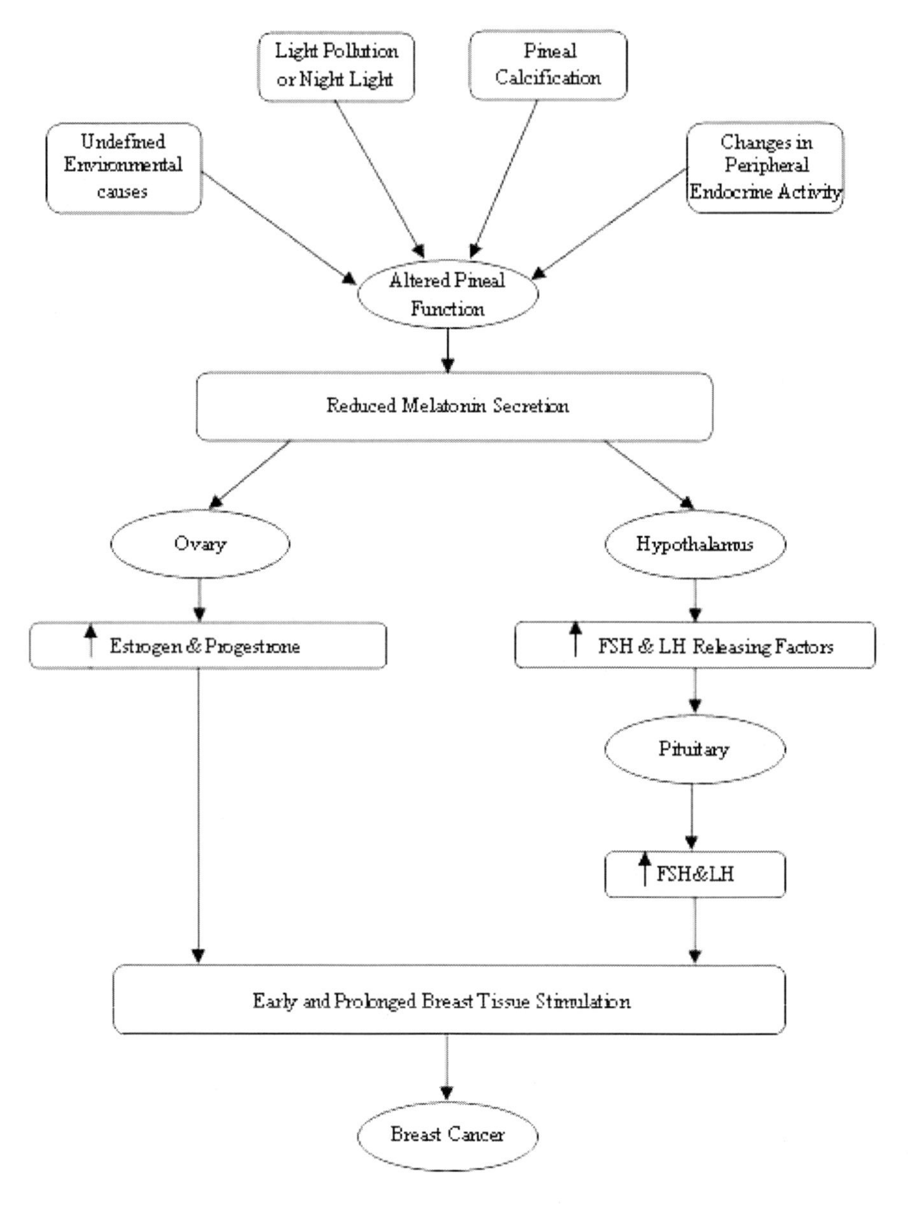

Figure 1. Schematic representation of mechanism involved in development of breast cancer from altered pinealgland function.

Schernhammer *et al.* (2006) examined the relationship between breast cancer and working on rotating night shifts during 10 years of follow-up in 78,562 women from the Nurses' Health Study. Information was ascertained in 1988 about the total number of years during which the nurses had worked rotating night shifts with at least three nights per month. From June 1988 through May 1998, they documented 2441 incident breast cancer cases. The authors observed a moderate increase in breast cancer risk among the women who worked 1–14 years or 15–29 years on rotating night shifts (multivariate adjusted relative risk (RR) = 1.08 [95% CI = 0.99 to 1.18] and RR = 1.08 [95% CI = 0.90 to 1.30] respectively). The risk was further increased among women who worked 30 or more years on the night shift (RR = 1.36; 95% CI = 1.04 to 1.78). The study concluded that women who work on rotating night shifts with at least three nights per month, in addition to days and evenings in that month, appear to have a moderately increased risk of breast cancer after extended periods of working rotating night shifts. Recently, Wise (2009) reported that Danish night shift workers suffering with breast cancer were awarded compensation.

Another study shows that using blue light or 'low wavelength light' (between 470 and 525 nm) in the night interferes with the bodies' ability to synthesis melatonin. This may be because blue light, the color of the day time sky reset the body clock back to day. These chances are more with people used to bed lamp, nightlight or a television switched on during sleep. The concept of light pollution and suppression of melatonin affects other animal species too. Rats are considered to be more sensitive to smaller changes of light than humans, once their melatonin levels drop due to a bright light source, the level stays down for the rest of the night (Kayumov, 2005).

Recently, Stevens (2009) stated that one should not just take melatonin as a supplement either. While supplements can be beneficial to slow down the growth rate of existing tumor, melatonin can cause a phase shift to their sleeping cycle, thus causing a circadian disruption instead of alleviating it. This may be a small price to pay for an increased protective effect of melatonin, so it would be better to seek the advice of a physician.

Blind Woman Model for Breast Cancer Management

The Pineal gland shows a marked difference between sighted and blind population. Earlier, our work showed that there is an inverse relationship between melatonin and estrogen (hormones of the pineal and the ovary). In menopause the control of estradiol and progesterone secretion is disturbed and therefore postmenopausal women are more vulnerable to breast cancer. Lehrer (1982) has shown that the average age of menarche in blind women is 2/3 years earlier than the sighted women. This indicates that they are exposed to estrogen for longer time, nevertheless they do not have high incidence rate of breast cancer. A correlation between prevalence of breast cancer with blindness was attempted in India by Pushkala and Gupta, 2009. In this pilot study consisting of blind menopausal (high risk age group) women (n = 204) from Chennai have shown that the ratio at risk of developing breast cancer is very much lower (1:100) compared to sighted women in the similar age group. The risk of developing breast cancer is 1:78 (Cumulative Risk 35 - 64 age), among sighted women. Statistical analysis provide enough evidence that blind women who are > 40 years of age had 13% greater risk of breast cancer compared with those in the age group < 40 years (RR =

1.125; 95% CI = 0.07 to 17.74).The susceptibility to develop the disease among partially blind women is almost twice than that of totally blind women (RR = 2.14; 95% CI = 0.14 to 33.68). Similarly menopausal stage of a woman has more risk of developing breast cancer than pre-menopausal stage (RR = 5.18; 95% CI = 0.33 to 80.75). Vision loss after menarche also indicates an increased risk (RR = 8.27; 95% CI = 0.54 to 127.6). The topographical location of India close to the equator and life style pattern of the people could be the major reasons for the very low prevalence of breast cancer in Chennai compared to earlier register based data from Finland (Pukkala *et al.,* 1999; Verkasalo *et al.,* 1999), Norway (Kliukiene *et al.,* 2001). None of the other high or low risk factors were found to be influencing blind women to develop breast cancer. The relationship between visible light and breast cancer can be studied by taking blind menopausal women as a model (Feychting *et al.,* 1998; Pushkala and Gupta, 2009).

Circadian Clock and Cancer Gene Regulation

Organisms exhibit behavior with 24-hour periodicity driven by cyclical environmental stimuli. Numerous activities, however, are expressed circadian way (with a periodicity of about -circa, a day- dies) in constant conditions, uncovering a further layer of temporal regulation, which is controlled by endogenous timekeepers (Crosthwaite, 2004). Circadian clocks are molecular time-keeping systems that underlie daily fluctuations in multiple physiological and biochemical processes. The circadian rhythm can be entrained by light and dark, kept precisely at 24 hour by the cycle of exposure to sunlight. In modern electrically lit societies, however, many, if not most, people suffer some degree of disruption of the circadian rhythms by exposure to LAN and by inadequate exposure to sunlight, especially in the morning (Schmutz *et al.,* 2010). It is well recognized now that dysfunctions of the circadian system (both genetically and environmentally induced) are associated with the development of various pathological conditions (Antocha and Chernov, 2009).

The rhythms or "hands" of the clock are preceded by changes in biochemistry and gene expression (Crosthwaite, 2004). The mammalian circadian system is hierarchically organized by central and peripheral oscillators. The circadian timing system is composed of a central pacemaker, the suprachiasmatic nuclei (SCN) in the hypothalamus, which coordinates molecular clocks in each cell through the generation and/or control of behavior, hormone, and neuromediator rhythms (Levi *et al.,* 2007). To obtain healthy rhythms with a periodicity of a day, the clocks use molecular oscillators consisting of two interlocked feedback loops. The core loop generates rhythms by transcriptional repression via the Period (PER) and Cryptochrome (CRY) proteins, whereas the stabilizing loop establishes roughly antiphasic rhythms via nuclear receptors. Nuclear receptors also govern many pathways that affect metabolism and physiology (Gupta and Pushkala, 1999). A dozen specific clock genes constitute the core of the molecular clock in mammals. These genes are involved in transcriptional and post-transcriptional activation and inhibition regulatory loops that result in the generation of the circadian oscillation in individual mammalian cells. In particular, the CLOCK: BMAL1 or NPAS2:BMAL1 protein dimers play a key role in the operation of the molecular clock. This protein dimer also exerts negative control on the cell cycle, both through the repression of c-myc and p21 and through the activation of p53 and weel. In proliferating cells, this results in the circadian control of the cell cycle stages, *e.g.*, the transition from G1 to S and from G2 to M. The automation switches sequentially between the phases G1, S, G2, and M after which the automation cell divides and two cells enter a new G1 phase (Kliukiene *et al.,* 2001).

In the mouse housed under a light–dark cycle of 12 hours of light followed by 12 hours of darkness (12:12 light–dark cycle), the WEE1 protein level rises during the second part of the dark phase, *i.e.*, at the end of the activity phase. Generally, human beings adhere to a routine of 16 hours of diurnal activity alternating with 8 hours of nocturnal sleep, 16:8 light–dark cycle (16 hours of light, from 8 a.m. to 12 p.m., followed by 8 hours of darkness, from 12 p.m. to 8 a.m.). The rise in WEE1 should occur at the end of the activity phase, *i.e.,* from 8 to 12 p.m. The decline in WEE1 activity is followed by a rise in the activity of the kinase Cdk1, which enhances the probability of transition to the M phase (Altinok *et al.,* 2007). Wee1 is expressed in-phase with mPer1 in liver and exhibits high-amplitude, robust oscillations.

Recent data reveal that major biological pathways, including those critical to cell division, that cell-cycle progression, response to genotoxic stress, aging and DNA-damage-response pathways are under circadian control (Gery *et al.*, 2006; Terazono *et al.,* 2008; Antocha and Chernov, 2009). Per1 provides an important link between the circadian system and the cell cycle system (Schmutz *et al.,* 2010). Over expression of Per1 sensitized human cancer cells to DNA damage-induces apoptosis, on other side, inhibition of Per1 in similarly treated cells blunted apoptosis. The apoptotic phenotype was associated with altered expression of key cell cycle regulators (Gupta and Saumyaa, 2008). Ectopic expression of Per1 in human cancer cell lines led to significant growth reduction. It was found that Per1 levels were reduced in human cancer patient samples (Grey *et al.,* 2006).

If circadian disruption can cause cancer, then one obvious avenue of investigation is the association of markers of circadian gene function and risk. Zhu *et al.* (2006) speculated that not only polymorphisms in circadian genes may have differential effects on cancer risk, but that promoter methylation of circadian genes may alter gene function in such a way, as to increase risk. If this is the case, then identifying what environmental factors change the promoter methylation status of circadian genes would become a rich new area for research. There is evidence for altered circadian gene expression in mammary tumor tissue compared to normal adjacent tissue, but this may be a result of tumor progression and not of etiologic significance (although it might still offer advances in cancer treatment).

In mammals, the route for circadian entrainment by light uses the retinohypothalamic tract, which connects directly to the central clock located in the suprachiasmatic nucleus (SCN). There are two prominent circadian photopigments, melanopsin and cryptochromes, both of which are expressed in retinal ganglion cells. Some evidence suggests melanopsin as the photoreceptor; other studies exhibit clock gene induction by light in the SCN, indicating that cryptochromes also play a critical role in photoreception (Chaurasia and Gupta, 1999; Cermakian *et al.,* 2002).

Cryptochromes are intimately linked to appropriate cell cycle progression *in vivo*. The disruption of circadian rhythmicity leads to deregulation of cell division and hence tumor growth (Figure 2). The circadian physiology is recognized by interactions between the central circadian oscillator of the SCN and the SCN-orchestrated activity of local, tissue-based circadian clocks. Tumors also contain effective circadian clockwork entrained to the host circadian system, which in turn drives elements of the tumor cell cycle.

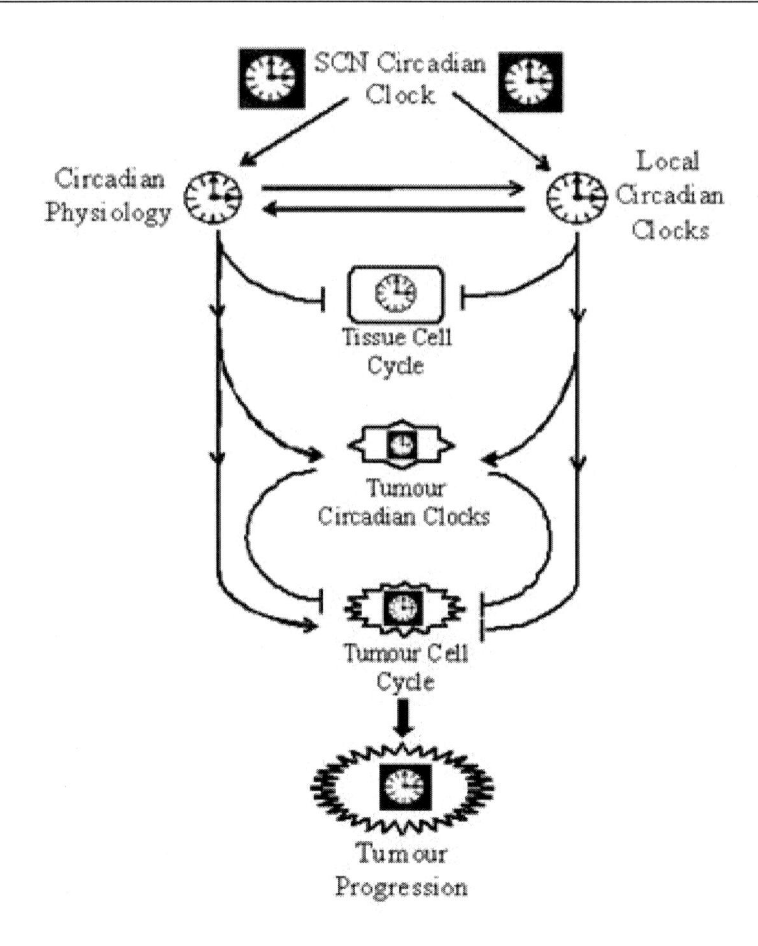

Figure 2. Schematic view of potential mechanisms underlying circadian regulation of cell division cycle.

In addition, the tumor cell cycle will be subject to circadian influences from the host, mediated by systemic and local paracrine signals. Environmental and genetic factors that disturb systemic and/or local circadian function may compromise temporal regulation of cell division and thereby promote tumor progression (Reddy *et al.,* 2005).

FOOD

Human beings take 2 types of food *viz.*, natural and processed. Generally, individual food may not support human health, because all the nutrients required for maintenance of metabolism are not present in a single food item and therefore for balancing the nutritional requirement we have evolved the nutritious and wholesome balanced diet. Nutrition is a complex requirement, which may vary as well as depend on the individual's genetic makeup, environment, and health status. An unhealthy diet is a major risk factor for a number of chronic diseases often called “life style diseases” which includes: high blood pressure, diabetes, abnormal blood lipids, overweight/obesity, cardiovascular diseases, and cancer.

Change in the Structure of Food

The food we consume, especially protein, will denature after it is cooked, baked, boiled, and fried. The word denature means, to change the structure (for protein the amino-acids change their structure) or reshape the food particle. Unfortunately when the food is denatured, some of its components become mutagenic. As a result, our body has a difficult time trying to break down these mutagenic food particulates. At this point, hopefully there are antioxidants in place that can pick some of these free radicals that are released from these mutagenic particulates. Free radicals are responsible for starting the process of cancer by damaging our cells.

Healthy Diet

The healthy diet is the one which is free from pathogens, poisonous heavy metals and carcinogenic substances (*e.g.* benzene); and can provide sufficient calories to maintain metabolic activities by having sufficient essential amino acids, carbohydrates, lipids, micronutrients and minerals. It should keep a good ratio (4:1) between carbohydrates and lipids but not excessive saturated high density (HDL) fat. There are certain foods which may be beneficial in small quantities and can be used occasionally, such as;

- foods or substances with directly toxic properties at high chronic doses (*e.g.* ethyl alcohol),
- foods that may interfere at high doses with other body processes (*e.g.* refined salts and sugars),

Diversity in Diet

A significant environmental effect is likely responsible for the different rates of breast cancer incidence between countries with different dietary customs. Researchers have long measured that breast cancer rates in an immigrant population soon come to resemble the rates of the host country after a few generations. The reason for this is speculated to be immigrant uptake of the host country diet. The prototypical example of this phenomenon is the changing rate of breast cancer after the arrival of Japanese immigrants to America (Deapen *et al.*, 2002).

The Ayurvedic Diet

Some foods are good for you, some are bad. But which are which? The answers, according to the latest nutritional science, are not the same as previously thought. While some age-old advice like "eat your vegetables" still holds true, many early assumptions have turned out to be wrong.

Scientists have learned much more about why some foods help prevent disease and why others promote it by studying the time tested Ayurvedic diet. The *Healthy Eating* has been suggested by modern nutritionists the food-health connection and takes on Holistic view of Ayurveda. Ayurveda has always classified food in terms of its positive and negative attributes, which are now being also recognized by modern nutrition studies. Ayurveda food; also influences" lifestyle", one's thoughts, behavior, physical appearance and actions.

That is why it is said YOU ARE WHAT YOU EAT. Food is seen as prana, the carrier of life force, and is judged according to how it affects the conscious self. Based on its inherent qualities, food is categorized into Sattva (purity), Rajas (activity, passion, the process of change), and Tamas (darkness, inertia).

Sattvic Food

This kind of food is always freshly cooked, simple, can be digested easily, help to build immunity, improves the health of those who are ill and puts one's mind in a state of balance. Foods closest to their natural forms like milk and milk products, fresh and dried fruits, and fresh vegetables (except onion and garlic), all whole grain cereals, most lentils, sprouts, natural sweeteners e.g. jaggery and honey, and natural oils such as homemade ghee, butter and cold-pressed oils are considered sattvic. Such food is lightly cooked with moderate spices (no chillies and black pepper) and consists of low fat content.

The spices commonly used in sattvic cooking are turmeric, ginger, cinnamon, coriander, fennel seeds and cardamom. Bear in mind, it's important to note that raw foods are not sattvic as they harbour parasites and microbes and are believed to weaken digestion and reduce ojas (vital energy) on which the proper functioning of the mind and spiritual development depends.

Though ghee is allowed (it helps the development of the mind), oily and fatty foods should be eaten sparingly. People with sattvic personality follow the sattvic way of eating, usually avoids alcohol, stimulants like tea, coffee, tobacco and non-vegetarian food and known to be clear minded, balanced, and spiritually aware.

Rajasic Food

Freshly cooked heavy nutritious food including non-vegetarian food like meat, fish, eggs, and chicken, all whole pulses and dals (not sprouted), hot spices like chillies, pepper and all vegetables including onion and garlic. Compared to sattvic food, rajasic food contain a little more oil and spices.

It benefits those who believe in action and aggression in a positive way such as business persons, politicians, and athletes.

The rajasic personality is linked to sensual stimulation. Rajasic people are usually aggressive (in a positive way) and full of energy, interested in the four Ps — power, prestige, position and prosperity. But they are quite in control of their lives and aren't obsessed by any of the above. They are go-getters and know how to enjoy life.

Tamasic Food

Not fresh, overcooked, stale and processed foods made from refined flour (maida), pastries, pizzas, burgers, chocolates, soft drinks, rumali roti, naan, tea, coffee, tobacco, alcohol, canned and preserved foods like jams, pickles and fermented foods, fried foods, sweets made from sugar, ice creams, puddings and most 'fun foods' are included in this list. All spicy, salty, sweet and fatty foods form part of the tamasic diet. Overeating is a tamasic trait. The irony is that overeating sattvic food too brings on tamas in an individual. The tamasic personality brings about stagnation leading to degeneration of people's health, normally lead a sedentary lifestyle. Such individuals suffer from intense mood swings, insecurity, desires, and cravings. These individuals are unable to deal with others in a balanced way.

They have little regard for the welfare of others and normally tend to be very self-centered. Their nervous system and heart do not function optimally and such individuals age fast and usually suffer from conditions like cancer, heart disease, diabetes, arthritis, and chronic fatigue. Switchover to sattvic diet in these people would have a positive benefit on the health status, especially in old age. Others, who live life in moderation and are into politics, business, and defense or into athletics, would do well by following the Rajasic way of life. Sattvic, rajasic and tamasic are more than just qualities in food, they are also a way of life.

Mediterranean Diet

Consuming a Mediterranean-style diet, rich in vegetables, fruits and nuts, olive oil, and legumes, may lead to longer life, less heart disease, and protection against some cancers. cereals and dairy products didn't make a major contribution, while moderate alcohol consumption and low meat consumption were linked to longer life.

The diet's main nutritional components include beta-carotene, vitamin C, tocopherols, polyphenols, and essential minerals. The individual nine components in the diet are found to contribute to the benefit of health, including moderate alcohol consumption, low meat and meat product consumption, high consumption of vegetables, fruit, legume, olive oil, and nuts, and a high monounsaturated to saturated lipid ratio. However, cereal and dairy consumption had only minimal effects, while increased fish and seafood consumption was linked to a non-significant increase in mortality.

Western and Mediterranean-Style Diet

Compounds produced by frying, grilling, or pasteurizing may be driving inflammation and ageing. Advanced glycation end products are toxic substances reportedly produced enough in western diets, as a result of heating, pasteurisation, drying, smoking, frying and/or grilling. The highly processed foods and meats in the western diet have been linked to a range of conditions, from obesity to colorectal cancer. Low intake of compounds called Advanced Glycation End Products (AGEs) may reduce inflammation and help boost the body's natural defense (Vlassara *et al.,* 2009).

INFLUENCE OF DIETARY CONSTITUENTS ON BREAST CANCER

1. Fats

Dietary influences have been proposed and examined, and recent research suggests that low-fat diets may significantly decrease the risk of breast cancer as well as the recurrence of breast cancer (Chlebowski *et al.,* 2006). Another study showed no contribution of dietary fat intake on the incidence of breast cancer in over 300,000 women (Hunter *et al.,* 1996). A randomized controlled study of the consequences of a low-fat diet, the Women's Health Initiative, failed to show a statistically significant reduction in breast cancer incidence in the group assigned to a low-fat diet, although the authors did find evidence of a benefit in the subgoup of women who followed the low-fat diet in a strict manner (Prentice *et al.,* 2006). A prospective cohort study, the Nurses' Health Study II, found that increased breast cancer incidence in premenopausal women with higher intake of animal fat, but not vegetable fat. Taken as a whole, these results point to a possible association between dietary fat intake and breast cancer incidence, though these interactions are hard to measure in large groups of women.

A. Omega-6 Polyunsaturated Fatty Acids

Modern Western diets typically have ratios of omega−6 to omega−3 in excess of 10 to 1, some as high as 30 to 1. The optimal ratio is thought to be 4 to 1 or lower. Excess n−6 fats interfere with the health benefits of n−3 fats, in part because they compete for the same rate-limiting enzymes (Simopoulos, 2003).A high consumption of omega-6 polyunsaturated fatty acids (PUFAs), which are found in most types of vegetable oil, may increase the likelihood that postmenopausal women will develop breast cancer (Sonestedt *et al.,* 2008). Similar effect was observed on prostate cancer (Berquin *et al.,* (2007). Polyunsaturated fatty acids (PUFAs), which are found in most types of vegetable oil (e.g. soybean oil, corn oil - the most consumed in USA, sunflower oil, etc.), may increase the likelihood that postmenopausal women will develop breast cancer (Sonestedt *et al.,* 2008).

B. Mufas on Erythrocyte Membrane and Derived Discussion About Dietary Fatty Acids

Pala *et al.* (2001) found higher levels of monounsaturated fatty acids MUFAs (especially oleic acid) in the erythrocyte membranes of postmenopausal women who developed breast cancer. A diet high in MUFAs is not the major determinant of erythrocyte membrane MUFAs, where most oleic acid in mammalian tissue is derived from the saturated stearic acid residue.

The key conversion is controlled by the Delta 9-desaturase, which also regulates the transformation of the other common saturated fatty acids (SFAs) (myristic and palmitic). The study discussed that fat content of the diet has an important effect on Delta 9-d activity, while high levels of SFAs increase Delta9-d activity by twofold to threefold, whereas polyunsaturated fatty acids (PUFAs) decrease (Pala *et al.,* 2001). This conclusion was partially contradicted by a latter study, which showed a direct relation between very high consumption of omega-6 (PUFAs) and breast cancer.

2. Phytoestrogens

Phytoestrogens have been extensively studied in animal and human *in vitro* and epidemiological studies. The actions of phytoestrogens on estrogen receptors and key enzymes that convert androgens to estrogens in relation to the growth of breast cancer cells has been reviewed extensively. In addition, comparison has been done to find out the experimental and epidemiological evidence pertinent to the potential, beneficial or harmful effects of phytoestrogens in relation to the incidence/progression of breast cancer and their efficacy as natural alternatives to conventional hormone replacement therapy during menopause (Helferich *et al.,* 2008). A novel anti-breast cancer agent, indole-3-carbinol (I3C) found naturally in vegetables such as broccoli and cabbage and can target breast cancer from a unique direction.

MCF-7 breast cancer cells treated with I3C showed a drastic decrease in the proliferation rate. The production and/or activity of key cell cycle regulatory proteins, CDK6 and CDK2 are also found to be decreased upon I3C treatment. Importantly, these effects were not related to the estrogen receptor status of the cell (Cover *et al.,* 1998). Estrogen-responsive breast cancer cells, such as MCF7 and T47D cells, express both estrogen receptor (ER)-α (ERα) and ERß. Indole-3-carbinol (I3C) strongly down-regulated ERα protein and transcript levels without altering the level of ERß protein in both cell lines. . Using an *in vitro* ERE binding assay, I3C was shown to inhibit the level of functional ERα and stimulated the level of ERE binding ERß even though the protein levels of this receptor remained constant. In ERα–/ERß+ MDA-MB-231 breast cancer cells, I3C treatment stimulated a 6-fold increase in binding of ERß to the ERE. Taken together, our results demonstrate that the expression and function of ERα and ERß can be uncoupled by I3C with a key cellular consequence being a significantly higher ERß: ERα ratio that is generally highly associated with antiproliferative status of human breast cancer cells (Sundar *et al.,* 2006). Pomegranates and mango are two more fruits found to have anticancerous principles. Periodically the list is growing as an outcome of scientific validation of many herbal products for its efficacy to curtail either the progression or the regression of breast cancer through the world. Brassica vegetable intake (broccoli, cauliflower, cabbage, kale and Brussels sprouts) was inversely related to breast cancer development. It has been estimated that women who consumed around 1.5 servings of Brassica vegetables per day had 42% less risk of developing breast cancer than those who consumed virtually none (Messina *et al.,* 2006). The literature support the following conclusions: Plant estrogen intake in early adolescence may protect against breast cancer later in life.

3. Vitamin D

Vitamin D is a group of prohormones, the two major forms of which are D2 and D3 of which .Vitamin D is related to reduced risk of breast cancer. Vit D's potential role in preventing cancer lies in a wide range of cellular mechanisms central to the development of cancer. These effects may be mediated through vitamin D receptors expressed in cancer cells. Polymorphisms of the vitamin D receptor (VDR) gene have been associated with an increased risk of breast cancer. Women with mutations in the VDR gene had an increased risk of breast cancer (Speeckaert *et al.,* 2010).

Geographical distribution pattern of breast cancer death rates clearly indicate that in white women the rate rises with distance from the equator and are highest in areas with long winters. Adult humans need much more vitamin D than the amount that used to be recommended (400 IU) probably somewhere around 3000-5000 IU daily. Diet alone cannot get enough vitamin D, therefore sun exposure is the preferred source of vitamin D in addition to dietary source. In women, an intake of approximately 1,000 mg of calcium per 1,000 kcal of energy with 800 IU of vitamin D would be sufficient to keep the rate under control. In observational studies, the source of approximately 90% of the calcium intake was vitamin D-fortified milk and may also be obtained from fish. Epidemiological data suggest that intake of 800 IU/day of vitamin D may be associated with enhanced survival rates among breast cancer cases (Garland *et al.,* 1999).

Some mushrooms offer an anti-cancer effect, which is thought to be linked to their ability to up-regulate the immune system. Some mushrooms known for this effect include, *Reishi, Agaricus blazei, Maitake, and Trametes versicolor.* Research suggests the compounds in medicinal mushrooms most responsible for up-regulating the immune system and providing an anti-cancer effect are a diverse collection of polysaccharide compounds, particularly beta-glucans. Beta-glucans are known as "biological response modifiers", and their ability to activate the immune system is well documented. Specifically, beta-glucans stimulate the innate branch of the immune system. Research has shown beta-glucans have the ability to stimulate macrophage, NK cells, T cells, and immune system cytokines. The mechanisms in which beta-glucans stimulate the immune system is only partially understood. One mechanism, in which beta-glucans are able to activate the immune system, is by interacting with the Macrophage (CD18) receptor on immune cells.

A highly purified compound isolated from the medicinal mushroom Trametes versicolor, known as Polysaccharide-K, has become incorporated into the health care system of a few countries. Japan's Ministry of Health, Labour and Welfare approved the use of Polysaccharide-K in the 1980s, to stimulate the immune systems of patients undergoing chemotherapy. In 2009, a case-control study of the eating habits of 2,018 women suggested that women who consumed mushrooms had an approximately 50% lower incidence of breast cancer. Since, mushrooms contain large amounts of vitamin D2, which when exposed to UV light; Mushrooms are the only known crop that naturally contains vitamin D (Lee *et al.,* 2009). Women who consumed mushrooms and green tea had a 90% lower incidence of breast cancer. A case control study of 362 Korean women also reported an association between mushroom consumption and decreased risk of breast cancer (Hong *et al.,* 2008).

CONCLUSION

Breast cancer is one of the most dreadful diseases; nevertheless, by change of lifestyle it can be prevented easily. Out of many risk factors, which aids in development of breast cancer in these review, only two very important factors have been discussed *viz.* Light and Food. It is a well established fact that incidence of breast cancer is higher in Scandinavian countries is higher than that of countries near equator, the reason being light conditions. This is further evidenced by higher incidence of breast caner in night shift workers and low incidence in blind subjects. Similarly some diets are better then others in prevention of breast cancers. For prevention, not only individuals but governments should adopt certain measures for rotation of shift duties restriction on food additives encourage people to eat less processed food and whole grains.

All these can be done only after adequate knowledge of molecular mechanism of action of light and food on DNA, cells, organs and body in a holistic way.

ACKNOWLEDGMENTS

Authors acknowledge their appreciation with thanks to Mr. T.V. Unnikrishnan, Technical writer, Comtech IT Solutions for his help for the preparation of the chapter.

REFERENCES

Altinok, A., Levi, F. and Goldbeter, A. 2007. A cell cycle automaton model for probing circadian patterns of anticancer drug delivery, *Adv. Drug Deliv. Rev.* 59: 1036–1053.

Antocha, M.P. and Chernov, M.V. 2009. Pharmacological modulators of the circadian clock as potential therapeutic drugs, *Mutation Res.* 680 : 109–115.

Barnett, G C. Shah, M., Redman, K., Easton, D.F., Ponder, B. A.J. and Pharoah, P. D. P. 2008. Risk Factors for the Incidence of Breast Cancer: Do They Affect Survival From the Disease?, *J. clin. Oncoy.* 26 :3310-3316.

Berquin, I.M., Min Y., Wu R.,WuJ., Perry P., Cline J.M., Thomas M.J., Thornburg T., Kulik G., SmithA., Edwards I.J., Agostino R.D., Zhang H., Wu H., Kang J.X. and Yong Chen Q. 2007.Modulation of prostate cancer genetic risk by omega-3 and omega-6 fatty acids, *J. Clin. Invest.* 117:1866-1875.

Blask, D.E., Dauchy R.T., Brainard, G.C., Hanifin, J.P. 2009. Circadian Stage-Dependent Inhibition of Human Breast Cancer Metabolism and Growth by the Nocturnal Melatonin Signal: Consequences of Its Disruption by Light at Night in Rats and Women, *Integr.Cancer Ther.* 8:341-353.

Brown, D. L., Feskanich D., Sanchez B. N., Rexrode K. M., Schernhammer E. S. and Lisabeth L. D. 2009. Rotating Night Shift Work and the Risk of Ischemic Stroke, *Am. J. Epidemiology.* 169: 1370-1377.

Cermakian, N., Pando, M.P., Thompson, C.L., Pinchak, A.B., Selby, C.P., Gutierrez, L., Wells D.E., Cahill, G.M., Sancar, A. and Sassone-Corsi, P. 2002. Light Induction of a Vertebrate Clock Gene Involves Signaling through Blue-Light Receptors and MAP Kinases, *Current Biol.* 12: 844-848.

Chaurasia, S.S. and Gupta P.D. 1999. Cryptochromes: The novel circadian photoreceptors, *Curr. Science.* 77: 55.

Chlebowski, R.T., Blackburn, G.L., Thomson, C. A., Nixon, D.W., Shapiro, A., Hoy, M.K., Goodman, M.T., Giuliano, A.E., Karanja, N., McAndrew, P., Hudis, C, Butler, J., Merkel, D., Kristal, A., Caan, B., Michaelson, R., Vinciguerra, V., Del Prete, S., Winkler, M., Hall, R., Simon, M., Winters, B.L. and Elashoff, R.M. 2006. Dietary fat reduction and breast cancer outcome: interim efficacy results from the Women's Intervention Nutrition Study (WINS) *J. Natl. Cancer Inst.* 98: 1767–1776.

Costa, G. 1996.The impact of shift and night work on health, *Applied Ergonomics.* 27: 9–16.

Cover, C.M., Hsieh, S.J., Tran, S.H., Hallden, G., Kim, G.S., Bjeldanes, L.F., and Firestone, G.L.1998. Indole-3-carbinol Inhibits the Expression of Cyclin-dependent Kinase-6 and Induces a G_1 Cell Cycle Arrest of Human Breast Cancer Cells, *J. Biol. Chem* 273: 3838-3847.

Crosthwaite, S.K. 2004. Circadian clocks and natural antisense RNA, *FEBS letters* 5671: 49-54.

Deapen, D., Lihua Liu, L., Perkins, C., Bernstein, L., Ross, R.K. 2002. Rapidly rising breast cancer incidence rates among Asian-American women, *Intl. J. Cancer* 99: 747-750.

Feychting, M., Osterlund, B. and Ahlbom, A. 1998. Reduced cancer incidence among the blind, *Epidemiology.* 9:490–494.

Garland, C.F., Garland, F.C and Gorham, E.D. 1999. Calcium and vitamin D. Their potential roles in colon and breast cancer prevention, *Ann. N. Y. Acad. Sci.* 889:107-119.

Gery, S., Komatsu, N., Baldjyan, L., Yu, A., Koo, D. and Koeffler, H.P. 2006. The circadian gene perl plays an important role in cell growth and DNA damage control in human cancer cells, *Mol. Cell.* 22: 375–382.

Gupta, P. D., Nayak U.Y. and Pushkala K. 2010. Darkside of the night light: implication in breast cancer, *J. Cell Tissue Res.* 10: 2173-2184.

Gupta, P.D. and Pushkala K. 1999. Importance of the role of calcium in programmed cell death: a review, *Cytobios.* 99: 83-95.

Gupta, P.D. and Pushkala K. 2006. Age dependent changes in steroid hormones level modulate progression and regression of breast cancer, *J. Cell Tissue Res.* 6: 825-836.

Gupta, P.D. and Saumyaa 2008. Lamin polymerization: a regulatory process for programmed cell survival, *Adv. Med. Dent. Sci.* 2: 34-39.

Helferich, W.G., Andrade. J.E., Hoagland. M.S. 2008. Phytoestrogens and breast cancer: a complex story, *Inflammopharmacology* 16 : 219-26.

Hong, S.A., Nam, K. and Kim, K. 2008. "A case-control study on the dietary intake of mushrooms and breast cancer risk among Korean women", *International journal of cancer. J. Intl. Cancer.* 122 : 919–923.

Hunter, D.J., Spiegelman, D., Adami, H.O., Beeso, L., van den Brand, P.A., Folsom, A.R., Fraser, G.E., Goldbohm, R.A., Graham, S., Howe, G.R., Kushi, L.H., Marshall, J.R., Mc Dermot, A., Miller, A.B., Speizer, F.E., Yaun, W.A . and Willett, W. 1996. "Cohort studies of fat intake and the risk of breast cancer--a pooled analysis", *New. Engl. J. Med.* 334 356–61.

Kayumov, L., Casper, R.F., Hawa, R.J, Perelman, B., Chung, S.A., Sokalsky, S. and Shapiro, C.M. 2005. Blocking low-wavelength light prevents nocturnal melatonin suppression with no adverse effect on performance during simulated shift work, *J. Clin. Endocrinol. Metab.* 90: 2755-2761.

Kliukiene, J., Tynes, T. and Andersen, A. 2001. Risk of breast cancer among Norwegian women with visual impairment, *Brit J. Cancer.* 84:397–399.

Lee, G.S., Byun, H.S., Yoon, K.H., Lee, S., Choi, K.C., Jeung, E.B. 2009. "Dietary calcium and vitamin D2 supplementation with enhanced Lentinula edodes improves osteoporosis-like symptoms and induces duodenal and renal active calcium transport gene expression in mice", *Eur. J. Nur.* 48 : 75–83.

Lehrer, S. 1982. Fertility of blind women, *Fertil. Steril.* 38:751-752.

Levi, F., Focan, C., Karaboue, A., Valette, V., Focan-Henrard, D., Baron, B., Kreutz, F. and Giacchetti, S. 2007. Implication of circadian clocks for rhythmic delivery of cancer therapeutics, *Adv. Drug Del. Rev.* 59: 1015–1035.

Lewy, A.J., Wehr, TA., Goodwin, F.K., Newsome, D.A., Markey, S.P.1980. Light suppresses melatonin secretion in humans, *Science.* 12: 210, 1267-1269.

Messina, M., McCaskill-Stevens, W. and Lampe, J.W. 2006. Addressing the soy and breast cancer relationship: review, commentary, and workshop proceedings, *J. Natl. Cancer Inst.* 98 : 1275–84.

Pala, V., Krogh, V., Muti, P., Chajès, V., Riboli, E., Micheli, A., Saadatian, M., SIPRI, S., Berrino, F. 2001. Erythrocyte Membrane Fatty Acids and Subsequent Breast Cancer: a Prospective Italian Study, *JNCL.* 93 : 1088.

Prentice, R.L., Caan, B., Chlebowski, R.T., Patterson, R., Lewis, H., Kuller., Ockene, J.K., Margolis, K.L., Limacher, M.C., Manson, J.E., Parker, L.M., Paskett, E., Phillips, L., Robbins, J., . Rossouw, J.E., Sarto, G.E., Shikany, J.M., Stefanick, M.L., Thomson, C.A. Horn, L.V., Vitolins, M.Z., Wactawski-Wende, J., Wallace, R.B., Wassertheil-Smoller, S., Whitlock, E., Yano, K., e Adams-Campbell, L., Anderson, G.L., Assaf, A.R., Beresford, S.A.A., Black, H.R., Brunner, R.L., Brzyski, R.G., Ford, L., Gass, M., Hays, J., Heber, D., Heiss, G., Hendrix, H.L., Hsia, J., Hubbell, F.A., Jackson, R.D., Johnson, K.C., Kotchen, J.M., LaCroix, A.Z., Lane, D.S., Langer, R.D., Lasser, N.L., Henderson, M.M. 2006. "Low-fat dietary pattern and risk of invasive breast cancer: the Women's Health Initiative Randomized Controlled Dietary Modification Trial", *JAMA.* 295 : 629–642.

Pukkala, E., Verkasalo, P.K., Ojamo, M., and Rudanko, S.L. 1999. Visual impairement and cancer: a population based cohort study in Finland, *Cancer Causes Con.* 10: 13-20.

Pukkala, E., Aspholm R., Auvinen A., Eliasch H., Gundestrup M., Haldorsen T., Hammar N., Hrafnkelsson J., Kyyrönen P., Linnersjö A., Rafnsson V., Storm H. and Tveten U. 2003. Cancer incidence among 10,211 airline pilots: a Nordic study, *Aviat. Space Environ. Med.* 74: 699-706.

Pukkala, E., Ojamo M., Rudanko S.L., Stevens R.G. and Verkasalo P.K. 2005. Does incidence of breast cancer and prostate cancer decrease with increasing degree of visual impairment, *Cancer Causes Control.* 17; 573-576.

Pushkala, K. and Gupta P.D. 2009. Prevalence of breast cancer in menopausal blind women, *Int. J. Med. Med. Sc.* 1 : 425-431.

Pushkala, K. and Gupta P.D. 2001. Steroid hormones regulate programmed cell death: a review, *Cytobios.* 106: 202-217.

Pushkala, K. and Gupta, P.D. 2011. Dark side of the night light Monograph. LAMBERT Academic Publishing GmbH and Co., Germany.

Reddy, A.B., Wong, G.K.Y., O'Neill J., Maywood E.S. and Hastings M.H. 2005.Circadian clocks: Neural and peripheral pacemakers that impact upon the cell division cycle, *Mutation Res.* 574: 76–91.

Reiter, R.J., Tan D.X., Korkmaz, A., Erren, T.C., Piekarski, C., Tamura, H. and Manchester, L.C. 2007. Light at night, chronodisruption, melatonin suppression, and cancer risk: a review, *Crit. Rev. Oncog.* 13: 303–328.

Schernhammer, E.S., Kroenke C.H., Laden F. and Hankinson S.E. 2006. *Epidemiology.* 17:108-11.

Schmutz, I., Ripperger, J.A., Baeriswyl-Aebischer, S. and Albrecht, U. 2010. The mammalian clock component PERIOD2 coordinates circadian output by interaction with nuclear receptors, *Genes Dev.* 24: 345-57.

Simopoulos, A. P. 2003. "Importance of the ratio of omega-6/omega-3 essential fatty acids: evolutionary aspects". In: Simopoulos, A.P. and L.G. Cleland , *eds., World Review of Nutrition and Dietetics* (Karger) 92 (Omega-6/Omega-3 Essential Fatty Acid Ratio: The Scientific Evidence. World Rev Nutr Diet. Basel, Karger,, vol 92, pp 1-22.

Sonestedt, E., Ericson, U., Gullberg, B., Skog, K., Olsson, H., Wirfält, E. 2008. Do both heterocyclic amines and omega-6 polyunsaturated fatty acids contribute to the incidence of breast cancer in postmenopausal women of the Malmö diet and cancer cohort?, *Intl J Cancer* (*UICC International Union Against Cancer) 123* : 1637–1643.

Speeckaert, M.M., Taes, Y.E., De Buyzere, M.L., Christophe, A.B., Kaufman, J.M., Delanghe, J.R. 2010. "Investigation of the potential association of vitamin D binding protein with lipoproteins", *Annals of Clinical Biochemistry* 47 : 143–150.

Stevens, R.G., 2009. Light-at-night, circadian disruption and breast cancer: assessment of existing evidence, *Int. J. Epidemiolog.* 38 (4): 963-970.

Sundar, S.N., Kerekatte, V., Equinozio, C.N., Victor, B. Doan, V.B., Bjeldanes, L.F. and Firestone, G.L. 2006. Indole-3-Carbinol Selectively Uncouples Expression and Activity of Estrogen Receptor Subtypes in Human Breast Cancer Cells, *Mol. Endocrin.* 20 : 3070-3082.

Terazono, H., Hamdan, A., Matsunaga, N., Hayasaka, N., Kaji, H., Egawa, T., Makino, K., Shigeyoshi, Y., Koyanagi, S. and Ohdo, S. 2008. Modulatory effects of 5-fluorouracil on the rhythmic expression of circadian clock genes: A possible mechanism of chemotherapy-induced circadian rhythm disturbances*, Biochem. Pharmacol.* 75: 1616 – 1622.

Verkasalo, P.K., Pukkala, E., Stevens, R.G., Ojamo, M. and Rudanko, S.L. 1999. Inverse association between breast cancer incidence and degree of visual impairment in Finland, *Brit. J. Cancer.* 80:1459–1460.

Vlassara, H., Cai, W., Goodman, S., Pyzik, R., Yong, A., Chen, X., Zhu, L., Neade, T., Beeri, M., Silverman, J.M., Ferrucci, L., Tansman, L., Striker, G.E. and Uribarri, J. 2009. Protection against loss of innate defenses in adulthood by low advanced glycation and products (AGE) intake: Role of the Antiinflammatory AGE Receptor-1, *J. Clinl. Endocrin. Metabol.* 94: 4483-4491.

Wise, J. 2009. Danish night shift workers with breast cancer awarded Compensation, *Brit. Med. J.* 338: b1152.

Zhu, Y., Tongzhang, Z., Stevens, R.G., Zhang, Y. and Boyle, P. 2006. Does"Clock" Matter in Prostate Cancer?, *Cancer Epidemiol. Biomarkers Prev.* 15 : 3-5.

WHO/World Cancer Day. 4 February *2009. Cancer* is a leading cause of death around the *world.* WHO estimates that 84 million people will die of *cancer* between 2005 www.who.int/mediacentre/events/.../world_cancer.../index.html

In: Natural Products and Their Active Compounds ... ISBN: 978-1-62100-153-9
Editors: M. Essa, A. Manickavasagan, and E. Sukumar

Chapter 6

OMEGA 3 FATTY ACIDS AND ITS POTENTIAL HEALTH BENEFITS

N. Guizani[1*], Mohamed M. Essa[1,2,3], Z. Al-Kharousi[1], V. Singh[1], B. Soussi[4] and G. J. Guillemin[2]

[1]Dept of Food Science and Nutrition,
College of Agriculture and Marine Sciences,
Sultan Qaboos University, Oman
[2]Neuropharmacology group, Dept of Pharmacology,
College of Medicine, University of New South Wales,
Sydney, Australia
[3]Developmental Neuroscience Lab, NYSIBR,
Staten Island, NY, US
[4]Dept of Marine Sciences, College of Agriculture and Marine Sciences,
Sultan Qaboos University, Oman

ABSTRACT

Omega-3 fatty acids, eicosapentaenoic acid (EPA) and docosahexaenoic acid (DHA), are one of the essential fatty acids. As these are necessary for human health but the human body can't synthesize it. It should be supplied through our diets. Omega-3 fatty acids (FAs) found mainly in fish, such as salmon, tuna, and halibut and seafood including algae and krill, some plants, and nut oils. It has been drawn the attention in the scientific community due to their capability to affect several processes in the body such as neurological, cardiovascular, and immune functions, and cancer. The American Heart Association recommends eating fish (particularly fatty fish such as mackerel, lake trout, herring, sardines, albacore tuna, and salmon) at least 2 times a week. The 2010 Dietary Guidelines for Americans recommends 8 oz. of a variety of seafood per week providing an average daily consumption of 250 mg EPA+DHA. Omega-3 fatty acids are highly concentrated in the brain and appear to be important for cognitive (brain memory and performance) and behavioral function. In fact, those infants are at risk for developing

* Corresponding author:guizani@squ.edu.om

vision and nerve problems that do not get enough omega-3 fatty acids from their mothers during pregnancy. Important deficiency symptoms of omega-3 FAs include fatigue, poor memory, dry skin, heart problems, mood swings or depression, and poor circulation. Further, it is necessary to maintain the balance between of omega-6 /omega-3 ratio in our diets. As omega-6 FAs increase the inflammatory responses in the body.

Keywords: Omega-3 fatty acids (Omega-3 FAs), Cardiovascular diseases (CVDs), Eicosapentaenoic acid (EPA) and Docosahexaenoic acid (DHA), cancer, pregnancy

1. Introduction

The successes of the Human Genome Project and the powerful advances that have been made in molecular biology have enrolled us in a new era of nutrition and medicine (Gillies, 2003). Rucker and Tinker (1986) described in their review how the tools of molecular biology can be used in the study of food components and essential nutrients as factors in the control of gene expression. In particular, they described the effects of various nutrients on some cellular events such as transcription, post-transcription, translation, and post-translational modification. Among these nutrients, long chain omega-3 polyunsaturated fatty acids that have been shown to regulate many diseases-related processes in the body such as neurological, cardiovascular, and immune functions, and cancer (Biondo*et al.*, 2008).The accumulating large numbers of studies that correlate consumption of foods high in omega-3 fatty acids to the increased health benefits have attracted even more scientists to unravel the molecular mechanisms involved in this process. The voluminous research data on omega-3 ingredients has given the marine and algae oil omega-3 ingredients a definite edge in the functional foods market. The 14,000 papers published about omega-3 eicosapentaenoic acid/ docosahexaenoicacids (EPA/DHA) have greatly improved the credibility of these polyunsaturated fatty acids (PUFAs) as vital functional ingredients (SAEMAO, 2010).Omega-3 FAs are a family of naturally occurring polyunsaturated fatty acids (PUFAs). Humans do not possess the necessary metabolic pathways to synthesize the precursor FA (linolenic acid), which is essential for the production of the longer bioactive omega 3 FAs. Therefore, these long-chain PUFAs must be obtained either from plant sources or by direct intake of EPA and DHA from marine or industrial products. EPA and DHA are mostly found in seafood, but fish do not actually produce these fatty acids. In fact, these compounds are produced by single-cell marine organisms that are consumed by fish.

The aim of this chapter is to review the recent advancements in pharmacological studies that have been conducted on dietary omega-3 fatty acids. To achieve this goal, a general view about fatty acids, their metabolism and health benefits will be discussed here.

2. Classification and Metabolism of Fatty Acids

PUFAs are classified into four families (n-9, n-7, n-6 and n-3) according to the position of the first double bond from the methyl end of the molecules (Jiang *et al.*, 1998). The n-carbon is also called ω-carbon. The n-3 and n-6 PUFAs are called essential fatty acids because they cannot be synthesized by mammals and must be obtained from the diet (Rose and Connolly, 1999).

PUFAs are important components of membrane phospholipids that determine membrane properties such as deformability and fluidity (Sellmayer*et al.*, 1997). The increased incorporation of PUFAs in cell membrane increases the fluidity of membranes because PUFAs acyl chains are very flexible and thus can alter conformational states. However, the flexibility of the acyl chain differs between omega-6 and omega-3 fatty acids and according to the number of double bonds. Therefore, omega-6 and omega-3 fatty acids composition of biological membranes influence physical properties of the membranes and so alter protein function and trafficking and vesicle budding and fusion (Schmitz and Ecker, 2008). The other physiological roles of PUFAs include regulation of gene expression, cell signaling, and energy provision (Diggle, 2002).

Omega-3 fatty acids are represented by α-linolenic acid (LNA; 18:3, n-3) and omega-6 fatty acids by linoleic acid (LA; 18:2, n-6). Both of these essential fatty acids are metabolized to longer chain fatty acids of 20 and 22 carbon atoms. LA is metabolized to arachidonic acid (AA; C20:4, n-6) and LNA to eicosapentaenoic acid (EPA; 20:5, n-3) and docosahexaenoic acid (DHA; 22:6, n-3) (Simopoulos, 1991). The essential fatty acids are the precursors of a group of compounds called eicosanoids (thromboxane, prostaglandin, and leukotriene) that have diverse biological effects like platelet aggregation and chemotaxis (Krey*et al.*, 1997). LNA is found in walnuts, leafy vegetables, canola oil, soybean oil and flaxseed while EPA and DHA are found mainly in cold-water fatty fish. Humans are able to produce EPA from LNA through chain elongation and desaturation but this conversion is limited such that EPA and DHA are obtained mainly from fish (Biondo*et al.*, 2008).

3. The Opposing Effects of Omega-3 and Omega-6 Fatty Acids

Recently, it has been shown that human beings consumed diets with a ratio of omega-6 to omega-3 fatty acids of about 1. Now, this ratio in western diets is about 15/1 to 16.7/1 which means that western diet is deficient in omega-3 fatty acids as compared to the diet on which human genetic patterns were established (Simopoulos, 2008). The story started when Americans decreased the use of saturated fatty acids in an attempt to increase heart-healthy fats incorporation into their diets. However, they substituted corn, safflower, and soybean oils for frying and baking purposes. These oils contain high amounts of omega-6 fatty acids (LA) and very little amounts of omega-3 fatty acids (ALA) (Hardman *et al.*, 2002). This misunderstanding happened because, until 1970, the emphasis was on the lipid hypothesis to lower serum cholesterol as a key factor in controlling cardiovascular disease. However, in 1970, the important roles of inflammation and thrombosis in the development of cardiovascular disease were emphasized by some researchers (Simopoulos, 1991) who noticed that Greenland Eskimos who consume a high seafood diet rich in omega-3 fatty acids, EPA and DHA, have low incidences of type 2 diabetes mellitus, coronary heart disease, multiple sclerosis, and asthma. Diets rich in omega-6 fatty acids contents result in increased amounts of eiosanoid metabolic products from arachidonic acid and large concentrations of them can contribute to the formation of atheromas, thrombus, inflammatory and allergic disorders and to proliferation of cells (Simopoulos, 2008). In general, omega-6 fatty acids and their derivatives enhance the production of proinflammatory eicosanoids whereas omega-3 fatty acids suppress this action (Pauwels and Kairemo, 2008). Omega-3 fatty acids lower serum cholesterol, increase bleeding time and possess antithrombotic actions (Simopoulos, 1991). The three major omega-3 fatty acids; LNA, EPA, and DHA suppress the production of AA from LA by competing more successfully than LA for the activity of Δ5 and Δ6 desaturases (Hardman, 2002). Moreover, omega-3 fatty acids can replace omega-6 fatty acids in the membranes of most cells (Simopoulos, 1991).

Bagga*et al*., (1997) tested the change in the ratio of omega-6 to omega-3 fatty acids in the plasma and adipose tissue of breast and buttocks of 25 women with breast cancer after consumption of a low-fat diet and fish oil supplements over a 3-months period. They found a significant decrease in omega-6 to omega-3 PUFAs ratios in plasma and breast adipose tissues after this short-term dietary intervention.

4. Biological and Pharmacological Activities

Since 2003, there have been introduction of thousands of foods and beverages enriched with omega fatty acids worldwide. Until late 2004, Food and Drug Administration (FDA) moved to allow a number of nutrient content claims for Omega fatty acids. Only after that, marketers try to sell the foods enriched in omega fatty acids.Therefore, the use of natural products is increasing due to their beneficial effects to prevent or delays many diseases and adverse health conditions such as cardiovascular disease, diabetes, multiple sclerosis etc.

4.1. Omega 3- FAs and Cardiovascular Disease (CVD)

Best way to prevent heart diseases is to eat low saturated fat and high monounsaturated and polyunsaturated fats (including omega-3 fatty acids) diet. 50 to 60% of deaths from cardiovascular disease (CVD) result from sudden cardiac death or sustained ventricular arrhythmias. Numerous epidemiological studies indicating that a high intake of omega 3 FAs is associated with a reduced risk for CVD mortality, MI and sudden death. Clinical evidence supports that EPA and DHA helps in the reduction in risk factors for heart disease, including high cholesterol and high blood pressure (Micallef and Garg, 2009). The major sources of EPA and DHA levels are dietary intake of cold water fish and less likely other seafood products (Kris et al., 2000). Consumption of fish oil (rich in Omega 3-FAs) reduces the cardiovascular mortality (Gruppo 1999), inflammatory and atherosclerotic processes (Caterina and Zampolli, 2004), and improves blood pressure and lipid status (Appel et al., 1993; Harris 1997).Combination of n-3 polyunsaturated fatty acid (PUFA) and plant sterol intake reduces systemic inflammation in hyperlipidemic individuals, hence are cardio-protective (Micallefa and Garg, 2009). According to some epidemiological studies, there is an inverse relationship between fish intake and the incidence of stroke (Gebska, 2008). There are many mechanisms behind the beneficial effects of PUFAs in cardioprotection. PUFA improves the vascular reactivity, decrease platelet aggregation, lowers the plasma tri-glycerides, decrease the blood pressure, prevents arrhythmias and reduces the inflammation

(Balk et al., 2006), an effect on atherothrombosis and on cardiac arrhythmias (Massaro et al., 2008). Evidence from recent studies indicates the consumption of combination of phytosterols and omega-3 fatty acids may reduce cardiovascular risk in a complementary and synergistic way. Further, intake of omega-3 FAs alone is also significantly associated with reduction in the risk factors for CVDs including hypotriglyceridemic effects (Micallef and Garg, 2009). Further, supplementation of combination of omega-3 FAs, oleic acid and vitamins are useful for reducing the risk factors for CVDs (Carrero et al., 2004). Clinical trials also support the role of omega 3 FAs in the significant reductions of cardiovascular events (GISSI, 1999; Yokoyama and Origasa, 2003; Yokoyama, 2005). Lower dose intake (1 g/day) resulted in reduction in total mortality, deaths due to cardiovascular disease, and fatal MI which includes sudden death while higher-dose consumption resulted in plaque stabilization, with observed reductions of 24% in unstable angina (RR, 0.76; 95% CI, 0.62 to 0.95) and 19% in nonfatal events (RR, 0.81; 95% CI, 0.68 to 0.96). Eating at least 2 servings of fish per week can reduce the risk of stroke by as much as 50%. However, high doses intake of fish oil/ omega-3 FAs may increase the risk of bleeding (GISSI, 1999; Yokoyama and Origasa, 2003; Yokoyama, 2005). While intake of very high intake (3 g/day) which is equivalent to 3 servings of fish per day may have higher risk for hemorrhagic stroke in which an artery in the brain leaks or ruptures. Overall, these results suggest that there was a significant reduction in total mortality and CAD death due to intake of omega-3 fatty acids which may be through a strong antiarrhythmic effect.These trials indicate the supplementation of EPA and DHA, if it ranged from 0.85 to 4.0 g/day will be most significantly benefitted. However, still more research is needed in this regard to estabilish the optimal doses (GISSI, 1999; Yokoyama and Origasa, 2003; Yokoyama, 2005).

4.2. Omega 3- FAs and Hypertension

Several clinical studies suggest that dietary supplementation of omega-3 FAs moderately decrease the blood pressure in people with hypertension (Ventura et al., 1993; Knapp et al., 1996). Appel et al. (1993) did meta analysis of 17 clinical trials using omega 3 FAs as a fish oil form and they found that taking more than 3 g of fish oil daily may reduce blood pressure in people with untreated hypertension. However, long-term efficacy and patient acceptability of lower doses is needed to confirm the use of omega-3 FAs as antihypertensive therapy.

4.3. Omega 3- FAs and Antihyperlipidemia

Supplementation of omega-3 fatty acid is associated with significant hypotriglyceridemic effects. Omega-3 FAs decreased plasma triglyceride levels in patients with hypertriglyceridemia. Moreover, there was not only decrease in very low-density lipoprotein (VLDL) levels but also there was a slight increase in LDL cholesterol levels. However, total and high-density lipoprotein (HDL) cholesterol levels were unaltered (Harris, 1997).

Further, supplementation of omega 3 FA also involves changes associated with the risk factors of CVDs like platelet function and pro-inflammatory mediators. Moreover, intake of combination of phytosterols and omega-3 fatty acids may reduce cardiovascular risk in both complementary as well as in a synergistic way. Also, there are many health benefits of phytosterols and omega-3 fatty acids, alone or in combination with statins which includes treatment and management of hyperlipidemia (Micallef and Garg, 2009).

4.4. Omega 3- FAs and Diabetes

There are many risk factors increasingly recognized and identified for the type 2 diabetes like food intake. Therefore, by following the healthy life style it can be prevented in a people with with impaired glucose tolerance (Ramachandran et al., 2006; Pan et al., 2007). An uncontrolled study by Sartorelli et al., 2010 supported the protective effect of consumption of omega 3 FA on the glucose metabolism in high-risk individuals. And they recommended further clinical studies to establish this fact clearly which helps in the development of strategies for the treatment of patients. The supplementation of omega-3 fatty acid (3g/day) for the period of 2 months significantly decreases the levels of homocysteine in diabetic patients without any change in fasting blood sugar (FBS), malondialdehyde (MDA) and C-reactive protein (CRP) levels (Pooya et al., 2010).

4.5. Omega 3- FAs and Cancer

MacLean *et al.*, (2006) reviewed 38 articles (from 1966 to 2005) that describe the effects of omega-3 fatty acids on the incidence of different types of cancer in human and concluded that dietary omega-3 fatty acids or their supplementation are unlikely to prevent cancer. This conclusion might be because the rare confirmation of the preclinical results of the beneficial effects of omega-3 fatty acids in reviews and meta-analysis of the epidemiological data published before 2005 (Pauwels and Kairemo, 2008). Moreover, Surette, 2008, emphasized that the diverse effects of omega-3 fatty acids should be viewed as a gentle shift in molecular mechanisms underlying an organism's functions toward a phenotype of lessened reactivity against environmental stimuli including those involved in carcinogenesis. In a review, Hardman (2002), summarized how omega-3 fatty acids can augment cancer therapy, he concluded that the bulk of the available evidence until that time indicated that increasing the amount of omega-3 fatty acids in the diet would be beneficial to cancer patients' survival. Likewise, Chapkin*et al.*, (2007) described how omega-3 fatty acids exert pleiotropic effects to reduce the development of colon cancer as compared to omega-6 fatty acids that enhance colon cancer development.

It is accepted that a high fat and thus a high calorie intake are risk factor in cancer development. However, the type of fat is very important because omega-6 fatty acids Arachidonic acid (AA) and the cyclooxygenase-2 (COX-2) enzyme are both found in abundance in most cancer tissues (Koki et al., 2002) and are thought to be positively correlated with cancer development while monounsaturated and omega-3 fatty acids like DHA and EPA are usually in low amount (Kokoglu et al., 1998) and it may be protective against cancer development or their obvious correlation with cancer is absent (Diggle, 2002). As the sources of these essential fatty acids are food intake therefore changes in the dietary habits are responsible for the change in the ratio of omega6/ omega 3 FAs. Following this, omega-6 fatty acid consumption should be reduced and that consumption of omega-3 fats should be increased. Moreover, to reduce total fat consumption, our diet should contain less meat and incorporate more fruit, vegetables and whole grains. It was also shown that omega-3 PUFAs sensitize cancer cells to effects of anti-cancer drugs and thus can improve their response to chemotherapy. However, currently, there are no specific nutrition guidelines for cancer patients undergoing chemotherapy (Biondo*et al.*, 2008). Moreover, many types of foods that are not traditional sources of omega-3 fatty acids, like dairy and bakery products, are now being fortified with omega-3 fatty acids (Surette, 2008). The International Society for the Study of Fatty Acids and Lipids recommends an intake of 500 mg/day of EPA and DHA. The American Heart Association recommends an intake of 2 fish meals weekly (at least 300 mg of omega-3 fatty acids) for people without coronary heart disease (CHD) and 1000 mg/day for patients with CHD (Surette, 2008). Incidences of certain type of cancers like breast cancers have increased with high consumption of westernized diet and unhealthy lifestyle. Many studies supported that consumption of omega-3 fatty acid is related with decreased risk of cancers of kidneys (Wolk et al., 2006), colon (Courtney et al., 2007), breast (Thiebaut et al., 2009), prostate (Fradet et al., 2009), and liver (Lim et al., 2009).

4.6. Omega 3- FAs and Neurological Effects

Many epidemiological studies suggested that higher intake of omega 3 FA rich foods like fish intake helps in reducing dementia/ Alzheimer's disease (AD)(Kalmijn et al., 1997a; Huang et al.,2005).But later, Arendash et al., in 2007 reported that dietary intake of omega 3 FAs including fish oil supplements will not protect the AD in a high-risk individuals but in a normal individuals may be cognitively benefitted. Omega 3 FAsintake (Dietary/ in capsules) is necessary for optimal cerebral and cognitive development of the infant (Bourre, 2007). ALA, DHA, EPA are important in the prevention or reduction in the risk of postpartum depression, manic-depressive psychosis, dementias (Alzheimer's disease and others) (Bourre, 2007). Supplementation of omega-3 fatty acids may improve general clinical function in patients with mild or moderate AD and mild cognitive impairment (MCI), but not their cognitive function. But its cognitive effects might be favored in patients with MCI rather than those with AD (Chiu et al., 2008).However, during pregnancy and lactation, consumption of excessive rich/ deficient diet in omega 3 FA may have adverse effects on brain development and sensory function (Church et al., 2009).

Attention-deficit hyperactivity disorder (ADHD) is a neurodevelopmental disorder affecting approximately 5–13% of children and adolescents (Faraone et al., 2003). As low levels of omega-3 FAs have been linked to the behavioral and mood disorders including ADHD. Gow et al., 2009 find that supplementation of EPA and DHA may influence processing in ADHD and also explain the conflicting findings of the literature about the EPA supplementation in ADHD and depression.

4.7. Omega 3- FAs and Metabolic Syndrome (MS)

Benito et al., 2006 did randomized, placebo-controlled and open clinical trial, among 72 patients with MS for 3 months. Control group were consumed semi-skimmed milk (500 cm^3 per day) while test group were consumed same quantity of enriched milk (5.7 g of oleic acid, 0.2 g of o-3 fatty acid, 150 mg of folic acid and 7.5 mg of vitamin E). They found a decrease in serum total cholesterol (-6.2%, p=0.006), LDL cholesterol (-7.5%, p=0.032), triglycerol (-13.3%, p= 0.016), Apo B (-5.7%, p=0.036), glucose (-5.3%, p= 0.013), and homocysteine (-9.5%, p=0.00) in a test group as compared to control. Hence, enriched milk diet supplementation reduces several cardiovascular risk factors in patients with MS.

4.8. Omega 3- FAs and Pregnancy

Dietary intake of omega-3 fatty acid will ensures that a woman's adipose tissue contains a reserve of these fatty acids for the developing fetus and newborn infant. During pregnancy and lactation, consumption of omega-3 FAs is beneficial for the fetal and infant brain development(Church 2009). However, diets relatively rich in omega-3 FAs can adversely affect fetal and infant development and the auditory brainstem response (ABR), a measure of brain development and sensory function. Infant/young adulthood that received excess/ deficient amount had postnatal growth retardation, poor hearing acuity and prolonged neural transmission times as evidenced by the ABR (Church 2009).

4.9. Omega 3- FAs and Hepatic Disease

Flax and pumpkin seed mixture (source of omega-3 FAs) had anti-atherogenic and hepatoprotective effects which are due to presence of unsaturated fatty acids in seed mixture (Makni et al., 2008).

Nonalcoholic fatty liver disease (NAFLD) patients were treated with EPA for a year. In the treatment group, hepatic steatosis and serum ALT, AST and γ-GT were reduced as compared to control group. Further, there was improvement in liver perfusion by the EPA supplementation (Capanni et al., 2006). Later researches were shown that omega 3 FAs supplementation can decrease nutritional hepatic steatosis in adults. Moreover, some clinical trials also suggested that the consumption of omega 3 FAs may be beneficial in NAFLD(Shapiro et al., 2011).

4.10. Omega 3- FAs and Kidney Disease

Many *in-vitro* and *in-vivo* studies support the omega-3 FAs efficiency on inflammatory pathways involved with the progression of kidney disease (Caterina et al., 1993). Recently, many renal diseases like lupus nephritis, polycystic kidney disease, and other glomerular diseases have been investigated. Clinical trials on immunoglobulin A (IgA) nephropathy reported conflicting results for the efficacy of omega-3 FAs in IgA nephropathy(Fassett et al., 2010). This may be due to varying doses, proportions of EPA and DHA, duration of therapy, and sample size of the study populations. Further, the quality of available meta-analysis of clinical trials in IgA nephropathy is the main limitation. Blood pressure which can intensify the renal disease decreased by the intake of omega 3 FAs. Still there is a need to further investigate the potential benefits of omega-3 polyunsaturated fatty acids on the progression of kidney disease through the well-designed, adequately powered, randomized, controlled clinical trials (Fassett et al., 2010).

4.11. Omega 3- FAs and Rheumatoid Arthritis (RA)

Rheumatoid arthritis (RA) is an autoimmune disease that causes inflammation in the joints. In one randomized, placebo-controlled, double-blind, fish oil versus olive oil study by Geusens et al. (1994) found better clinical scores and reduction in their concomitant antirheumatic medications. A number of clinical studies found that supplementation of fish oil helps in the reduction of clinical symptoms of RA which includes joint pain and morning stiffness. Moreover, there was significant decrease in levels of IL-1 beta and subjects consuming fish oil were able to cease their dose of non-steroidal anti-inflammatory drugs (NSAIDs) (Kremer et al., 1995; Kremer 1996).

4.12. Omega 3- FAs and Inflammatory Bowel Disease (IBD)

Many clinical studiesare available for the dietary intake omega 3 FAs and improvements in the Inflammatory bowel disease (IBD). The blind cross-over trial of fish oil versus vegetable oil found reductions in rectal dialysate leukotriene B4 levels, improvements in histologic findings, and weight gain in patients with IBD (Stenson et al., 1992).Another double-blind, placebo (maize oil) controlled trial found that dietary intake of omega-3 FAs resulted in only delay in relapse but no prevention (Loeschke et al., 1996). Some studies suggest that omega-3 fatty acids may help when added to medication, such as sulfasalazine (a standard medication for IBD), while other studies don't support it.Aslan and Triadafilopoulos, (1992) studied the effect of intake of fish oil supplementation in active ulcerative colitis. They found a reduction in the disease activity index and marked reduction/elimination in anti-inflammatory drugs in patients with ulcerative colitis. But, further studies are needed to elucidate the mechanism of action and optimal dose and duration of fish oil supplementation in IBD.

4.13. Omega 3- FAs and Eye

The retina is an extension of the optic nerve, leading electrical impulses to the posterior part of the brain. In the retina, light energy is transformed into electrical impulses by conformational changes of the light-sensitive protein, rhodopsin. The omega-3 fatty acid, docosahexaenoic acid (DHA), plays a part in the generation of these impulses. DHA is major structural lipid of retinal photoreceptor outer segment membranes (Giovanni and Chew, 2005). It may affect photoreceptor membrane function by altering permeability, fluidity, thickness, and lipid phase properties due to its bio-physical and chemical properties. It involves in signaling cascades to enhance activation of membrane-bound retinal proteins and may also be involved in rhodopsin regeneration. Insufficient DHA resulted in changes in retinal function. DHA activates a many nuclear hormone receptors that act as transcription factors for molecules which amend reduction-oxidation-sensitive and proinflammatory genes(Giovanni and Chew, 2005). These include the peroxisome proliferator-activated receptor-a (PPAR-a) and the retinoid X receptor. The DHA content in the retina is the highest of any tissue in the human body, and it is located in the membranes of the photoreceptors, especially in the rods. Premature babies fed with infant formula without marine omega-3 fatty acids do not have the same visual acuity as babies who are given mother's milk containing DHA(Giovanni and Chew, 2005). Further, intake of omega -3 and omega- 6 FAs intake have therapeutic advantage in patients suffering from ocular dryness (Garcher et al., 2011). Patients with dry eye syndrome can be given the supplementation of essential fatty acids (omega-3 FAs) as a novel treatment (Rand and Asbell, 2011). DHA supplementation helps in significant improvement in better right eye visual acuity in healthy and older people (Stough et al., 2011).

CONCLUSION

Epidemiological and clinical studies on omega-3 FAs suggest its important function in the prevention of many diseases including cardiovascular disease, cancers, neurological diseases (like Alzheimer's disease), diabetes, hypertension, inflammatory disorders including arthritis, liver, kidney diseases etc. Particularly, it plays an important role, starting from birth of infant till end of life. Furthermore, it has great result for healthy vision. Based on the above review, it is necessary to include the fish (atleast 1-2 servings/ week) or seafood or its other sources in our diet for the prevention of the diseases as well as to maintain our health. But, there should be a need for complete and comprehensive clinical studies to prove the above said effects.

REFERENCES

Appel, L.,J., Miller, E.,R., Seidler, A.,J., Whelton, P.,K. 1993. Does supplementation of diet with ''fish oil'' reduce blood pressure? A meta-analysis of controlled clinical trials. *Arch. Intern. Med.* 153: 1429.

Balk, E., M., Lichtenstein, A.,H., Chung, M., Kupelnick, B., Chew,P., Lau, J. 2006. Effects of omega- 3 fatty acids on serum markers of cardiovascular disease risk: a systematic review. *Atherosclerosis.* 189: 19–30.

Benito, P., Caballerob, J., J., Morenoa, J., Alcantaraa, C., G., Munozb, C, Rojob, G., Garciab,S, Soriguerb, F,C. 2006. Effects of milk enriched with *x*-3 fatty acid, oleic acid and folic acid in patients with metabolic syndrome. Clinical Nutrition. 25: 581–587.

Biondo, P.,D., Brindley, D., N., Sawyer, M.,B., Field, C.,J. 2008. The potential for treatment with dietary long-chain polyunsaturated n-3 fatty acids during chemotherapy. *Journal of Nutritional Biochemistry.* 19: 787-796.

Bryhn, M., Silentia, A., S., and Svelvik.Can Omega-3 Fatty Acids Prevent Blindness? *EPAX Marine omega 3 formula.*

Capanni, M., Calella, F., Biagini, M.,R., Genise, S., Raimondi, L., Bedogni, G., et al. Prolonged n-3 polyunsaturated fatty acid supplementation ameliorates hepatic steatosis in patients with non alcoholic fatty liver disease: a pilot study. *Aliment Pharmacol. Ther.* 23: 1143-51.

Carrero, J., J., Baro, L., Fonolla, J., Santiago,M., G., Ferez, A., M., Castillo,R., Jimenez, J., Boza, J., J., Huertas, E., L. 2004. Cardiovascular effects of milk enriched with omega-3 polyunsaturated fatty acids, oleic acid, folic acid, and vitamins e and b6 in volunteers with mild hyperlipidemia. *Nutrition.* 20: 521–527.

Caterina, R., Caprioli, R., Giannessi, D., et al. 1993. n-3 fatty acids reduce proteinuria in patients with chronic glomerular disease. *Kidney International.* 44: 843.

Caterina, R., Zampolli, A. 2004. From asthma to atherosclerosis 5-Lipoxygenase.leukotrienes, and inflammation. *N. Engl. J. Med.* 350: 4-7.

Chiu, C., C., Su, K.,P., Cheng, T., C., Liu, H., C., Chang, C., J., Dewey, M., E., Stewart, R., Huang, S.,Y., 2008. The effects of omega-3 fatty acids monotherapy in Alzheimer's disease and mild cognitive impairment: A preliminary randomized double-blind placebo-controlled study. *Progress in Neuro-sychopharmacologyand Biological Psychiatry.* 32: 1538–1544.

Church, M.,W., Jen, K.,L.,C., Jackson, D.,A., Adams, B.,R., Hotra, J.,W. 2009. Abnormal neurological responses in young adult offspring caused by excess omega-3 fatty acid (fish oil) consumption by the mother during pregnancy and lactation. *Neurotoxicology and Teratology.* 31: 26–33.

Courtney, E., D., Matthews, S., Finlayson, C., Di Pierro, D., Belluzzi, A., Roda, E., Kang, J.,Y., Leicester, R.,J. 2007. Eicosapentaenoic acid (EPA) reduces crypt cell proliferation and increases apoptosis in normal colonic mucosa in subjects with a history of colorectal adenomas, *Int. J. Colorectal Dis.* 22: 765–776.

Diggle, C., P. 2002. In vitro studies on the relationship between polyunsaturated fatty acids and cancer: Tumour or tissue specific effects? *Progress in Lipid Research.* 41: 240-253.

Etherton, P.,M.,K., Taylor, D.,S., Poth, S., Y. et al. 2000. Polyunsaturated fatty acids in the food chain in the United States. Am. J. Clin. Nutr. 71: S179-S188 (suppl 1).

Fassett, R., G., Gobe, G., C., Peake, J., M., Coombes, J., S. 2010. Omega-3 Polyunsaturated Fatty Acids in the Treatment of Kidney Disease. *American Journal of Kidney Diseases.* 56: 728-742.

Faraone, S., V., Sergeant, J., Gillberg, C., Biederman, J. 2003. The worldwide prevalence of ADHD: is it an American condition? *World Psychiatry.* 2: 104–113.

Fradet, V., Cheng, I., Casey, G., Witte, J.,S. 2009. Dietary omega-3 fatty acids, cyclooxygenase-2 genetic variation, and aggressive prostate cancer risk. *Clin. Cancer Res.* 15: 2559–2566.

Garcher, C., C., Baudouin, C., Labetoulle, M., Pisella, P.,J,, Mouriaux, F., Ouertani, A.,M., Matri, L., Khairallah, M., Baudouin, F., B. 2011. Efficacy assessment of Nutrilarm(®), a per os omega-3 and omega-6 polyunsaturated essential fatty acid dietary formulation versus placebo in patients with bilateral treated moderate dry eye syndrome. *J. Fr. Ophtalmol.* 2011 Jun 20. [Epub ahead of print]

Gebska, M., A. 2008. Are All Fish Equally Close to the Heart? *Mayo Clin. Proc.* 83: 723-30.

Geusens, P., Wouters, C., Nijs, J., et al. 1994. Long term effect of omega-3 fatty acid supplementation in active rheumatoid arthritis. A 12 month, double blind, controlled study. *Arthritis Rheum.* 37: 824.

Giovanni, J., P.,S., Chew, E. Y. 2005. The role of omega-3 long-chain polyunsaturated fatty acids in health and disease of the retina. *Progress in Retinal and Eye Research.* 24: 87–138.

Gow, R., V., Matsudaira, T., Taylor, E., Rubia, K.,Crawford, M., Ghebremeskel, K., Ibrahimovic, A., Tourangeau, F., V., Williams, L., M., Sumich, A. 2009. Total red blood cell concentrations of ω-3 fatty acids are associated with emotion-elicited neural activity in adolescent boys with attention-deficit hyperactivity disorder. *Prostaglandins, Leukotrienes and Essential Fatty Acids.* 80: 151–156.

GruppoItaliano per lo Studio dellaSopravvivenzanell'Infartomiocardico (GISSI)-Prevenzione Investigators. 1999. Dietary supplementation with n-3 polyunsaturated fatty acids and vitamin E after myocardial infarction: results of the GISSI-Prevenzione trial. *Lancet.* 1999;354: 447– 455.

Harris, W., S. 1997.n-3 Fatty acids and serum lipoproteins: Human studies. *Am. J. Clin. Nutr.* 65: S1645-S1654. (suppl 5).

Knapp, H., R. 1996. n-3 fatty acids and human hypertension. *Curr. Opin. Lipidology.* 7: 30.

Kalmijn, S., Launer, L., Ott ,A., Witteman, J., Hofman, A., Breteler, M. 1997a. Dietary fat intake and risk of incident dementia in the Rotterdam Study. *Ann. Neurol.* 42: 776–782.

Kokoglu, E., Tuter, Y., Yazici, Z., Sandikci, K., S., Sonmez, H., Ulakoglu, E., Z., Ozyurt, E. 1998. Profiles of the fatty acids in the plasma membrane of human brain tumors, *Cancer Biochem. Biophys.* 16: 301–312.

Koki, A.,T., Masferrer, J.,L. 2002. Celecoxib: a specific COX-2 inhibitor with anticancer properties. *Cancer Control.* 9: 28–35.

Kremer, J., M., Lawrence, D., A., Petrillo, G.,F., et al. 1995. Effects of high-dose fish oil on rheumatoid arthritis after stopping nonsteroidal anti-inflammatory drugs. Clinical and immune correlates. *Arthritis Rheum.* 38: 1107.

Kremer, J., M. 1996. Effects of modulation of inflammatory and immune parameters in patients with rheumatic and inflammatory disease receiving dietary supplementation of n-3 and n-6 fatty acids. *Lipids.* 31: S243.

Krey, G., O., Braissant, F., Horset, E., Kalkhoven, M., Parker, P., M., G., Wahli, W. 1997. Fatty acids, eicosanoids, and hypolipidemic agents identified as ligands of peroxisome proliferators-activated receptors by coactivator-dependent receptor ligand assay. *Mol. Endo.* 11: 779-791.

Loeschke, K., Ueberschaer, B., Pietsch, A. 1996. n-3 fatty acids only delay early relapse of ulcerative colitis in remission. *Dig. Dis. Sci.* 41: 2087.

Lim, K., Han, C., Dai, Y., Shen, M., Wu, T. 2009. Omega-3 polyunsaturated fatty acids inhibit hepatocellular carcinoma cell growth through blocking beta-catenin and cyclooxygenase-2. *Mol. Cancer Ther.* 8: 3046-55.

Makni, M., Fetoui, H., Gargouri, N., K., Garoui, E., M., Jaber, H., Makni, J., Boudawara, T., Zeghal, N., 2008. Hypolipidemic and hepatoprotective effects of flax and pumpkin seed mixture rich in x-3 and x-6 fatty acids in hypercholesterolemic rats. *Food and Chemical Toxicology.* 46: 3714–3720.

Massaro, M., Scoditti, E., Carluccio, M, A, Caterina, R., 2008.Basic mechanisms behind the effects of n-3 fatty acids on cardiovascular disease. *Prostaglandins, Leukotrienes and Essential Fatty Acids.* 79: 109–115.

Micallef, M, A., Garg, M., A. 2009. Beyond blood lipids: phytosterols, statins and omega-3 polyunsaturated fatty acid therapy for hyperlipidemia. *Journal of Nutritional Biochemistry*. 20: 927–939.

Micallefa, M., A., Manohar L. Garg, M., L. 2009.Anti-inflammatory and cardioprotective effects of n-3 polyunsaturated fatty acids and plant sterols in hyperlipidemic individuals. *Atherosclerosis*. 204: 476–482.

Pan, X.,R, Li, G.,W., Hu, Y.,H., Wang, J.,X., Yang, W.,Y., An, Z.,X., et al. 1997. Effect of diet and exercise in preventing NIDDM in people with impaired glucose tolerance. *Diabetes Care*. 20: 537–44.

Pooya, S., Jalali, M.,D.,J., Jazayery, A., D., Saedisomeolia, A., Eshraghian, M., R., Toorang, F. 2010. The efficacy of omega-3 fatty acid supplementation on plasma homocysteineand malondialdehyde levels of type 2 diabetic patients. *Nutrition, Metabolism and Cardiovascular Diseases*. 20: 326-331.

Ramachandran, A., Snehalatha, C., Mary, S., Mukesh, B., Bhaskar, A.,D., Vijay, V., et al. 2006. The Indian Diabetes Prevention Programme shows that lifestyle modification and metformin prevent type 2 diabetes in Asian Indian subjects with impaired glucose tolerance (IDPP-1). *Diabetologia*. 49: 289–97.

Rand, A., L., Asbell, P., A. 2011. Nutritional supplements for dry eye syndrome. *Curr. Opin. Ophthalmol*. 22: 279-82.

Rose, D.,P., and Connolly, J., M. 1999. Omega-3 fatty acids as cancer chemopreventive agents. *Pharmacology and Therapeutics*. 83: 217-244.

Sartorelli, D., S., Damiao, R., Chaim, R., Hirai, A., Gimeno, S., G., A., and Sandra R., G., Ferreira, S., R., G. 2010. Dietary u-3 fatty acid and ω-3: ω-6 fatty acid ratio predict improvement in glucose disturbances in Japanese Brazilians. *Nutrition*. 26: 184–191.

Schmitz, G., and Ecker, J. 2008. The opposing effects of n-3 and n-6 fatty acids. *Progress in Lipid Research*. 47: 147-155.

Sellmayer, A., Danesch, U., Weber, P., C. 1997. Modulation of the expression of early genes by polyunsaturated fatty acids. *Prostaglandins, Leukotrienes and Essential Fatty Acids*. 57: 353-357.

Simopoulos, A., P. 1991. Omega-3 fatty acids in health and disease in growth and development. *Am. J. Clin. Nutr*. 54: 438-463.

Stenson, W.,F., Cort, D., Rodgers, J., et al. 1992. Dietary supplementation with fish oil in ulcerative colitis. *Ann. Intern. Med*. 116: 609.

Strategic Analysis of the European Marine and Algae Oil (SAEMAO) 2010. http://www.companiesandmarkets.com/Market-Report/strategic-analysis-of-the-european-marine-and-algae-oil-omega-3-ingredients-market-349953.asp.

Stough, C., Downey, L., Silber, B., Lloyd, J., Kure, C., Wesnes, K., Camfield, D. 2011. The effects of 90-day supplementation with the Omega-3 essential fatty acid docosahexaenoic acid (DHA) on cognitive function and visual acuity in a healthy aging population. *Urobiol Aging*. 2011 Apr 30. [Epub ahead of print]

Surette, M., E. 2008. The science behind dietary mega-3 fatty acids. *CMAJ*. 178: 177-180.

Thiebaut, A., C., Chajes, V., Gerber, M., Boutron-Ruault, M., C., Joulin, V., Lenoir, G., Berrino, F., Riboli, E., Benichou, J., Chapelon, F., C. 2009. Dietary intakes of omega-6 and omega-3 polyunsaturated fatty acids and the risk of breast cancer. *Int. J. Cancer*. 124: 924–931.

Ventura, H.,O., Milani, R.,V., Lavie, C.,J., et al. 1993. Cyclosporine-induced hypertension. Efficacy of omega-3 fatty acids in patients after cardiac transplantation. *Circulation.* 88: 281

Wolk, A., Larsson, S.,C., Johansson, J., E. , Ekman, P. 2006.Long-term fatty fish consumption and renal cell carcinoma incidence in women. *JAMA.* 296: 1371–1376.

Yokoyama, M., Origasa, H., for the JELIS Investigators. 2003. Effects of eicosapentaenoic acid on cardiovascular events in Japanese patients with hypercholesterolemia: rationale, design, and baseline characteristics of the Japan EPA Lipid Intervention Study (JELIS). *Am. Heart J.* 146: 613– 620.

Yokoyama, M. 2005. Effects of eicosapentaenoic acid (EPA) on major cardiovascular events in hypercholesterolemic patients: the Japan EPA Lipid Intervention Study (JELIS). Presented at the American Heart Association Scientific Sessions. November 14, Dallas, TX.

In: Natural Products and Their Active Compounds ... ISBN: 978-1-62100-153-9
Editors: M. Essa, A. Manickavasagan, and E. Sukumar

Chapter 7

NEUROPREVENTIVE ROLE OF *WITHANIA SOMNIFERA* ROOT POWDER AND *MUCUNA PRURIENS* SEED POWDER IN PARKINSONIC MICE

***K. Tamilselvam and T. Manivasagam**[*]
Department of Biochemistry and Biotechnology
Faculty of Science
Annamalai University
Tamilnadu, India

INTRODUCTION

Parkinson's disease (PD) is the second most common neurodegenerative disorder, after Alzheimer's disease, affecting 1% of the population by the age of 65 years and 4–5% of the population by the age of 85 years (Eriksen and Petrucelli, 2004). PD is found to have either familial or non-familial etiology (Przedborski *et al*., 2001). The interactions between external toxins (which arise from environmental, dietary and lifestyle factors), internal toxins arising from normal metabolism and the genetic (nuclear genes) and epigenetic (mitochondria, membranes, and proteins) components of neurons occur continuously and could initiate degeneration in DA neurons (Prasad *et al.,* 1999). The motor disabilities characterizing PD are primarily due to the loss of dopaminergic neurons in the substantia nigra and depletion of dopamine in corpus striatum.

Symptoms and Pathophysiology

PD has major adverse impact on patients' lives. Patients' symptoms such as tremor, hypokinesia, rigidity, hypophonic voice, painful dystonia, postural abnormalities, gait disorders, sleep disturbances, depression and drug related problems may progressively lead to

[*] Corresponding author: E-mail: mani_pdresearchlab@rediff.com

falls, social embarrassment, loneliness and increasing dependence on others for everyday activities (Behari *et al.,* 2005). Different combinations of these symptoms result in a wide range of abnormalities of gait. The disease progresses slowly and may ultimately produce complete akinesia (Jankovic, 2008).

The major pathological change in patients with PD is the loss of melanin containing dopaminergic neurons in zona compacta (SNc) of the substantia nigra (Hassler, 1938). These pigmented neurons have been identified as nigrostriatal dopamine neurons (Ehringer and Hornykiewiez 1960) and loss of these neurons results in decrease of dopamine content in striatum (Ellis *et al.,* 1992). Clinical symptoms appear only when dopaminergic neuronal death exceeds a critical threshold 70-80% of striatal nerve terminals (Bezard, 1998).

Oxidative damage to proteins, lipids, and nucleic acids has been found in the SN of patients with PD (Dexter *et al*., 1994). Both overproduction of reactive species and failure of cellular protective mechanisms appear to be operative in PD. Dopamine metabolism promotes oxidative stress through the production of quinones, peroxides, and other reactive oxygen species (ROS) (Hastings *et al*., 1996). Mitochondrial dysfunction is another source for the production of ROS, which can then further damage mitochondria. Complex I activity is diminished in the SN of PD patients (Schapira *et al.,* 1989), while inhibitors of complex I, such as MPP^+ and rotenone, cause a Parkinsonian syndrome in animal models (Betarbet *et al.,* 2000). Increased iron levels seen in the SN of PD patients (Riederer *et al.,* 1989) and also promote free radical damage, particularly in the presence of neuromelanin. The pathogenesis of neurodegenerative diseases such as Alzheimer's or Parkinson's diseases is multifactorial with toxic reactions including inflammation, the glutamatergic toxicity, the dysfunction of mitochondrial activity and of the ubiquitin/ proteasome system, the activation of apoptosis pathways, the elevation of iron and nitric oxide, the alteration of the homeostasis of antioxidants/oxidation are involved.

Treatment

Current pharmacological therapies for the disease are inadequate; these are only able to provide symptomatic relief and after long use produce stern side effects and even worsen the condition. For these reasons there has been an increased tendency to search for other strategies such as neural transplantation and stem cell transplantation. Unfortunately, most of these novel approaches remains in experimental stage. As the knowledge of the evolution of the pathology of PD is still fragmentary, this can be overcome only through basic science research.

Experimental PD

For better understanding of pathogenesis of PD mechanism and to develop new therapies for PD different animal models treated with a number of neurotoxins which are mitochondrial electron transport chain inhibitors such as 6-hydroxy dopamine, 1-methyl-4-phenyl-1,2,3,6 tetrahydropyridine (MPTP), rotenone and paraquat (pyridine containing herbicides) (Betarbet *et al.,* 2002).

MPTP

The compound, 1-methyl, 4-phenyl-1,2,3,6-tetrahydropyridine (MPTP) is a selective neurotoxin of dopaminergic neurons in the substantia nigra and striatum and has been shown to induce PD in various experimental animals including monkeys, mice, cats, dogs, rats and goldfish (Gerlach and Riederer, 1996). It induced neurochemical, behavioural and histopathological alterations very closely replicate the clinical symptoms of PD patients mainly mediated by loss of dopamine, oxidative stress and mitochondrial dysfunction (Schmidt and Ferger, 2001).

Various behavioral tests such as rotarod test, hang test and open field test are used as indices to measure the movement (motor) impairments in MPTP-induced animal models (Rozas *et al.,* 1998). The animal to balance and walk on a rotating rod is widely used to measure the coordinated motor skills (Rozas *et al.,*1998). Regarding, open field test, it mainly indicates mental stress, acclimatization ability and motor activity assessment tool (Fredriksson and Archer 1994; Crabbe *et al.,* 2003). Peripheral square crossing is an indication of general motor performance and acclimatization attempt. Central square crossing is an indication of exploratory behaviour (Sedelis *et al.,*2001). The narrow beam walking, used to test the balance, vestibular integrity and muscular co-ordination (Crabbe *et al.,*2003: Fleming *et al.,*2004) was significantly altered by MPTP treatment. The measures of forepaw step length reflected the persistent dopaminergic loss in mice exposed to MPTP and these behavioural deficits were reversed following L-dopa treatment (Tillerson and Miller, 2003). MPTP induced mice displayed a reduced mean hind limb stride length compared with forelimbs, which resembles reduced stride length in PD gait (Fernagut *et al.,*2002). Typically, PD patients present a shortened stride length with a shuffling gait and reduced velocity (gait hypokinesia) (Moriss *et al.,* 1998).

Ayurveda and PD

Ayurveda is perhaps, the most ancient of all medicinal traditions is probably older than the traditional Chinese medicine. Ayurveda is derived from the Indian words '*Ayar'* (Life) and '*veda*' (Knowledg or Science) and hence means the Science of Life. Ayurvedic texts describe Kampavata, a nervous malady bearing similarities to Parkinson's syndrome and some of the important plant ayurveda drugs used to treat PD were (i) Aswagandha (root of *Withania somnifera*, (ii) Atmagupta (seed of *Mucuna prureins*, (iii) Bala (root of Sida cordifolia) and (iv) Paraseekayavanee (dry fruit of *Hyocyamus reticulatus*) (Gourie *et al.*, 1991).

Withania Somnifera

Withania somnifera (WS) is popularly known as Ashwagandha or Winter Cherry (Andallu and Radhika, 2000). The practitioners of the traditional system of medicine in India regard *W. somnifera* as the "Indian Ginseng" (Singh et al., 2001). It is classified in the ancient Indian system of medicine (Ayurveda and Siddha) as a rasayana, a group of plant derived drugs that improve overall physical and mental health and put off diseases by rejuvenating the incapacitated conditions. It is an official drug and is mentioned in Indian Pharmacopoeia (1985). *Withania somnifera* is known to have antitumour, cardioprotective, antioxidant and antialzheimeric (Bhattacharya et al., 1995; Dhuley, 2000) properties and has positive effects on the endocrine (Andallu and Radhika, 2000) and central nervous systems (Dhuley, 2000). It is in use for a long time for all age groups and both sexes and even during pregnancy without any side effects (Ziauddin et al., 1996).

Mucuna Pruriens

Mucuna pruriens (MP) (also known as ''the cowhage'' or 'velvet'' bean; and ''atmagupta'' in India) is aclimbing legume endemic in India and in other parts of the tropics including Central and South America. Charakasamhitha'', the classical text in Ayurveda, a concoction in cow's milk of powdered *Mucuna pruriens* prescribed for treating Parkinson's disease and is reported to contain L-DOPA as one of its constituents (Chaudhri et al., 1996). Seeds of Mucuna are prescribed as powder to treat leucorrhoea, spermatorrhoea and wherever aphrodisiac action required (Nadkarni 1982). Seeds possess anabolic, analgesic (pain-relieving), anti-inflammatory, anti-Parkinson's, hormonal, immunomodulator, nervine (nerve balancing), neurasthenic (nerve pain relieving), central nervous system stimulant, diuretic, hypotensive (blood pressure lowering), menstrual stimulant, uterine stimulant and vermifuge (Bhat et al., 2007). *Mucuna pruriens* seeds are well known to contain levodopa (Bell and Janzen, 1971; Damodaran and Ramaswamy, 1937).

In the present study, we evaluated the synergestic effect of *Withania somnifera* root powder and *Mucuna pruriens* seed powder on biochemical, behavioural and neurochemical variables against MPTP induced experimental Parkinsonism.

Materials and Methods

All the experiments were carried out in Male Swiss albino mice (*Mus musculus*) of body weight ranging from 25 to 35 gm. Albino mice of 3-4 months old were obtained from Central Animal House, Annamalai University. Mice were provided with food and water ad libitum.

1-methyl 4-phenyl 1,2,3,6-tetrahydropyridine hydrochloride (MPTP), thiobarbituric acid (TBA), reduced glutathione and 3,5-dithio-bis-nitrobenzoic acid (DTNB) were purchased from Sigma chemical Company, Bangalore, India. The commercially available powdered root of *Withania somnifera* and powdered seeds of *Mucuna pruriens* were obtained from Indian Medical Practitioners Cooperative Society (IMCOPS), Adyar, Chennai, India. All other chemicals were of analytical grade.

The mice were randomized and divided into four groups of twelve animals each. Group I are of Normal mice with saline served as control. Group II mice received intraperitoneal injection of MPTP (20 mg/kg body wt) (Bezard *et al.,* 2000; Lin *et al.,* 2004).Group III are treated with WsRP (*Withania somnifera* root powder) (100mg/kg body wt) and MpSP (*Mucuna pruriens* seed powder) (50mg/kg body wt) orally and then received MPTP for 4 consecutive days (from 4th day to 7th day of WsRP and MpSP administration). MPTP was given after 1 hr interval of WsRp and MpSp administration to this group.

Phase I

At the end of the experiment (8th day), the animals were analysed for behavioural studies such as open field test (number of rearings, grooming and number of visits to the central and peripheral compartments were calculated during a single 5-min session) (Fernagut *et al.*, 2002) and narrow beam walking (Crabbe *et al.*, 2003: Fleming *et al.*, 2004) were performed. Following the behavioural analysis the mice were sacrificed by cervical dislocation and the brain was dissected to procure corpus striatum and midbrain by the method described by Glowinski and Iversen (1966) for the analysis of following biochemical parameters.

Analysis of Dopamine and Its Metabolites

At the end of the experiment (8th day), striatal dopamine and its metabolites 3,4-dihydroxyphenylacetic acid (DOPAC) and homovanillic acid (HVA) were analysed using HPLC combined with electrochemical detection under isocratic conditions (Muralikrishnan and Mohanakumar, 1998).

Estimation of TBARS, Reduced Glutathione and Glutathione Peroxidase

For biochemical estimations such as TBARS (Utley *et al.*,1967), GSH (Jollow *et al.*,1974) and GPx (Mohandas *et al.*, 1984), midbrain were homogenized (10% w/v) in 0.01 M phosphate buffer (pH 7.0) and centrifuged at 10500 g for 20 min at 4°C.

Phase II

At the end of the experiment (8th day), behavioural studies such as Rotarod test (Rozas *et al.*,1998) and stride length measurement (Fernagut *et al.*, 2002) were performed in all the groups.

Determination of Superoxide Dismutase and Catalase Activities

For enzymatic determinations such as SOD (Beauchamp and Fridovich, 1971) and catalase (Claiborne, 1985), midbrain were homogenized (10% w/v) in 0.01 M phosphate buffer (pH 7.0) and centrifuged at 10500 g for 20 min at 4°C.

RESULTS

Figure 1 (a, b, c) summarizes the levels of dopamine and its metabolites DOPAC and HVA in mice striata. They are significantly decreased in the MPTP intoxicated mice (group II) compared to the control mice (group I).

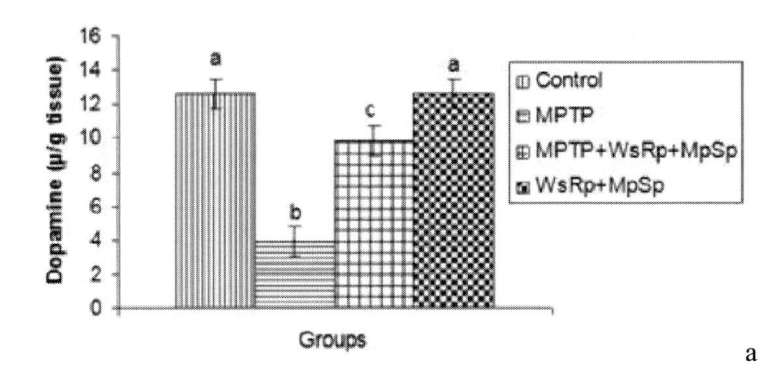

a

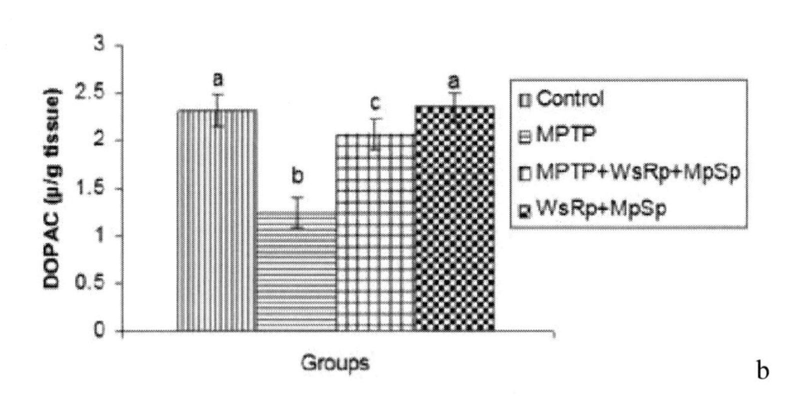

b

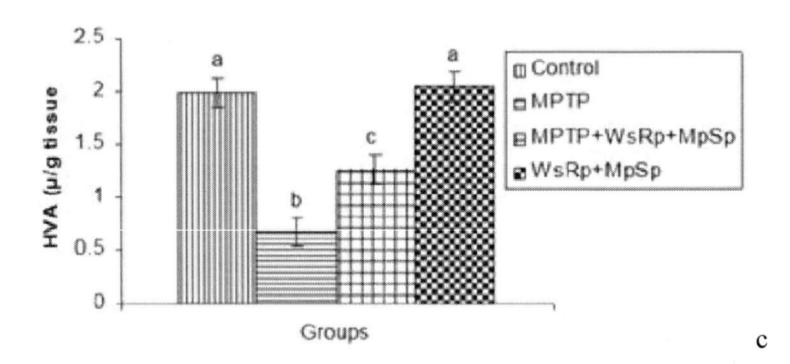

c

Figure 1. (a, b, c) depicts the levels of Dopamine and its metabolites DOPAC and HVA were decresased in PD mouse compared to control. The WsRp and MpSp pretreated and MPTP administered mice (group III) showed a significant increase in level of dopamine and its metabolites compared to MPTP intoxicated mice group II.

The WsRP and MpSP pretreated and MPTP administered mice (group III) showed a significant increase in level of dopamine and its metabolites compared to MPTP intoxicated mice group II. There is no significant change in the dopamine and its metabolites level in the WsRP and MpSP alone pretreated mice (group IV) compared to control mice (group I).

Figure 2 (a,b,c,d) elucidates the open field test. Significant reduction ($p<0.05$) in pheripheral movements and central movements along with rearing and grooming in MPTP injected mice (group II) compared to controls mice (group I). Pretreatment of WsRP and MpSP to MPTP administered in mice (group III), made them to exhibit increased peripheral movements and central movements along with rearing and grooming significantly ($p<0.05$).

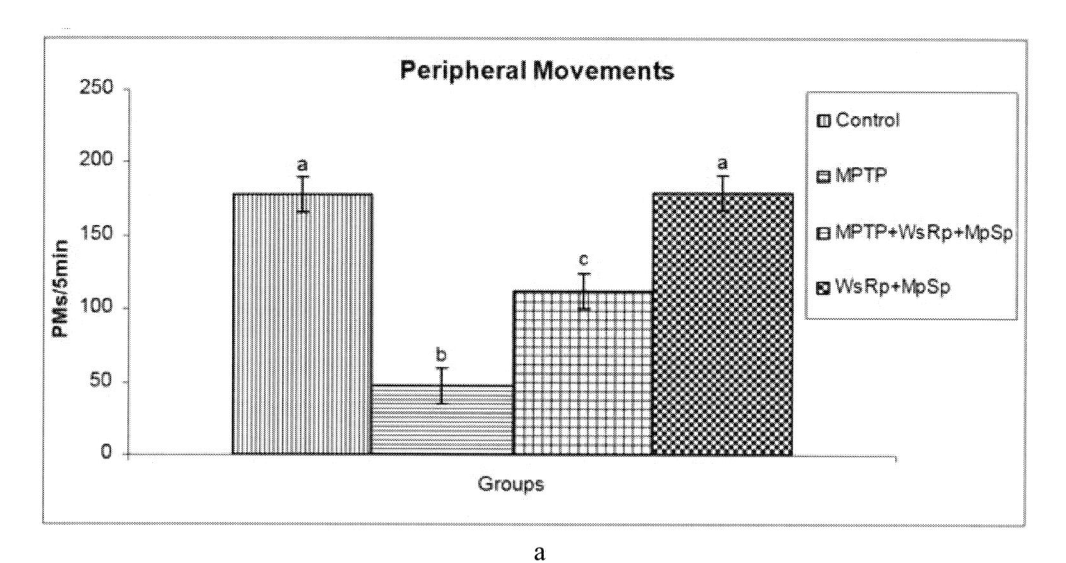

a

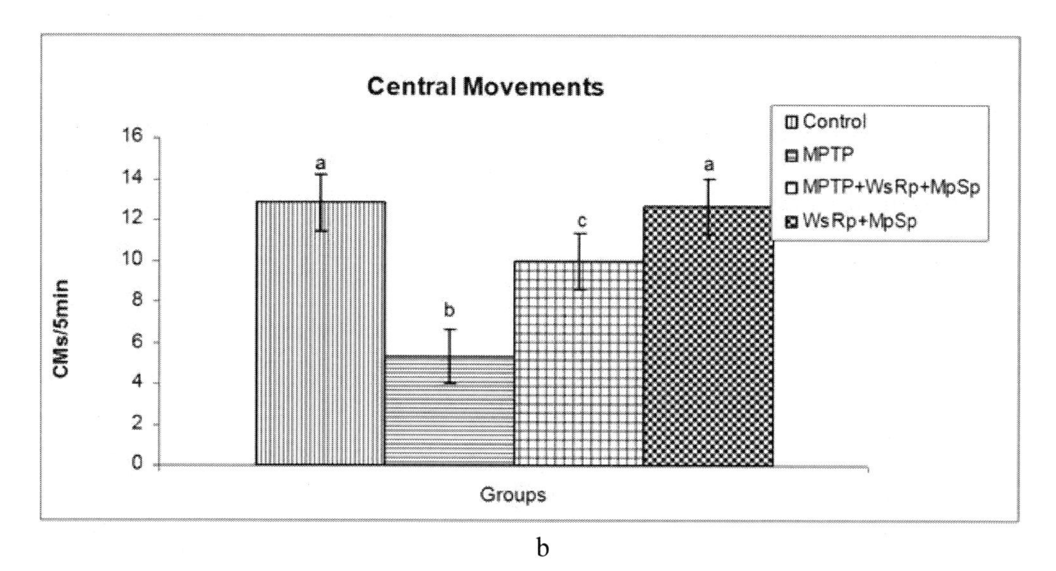

b

Figure 2. (a and b) depicts the (Open field test) peripheral movements, central movements; were decreased in PD mouse compared to control. WsRp and MpSp significantly improved this behaviors compared to PD mouse (< 0.05).

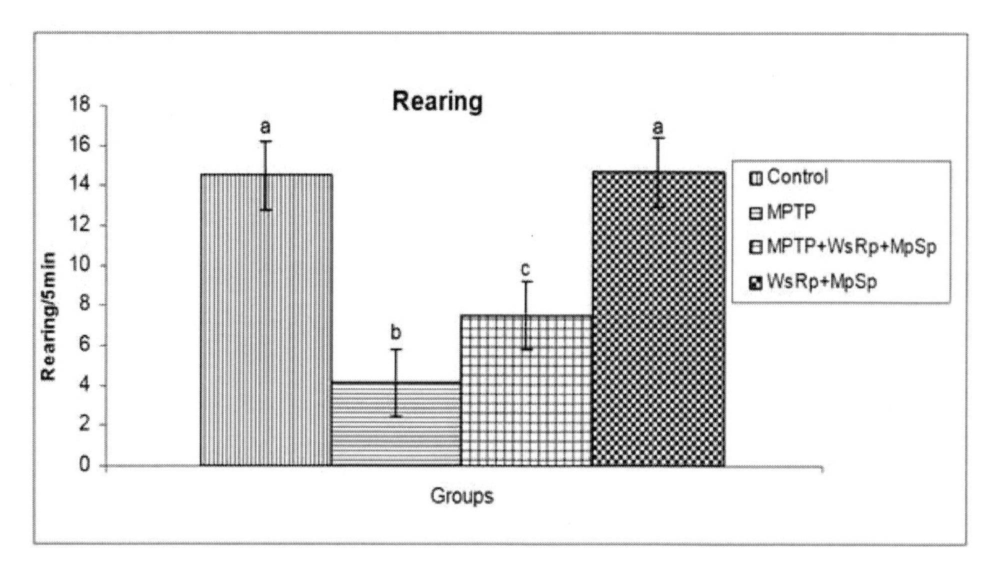

c

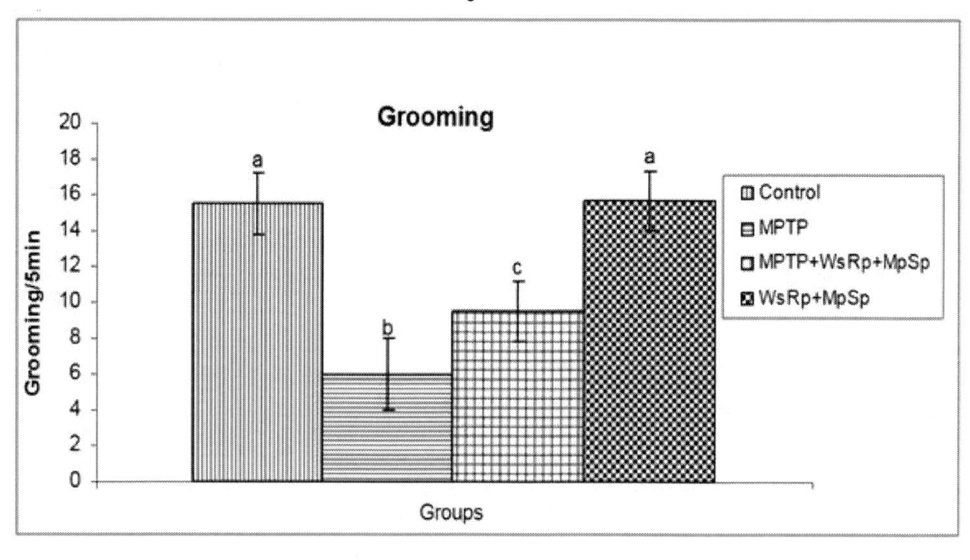

d

Figure 2. (c and d) depicts the (Open field test) grooming and rearing were decreased in PD mouse compared to control. WsRp and MpSp significantly improved this behaviors compared to PD mouse (< 0.05).

Figure 3 shows the narrow beam walking test, MPTP treated mice (group II) were observed to have a significant increase in the duration to cross the beam and exhibit more foot slip errors compared with control mice (group I). WsRP and MpSP pretreated and MPTP lesion mice (group III) significantly ($p<0.05$) consumed a reduced time to cross the beam and showed a decrease in the frequency of foot slip errors.

Figure 4 shows the mean stride length differences for the fore and hind limbs in the control and experimental groups. The stride length in fore and hind limb significantly decreased in the MPTP-treated mice (group II) relative to the control mice (group I).

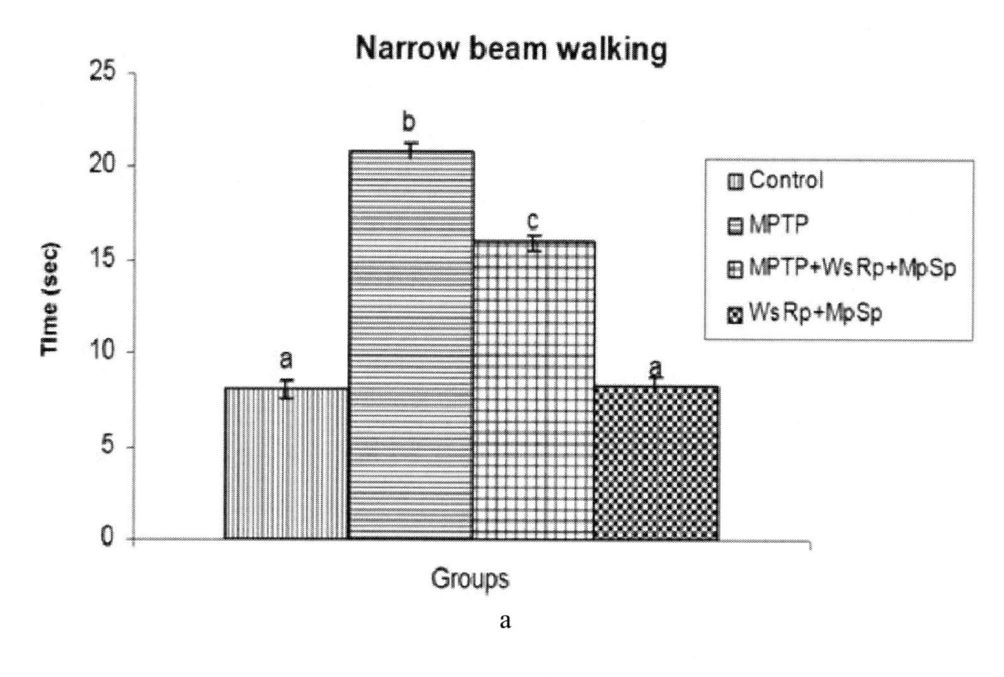

a

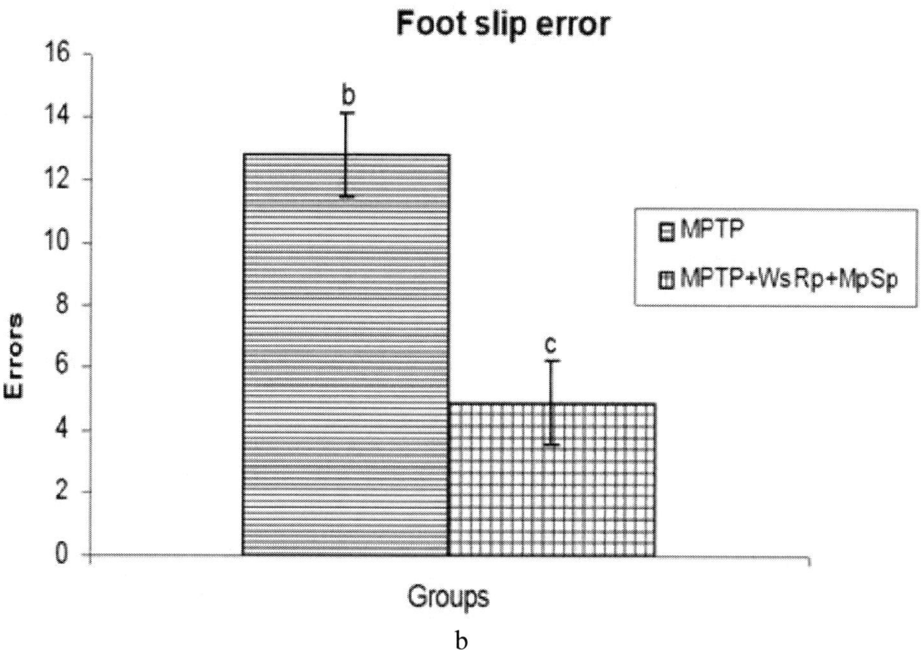

b

Figure 3. (a and b) depicts the narrow beam walking and foot slip errors test, the time consumption to cross the beam and foot slip errors increased in PD mouse compared to control. WsRp and MpSp significantly reduced this compared to PD mous (< 0.05).

WsRP and MpSP Pretreated and MPTP lesion (group III) showed significant increased stride length relative to the MPTP-intoxicated mice (group II). In all behavioral analysis there is no significant altered behavior in the WsRP and MpSP alone pretreated (group IV) mice compared to control mice (group I).

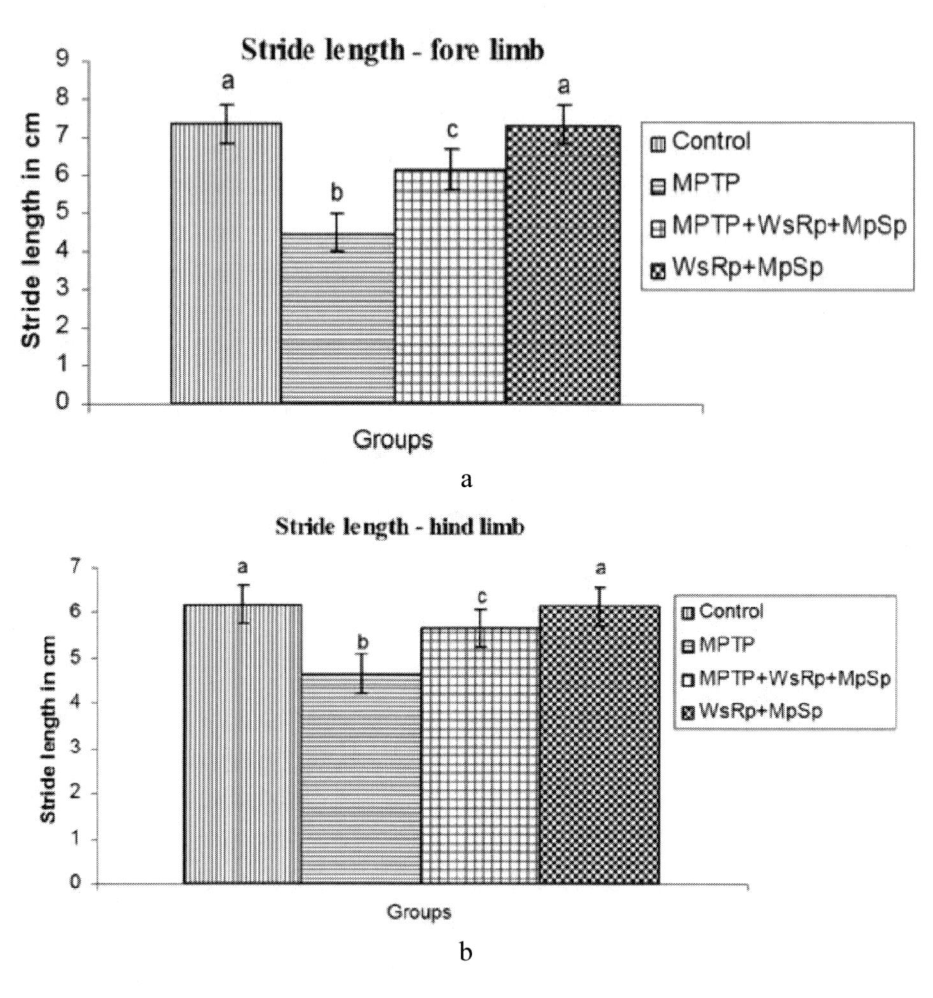

Figure 4. Elucidates the changes in the forelimbs and hind limbs stride length in control and experimental mice groups. Group II PD mice showed reduced stride length in both forelimb and hind limb stride length measurements. WsRp and MpSp pretreated group III mice shown increase stride length in both forelimb and hind limb measurements.

In the rotarod analysis (Figure 5), the MPTP lesioned (group II) shows reduced retention time on the rotarod compared to control (group I). The WsRP and MpSP pretreated MPTP lesion mice (group III) were withstand on the rotarod for appreciable time compared to MPTP intoxicated group II.

Table 1 depicts the levels of TBARS and GSH in midbrain of mice of the control and experimental groups. The lipid peroxidation levels were significantly elevated in MPTP-treated animals (group II) relative to the controls (group I). The prior treatment of WsRP and MpSP to MPTP-administered mice (group III) significantly reduced the lipd peroxidation. The levels of GSH were significantly decreased in MPTP treated group II compared to the control mice (group I).

The prior treatment of WsRP and MpSP to MPTP-administered mice (group III) significantly increased the GSH level. No significant changes in the levels of TBARS and GSH were found in mice treated with WsRP and MpSP alone (group IV) compared to control.

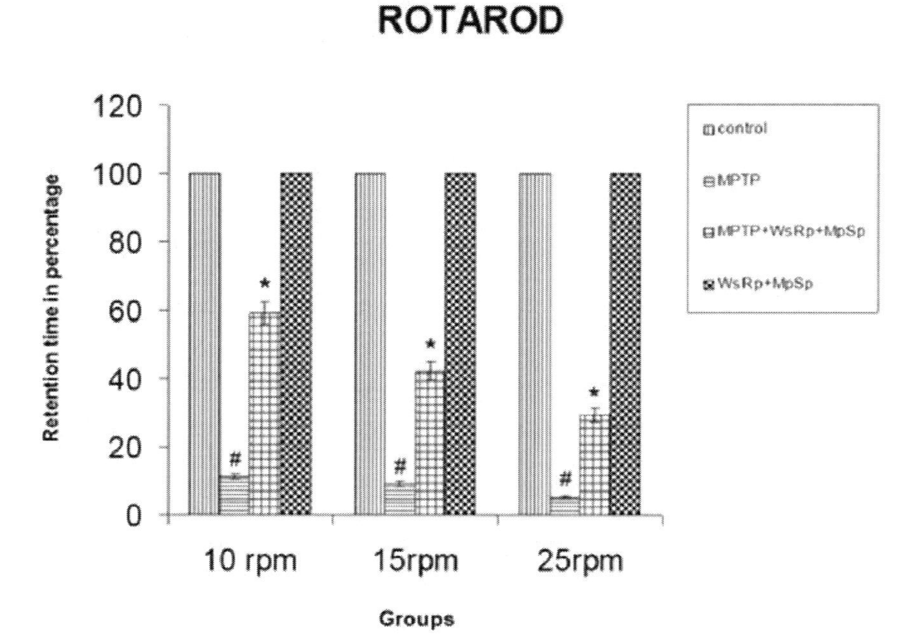

Figure 5. Depicts the rotarod analysis in control, PD and WsRp and MpSp treated mice groups and the significantly improved balance and muscular coordination in WsRp and MpSp pretreated group III compared to PD mice group II (< 0.05).

Table 1. Changes in the levels of TBARS and GSH in control and experimental mice midbrain

Groups	TBARS (nmoles/g)	GSH (µg/ gram tissue)
Control	1.37± 0.09[a]	0.48 ± 0.03[a]
MPTP	3.42 ± 0.24[b]	0.19 ± 0.02[b]
MPTP+WsRp+MpSp	2.01 ± 0.14[c]	0.38 ± 0.04[c]
WsRp+MpSp	1.39± 0.10[a]	0.50±0.06[a]

Values not sharing a common superscript letter differ significantly at $p<0.05$ (DMRT).

Table 2 depicts the activities of SOD, Catalase, and GPx in mid brain regions of mice. The activities of SOD and Catalase in the both the regions were significantly elevated in MPTP-treated animals (group II) relative to the controls (group I). The pre- WsRP and MpSP treated and MPTP lesioned mice (group III) tends to decrease the activities SOD and catalase compared to the MPTP treated mice (group II).

The activity of GPx is reduced in the MPTP treated mice (group II) compared to control mice (group I). The pretreatment of WsRP and MpSP significantly increase the GPx activity in MPTP lesion mice (group III). There is no significant changes in the activities of enzymatic antioxidants were found in mice treated with WsRP and MpSP alone (group IV).

Table 2. Changes in the activities of SOD, Catalase and GPX in control and experimental mice midbrain

Groups	SOD U[A]/mg protein)	CAT (U^B/mg protein)	GPx (U^c/mg protein)
Control	0.99 ± 0.09^a	0.38 ± 0.03^a	4.93 ± 0.39^a
MPTP	2.07 ± 0.15^b	0.98 ± 0.07^b	3.27 ± 0.22^b
MPTP+WsRp+MpSP	1.68 ± 0.11^c	0.56 ± 0.04^c	4.50 ± 0.35^c
WsRp+MpSP	1.01 ± 0.07^a	0.41 ± 0.03^a	5.11 ± 0.36^a

[A]= amount of enzyme required to inhibit 50% of NBT reduction.
[B]=μmoles of H_2O_2 consumed /min/mg protein.
[C]= amount of glutathione utilized/minute.
Values not sharing a common superscript letter differ significantly at $p<0.05$ (DMRT).

DISCUSSION

Outcome of Neurochemical Variables

MPTP when administered in mice was found to decrease striatal dopamine level and its metabolite 3,4 dihydroxyphenylacetic acid (DOPAC) contents and locomotor activity (Sundstrom *et al.*,1990). The loss of dopaminergic neurons in the SN and the concomitant decrease in DA levels in the striatum have been reported in many species, including monkeys, mice and dogs (Rapisardi *et al.*,1990; Jackson-Lewis *et al.*,1995) during MPTP treatment. Administration ayurvedic plant extracts elevated the levels of dopamine and its metabolites in MPTP intoxicated rats.

Outcome of Behavioural Analysis

Various behavioural tests are used as indices to measure the movement impairments in MPTP induced animal models (Rozas *et al.*, 1998) and they closely mimic the clinical symptoms of PD patients such as akinesia, rigidity and gait disturbance. Reduction in the open field spontaneous activity has been reported in MPTP treated mice and is inhibited following treatment with the WsRP and MpSP. Reduced performance of lesioned animals in crossing peripheral squares and central squares may be due to reduced motor performance associated with mental stress. Similarly rearing and grooming activities are also indicators of stress. Rearings are known to be highly sensitive to either SNc (MPTP) or striatal lesions (Sedelis *et al.*, 2001).

The reduction in terms of rearing and grooming were found in MPTP treated mice and is inhibited following treatment with the WsRP and MpSP.

In narrow beam walking, the lesioned animals (MPTP) showed more deficits in crossing time and foot slip errors on 8th day. In lesioned and WsRP and MpSP pretreated animals, although the deficits were more when compared with normal animals, the overall deficit was significantly less than that of group II animals. The narrow beam walking, used to test the balance, vestibular integrity and muscular co-ordination (Crabbe *et al.,* 2003: Fleming *et al.,* 2004) was significantly altered by MPTP treatment.

The measures of forepaw step length reflected the persistent dopaminergic loss in mice exposed to MPTP and these behavioural deficits were reversed following L-dopa treatment (Tillerson and Miller, 2003). MPTP induced mice displayed a reduced mean hind limb stride length compared with forelimbs, which resembles reduced stride length in PD gait (Fernagut *et al.,* 2002) which is corroborated with our present results. Typically, PD patients present a shortened stride length with a shuffling gait and reduced velocity (gait hypokinesia) (Moriss *et al.,* 1998). Results of forepaw stride length during walking were consistent with the findings of Fernagut *et al*., (2002) and Tillerson and Miller, (2003). Since stride length was established as a sensitive index of nigrostriatal pathway function (Fernagut *et al*., 2002), tends to normalization of stride length in treated animals was therefore significant and indicative efficacy of WsRP and MpSP.

Basal ganglia (striatum) are involved in the control of automatic movements such as gait, mainly through their interaction with cortical motor areas (i.e. supplementary motor area) and disruption of such a system may result in gait disorders such as in Parkinson's disease (Hanakawa *et al*., 1999). Treatment with WsRP and MpSP improves the behavioural status of MPTP intoxicated mice may be due to their pharmacological properties.

Outcome of Biochemical Estimations

In the present study, the levels of lipid peroxidation and the activities of SOD and CAT were increased in midbrain of lesioned animals which are an indication of release of more free radicals. Interestingly GSH and its dependent enzyme (GPx) activities were correspondingly decreased in lesion groups. Increase of SOD and CAT in lesioned animal must be a natural response for the increase of LPO.

MPTP has been reported to cause its dopaminergic toxicity by the process of lipid peroxidation and oxidative stress. Excessive free radical formation or antioxidant enzyme deficiency can result in an oxidative stress, a mechanism proposed in the toxicity of MPTP and in the pathogenesis of Parkinson's disease (Muralikrishnan and Mohanakumar, 1998). Several lines of evidence from both *in vivo* and *in vitro* studies suggest that the damaging effects of MPTP are at least partly due to the formation of free radicals especially hydroxyl radicals during MPTP metabolism (Muralikrishnan and Mohanakumar, 1998). In the present study MPTP treated animals showed marked elevation in the production of TBARS and depletion of GSH in the mid-brain. This is in support of previous findings in different strains of mice (Muralikrishnan and Mohanakumar, 1998; Munoz *et al.,* 2006; Chen *et al.,* 2005). The activities of the SOD and CAT were elevated significantly in MPTP induced Parkinson's disease group animals which were correlated with previous experiments (Muralikrishnan and Mohanakumar, 1998; Chen *et al.,* 2005). The activity of GPx was found to be reduced in our experiment, which is corroborated with previous experiment (Genc *et al.,* 2002).

Increased levels of malondialdehyde (MDA) and lipid hydroperoxides (Dexter *et al.,* 1989; 1994a), markers of oxidative damage were reported in midbrain of the PD patients. Rojas and Rios (1993), found enhanced lipid peroxidation after MPP^{+} (metabolite of MPTP) administration to mice, a process dependent on over production of free radicals. Similarly the formation of superoxide anions in the presence of MPP^{+} (Adams *et al.,* 1993), results in increased activity of SOD in SN (Muralikrishnan and Mohanakumar, 1998). It is still unclear that the oxidative damage in neurodegenerative disorders results predominantly from an increase in free radical production, a decrease in protective mechanisms, or equally from both (Halliwell, 2001).

In treated and lesioned animals both LPO and anti-oxidants remain close to normal values and thus from the data it appears that root powder of *Withania somnifera* and seed powder of *Mucuna prureins* nullify / reduce the toxicity of MPTP primarily by blocking the release of free radicals rather than normalising the anti-oxidants levels.

The Efficacy of *Withania Somnifera* and *Mucuna* - Possible Mechanism

Pre treatment with WsRP and MpSP has significantly reversed the toxic effects of MPTP in experimental animals. *W. somnifera* contains biologically active alkaloids, steroids, lactones, saponins and withanolides (Mishra *et al.,* 2000). Ws has been reported to be a potent enhancer of cellular antioxidants and possesses a significant free radical scavenging activity in various disease models (Bhattacharya *et al.,* 2001; Davis and Kuttan, 2001). Bhattacharya *et al.,* (1997) reported that glycowithanolides, the active principles of Ws were found to increase the cortical and striatal concentrations of the antioxidant enzymes, SOD, CAT and GPx of adult Wistar rats.

It is of interest to note that, although *W. somnifera* has been reported to increase rat brain SOD concentrations per se (Bhattacharya *et al*., 1997), it tended to normalise chronic stress-induced augmentation of SOD activity in the frontal cortex and striatum which is corroborate with our study. It was described that thirty days treatment of *Withania somnifera* root produced a significant decrease in LPO, and an increase in both SOD and CAT thus indicating that Ashwagandha root powder possesses free radical scavenging activity (Panda and Kar, 1997). *Withania somnifera* has also been used in convulsive disorders, tremors and in diseases, which closely resemble the modern concept of neurodegenerative disorders (Weiner and Weiner, 1994).

Nagashayana *et al.,* (2000) have demonstrated that the *Mucuna pruriens* content of L-DOPA the precursor of dopamine, and a drug of choice for PD is present in MpSP. In 1978, Vaidya *et al.* reported a study of patients with Parkinson's disease who were initially treated with synthetic levodopa followed by treatment with a powder made from the whole bean of *Mucuna*, wherein the patients experienced a decrease in the incidence of adverse effects during treatment with the *Mucuna* powder compared to the synthetic levodopa treatment (Vaidya *et al.* 1978). This report supports our results with mice in the prevention of the behavioral pattern and dopamine level. Dhanasekaran *et al*., 2008 reported the *Mucuna pruriens* cotyledon powder's antioxidant and metal chelating activity. Their report agrees with our work. SanjayKasture *et al*., 2008 reported the symptomatic and neuroprotective efficacy of *Mucuna pruriens* seed extract in rodent model of Parkinson's disease their results are corroborated with our results.

In a nutshell, the present study demonstrates MPTP induced stress in brain of mice is nullified by the significant potency of WsRP and MpSP, due to the presence of number active principle components to a significant level and further study is necessary to establish their neuroprotective role.

REFERENCES

Andallu, B. and Radhika B. 2000. Hypoglycemic diuretic and hypocholesterolemic effect of winter cherry (*Withania somnifera* Dunal) root, *Indian J. Exp. Biol. 38*:607– 609.

Beauchamp, C. and Fridovich I. 1971. Superoxide dismutase: Improved assays and an assay applicable to acrylamide gels. *Anals of Biochemistry. 44*: 276 – 87.

Behari,M., Achal K. and Srivastavaa P. 2005. Quality of life in patients with Parkinson's disease. *Parkinson. Rel. Disorders. 11*:221 – 226.

Bell, E.A. and Janzen D.H. 1971. Medical and ecological considerations of L-dopa and 5-HT in seeds. *Nature. 229*: 136 – 137.

Betarbet, R., Sherer T.B. and Greenamyre J.T. 2002. Animal models of Parkinson's disease. *Bioessays. 24*:308 – 318.

Betarbet, R., Sherer T.B. and MacKenzie G. 2000. Chronic systemic pesticide exposure reproduces features of Parkinson's disease. *Nat. Neurosci. 3:* 1301–1306.

Bezard, E., Boraud T. and Imbert C. 1998. Towards a dynamic approach of experimental parkinsonism. *Prog. Neuropsychopharmacol. Biol. Psychiatry. 23*: 1317–1329.

Bezard, E., Jaber M. and Gonon F. 2000. Adaptive changes in the nigrostriatal pathway in response to increased 1- methyl- 4- phenyl-1,2,3,6- tetrahydropyridine-induced neurodegeneration in the mouse. *Eur. J. Neurosci. 12*: 2892 – 2900.

Bhat, R., Sridhar K.R. and Yokotani K.T. 2007. Effect of ionizing radiation on antinutritional features of velvet Bean seeds (*Mucuna pruriens*). *Food Chemistry. 103*: 860 – 866.

Bhattacharya, A., Ghosal S. and Bhattacharya S.K. 2001. Anti-oxidant effect of *Withania somnifera* glycowithanolides in chronic footshock stress-induced perturbations of oxidative free radical scavenging enzymes and lipid peroxidation in rat frontal cortex and striatum. *Journal of Ethnopharmacology. 74*: 1– 6.

Bhattacharya, S.K., Kumar A. and Ghosal S. 1995. Effects of glycowithanolides from *Withania somnifera* on an animal model of Alzheimer's disease and perturbed central cholinergic markers of cognition in rats. *Phytother. Res. 9*:110 – 113.

Bhattacharya, S.K., Satyan K.S. and Ghosal S. 1997. Antioxidant activity of glycowithanolides from *Withania somnifera*. *Indian J. Exp. Biol*. 35: 236 –239.

Chaudhri R.D. 1996. Herbal drug industry a practical approach to industrial pharmacognosy, New Delhi*: Eastern Publishers*

Chen, X., Zhou Y. and Chen Y. 2005. Ginsenoside Rgz 1 reduces MPTP induced substantia nigra neuron loss by suppressing oxidative stress. *Acta Pharmacol. Sin. 26*: 56 – 62.

Claiborne, A. Catalase activity. In Green Wald RA ed.1985. CRC Handbook of methods for oxygen radical research. Boca Raton, FL: CRC Press. 283 – 284.

Crabbe, J.C., Metten P. and Yu C.H. 2003. Genotypic differences in ethanol sensitivity in two tests of motor incoordination. *J. Appl. Physiol. 95*: 1338 – 1351.

Damodaran M. and Ramaswamy R. 1937. Isolaion of l-3:4-dihydroxyphenylalanine from the seeds of *Mucuna pruriens*. *Biochem. J. 31*:2149 – 2152.

Davis L. and Kuttan G. 2001. Effect of *Withania somnifera* on DMBA induced arcinogenesis. *J. Ethnopharmacol. 75*: 165 – 168.

Dexter, D.T., Carter C.J. and Wells F.R. 1989. Basal lipid peroxidation in substantia nigra is increased in Parkinson's disease. *J. Neurochem. 52*:381–389.

Dexter, D.T., Holley A.E. and Flitter W.D. 1994. Increased levels of lipid hydroperoxides in the parkinsonia substania nigra: an HPLC and ESR study. *Mov. Disord. 9*:92–97.

Dhanasekaran, M., Tharakan B. and Manyam B.V. 2008. Antiparkinson drug--*Mucuna pruriens* shows antioxidant and metal chelating activity. *Phytother Res. 22*: 6 –11.

Dhuley J.N. 2000. Adaptogenic and cardioprotective action of ashwagandha in rats and frogs. *J. Ethnopharmacol. 70*:57 – 63.

Ehringer H. and Hornykiewicz O. 1960. Verteilung von Nonadrenalin und dopamine (3-hydroxytyramin) im Gerhim des Menschen und ihr verhalten bei Erkankungen des extrapyramidalen systems. In : Bezard E. Brotchie J.M. and Gross C.E. 2001. Pathophysiology of levodopa- induced dyskinesia: potential for new therapies. *Nat. Rev. Neurosci. 2*: 577 – 588.

Ellis, J.E., Byrd L.D. and Bakay R.A.E. 1992. A method for quantitating motor deficits in a nonhuman primate following MPTP induced hemiparkinsonism and co-grafting. *Exp. Neurol. 115*: 376 – 387.

Eriksen, J.L. and Petrucelli L. 2004. Parkinson's disease - molecular mechanisms of disease. *Drug Dis Today: Dis Mech. 1*: 399 – 405.

Fernagut, P.O., Diguet E. Labattu B. and Tison F. 2002. A simple method to measure stride length as an index of nigrostriatal dysfunction in mice. *J. Neurosci. Meth. 113*: 123 – 130.

Fleming, SM., Salcedo J. Fernagut P.O. 2004. Early and Progressive Sensorimotor Anomalies in Mice Overexpressing Wild-Type Human α-Synuclein. *J. Neurosci. 24*: 9434 – 9440.

Fredriksson A. and Archer T. 1994. MPTP-induced behavioural and biochemical deficits: a parametric a1994nalysis. *J. Neural Transm. 7*:123 – 132.

Genc, S., Akhisaroglu M. Kuralay F. and Genc K. 2002. Erythropoietin restores glutathione peroxidase activity in 1-methyl-4-phenyl-1,2,5,6-tetrahydropyridine-induced neurotoxicity in C57BL mice and stimulates murine astroglial glutathione peroxidase production *in vitro. Neurosci. Lett. 321*: 73 – 76.

Gerlach, M., Russ H. Winker J. Witzmann K. Traber J. Stasch J.P. Riederer P. and Przuntek H. 1993. Effects of nimodipine on the I-methyl4- phenyl-1,2,3,6-tetrahydropyridine-induced depletions in the biogenic amine levels in mice. Arzneimittelforschung. *43*: 413 – 415.

Glowinski, J. and Iversen L.L. 1966. Regional studies of cateholamines in the rat brain-I. *J. Neurochem. 13*: 655 – 669.

Gourie-Devi, M., Ramu M.G. and Venkataram B.S. 1991. Treatment of Parkinson's disease in Ayurveda (ancient Indian system of medicine): discussion paper. *J. Roy. Soc. Med. 84*:491 – 492.

Halliwell B. 2001. Role of free radicals in the neurodegenerative disease therapeutic implications of antioxidant treatment. *Drug aging. 18*:685 – 716.

Hanakawa, T., Fukuyama H. and Katsumi Y. 1999. Enhanced lateral premotor activity during paradoxical gait in Parkinson's disease. *Ann. Neurol. 45*:329 – 336.

Hassler R. 1938. Zur Pathologie der Paralysis agitans und des postenzephalitischen Parkinsonismus. *J. Psychol. Neurol. 48*: 387 – 476.

Hastings, T.G., Lewis D.A.and Zigmond M.J. 1996. Reactive dopamine metabolites and neurotoxicity: implications for Parkinson's disease. *Adv. Exp. Med. Biol. 387*:97–106.

Jackson-Lewis, V., Jokowec M. Burke R.E. and Przedborski S. 1995. Time course and morphology of dopaminergic neuronal death caused by the neurotoxin 1-methyl 4-phenyl 1,2,3,6 tetrahydropyridine. *Neurodegeneration. 4*:257 – 269.

Jankovic J. 2008 Parkinson's disease: clinical features and diagnosis. *J. Neurol. Neurosurg. Psychiatr. 79*: 368 – 376.

Jollow, D.J., Mitchell J.R. Zampagloine N. and Gillete J.R. 1974. Bromobenzene-induced liver necrosis: protective role of glutathione and evidence for 3, 4 bromobenzeneoxide as the hepatotoxic intermediate. *Pharmacology. 11*: 151–169.

Lin, AM., Yang C.H. Ueng Y.F. Luh T.Y. Liu T.Y. Lay Y.P. and Ho L.T. 2004. Differential effects of carboxyfullerene on MPP^+/MPTP-induced neurotoxicity. *Neurochem Int. 44*: 99 –105.

Mishra, L.C., Singh B.B. and Dagenais S. 2000. Scientific basis for the therapeutic use of *Withania somnifera* (ashwagandha). *Altern. Med. Rev. 5*:334 –346.

Mohandas, J., Marshall J.J. Duggin G.G. Horvath J.S. and Tiller D. 1984. Differential distribution of glutathione and glutathione related enzymes in rabbit kidneys: Possible implication in analgesic neuropathy. *Cancer Research. 44*: 5086 – 5091.

Moriss, M., Iansek R. Matyas T. and Summers J. 1998. Abnormalities in the stride length cadence relation in Parkinsonian gait. *Mov. Disord. 13*:61 – 69.

Munoz, A., Rey P. and Guerra M.J. 2006. Reduction of dopaminergic degeneration and oxidative stress by inhibition of angiotensin converting enzyme in a MPTP model of parkinsonism. *Neuropharmacol. 51:*112 – 120.

Muralikrishnan D. and Mohanakumar K.P. 1998. Neuroprotection by bromocriptine against 1-methyl-4-phenyl-1,2,3,6-tetrahydropyridine induced neurotoxicity in mice. *FASEB J. 12:* 905 – 912.

Nadkarni A.K. 1982. The Indian Materia Medica. Volume 1, Popular Prakashan Privte Limited, Bombay, India. .

Nagashayana, N., Sankarankutty P. Nampoothiri M.R.V. Mohan P.K. and Mohankumar K.P. 2000. Association of L-DOPA with recovery following Ayurveda medication in Parkinson's disease. *J. Neurol. Sci. 176*:124 –127.

Panda S. and Kar A. 1997. Evidence for free radical scavenging activity of Ashwagandha root powder in mice. *Indian J. Physiol. Pharmacol. 41*: 424 – 426.

Prasad. KN., Cole W.C. and Kumar B. 1999 Multiple antioxidants in the prevention and treatment of Parkinson's disease. *Journal of the American College of Nutrition. 18*: 413– 423.

Przedborski, S., Chen Q. Vila M. Giasson B.I. Djaldatti R. Vukosavic S. Souza J.M. Jackson-Lewis V. Lee V.M. Ischiropoulos H. 2001. Oxidative post translational modifications of alpha synuclein in the 1-methyl 4-phenyl 1,2,3,6 tetrahydropyridine (MPTP) mouse model of Parkinson's disease *J. Neurochem. 76*: 637–640.

Rapisardi, S.C., Warrington V.O. and Wilson J.S. 1990. Effects of MPTP on the fine structure of neurons in substantia nigra of dogs.*Brain. Res. 512*: 147–54.

Riederer, P., Konradi C. Hebenstreit G. and Youdim M.B.H. 1989. Neurochemical perspectives to the function of monoamine oxidase. *Acta Neurol. Scand. 126*: 41–45.

Rojas P. and Rios C. 1993. Increased striatal lipid peroxidation after inter- cerbroventricular MPP^+ administration to mice. *Pharmacol. Toxicol. 72*:364–368.

Rozas, G., Lopez-Martin E. Guerra M.J. and Labanderia-Garcia J.L. 1998. The overall rod performance test in the MPTP –treated moue model of parkinsonism. *J. Neurosci. Meth. 83*:165–175.

Kasture, S., Pontis S. Pinna A. Schintu N. Spina L. Longoni R. Simola N. Ballero M. and Morelli M. 2009. Assessment of symptomatic and neuroprotective efficacy of *Mucuna pruriens* seed extract in rodent model of Parkinson's Disease. *Neurotox Res. 15*:111–122.

Schapira, A.H., Cooper J.M. Dexter D. Jenner P. Clark J.B. and Marsden C.D. 1989. Mitochondrial complex I deficiency in Parkinson's disease, Lancet 1

Schmidt N. and Ferger B. 2001. Neurochemical findings in the MPTP model of Parkinson's disease. *J. Neural. Transm. 108*:1263–1282.

Schulz, J.B., Matthews R.T. Muqit M.M. Browne S.E. Beal M.F. 1995. Inhibition of neuronal nitric oxide synthase by 'I-nitroindazole protects against MPTP-induced neurotoxicity in mice. *J Neurochem. 64*:936–939.

Sedelis, M., Schwarting R.K. and Huston J.P.I. 2001. Behavioural phenotyping of the MPTP mouse model of Parkinson's disease. *Brain Res. 125*:109 – 125.

Singh, B., Saxena A.K. and Chandan B.K. 2001. Adaptogenic activity of a novel withanolide-free aqueous fraction from the roots of *Withania somnifera* Dun. *Phytother Res. 15*: 311–8.

Sundstrom, E., Fredriksson A. and Archer T. 1990. Chronic neurochemical and behavioral changes in MPTP-lesioned C57BL/6 mice: a model for Parkinson's disease. *Brain Res. 528*: 181–188.

Szirakin, I., Kardos V. Patthy M. Pátfalusib M. and Budaia G. 1994. Methamphetamine protects against MPTP neurotoxicity in C57 BL mice. *Eur J Pharmacol. 251*:311–314.

Tillerson J.L. and Miller G.W. 2003. Grid performance test to measure behavioral impairment in the MPTP-treated mouse model of parkinsonism. *J. Neurosci. Meth. 123*:189–200

Turski, L., Bressler K. Rettig K.J. Löschmann P.A. and Wachtel H. 1991. Protection of substantia nigra from MPP+ neurotoxicity by N-methyl-naspartateantagonists. *Nature. 349*:414 – 418.

Utley, H.C., Bernheim F. and Hochslein P. 1967. Effect of sulfhydryl reagent on peroxidation in microsome. *Archives of Biochemistry and Biophyics. 118*: 29–32.

Vaidya, R.A., Sheth A.R. Aloorkar S.D. Rege N.R. Bagadia V.N. Devi P.K. and Shah L.P. 1978. The inhibitory effect of the cowhage plant-*Mucuna pruriens*-and L-dopa on chlorpromazine-induced hyperprolactinaemia in man. *Neurol India 26*:177–178.

Weiner M.A. and Weiner J. 1994. Ashwagandha (Indian ginseng). In: Herbs that Heal, Quantum Books, Mill Valley, California. 70–72.

Ziauddin, M., Phansalkar N. and Patki P. 1996. Studies on the immunomodulatory effects of Ashwagandha. *J. Ethnopharmacol. 50*: 69–76.

In: Natural Products and Their Active Compounds ... ISBN: 978-1-62100-153-9
Editors: M. Essa, A. Manickavasagan, and E. Sukumar © 2012 Nova Science Publishers, Inc.

Chapter 8

Role of Phytochemicals on Radiation-Modification in Normal and Cancer Cells

*N. Rajendra Prasad** *and G. Kanimozhi*

Department of Biochemistry and Biotechnology,
Annamalai University, Annamalainagar,
Tamilnadu, India

Abstract

Introduction: The main objective of any anti-cancer therapy is to provide maximal cancer cell killing while at the same time minimizing the damage to normal cells and tissues. Herbs and dietary phytochemicals have been increasingly recognized in prevention and treatment of human diseases including cancer. There exist enormous prospect for the evaluation of phytochemicals to increase radioprotection of normal cells during cancer radiotherapy.

Our work have focused on the mechanism of variety of dietary phytochemicals, namely, ferulic acid, curcumin, sesamol, lycopene and quercetin etc., on normal cells with view to design effective protocols in practical radioprotection. Recent reports have shown that phytochemicals induces radiosensitization of tumor cells by generating reactive oxygen species (ROS). Further, phytochemicals modulate G2/M checkpoint, thus leaving the cancer cells less time to repair the radiation-induced DNA damage and eventually leads them to apoptotic processes.

Further it is emphasized that modulation of the apoptotic and signaling pathways may help to achieve efficient destruction of cancer cells, this may provide a new approach in developing effective treatment for cancer. Phytochemicals can therefore be considered as adjunct in radiation therapy. *Conclusion:* This chapter critically explains the role phytochemicals as radioprotectors in normal cells and its suitability as radiosensitizers in cancer cells.

* E-mail: drprasadnr@gmail.com; Ph : 91 + 954144 – 238343; Fax : + 91 4144 - 239141

1. Introduction

Eighty percent of cancer patients need radiotherapy at some time or other, either for curative or palliative purpose [1]. As the tumor cells proliferate very rapidly, they usually overgrow their vascular blood supply, resulting in centrally necrotic and hypoxic regions, rendering radiation ineffective in these areas [2]. To overcome this problem, higher doses of radiation must be delivered to control the tumor. This is clinically not feasible, since the normal tissues surrounding the tumor are well perfused, vascularised and remain oxygenated, and are therefore more prone to radiation damage [3]. Another problem associated with cancer therapy is most of the cancer cells are modestly responsive or non-responsive to radiotherapy [4]. The success of radiotherapy, therefore, depends on increasing the sensitivity of the cancer cells to radiation [5]. Improvement of the radiotherapy can be accomplished in two different ways: enhancement of tumor radioresponse or protection of normal tissues [6]. It is quite uncommon to pursue both of these strategies with just one compound given at the time of radiotherapy, although discovery of such a compound would be a major breakthrough. Many scientists have hypothesized that various dietary modulators and phytochemicals can work as excellent adjuvants to radiation therapy in a variety of cancers. So, this present chapter critically explores the role of phytochemicals in radioprotection of normal cells and radiosensitization of cancer cells.

2. Radioprotectors

Radioprotectors are the substances that protect normal tissues during radiation exposure. Clinical gain can be obtained either by a reduction in morbidity if the effects are confined to normal tissues, or by exploiting the hoped-for reduced radiosensitivity of normal tissues to deliver higher radiation doses and, thus, enhanced tumour cell kill, the latter strategy obviously not without risk [7]. Among the molecular radioprotectors, WR-2721 (S-2- (3-aminopropyl-amino) ethyl phosphorothioic acid), also known as amifostine is the most thoroughly investigated radioprotective drug, initially developed at the Walter Reed Army Research Institute, USA under the Antiradiation Drug Development Program of the US Army Medical Research and Development Command [8]. However, the radioprotective effect of synthetic phosphorothioate compounds, including amifostine, is short term, and is associated with severe side effects (e.g. nausea, vomiting, diarrhea, hypotension, hypocalcaemia, nephro- and neuro-toxicity) at clinically effective doses [9]. These limitations have greatly restricted their clinical use.

Despite its drawbacks, amifostine is the only radioprotective drug that has been approved by the Food and Drug Administration (FDA), USA. Amifostine is being used clinically for ameliorating the incidence of xerostomia (dry mouth) in patients undergoing radiotherapy for the treatment of head and neck cancer [10]. Hence, the success with these compounds has also been limited. The fact remains that to date there is no single radioprotective agent available which meets all the prerequisites of an ideal radioprotector, i.e. produces no cumulative or irreversible toxicity, offers effective long-term protection, possesses a shelf life of 2–5 years, and can be easily administered. In view of this, the search for newer, less toxic and more effective radioprotective drugs continues.

3. PHYTOCHEMICALS ACT AS RADIOPROTECTORS

Vegetables and fruits are excellent sources of cancer preventive substances. Phytochemicals are non-nutritive components present in the plant-based diet that possess substantial anticarcinogenic, antioxidant and antimutagenic properties [11]. As folklore medicine has an enormous heritage of vast natural dietary and time tested medicinal resources it is worth exploring the possibility of developing efficient, economically viable and clinically acceptable radiomodifiers for human application from these resources. Numerous phytochemicals derived from edible plants have been reported to interfere with a specific stage of the carcinogenic process (figure 1). Many mechanisms have been shown for the anticarcinogenic actions of dietary constituents, but attention has recently been focused on the modulation of radiation response signaling elements for the treatment of various tumors [12].

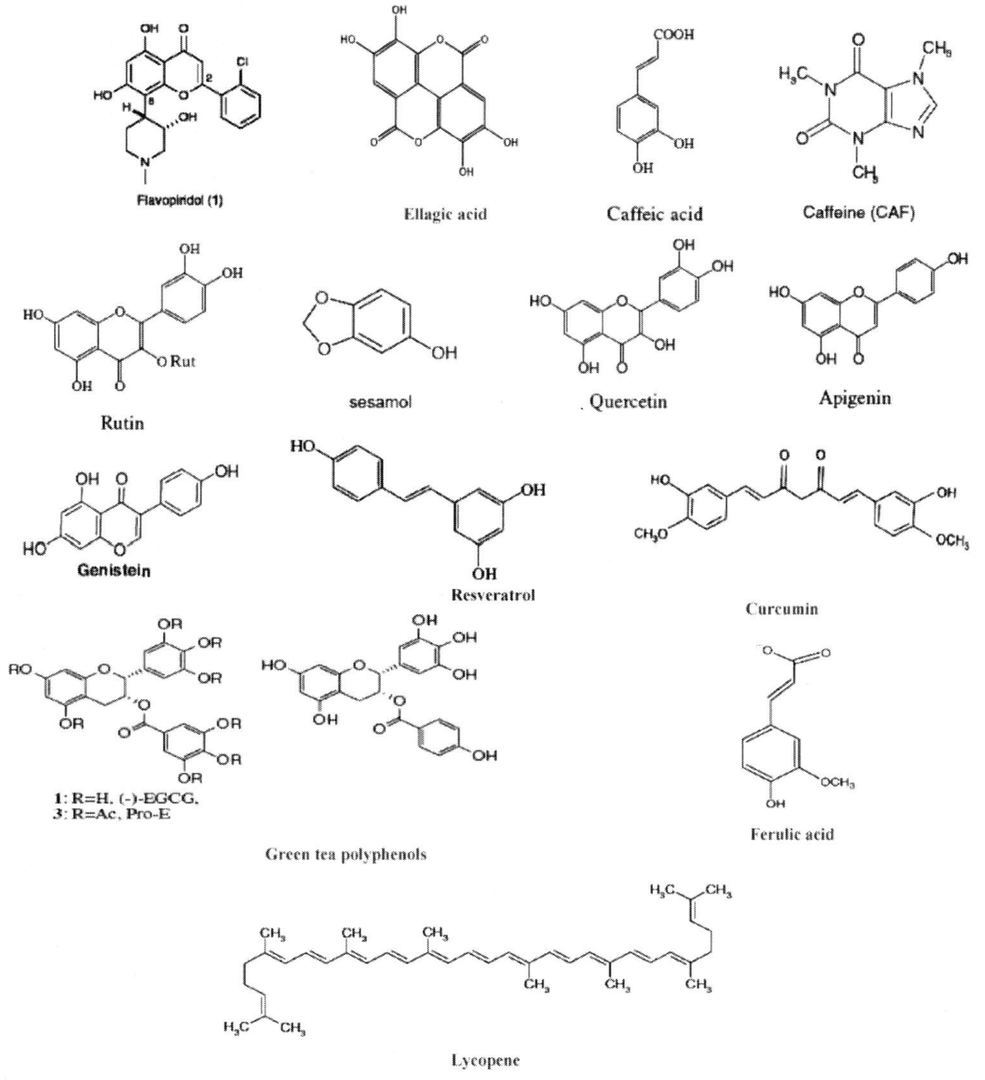

Flavopiridol (1)

Ellagic acid

Caffeic acid

Caffeine (CAF)

Rutin

sesamol

Quercetin

Apigenin

Genistein

Resveratrol

Curcumin

1: R=H. (-)-EGCG.
3: R=Ac, Pro-E

Green tea polyphenols

Ferulic acid

Lycopene

Figure 1. Structure of some phytochemicals discussed in this review.

Phytochemicals are known to possess antioxidant properties [13]. When compare to amifostine, a conventional radioprotector, phytochemicals may likely to serve as an alternative sources as non-toxic radioprotectors. A large number of plants contain antioxidant phytochemicals reported to be radioprotective in various model systems (Table 1, figure 2). Protection against radiation induced genetic instability by non-toxic dietary ingredients form an attractive proposition to prevent radiation-induced carcinogenesis [14]. Number of flavonoids reduces the frequency of micronucleated reticulocytes in the peripheral blood of irradiated mice [50]. Procyanadins from grape seed extract, including rutin, were radio-protective as measured by a decrease in frequency of micronucleated erythrocytes from bone marrow of irradiated mice [15]. Tea polyphenols protects radiation-induced DNA damage [51]. Curcumin in the diet protects radiation-induced and diethylstilbestrol-promoted tumors in mice [52]. Curcumin treatment before irradiation was found to inhibit radiation-induced chromosomal damage [53, 17]. The isoflavone genistein, found in soybeans, is of interest because of its potential radioprotective effects [54]. Genistein is radioprotective in mice when administered in a single dose 24 h before γ-irradiation [18]. Sesamol, simple phenolics present in the roasted sesame oil, possess potent radioprotective property against γ-irradiation induced lipid peroxidation and DNA damage [19]. Ferulic acid, a dietary phenolic phytochemical, protects cultured human lymphocytes and rat hepatocytes from γ-radiation induced chromosomal aberrations [21, 17]. Under *in vitro* condition lycopene act as a natural protector against γ-radiation induced DNA damage, lipid peroxidation and protects antioxidant status in primary culture of isolated rat hepatocytes [23]. Protective effect of apigenin on radiation-induced chromosomal damage in human lymphocytes was studied, the frequency of micronuclei was decreased as the concentration of apigenin increased, suggesting the radioprotective effect of apigenin [24].

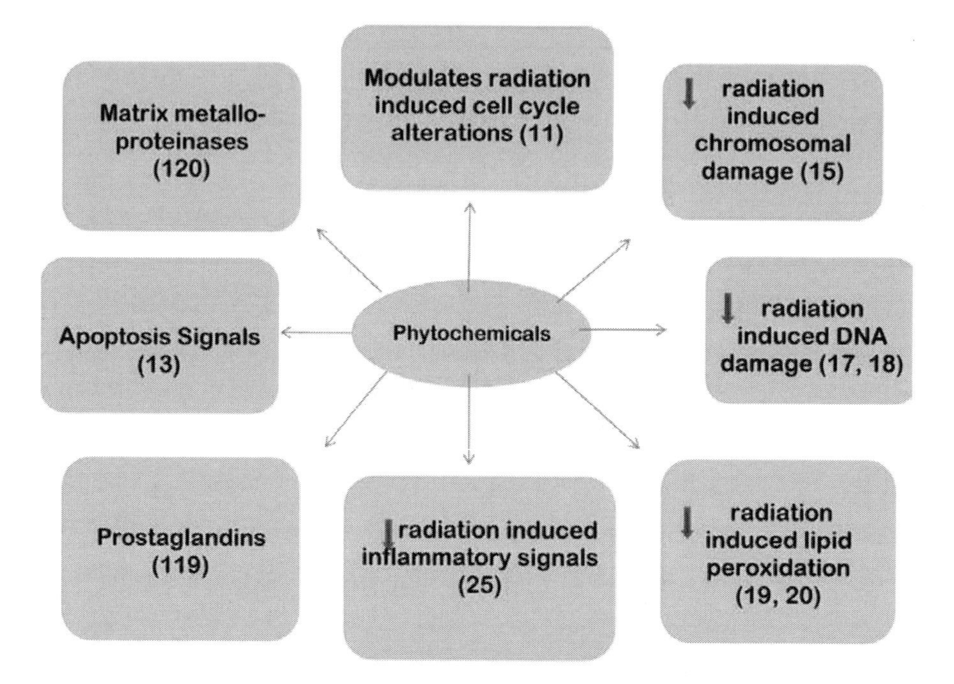

Figure 2. Schematic representation showing radioprotective properties of phytochemicals.

Table 1. Effect of phytochemicals on radiation protection

S.No	Name of the phytochemicals	Concentration of the phytochemicals	Radiation dose	Radioprotective effect	References
1	Rutin	3.3 μM	2 Gy	Antioxidant activity and radioprotective effects against chromosonal damage induced *in vivo* by X-rays.	[15]
2	Tempol	275 mg/kg	15 Gy	Evaluate the effect of different doses and times of administration, the heads of C3H mice were exposed to a single irradiation dose of 15 Gy, with i.p. tempol injection.	[16]
3	Curcumin	40 μg/mL	4 Gy	Protective effect of curcumin on γ-radiation induced DNA damage and lipid peroxidation in cultured human lymphocytes.	[17]
4	Genistein	150 μM	2 Gy	Genistein treatment protects mice from ionizing radiation injury.	[18]
5	Sesamol	10 μg/mL 25-100 mg	4 Gy, 4.5 Gy	(i) Radioprotective effect of sesamol on γ-radiation induced DNA damage, lipid peroxidation and antioxidants levels in cultured human lymphocytes. (ii) Radioprotective effect of sesamol on cytotoxicity radiation-in Swiss albino mice.	[19,20]
6	Ferulic acid	10 μg/mL	4 Gy	Protective effect of ferulic acid on γ-radiation-induced micronuclei, dicentric aberration and lipid peroxidation in human lymphocytes. Radiation protection of DNA by ferulic acid under *in vitro* and *in vivo* conditions. The DMF value is 2.0.	[21,22]
7	Lycopene	18.62 μM	4 Gy	Lycopene as a natural protector against γ-radiation induced DNA damage, lipid peroxidation and antioxidant status in primary culture of isolated rat hepatocytes *in vitro*.	[23]

Table 1. (Continued)

S.No	Name of the phytochemicals	Concentration of the phytochemicals	Radiation dose	Radioprotective effect	References
8	Apigenin	25 μg/mL	2 Gy	Protective effect of apigenin on radiation-induced chromosomal damage in human lymphocytes.	[24]
9	Glycyrrhizic acid	4 mM	1.25 Gy	Protect against γ-radiation induced DNA damage to plasmid pBR322 *in vitro*, human peripheral blood leukocytes and bone marrow cell *in vivo*.	[25]
10	Mangiferin	50 μg/mL	3 Gy	Reduces the radiation induced damage and enhance the repair of DNA double strand breaks in human peripheral blood lymphocytes.	[26]
11	Naringin	2 mg/b.wt	4 Gy	Naringin protects mouse liver and intestine against the radiation induced damage by elevating the antioxidant status and reducing the lipid peroxidation.	[27]
12	flavonone,	7.5 mg/kg b.wt	1-5 Gy	Aphanamixis polystachya (EAP) in the presence of flavonone, Naringin protects mouse bone marrow cells against radiation-induced chromosomal aberrations and lipid peroxidation	[28]
13	Quercetin	1 mg/b.wt	6–8 Gy	Radioprotective effects of propolis and quercetin in γ - irradiated mice evaluated by the alkaline comet assay.	[29]
14	5,7- Di hydroxy chromone-2-carboxylic acid	1 mmol/kg b.wt	8.2 Gy	5,7-Dihydroxychromone-2-carboxylic acid and it's transition-metal (Mn and Zn) chelates as non-thiol radioprotective agents in mice.	[30]
15	4-hydroxy-3,5-dimethoxy benzaldehyde	20 μmol	2-8Gy	Radioprotection of 4-hydroxy-3,5-dimethoxybenzaldehyde (VND3207) in culture cells is associated with minimizing DNA damage.	[31]
16	Epicatechin	2 mM	50 Gy	Protection of cellular DNA was analyzed in peripheral blood leucocytes of whole body irradiated mice.	[32]

S.No	Name of the phytochemicals	Concentration of the phytochemicals	Radiation dose	Radioprotective effect	References
18	Paeoniflorin	200 μg/ml	4 Gy	Radioprotective effect of paeoniflorin on irradiation-induced cell damage involved in modulation of reactive oxygen species and the mitogen-activated protein kinases in thymocytes.	[34]
19	Dehydrozingerone	100 mg/kg b.wt	10 Gy	Free radical scavenging and radioprotective activity of dehydrozingerone against whole body gamma irradiation in Swiss albino mice. The DMF value is 1.09.	[35]
20	Tannins from *Pinus caribaea*	2 mg/ml	150 Gy	Tannins from barks of *Pinus caribaea* protect *Escherichia coli* cells against DNA damage induced by γ-rays	[36]
21	Famotidine	10 μg/mL	12 Gy	Modulation of gamma-ray-induced apoptosis in human peripheral blood leukocytes by famotidine and vitamin C	[37]
22	Luteolin	10 μmol/kg b.wt	6 Gy	Radioprotective effects of antioxidative plant flavonoids in mice	[38]
23	Lignans from Myristica fragrans	500μg/mL	4.6 Gy	Immunomodulatory and radioprotective effects of lignans derived from fresh nutmeg mace (*Myristica fragrans*) in mammalian splenocytes	[39]
24	Melatonin	10 mg/kg b.wt	4 Gy	Ameliorative effect of melatonin against gamma-irradiation-induced oxidative stress and tissue injury	[40]
25	Pilocarpine	4 mg/kg b.wt	15 Gy	Optimum dose range for the amelioration of long term radiation-induced hyposalivation using prophylactic pilocarpine treatment	[41]
26	Hesperidin	100 mg/kg b.wt	5 Gy	Hesperidin a flavanoglycone protects against γ-irradiation induced hepatocellular damage and oxidative stress in Sprague-Dawley rats	[42]
27	Vitamin-E	600 mg/kg b.wt	15 Gy	Evaluate the radioprotective effect of vitamin E in salivary disfunction in irradiated male rats.	[43]
28	Podophyllotoxin	2.5 μg/mL	400 Gy	Radioprotective action of podophyllotoxin in *Saccharomyces cerevisiae* yeast	[44]

Table 1. (Continued)

S.No	Name of the phytochemicals	Concentration of the phytochemicals	Radiation dose	Radioprotective effect	References
29	Isoflavone genistein-8-c-glycoside	75 mg/kg b.wt	1 Gy	Antioxidant and radioprotective effects of genistein-8-C-glicoside (G8CG), on rat blood plasma and liver microsomal membrane lipid peroxidation.	[45]
30	Catechines	200 mg/kg b.wt	6 Gy	Catechines have definite radioprotective effect against radiation injury and in mice body weight.	[46]
31	Paeoniflorin	50-200 μg/mL	4 Gy	Protective effect of paeoniflorin on irradiation-induced cell damage involved in modulation of reactive oxygen species and the mitogen-activated protein kinases	[47]
32	Vanillin	50 mg/kg b.wt	4 Gy	Vanillin 0.5mM has a dose-modifying factor (DMF) of 6.75 for 50% inactivation of ccc form. Exposure of human peripheral blood leucocytes in vitro to γ-radiation causes strand breaks in the cellular DNA.	[48]
33	Orientin and Vicenin	17.5 μM	4 Gy	Orientin (50 μg/ kg body weight) is having DMF value 1.6 for stem cell survivel, exogenous spleen colon (CFU-S). The DMF value is 1.6.	[49]

4. Phytochemicals Modulates Radiation Effect in Experimental Animals

Supplementation of phytochemicals to improve the efficacy of radiotherapy is today's proposed strategy. Certain phytochemicals selectively inhibit the growth of tumor cells, induce cellular differentiation, and alter the intracellular redox state thereby may enhance efficacy of radiotherapy *in vivo* [55]. Dehydrozingerone (DZ) is an important constituent of ginger [56], has the structure corresponding to half an analog of curcumin. The free radical scavenging and radioprotective activity of dehydrozingerone against whole body γ-irradiation in Swiss albino mice has been studied [35]. Pretreatment with DZ 100 mg/kg bwt reduced the mucosal erosion and maintains villus height, the improvement in crypt cells proliferation has been observed on DZ pretreatment upon radiation treatment. [35]. Positive results obtained on γ-irradiated mice given propils and quercetin, on survival of irradiated mice, indicate that these compounds could be considered effective non-toxic radioprotectors [29]. Nuclear enlargement in epithelial cells of jejunum was lower in morin (2-(2, 4-dihydroxyphenyl)-3, 5,

7 trihydroxy-4H-1-benzopyran-4-one) a flavonoid constituent of many herbs and fruits) treated mice compared to radiation control. Morin (100 mg/kg bwt) also significantly elevated the endogenous antioxidant enzymes viz. glutathione S transferase (GST), superoxide dismutase (SOD) and reduced glutathione (GSH), in normal mice at 2, 4 and 8 h post treatment. Drastic decrease in endogenous enzymes (GSH, GST, catalase and SOD) and total thiols has been observed in irradiated mice at 2, 4 and 8 h post irradiation, while pretreatment with morin (100 mg/kg bwt) prevented this decrease. Morin (100 mg/kg bwt) also elevated radiation LD_{50} from 9.2 to 10.1 Gy, indicating a dose modifying factor (DMF) of 1.11. [57]. An intraperitoneal injection of 5-aminosalicilic acid at a dose of 25 mg/kg bwt 30 min before lethal radiation increased survival, giving a DMF of 1.08. Injection of 5ASA (25 mg/kg b wt.) 60 or 30 min before or within 15 min after 3Gy whole body radiation resulted in a significant decrease in the radiation-induced aberrant metaphases, at 24 h post-irradiation [58]. Earlier study shows that radiation increased the micronuclei (MN) frequency linearly (r2 = 0.99) with dose. Pretreatment with 5ASA significantly reduced the MN counts to 40–50% of the radiation (RT) alone values, giving a DMF of 2.02 (MPCE) and 2.53 (MNCE) [59].

The radio protective ability of sesamol administered intraperitoneally 30 min prior to 9.5 Gy whole-body γ-irradiation has been studied in Swiss albino mice. Sesamol pretreatment with 50 mg/kg bwt was found to be the most effective dose in maintaining body weight and in reducing the percentage mortality, while 100 mg/kg bwt was found to be more effective in maintaining the spleen index and in stimulation of endogenous spleen colony-forming units. Pretreatment with sesamol (50 mg/kg bwt) in mice irradiated with 15 Gy significantly reduced dead, inflammatory, mitotic and goblet cells in irradiated jejunum. Sesamol at 50 mg/kg bwt also increased crypt cells, to improvement villus height and prevented mucosal erosion irradiated mice. Nuclear enlargement in epithelial cells has been found less in sesamol-treated mice compared with the irradiated control [20].

Eckol, a component of the seaweed *Ecklonia cava*, protected the lymphocytes' viability and rescued intestinal cells from radiation-induced apoptosis by decreasing the amount of pro-apoptotic p53 and Bax and increasing that of anti-apoptotic bcl-2. Eckol modulates p53-dependent pathways, thereby blocking radiosensitive cells from entry into apoptosis [60]. Both synthetic and natural radioprotective agent such as WR-2721, silibinin, and aloe polysaccharides have been reported to confer radioprotection by preventing destruction of blood cells and intestinal crypt cells through the modulation of the p53-dependent pathways [61]. The effect of genistein and radiation combined therapeutic approach has been tested *in vivo*, using an orthotopic PC-3/nude mouse xenograft Pca tumor model [62]. The combined genistein and radiation treatment caused significant prostate tumor growth inhibition and controlled spontaneous metastasis to para-aortic lymph nodes compared each modality alone.

A significant increase in mouse survival has been observed. Genistein pretreatment and radiation exposure shows an increase in giant cells, apoptosis and inflammatory cells and fibrosis compared to each modality alone [62]. Enhanced tumor growth inhibition and metastasis by treatment with genistein and primary tumor irradiation has been observed in two additional metastatic tumor models, the syngeneic murine RM-9 orthotopic prostate tumor model in immuno-competent C57BL/6 mice [63] and the xenograft orthotopic metastatic KCI-18 renal cell carcinoma (RCC) model [64]. Hence, *in vivo* studies in metastatic tumor models demonstrate that the combination of genistein with primary tumor irradiation is an effective therapeutic approach for cancer therapy. Considering that the recovery of intestinal stem cell is a major complication of radiation exposure, thereby facilitating opportunistic gastrointestinal infections [65], the augmentation of the intestinal stem cell protection is an important factor of radioprotective efficacy. These effects may help to prevent the death of irradiated animals associated with gastrointestinal malabsorption, bleeding and tissue destruction.

5. PHYTOCHEMICALS ACT AS RADIOSENSITIZERS IN CANCER CELLS

There appears to be a differential effect of phytochemicals on normal and malignant cells. Phytochemicals can act as a radiosensitizers under *in vitro* and *in vivo* conditions in a variety of cancer cell lines (Table 2).

Table 2. Effect of phytochemicals on radiation sensitization

S.No	Phytochemicals	Concentration of the phytochemical	Radiation dose	Effect	References
1	Hypericin	5 mM	8 Gy.	Enhancement of Radiosensitivity in Human Malignant Glioma Cells by Hypericin in Vitro. DMF value is 1.45.	[66]
2	Protohypericin	5 mM	8 Gy.	Cytotoxicity and antiproliferative effect of hypericin and derivatives after photosensitization.	[67]
3	Epigallocate chin-galate	20 µ0 n-	3 Gy	Study of the combined effect of X-irradiation and epigallocatechin-galate on the growth inhibition and induction of apoptosis in human cancer cell lines.	[51]
4	Ellagic acid	100 µ00 gi	6 Gy	Ellagic acid enhanced radiation-induced oxidative stress and subsequent cytotoxicity in tumor cells.	[68]
5	Tea polyphenols	25 µ5	10- 30 Gy	The Survivin-mediated radioresistant phenotype of glioblastomas is regulated by RhoA and inhibited by the green tea polyphenol	[51]
6	Caffeic acid	10 mg/kg b.wt	8 Gy	Caffeic acid phenethyl ester (CAPE) sensitizes CT26 colorectal adenocarcinoma to ionizing radiation	[69]
7	Genistein	200 mM b.wt	5 Gy	Genistein potentiates the effects of irradiation in a human prostate carcinoma cell line.	[70]
8	Carvacrol	1%	14.42 kGy	The influence of atmosphere conditions on Escherichia coli and Salmonella typhi radiosensitization in irradiated ground beef containing carvacrol and tetrasodium pyrophosphate	[71]

S.No	Phytochemicals	Concentration of the phytochemical	Radiation dose	Effect	References
9	Chlorophyllin	5 %	20 Gy	Chlorophyllin as a protector of mitochondrial membranes against g-radiation and photosensitization.	[72]
10	Resveratrol	3 mM	8 Gy	Resveratrol sensitization of DU145 prostate cancer cells to ionizing radiation is associated to ceramide increase	[73]
11	Flavopiridol	500 nM	40 Gy	Antisense inhibition of cyclin D1 expression is equivalent to flavopiridol for radiosensitization of zebrafish embryos	[74]
12.	Curcumin	10 µ0 cu	4 Gy	Curcumin: An ''old-age'' disease with an ''age-old'' solution to act radiosensitization in cancer cells.	[75]
13	Gossypol	2 µ	6 Gy	Radiosensitization of tumour cell lines by the polyphenol Gossypol results from depressed double-strand break repair and not from enhanced apoptosis	[76]

Presence of phytochemicals during irradiation amplifies their effects by multi-factorial mechanisms including (i) toxic reactions of free radicals, (ii) overriding cell cycle arrest and (iii) by inducing apoptosis [77].

5.1. Phytochemicals Generates Free Radicals in Cancer Cells

The presence of phytochemicals during irradiation amplifies radiation effects by inducing toxic reactions of free radicals. Free radicals are toxic entities known to cause cellular damage and to mediate the effect of ionizing radiation [78-80]. Phytochemicals have been reported to interfere with a specific stage of the carcinogenic process [81]. A large number of natural compounds have shown cytotoxic effects in a variety of pathological situations either alone or together with radiation [82]. Phytochemicals found in herbals has been reported to possess antioxidant properties, which protects normal cells from oxidative stress. On the other hand, they exhibit prooxidant activity, which contributes cancer cell killing (Figure 3). Flavonoids autooxidize in aqueous medium and may form highly reactive hydroxyl radicals. Polyphenols may act as substrate for peroxidases and other metalloenzymes, yielding quinine or quinomethide type prooxidants [83]. It has been shown that curcumin potentiates the effect of γ-radiation on hamster ovary cells [84]. Sahu and Washington, 1992 showed that the effect of curcumin on irradiated cells and under certain conditions it may serve as a prooxidant and stimulates radiation damage [85].

High concentration of intracellular thiol is an important factor of resistance to ionizing radiation in cancer cells [86]. Some of the phytochemicals depletes intracellular antioxidants levels thereby enhance radiation effects in cancer cells. Caffeic acid phenethyl ester (CAPE) sensitizes CT26 colorectal adenocarcinoma to ionizing radiation, which may be via depleting GSH and inhibiting NF-κB activity, without toxicity to normal cells. Cells treated with CAPE had a lower ratio of GSH: GSSG. Glutathione reductase is the principal enzyme involved in the regeneration of GSH from GSSG, it is suggested that CAPE may deplete GSH and inhibit the recycling of GSH in CT26 cells. Because CAPE depletes GSH via increasing oxidation of GSH and inhibiting GSH recycling, it is implicated that CAPE might act synergistically with inhibitors of *de novo* synthesis of GSH to sensitize tumor cells to radiation [69].

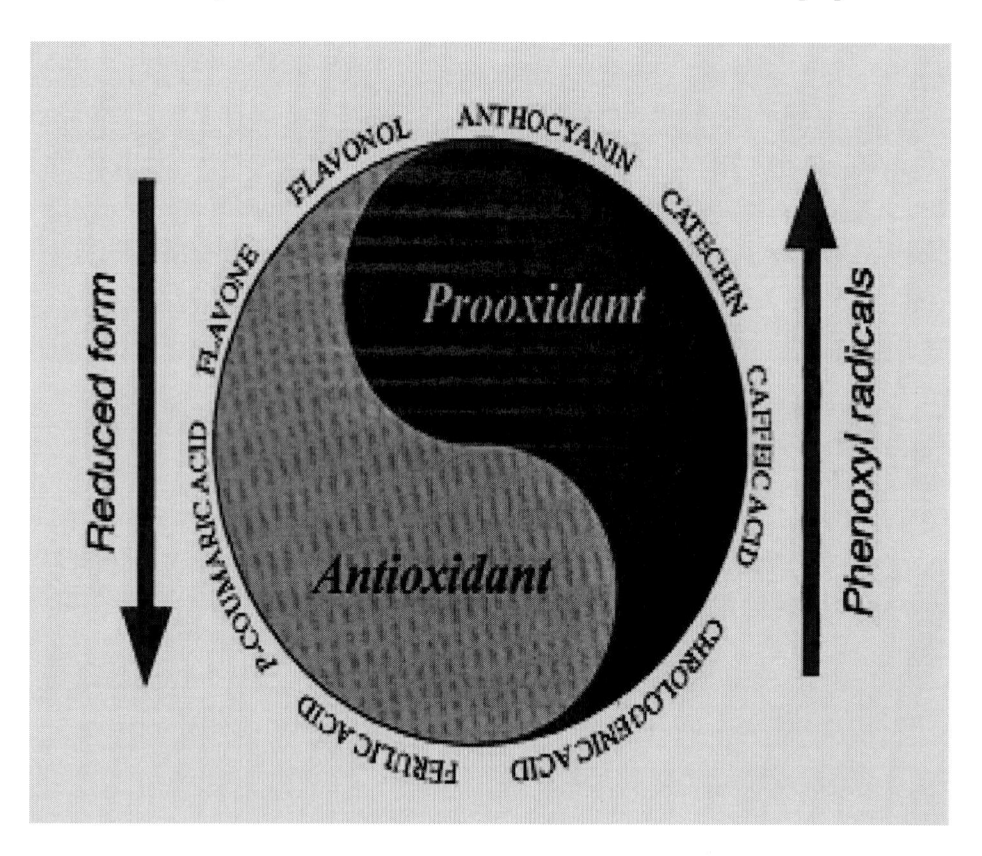

Figure 3. The Yin-Yang symbol represents the balance between antioxidant and prooxidant characteristics of phytophenolics. Although the reduced forms of phytochemicals act as antioxidants, the oxidized form, phenoxyl radicals, can exhibit prooxidant activities under conditions that prolong the radical lifetime.

Many studies suggested that antioxidant systems are critical in protecting against tumor promoting agents. Interestingly, cell malignancy or transformation is often accompanied by a decrease in activity of antioxidant enzymes like superoxide dismutase (SOD), catalase (CAT) and glutathione peroxidase (GPx), which increases the cell sensitivity to prooxidant compounds [87]. As a result, the susceptibility of tumor cells to radiation is associated with decreased levels of antioxidants [88]. Bhosle et al showed that ellagic acid decreases antioxidant enzymes like SOD, CAT and GPx in mice treated with radiation. On the other hand ellagic acid shows protection against radiation induced ROS in normal cells. This shows differential action of radiation and ellagic acid in tumor cells and its normal counterpart [68]. Thus, it can be thought that a single molecule can differentially act in cancer and normal cells.

5.2. Significance of ROS in the Inhibition of Cancer Cell Proliferation by Phenolic Phytochemicals and Isothiocyanates

Numerous studies have established that phenolic phytochemicals inhibit the proliferation of cancer cells by inducing cell cycle arrest and/or apoptosis, which might be due to scavenging by the phenolic phytochemicals of hydrogen peroxidase (H_2O_2) or other Reastive oxygen spieces (ROS) needed by cancer cells for their vitality or viability (Figure 4). For example, the phenolic phytochemical, apigenin, induced growth inhibition of human anaplastic thyroid carcinoma cells that was later followed by apoptosis [89].

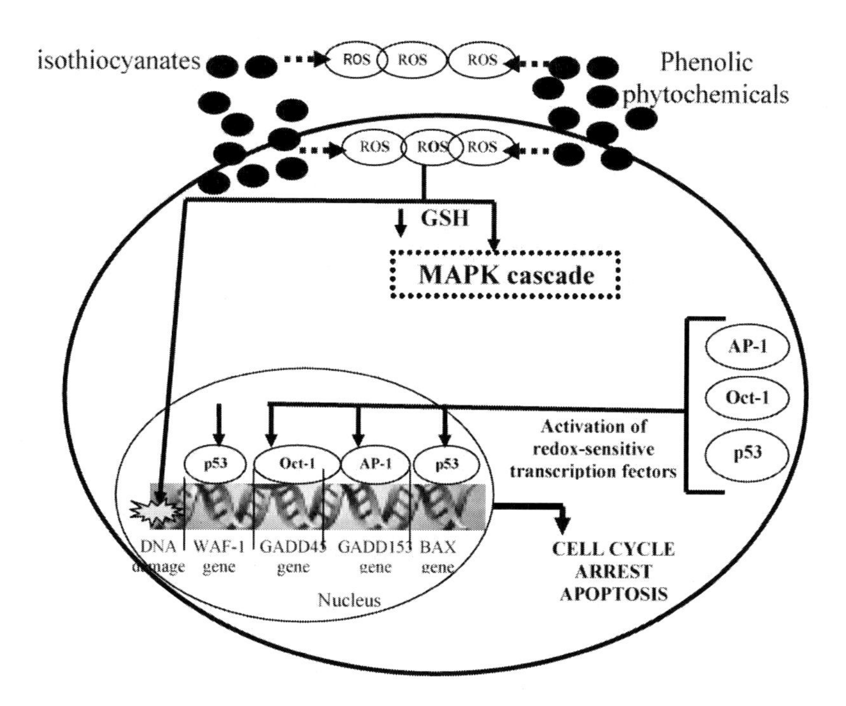

Figure 4. Concept of how phenolic phytochemicals inhibit cancer cell proliferation by inducing the formation of intolerable amounts of reactive oxygen species (ROS). ROS cause irrepearable DNA gamage as well as activation of stress/survival/death genes that result in cell cycle arrest and/or apoptosis.

There was inhibition of EGFR tyrosine autophosphorylation and downstream phosphorylation of MAPK. Thus, in exerting its effects, apigenin may have directly inhibited the protein phosphorylation, or alternatively, scavenged H_2O_2 that activates the protein kinases. Resveratrol induced G1-phase cell cycle arrest in human epidermoid carcinoma A431 cells [90].

There was decreased protein expression of cyclins and cyclin-dependent kinases (CDK) along with reduced CDK activity that normally mediate cell cycle progression. Consequently, expression of the CDK inhibitor, p21, increased so that apoptosis was the end result. It has been suggested that resveratrol induces apoptosis via JNK activation that leads to phosphorylation and hence activation of p53 [91], which is a transcription factor that activates the Waf-1 gene, which codes for p21, or the bax gene, whose translated product is known to promote apoptosis. Other phenolic phytochemicals such as quercetin and genistein induced apoptosis in pancreatic carcinoma cells, where mitochondrial depolarization, cytochrome c release, and caspase activation were early apoptotic events [92]. Therefore, since the phenolic phytochemicals have antioxidant activity, it is logical to think that they induced cell cycle arrest and apoptosis in cancer cells by scavenging H_2O_2, thereby depriving them of an essential molecule needed for their existence.

5.3. Phytochemicals Override Cell Cycle Arrest in Cancer Cells

The basis of many anti-cancer therapies is the use of genotoxic agents that damage DNA and thus kill dividing cells. Agents that cause cells to override the DNA-damage checkpoint are predicted to sensitize cancer cells. Phytochemicals accelerate the traverse of the G2/M checkpoint, thus leaving the cancer cells less time to repair radiation-induced DNA damages and eventually leads them to apoptotic processes. Normal maintenance of cell cycle requires the coordinated activity of the CDK, controlled by their association with cyclins, as well as their state of phosphorylation [93]. Two major regulators of G1 progression include three D-type cyclins (D1, D2, and D3) that combine with CDK4/6, and cyclin E that combines with CDK2. Cyclin D-CDK4/6 and cyclin E-CDK2 complexes sequentially phosphorylate the product of the Rb to initiate the G1-S transition [94]. Cyclin A-CDK2 and cyclin B-CDK1 complexes are required for S-phase progression and G2-M transition, respectively [95]. Studies shows phytochemicals act as CDK inhibitor therefore they may act as a potential anticancer agent. One such compound, caffeine, uncouples cell-cycle progression from the replication and repair of DNA [96]. Studies shows caffeine is known to reduce the radiation survival of many cell lines. Caffeine, is a G2 checkpoint inhibitor, releases radiation induced G2 block [97]. Flavopiridol is a novel flavone, structurally related to a natural alkaloid originally purified from the extract of the stem bark *Dysoxylum binectariferum*, a plant native to India [98, 99]. It has been shown that resveratrol, a polyphenolic compound alters cell cycle progression and show cytotoxic response to ionizing radiation in cervical cancer cell lines [100, 77]. It is also known that curcumin confers radiosensitizing effects in prostate cancer cell line by inhibiting the growth of human prostate PC-3 cancer cells and down regulates radiation-induced prosurvival factors [101]. Genistein can act as a radiosensitizers under *in vitro* and *in vivo* conditions in a variety of cancer cell lines. For example, genistein potentiates the effects of irradiation in a human prostate carcinoma cell line [70] and acted as a radiosensitizers in leukemic cells [102] and cervical cancer cells [103] exposed to radiation. Using an orthotopic xenograft in mice, it was demonstrated that genistein combined with irradiation caused greater inhibition of primary tumor growth than irradiation alone [104]. Genistein has a heterocyclic diphenolic structure and has been shown to inhibit tyrosine protein kinases, topoisomerase I and II, protein histidine kinase and 5a-reductase and induces G2/M cell cycle arrest in some cells that leads to cell growth inhibition [105]. Effect of genistein on disruption of the cell cycle may be the likely mechanism that results in the augmentation of cell killing induced by radiation.

5.4. Phytochemicals Modulates Radiation-Induced Apoptosis

Results demonstrates that treatment of HL-60 human promyelocytic leukemia cells lacking p53, a genetic abnormality frequently found in tumors, with radiation doses of 2–6 Gy induces only a insignificant apoptotic response. However, if cells were pre-incubated with some of the phytochemicals prior to radiation effectively induces apoptosis [106]. The enhancement of radiation-induced apoptosis by phytochemicals showed an additive effect according to the model of Steel and Peckham [107]. Phytochemicals potently activates the apoptosome by the release of Cyt-C from mitochondria and the activation of caspases 9, 8, 7 and 3 [108].

There is a considerable void in the information with regard to the status on signal transduction and oncogene expression in response to ionizing radiation. Agents designed to inhibit the activation of ras or protein kinase-C (PKC) could achieve this objective, since these are the primary signal transduction elements that impart radioresistance. Inhibition of ras activity in tumors is difficult to achieve, and since activation of ras can be mediated through phosphorylation of GAP [109], modulation of PKC would be the most appropriate strategy to increase the apoptotic potential of cell [110]. Most PKC inhibitors are toxic to normal cells. It is important to screen natural compounds, which can selectively inhibit PKC activation. Nicotinamide, an endobiotic, has recently been shown confer radiosensitivity in cancer cells [111]. Curcumin has been observed to induce apoptosis in human leukemic cell lines [112, 113]. Rutin, a naturally occurring glycoside is the effective inhibitors of PKC [114]. Lycopene, carotenoid abundantly present in tomato, also has been found recently to inhibit activation of PKC [115]. Studies could be extended to screen other PKC modulators from natural origin and to evaluate their efficacy to enhance radiation-induced apoptosis in tumor cells. This could form a rational strategy for the treatment of resistant tumors.

The proliferative and apoptotic responses to irradiation after genistein treatment were monitored. The gene expression and cell cycle effects reported for genistein have been shown to be different for normal compared to cancer cells [116]. For example, genistein increased the cell-cycle regulator CDKN1a in normal breast epithelial cells but showed little effect on CDKN1a in malignant breast cell. Several studies have reported that genistein induces apoptosis in cancer cells through a shift in the bax/bcl2 protein ratios [117]. In addition, Sarkar and Li [118] observed an increase in bax and a decrease in bcl-2 proteins in cancer cells tested after treatment with genistein for approximately 24 h. This resulted in a significant elevation in the bax/bcl-2 ratio, and an increase in apoptosis in these cells [119, 120]. This could form a rational strategy for the treatment of resistant tumors [121]. Resveratrol sensitization of DU145 prostate cancer cells to ionizing radiation is associated to ceramide increase. To investigate whether an increase of intracellular ceramide is associated with increased radio-sensitivity of cancer cells acquired after pretreatment with phytochemicals, it is necessary to perform ceramide status of the cell [122]. Flavopiridol also has been reported to enhance the effect of radiation in malignant glioma cells by enhancing apoptosis [122]. Since, phytochemicals has been shown to modulates cell cycle progress their radiosensitizing potential have to be explored both *in vitro* and *in vivo*, and in clinical trials.

CONCLUSION

By modulating cell cycle control, apoptosis, and DNA repair pathways, phytochemicals may alter radiation response and prepare the cancer cells to more apoptotic death on the other hand phytochemicals protects the normal cells from radiation effects mainly through their antioxidant proportion. The future perspectives lie in identifying more such compounds and elucidating the mechanism through which they act for developing effective protocol for cancer radiotherapy.

REFERENCES

[1] Poulsen MG, Riddle B, Keller J, Porceddu SV, and Tripcony L. (2008). Predictors of acute grade 4 swallowing toxicity in patients with stages III and IV squamous carcinoma of the head and neck treated with radiotherapy alone. *Radiother and Oncol. 87,* 253–259.

[2] Vaupel P. (2004). Tumor microenvironmental physiology and its implications for radiation oncology. *Seminars in Radiation Oncology. 14,* 98-206.

[3] Sweeney TR. A survey of compounds from the anti-radiation drug development program of the U.S. Army Medical Research Development Command, Walter Reed Army Institute of Research, Washington, DC, 1979.

[4] Gordon AT, and Mcmillan TJ. (1997). A role for molecular radiobiology in radiotherapy. *Clin. Oncol. 9,* 70-78.

[5] Wardman P. (2007). Chemical Radiosensitizers for use in radiotherapy. *Clin. Oncol. 19,* 397-417.

[6] Girdhani S, Bhosle SM, Thulsidas SA, Kumar A, and Mishra KP. (2005). Potential of radiosensitizing agents in cancer chemo-radiotherapy. *J. of Can. Res. Ther. 3,* 129-131.

[7] Maurya DK, Devasagayam TPA, and Nair CKK. (2006). Some novel approaches for radioprotection and the beneficial effect of natural products. *Inter. J. of Exper. Biol. 44,* 93-114.

[8] Simone N L, Nard C M, Soule B P, Albert PS, Guion P, Smith S, Godette D, Crouse NS, Sciuto LC, Cooley-Zgela T, Camphausen KC, Coleman N, and Singh AK. (2008). ntrarectal amifostine during external beam radiation therapy for prostate cancer produces significant improvements in quality of life measured by epic score. *Inter. J. of Radia. Oncol. Biol. and Phy. 70(1),* 90–95.

[9] Landauer MR, Davis HD, Dominitz JA, and Weiss JF. (1987). Doss and time relationships of the radioprotector WR-2721 on locomator activity in mice. *Pharmaco. Biochem. Behav. 27,* 573–576.

[10] Brizel DM, Wasserman TH, and Henke M. (2000). Phase III randomized trial of amifostine as a radioprotector in head and neck cancer. *J. Clin. Oncol. 18,* 3339–3345.

[11] Surh YJ. (2003). Cancer chemoprevention with dietary phytochemicals. *Nature s|ca. 3,* 768-780.

[12] Comini LR, Nunez Montoya SC, Sarmiento M, Cabrera JL, and Arguello GA. (2007). Characterizing some photophysical, photochemical and photobiological properties of photosensitizing anthraquinones. *J. Photochem. and Photobiol. A: Chem. 188 (2-3),* 185-191.

[13] Soobrattee MA, Neergheen VS, Luximon-Ramma A, Aruoma O I, and Bahoruna T. (2005). Phenolics as potential antioxidant therapeutic agents: Mechanism and actions. *Mut. Res. 579,* 200-213.

[14] Arora R, Gupta G, Chawla R, and Sharma RK. (2005). Radioprotection by plant products: present status and future prospects. *Phytomed. Res. 19,* 1-22.

[15] Castillo J, Benavente-Garcia O, Lorente J, Alcaraz M, Redondo A, Ortuno A, and Del Rio JA. (2000). Antioxidant activity and radioprotective effects against chromosonal damage induced *in vivo* by X-rays of flavan-3-ols (procyanidins) from grape seeds (*Vitis vinifera*), Comparative study versus other phenolic and organic compounds. *J. Agricul. and Food Chem. 48,* 1738-1745.

[16] Cotrim AP, Sowers AL, Lodde BM, Vitolo JM, Kingman A, Russo A, Mitchell JB, and Baum BJ. (2005). Kinetics of tempol for prevention of xerostomia following head and neck irradiation in a mouse model. *Clin. Cancer Res. 11(20),* 7564-7568.

[17] Srinivasan M, Rajendra Prasad N, and Menon VP. (2006). Protective effect of curcumin on γ-radiation induced DNA damage and lipid peroxidation in cultured human lymphocytes. *Mut. Res. 611,* 96–103.

[18] Landauer MR, Srinivasan SV, and Seed TM. (2003). Genistein treatment protects mice from ionizing radiation injury. *J. of App. Toxicol. 23,* 379-385.

[19] Rajendra Prasad N, Menon VP, Vasudev V, and Pugalendi KV. (2005). Radioprotective effect of sesamol on γ-radiation induced DNA damage, lipid peroxidation and antioxidants levels in cultured human lymphocytes. *Toxicol. 209,* 225–235.

[20] Parihar, Prabhakar KR, Veerapur VP, Sudheer Kumar M, Rosi Reddy Y, Ravi Joshi, Unnikrishnan MK, and Mallikarjuna Raoa C. (2006). Effect of sesamol on radiation-induced cytotoxicity in Swiss albino mice. *Mut. Res. 611,* 9–16.

[21] Rajendra Prasad N, Srinivasan M, Pugalendi KV, and Menon VP. (2006). Protective effect of ferulic acid on γ-radiation-induced micronuclei, dicentric aberration and lipid peroxidation in human lymphocytes. *Mut. Res. 603,* 29–134.

[22] Maurya D K, Salvi VP, and Nair CK. (2005). Radiation protection of DNA by ferulic acid under *in vitro* and *in vivo* conditions. *Mol. Cell Biochem. 280(1-2),* 209-217.

[23] Srinivasan M, Ram Sudheer A, Raveendran Pillai K, Raghu Kumar P, Sudhakaran PR, and Menon VP. (2007). Lycopene as a natural protector against γ-radiation induced DNA damage, lipid peroxidation and antioxidant status in primary culture of isolated rat hepatocytes *in vitro*. *Biochim. Ethno. Biophy. Acta. 1770(4),* 659-665.

[24] Rithidech KN, Tungjai M, and Whorton EB. (2005). Protective effect of apigenin on radiation-induced chromosomal damage in human lymphocytes. *Mut. Res. 585(1-2),* 96-111.

[25] Gandhi NM, Maurya DK, Salvi V, Kapoor S, Mukherjee T, and Nair CKK. (2004). Radioprotection of DNA by glycyrrhizic acid through scavenging free radicals. *J. Radiat. Res. 45:* 461.

[26] Venkatesha VA, and Jagetia GC. (2004). Reduction in the radiation-induced DNA damage and repair enhancement by mangiferin in culture human peripheral blood lymphocytes. *Indian J. of Radia Res. 1,* 21.

[27] Jagetia TGC, and Reddy TK. (2005). Modulation of radiation-induced alteration in the antioxidant status of mice by naringin. *Life Scie. 77,* 780–794.

[28] Jagetia GC, Venkatesha VA. (2006). Treatment of mice with stem bark extract of Aphanamixis polystachya reduces radiation-induced chromosome damage. *Int. J. Radiat. Biol. 82(3),* 197-209.

[29] Benkovic V, Horvat Knezevic A, Dikic D, Lisicic D,. Orsolic N, Basic I, Kosalec I, and Kopjar N. (2008). Radioprotective effects of propolis and quercetin in γ-irradiated mice evaluated by the alkaline comet assay. *Phytomed. x,* xx–xx.

[30] Hosseinimehr, Emami S, Taghdisi SM, and Akhlaghpoor S. (2008). 5,7-Dihydroxychromone-2-carboxylic acid and it's transition-metal (Mn and Zn) chelates as non-thiol radioprotective agents. *Europ. J. of Med. Chem. 43,* 557-561.

[31] Zheng H, Chen ZW, Wang L, Wang SY, Yan YQ, Wu K, Xu QZ, Zhang SM, Zhou PK. (2008). Radioprotection of 4-hydroxy-3,5- imethoxybenzaldehyde (VND3207) in culture cells is associated with minimizing DNA damage and activating Akt. *Europ. J. of Pharmaceut. Scie. 33,* 52–59.

[32] Nair CK, and Salvi V P. (2008). Protection of DNA from γ-radiation induced strand breaks by Epicatechin. *Mut. Res. 650,* 48–54.

[33] Parihar, Prabhakar KR, Veerapur VP, Priyadarsini KI, Unnikrishnan MK, and Mallikajuna Rao C. (2007). Anticlastogenic activity of morin against whole body γ-irradiation in Swiss albino mice. *Europ. J. of Pharm. 557,* 58–65.

[34] Li CR, Zhou Z, Zhu D, Sun YN, Dai JM, and Wang SQ. (2007). Protective effect of paeoniflorin on irradiation-induced cell damage involved in modulation of reactive oxygen species and the mitogen-activated protein kinases. *Interl. J. Biochem. and Cell Biol. 39,* 426–438.

[35] Parihar, Dhawan J, Kumar S, Manjula SN, Subramanian G, Unnikrishnan, MK, and Mallikarjuna Rao C. (2007). Free radical scavenging and radioprotective activity of dehydrozingerone against whole body γ-irradiation in Swiss albino mice. *Chem-Biol. Interac. 170,* 49–58.

[36] Fuentes JL, Vernhe M, Cuetara EB, Sánchez-Lamar A, Santana JL, and Llagostera M. (2006). Tannins from barks of *Pinus caribaea* protect Escherichia coli cells against DNA damage induced by γ-rays. *Fitoterapia. 77,* 116–120.

[37] Mozdarani H, and Ghoraeian P. (2008). Modulation of γ-ray-induced apoptosis in human peripheral blood leukocytes by famotidine and vitamin C. *Mut. Res. 649,* 71–78.

[38] Shimoi K, Masuda S, Shen B, Furugori M, and Kinae N. (1996). Radioprotective effects of antioxidative plant flavonoids in mice. *Mut. Res. 350,* 153-161.

[39] Checker R, Chatterjee S, Sharma D, Gupta S, Variyar P, Sharma A, and Poduval TB. (2008). Immunomodulatory and radioprotective effects of lignans derived from fresh nutmeg mace (Myristica fragrans) in mammalian splenocytes. *Inter. Immunopharm.* 8, 661–669.

[40] El-Missiry MA, Fayed TA, El-Sawy MR, and El-Sayed AA. (2007). Ameliorative effect of melatonin against γ-irradiation-induced oxidative stress and tissue injury. *Ecotoxicol. and Environ. Saf. 66,* 278–286.

[41] Burlage FR, Roesink JM, Hette Faber, Vissink A, Langendijk JA, Kampinga HH, and Coppes RP. (2008). Optimum dose range for the amelioration of long term radiation-induced hyposalivation using prophylactic pilocarpine treatment. *Radiother. and Oncol. 86,* 347–353.

[42] Pradeep K, Park SH, Cheol Ko K, and Choi SM. (2008). Hesperidin a flavanoglycone protects against γ-irradiation induced hepatocellular damage and oxidative stress in Sprague-Dawley rats. *Europ. J. of Pharmacol.* xx, xx-xx.
[43] Ramos FM, Pontual ML, de Almeida SM, Boscolo FN, Tabchoury CP, and Novaes PD. (2006). Evaluation of radioprotective effect of vitamin E in salivary dysfunction in irradiated rats. *Arch. Oral Biol. 51(2),* 96-101.
[44] Bala M, Goel HC. (2004). Radioprotective effect of podophyllotoxin in Saccharomyces cerevisiae. *J. Environ. Pathol. Toxicol. Oncol. 23(2),* 139-44.
[45] Zavodnik LB. (2003). Isoflavone genistein-8-c-glycoside prevents the oxidative damages in structure and function of rat liver microsomal membranes. *Radiats Biol. Radioecol. 43(4),* 432-438.
[46] Shi WM, Zhao XP, and Lu T. (2006). Radioprotective effect of catechines against radiation injury in mice. Nan Fang Yi, Ke Da Xue, Xue Bao. 26(11): 1621-2.
[47] Li CR, Zhou Z, Zhu D, Sun YN, Dai JM, and Wang SQ. (2007). Protective effect of paeoniflorin on irradiation-induced cell damage involved in modulation of reactive oxygen species and the mitogen-activated protein kinases. *Int. J. Biochem. Cell Biol.* 39(2), 426-438.
[48] Maurya DK, Adhikari S, Nair CKK, and Devasagayam TP. (2007). DNA protective properties of vanillin against γ-radiation under different conditions: Possible mechanisms. *Mut. Res. 634,* 69–80.
[49] Nayak V, and Devi UP. (2005). Protection of mouse bone marrow against radiation induced chromosome damage and stem cell death by the Ocimum Flavonoids orientin and vicenin. *Radiat. Res. 163,* 165-171.
[50] Shimoi K, Masuda S, Furogori M, Esaki S, and Kinae N. (1994). Radioprotactive effect of antioxidative flavonoids in γ-ray irradiated mice. *Carcinogen. 15,* 2669-2672.
[51] Baatout S, Jacquet P, Derradji H, Ooms D, Michaux A, and Mergeay M. (2004). Study of the combined effect of X-irradiation and epigallocatechin-galate (a tea component) on the growth inhibition and induction of apoptosis in human cancer cell lines. *Oncol. Res. 12,* 159-167.
[52] Inano H, and Onoda M. (2002). Radioprotective action of curcumin extracted from *Curcuma longa* Linn: Inhibitory effect of formation of urinary 8-hydroxy-2'-deoxyguanosine, tumorigenesis, but not mortality induced by γ-ray irradiation. *Inter. J. of Radiat. Oncol. Biol. Phys. 53,* 735-743.
[53] Abraham SK, Sarma L, and Kesavan PC. (1993). Protective effects of chorogenic acid, curcumin and beta-carotene against γ-radiation-induced *in vivo* chromosomal damage. *Mutat. Res. 303,* 109-112.
[54] Landauer MR, Srinivasan V, Shapiro A, Takimote C, and Seed TM. (2000). Protection against lethal irradiation by genistein. *Inter. J. of Toxicol. 19,* 37-40.
[55] Conklin KA. (2002). Dietary polyunsaturated fatty acids: impact on cancer chemotherapy and radiation. *Altern. Med. Rev. 7,* 4–21.
[56] Rajakumar DV, and Rao MN. (1993). Dehydrozingerone and isoeugenol as inhibitors of lipid peroxidation and as free radical scavengers. *Biochem. Pharmacol. 46,* 2067–2072.
[57] Zhang R, Kang KA, Kang SS, Park JW, Hyun JW. (2011). Morin (2′,3,4′,5,7-pentahydroxyflavone) protected cells against γ-radiation-induced oxidative stress. *Basic Clin. Pharmacol. Toxicol. 108(1),* 63–72.

[58] Mantena SK, Unnikrishnan MK, Joshi R, Radha V, Uma Devi P, and Mukherjee T. (2008). *In vivo* radioprotection by 5-aminosalicylic acid. *Mut. Res. 650,* 63–79.

[59] Sudheer Kumar M, Unnikrishnan MK, and Uma Devi P. (2003). Effect of 5-aminosalicylic acid on radiation-induced micronuclei in mouse bone marrow. *Mut. Res. 527,* 7–14.

[60] Park E, Lee NH, Joo HG, and Jee Y. (2008). Modulation of apoptosis of eckol against ionizing radiation in mice. *Biochem. Biophys. Res. Com. 372,* 792–797.

[61] Wang ZW, Zhou JM, Huang ZS, Yang AP, Liu ZC, Xia YF, Zeng YX, and Zhu XF. (2004). Aloe polysaccharides mediated radioprotective effect through the inhibition of apoptosis. *J. Radiat. Res. 45,* 447–454.

[62] Hillman GG, Wang Y, Kucuk O, Che M, Doerge DR, Yudelev M. Joiner MC, Marples B, Forman JD, and Sarkar FH. (2004). Genistein potentiates inhibition of tumor growth by radiation in a prostate cancer orthotopic model. *Mol. Can. Ther. 3,* 1271-1279.

[63] Wang Y, Raffoul JJ, Che M, Doerge DR, Joiner MC, Kucuk O, Sarkar FH, and Hillman GG. (2006). Prostate cancer treatment is enhanced by genistein *in vitro* and *in vivo* in a syngeneic orthotopic tumor model. *Radiat. Res. 166,* 73-80.

[64] Hillman GG, Wang Y, Che M, Raffoul JJ, Yudelev M, Kucuk O, Sarkar FH. (2001). Progression of renal cell carcinoma is inhibited by genistein and radiation in an orthotopic model. *BMC Can. 7,* 4.

[65] Raffoul JJ, Sarkar FH, and Hillman GG. (2007). Radiosensitization of Prostate Cancer by Soy Isoflavones. *Current Can. Drug Targ. 7,* 759-765.

[66] Zhang W, Anker L, Law RE, Hinton DR, Gopalakrishna R, Qian Pu, Gundimeda U, Weiss MH, and Couldwell WT. (1996). Enhancement of Radiosensitivity in Human Malignant Glioma Cells by Hypericin *in Vitro. Clin. Can. Res. 2,* 843-846.

[67] Vanden bogaerde AL, Delaey EM, Vantieghem AM, Himpens BE, Merlevede WJ, and de Witte PA. (1998). Cytotoxicity and antiproliferative effect of hypericin and derivatives after photosensitization. *J. of Photochem. and Photobiol. A: Chem. 67,* 119–125.

[68] Bhoslea SM, Huilgola NG, and Mishra KP. (2005). Enhancement of radiation-induced oxidative stress and cytotoxicity in tumor cells by ellagic acid. *Clin. Chim. Acta. 359,* 89–100.

[69] Chen YJ, Liao HF, Tsai TH, Wang SY, and Shiao MS. (2005). Caffeic acid phenethyl ester preferentially sensitizes CT26 colorectal adenocarcinoma to ionizing radiation without affecting bone marrow radioresponse. *Inter. J. Radia. Oncol. Biol. Phy. 63(4),* 1252-1261.

[70] Hillman GG, Raffoul JJ, Wang Y, Kucuk O, Joiner MC, Yudelev M, Forman JD, Che M, and Sarkar FH. (2006). Radiosensitization of prostate cancer by genistein *in vitro* and in an orthotopic model *in vivo. Radiother. and Oncol. 78,* S48-S49.

[71] Lacroixa M, Borsab J, Chiassona F, and Ouattaraa B. (2004). The influence of atmosphere conditions on Escherichia coli and Salmonella typhi radiosensitization in irradiated ground beef containing carvacrol and tetrasodium pyrophosphate. *Radia. Phy. and Chem. 71,* 59–62.

[72] Boloor KK, Kamat JP, and Devasagayam TPA. (2000). Chlorophyllin as a protector of mitochondrial membranes against γ-radiation and photosensitization. *Toxicol. 155,* 63–71.

[73] Scarlatti F, Sala G, Ricci C, Maioli, C, Milani F, Minella M, Botturi M, and Ghidoni R. (2007). Resveratrol sensitization of DU145 prostate cancer cells to ionizing radiation is associated to ceramide increase. *Can. Lett. 253,* 124–130.

[74] McAleer, Duffy KT, Davidson WR, Kari G, Dicker AP, Rodeck U, and Wickstrom E. (2006). Antisense inhibition of cyclin D1 expression is equivalent to flavopiridol for radiosensitization of zebrafish embryos. *Inter. J. of Radia. Oncol. Biol. Phy. 66(2),* 546-551.

[75] Preetha A, Chitra S, Sonia J, Kunnumakkara AB, and Aggarwa BB. (2008). Curcumin and cancer: An ''old-age'' disease with an ''age-old'' solution. *Can. Let. 267, 133–164.*

[76] Kasten-Pisula U, Windhorst S, Dahm-Daphi J, Mayr G, and Dikomey E. (2007). Radiosensitization of tumour cell lines by the polyphenol Gossypol results from depressed double-strand break repair and not from enhanced apoptosis. *Mol. Radiobiol. 83,* 296–303.

[77] Zoberi I, Bradbury CM, Curry HA, Bisht KS, Goswami PC, and Roti JL. (2002). Radiosensitizing and antiproliferative effects of resveratrol in two human cervical tumor cell lines. *Can. Let. 175,* 65–173.

[78] Sandhya T, and Mishra KP. (2006). Cytotoxic response of breast cancer cell lines, MCF 7 and T 47 D to triphala and its modification by antioxidants. *Can. Lett. 238,* 304–313.

[79] Pandey BN, and Mishra KP. (2004). Modification of thymocytes membraneoxidative damage and apoptosis by eugenol. *J. of Environ. Pathol. Toxicol. Oncol. 23,* 117–122.

[80] Coleman CN, Stone HB, Modulder JE, and Pellmar TC. (2004). Modulation of radiation injury. *Scie. 304,* 693-694.

[81] Yamamoto T, Hsu S, Lewis J, Wataha J, Dickinson D, and Singh B. (2003). Green tea polyphenol causes differential oxidative environments in tumor verses normal epithelial cells. *J. Pharmacol. and Exper. Ther.* 307, 230–236.

[82] Vermund H, and Gollin FF. (1968). Mechanisms of action of radiotherapy and chemotherapeutic adjuvants (A). *Can. 21,* 58-76.

[83] Metodiewa D, Jaiswal AK, Cenas N, Dickancaite E, and Aguilar JS. (1999). Quercetin may act as a cytotoxic prooxidant after its metabolic activation to semiquinone and quinoidal product. *Free Rad. Biol. Med. 26,* 107–16.

[84] Srinivasan M, Ram Sudheer A, Rajasekaran KN, and Menon VP. (2008). Effect of curcumin analog on γ -radiation induced cellular changes in primary culture of isolated rat hepatocytes *in vitro. Chemico-Biol. Interact. x,* xx-xx.

[85] Sahu SC, and Washington MC. (1992). Effect of ascorbic acid and curcumin on quercetin-induced nuclear DNA damage, lipid peroxidation and protein degradation. *Can. Let. 63(3),* 237-241.

[86] Chen MF, Keng PC, Lin PY, Yang CT, Liao SK, and Chen WC. (2005). Caffeic acid phenethyl ester decreases acute pneumonitis after irradiation *in vitro* and *in vivo. BMC Cancer. 5,* 158.

[87] Sergediene E, Jonsson K, Szymusiak H, Tyrakowsa B, Rietjensi M, and Cenas N. (1999). Prooxidant toxicity of polyphenolic antioxidants to HL-60 cells description of quantitative structure activity relationships. *FEBS Let. 462,* 392– 406.

[88] Dal-Pizzol F, Ritter C, Klamt F, Andrades M, DA Frota JR, and Diel C. (2003). Modulation of oxidative stress in response to γ-radiation in human glioma cell lines. *J. of Neurooncol. 61(2),* 89 – 94.

[89] George Loo. (2003). Redox-sensitive mechanisms of phytochemical-mediated inhibition of cancer cell proliferation1 (). *J. Nut. Biochem. 14,* 64–73.
[90] Ahmad N, Adhami VM, Afaq F, Feyes DK, and Mukhtar H. (2001). Resveratrol causes WAF-1/p21-mediated G1-phase arrest of cell cycle and induction of apoptosis in human epidermoid carcinoma A431 cells. *Clin. Cancer Res. 7,* 1466–1473.
[91] She QB, Huang C, Zhang Y, and Dong Z. (2002). Involvement of c-jun NH2-terminal kinases in resveratrol-induced activation of p53 and apoptosis. *Mol. Carcinog. 33,* 244–250.
[92] Mouria M, Gukovskaya AS, Jung Y, Buechler P, Hines OJ, Reber HA, and Pandol, SJ. (2002). Food-derived polyphenols inhibit pancreatic cancer growth through mitochondrial cytochrome C release and apoptosis. *Int. J. Cancer. 98*, 761–769.
[93] Maclachlan TK, Sang N, and Giordano A. (1995). Cyclins, cyclin-dependent kinases and CDK inhibitors: implications in cell cycle control and cancer. *Crit. Eukary Gene Expre. 5,* 127–156.
[94] Sherr CJ. (1994). G_1 phase progression: cycling on cue. *Cell. 79,* 551–555.
[95] Pines J. (1991). Cyclins: wheels within wheels. *Cell Growth Diff. 2,* 305–310.
[96] Valenzuela MT, Mateos S, Almodoavar JMR, and Mcmillan TJ. (2000). Variation in sensitizing effect of caffeine in human tumour cell lines after γ-irradiation. *Radiother. Oncol. 54,* 261-271.
[97] Higuchi K, Mitsuhashi N, Saitoh J, Maebayashi K, Sakurai H, Akimoto K, and Niibe H. (2000). Caffeine enhanced radiosensitivity of rat tumor cells with a mutant type p53 by inducing apoptosis in a p53-independent manner. *Can. Lett. 152,* 57-162.
[98] Kim JC, Saha D, Qianwen Cao, and Hak Choy. (2004). Enhancement of radiation by combined docetaxel and flavopiridol treatment in lung cancer cell. *Radiother. and Oncol. 71,* 213-221.
[99] Bible KC, and Kaufmann SH. (1997). Cytotoxic synergy between flavopiridol (NSC 649890, L86-8275) and various antineoplastic agents: the importance of sequence of administration. *Can. Res. 57,* 3375–3380.
[100] Scarlatti F, Sala G, Ricci C, Maioli C, Milani F, Minella M, Botturi M, and Ghidoni R. (2007). Resveratrol sensitization of DU145 prostate cancer cells to ionizing radiation is associated to ceramide increase. *Can. Let. 253,* 24–130.
[101] Dorrah Deeb, Yong X, Hao XU, Jiang Xiaohua GAO, Nalini J, Robert A, Chapman, Subhash C, and Gautam. (2003). Curcumin (Diferuloyl-Methane) Enhances Tumor Necrosis Factor-related Apoptosis-inducing Ligand-induced Apoptosis in LNCaP Prostate Cancer Cells. *Mol. Can. Therap. 2,* 95–103.
[102] Oki T, Sowa Y, Hirose T, Takagaki N, Horinaka M, Nakanishi R, Yasuda C, Yoshida T, Kanazawa M, Satomi Y, Nishino H, Miki T, and Sakai T. (2004). Genistein induces Gadd45 gene and G_2/M cell cycle arrest in the DU145 human prostate cancer cell line. *FEBS Let. 577,* 5–59.
[103] Yashar CM, Spanos WJ, Taylor DD, and Gercel-Taylor C. (2005). Potentiation of the radiation effect with genistein in cervical cancer cells. *Gynecological. Oncol. 99,* 199–205.
[104] Zhivotovsky B, Joseph B, and Orrenius S. (1999). Tumor radiosensitivity and apoptosis. *Experimen. Cell Res. 248,* 10–17.

[105] Hongzhong, Wang, Yu Zhu, Chonghua Li, Liping Xie, Guang Chen, Yancheng Nie, and Rongqing Zhang. (2007). Effects of genistein on cell cycle and apoptosis of two murine melanoma cell lines. *Tsinghual. Scie. and Technol. 2(4),* 372-380.

[106] Baust H, Schoke A, Brey A, Gern U, Marek Los, Schmid RM, Ttinger EMR, and Seufferlein T. (2003). Evidence for radiosensitizing by gliotoxin in HL-60 cells: implications for a role of NF-kB independent mechanisms. *Oncogen. 22,* 8786–8796.

[107] Steel GG, and Peckham MJ. (1979). Internatioinal Journal of Radiation Oncology and Biology Physics. 5, 85–91.

[108] Shimizu M, and Weinstein IB. (2005). Modulation of signal transduction by tea catechins and related phytochemicals. *Mut. Res. 591(1-2),* 147-160.

[109] Zhu AZ, and Willett CG. (2003). Chemotherapeutic and biologic agents as radiosensitizers in rectal cancer. *Sem. In. Rad. Oncol. 13(4),* 54-468.

[110] Varadkar P, Dubey P, and Krishna M. (2001). Modulation of radiation induced protein kinase C activity by phenolics. *J. Radiol. Prot. 21(4),* 361-70.

[111] Horsmana MR, Siemannb DW, Chaplin DJ, and Overgaarda J. (1997). Nicotinamide as a radiosensitizer in tumours and normal tissues:the importance of drug dose and timing. *Radiother. and Oncol. 45,* I67-174.

[112] Thangapazham RL, Sharma A, and Maheshwari RK, (2006). Multiple molecular targets in cancer chemoprevention by curcumin. *The AAPS J. 8(3),* 43-449.

[113] Krishnan S, Sandur SK, Shentu S, Aggarwal BB. (2006). Curcumin. *Inter. J. of Radia Oncol. Biol. Phy. 66,* S547.

[114] Webster RP, Gawde MD, and Bhattacharya RK. (1996). Protective effect of rutin, a flavonol glycoside, on the carcinogen-induced DNA damage and repair enzymes in rats. *Can. Let. 109(1-2),* 185-91.

[115] Scolastici C, Alves De Lima RO, Barbisan LF, Ferreira AL, Ribeiro AD, and Salvadori DMF. (2007). Lycopene activity against chemically induced DNA damage in Chinese hamster ovary cells. *Toxicol. in Vitro. 21,* 840–845.

[116] Grace MB, Blakely WF, and Landauer MR. (2007). Genistein-induced alterations of radiation-responsive gene expression. *Rad. Measurements. 42(6-7),* 1152-1157.

[117] Park OJ, Shin J I. 2004. Proapoptotic potentials of genistein under growth stimulation by estrogen. *Annuvel. New York Acad. Scie. 1030,* 410–418.

[118] Sarkar FH, and Li Y. (2002). Mechanisms of cancer chemoprevention by soy isoflavone genistein. *Can. Metast. 21,* 265–280.

[119] Sarkar FH, and Li Y. (2004). The role of isoflavones in cancer chemoprevention. *Front. Biosci. 9,* 2714–2724.

[120] Bionda C, Hadchity E, Alphonse G, Chapet O, Rousson R, Rodriguez-Lafrasse C, and Ardail D. (2007). Radioresistance of human carcinoma cells is correlated to a defect in raft membrane clustering. *Free Radical. Biol. and Med. 43(5),* 681-694.

[121] She QB, Bode AM, Ma WY, Chen NY, and Dong Z. (2001). Resveratrol-induced Activation of p53 and Apoptosis Is Mediated by Extracellular-Signal-regulated Protein Kinases and p38 Kinase. *Can. Res. 61(15),* 604–1610.

[122] Stenzel RA, Rajesh D, and Howard SP. (2002). Flavopiridol as a radiosensitizer in malignant gliomas (abstract). *Proc. Am. Ass. Can. Res. 43,* A975.

In: Natural Products and Their Active Compounds ... ISBN: 978-1-62100-153-9
Editors: M. Essa, A. Manickavasagan, and E. Sukumar © 2012 Nova Science Publishers, Inc.

Chapter 9

NATURAL PRODUCTS IN THE PREVENTION AND TREATMENT OF INVASIVE HUMAN DISEASES

Majekodunmi O. Fatope[*] and Saleh N. Al Busafi
Chemistry Department, College of Science,
Sultan Qaboos University, Al Khod, Muscat, Oman

ABSTRACT

The diseases confronting human race are many and a number of them, which are invasive, can be grouped into three major categories - degenerative, parasitic and microbial. Historically, herbal preparations have been used to manage invasive human diseases. At present, the majority of chemical substances commercially available for treating invasive diseases are dietary supplements, natural products, and synthetic drugs modeled after natural products.

This chapter provides an outlook for the role and mode of action of natural products and selected herbs used to treat invasive diseases linked to oxidative stress, cancer, malaria, onchocerciasis, leishmaniasis, and microbes. Considering the therapeutic spectrum and potency of natural products currently in use for the treatment of invasive diseases, a combination of dietary supplements, herbs, natural products, synthetic drugs and life style will continue to play important roles in the treatment of invasive diseases.

1. INTRODUCTION

Natural products are chemical entities synthesized by living organisms through complex enzyme-mediated biochemical reactions [1]. They have nutritional, medicinal or unknown human benefits. Historically, humans have learned through careful observation of nature to use natural products in forms of herbs, dietary supplements, spices, and alcoholic beverages to prevent or cure invasive human diseases [2, 3]. Arthropods, reptiles and amphibians also use natural products [4] as chemical weapons against predators; and plants have also been

[*] Correspondence author: Tel: + 968 24141491; fax: + 969 24141469. E-mail address: majek@squ.edu.om

observed to use natural products to ward off invasive organisms or suppress other plants [5]. Natural products produced by microbes play important roles in the modulation of the microenvironment of the organisms that produce them [6].

Since the mid-19th century, natural products with extraordinary chemical structures and biological activities have been isolated from plants, fungi, bacteria [7] marine organisms [8] and bugs [9] and tested for anti-cancer, antioxidant, antimicrobial and anti-parasitic activities. Some of these chemical entities have been clinically approved, and regularly modified and used to treat invasive human diseases.

1.1. Invasive Human Diseases

Invasive human diseases are caused by multiple factors and several organisms, and once established in the body, they can spread to different parts of the body and to other organisms in a non-contagious or contagious manner. The majority of invasive and non-contagious diseases are degenerative, cancerous, or parasitic . The causes of neurodegenerative diseases are not clearly understood but their expressions are clinically unmistakable. Neuro-degenerative diseases like Alzheimer's, Huntington's, Parkinson's, and Prion's have complex pathogenesis but radicals, trace metals and life style are generally accepted to play crucial roles in their development and complications [10]. Cancerous diseases are non-contagious. The genetic disposition of individuals, diet, and environmental factors play key roles in their development. Cancerous growth usually starts at a primary site such as the kidney, skin, lung, breast, prostrate, testis, and gastrointestine and invades a secondary site such as the bone, liver, brain, ovary, thyroid and bowl. The pathogenesis and metastasis of cancer are fundamentally known, but the discovery of target-specific and non-toxic natural product for cancer chemotherapy has so far been elusive.

Malaria, onchocerciasis, schitosomiasis, trypanosomiasis, leishmaniasis are invasive diseases caused by parasitic protozoans. They are non-contagious and are spread by insects or mollusks. However, invasive diseases such as Severe Acute Respiratory Syndrome (SARS), Acquired Immune-Deficient Syndrome (AIDS), Sexually Transmitted Diseases (STDs), Legionnaires' disease, Tuberculosis (TB), leprosy, skin diseases are caused by microbial organisms. They invade the body and multiply once they are inside and are spread through physical or sexual contact.

In general, natural products in forms of herbs, spices, drugs or dietary supplements have been used to delay the progression of neurodegenerative diseases, kill or inhibit cancerous growths and eliminate parasitic protozoan.

2. Natural Chemo-Preventive Chemical Entities from Dietary Sources

The ability of natural chemical substances to interact with biological systems and prevent, suppress or reverse the development of invasive diseases is broadly referred to as chemoprevention. Natural foods have been speculated to contain 5000-25,000 phytochemicals which interact in a complex but little understood pathways to delay or prevent the progression of invasive diseases linked to oxidative stress and cancer. They have been demonstrated to produce therapeutic effects, which are superior to single bioactive compounds in animal model experiments of degenerative diseases, and to play important roles in the prevention or progression of cancer [11, 12, 13]. This knowledge is not new to traditional healers who think of food as medicinal. Evidences from the teachings of Hippocrates and his followers in Greece as far back as 420 BC, Ayurvedic medicine and beliefs of Indians and Sri Lankans, Chinese traditional medicine, herbal medicine of Africans and Arabs, and publications of holistic nutritionist, Mike Adam [14] have attested to the potential healing power of food. Scientists have now discovered that the consumption of natural foods [15, 16] can decrease the risk of neurodegenerative or cancerous diseases.

For instance, vitamins A, C, and E, which occur naturally in fruits and vegetables, have been demonstrated to have antioxidant and other chemo-preventive activities *in vitro* [17]. Cruciferous vegetables such as broccoli, watercress, cabbage, radish, turnip, garden cress, contain isothiocyanates in the form of glucosinolates [18]; sulforaphane, an isothiocyanate from broccoli [19] is endowed with strong cancer chemo-preventive activities. Apples (*Malus* sp.,) (Rosaceae) are rich in oligomeric procyanidins; green or black tea, grape, berry and red wine contain polyphenols, epigallocatechin, and anthocyanidines which exert cancer chemo-preventive [19, 20] and antioxidant [17] properties through multiple mechanisms of action. Flavonol in fruits and vegetables, flavanone in citrus fruits, flavanol in tea leaves, isoflavanone in legumes, and anthocyanin in berries and red grapes have all been demonstrated to be effective in aged male rats' model of Alzheimer's disease [21, 22]. Of the commonly eaten vegetables, the order of decreasing antioxidant capacity, based on laboratory tests is garlic, asparagus, spinach, beet, pepper, mushroom, broccoli, cabbage, corm, onion, beans, carrot, cauliflower, sweet potato, tomato, lettuce, squash, celery and cucumber. Pomegranate and berries [23] have the highest cellular antioxidant activity [24] amongst the 25 fruits commonly consumed in the United States.

Edible African walnut, *Tetracarpidium conophorum* Mull. or *Ricinodendron heutelotii* (Bail.)(Euphorbiaceae)('ukpa' in Ibo and 'awusa' or 'asala' in Yoruba) is rich in oil with high contents of linolenic, α-linolenic, and α-elaeostearic acids [25]; and these acids and oil have been demonstrated to lower the cholesterol level of blood in rats, thus suggesting some role in the prevention of invasive diseases of the cardiovascular system.

2.1. Natural Chemo-Preventive Agents: Mode of Action

Fundamental knowledge about the causes and development of degenerative diseases is important to our appreciation of the mode of action of chemo-preventive agents. The etiology of degenerative diseases like Alzheimer's, Parkinson's, arthritis, cancer, atherosclerosis and AIDS are complex, however, oxidative stress, mitochondria dysfunction, aggregation and expression of mutant genes, are believed to play some roles in their initiation, progression and associated complications. Once established, these diseases develop in stages that include incubation or initiation, progression, and malignant or virulent segments [26]. The incubation period depends on the type of disease and during this period the disease is difficult to detect. Natural chemo-preventive agents can intervene through multiple modes of action at different phases to delay the progression of disease.

The majority of natural chemo-preventive agents are phytochemicals [27] with anti-oxidant or antiradical [28, 29] properties. Antioxidants can prevent degenerative diseases and cancer by inhibiting the production of radicals, chelating pro-oxidant cations, and quenching free radicals through hydrogen or electron transfer processes. Free radicals have short half-life ($t_{1/2} = 10^{-9}$ s) and are highly reactive chemical species. They can "steal electrons" from lipids, organs and tissues, causing them to malfunction [28].

Lipids are the main component of the cell membrane that regulates what enters and passes out of the cell. Damage to lipids can impair the structural function of cells, organs and tissues. Free radicals in a cell system can also damage DNA and proteins, leading to cleavage of molecular bonds [30]. Reactive oxygen species (ROS) such as singlet oxygen ($^{\cdot}O_2$), peroxy ($ROO^{\cdot}$), alkoxy ($RO^{\cdot}$), hydroxyl ($^{\cdot}OH$) radicals, superoxide anion ($O_2^{\cdot-}$) and hydrogen peroxide (H_2O_2) are formed as by products of normal metabolic activities during cell division and production of ATP [31]. ROS and reactive nitrogen species (RNS) ($NO^{\cdot}$) can also be generated in the body in response to chemical stress from polluted environment; and they can abstract hydrogen radicals from the hydrocarbon chain of lipids thus generating new free radical sites which are susceptible to peroxidation [32] and damage. The amount of ROS and RNS produced in the body are regulated by enzymes [32] such as gluthatione peroxidase, superoxide dismutase, glutathione peroxidase and catalase, and by chemical substances like albumin, gluthathione, histidine peptides, iron binding proteins, hormones, reduced CoQ10 factor, urates and natural products present in diet [33].

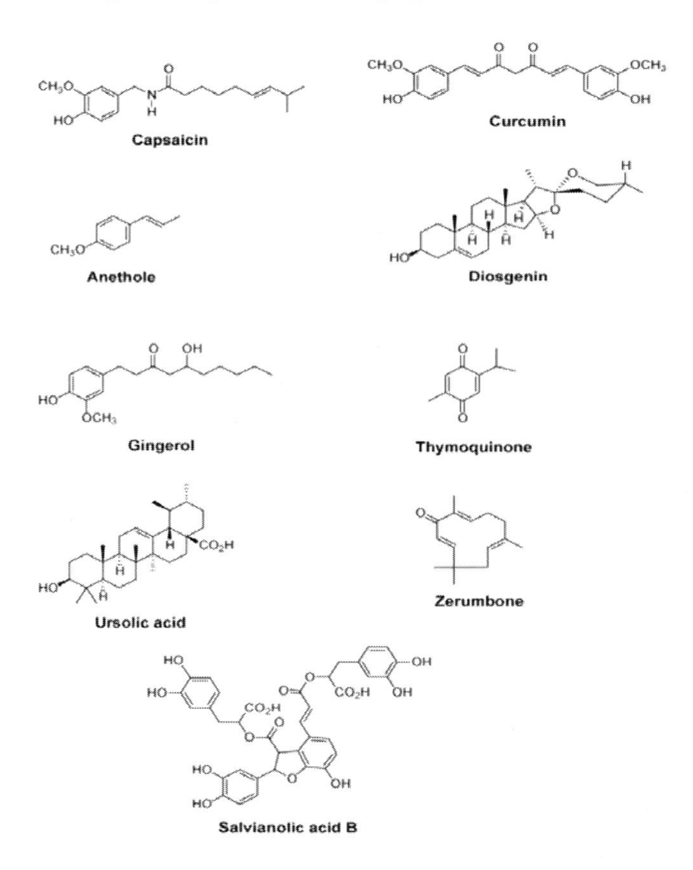

Figure 1. Structures of antioxidants derived from spices.

For instance, flavonoids, isoflavones, flavanols, phenols, vitamins A, G and E found in vegetable and fruits can break the chain reactions initiated by free radicals [15, 34] by using carbonyl and hydroxyl functional groups to chelate harmful zinc and copper trace metals, and conjugated double bonds systems to divert radicals from susceptible targets. Flavonoids have lower redox potentials than ROS and RON [35] and can thus reduce them to harmless products. The chemical structures of natural products isolated from spices, vegetables and fruits [36-38] which have been shown in laboratory experiments to exert antioxidant and cancer-chemo preventive actions [39, 40, 41] are shown in Figures 1 and 2.

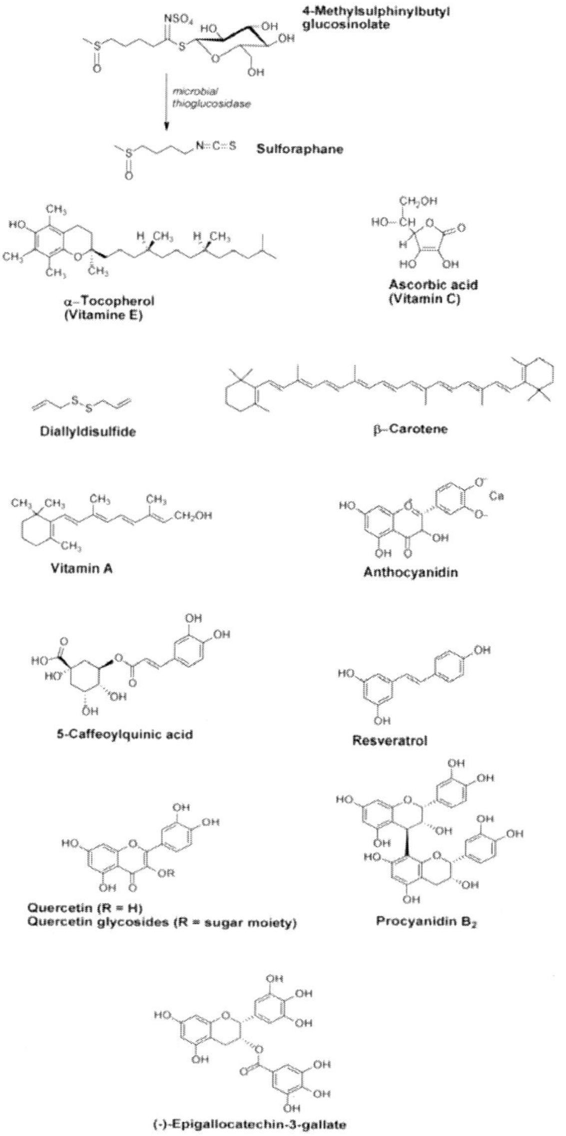

4-Methylsulphinylbutyl glucosinolate

microbial thioglucosidase

Sulforaphane

α-Tocopherol (Vitamine E)

Ascorbic acid (Vitamin C)

Diallyldisulfide

β-Carotene

Vitamin A

Anthocyanidin

5-Caffeoylquinic acid

Resveratrol

Quercetin (R = H)
Quercetin glycosides (R = sugar moiety)

Procyanidin B_2

(-)-Epigallocatechin-3-gallate

Figure 2. Structures of antioxidants from vegetables and fruits.

At present, there is no proven chemical entity that can be prescribed for the general population as a cure for degenerative diseases as fluoride is to dental caries. Epidemiological and animal model investigations have established an inverse relationship between vegetable chemicals and degenerative diseases but clinical evidence in human is limited and often controversial [42].

3. Natural Anti-Cancer Agents and Their Mode of Action

Cancer is a group of diseases developed from uncontrolled proliferation of cells. This condition can be initiated by inherited mutated genes, exposure to pro-oxidant chemicals, microbial toxins, UV irradiation and smoking. Anti-cancer agents are chemical entities used to stop the proliferation of cancerous cells.

According to recent analysis, about 67% of drugs that are effective in treating various forms of cancer are derived from plants [43]. Amongst the most effective cancer chemotherapy agents are vinblastine, vincristine, taxol (paclitaxel), etoposide, teniposide and camptothecin (Figure 3).

Vinblastine and Vincristine

The vinca alkaloids, vinblastine and vincristine, were isolated from *Catharanthus roseus* L. G. Don (Apocynacea), a Madagascar periwinkle [44]. They are used to treat cancer of the lymph glands, spleen and liver.

Taxol (Paclitaxel)

Taxol was originally obtained from the bark of *Taxus brevifolia* Nutt (Taxaceae), a yew tree from the Pacific Northwest [45]. It exerts anti-cancer properties by binding to microtubules during cell division thus disrupting normal mitosis and blocking the proliferation of cell [46].

Etoposide And Teniposide

These are semi-synthetic products, derived from podophyllotoxin, a natural product isolated from the root of *Podophyllum* spp (Berberidacea) [47]. They act as anti-cancer agent by inhibiting the enzyme topoisomerase II, which is responsible for the synthesis and replication of DNA.

Camptothecin

Camptothecin was isolated from the bark of *Camptotheca acuminate* Decne (Nyssaceae), a tree that is endemic to Tibet and West China. Camptothecin is used to treat ovarian cancer and acts by inhibition of the enzyme topoisomerase I involved in the reassembly and replication of DNA.

Several anti-cancer chemical entities from plants [7] and marine organisms [8], which are potentially useful for the treatment of cancer, have reached clinical trials [48]. Plants from the families Annonceae, Apocynaceae, Clusiaceae, Euphorbiaceae, Fabaceae (Leguminaceae) Flacourtiaceae, Meliaceae, Myrsinaceae, Myrtaceae, Rubiaceae, Sapindinacea and Solanaceae have been recognized as "hot" terrestrial plants from which potent anti-cancer chemical entities [7] can be obtained based on laboratory tests.

Some anti-tumor drugs have also been obtained from microbes. Adriamycin (Doxorubicin) is a secondary metabolite produced by the cultures of *Streptomyces peucetius* var *caesius*. It has a broad spectrum of anti-tumor activities and is clinically used to treat a variety of solid tumors, acute leukaemias and lymphomas. It acts by intercalating between the base pairs of DNA, thus inhibiting enzyme topoisomerase II that is responsible for cleaving and reassembling double stranded DNA during replication.

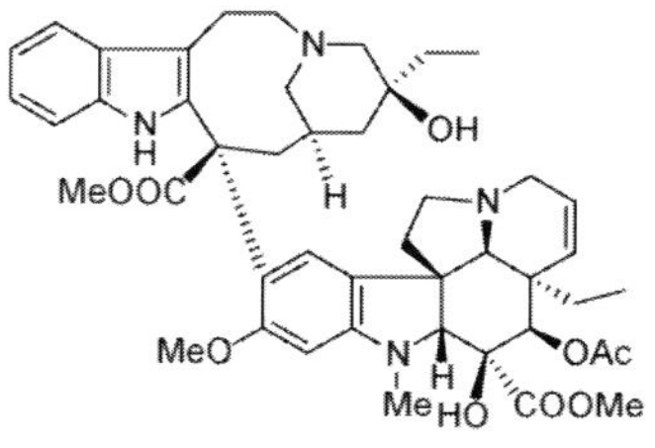

Vinblastine

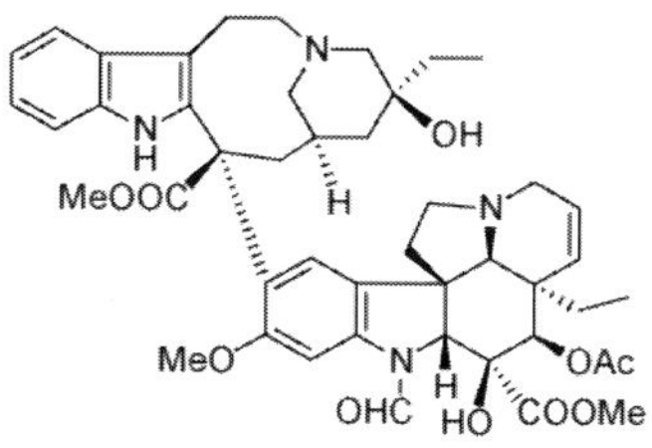

Vincristine

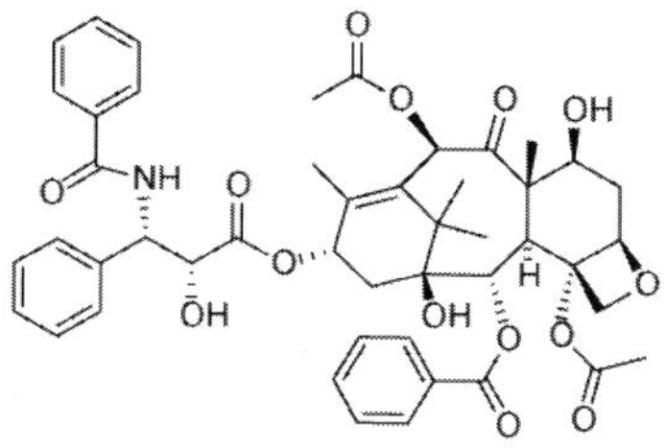

Paclitaxel

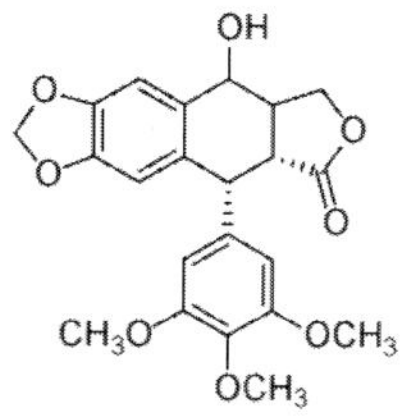

Podophyllotoxin

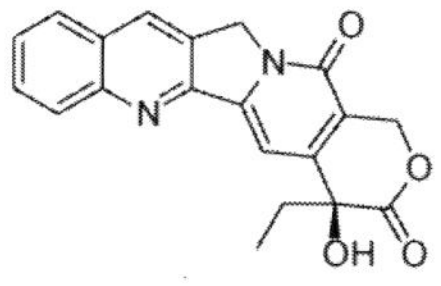

Camptothecin

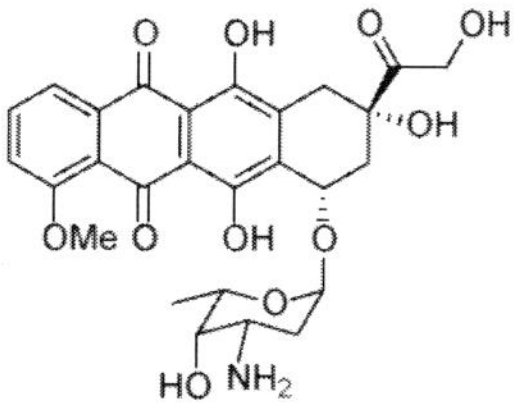

Doxorubicin (Adriamycin)

Figure 3. Structures of natural products used to treat cancer.

4. NATURAL ANTI-PARASITES

Parasitic diseases are a scourge that incapacitates the inhabitants of the developing countries. The most prominent amongst them in the tropics are malaria, African trypanosomiasis, Chagas disease, schistosomiasis, onchocerciasis and leshmaniasis. Several of these diseases are transmitted by insects, mollusks, and other vectors. They are the so called 'neglected diseases' and remain a major health problem in the poorest regions of the world.

4.1. Natural Products in Malaria Chemotherapy

Malaria is caused by protozoal parasites, *Plasmodium falciparum. P. vivax, P. ovale. P. malriae*, and it is the most important parasitic disease in the tropics. Malaria is endemic to about 100 countries and transmitted through bites of female *Anopheles* mosquitoe. *P. faciparum* has devastated several communities in Sub-Saharan West Africa, South East Asia and South America, putting about 2.2 billion people at risk [49, 50]. Medicinal plants have been used to treat malaria for centuries in West Africa [51, 52], Brazil and sub-regions [53], East Africa, Turkey, Cuba, South Vietnam, Ethiopia, Puerto Rico and Bolivia [54]. Two of the world's most valuable anti-malarial drugs, quinine and artemisinin (Figure 4), are natural products from plants.

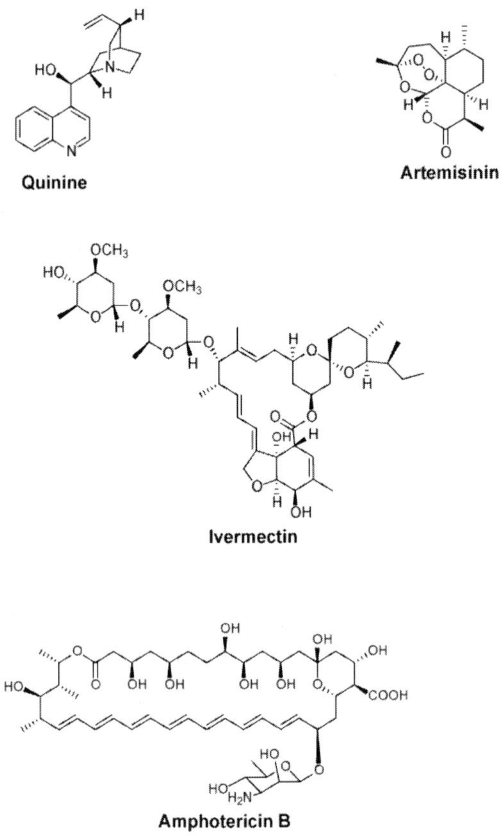

Quinine

Artemisinin

Ivermectin

Amphotericin B

Figure 4. Structures of natural products used to treat parasites.

Quinine

Quinine was initially isolated from the bark of the Cinchona tree, *Cinchona succiruba* Pavon (Rubiaceae) and it has been used against malaria since the Second World War. It is the first effective plant-derived anti-malarial chemical entity. *Cinchona* tree is native to South America and grows abundantly in the Aedes Mountains stretching from Bolivia to Venezuela. The antimalarial property of the *Cinchona* tree was discovered by chance when it was recognized that herbal preparations from the bark of the tree is an effective herbal remedy for people that contacted certain fever which was later shown to be caused by malaria parasites.

Artemisinin

Artemisinin was isolated in 1972 from *Artemisia annua* L. (Asteraceae) and proved to be effective against chloroquine-resistant parasites [55]. Artemisinin is a non-polar compound and several derivatives including ethers (artemether, arteether) and salts (sodium artesunate, sodium artelinate) have been synthesized to improve its formulation characteristics, potency, and pharmacokinetics and to serve as alternatives to quinine [56]. Artemisinin and its commercial derivatives are generally of low toxicity and they reduce the transmission of malaria by acting on the gametocytes of the parasite [57]. Some derivatives of artemisinin have short half life and as a result long periods of treatment may be required to prevent recrudescence problem [58].

4.2. Natural Products in Onchocerciasis and Leishmaniasis Chemotherapy

Onchocerciasis otherwise known as river blindness is caused by *Onchocerca volvulus*, a species of filarial nematode. It afflicts mostly the inhabitants of Africa, putting about 100 million at risk and about 300,000-500,000 are either blind or visually impaired by this disease [59]. It is transmitted by *Simulium* black flies which live near fast moving rivers. Onchocerciasis is a major cause of blindness in the areas of sub-Saharan African countries outside those monitored by the Onchocerciasis Control Program. Ivermectin (Mectizan MK-933) is an efficient microfilaricidal drug used for mass treatment of onchocerciasis [60, 61] and it was donated free in 1967 by Merck to all who need it. Ivermectin is a semi-synthetic derivative of one of the avermectins, a natural product previously developed as a veterinary drug against worms in horses, and accidentally discovered through screening programs to be effective against human parasitic diseases. Ivermectin binds to the GABA receptor-chloride channel complex of the peripheral nervous system (neuromuscular junction), causing inflow of chloride ions, decrease in nerve and paralysis of worm. *Leishmaniasis* is another disease caused by protozoal parasites of the *Leishmania* genus, *L. infantum* and *L. dovovania* and is transmitted by sandfly bites. It is one of the neglected tropical diseases. Infections vary from simple cutaneous to visceral leishmaniasis. Very limited drugs are available for treating the disease. A polyene antibiotic, amphotericin (see Figure 4) produced by rare actinomycetes species, has been demonstrated to eliminate intracellular amastigotes stages of this disease. It is however extremely toxic.

5. NATURAL ANTIBIOTICS FROM MICROBES

Fungi and filamentous bacteria are the most common sources of antibiotic drugs used for treating invasive and contagious microbial diseases. Antibiotics are believed to play important roles in the regulation of metabolic functions of the microbial communities that produce them. Antibiotics are marketed under different trade names as pure natural products or semi-synthetic derivatives. The synthetic derivatives have better potency, activity profile, stability and pharmacokinetic properties. Of the 90 or so antibiotics marketed between 1982 and 2002, about 70 were derived from natural products [8]. In the 1920s, a low yield penicillin-producing mould was serendipitously discovered, and through strain improvement processes other *Penicillium* spp were discovered which could produce more than 50 g of penicillin per liter of culture, thus allowing commercial production of penicillin to start in the 1940s. Penicillin is not only produced by *Penicillium* spp, but also by microbes of the genera *Aspergillus, Trichophyton* and *Epidermophyton*.

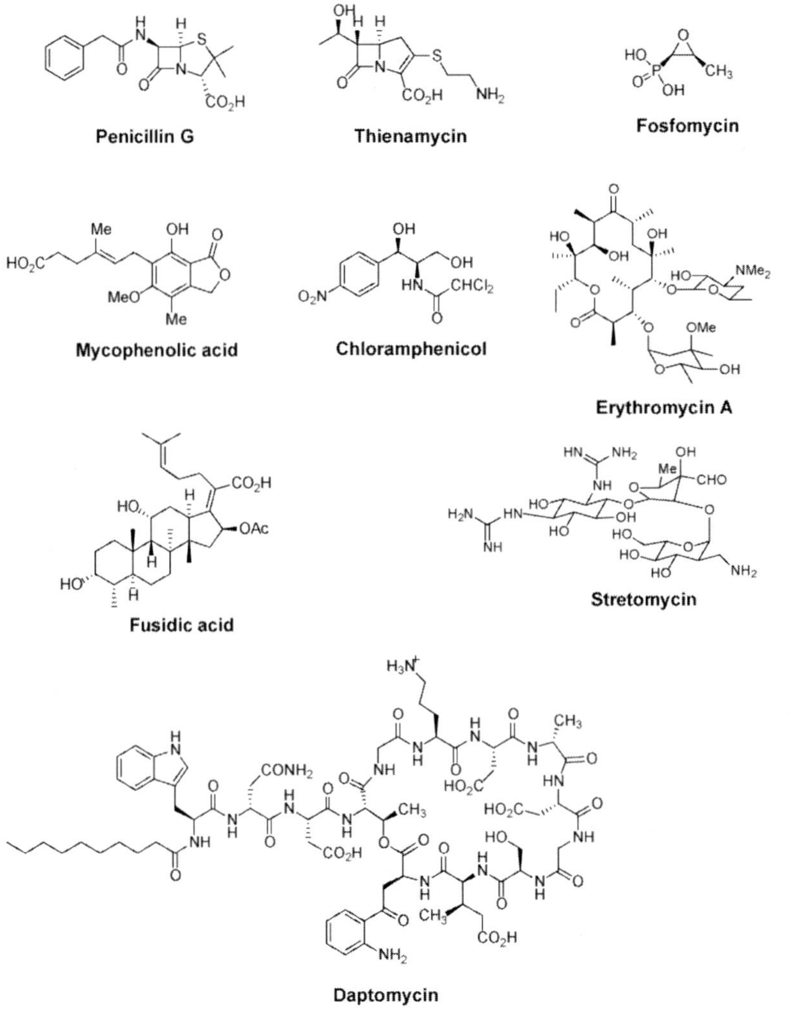

Figure 5. Structures of natural antibiotics.

The following antibiotics (Figure 5) (name of producing organism) from natural sources are currently available in the market: Penicillins (*Penicillium* spp., *Aspergillus* spp.), Cephalosporins (*Acremonium* spp., *Emericellopsis* spp, *Amycolatopsis lactamdurans*, *Streptomyces clavuligerus*), Thienamycin (*Streptomyces cattleya*), Erythromycin (*Saccharopolyspora erythraea*), Vancoycin (*Streptomyces orientalis*), Chloramphenicol (*Streptomyces venezuelae*), Fosfomycin (*Streptomyces fradiae)*, Mupirocin (Pseudomonic Acid) *(Pseudomonas fluorescens*), Fusidic Acid (*Fusidium griseum*), Tetracycline

(*Streptomyce* spp*),* Streptogramins (*Streptomyces pristinaespiralis)* and Daptomycin *Streptomyces roseosporus)* [62].

Over the past two decades, Streptogramins and Daptomycin are the only two natural products antibiotics that have reached the market.

5.1. Natural Antibiotics: Mode of Action and Resistance

Antibiotics have different modes of action depending on their structure: The β-lactams (penicillin G, cephalosporins, thienamycins), glycopeptides (vancomycins), and lipopeotides (daptomycin) block the synthesis of bacterial cell wall; the macrolides (erythromycin A), polyketides (tetracyclines), aminoglycosides (streptomycin) and oxazolidines block the synthesis of proteins in bacteria; and the synthetic fluoroquinolones and ciprofloxacin for example block the replication and repair of DNA [63]. Given the potency of theses chemical entities and their semi-synthetic derivatives, victory was declared at a time on microbial diseases, and pharmaceutical companies disinvested in the development of new antibiotics by the 1980s [64, 65]. This optimism did not last because of the emergence of TB, STDs, SARS, diarrhea, 'flesh-eating' bacteria, Legionnaire's disease-causing microbes which are resistant to previously effective antibiotics.

The mechanisms of resistance of bacteria to various classes of antibiotics are fundamentally known. Some bacteria pump antibiotics from bacterial targets using multiple efflux pump systems [66], and others destroy the 'chemical war head' of antibiotics. For example, the hydrolysis of the β-lactam ring of penicillin or cephalosporin by resistant bacteria leads to loss of antibacterial potency. Some bacteria develop resistance to antibiotics by reprogramming their molecular architecture to camouflage targets from antibiotic missiles.

6. Herbs in Indigenous Skin Care

The skin as an external organ is susceptible to invasive diseases caused by microbes. More than 80% of the world's population depends on herbs for skin care and treatment of skin diseases [67]. The herbs used for skin care contain antibacterial or antioxidant phytochemicals [68, 69].

Skin care is an important cultural practice in Oman as evidenced from the diversity of plants traditionally used to stimulate wound-healing and ward off bacterial and fungal infections of the skin [70].

6.1. Wound-Healing and Antimicrobial Herbs

Wound-healing is a complex process requiring enhanced activities of antioxidant and antimicrobial agents and stimulation of the growth of fibroblast cells. *Staphylococcus aureus*, particularly the methicillin-resistant strain, has severally been implicated in wound-related infection. It has the ability to thrive on skin hair and salty microenvironments.

Herbs with wound-healing properties have been recognized [71] and in some instances formulated into products used as lotions, liniments, ointments, tinctures, shampoos, effervescent granules and bath soaps which are applied directly to the skin to prevent itching or to cure microbial diseases such as impetigo, shingles, ringworm, eczema and acne.

The juice of *Aloe dhofarensis* (Asphodelaceae) (a species of Aloe that is regionally endemic to Oman) for instance, is mixed with hot water and painted on skin to cure rashes and eruptions caused by measles and chickenpox. The *Aloe* plants have been used for thousand of years to treat burns and skin disorders and over 400 species of *Aloe* are known [72]. The plants have been credited with wound-healing, antifungal, anti-cancer, antioxidants, and immunomodulatory properties [73]. The skin care benefits of *Aloe* have been attributed in part to the polysaccharides present as gel in the leaves. The moisturizing and soothing properties of the gel are important to commercial exploitation of *Aloe* extracts for skin and hair care products.

The latex from *Jatropha dhofarica* Radcliff-Smith (Euphorbiaceae) is painted on fresh wound and other cutaneous eruptions as an antiseptic [70]. *Jatropha* spp are rich in diterpenoids, some of which have antibacterial [74] and anti-tumor properties. Nonyl ferulate has been isolated from *J. dhofarica* [75] and the ferulates are active components of skin lotions [76]. The therapeutic biology of *Jatropha* spp has been reviewed by Thomas and Sharma [77].

The root bark of *Calotropis procera* is used to promote secretions and treat skin infections. Proceragenin, a cardenolide isolated from *C. procera*, showed antibacterial activity against opportunistic Gram positive and Gram negative bacteria that can damage the skin [78].

The juice expressed from the leaf of *Sansevieria ehrenbergii* Schweinf. (Agavaceae) is used to treat skin eruptions [70]. The extracts of the leaves showed strong antimicrobial and cytotoxic activities [79]. The powdered frankincense from *Boswellia sacra* (Aiton) W. T. Aiton (Burseraceae) is used traditionally in Oman to treat skin lesions and eruptions [70]. Isolates from frankincense are used to treat skin conditions [80] and to improve psoriasis [81].

The leaf extracts of *Ziziphus spina-christi* (L.) Willd (Rhamnaceae) are rich in saponins and have been shown to possess antibacterial, deodorant, and moisturizing properties [78] when formulated into cosmetics.

The leaves of *Plectranthus barbatus* Andr. (Lamiaceae) and *P. cylindraceus* Hoechst. Ex Benth (syn *P. montanus* (Lamiaceae) are rubbed between the palms and applied over the body as fragrant disinfectant. The volatile oil of *P. cylindraceus* and the leaves of *P. barbatus* and *P. ecklonii* have strong anti-bacterial and anti-fungal activities [82, 83].

Two other plants also used in Oman as wound-healing and excellent emollient lotions are *Moringa peregina* (Forssk) Fiori (Moringaceae) and *Lawsonia inermis* L. (Lythraceae).

The seed oil of *M. peregrina* is used to treat pustules on the face and relief itching of the skin [70]. The oil is rich in oleic acid and lacks antibacterial properties.

Natural products from plants that can inhibit the growth of *Staphylococci* at minimum inhibitory concentration (mic) values below 64 μg/ml or myco-bacteria at mic values lower than 50 μg/ml *in vitro* have been reviewed [84, 85]. However, in spite of the huge potentials of natural products from plants with antibacterial properties, there has been no new antibacterial product developed in the past two decades from plants for general use as topical disinfectants [86].

6.2. Herbal Shampoos

Several plants are used to prepare antibacterial shampoos. The boiled leaves of *Z. spina-christi* are ground into a paste and the fluid strained from it is used as a lotion for sores [70] and a shampoo for hair and scalp in Oman. The crushed leaves of *Becium dhofarense* (Baker) Sebald (Labiatae = Lamiaceae) are mixed with water to make shampoos for washing the skin and hair. *B. dhofarense* is rich in flavonoids and triterpenoids, and showed high anti-oxidant capacity [87]. The seeds of *Ipomoea nil* (L.) Ker-Gawl (Convolvulaceae) are ground and mixed with water to produce shampoo used as body cleanser. The shampoo is believed to stimulate the growth of hair, giving it a lustrous and glossy appearance, and to kill head lice and disinfect the scalp.

Conclusion and Outlook

Natural products have historically played important roles in the discovery of drugs for treating invasive human diseases. Several of the new chemical entities routinely developed in the pharmaceutical industries today revolve around natural products. Initially, natural products were isolated from plants to satisfy the curiosity of organic chemists for structurally novel compounds present in medicinal plants, vegetables and fruits. In the last few decades, research interests in biological interactions between natural products and life forms have led to the discovery of new chemical entities that have useful medicinal applications. At present a number of natural products from herbs, spices, fruits, vegetables, medicinal plants and microbes have been discovered that can control invasive diseases of microbial, parasitic and cancer origin. Medicinal plants seem to have potential applications as topical rather than systemic antimicrobial agents. In the future, ethno-medical information about herbs, isolation, structural analysis and bioassay methods will remain the primary tools needed to discover natural products with useful anti-invasive activities. Research collaborations between natural products chemists and molecular biologists will also play important roles in our understanding of the mode of action of natural products that can be used to treat invasive human diseases.

References

[1] Dewick, P. M. (2009). *Medicinal Natural products: A Biosynthetic Approach.* (3rd Ed.) Wiley and Sons, Ltd; pp 539.

[2] Plotkin, M. J. (2000) *Medicine quest in search of nature's healing secrets*, Viking, New York, American chemical Society, Washington DC, pp 224.

[3] Harvey, A. Strategies for discovering drugs from previously unexplored natural products. *Drugs Discovery Today (DTT)*, 2000, 5, 294-300.

[4] ACS Symposium Series 449, 1991. *Naturally Occurring Pest Bioregulators.* Hedin PA (Ed), pp 456.

[5] Belz, RG; Allelopathy in crop/weed interactions - an update. *Pest Manag. Sci.*, 2007, 63, 308-326.

[6] Harborne, J. B. (1999). Classes and functions of secondary products from plants. In *Chemicals from plants. Perspectives on plant secondary metabolites*. Walton NJ, Brown, E (Eds), Imperial College Press, London, page 1-25.
[7] Cragg, GM; Newman DJ; Yang SS. Natural products of plant and marine origin having antileukemia potential. The NCI Experience, *J. Nat. Prod.*, 2006, 69, 488-498.
[8] Newman, DJ; Cragg, GM. Marine natural products and related compounds in clinical and advanced trials. *J. Nat. Prod.*, 2004, 67, 1216-1238.
[9] Jones, OT. The commercial exploitation of pheromones and other semiochemicals, *Pesticide Science*, 1998, 54, 293-296.
[10] Metal-Based Neurodegeneration: From Molecular Mechanisms To Therapeutic Strategies, John Wiley and Sons 2005, pp238.
[11] Milner, JA. Nutrition and cancer: Essential elements for a roadmap. *Cancer Lett.*, 2008, 269, 189-198.
[12] Liu, RH. Health benefits of fruits and vegetables and from additive and synergistic combinations of phytochemicals. *Amer. J. Clin. Nutr.*, 2003, 78, 717s-520s.
[13] Wattenberg, LW. Inhibitors of chemical carcinogenesis. *Adv. Cancer Res.*, 1978, 26, 197-226.
[14] Adams, M. (2009). *Secret sources of healing foods and natural medicines that can save your life*. (2nd Ed), Truth publishing international Ltd, Arizona, pp 57.
[15] Wolfe, KL; Kang, X; He, X; Dong, M; Zhang, Q; Liu, RH. Cellular antioxidant activity of common fruits. *J. Agric. Food Chem.,* 2008, 56, 8418-8426.
[16] Ursell A. (2000). *The complete guide healing foods*. Dorling Kindersley Limited, London, pp 256.
[17] Wang, H; Cao, G; Prior, RL. Total antioxidant capacity of fruits. *J. Agric. Food Chem.* 1996, 44, 701-705.
[18] Pham NA; Jacobberger, JW; Schimmer, AD; Cao, P; Gronda, M; Hedley, DW. The dietary isothiocyanate sulforaphane targets pancreatic cancer cells and inhibits tumor growths in severe combined immunodeficient mice. *Mol. Cancer Ther.* 2004, 3, 1239-1248.
[19] Gerhauser, C. Cancer chemopreventive potential of apples, apple juice and apple components, *Planta Med*, 2008, 74, 1608-1624.
[20] Jang, M; Cai, L; Udeani, G. O.; Slowing, KV; Thomas, CF; Beecher, CWW; Fong HHS; Farnsworth, NR; Kinghorn, D; Mehta, RG; Moon, RC; Pezzuto, JM. Cancer chemopreventive activity of resveratrol, a natural product derived from grapes. *Science,* 1997, 275, 218-220.
[21] Musialik, M; Kuzmicz, R; Pawlowski, TS; Litwinienko, G. Acidity of hydroxyl groups: An overlooked influence on antiradical properties of flavonoids. *J. Org. Chem.,* 2009, 74, 2699-2709.
[22] Farbood, Y; Sarkaki, A; Badavi, M. Preventive effect of Grape Seed hydroalcholic extract on dementia type of Alzheimer's disease in aged male rats, *Intern. J. of Pharmacol*, 2009, 5, 257-262.
[23] Schreckinger, ME; Lotton, J; Lila, MA; Gonzalez de Mejia E; Berries from South America: Comprehensive review on chemistry, health potential, and commercialization. *J. Med. Food*, 2010, 13, 233-246.
[24] Negi PS; Jayaprakasha, GK; Jena, BS. Antioxidant and antimutagenic activities of pomegranate peel extracts. *Food Chemistry*, 2003, 80, 393-397.

[25] Carine, B; Leudeu, T; Tchiegang, C; Barbe, F; Nicolas, B; Gueant, J-L; *Ricinodendron heutelotii* (Bail.) or *Tetracarpidium conophorum* Muell. oils fed to male rats lower blood lipids *Nutr. Res.,* 2009, 29, 503-509.
[26] Poste, G., and Fidler I. J. The pathogenesis of cancer metastasis. Nature, 283, 139-146.
[27] Aggarwal, B. B; Kumar, A; Aggarwal, M. S; Shishodia, S. (2004) In *Phytopharmaceuticals in Cancer Chemoprevention*. Bagchi, D and Preus HG. (Eds) CRC press: Boca Raton, FL, Chapter 23, 349-387.
[28] Halliwell, B; Gutteridge, J. M. C. (1998) In *Free Radicals in Biology and Medicine*; Oxford University Press, Oxford.
[29] Valko, M; Leibfritz D; Moncol, J; Cronin, MT; Mazur, M.; Telser, J. Free radicals and antioxidants in normal physiological functions and human disease. *Int. J Biochem. Cell Biol.* 2007, 39, 44-84.
[30] Poulsen, HE Oxidative DNA modifications. *Exp. Toxicol. Pathol*. 2005, 57, 161-169.
[31] Halliwell, B. Mechanisms involved in the generation of free radicals. *Pathologie Biologie*, 1996, 44, 6-13.
[32] Gutteridge, JMC. Lipid peroxidation and antioxidants as biomarkers of tissue damage. *Clin. Chem*., 1995, 41, 1819-1828.
[33] Antosiewiz J; Ziolkowski W; Kar S; Powolny AA; Singh, SV. Role of reactive oxygen intermediates in cellular responses to dietary cancer chemopreventive agent, *Planta Med*. 2008, 74, 1570-1579.
[34] Pier-Giorgio P. Flavonoids as antioxidants. *J. Nat. Prod*, 2000, 63, 1035-1042.
[35] Pietta, P. G. and Simonetti, P. (1999). In *Antioxidant Food Supplements in Human Health*; Packer, L., Hiramatsu, M., Yoshikawa, T. (Eds.) Academic Press, New York.
[36] Juturu, V. Resveratrol and cardiovascular disease. *Current Nutrition and Food Science*, 2008, 4, 227-230.
[37] Cucciola, V; Borriello; Oliva, A; Galletti, P; Zappia, V; Della Ragione, F. Resveratrol: From basic science to the clinic. *Cell Cycle*, 2007, 6, 2495-2510.
[38] Pamela M. Nutrition Isoflavones, *The Pharmaceutical Journal*, 2001, 266, 16-19.
[39] Aggarwal, BB; Kunnumakkara, AJ; Harikumar KB; Tharakan, ST; Sung, B; Anand, P. Potential of spice-derived phytochemicals for cancer prevention, *Planta Med*. 2008, 74, 1560-1569.
[40] Pratchayasakul W; Pongruangporn, M; Chattipakorn, N; Chattipakorn S. *Current Topics in Nutraceutical Research,* 2009, 7, 11-26.
[41] Yinrong, Lu; and Yeap Foo, L. Polyphenols of *Salvia*- a review. *Phytochemistry*, 2002, 59, 117-140.
[42] Jensen, GS; Wu, X; Patterson, KM; Barnes, J; Carter, SG; Scherwitz, L; Beaman, R; Endres, JR; Schauss, AG. *In vitro* and *in vivo* antioxidant and anti-inflammatory capacities of an antioxidant-rich fruit and berry juice blend. Results of a pilot and randomized, double-blinded, placebo-controlled cross-over study. *J. Agric. Food Chem.,* 2008, 56, 8326-8333.
[43] Cragg, G. M; Kingston, D. G. I; Newman, G. J. Eds. (2005). In *Anticancer Agents from natural Products*; Taylor and Francis Group, Boca Raton FL, 2005.
[44] Gueritte, F., Fahy, J. (2005). In *Anticancer Agents from natural Products*. Taylor and Francis Group; Boca Raton FL, 123-136.
[45] Kingston, D. G. I. (2005) In *Anticancer Agents from natural Products*; Taylor and Francis Group, Boca Raton FL, 99-122.

[46] Amos, LA; Lowe, J. How Taxol stabilizes microtubule structure. *Chem and Bio*, 1999, 6, R65-R69.

[47] Lee K.-H; Xiao Z. (2005). In *Anticancer Agents from natural Products*; Taylor and Francis Group, Boca Raton FL, 71-88.

[48] Newman, DJ; Cragg, GM; Snader, KM. Natural products as sources of new drugs over the period 1981–2002. *J. Nat. Prod.*, 2003, 66, 1022-1037.

[49] WHO, Africa Malaria Report, World Health Organisation, Geneva, WHO/CDS/MAL/2003.1093, 2003.

[50] WHO, World Health Report, Changing History, World Health Organisation, Geneva, 2004.

[51] Soh, PN; Benoit-Vical, F. Are West African plants a source of future antimalarial drugs? *J. Ethnopharmaco*logy, 2007, 114, 130-140.

[52] Mohammed, M. S. A. (2009). In *African Natural Plant Products: New discoveries and Challenges in Chemistry and quality*, Juliani, H. R; Simon, J. E; and Ho, C-T. (Eds.) ACS Symposium Series, Washington DC.

[53] Braga de Oliveira, A; Dolabela, MF; Braga, FC; Jacome, RLRP; Varotti, FP; Povoa, MM. Plant-derived antimalarial agents: new leads and efficient phytomedicines. Part I. alkaloids *Anais da Academia Brasileira de Ciencias*, 2009, 81, 715-714.

[54] Kaur K; Jain, M; Kaur, T; Jain, R. Antimalarials from nature, *Bioorganic and Medicinal Chemistry*, 2009, 17, 3229-3256.

[55] Klayman, DL. Qinghaosu (Artemisinin): An antimalarial drug from China. *Science*, 1985, 228, 1049-1054.

[56] Wilairatana P and Looareesuwan S. (2002). *The clinical use of artemisinin and its derivatives in the treatment of malaria* In: Artemisia. Wright C.W. (Ed), Taylor and Francis London, 289-307.

[57] Price, RN; Nosten, F; Luxemberger, C; Kuile FO; Paiphun, L; Chongsuphajaisiddhi, T; White, NJ. Effects of artemisinin derivatives on malaria transmissibility. *Lancet*, 1996, 347, 1654-1658.

[58] Wright C. W. and Warhurst, D. C. (2002). The mode of action of artemisinin and its derivatives: In: *Artemisia.* Wright C.W. (Ed), Taylor and Francis, London, 249-288.

[59] Thylefors, B; Negrel, AD; Pararajasegaram, R; Dadzie, KY. Global data on blindness. *Bull World Health Organisation*, 1995, 73, 115-121.

[60] Brieger, WR; Awedoba, AK; Eneanya, CI; Hagan, M; Ogbuagu, KF; Okello, DO; Ososanya, OO; Ovuga, EB; Noma, LM; Kale,OO; Burnham, GM; Remme, JHF. The effects of ivermectin on onchocercal skin and severe itching: results of a multicentre trial. *Tropical Medicine and International Health*, 1998, 3, 951-961.

[61] Abiose, A. Onchocercal eye disease and the impact of Mectizan treatment. *Annals of Tropical Medicine and Parasitology*, 1998, 92, Supplement No 1, S11-S22.

[62] Pelaez. F. The historical delivery of antibiotics from microbial natural products- can history repeat? *Biochemical Pharmacology* 2006, 981-990.

[63] Walsh, C. Molecular mechanisms that confer antibacterial drug resistance. *Nature*, 2000, 406, 778-781.

[64] Culotta E. Funding cut? Hobbles antibiotic resistance research, *Science* 1994, 264, 362-363.

[65] Bax RP. Antibiotic resistance: A view from pharmaceutical Industry. *Clin. Infect. Dis.* 1997, 24 (Suppl), 5151-5153.

[66] Marshall, NJ and Piddock, LJ. Antibacterial efflux systems. *Microbiologia*, 1997, 13, 285-300.

[67] SteenKamp, V; Mathivha, E; Gouws, M. C; van Rensburg C. E. J. Studies on antibacterial, antioxidant and fibroblast growth stimulation of wound healing remedies from South Africa. *J. Ethnopharmacol.*, 2004, 95, 353-357.

[68] Dinkova-Kostova, AT.Phytochemicals as protectors against ultraviolet radiation: Versatility of effects and mechanism. *Planta Med.* 2008, 74, 1548-1559.

[69] Jorge ATS; Arroteia, KF; Lago, JC, de Sa-Rocha, VB; Gesztesi J; Moreira, PL A new potent natural antioxidant mixture provides global protection against oxidative skin cell damage. *International Journal of Cosmetic Science*, 2010, 1-9, doi:10.1111/j.1468-2494.2010.00595.x

[70] Miller, A. G., Morris, S. M., Stuart, S. (1988). *Plants of Dhofar The Southern Region of Oman: Traditional, Economic and Medical Uses*. The Office of the Advisor for Conservation of the Environment, diwan of Royal Court, Sultanate of Oman, pp361.

[71] Bodeker G. amd Hughes, M. A. (1998). Wound healing, traditional treatments, and research policy. In *Plants for Food and Medicinal Prendergast*, H. D. V., Etkin, N. L., Harris, D. R., Houghton P.J., (Eds). Royal Botanic Gardens, Kew, London.

[72] Dagne, E; Bistrat, D; Viljoen, A; Van Wyk, BE. Chemistry of *Aloe* species. *Current Organic Chemistry*, 2000, 4, 1055-1078.

[73] Singh, S; Sharma, PK; Kumar, N; Dudhe, R. Biological activities of *Aloe vera*. *International Journal of Pharmacy and Technology*, 2010, 2, 259-280.

[74] Viswanathan, MB; Ramesh, N; Ahilan, A; Lakshmanaperumalsamy, P. Phytochemical *Medicinal Chemistry Research,* 2004, 13, 361-368.

[75] Mahfouda K. Al-Busaidi, BSc Thesis, Chemistry Department Sultan qaboos University, Oman December 2009.

[76] Taniguchi, H; Nomura, E; Tsuno, T; and Minami S. Ferulic acid esters antioxidants/UV absorbance, *United States Patent*, June 1999, 5 908 615.

[77] Thomas, R; Sah, NK; Sharma, PB. Therapeutic biology of *Jatropha curcas*: A mini review. *Current Pharmaceutical Biotechnology*, 2008, 9, 315-324.

[78] Akhtar, N; Malik, A; Ali, SN; Kazmi, SU. Proceragenin, an antibacterial cardenolide from *Calotropis procera*. *Phytochem.,* 1992, 31, 2821-2824.

[79] Mohamed, A; Martina, W; Gudrun, S; Ulrike, L. Antioxidant, antimicrobial and cytotoxic activities of selected medicinal plants from Yemen *J. Ethnopharmacol.* 2007, 111, 657-66.

[80] Kondo, C; Suenobu, N; Shishido, M; Oikawa, M; Sakata, A. Skin condition restoration stimulators containing essential oils from *Burseraceae* plants, and topical preparations containing them. Jpn. Kokai Tokkyo Koho 2010, JP 2010195723 A 20100909.

[81] Legg, K. Frankincense-derived triterpenoid improves psoriasis. *Nature Reviews Rheumatology*, 2009, 5, 654.

[82] Figueiredo, NL; de Aguiar, SRMM; Fale, PL; Ascensao, L; Serralheiro, MLM; Lino, ARL. The inhibitory effect of *Plectranthus barbatus* and *Plectranthus ecklonii* leaves on the viability, glucosyltransferase activity and biofilm formation of *Streptococcus sobrinus* and *Streptococcus mutans*. *Food Chemistry*, 2010, 119, 664-668.

[83] Marwa, RG; Fatope, MO; Deadman, DL; Ochei, JE and Al-Saidi SH. Antimicrobial activity and major componenets of the essential oil of *Plectrantus cylindraceus*, *J. Applied Microbiology*, 2007, 103, 1220-1226.

[84] Gibbons S. Anti-staphylococcal plant natural products. *Nat. Prod. Rep.* 2004, 21, 263-267.

[85] Gibbons, S. Phytochemicals for resistance – strengths, weaknesses and opportunities *Planta Med.* 2008, 74, 594-602.

[86] Okunade, AL; Elvin-Lewis, MPF; Lewis, WH. Natural antimycobacterial metabolites: current status. *Phytochemistry*, 65 (2004), 1017-1032.

[87] Marwah, RG; Fatope, MO; Al Mahrooqi, R; Varma, GB; Al Abadi, H; Al-Burtamani, SKS. Antioxidant*Food Chemistry*, 2007, 101, 465-470.

In: Natural Products and Their Active Compounds ... ISBN: 978-1-62100-153-9
Editors: M. Essa, A. Manickavasagan, and E. Sukumar © 2012 Nova Science Publishers, Inc.

Chapter 10

PHYTOCHEMICALS IN CIGARETTE/TOBACCO SMOKE-MEDIATED INFLAMMATION

Saravanan Rajendrasozhan[1*] and Irfan Rahma[2]
[1]Department of Chemistry, Faculty of Science, University of Hail, Hail, Saudi Arabia
[2]Department of Environmental Medicine, Lung Biology and Disease Program, University of Rochester Medical Center, Rochester, NY, US

ABSTRACT

Cigarette/tobacco smoking is associated with an increased incidence and severity of lung inflammation, and impaired lung function. It is a major etiological factor for chronic obstructive pulmonary disease (COPD), a leading cause of mortality and morbidity worldwide. Corticosteroids are considered as the most effective anti-inflammatory agent available in clinical use at present. However, smokers and patients with COPD develop insensitivity to corticosteroid treatment, which leads to considerable management problems. Therefore, there is a great need for the novel anti-inflammatory agent without any secondary side effects. The use of phytochemicals to control cigarette/tobacco smoke-induced lung inflammation may have a potential benefit and is reviewed in this chapter. Plant-derived chemical compounds, such as curcumin, resveratrol, epigallo-catechin-3-gallate, ursolic acid, zerumbone, parthenolide, helenalin, baicalin, caffeic acid phenethylester, garcinol, vitamin C, vitamin E, β-carotene, sulforaphane and apocynin showed anti-inflammatory effect against cigarette/tobacco smoke exposure. Depending on the molecular target, the phytochemicals were classified into NF-κB inhibitors (which reduce NF-κB-mediated pro-inflammatory gene transcription), histone acetyltransferase inhibitors or histone deacetylase activators (which reduce transcription factor-DNA binding and thereby reduce pro-inflammatory gene transcription), and antioxidants (which scavenge reactive oxygen species-induced pro-inflammatory gene transcription and tissue damage). Since a variety of pathways are involved in the pathogenesis of lung inflammation and COPD, therapeutic administration of a phytochemical with multiple molecular targets or mixture of multiple phytochemicals will be effective in management of lung inflammation and COPD.

* For communication: Dr. Saravanan Rajendrasozhan; Assistant Professor of Biochemistry; Department of Chemistry; Faculty of Science; Floor 3, Room No. 2, New Campus; PO Box 2440; University of Hail; Hail, Saudi Arabia. Mobile: +966 535694288; Fax: +966 5310228; E-mail: s.rajendrasozhan@uoh.edu.sa

Keywords: Cigarette, lung inflammation, COPD, polyphenols, NF-κB, Nrf2, HDAC, antioxidants

ABBREVIATIONS

BALF	Bronchoalveolar lavage fluid
CAPE	Caffeic acid phenethyl ester
COPD	Chronic obstructive pulmonary disease
COX-2	Cyclooxygenase-2
CS	Cigarette smoke
EGCG	Epigallocatechin gallate
GCL	Glutamate-cysteine ligase
GMC-SF	Granulocyte-Macrophage Colony Stimulating Factor
H_2O_2	Hydrogen peroxide
HAT	Histone acetyltransferases
HDAC	Histone deacetylase
ICAM-1	Intercellular adhesion molecule-1
IKK	I-kappaB kinase
IL	Interleukin
iNOS	Inducible nitric oxide synthase
IκBα	Inhibitor kappa B alpha
Keap1	Kelch-like ECH-associated protein 1
MCP-1	Monocyte chemotactic peptide-1
MEK	Mitogen-activated kinase
MIP-1α	Macrophage inflammatory protein-1 alpha
NADPH	Nicotinamide adenine dinucleotide phosphate
NF-κB	Nuclear factor-kappa B
NIK	NF-κB inducing kinase
NO_2	Nitrogen dioxide
Nrf2	Nuclear factor erythroid 2-related factor 2
$^{\bullet}OH$	Hydroxyl radical
$O_2^{\bullet -}$	Superoxide anion
$OONO^-$	Peroxynitrite radical
RNS	Reactive nitrogen species
ROS	Reactive oxygen species
SIRT1	Sirtuin 1
SOD	Superoxide dismutase
TNF-α	Tumor necrotic factor-alpha

INTRODUCTION

Cigarette smoke (CS) is a toxic cocktail of more than 4000 chemicals including high concentrations of oxidants (10^{14} molecules per puff) generated from the combustion of tobacco leaves. It contains a high concentration of reactive oxygen species (ROS), reactive nitrogen species (RNS), pro-inflammatory agents, and carcinogens. Active cigarette smoking causes a broad spectrum of human pathologies, including lung diseases, cardiovascular diseases, atherosclerosis, cancer, infertility, and oral cavity pathologies [1, 2]. However, lung is the major organ which is affected by CS because of its large surface area and direct contact with inhaled toxicants. Prolonged exposure to CS is associated with an increased incidence and severity of lung airway inflammation and impaired lung function. CS-induced lung inflammation is the major known risk factor for the development of chronic obstructive pulmonary disease (COPD).

COPD is an inflammatory disease characterized by an irreversible airflow limitation in lung. Inflammation in COPD leads to progressive destruction of lung parenchyma, airway remodeling and pulmonary emphysema [3, 4]. According to the World Health Organization, about 80% of COPD is caused by cigarette/tobacco smoking [5]. COPD is the fifth leading cause of mortality worldwide, and is expected to become the third leading cause of mortality by 2020 [6]. Since cigarette smoking and COPD impose a huge public health burden, treatment and management strategy have become more important.

PHARMACOTHERAPY

Targeting airway and lung inflammation is vital in the management of inflammatory lung diseases, such as COPD. Corticosteroids (glucocorticoids) are considered as the most effective anti-inflammatory drug available in clinical use at present. However, in most clinical studies, corticosteroid has not been able to reduce the airway inflammation in patients with COPD and asthmatics who smoke [7, 8]. As there are only limited options available for the anti-inflammatory therapy, insensitivity to corticosteroid treatment leads to considerable management problems in COPD and severe asthmatics. Therefore, there is a great need to identify novel anti-inflammatory agents without any apparent toxic side effects to combat airway/lung inflammation.

NATURAL PRODUCTS

The use of natural health products is a popular form of complementary medicine, and is becoming an attractive approach for the treatment of various inflammatory disorders. Understanding the anti-inflammatory properties of the active components of medicinal plants (phytochemicals) will facilitate the treatment strategy for lung inflammation. In this chapter, the effect of various phytochemicals on CS/tobacco smoke-induced lung inflammation, a major contributory factor in the pathogenesis of COPD, is discussed along with their possible mechanism of therapeutic action.

1) NF-κB Inhibitors

NF-κB (nuclear factor kappaB) is a transcription factor that regulates the expression of a wide spectrum of genes encoding pro-inflammatory cytokines (IL-1, IL-2, IL-6, and TNF-α), chemokines (IL-8, MIP-1α, and MCP-1), and inducible enzymes (COX-2 and iNOS), all of which play a critical role in inflammatory process. NF-κB represents a group of five proteins namely RelA (p65), RelB, c-Rel, NF-κB1 (p50) and NF-κB2 (p52). It exists as a heterodimer, predominantly composed of p65 and p50 subunits. In unstimulated cells, NF-κB resides in the cytoplasm in an inactive state via physical association with NF-κB inhibitory proteins, called IκBs. Activation of NF-κB is tightly controlled by IκB kinase (IKK)-mediated IκBα phosphorylation, which leads to polyubiquitination and proteasomal degradation of IκBα. In the absence of inhibitory IκBα, NF-κB is translocated into the nucleus, bind to κB DNA elements and induce transcription of pro-inflammatory genes.

NF-κB activation occurs both *in vitro* (cell culture) and *in vivo* (animal models) in response to CS exposure. NF-κB activation was also increased in the lungs of smokers and patients with COPD [9]. CS activates NF-κB through a series of events, such as IKK activation, IκBα phosphorylation, IκBα degradation, nuclear translocation of NF-κB, and NF-κB-DNA binding, which leads to pro-inflammatory gene transcription. Inhibition of NF-κB activation using kinase inhibitors and dietary polyphenols (see below) represent an exceptional strategy to control CS-mediated lung inflammation.

I) Curcumin

Curcumin (1,7-bis(4-hydroxy-3-methoxyphenol)-1,6heptadiene-3,5-dione) is a polyphenol derived from the rhizome of *Curcuma longa*. It has a remarkable range of protective effects in various diseases due to its anti-inflammatory, antioxidant, chemo-preventive, and chemotherapeutic properties. A considerable number of studies suggest that curcumin emerged as the powerful anti-inflammatory drug, especially due to its ability to block or down-regulate NF-κB activation [10].

Curcumin has been shown to inhibit CS-induced NF-κB activation in human lung epithelial cells and monocytes [11-13], which resulted in a decreased pro-inflammatory mediator secretion. The molecular mechanisms whereby curcumin inhibits NF-κB is due to IκB kinase activation, IκB phosphorylation and degradation, NF-κB nuclear translocation, and DNA binding of NF-κB. Curcumin-mediated inhibition of NF-κB activation correlated with the suppression of CS-induced NF-κB-dependent pro-inflammatory gene expression [12]. Oral administration of curcumin attenuates CS-induced inflammatory cell influx in bronchoalveolar lavage fluid (BALF) and emphysema (airspace enlargement) in mice [14]. Dietary supplementation of curcumin also inhibits COPD-like airway inflammation and lung cancer progression in mice [15].

Spurred by the encouraging findings on the anti-inflammatory effects of curcumin in experimental models, a few clinical studies have been conducted that showed the protective effect of curcumin in inflammatory diseases [16]. In human, bioavailability of curcumin from oral administration is low because of its lipophilic nature and poor absorption. Nevertheless, curcumin has been reported to demonstrate biological activity in a number of human clinical trials. Improvement of curcumin bioavailability by coingestion with piperine (an alkaloid from black pepper) [17] or liposome encapsulation [18] might contribute to a better pharmacological effect of curcumin. As curcumin intake did not show any apparent toxic side effects in human, it warrants further research and clinical trials on CS-induced human lung diseases.

II) Resveratrol

Resveratrol (3,4',5-trihydroxystilbene) is a polyphenoic phytoalexin found in grapes, red wine, peanuts, and some berries. It has a wide spectrum of therapeutic properties which includes antioxidant, anti-inflammatory, chemopreventive, and immunomodulatory activities. Resveratrol exhibit anti-inflammatory effect by reducing the release of pro-inflammatory mediators, IL-8 and GMC-SF (that are regulated by NF-κB), from lung epithelial cells and monocytes [19, 20]. In alveolar macrophage from smokers and patients with COPD, resveratrol inhibits inflammatory cytokine release at basal condition as well as in response to CS-mediated pro-inflammatory stimuli [21]. In consistent with the above finding, resveratrol has been reported to reduce the release of inflammatory cytokines, such as IL-8, GM-CSF, from human airway smooth muscle cells in COPD [22]. The anti-inflammatory effects of resveratrol are, at least partially, based on their ability to suppress the activation of NF-κB. Resveratrol inhibits NF-κB activation and NF-κB-dependent gene expression through its ability to inhibit IκB kinase activity, the key regulator in IκBα degradation and NF-κB activation [23].

Resveratrol is a lipid-soluble compound. It is well-absorbed by humans when taken orally, but its bioavailability is relatively low due to its rapid metabolism and elimination [24]. The absorption and bioavailability of resveratrol can be improved by using micronized resveratrol (particle size of 1 to 4 microns) and coingestion with tween-80, a commonly used emulsifier and surfactant [25]. Inhalation of atomized resveratrol has been shown to reduce inflammatory mediators and alleviate lung injury in COPD rat model (exposed to bleomycin and CS) [26]. Clinical trials with resveratrol indicate no adverse effects in human even at a dose of 5 grams [27].

III) Gallate

Epigallocatechin-3-gallate is the ester of epigallocatechin and gallic acid found in green tea (*Camellia sinensis*). Regular consumption of green tea has been reported to reduce inflammation and incidence of carcinogenesis [28]. Green tea polyphenols decreased the production of pro-inflammatory mediator TNF-α in response to inflammatory stimuli in mice lung [29]. The green tea polyphenols include epicatechin, epigallocatechin, epicatechin-3-gallate, and epigallocatechin-3-gallate (EGCG). EGCG generally accounts for more than 40% of the total polyphenols present in green tea [30]. It has potent antioxidant and anti-inflammatory properties. ECGC showed anti-inflammatory activity in macrophage cell line and peritoneal macrophages by blocking NF-κB activation [29]. Among the various green tea polyphenols, EGCG is the most potent inhibitor of IKK, and it blocks the phosphorylation of IκBα, which leads to inhibition of NF-κB activation. The gallate group plays a major role for inhibiting IKK, and hence NF-κB activation [28]. Pretreatment of human bronchial epithelial cells with EGCG suppressed CS-induced phosphorylation of IκB-α, activation as well as nuclear translocation of NF-κB [31]. Interestingly, it has been reported that EGCG-mediated inhibition of NF-κB and anti-inflammatory action is independent of its antioxidant property

[28]. Although EGCG demonstrated anti-inflammatory effect against CS exposure, it did not decrease CS-induced pulmonary emphysema severity in mice [32].

EGCG possesses a longer half-life and slower rate of elimination as compared with epigallocatechin and epicatechin in rats [33]. In human, oral bioavailability of EGCG can be improved by using peracetate protected EGCG or nanolipidic EGCG [34, 35]. Clinical trials with EGCG in healthy participants (400 mg to 800 mg per day for 4 weeks) or cancer patients (400 mg to 2000 mg, twice a day for a month) demonstrated that EGCG consumption was well tolerated and had no serious adverse effects [36, 37].

IV) Ursolic Acid

Ursolic acid (3β-hydroxy-urs-12-en-28-oic acid) is a pentacyclic triterpenoid found in a wide range of plants, such as *Rosemarinus officinalis*, *Eriobotrya japonica*, *Calluna vulgaris*, *Ocimum sanctum*, and *Eugenia jumbolana* in the form of aglycones or as the free acid. Ursolic acid has been recognized to have anti-inflammatory properties in laboratory animals [38]. Ursolic acid is used as an anti-inflammatory agent as it suppresses NF-κB activation induced by CS. Ursolic acid inhibits IKK, leading to inhibition of CS-induced IκBα phosphorylation, IκBα degradation, NF-κB nuclear translocation, and ultimately resulting in abrogation of NF-κB-DNA binding and gene transcription [39]. Interestingly, some laboratory studies have recently revealed that the effects of ursolic acid on normal cells and tissues during inflammatory response [40]. Ursolic acid is relatively non-toxic, and has been used in health products and cosmetics [38].

V) Zerumbone

Zerumbone is a sesquiterpene found in subtropical ginger, *Zingiber zerumbet.* It has been implicated as a promising anti-inflammatory agent on acute and chronic inflammation models in mice [41]. Zerumbone suppressed NF-κB activation induced by CS condensate, tumor necrosis factor (TNF), okadaic acid, phorbol myristate acetate, and H_2O_2. The α, β-unsaturated carbonyl group in zerumbone may play a pivotal roles in inhibiting NF-κB and its anti-inflammatory action. NF-κB inhibition by zerumbone correlated with sequential suppression of the IκBα kinase activity, IκBα phosphorylation, IκBα degradation, p65 phosphorylation, p65 nuclear translocation, and p65 acetylation [42].

VI) Parthenolide

Parthenolide is a sesquiterpene lactone found in the medicinal plant, feverfew (*Tanacetum parthenium*). Parthenolide is one of the potent inhibitor of NF-κB activation by preventing the degradation of IκBα [43]. The multisubunit IκB kinase complex (IKKα and IKKβ) is the molecular target for the NF-κB-inhibiting activity of parthenolide. This compound also prevents activation and DNA binding of NF-κB induced by MEK kinase (MEKK) and NF-κB-inducing kinase (NIK) [43, 44]. Parthenolide has an exomethylene group and an epoxide ring that are responsible for its inhibitory effects on target molecule(s), probably by reacting with exposed cysteines [45]. The blockade of NF-κB activation by parthenolide prevented the upregulation of pro-inflammatory mediator (IL-8) secretion in response to CS exposure in human monocytes [11].

VII) Helenalin

Helenalin, derived from *Arnica Montana*, is a potent anti-inflammatory compound with a unique molecular mechanism of action. It inhibits NF-κB activity by directly alkylating the p65 subunit of NF-κB, which leads to blockade of NF-κB binding to DNA. Neither IκB degradation nor NF-κB nuclear translocation are inhibited by helenalin [46]. Alkylation of NF-κB may be specific for the highly reactive helenalin, which causes the inactivation of multiple target enzymes. Helenalin reduced TNFα-induced inflammatory cytokine expression in human airway smooth muscle cells [47]. Increase of NF-κB activity stimulated by CS extract was attenuated by selective inhibition of NF-κB by helenalin *in vitro* [48]. Thus, helenalin would ameliorate the NF-κB-dependent pro-inflammatory mediator synthesis in response to CS exposure.

VIII) Baicalin

Baicalin (7-glucuronic acid,5,6-dihydroxy-flavone) is a flavonoid derived from the roots of *Scutellaria baicalensis*. Baicalin has antioxidant, anti-inflammatory and anti-microbial activities [49]. Baicalin inhibits the release of pro-inflammatory cytokines (TNF-α, IL-1β, and IL-6), and also enhances the production of anti-inflammatory cytokine (IL-10), thus resulting in an overall attenuation of the pro-inflammatory/anti-inflammatory cytokine ratio, which may contribute to its anti-inflammatory effect [50]. Baicalin has anti-inflammatory effect on CS-induced COPD in rat model by reducing the expression of pro-inflammatory cytokines (IL-8, IL-6 and TNF-α) in plasma and BALF in the lungs. Decrease in CS-induced IκB and p65 phosphorylation, and NF-κB DNA-binding activity by baicalin indicates that the anti-inflammatory effect is likely to be achieved by inhibiting the NF-κB pathway [51].

IX) Caffeic Acid

Caffeic acid phenethyl ester (CAPE) is plant polyphenol found in pears, basil, thyme, and propolis from honeybee hives. It is known to have antioxidant, anti-inflammatory and anti-carcinogenic properties [52-54]. It is a potent and specific inhibitor of NF-κB activation. The sulfhydryl groups in CAPE play a crucial role in its NF-κB inhibitory activity [55]. Previous reports have shown that the CAPE exerts its protective effect by inhibiting NF-κB activation and reducing inflammation [56-58]. CAPE exerts protective effect in the lung against CS-induced inflammation, emphysematous and other pathological changes in rabbit [59].

CAPE showed anti-inflammatory effect by reducing the expression of pro-inflammatory mediator (TNF-α), and production of IL-8 in response to pro-inflammatory stimuli. CAPE exerts its anti-inflammatory effect possibly through the suppression of IκB-α degradation and NF-κB inactivation [58].

In contrast, CAPE has been reported to have no significant effect on TNF-induced IκB-α degradation, but prevents the nuclear translocation of NF-κB, delays IκB-α resynthesis and specific inhibition of NF-κB binding to the DNA [55]. In addition, CAPE-mediated improvement in tissue antioxidant *status in vivo* against CS exposure [52-54] may play a supportive role in its anti-inflammatory activity.

2) Protein Deacetylase Activators and Acetyltransferase Inhibitors

It is estimated that the human body contains 100 trillion meters of DNA which are tightly coiled around histone proteins to form compact DNA-protein complex, called chromatin [60]. Tightly coiled DNA cannot bind to transcription factors, whereas uncoiled DNA can bind to specific transcription factors to induce gene transcription. Coiling and uncoiling of DNA is controlled by histone deacetylation and acetylation, respectively. Coiling of DNA is facilitated by protein/histone deacetylases (HDACs). Uncoiling of DNA is facilitated by protein/histone acetyltransferases (HAT). Furthermore, HAT also mediate acetylation of transcription factors such as NF-κB, leading to increased NF-κB-DNA binding and pro-inflammatory gene transcription, which can be reversed by HDACs. Decreased level and activity of HDACs (especially HDAC2 and SIRT1) has been reported in smokers and patients with COPD [9, 61]. It is associated with increased histone and NF-κB acetylation as well as increased pro-inflammatory cytokine levels. As HDAC/HAT balance is a key regulator of gene transcription, HAT inhibitors and HDAC activators can play a pivotal role in reducing CS-induced pro-inflammatory gene transcription.

I) Curcumin

The anti-inflammatory effect of curcumin is not only mediated through the inhibition of NF-κB activation, but also through the change of HDAC/HAT balance in favor of greater histone deacetylation. This condition may leads to reduction in transcription factor-DNA binding and pro-inflammatory gene expression. It is supported by the finding that curcumin treatment inhibited the HAT activity both *in vivo* and *in vitro* resulted in a comparable inhibition of histone acetylation [62]. Moreover, CS and oxidant (H_2O_2)-mediated decrease in HDAC2 activity and expression was also increased by curcumin treatment *in vitro* [63].

HDAC-2 plays a pivotal role in the activity of corticosteroid, a commonly used anti-inflammatory drug. Corticosteroids suppress inflammatory genes by inhibiting HAT and, in particular, by recruiting HDAC2 to the NF-κB-activated inflammatory gene complex [62]. Recruitment of HDAC2 leads to histone deacetylation and reduced binding of NF-κB with DNA results in decreased pro-inflammatory cytokine synthesis. In smokers and patients with COPD, HDAC2 level and activity were reduced due to the increased oxidative stress, thus preventing corticosteroids from suppressing inflammation [64]. Curcumin acts at a post-translational level to maintain both HDAC2 level and activity, thereby reversing steroid insensitivity induced by either CS extract or oxidative stress [63]. Thus, curcumin showed a concentration-dependent restoration of corticosteroid-mediated suppression of pro-inflammatory cytokine release in response to CS.

II) Resveratrol

Resveratrol is a potent activator of SIRT1. Resveratrol has the potency to increase SIRT1 activity by increasing affinity for acetylated substrate [65]. Resveratrol-mediated SIRT1 activation requires the presence of a fluorophore covalently attached to the acetylated substrate [65]. It was found to be the most potent activator of SIRT1 among 15 small molecular plant phenols, including quercetin and piceatannol [66].

SIRT1 regulates the transcriptional activity of NF-κB. SIRT1 physically interacts with the RelA/p65 subunit of NF-κB, and inhibits transcription by deacetylating p65 subunit at lysine 310 [67]. CS-induced upregulation of inflammatory markers (ICAM-1, iNOS, IL-6, and TNF-α) is abrogated by resveratrol treatment. Resveratrol inhibited CS extract-induced NF-κB activation, inflammatory gene expression and eNOS activity (endothelial function) in endothelial cells [68]. The NF-κB inactivation and anti-inflammatory activities of resveratrol were abolished by knockdown of SIRT1, which shows that resveratrol require SIRT1 for its protective effect in against inflammation and autophagy [69, 70]. Resveratrol is also reported to inhibit CS extract-mediated proinflammatory cytokine release in macrophages [71]. Taken together, resveratrol, likely via inducing SIRT1-dependent deacetylation of NF-κB, reduced the pro-inflammatory cytokine release and abrogated the pro-inflammatory effects of cigarette smoking.

III) Garcinol

Garcinol, also called as camboginol, (tri-isoprenylated chalcone) is derived from *Garcinia indica.* Due to the presence of one phenolic catechol group and 3 beta-diketone groups, it acts as a potent inhibitor of HAT (p300). It binds to histone and acetyl-CoA binding sites of HAT to inhibit the acetyltransferase activity [72]. Blockade of HAT by garcinol may attenuate inflammatory responses in airway diseases [73]. CS extract-mediated increase in the expression of pro-inflammatory enzyme COX-2 was reduced by pretreatment with garcinol, probably by disrupting the interaction between HAT and NF-κB [74]. The anti-inflammatory activities of garcinol may also due to its ability to scavenge reactive oxygen species and inhibit NF-κB activation [75, 76].

IV) Gallate

EGCG is a novel inhibitor of HAT with global specificity for the majority of HAT enzymes but with no activity toward histone deacetylase enzymes including HDAC and SIRT1. EGCG abrogates p300-mediated NF-κB (p65 subunit) acetylation *in vitro* and *in vivo*, and thereby suppresses NF-κB activation in response to pro-inflammatory stimuli [77]. However, further research is warranted to find out whether the same phenomenon happens in response to CS exposure.

3) Antioxidants

CS/tobacco smoke is a well-proven source of oxidative stress. Each puff of smoke contains more than 10^{14} molecules per puff reactive oxygen and reactive nitrogen species, which includes $O_2^{\bullet -}$, NO_2, $^{\bullet}OH$, $OONO^-$, and H_2O_2. In addition, cigarette smoking induces the release of free radicals from activated inflammatory cells [4, 78, 79]. The level of anti-oxidants in blood is significantly reduced in smokers. Thus, excessive generation of reactive oxygen species and insufficient antioxidant defense resulted in oxidant/antioxidant imbalance and oxidative stress. Given the importance of oxidative stress in CS-mediated lung inflamemation and the pathogenesis of COPD, one rational approach would be antioxidants targeted against oxidative stress that may be useful in the management of CS-mediated inflammation [80]. Plants are considered as a good source of natural antioxidants as they have a variety of phenolic compounds, which possess antioxidant activity [81]. These polyphenolic compounds attenuate CS-mediated oxidative stress and inflammation.

I) Vitamin C, E, β-Carotene

Non-enzymic antioxidants, such as vitamin C (ascorbic acid), vitamin E (α-tocopherol) and β-carotene are involved in decomposition of ROS and play an important role in the antioxidant defense of the lung. Smokers and patients with COPD showed decreased plasma levels of vitamin C, vitamin E and β-carotene, which is associated with CS-induced inflammatory response. Oral supplementation of vitamin C, vitamin E and β-carotene efficiently reduced CS-induced inflammatory process in animal experimental model [82, 83]. In human, intake of antioxidant vitamins is associated with decreased oxidative damage and improved lung function [84]. Daily intake of vitamin C over the recommended daily allowance (60 mg/day among nonsmokers and 100 mg/day among smokers) may have protective effect against COPD [85]. Unfortunately, inconsistent results are obtained during the clinical trials with the supplementation of vitamin C, vitamin E, and β-carotene in smokers and patients with COPD [86, 87].

II) Resveratrol

Resveratrol is an effective antioxidant because of its ability to scavenge reactive oxygen species inhibit low density lipoprotein oxidation [88, 89]. In humans, the oral bioavailability of resveratrol is low because of their rapid conversion to resveratrol metabolites, which has lower antioxidant capacity than resveratrol. However, resveratrol has been reported to increase the antioxidant capacity and protect against oxidative damage *in vitro* and *in vivo* [90, 91]. It seems that the antioxidant activity of resveratrol is mainly due to its ability to activate nuclear factor erythroid 2-related factor 2 (Nrf2), even at low concentration [92, 93]. Nrf2 is a redox-sensitive transcription factor involved in transcriptional regulation of many antioxidant genes.

The lungs of smokers and patients with COPD showed decline in the levels of Nrf2 protein and Nrf2-regulated antioxidant enzymes [94]. Nrf2-deficiency is associated with increased susceptibility to oxidative stress and CS-induced lung injury [92]. Resveratrol treatment induced the nuclear translocation of Nrf2 in response to CS exposure by inhibiting oxidant mediated post-transcriptional modifications. It also increased the antioxidant defense against CS through the activation of Nrf2 leading to increased glutamate-cysteine ligase (GCL) and glutathione synthesis in human lung epithelial cells [92]. Resveratrol-induced Nrf2 antioxidant defense pathway is gaining importance, because resveratrol has reported to reduce the levels of pro-inflammatory cytokines in rodent model of airway inflammation via NF-κB-independent mechanism [92]. Thus, antioxidant capacity of resveratrol may play a pivotal role in its anti-inflammatory properties.

III) Sulforaphane

Sulforaphane, 1-isothiocyanate-(4R)-(methylsulfinyl)butane, is present in many cruciferous vegetables, especially at high concentrations in broccoli (*Brassica oleracea*). It exerts its antioxidant effect by activating Nrf2 and thereby inducing phase 2 enzymes. Sulforaphane-induced activation of Nrf2 is mediated through specific modification in Kelch-like ECH-associated protein 1 (Keap1) that inhibit the activity of Nrf2 by sequestering it in the cytoplasm. Sulforaphane can react with thiol groups present in Keap1 to form thionoacyl adduct. Thus, the modification in Kelch domain of Keap1 leading to increased nuclear translocation of Nrf2 and expression of antioxidant genes [95]. Sulforaphane has been reported to induce the expression of Nrf2-dependent genes, such as NADPH quinone oxidoreductase-1, heme oxygenase-1, and glutamate cysteine ligase modulatory subunit in response to cigarette smoke exposure in human epithelial cells. The increased phase2 enzyme synthesis is associated with decreased *de novo* synthesis of pro-inflammatory mediators, IL-8 and MCP-1 [96]. Oral administration of sulforaphane reduced lung inflammation in chronic cigarette smoke-exposed mice [97], as well protected endothelial cells in arteries in response to inflammatory stimuli [98]. The ability of sulforaphane to inhibit NF-κB activation, by binding to the cysteine residues of NF-κB, may be responsible for its anti-inflammatory activity [99, 100]. However the anti-inflammatory activity of sulforaphane was not seen in Nrf2-deficient mice, which shows that sulforaphane exerts its anti-inflammatory effect via activation of Nrf2 [97].

IV) Curcumin

Curcumin possess a strong antioxidant activity and is more potent than vitamins C and E, because of the presence of phenolic groups and conjugated double bonds. It is also a potent inducer of Nrf2-related antioxidant enzymes under various pro-inflammatory conditions [101-103]. Oral administration of curcumin has been reported to up-regulate gene expression of Nrf2-regulated antioxidants, such as glutamate-cysteine ligase and glutathione reductase, and to attenuate CS-induced pulmonary inflammation and emphysema in mice [104]. Thus, the antioxidant activity of curcumin may be one of the multiple mechanisms whereby curcumin regulate CS-induced pro-inflammatory gene expression.

V) Gallate

EGCG has potent antioxidant activity which is 25 to 100 times more effective than vitamins C and E [105]. The consumption of tea and green tea catechins leads to modest transient increases in plasma antioxidant capacity in human [106]. Green tea (which has EGCG as its main active ingredient) have the ability to suppress CS-induced oxidative stress by inducing the activities of lung superoxide dismutase (SOD) and catalase that leads to protection against lung injury [107]. Recently, it has been shown that EGCG is an activator of Nrf2 pathway, and is associated with increased antioxidant defense and decreased nuclear NF-κB levels [108]. Thus, the antioxidant efficacy of EGCG may be helpful to execute its anti-inflammatory properties.

VI) Apocynin

Apocynin, also known as acetovanillone, (4-hydroxy-3-methoxy-acetophenone) is derived from the roots of *Apocynum cannabinum* and *Picrorhiza kurroa*. It is a selective inhibitor of nicotinamide adenine dinucleotide phosphate (NADPH) oxidase, a principle ROS-generating enzyme in inflammatory cells. Apocynin is effective in preventing NADPH oxidase-mediated production of the superoxide radicals. The mechanism of action involves metabolic activation of apocynin in a (myelo)-peroxidase-dependent reaction, followed by prevention of the assembly of NADPH:O_2 oxidoreductase by conjugation to essential thiol groups [109]. CS has been reported to induce NADPH oxidase activation and superoxide anion formation, which are associated with CS-induced lung inflammation and injury [110].

Apocynin showed anti-inflammatory effect in a variety of cell and animal models of inflammation [111], which may be mediated by reducing the activity of NF-κB [112]. CS extract-induced NF-κB activity and pro-inflammatory mediator (iNOS, TNF-α, IL-1, and IL-6) expression were reduced by apocynin in endothelial cells [113]. The mechanism of anti-inflammatory effect of apocynin may be via the reduction NADPH oxidase-derived ROS production leading to decreased NF-κB activation. Thus, apocynin may be used for redressing the oxidant/antioxidant imbalance and reducing inflammation in COPD. Oral administration of apocynin showed very low toxicity in mice (LD50: 9 g/kg). In contrast, genetic ablation of NDAPH leads to increased inflammatory response to CS, which was accompanied by the development of distal airspace enlargement and alveolar destruction [110]. The reason for the contrary is not known, however it shows that the extent of NADPH oxidase inhibition is very critical.

Table I. Phytochemicals in cigarette smoke-mediated inflammation

Compound	Natural source	Mode of action
Curcumin	Turmeric	Blocks NF-κB activation Increases HDAC2 activity Scavenges free radicals Activates Nrf2
Resveratrol	Grapes, red wine, peanuts, and some berries	Blocks NF-κB activation Increases SIRT1 activity Scavenges free radicals Activates Nrf2
EGCG	Green tea	Blocks NF-κB activation Inhibits HAT Scavenges free radicals
Ursolic acid	Rosemary, loquat, Heather, tulsi, and black plum	Blocks NF-κB activation
Zerumbone, gingerol	Ginger	Block NF-κB activation
Parthenolide	Feverfew	Blocks NF-κB activation
Helenalin	Leopard's bane	Alkylation of NF-κB
Baicalin	Baikal scullcap	Blocks NF-κB activation
Caffeic acid phenethyl ester	Pears, basil, and thyme	Blocks NF-κB activation
Garcinol	Kokum	Inhibits HAT (p300) activity
Vitamin C, E, β-carotene	Citrus fruits, seed oils, and carrots, respectively.	Scavenges free radicals
Sulforaphane	Broccoli	Activates Nrf2
Apocynin	Indian hemp, and Kutki	Inhibits NADPH oxidase

4) Other anti-inflammatory agents

Number of phytochemicals such as quercetin, capsaicin, gingerol, diallyl sulphide, lycopene, magnolol, kaempferol, scutellarin, apigenin, luteolin, oleanolic acid, ellagic acid, genistein and puerarin showed inhibitory effect on NF-κB activation and anti-inflammatory effects against various pro-inflammatory stimuli. However, detailed research is needed to evaluate their effect on CS-mediated lung inflammation and mode of action.

CONCLUSION

This chapter provides the current knowledge and update on potential therapeutic applications of phytochemicals in CS/tobacco smoke-mediated lung inflammation. Among the number of anti-inflammatory phytochemicals, curcumin, resveratrol and EGCG are of high interest because of their multiple mode of action, to inhibit pro-inflammatory gene transcription, including inhibition of NF-κB activation, modification of HAT/HDAC balance, and improvement in antioxidant defense. Potential protective effects of phytochemicals in *in vitro* and *in vivo* experimental models of inflammation suggest that phytochemicals can be considered as an alternative non-steroidal anti-inflammatory drug for inflammatory lung diseases, such as asthma and COPD.

REFERENCES

[1] Sherman, C. B. Health effects of cigarette smoking. *Clin. Chest. Med.* 1991, 12, 643-658.

[2] Mallampalli, A.; Guntupalli, K. K. Smoking and systemic disease. *Med. Clin. North Am.* 2004, 88, 1431-1451.

[3] Tuder, R. M.; Yoshida, T.; Arap, W.; Pasqualini, R.; Petrache, I. State of the art. Cellular and molecular mechanisms of alveolar destruction in emphysema: an evolutionary perspective. *Proc. Am .Thorac. Soc.* 2006, 3, 503-510.

[4] Rahman, I.; Adcock, I. M. Oxidative stress and redox regulation of lung inflammation in COPD. *Eur. Respir. J.* 2006, 28, 219-242.

[5] McNabola, A.; Gill, L. W. The control of environmental tobacco smoke: a policy review. *Int. J. Environ. Res. Public Health.* 2009, 6, 741-758.

[6] Murray, C. J.; Lopez, A. D. Alternative projections of mortality and disability by cause 1990-2020: Global Burden of Disease Study. *Lancet.* 1997, 349, 1498-1504.

[7] Barnes, P. J. Corticosteroids: the drugs to beat. *Eur. J. Pharmacol.* 2006, 533, 2-14.

[8] Culpitt, S. V.; Rogers, D. F.; Shah, P.; De Matos, C.; Russell, R. E.; Donnelly, L. E.; Barnes, P. J. Impaired inhibition by dexamethasone of cytokine release by alveolar macrophages from patients with chronic obstructive pulmonary disease. *Am. J. Respir. Crit. Care Med.* 2003, 167, 24-31.

[9] Rajendrasozhan, S.; Yang, S. R.; Kinnula, V. L.; Rahman, I. SIRT1, an antiinflammatory and antiaging protein, is decreased in lungs of patients with chronic obstructive pulmonary disease. *Am. J. Respir. Crit. Care Med.* 2008, 177, 861-870.

[10] Bengmark, S. Curcumin, an atoxic antioxidant and natural NFkappaB, cyclooxygenase-2, lipooxygenase, and inducible nitric oxide synthase inhibitor: a shield against acute and chronic diseases. *JPEN J. Parenter Enteral. Nutr.* 2006, 30, 45-51.

[11] Lerner, L.; Weiner, D.; Katz, R.; Reznick, A. Z.; Pollack, S. Increased pro-inflammatory activity and impairment of human monocyte differentiation induced by in vitro exposure to cigarette smoke. *J. Physiol. Pharmacol.* 2009, 60 Suppl 5, 81-86.

[12] Shishodia, S.; Potdar, P.; Gairola, C. G.; Aggarwal, B. B. Curcumin (diferuloylmethane) down-regulates cigarette smoke-induced NF-kappaB activation through inhibition of IkappaBalpha kinase in human lung epithelial cells: correlation with suppression of COX-2, MMP-9 and cyclin D1. *Carcinogenesis.* 2003, 24, 1269-1279.

[13] Biswas, S. K.; McClure, D.; Jimenez, L. A.; Megson, I. L.; Rahman, I. Curcumin induces glutathione biosynthesis and inhibits NF-kappaB activation and interleukin-8 release in alveolar epithelial cells: mechanism of free radical scavenging activity. *Antioxid Redox Signal.* 2005, 7, 32-41.

[14] Sandelowski, M. Telling stories: narrative approaches in qualitative research. *Image J. Nurs. Sch.* 1991, 23, 161-166.

[15] Moghaddam, S. J.; Barta, P.; Mirabolfathinejad, S. G.; Ammar-Aouchiche, Z.; Garza, N. T.; Vo, T. T.; Newman, R. A.; Aggarwal, B. B.; Evans, C. M.; Tuvim, M. J.; Lotan, R.; Dickey, B. F. Curcumin inhibits COPD-like airway inflammation and lung cancer progression in mice. *Carcinogenesis.* 2009, 30, 1949-1956.

[16] White, B.; Judkins, D. Z. Clinical Inquiry. Does turmeric relieve inflammatory conditions? *J. Fam. Pract.* 2011, 60, 155-156.

[17] Shoba, G.; Joy, D.; Joseph, T.; Majeed, M.; Rajendran, R.; Srinivas, P. S. Influence of piperine on the pharmacokinetics of curcumin in animals and human volunteers. *Planta Med.* 1998, 64, 353-356.

[18] Li, L.; Braiteh, F. S.; Kurzrock, R. Liposome-encapsulated curcumin: in vitro and in vivo effects on proliferation, apoptosis, signaling, and angiogenesis. *Cancer.* 2005, 104, 1322-1331.

[19] Shen, F.; Chen, S. J.; Dong, X. J.; Zhong, H.; Li, Y. T.; Cheng, G. F. Suppression of IL-8 gene transcription by resveratrol in phorbol ester treated human monocytic cells. *J. Asian Nat. Prod. Res.* 2003, 5, 151-157.

[20] Donnelly, L. E.; Newton, R.; Kennedy, G. E.; Fenwick, P. S.; Leung, R. H.; Ito, K.; Russell, R. E.; Barnes, P. J. Anti-inflammatory effects of resveratrol in lung epithelial cells: molecular mechanisms. *Am. J. Physiol. Lung Cell Mol. Physiol.* 2004, 287, L774-783.

[21] Culpitt, S. V.; Rogers, D. F.; Fenwick, P. S.; Shah, P.; De Matos, C.; Russell, R. E.; Barnes, P. J.; Donnelly, L. E. Inhibition by red wine extract, resveratrol, of cytokine release by alveolar macrophages in COPD. *Thorax.* 2003, 58, 942-946.

[22] Knobloch, J.; Sibbing, B.; Jungck, D.; Lin, Y.; Urban, K.; Stoelben, E.; Strauch, J.; Koch, A. Resveratrol impairs the release of steroid-resistant inflammatory cytokines from human airway smooth muscle cells in chronic obstructive pulmonary disease. *J. Pharmacol. Exp. Ther.* 2010, 335, 788-798.

[23] Holmes-McNary, M.; Baldwin, A. S., Jr. Chemopreventive properties of trans-resveratrol are associated with inhibition of activation of the IkappaB kinase. *Cancer Res.* 2000, 60, 3477-3483.

[24] Walle, T.; Hsieh, F.; DeLegge, M. H.; Oatis, J. E., Jr.; Walle, U. K. High absorption but very low bioavailability of oral resveratrol in humans. *Drug Metab. Dispos.* 2004, 32, 1377-1382.

[25] Smith, J. J.; Kenney, R. D.; Gagne, D. J.; Frushour, B. P.; Ladd, W.; Galonek, H. L.; Israelian, K.; Song, J.; Razvadauskaite, G.; Lynch, A. V.; Carney, D. P.; Johnson, R. J.; Lavu, S.; Iffland, A.; Elliott, P. J.; Lambert, P. D.; Elliston, K. O.; Jirousek, M. R.; Milne, J. C.; Boss, O. Small molecule activators of SIRT1 replicate signaling pathways triggered by calorie restriction in vivo. *BMC Syst. Biol.* 2009, 3, 31.

[26] Zhou, M.; He, J. L.; Yu, S. Q.; Zhu, R. F.; Lu, J.; Ding, F. Y.; Xu, G. L. Effect of resveratrol on chronic obstructive pulmonary disease in rats and its mechanism. *Yao Xue Xue Bao.* 2008, 43, 128-132.

[27] Boocock, D. J.; Faust, G. E.; Patel, K. R.; Schinas, A. M.; Brown, V. A.; Ducharme, M. P.; Booth, T. D.; Crowell, J. A.; Perloff, M.; Gescher, A. J.; Steward, W. P.; Brenner, D. E. Phase I dose escalation pharmacokinetic study in healthy volunteers of resveratrol, a potential cancer chemopreventive agent. *Cancer Epidemiol. Biomarkers Prev.* 2007, 16, 1246-1252.

[28] Yang, F.; Oz, H. S.; Barve, S.; de Villiers, W. J.; McClain, C. J.; Varilek, G. W. The green tea polyphenol (-)-epigallocatechin-3-gallate blocks nuclear factor-kappa B activation by inhibiting I kappa B kinase activity in the intestinal epithelial cell line IEC-6. *Mol. Pharmacol.* 2001, 60, 528-533.

[29] Yang, F.; de Villiers, W. J.; McClain, C. J.; Varilek, G. W. Green tea polyphenols block endotoxin-induced tumor necrosis factor-production and lethality in a murine model. *J. Nutr.* 1998, 128, 2334-2340.

[30] Hara, Y. Influence of tea catechins on the digestive tract. *J. Cell Biochem. Suppl.* 1997, 27, 52-58.

[31] Syed, D. N.; Afaq, F.; Kweon, M. H.; Hadi, N.; Bhatia, N.; Spiegelman, V. S.; Mukhtar, H. Green tea polyphenol EGCG suppresses cigarette smoke condensate-induced NF-kappaB activation in normal human bronchial epithelial cells. *Oncogene.* 2007, 26, 673-682.

[32] March, T. H.; Wilder, J. A.; Esparza, D. C.; Cossey, P. Y.; Blair, L. F.; Herrera, L. K.; McDonald, J. D.; Campen, M. J.; Mauderly, J. L.; Seagrave, J. Modulators of cigarette smoke-induced pulmonary emphysema in A/J mice. *Toxicol. Sci.* 2006, 92, 545-559.

[33] Chen, L.; Lee, M. J.; Li, H.; Yang, C. S. Absorption, distribution, elimination of tea polyphenols in rats. *Drug Metab. Dispos.* 1997, 25, 1045-1050.

[34] Chow, H. H.; Hakim, I. A.; Vining, D. R.; Crowell, J. A.; Ranger-Moore, J.; Chew, W. M.; Celaya, C. A.; Rodney, S. R.; Hara, Y.; Alberts, D. S. Effects of dosing condition on the oral bioavailability of green tea catechins after single-dose administration of Polyphenon E in healthy individuals. *Clin. Cancer Res.* 2005, 11, 4627-4633.

[35] Smith, A.; Giunta, B.; Bickford, P. C.; Fountain, M.; Tan, J.; Shytle, R. D. Nanolipidic particles improve the bioavailability and alpha-secretase inducing ability of epigallocatechin-3-gallate (EGCG) for the treatment of Alzheimer's disease. *Int. J. Pharm.* 2010, 389, 207-212.

[36] Shanafelt, T. D.; Call, T. G.; Zent, C. S.; LaPlant, B.; Bowen, D. A.; Roos, M.; Secreto, C. R.; Ghosh, A. K.; Kabat, B. F.; Lee, M. J.; Yang, C. S.; Jelinek, D. F.; Erlichman, C.; Kay, N. E. Phase I trial of daily oral Polyphenon E in patients with asymptomatic Rai stage 0 to II chronic lymphocytic leukemia. *J. Clin. Oncol.* 2009, 27, 3808-3814.

[37] Chow, H. H.; Cai, Y.; Hakim, I. A.; Crowell, J. A.; Shahi, F.; Brooks, C. A.; Dorr, R. T.; Hara, Y.; Alberts, D. S. Pharmacokinetics and safety of green tea polyphenols after

multiple-dose administration of epigallocatechin gallate and polyphenon E in healthy individuals. *Clin. Cancer Res.* 2003, 9, 3312-3319.

[38] Liu, J. Pharmacology of oleanolic acid and ursolic acid. *J. Ethnopharmacol.* 1995, 49, 57-68.

[39] Shishodia, S.; Majumdar, S.; Banerjee, S.; Aggarwal, B. B. Ursolic acid inhibits nuclear factor-kappaB activation induced by carcinogenic agents through suppression of IkappaBalpha kinase and p65 phosphorylation: correlation with down-regulation of cyclooxygenase 2, matrix metalloproteinase 9, and cyclin D1. *Cancer Res.* 2003, 63, 4375-4383.

[40] Ikeda, Y.; Murakami, A.; Ohigashi, H. Ursolic acid: an anti- and pro-inflammatory triterpenoid. *Mol. Nutr. Food Res.* 2008, 52, 26-42.

[41] Sulaiman, M. R.; Perimal, E. K.; Akhtar, M. N.; Mohamad, A. S.; Khalid, M. H.; Tasrip, N. A.; Mokhtar, F.; Zakaria, Z. A.; Lajis, N. H.; Israf, D. A. Anti-inflammatory effect of zerumbone on acute and chronic inflammation models in mice. *Fitoterapia.* 2010, 81, 855-858.

[42] Takada, Y.; Murakami, A.; Aggarwal, B. B. Zerumbone abolishes NF-kappaB and IkappaBalpha kinase activation leading to suppression of antiapoptotic and metastatic gene expression, upregulation of apoptosis, and downregulation of invasion. *Oncogene.* 2005, 24, 6957-6969.

[43] Saadane, A.; Masters, S.; DiDonato, J.; Li, J.; Berger, M. Parthenolide inhibits IkappaB kinase, NF-kappaB activation, and inflammatory response in cystic fibrosis cells and mice. *Am. J. Respir. Cell Mol. Biol.* 2007, 36, 728-736.

[44] Hehner, S. P.; Hofmann, T. G.; Droge, W.; Schmitz, M. L. The antiinflammatory sesquiterpene lactone parthenolide inhibits NF-kappa B by targeting the I kappa B kinase complex. *J. Immunol.* 1999, 163, 5617-5623.

[45] Hehner, S. P.; Heinrich, M.; Bork, P. M.; Vogt, M.; Ratter, F.; Lehmann, V.; Schulze-Osthoff, K.; Droge, W.; Schmitz, M. L. Sesquiterpene lactones specifically inhibit activation of NF-kappa B by preventing the degradation of I kappa B-alpha and I kappa B-beta. *J. Biol. Chem.* 1998, 273, 1288-1297.

[46] Lyss, G.; Knorre, A.; Schmidt, T. J.; Pahl, H. L.; Merfort, I. The anti-inflammatory sesquiterpene lactone helenalin inhibits the transcription factor NF-kappaB by directly targeting p65. *J. Biol. Chem.* 1998, 273, 33508-33516.

[47] Kelsey, N. A.; Wilkins, H. M.; Linseman, D. A. Nutraceutical antioxidants as novel neuroprotective agents. *Molecules.* 2010, 15, 7792-7814.

[48] Cheng, S. E.; Luo, S. F.; Jou, M. J.; Lin, C. C.; Kou, Y. R.; Lee, I. T.; Hsieh, H. L.; Yang, C. M. Cigarette smoke extract induces cytosolic phospholipase A2 expression via NADPH oxidase, MAPKs, AP-1, and NF-kappaB in human tracheal smooth muscle cells. *Free Radic Biol. Med.* 2009, 46, 948-960.

[49] Li-Weber, M. New therapeutic aspects of flavones: the anticancer properties of Scutellaria and its main active constituents Wogonin, Baicalein and Baicalin. *Cancer Treat Rev.* 2009, 35, 57-68.

[50] Chou, T. C.; Chang, L. P.; Li, C. Y.; Wong, C. S.; Yang, S. P. The antiinflammatory and analgesic effects of baicalin in carrageenan-evoked thermal hyperalgesia. *Anesth. Analg.* 2003, 97, 1724-1729.

[51] Lixuan, Z.; Jingcheng, D.; Wenqin, Y.; Jianhua, H.; Baojun, L.; Xiaotao, F. Baicalin attenuates inflammation by inhibiting NF-kappaB activation in cigarette smoke induced inflammatory models. *Pulm. Pharmacol. Ther.* 2010, 23, 411-419.
[52] Pekmez, H.; Ogeturk, M.; Ozyurt, H.; Sonmez, M. F.; Colakoglu, N.; Kus, I. Ameliorative effect of caffeic acid phenethyl ester on histopathological and biochemical changes induced by cigarette smoke in rat kidney. *Toxicol. Ind. Health.* 2010, 26, 175-182.
[53] Pekmez, H.; Kus, I.; Colakoglu, N.; Ogeturk, M.; Ozyurt, H.; Turkoglu, A. O.; Sarsilmaz, M. The protective effects of caffeic acid phenethyl ester (CAPE) against liver damage induced by cigarette smoke inhalation in rats. *Cell Biochem. Funct.* 2007, 25, 395-400.
[54] Ozyurt, H.; Pekmez, H.; Parlaktas, B. S.; Kus, I.; Ozyurt, B.; Sarsilmaz, M. Oxidative stress in testicular tissues of rats exposed to cigarette smoke and protective effects of caffeic acid phenethyl ester. *Asian J. Androl.* 2006, 8, 189-193.
[55] Natarajan, K.; Singh, S.; Burke, T. R., Jr.; Grunberger, D.; Aggarwal, B. B. Caffeic acid phenethyl ester is a potent and specific inhibitor of activation of nuclear transcription factor NF-kappa B. *Proc. Natl. Acad. Sci. U. S. A.* 1996, 93, 9090-9095.
[56] Jung, W. K.; Lee, D. Y.; Choi, Y. H.; Yea, S. S.; Choi, I.; Park, S. G.; Seo, S. K.; Lee, S. W.; Lee, C. M.; Kim, S. K.; Jeon, Y. J.; Choi, I. W. Caffeic acid phenethyl ester attenuates allergic airway inflammation and hyperresponsiveness in murine model of ovalbumin-induced asthma. *Life Sci.* 2008, 82, 797-805.
[57] Andrade-Silva, A. R.; Ramalho, F. S.; Ramalho, L. N.; Saavedra-Lopes, M.; Jordao, A. A., Jr.; Vanucchi, H.; Piccinato, C. E.; Zucoloto, S. Effect of NFkappaB inhibition by CAPE on skeletal muscle ischemia-reperfusion injury. *J. Surg. Res.* 2009, 153, 254-262.
[58] Song, J. J.; Cho, J. G.; Hwang, S. J.; Cho, C. G.; Park, S. W.; Chae, S. W. Inhibitory effect of caffeic acid phenethyl ester (CAPE) on LPS-induced inflammation of human middle ear epithelial cells. *Acta Otolaryngol.* 2008, 128, 1303-1307.
[59] Sezer, M.; Sahin, O.; Solak, O.; Fidan, F.; Kara, Z.; Unlu, M. Effects of caffeic acid phenethyl ester on the histopathological changes in the lungs of cigarette smoke-exposed rabbits. *Basic Clin. Pharmacol .Toxicol.* 2007, 101, 187-191.
[60] Annunziato, A. DNA Packaging: Nucleosomes and Chromatin. *Nature Education.* 2008, 1.
[61] Ito, K.; Ito, M.; Elliott, W. M.; Cosio, B.; Caramori, G.; Kon, O. M.; Barczyk, A.; Hayashi, S.; Adcock, I. M.; Hogg, J. C.; Barnes, P. J. Decreased histone deacetylase activity in chronic obstructive pulmonary disease. *N. Engl. J. Med.* 2005, 352, 1967-1976.
[62] Marwick, J. A.; Ito, K.; Adcock, I. M.; Kirkham, P. A. Oxidative stress and steroid resistance in asthma and COPD: pharmacological manipulation of HDAC-2 as a therapeutic strategy. *Expert Opin. Ther. Targets.* 2007, 11, 745-755.
[63] Meja, K. K.; Rajendrasozhan, S.; Adenuga, D.; Biswas, S. K.; Sundar, I. K.; Spooner, G.; Marwick, J. A.; Chakravarty, P.; Fletcher, D.; Whittaker, P.; Megson, I. L.; Kirkham, P. A.; Rahman, I. Curcumin restores corticosteroid function in monocytes exposed to oxidants by maintaining HDAC2. *Am. J. Respir. Cell Mol. Biol,*.2008, 39, 312-323.

[64] Barnes, P. J. Histone deacetylase-2 and airway disease. *Ther. Adv. Respir. Dis.* 2009, 3, 235-243.

[65] Borra, M. T.; Smith, B. C.; Denu, J. M. Mechanism of human SIRT1 activation by resveratrol. *J. Biol. Chem.* 2005, 280, 17187-17195.

[66] Howitz, K. T.; Bitterman, K. J.; Cohen, H. Y.; Lamming, D. W.; Lavu, S.; Wood, J. G.; Zipkin, R. E.; Chung, P.; Kisielewski, A.; Zhang, L. L.; Scherer, B.; Sinclair, D. A. Small molecule activators of sirtuins extend Saccharomyces cerevisiae lifespan. *Nature.* 2003, 425, 191-196.

[67] Yeung, F.; Hoberg, J. E.; Ramsey, C. S.; Keller, M. D.; Jones, D. R.; Frye, R. A.; Mayo, M. W. Modulation of NF-kappaB-dependent transcription and cell survival by the SIRT1 deacetylase. *Embo J.* 2004, 23, 2369-2380.

[68] Arunachalam, G.; Yao, H.; Sundar, I. K.; Caito, S.; Rahman, I. SIRT1 regulates oxidant- and cigarette smoke-induced eNOS acetylation in endothelial cells: Role of resveratrol. *Biochem. Biophys. Res. Commun.* 2010, 393, 66-72.

[69] Csiszar, A.; Labinskyy, N.; Podlutsky, A.; Kaminski, P. M.; Wolin, M. S.; Zhang, C.; Mukhopadhyay, P.; Pacher, P.; Hu, F.; de Cabo, R.; Ballabh, P.; Ungvari, Z. Vasoprotective effects of resveratrol and SIRT1: attenuation of cigarette smoke-induced oxidative stress and proinflammatory phenotypic alterations. *Am. J. Physiol. Heart Circ. Physiol.* 2008, 294, H2721-2735.

[70] Hwang, J. W.; Chung, S.; Sundar, I. K.; Yao, H.; Arunachalam, G.; McBurney, M. W.; Rahman, I. Cigarette smoke-induced autophagy is regulated by SIRT1-PARP-1-dependent mechanism: implication in pathogenesis of COPD. *Arch. Biochem. Biophys.* 2010, 500, 203-209.

[71] Yang, S. R.; Wright, J.; Bauter, M.; Seweryniak, K.; Kode, A.; Rahman, I. Sirtuin regulates cigarette smoke-induced proinflammatory mediator release via RelA/p65 NF-kappaB in macrophages in vitro and in rat lungs in vivo: implications for chronic inflammation and aging. *Am. J. Physiol. Lung Cell Mol. Physiol.* 2007, 292, L567-576.

[72] Arif, M.; Pradhan, S. K.; Thanuja, G. R.; Vedamurthy, B. M.; Agrawal, S.; Dasgupta, D.; Kundu, T. K. Mechanism of p300 specific histone acetyltransferase inhibition by small molecules. *J. Med. Chem.* 2009, 52, 267-277.

[73] Luo, S. F.; Chang, C. C.; Lee, I. T.; Lee, C. W.; Lin, W. N.; Lin, C. C.; Yang, C. M. Activation of ROS/NF-kappaB and Ca2+/CaM kinase II are necessary for VCAM-1 induction in IL-1beta-treated human tracheal smooth muscle cells. *Toxicol. Appl. Pharmacol.* 2009, 237, 8-21.

[74] Yang, C. M.; Lee, I. T.; Lin, C. C.; Yang, Y. L.; Luo, S. F.; Kou, Y. R.; Hsiao, L. D. Cigarette smoke extract induces COX-2 expression via a PKCalpha/c-Src/EGFR, PDGFR/PI3K/Akt/NF-kappaB pathway and p300 in tracheal smooth muscle cells. *Am. J. Physiol. Lung Cell Mol. Physiol.* 2009, 297, L892-902.

[75] Liao, C. H.; Sang, S.; Liang, Y. C.; Ho, C. T.; Lin, J. K. Suppression of inducible nitric oxide synthase and cyclooxygenase-2 in downregulating nuclear factor-kappa B pathway by Garcinol. *Mol. Carcinog.* 2004, 41, 140-149.

[76] Yamaguchi, F.; Saito, M.; Ariga, T.; Yoshimura, Y.; Nakazawa, H. Free radical scavenging activity and antiulcer activity of garcinol from Garcinia indica fruit rind. *J. Agric. Food Chem.* 2000, 48, 2320-2325.

[77] Choi, K. C.; Jung, M. G.; Lee, Y. H.; Yoon, J. C.; Kwon, S. H.; Kang, H. B.; Kim, M. J.; Cha, J. H.; Kim, Y. J.; Jun, W. J.; Lee, J. M.; Yoon, H. G. Epigallocatechin-3-

gallate, a histone acetyltransferase inhibitor, inhibits EBV-induced B lymphocyte transformation via suppression of RelA acetylation. *Cancer Res.* 2009, 69, 583-592.

[78] Rahman, I. Antioxidant therapies in COPD. *Int. J. Chron. Obstruct. Pulmon. Dis.* 2006, 1, 15-29.

[79] Rahman, I.; Kilty, I. Antioxidant therapeutic targets in COPD. *Curr. Drug Targets.* 2006, 7, 707-720.

[80] Rahman, I.; Biswas, S. K.; Kirkham, P. A. Regulation of inflammation and redox signaling by dietary polyphenols. *Biochem. Pharmacol.* 2006, 72, 1439-1452.

[81] Schubert, S. Y.; Neeman, I.; Resnick, N. A novel mechanism for the inhibition of NF-kappaB activation in vascular endothelial cells by natural antioxidants. *Faseb J.* 2002, 16, 1931-1933.

[82] Silva Bezerra, F.; Valenca, S. S.; Lanzetti, M.; Pimenta, W. A.; Castro, P.; Goncalves Koatz, V. L.; Porto, L. C. Alpha-tocopherol and ascorbic acid supplementation reduced acute lung inflammatory response by cigarette smoke in mouse. *Nutrition.* 2006, 22, 1192-1201.

[83] Pang, B.; Wang, C.; Weng, X.; Tang, X.; Zhang, H.; Niu, S.; Mao, Y.; Xin, P.; Huang, X.; Zhang, H.; Zhu, J. Beta-carotene protects rats against bronchitis induced by cigarette smoking. *Chin. Med. J. (Engl.),* 2003, 116, 514-516.

[84] Grievink, L.; Smit, H. A.; Ocke, M. C.; van 't Veer, P.; Kromhout, D. Dietary intake of antioxidant (pro)-vitamins, respiratory symptoms and pulmonary function: the MORGEN study. *Thorax.* 1998, 53, 166-171.

[85] Romieu, I.; Trenga, C. Diet and obstructive lung diseases. *Epidemiol. Rev.* 2001, 23, 268-287.

[86] Nadeem, A.; Raj, H. G.; Chhabra, S. K. Effect of vitamin E supplementation with standard treatment on oxidant-antioxidant status in chronic obstructive pulmonary disease. *Indian J. Med. Res.* 2008, 128, 705-711.

[87] Wu, T. C.; Huang, Y. C.; Hsu, S. Y.; Wang, Y. C.; Yeh, S. L. Vitamin E and vitamin C supplementation in patients with chronic obstructive pulmonary disease. *Int. J. Vitam. Nutr. Res.* 2007, 77, 272-279.

[88] Stojanovic, S.; Sprinz, H.; Brede, O. Efficiency and mechanism of the antioxidant action of trans-resveratrol and its analogues in the radical liposome oxidation. *Arch. Biochem. Biophys.* 2001, 391, 79-89.

[89] Vivancos, M.; Moreno, J. J. Effect of resveratrol, tyrosol and beta-sitosterol on oxidised low-density lipoprotein-stimulated oxidative stress, arachidonic acid release and prostaglandin E2 synthesis by RAW 264.7 macrophages. *Br. J. Nutr.* 2008, 99, 1199-1207.

[90] Fremont, L. Biological effects of resveratrol. *Life Sci.* 2000, 66, 663-673.

[91] Floreani, M.; Napoli, E.; Quintieri, L.; Palatini, P. Oral administration of trans-resveratrol to guinea pigs increases cardiac DT-diaphorase and catalase activities, and protects isolated atria from menadione toxicity. *Life Sci.* 2003, 72, 2741-2750.

[92] Kode, A.; Rajendrasozhan, S.; Caito, S.; Yang, S. R.; Megson, I. L.; Rahman, I. Resveratrol induces glutathione synthesis by activation of Nrf2 and protects against cigarette smoke-mediated oxidative stress in human lung epithelial cells. *Am. J. Physiol. Lung Cell Mol. Physiol.* 2008, 294, L478-488.

[93] Hasko, G.; Pacher, P. Endothelial Nrf2 activation: a new target for resveratrol? *Am. J. Physiol. Heart Circ. Physiol.* 2010, 299, H10-12.

[94] Malhotra, D.; Thimmulappa, R.; Navas-Acien, A.; Sandford, A.; Elliott, M.; Singh, A.; Chen, L.; Zhuang, X.; Hogg, J.; Pare, P.; Tuder, R. M.; Biswal, S. Decline in NRF2-regulated antioxidants in chronic obstructive pulmonary disease lungs due to loss of its positive regulator, DJ-1. *Am. J. Respir. Crit. Care Med.* 2008, 178, 592-604.
[95] Hong, F.; Freeman, M. L.; Liebler, D. C. Identification of sensor cysteines in human Keap1 modified by the cancer chemopreventive agent sulforaphane. *Chem. Res. Toxicol.* 2005, 18, 1917-1926.
[96] Starrett, W.; Blake, D. J. Sulforaphane inhibits de novo synthesis of IL-8 and MCP-1 in human epithelial cells generated by cigarette smoke extract. *J. Immunotoxicol.* 2011, 8, 150-158.
[97] Harvey, C. J.; Thimmulappa, R. K.; Sethi, S.; Kong, X.; Yarmus, L.; Brown, R. H.; Feller-Kopman, D.; Wise, R.; Biswal, S. Targeting Nrf2 Signaling Improves Bacterial Clearance by Alveolar Macrophages in Patients with COPD and in a Mouse Model. *Sci. Transl. Med.* 2011, 3, 78-132.
[98] Zakkar, M.; Van der Heiden, K.; Luong le, A.; Chaudhury, H.; Cuhlmann, S.; Hamdulay, S. S.; Krams, R.; Edirisinghe, I.; Rahman, I.; Carlsen, H.; Haskard, D. O.; Mason, J. C.; Evans, P. C. Activation of Nrf2 in endothelial cells protects arteries from exhibiting a proinflammatory state. *Arterioscler. Thromb. Vasc. Biol.* 2009, 29, 1851-1857.
[99] Kivela, A. M.; Makinen, P. I.; Jyrkkanen, H. K.; Mella-Aho, E.; Xia, Y.; Kansanen, E.; Leinonen, H.; Verma, I. M.; Yla-Herttuala, S.; Levonen, A. L. Sulforaphane inhibits endothelial lipase expression through NF-kappaB in endothelial cells. *Atherosclerosis.* 2010, 213, 122-128.
[100] Heiss, E.; Herhaus, C.; Klimo, K.; Bartsch, H.; Gerhauser, C. Nuclear factor kappa B is a molecular target for sulforaphane-mediated anti-inflammatory mechanisms. *J. Biol. Chem.* 2001, 276, 32008-32015.
[101] Balogun, E.; Hoque, M.; Gong, P.; Killeen, E.; Green, C. J.; Foresti, R.; Alam, J.; Motterlini, R. Curcumin activates the haem oxygenase-1 gene via regulation of Nrf2 and the antioxidant-responsive element. *Biochem. J.* 2003, 371, 887-895.
[102] Betsuyaku, T.; Hamamura, I.; Hata, J.; Takahashi, H.; Mitsuhashi, H.; Adair-Kirk, T. L.; Senior, R. M.; Nishimura, M. Bronchiolar chemokine expression is different after single versus repeated cigarette smoke exposure. *Respir. Res.* 2008, 9, 7.
[103] Nishinaka, T.; Ichijo, Y.; Ito, M.; Kimura, M.; Katsuyama, M.; Iwata, K.; Miura, T.; Terada, T.; Yabe-Nishimura, C. Curcumin activates human glutathione S-transferase P1 expression through antioxidant response element. *Toxicol. Lett.* 2007, 170, 238-247.
[104] Suzuki, M.; Betsuyaku, T.; Ito, Y.; Nagai, K.; Odajima, N.; Moriyama, C.; Nasuhara, Y.; Nishimura, M. Curcumin attenuates elastase- and cigarette smoke-induced pulmonary emphysema in mice. *Am. J. Physiol. Lung Cell Mol. Physiol.* 2009, 296, L614-623.
[105] Vinson, J. A.; Dabbagh, Y. A. Tea phenols: Antioxidant effectiveness of teas, tea components, tea fractions and their binding with lipoproteins. *Nutrition Research.* 1998, 18, 1067-1075.
[106] Higdon, J. V.; Frei, B. Tea catechins and polyphenols: health effects, metabolism, and antioxidant functions. *Crit. Rev. Food Sci. Nutr.* 2003, 43, 89-143.

[107] Chan, K. H.; Ho, S. P.; Yeung, S. C.; So, W. H.; Cho, C. H.; Koo, M. W.; Lam, W. K.; Ip, M. S.; Man, R. Y.; Mak, J. C. Chinese green tea ameliorates lung injury in cigarette smoke-exposed rats. *Respir. Med.* 2009, 103, 1746-1754.

[108] Sahin, K.; Tuzcu, M.; Gencoglu, H.; Dogukan, A.; Timurkan, M.; Sahin, N.; Aslan, A.; Kucuk, O. Epigallocatechin-3-gallate activates Nrf2/HO-1 signaling pathway in cisplatin-induced nephrotoxicity in rats. *Life Sci.* 2010, 87, 240-245.

[109] Hart, B. A.; Simons, J. M. Metabolic activation of phenols by stimulated neutrophils: a concept for a selective type of anti-inflammatory drug. *Biotechnol. Ther.* 1992, 3, 119-135.

[110] Yao, H.; Edirisinghe, I.; Yang, S. R.; Rajendrasozhan, S.; Kode, A.; Caito, S.; Adenuga, D.; Rahman, I. Genetic ablation of NADPH oxidase enhances susceptibility to cigarette smoke-induced lung inflammation and emphysema in mice. *Am. J. Pathol.* 2008, 172, 1222-1237.

[111] Stefanska, J.; Pawliczak, R. Apocynin: molecular aptitudes. *Mediators Inflamm.* 2008, 2008, 106507.

[112] Kim, J. H.; Na, H. J.; Kim, C. K.; Kim, J. Y.; Ha, K. S.; Lee, H.; Chung, H. T.; Kwon, H. J.; Kwon, Y. G.; Kim, Y. M. The non-provitamin A carotenoid, lutein, inhibits NF-kappaB-dependent gene expression through redox-based regulation of the phosphatidylinositol 3-kinase/PTEN/Akt and NF-kappaB-inducing kinase pathways: role of H(2)O(2) in NF-kappaB activation. *Free Radic. Biol. Med.* 2008, 45, 885-896.

[113] Orosz, Z.; Csiszar, A.; Labinskyy, N.; Smith, K.; Kaminski, P. M.; Ferdinandy, P.; Wolin, M. S.; Rivera, A.; Ungvari, Z. Cigarette smoke-induced proinflammatory alterations in the endothelial phenotype: role of NAD(P)H oxidase activation. *Am. J. Physiol. Heart Circ. Physiol.* 2007, 292, H130-139.

In: Natural Products and Their Active Compounds … ISBN: 978-1-62100-153-9
Editors: M. Essa, A. Manickavasagan, and E. Sukumar © 2012 Nova Science Publishers, Inc.

Chapter 11

ANTIMICROBIAL ACTIVITY OF NATURAL PRODUCTS

Nallusamy Sivakumar
Sultan Qaboos University, Oman

ABSTRACT

Natural products have been the most productive source of leads for the development of drugs. Natural products provide an unlimited opportunity for new drug leads because of the unparalleled chemical diversity. Many pharmaceutical agents have been discovered by screening natural products from plants, animals, marine organisms and microorganisms.

These biologically derived antibacterial compounds could have reasonable value in controlling antibiotic resistant pathogens also. The prevalence of antibacterial natural products drugs may be due to the evolution of secondary metabolites as biologically active chemicals that conferred selective advantages to the producing organisms. This chapter provides an overview of natural products, their sources, their role in increasing antibiotic susceptibility of drug resistant bacteria and the synergism between natural products and antibiotics.

INTRODUCTION

Infectious diseases are caused by organisms like bacteria, viruses, fungi and parasites. The disease is due to a complex interaction between the pathogen, host and environment. Natural products have been investigated and utilized to relieve from diseases since early human history. Before the synthetic era, in the early 1900s, 80% of all medicines were obtained from barks, leaves and roots. In recent times also natural products have continued to be significant sources of drugs. Approximately 60% of anticancer compounds and 75% of drugs for infectious diseases that are either natural products or natural product derivatives evidenced their leading role (Newman et al, 2003; Cragg et al., 2005). The application of antibiotics defended the human beings from the damage caused by these infectious agents.

The indiscriminate use of antibiotics led to the development of multidrug resistant pathogens. The impact of infectious diseases is high in developing countries due to the relative unavailability of medicines and the emergence of widespread drug resistance (Okeke et al., 2005). In most of the Asian countries around 70-80% of the *Staphylococcal aureus* strains are methycillin resistant (Chambers, 2001) but the same strains are penicillin resistant worldwide (Casal et al., 2005). Research on new antimicrobial substances must therefore be continued and all possible strategies should be explored.

Drugs derived from plant resources have been empirically used to treat various human infections in the form of traditional medicines and the current research on natural molecules and products mainly focuses on plants since they can be sourced more easily and be selected on the basis of their ethno-medicinal use (Verpoorte et al., 2005).Till date a number of natural products like aspirin, morphine and quinine are in use (Singh and Barrett, 2006). Invariably, plant derived antibacterial compounds are a good source of novel therapeutics. Of the 104 new drugs developed in the past 37 years, 60 derived from plants used in traditional medicine of China (Gen, 1986). Plants which are rich in a wide variety of secondary metabolites belonging to chemical classes such as alkaloids, polyphenols, tannins and terpenoids, are having a greater potential in anti-microbial activities (Cowan, 1999). In addition, the discovery of modern drugs such as quinine, vincristine, digoxin, digitoxin, emetine etc., from medicinal plants signify the high prospective that still exists for the production of large number of novel pharmaceuticals. Only a little portion of the available diversity among bacteria, fungi, marine fauna and flora and plants has been explored and copious provisions still ahead. Drugs produced from microbial fermentations also have played a significant role in modern discovery by revolutionized medicine and protecting human and animal lives. Advances in microbial isolation, fermentation and natural product chemistry techniques provide tremendous opportunities to adopt these tools to the discovery of novel antibiotics.

Discovery and Development of Natural Products

Historically, natural products have provided a never-ending resource of medicine. Plant-derived products have dominated the human pharmacopoeia for thousands of years almost unchallenged (Raskin and Ripoll, 2004). In 1897, Arthur Eichengrün and Felix Hoffmann, developed the first synthetic drug, aspirin (acetylsalicylic acid). Aspirin was synthesized from salicylic acid, an active component of analgesic herbal remedies. This achievement leads to an era dominated by the pharmaceutical industry. In 1928 penicillin was discovered by Alexander Fleming which added microbes as important sources of novel drugs.The introduction of sulphonamides in 1930s and penicillin in 1940s change radically the medicinal practice by considerably reducing the casualty rates associated with bacterial infections. The success of penicillin led to discover other compounds from natural sources for the treatment of bacterial infections resulting in nearly all novel classes of antibiotics from natural products through 1962. These were discovered simply by measuring inhibition zones of bacteria on agar plates by applying whole broth or extracts obtained from microbial fermentation. The wide spread use of natural product derived antibacterial drugs is due to the role of secondary metabolites as bioactive components which confers survival advantage to the producing organisms by penetrating cell membranes and interact with specific targets (Stone and

Williams, 1992). Analysis of FDA new-drug approvals from 1981 to 2002 reveals that natural products continued to play a vital role during that time, even if the industry started to practice other discovery strategies (Newman et al., 2000). These discoveries initiated the search for new antibacterial drugs and resulted in most of the antibacterial drug classes known today.

The time required for the commercialization of natural products ranges from few to many years. For example, in 1971, the chemical structure of paclitaxel (1) was identified and reported as the cytotoxic active component of *Taxus brevifolia* extracts (Wani et al., 1971). Taxol_ (1) was approved for marketing as a cancer chemotherapeutic agent at the end of 1992. On average, new pharmaceuticals require a decade for development and commercialization. More than 90% of present day therapeutic classes derived from natural products and even today, roughly two-thirds to three quarters of the world's population relies upon medicinal plants for its primary pharmaceutical care (World Health Organization, 2002). Between 2001 and 2005, 23 new drugs derived from natural products were introduced for the treatment of disorders such as bacterial and fungal infections, cancer, diabetes, dyslipidemia, atopic dermatitis, Alzheimer's disease and genetic diseases such as tyrosinaemia and Gaucher disease. Two drugs have been approved as immunosuppressive agents and one for pain management (Lam, 2007).

Natural Products from Plants

Plants have an ability to synthesize secondary chemical substances continuously, which may have a protective role against herbivores, pathogens and also an attractant towards pollinators or symbionts (Wink and Schimmer, 1999). Some of the secondary-derived compounds may therefore have beneficial effects in the treatment of microbial infections in animals and humans (Kuete et al., 2007). Recently, the interest in these metabolites has increased following searches for new antimicrobial agents from plant sources (Hostettmann et al., 2000).

It is generally estimated that there are approximately 300,000 species of higher plants on earth (Lawrence, 1951). Out of this, about 1% or approximately 3000, has been utilized for food. Of those 3000, about 150 have been commercially cultivated. On the other hand, approximately 10,000 of the world's plants have recognized for their medicinal use, significantly more than the plants that have been utilized for food. The utilization of plant materials in western medicine is found roughly 150–200. This is still a very small proportion and thus, there are potentially many more important discoveries in the plant kingdom to be exploited for pharmaceutical application (McChesney et al., 2007).

Selection of Plants

Pharmacological and phytochemical insights into several plants led either to the isolation of novel structures for the manufacture of new drugs or to templates that served for the production of synthetically improved therapeutic agent. Four standard approaches are available for selecting plants: The first one is random selection of plants followed by chemical screening i.e phytochemical approach. This search is made for the availability of secondary metabolites having antimicrobial activities. This approach is still very popular in developing countries because of the ease but false-positive tests often make results difficult to interpret. In the second approach, random selection is followed by antimicrobial assays. In spite of prior knowledge, all available parts of the plants are collected. This methodology is expensive and laborious and depends greatly on the pathogens tested and the criteria used. The third approach is based on the antimicrobial activity reports already available. In this, the enormous available sources and published reports on antimicrobial activities are utilized. However, sometimes, critical assessment of conflicting test results is warranted and prior confirmation of the published results remains prerequisite. The fourth approach is the follow-up of ethno medical or traditional uses of plants against infectious diseases (Fabricant and Farnsworth, 2001). In this, oral, written and information available in the organized traditional medical systems on the medicinal application of a plant forms the basis for selection.

Regardless of the plant collection strategy used, an important step is the processing of the plant material that will be used in the screening. Suitable measures should be taken to assure that the potential active constituents are not lost, altered or destroyed during the extract preparation.

Different screening methods are available to identify the primary pharmacological activity in chemical or natural products. The screening option largely depends on the specific nature of the targeted disease and the availability of practical and biological laboratory models.

Anti-Microbial Properties of Some Medicinal Palnts

A variety of natural products obtained from plants showed inhibitory activity against disease causing organisms. Effect of extracts of *Commiphora harveyi* (Engl.) Engl. (Burseraceae), *Combretum vendae* (A.E. van Wyk) (Combretaceae), *Khaya anthotheca* (Welm.) C.DC (Meliaceae), *Loxostylis alata* A. Spreng. ex Rchb. (Anacardiaceae) and *Protorhus longifolia* (Bernh.) Engl. (Anacardiaceae) *Kirkia wilmsii* Engl. (Kirkiaceae) and *Ochna natalitia* (Meisn.) Walp. (Ochnaceae) were tested against different pathogens such as *Staphylococcus aureus, Enterococcus faecalis, Pseudomonas aeruginosa, E.coli* and five fungal organisms *Aspergillus fumigatus, M. canis, Candida albicans, C. neoformans* and *S. schenckii*. The plant extracts tested had varying levels of activity against bacteria. The hexane extract of *K. anthotheca* had the highest antibacterial activity with lowest MIC against *S. aureus*. The acetone extract of *L. alata* had the highest antifungal activity with lowest MIC against *S. schenckii* (Suleiman et al., 2010).

Infectious diseases of microbial origin, such as *Neisseria gonorrhoea*, *S. aureus*, *Bacillus cereus*, *Shigella* spp., etc, constitute the major cause of moribidity. Geyid et al., (2005) analysed the activity of crude methanol, petroleum ether and aqueous extract of 67 plant species against 10 strains of bacteria (*S. aureus, Streptococcus pyogenes, S. pneumonia, N. gonorrhoea, E. coli. Bacillus cereus*, *Shigella dysentery* A, *S. flexineri* B, *Salmonella typhi* and *S. typhimurium*) and 6 fungal strains (*A. flavus*, *A. niger*, *C. albicans*, *Trichophyton mentagrophytes*, *Cryptococcus neoformas* and *Trichophyton violacum*). 44 species (66%) showed activity against one or more pathogenic bacteria. 23 species inhibited the growth of one or more organisms at dilution 250 µg/ml. The most potent was *Syzygium guineense*, effective against *N. gonorrhoea*, *S. aureus* and *S. dysentriae*. On the other hand, *Bersama abyssinica, Ferula communsis, Gardenia lutea* and *Combretum molle* showed activity against two pathogens. *Albizzia gummifera* inhibited the growth of all the 10 test organisms at dilutions of 500-2000 μg/ml. *Trichila emetica* and *Dovyalis abyssinica* inhibited the growth of four to five fungal strains at 100 μg/ml. Chemical screening conducted on the extracts of all the plants showed the presence of several secondary metabolites, mainly, polyphenols, alkaloids, tannins sterols/terpenes, saponins and glycosides. The plants containing more of these metabolites demonstrated stronger anti-microbial properties stressing the need for further investigations using fractionated extracts and purified chemical components.

Erycristagallin from *Erythrina variegate* interferes with bacterial uptake of metabolites (Sato et al., 2002). Kuwanon G isolated from root bark of *Morus alba* (Moraceae) exhibits antibacterial activity against food poisoning micro-organisms and *S. mutans* (Park et al., 2003). Malvidin-3,5-diglucoside (malvin) was identified as the active constituent of an ethanol extract of *Alcea longipedicellata* (Malvaceae) responsible for activity against oral streptococci (Esmaeelian et al., 2007). A plant compound, cubebin, naturally found derivative and three semi-synthetic derivatives showed bacteriostatic and fungicidal activity against a number of Gram-positive oral bacteria and the yeast *C. albicans* (Silva et al., 2006). Naringin is a polymethoxylated flavonoid commonly found in citrus fruit and it is a FDA-approved health supplement. It inhibits the growth of periodontal pathogens and other common oral microorganisms and particularly effective against anaerobic Gram negative bacteria *Actino-bacillus actinomycetemcomitans* and *Porphyromonas gingivalis* (Tsui et al, 2008). Bakuchiol isolated from the Chinese medicinal plant, *Psoralea corylifolia* (Fabaceae), has shown activity against numerous Gram-positive and Gram-negative oral pathogens and able to inhibit the growth of *S. mutans* (Katsura et al., 2001). Several components found in hops, *Humulus lupulus* (Cannabaceae), have been found to display antibacterial activity against *S. mutans, S. saliviarius* and *S. sanguis* in disc diffusion assays (Bhattacharya et al., 2003). Garlic extract showed excellent antibacterial activity against different serotypes of *E.coli*, *Salmonella* sp., *Aeromonas hydrophila* and *A. niger* (Indu et al., 2006; Sivakumar et al., 2005).

Anti-Microbial Natural Product from Betulin

Betulin is a well-known natural lupane triterpene with various pharmacological activities. It is widely distributed in nature and can be easily extracted from birch bark, a side product of forest industry. Betulin-derived compounds have been extensively studied for their anticancer activity. It has been proposed that betulin and its natural derivative, betulinic acid, inhibit the sPLA2 (secreted phospholipase A2, EC 3.1.1.4) enzyme (Bernard, 2001), intracellular isoforms of which have recently been studied as essential components for *Chlamydia* gaining lipids for its replication requirements. The derivatives of betulin have also shown a wide range of anti-viral, antiprotozoal and anti-fungal activities, in addition to their activities against Gram-positive bacteria. Betulin derivatives are active against intracellular pathogens like *Leishmania* and *Plasmodium* and alphaviruses, and thus, they are attractive compounds for antichlamydial screening(Alakurtti et al., 2010; Pohjala et al., 2009).

Chlamydia pneumoniae is a Gram-negative human respiratory pathogen that causes acute respiratory infections. It has been estimated that *C. pneumoniae* infection is the causative agent in 5–10% of community-acquired pneumonia, bronchitis and sinusitis cases. This intracellular bacterium also causes a chronic infection that has been associated with atherosclerosis, asthma, lung cancer and Alzheimer's disease (Salin et al., 2010). Antibiotics used against *C. pneumoniae* are azithromycin (macrolide), doxycycline (tetracycline) and rifampicin (rifamycin). Rifampicin is highly effective against Chlamydia in vitro, but clinical use of rifampicin is often limited to multidrug treatments of tuberculosis and other difficult infections (Kutlin et al., 2005). Doxycycline is widely used against chlamydia but may induce persistence instead of eradication if the cellular concentration remains subinhibitory (Gieffers et al., 2004).

Antichlamydial lead discovery is challenging, as the biology of *Chlamydia* is poorly understood. Because of the intracellular nature of the bacterium, the genetic modifications have not been successful mostly and this hampered the efforts to design new antichlamydial agents. Thirty-two betulin derivatives were assayed against *C. pneumoniae* using an acute infection model in vitro. Five compounds with potential characteristics were identified. Compound 24 (betulin dioxime) gave a minimal inhibitory concentration (MIC) of 1 mM against strain CWL-029 and showed activity in nanomolar concentrations, as 50% inhibition was achieved at 290 nM. The antichlamydial effect of compound 24 was confirmed with a clinical isolate CV-6, showing a MIC of 2.2 mM. The clear structure-activity relationships and the pharmacological profile suggested that betulin derivatives, especially compound 24, is highly potential candidate for the treatment of *C. pneumoniae* infection in vitro, with suitable lead compound properties (Salin et al., 2010)

Natural Product Production by Bioconversion

Bioconversion is a ''green'' technology that converts a substrate into entirely new chemical compounds with antimicrobial, industrial or biomedical properties. The bio-conversion reactions by bacterial species have been quoted extensively among microbial systems that produce several value-added products. Strain *P. aeruginoasa* PR3, isolated from waste water stream was found to convert oleic acid to a novel compound 7,10-dihydroxy-8(E)-octadecenoic acid, which inhibited the laboratory growth of *C. albicans* (Hou and Bagby, 1991). A number of microbially bioconverted products have been shown to possess antimicrobial activities. It was reported that bioconverted product of docosahexaenoic acid possessed strong antibacterial activity against food spoilage and food borne pathogenic bacteria and also a potent antifungal agent (Shin et al., 2007; Bajpai et al., 2009). Recently, production of antifungal agents through bioconversion by microorganisms has become a major focus of research.

C. albicans is the organism most often associated with serious fungal infection and it is showing increased resistance to traditional antifungal agents. Multiple resistant phenotypes of *C. albicans* have been found to coexist during oropharyngeal candidiasis in AIDS patients (Lopez-Ribot et al., 1999). There is an increasing rate of triazole resistance amongst *C. albicans* isolates worldwide. *Candida* overgrowth on mucous membranes can lead to inflammation, impaired tissue function and a variety of systemic orders. So there is an urgent need for novel and active anticandidal agents. Cabbage (*Brassica oleracea* var. *capitata*) was microbially bioconverted by a bacterial strain *Pectobacterium carotovorum* pv. *carotovorum* 21 and examined for the anticandidal effect against various isolates of *Candida* species including a clinical isolate. The bioconverted product showed potential anticandidal effect against *Candida albicans*, *C. geochares*, *C. saitoana* and *C. glabrata*. This finding indicates that bioconverted product of cabbage has potential therapeutic value of medicinal significance to control *Candida* species including clinical isolates. Cabbage is also a source of indole-3-carbinol, a compound used as an adjuvant therapy for recurrent respiratory papillomatosis, a disease of the head and neck caused by human papillomavirus that causes growths in the airway that can lead to death (Bajpai et al., 2010).

NATURAL PRODUCTS FROM MARINE ORGANISMS

The oceans cover more than 70% of the earth's surface and contain more than 300 000 species of plants and animals (Jemino, 2002; Pomponi, 1999). More than 12,000 novel chemicals were produced from a small number of marine plants, animals, and microbes and hundreds of new compounds still being discovered every year. These efforts have yielded several bioactive metabolites that have been developed by the pharmaceutical industry fruitfully (Faulkner, 2001). Marine organisms contain secondary metabolites different from their terrestrial counterparts in structure and activity has led to the hypothesis that marine organisms may contain efficient and novel antimicrobial compounds.

Dedemnin B isolated from the tunicate *Trididemnum solidum* was the first defined marine product to enter clinical trials for any major human disease (Zhang et al., 2005). Gambieric acids are effective antifungal metabolites isolated from a strain of the epiphytic marine dinoflagellate *Gambierdiscus toxicus*. These metabolites are the first antifungal representatives of the brevetoxin. Gambieric acid A inhibits the growth of *A.niger*. Jasplakinolide is the first example of a cyclodepsipeptide isolated from a sponge and was identified from a *Jaspis* sp Jasplakinolide, also named jaspamide, has selective in-vitro antimicrobial activity against *C.albicans* (Crews et al., 1986). Jorumycin is a dimeric isoquinoline alkaloid, isolated from the Pacific nudibranch *Jorunna funebris*, inhibits the growth of *B. subtilis*, *S. aureus* and various other Gram positive bacteria (Fontana, 2000).

Most malaria cases and deaths are caused by the parasite *Plasmodium falciparum*. Manzamine A exhibits potent in-vitro activity against *P falciparum* (Ang et al., 2000). Squalamine is the first aminosterol isolated from dogfish shark *Squalus acanthias* (squalidae). It has strong antimicrobial activity against *S. aureus* (Rao et al., 2000) in addition to antitumour properties. *Sphaerococcus coronopifolius*, a red alga present along the Atlantic coast of Morocco, contains the potent antibacterial diterpene, bromosphaerone (Etahiri, 2001). Cribrostatin 3, isolated from a blue sponge *Cribrochalina* sp, has potent inhibitory activity against *Neisseria gonorrheae*, with an MIC of 0·09 g/mL. It is also active against penicillin resistant *N. gonorrheae* with an MIC of 0·39 g/mL (Pettit, 2000). The resistance acquired by bacteria against current antibacterials continues to be a serious difficulty in the treatment of infectious diseases.Therefore the discovery and development of new antibiotics has become a main concern. In the continuing effort by the marine natural products community, many antibacterial agents have been identified.

The bioactive compounds or drug produced from marine organisms could also lead to toxicity because of the high potency, might be to overcome the dilution effect in marine environment. Many marine drugs could be effective at a low dosage if synergistic compounds are introduced simultaneously. This synergism offers a unique method for marine drug discovery. Bryostatin is likely to be produced by a bacterial symbiont and is now in combination therapy (Newman and Cragg, 2005). The microorganisms that can be cultured only represent a very little fraction of the microbial population and might not be the most common in the natural environment. Though already efforts are going on, much breakthrough is expected in the next decade to grow these less-culturable microbes.

Natural Products from Microorganisms

Production of any natural product requires the processing of biological sources which needs proper handling and specific equipments. As a result, it would take considerable time to explore natural products from all the resources. Focusing the attention to a most reliable source of natural products can minimize the time duration and maximize the success of the drug discovery process. Over 130 commercial drugs in present use are of microbial origin. presently, 67 microbial or microbially derived compounds are in various stages of clinical development in the areas of antibacterial, antifungal, antiparasitic, antiviral, antiinflammatory, anticancer, neurological, cardiovascular, metabolic and immunological diseases. In addition, advances in combinatorial biosynthesis and genomics can be effectively applied to the discovery of novel microbial natural products make microbial products the most attractive source for natural product drug discovery (Lam, 2007).

Searching for natural product from a microbiological material involves the collection of soil and environment samples from different geographical areas and habitats. The sample collection is one of the most critical steps for the discovery of novel natural products. Microorganisms from soil or other environmental samples are recovered by various methods. Classical methods for the isolation of producer organisms have favoured rapidly growing organisms. But recent improvements in the microbial isolation techniques allow the isolation of single strains of producer regardless of their growth rates. Specifically, the classical approach has been to plate microorganisms on an agar lawn with or without antibiotics and observed for the zone of inhibition for subsequent assessment of their ability to produce a bioactive compound (Connon, 2002). Antibiotics are secondary metabolites that are produced by microorganisms as defense mechanisms from either co-existing life forms. The changing environmental conditions, whether in nature or in the laboratory, select for the fittest microorganism that produce new antibiotics. New isolation techniques have been developed to aid in the isolation of microorganisms and it is believed that up to 99% of microorganisms have yet to be discovered (Davis et al., 2005).

During the last two decades, soil myxobacteria have become one of the major sources of microbial natural products. It produces large number of structurally unique compounds with biological activities that are different from actinomycetes and fungi (Bode and Muller, 2006). The terrestrial environment will continue to be the source of the majority of the new natural products for some time. During 2001 to 2005, at least 23 natural product and natural product-derived drugs have been launched in Europe, Japan and the United States markets. This includes antibacterials such as Thienamycin, Daptomycin, Carbapenem, Tetracycline and Erythromycin and anti fungals like Pneumocandin B and Micafungin (Lam, 2007). Furthermore, a total of 136 natural products and the drugs derived from them have undergone different stages of clinical development in all major therapeutic areas (Butler, 2005). It is interesting to note that all antibacterial and antifungal clinical candidate originated from natural product were derived from microorganisms.

In the anti-infective field there is a shortage of lead compounds making progress into clinical trials. New antibacterial templates have advantages over known antibacterial agents against multi-drug resistant bacteria. Numerous microbial natural products have been identified with prospectives for new classes of antibacterial agents. Arylomycin 1, GE81112 4, Muraymycin 7 and Platensimycin 10 from *Streptomyces* sp., Nocathiacins 8 Nocathiacin I from *Nocardia* sp., Ramoplanin 9 from *Actinoplanes* sp. have been identified as potential natural products of microbial origin in development as anti-infective agents (Lam, 2007).

NEW ANTIMICROBIAL TEMPLATES

Drugs with new antimicrobial templates and novel mechanisms of action have more advantages over known antibacterials in the battle against new pathogens and multi-drug resistant bacteria. According to Butler and Buss, in 2006, 14 of the 19 candidates undergoing antibacterial clinical evaluation were derivatives of known drugs. The remaining five (Bacterial peptide deformylase, actinonin 2, Pleuromutilin 16, ramoplanin 17 and Tiacumicin B 21) were of interest because they contain new antibacterial templates not found in the market previously. Bacterial peptide deformylase (PDF) is responsible for removing the N-formyl group from the N-terminal methionine following translation. PDF is an essential gene for bacterial survival and does not share close homology with any mammalian equivalent. It was reported that actinonin 2, a *Streptomyces*-derived antibiotic, was a potent inhibitor of PDF (Chen et al., 2000). A combinatorial chemistry approach to the lead optimisation of actinonin 2 led to the identification of several promising compounds, such as VRC3375, VRC4307 and LBM-415 1. Recent papers have reported that LBM-415 1 has excellent antibacterial activity and a low level of bacterial mutation rate (Bell et al., 2005; Fritsche et al., 2005). Pleuromutilin 16 is a fungal metabolite that exerts antimicrobial activity by binding to the 50S bacterial ribosome (Schlunzen et al., 2004). Ramoplanin is a lipopeptide antibiotic complex isolated from *Actinoplanes* sp. ATCC33076. It will bind to the peptidoglycan intermediate Lipid II and interrupt bacterial cell wall synthesis. In preclinical studies, ramoplanin 17 showed excellent activity against methicillin-resistant *S. aureus* (MRSA), vancomycin-resistant Enterococci (VRE) and *Clostridium difficile* (Farver et al., 2005). Tiacumicin B 21 is the major component of the tiacumicin antibiotic complex produced by *Dactylosporangium aurantiacum* spp. *hamdenensis* NRRL 18085. It inhibits RNA synthesis possesses broad-spectrum Gram-positive antibacterial activity and is especially active against various *Clostridium* species (Theriault et al., 1987).

The arylomycin antibiotic complex, effective against Gram-positive bacteria, was first reported from *Streptomyces* sp. Tu¨ 6075 (Schimana et al., 2002). Mannopeptimycins from *Streptomyces hygroscopicus* LL-AC98, which have activity against MRSA and VRE (He, 2005). The caprazamycins isolated from *Streptomyces* sp. MK730-62F2 possess activity against acid fast bacteria including *Mycobacterium tuberculosis* and *M. avium* (Igarashi et al., 2005). Similarly Nocathiacin I 27 from *Nocardia* sp. was identified by screening natural product extracts against a multiple drug resistant strain of *Enterococcus faecium* (Li et al., 2003).

Use of Natural Antimicrobials to Increase Antibiotic Susceptibility of Drug Resistant Bacteria

Certain strains of *E. coli*, the normal flora of human gut, cause infections and becoming resistant to antibiotics. From the late 1990s, multidrugresistant Enterobacteriaceae, mostly *E. coli*, that produce extended-spectrum β lactamases (ESBLs) have emerged within the community and became an important cause of urinary tract infections. Recent studies have also revealed ESBL-producing *E. coli* as a cause of bloodstream infections associated with these community onsets of UTI (Pitout and Laupland, 2008). A study on effect of ethno-medicinal plants used in folklore medicine in Jordan as antibiotic resistant inhibitors on *E. coli* reveals that methanolic extracts of the plant materials significantly improved the inhibitory effects of chloramphenicol, neomycin, doxycycline, cephalexin and nalidixic acid against both the standard strain and to a lesser extent the resistant strain of *E. coli*. The plant extracts were combined with antibiotics concentrations of half the MICs . The effects varied considerably according to the antibiotic, plant extract and the *E. coli* strain. Two edible plant extracts, *Gundelia tournefortii* L. and *Pimpinella anisum* L. generally improved the activity against resistant strain of *E. coli*.

Plant extracts from *Anagyris foetida* (Leguminosae) and *Lepidium sativum* (Umbelliferae) reduced the activity of amoxicillin against the standard strain but enhanced the activity against resistant strains. Three edible plants, *Gundelia tournefortii* L. (Compositae) *Eruca sativa Mill.* (Cruciferae) and *Origanum syriacum* L. (Labiateae) enhanced the activity of clarithromycin significantly against the resistant *E. coli* strain. The main mechanisms of resistance to antibiotics used in this study are active efflux and enzymatic inactivation (Olga and Will, 2001). Several studies have been performed to identify drugs interfering with these pumps. Plant products, as ethanol extracts of *Mentha arvensis*, are known to affect the efflux system of an *E. coli* multiresistant to aminoglycosides, inhibiting these resistance mechanisms (Gyongyi et al., 2000). This synergistic multi-target effects refers to the use of herbals and drugs in a multi targeted approach. Mono or multi-extract combinations affect several targets, cooperating in an synergistic way. This approach is not exclusive for extract combinations, but combinations between single natural products or extracts with chemosynthetic or antibiotics are possible too (Darwish and Aburjai, 2010).

Synergism between Natural Products and Antibiotics

The development of antibiotic resistance can be natural or acquired and this can be transmitted within same or different species of bacteria. Bacteria gains natural resistance by spontaneous gene mutation and the acquired resistance is through the transfer of DNA fragments. One strategy applied to overcome these antibiotic resistance mechanisms is the application of combination of natural products and other drugs. One example is the administration of β-lactamases inhibitors along with antibiotics as co-drugs. Entero-hemorrhagic *E. coli* (EHEC) produces Shiga-toxin (stx). EHEC infection causes hemolytic uremic syndrome, which is the cause of acute renal failure in children. It is already reported that Japanese green tea extract (JGTE) had protective effects in a gnotobiotic mouse with EHEC O157 infection (Isogai et al., 1998). In the mice group given with JGTE plus levofloxacin (LVFX group), stx levels were not detectable. Major JGTE catechins inhibit extracellular release of stx from EHEC and catechins have bactericidal and antitoxin activity against EHEC (Konishi et al., 1999). JGTE protected the mice from the organ damage, observed in LVFX-treated mice, indicates antibiotics could be used more safely when used in combination with JGTE (Isogai et al., 2001). Humans can be infected with antibiotic resistant bacteria of animal origin, for which limited therapeutic options are available, which may lead to increased chances of treatment failure. There is a growing interest in using natural antibacterial compounds such as extracts of spices and herbs for food preservation (Shan et al., 2007) and these could be an alternative source of novel therapeutics.

Individual activity of natural antimicrobials such as eugenol, thymol, carvacrol, cinnamaldehyde and allyl isothiocyanate and activity when paired with an antibiotic was studied tested against ampicillin, penicillin, erythromycin and bacitracin resistant *S. typhimurium* and *E. coli,* ampicillin, penicillin and bacitracin resistant *S. aureus* and erythromycin resistant *S. pyogenes*.

A synergistic interaction was found for carvacrol and thymol with all antibiotics used in the study. *S. pyogenes* was highly resistant to erythromycin but was more sensitive to the natural antimicrobials than the other microbes tested but a synergistic interaction was seen with thymol, carvacrol or allyl isothiocyanate. The results showed that natural antimicrobials decreased the MIC of antibiotics considerably. Similarly for *E. coli*, eugenol was synergistic with tetracycline at 0.31 mM. The same results were obtained when *S. typhimurium* treated with thymol and cinnamaldehyde. For *S. aureus* no interaction was found between bacitracin and eugenol, but a synergistic interaction was found between penicillin and 0.31 mM of eugenol, and with ampicillin and 1.25 mM eugenol. The natural antimicrobials tested were either synergistic or showed no interference with antibiotic activity. Gram positive bacteria were more sensitive to the natural antimicrobials than the Gram negative organisms when tested individually and in paired combination with antibiotics (Palaniappan and Holley, 2010).

The effect of combinations of the crude methanolic extract of the leaves of *Helichrysum pedunculatum* and eight antibiotics were investigated against a panel of bacterial strains concerned in wound infections. The plant extract showed appreciable antibacterial activities against the test bacteria and combination of the plant extract and the antibiotics resulted in reduction of bacterial counts. It was found that 60% of the interactions were synergistic and no antagonism was observed. So the leaf extracts of *Helichrysum pedunculatum* could be of relevance in combination therapy (Aiyegoro, et al., 2009).

It is also reported that natural products Tellimagrandin I and Rugosin B from red rose tree when administered along with β-Lactum antibiotics showed synergism against methicillin resistant *S. aureus* (Shiota et al., 2000). Similarly pomegranate extract showed synergism with ciprofloxacin, chloramphenicol, gentamycin, ampicillin, tetracycline and oxacillin against methicillin resistant *S. aureus* (Braga et al., 2005). The application of myricetin along with amoxicillin was effective against extended spectrum β-lactamase producing *Klebsiella pneumoniae* (Lin et al., 2005). Totarol from *Ferula communis* and plumbagin from *Plumbago zeylanica* in combination with isonicotinic acid hydrazide exhibited synergistic action against *Mycobacterium intracellulare, M. smegmatis, M. xenopei* and *M. chelonei* (Mossa et al., 2004).

CONCLUSION

In the search for novel antibiotics, it would be more difficult to visualize a more specific source of naturally occurring antimicrobials. Despite the success of antibiotic drug discovery, infectious diseases remain the second-leading cause of death worldwide. Discovery of novel antibiotics and clinically effective natural products faced a gradual decrease over the past few decades, in spite of one million cultures screened by pharmaceutical companies for the biological activities. The major challenge that still remains for natural product based drug discovery is the tendency of medicinal chemists to take up the task of optimizing the chemical scaffolds with a range of diverse functional groups. Increased antibiotic resistance by pathogens and the prolonged process of marketing new drugs simply compounds this turn down. Therefore it is needed to develop some innovative approaches to drug discovery.

The large number of DNA sequences in the public database signifies the beginning of the new era. Genome sequencing of actinomycetes and other microbes has disclosed a number of biosynthetic enzymes that cannot be related to known metabolites, indicating that the occurrence of natural products has been underrated by using conventional. The chemical diversity of natural products is well suited to provide the core scaffolds for future drugs. Hence, strategies should be developed to take an advantage of the wealth of genomic information using different high-throughput techniques.

REFERENCES

A.A.M., Martins, C.H. and Bastos, J.K. (2007). Evaluation of *Piper cubeba* extract, (-)-cubebin and its semi-synthetic derivatives against oral pathogens. *Phytotherapy Research. 21*, 420–422.

Aiyegoro OA, Afolayan AJ, Okoh AI. 2009. Synergistic interaction of Helichrysum pedunculatum leaf extracts with antibiotics against wound infection associated bacteria. *Biological Research. 42*, 327-38.

Alakurtti, S., Bergstro¨m, P., Sacerdoti-Sierra, N., Jaffe, C. L., Yli-Kauhaluoma, J. (2010). Antileishmanial activity of betulin derivatives. *The Journal of Antibiotics. 63*,123–126.

Ang, K. K. H., Holmes, M. J., Higa, T., Hamann, M. T. and Kara, U. A. K. (2000). In vivo antimalarial activity of the betacarboline alkaloid manzamine A. *Antimicrobial Agents and Chemotherapy. 44*, 1645–1649.

Bajpai, V. K., Kang, S. C., Heu, S., Shukla, S., Lee, S. and Baek, K.H. (2010). Microbial conversion and anticandidal effects of bioconverted product of cabbage (*Brassica oleracea*) by *Pectobacterium carotovorum* pv. *carotovorum* 21. *Food and chemical toxicology. 48*, 2719-2724.

Bajpai, V. K., Kim, H. R., Hou, C. T. and Kang, S.C. (2009). Microbial conversion and in vitro and in vivo antifungal assessment of bioconverted docosahexaenoic acid (bDHA) used against agricultural plant pathogenic fungi. *Journal of Industrial Microbiology and Biotechnology. 36*, 695–704.

Bell, J. M., Turnidge, J. D., Inoue, M., Kohno, S., Hirakata, Y., Ono, Y. and Jones, R. N. (2005). Activity of a peptide deformylase inhibitor LBM415 (NVP PDF-713) tested against recent clinical isolates from Japan. *Journal of Antimicrobial Chemotherapy. 55*, 276–278.

Bernard, P., Scior, T., Didier, B., Hibert, M. and Berthon, J. (2001). Ethnopharmacology and bioinformatic combination for leads discovery: application to phospholipase A2 inhibitors. *Phytochemistry. 58*, 865–74.

Bhattacharya, S., Virani, S., Zavro, M. and Haas, G. J. (2003). Inhibition of *Streptococcus mutans* and other oral Streptococci by hop (*Humulus lupulus* L.) constituents. *Economic Botany.* 57, 118–25.

Bode, H. B. and Muller, R. (2006). Analysis of myxobacterial secondary metabolism goes molecular. *Journal of Industrial Microbiology and Biotechnology. 33*, 577–588.

Braga, L. C., Leite, A. A. M., Xavier, K. G. S., Takahashi, J. A., Bemquerer, M. P., Chartone-Souza, E., Nascimento, A. M. A. (2005). Synergic interaction between pomegranate extract and antibiotics against Staphylococcus aureus. *Canadian Journal of Microbiology. 51*, 541–547.

Butler, M. S. (2005). Natural products to drugs: natural product derived compounds in clinical trials. *Natural Products Report. 22*, 162–195.

Casal, M., Vaquero, M., Rinder, H., Tortoli, E., Grosset, J., Rusch-Derdes, S., Gutierrez, J. and Jarlier, V. (2005). A Case-Control Study for Multidrug-Resistant Tuberculosis: Risk Factors in Four European Countries. *Microbial Drug Resistance. 11,* 62-67.

Chambers, H.F. (2001). The changing epidemiology of *Staphylococcus aureus. Emerging Infectious Diseases. 7,* 178-182.

Chen, D. Z., Patel, D. V., Hackbarth, C. J., Wang, W., Dreyer, G., Young, D. C., Margolis, P.S., Wu, C., Ni, Z.J., Trias, J., White, R.J. and Yuan, Z. (2000). Actinonin, a naturally occurring antibacterial agent, is a potent deformylase inhibitor. *Biochemistry.* 39, 1256–1262.

Connon, S. A. and Giovannoni, S. J. (2002). High-throughput methods for culturing microorganisms in very-low-nutrient media yield diverse new marine isolates. *Antimicrobial Agents and Chemotherapy. 68*, 3878–3885.

Cowan, M. M., 1999. Plant products as antimicrobial agents. *Clinical Microbiology Reviews. 12*, 564-582.

Cragg, G. M., Kingston, D. G. I., and Newman, D. J. (2005). Anticancer Agents from Natural Products. Boca Raton, FL :CRC Press, Taylor and Francis Group.

Crews, P., Manes, L. V., Boehler, M. (1986). Jasplakinolide a cyclodepsipeptide from the marine sponge *Jaspis* sp. *Tetrahedron Letters. 27*, 2797–2800.

Darwish, R.M., Aburjai, T.A., 2010. Effect of ethnomedicinal plants used in folklore medicine in Jordan as antibiotic resistant inhibitors on *Escherichia coli. BMC Complementary and Alternative Medicine. 10*, 1-9.

Davis, K. E. R., Joseph, S. J. and Janssen, P. H. (2005). Effects of growth medium, inoculum size, and incubation time on culturability and isolation of soil bacteria. *Applied and Environmental Microbiology. 71*, 826–834.

Esmaeelian, B., Kamrani, Y. Y, Amoozegar, M. A, Rahmani, S., Rahimi, M., Amanlou, M. (2007). Anti-cariogenic properties of malvidin- 3,5-diglucoside isolated from Alcea longipedicellata against oral bacteria. *International Journal of Pharmacology. 3*, 468–474.

Etahiri, S., Bultel-Ponce, V., Caux, C. and Guyot, M. (2001). New bromoditerpenes from the red alga *Sphaerococcus coronopifolius*. *Journal of Natural Products. 64,* 1024–1027.

Fabricant, D. S. and Farnsworth, N. R. (2001). The value of plants used in traditional medicine for drug discovery. *Environmental Health Perspectives. 109*, 69–75.

Farver, D. K., Hedge, D. D., Lee, S. C. (2005). Ramoplanin: a lipoglycodepsipeptide antibiotic. *The Annals of Pharmacotherapy. 39*, 863–868.

Faulkner, D. J. (2001). Marine natural products. *Natural Products Report. 18*, 1–49.

Fontana, A., Cavaliere, P., Wahidulla, S. and Naik, C. G. (2000). Cimino, G. A new antitumor isoquinoline alkaloid from the marine nudibranch *Jorunna funebris*. *Tetrahedron Letters. 56*, 7305–7308.

Fritsche, T.R., Sader, H. S., Cleeland, R. and Jones, R.N. (2005). Comparative antimicrobial characterization of LBM415 (NVP PDF-713), a new peptide deformylase inhibitor of clinical importance.*Antimicrobial Agents and Chemotherapy. 49*, 1468–1476.

Gen, X. P., 1986. Medicinal plants: The Chinese approach. *World Health Organization forum. 7*, 84-85.

Geyid, A., Abebe, D., Debella, A., Makonnen, Z., Aberre, F., Teka, F., Kebede, T., Urga, K., Yersaw, K., Biza, T., Mariam, B.H. and Guta, M. (2005). Screening of some medicinal plants of Ethiopia for their anti-microbial properties and chemical profiles. *J. Ethanopharmacology. 97*, 421-427.

Gieffers, J., Rupp, J., Gebert, A., Solbach, W. and Klinger, M. (2004). First-choise antibiotics at subinhibitory concentrations induce persistence of *Chlamydia pneumoniae*. *Antimicrobial Agents and Chemotherapy. 48*, 1402–1405.

Gyongyi, G., Noboru, M., Leonard, A., Sandor, F. and Joseph, M. (2000). Interaction between antibiotics and non-conventional antibiotics on bacteria. *International Journal of Antimicrobial Agents. 14*, 239-242.

He, H. (2005). Mannopeptimycins, a novel class of glycopeptides antibiotics active against Gram-positive bacteria. *Applied Microbiology and Biotechnology*. 67, 444–452.

Hostettmann, K., Marston, A., Ndjoko, K. and Wolfender, J.L. (2000). The potential of African plants as a source of drugs. *Current Organic Chemistry. 4*, 973–1010.

Hou, C. T. and Bagby, M. O. (1991). Production of a new compound, 7,10-dihydroxy-8(E)-octadecenoic acid from oleic acid by *Pseudomonas* sp. PR3. *Journal of Industrial Microbiology. 7,* 123–130.

Igarashi, M., Takahashi, Y., Shitara, T., Nakamura, H., Naganawa, H., Miyake, T. Akamatsu, Y. (2005). Caprazamycins, novel liponucleoside antibiotics, from *Streptomyces* sp. II. Structure elucidation of caprazamycins. *Journal of Antibiotics.* (Tokyo), *58*, 327–37.

Indu, M.N., Hatha, A.A.M., Abirosh, C., Harsha, U. and Vivekanandan, G. (2006). Antimicrobial activity of some of the south-Indian spices against serotypes of *Escherichia coli*, *Salmonella*, *Listeria monocytogenes* and *Aeromonas hydrophila.* Brazilian Journal of Microbiology. 37, 153-158.

Isogai, E., Isogai, H., Hara, Y. and Shimamura, T. (2001). Epigallocatechin gallate synergy with ampicillin/sulbactam against 28 clinical isolates of methycillin-resistant *Staphylococcus aureus. The Journal of Antimicrobial Chemotherapy. 48*, 361-364.

Isogai, E., Isogai, H., Takeshi, K. and Nishikawa, T. (1998) Protective effect of Japanese green tea extract on gnotobiotic mice infected with an *Escherichia coli* O157:H7 strain. *Microbiology and Immunology. 42*, 125–128.

Jimeno, J. M. (2002). A clinical armamentarium of marine derived anti-cancer compounds. *Anticancer Drugs. 13*, 15–19.

Katsura, H., Tsukiyama, R. I, Suzuki, A. and Kobayashi, M (2001). In vitro antimicrobial activities of bakuchiol against oral microorganisms. *Antimicrobial Agents and Chemotherapy. 45,* 3009–3013.

Konishi, S. Y., Kubo, H. Y., Amano, F., Okubo, T., Aoi, N., Iwaki, M. and Kumagai, S. (1999). Epigallocatechin gallate and gallocatechin gallate in green tea catechins inhibit extracellular release of Vero toxin from enterohemorrhagic *Escherichia coli.* O157:H7. *Biochimica et Biophysica Acta. 1472*, 42–50.

Kuete, V., Nguemeving, R. J., Beng, V. P., Azebaze, A. G. B., Etoa, F., Meyer, M., Bodo, B. and Nkengfack, E. (2007). Antimicrobial activity of the methanolic extract of Vismia laurentii De Wild (Guttiferae). *Journal of Ethnopharmacology. 109*, 372–379.

Kutlin, A., Kohlhoff, S., Roblin, P., Hammerschlag, M. and Riska, P. (2005). Emergence on resistance to rifampin and rifazil in *Chlamydophila pneumoniae* and *Chlamydia trachomatis*. *Antimicrobial Agents and Chemotherapy. 49*, 903–907.

Lam, K.S. (2007). New aspects of natural products in drug discovery. *TRENDS in Microbiology. 15*, 279-289.

Lawrence, G. H. M. (1951). The Taxonomy of Vascular Plants. New York: The Macmillan Company.

Lin, R. D., Chin, Y. P. and Lee, M. H. (2005). Antimicrobial Activity of Antibiotics in Combination with Natural Flavonoids against Clinical Extended-Spectrum β –Lactamase (ESBL)-producing *Klebsiella pneumoniae. Phytotherapy Research. 19*, 612–617.

Li, W., Leet, J. E., Ax, H. A., Gustavson, D. R., Brown, D. M., Turner, L., Brown, K. Y., Clark, J., Yang, H., Fung-Tomc, J. and Lam, K. S. (2003). Nocathiacins, new thiazolyl peptide antibiotics from Nocardia sp. I. Taxonomy, fermentation and biological activities. *The Journal of Antibiotics.* (Tokyo), 56, 226–231.

Lopez-Ribot, J. L., McAtee, R. K., Perea, S., Kirkpatrick, W. R., Rinaldi, M. G. and Patterson, T. F. (1999). Multiple resistance phenotypes of *Candida albicans* coexist during episodes of eosopharyngeal candidiasis in human immunodeficiency virusinfected patients. *Antimicrobial Agents and Chemotherapy. 43*, 1621–1630.

McChesney, J. D., Venkataraman, S. K. and Henri, J.T. (2007). Plant natural products: Back to the future or into extinction?. *Phytochemistry. 68*, 2015–2022.

Mossa, J. S., El-Feraly, F. S. and Muhammad, I. (2004). Antimycobacterial constituents from Juniperus procera, Ferula communis and Plumbago zeylanica and their in vitro synergistic activity with isonicotinic acid hydrazide. *Phytotherapy Research. 2*, 934–937.

Newman, D. J, and Cragg, G.M. (2005). The discovery of anticancer drugs from natural sources. In: Zhang, L. and Demain, A. (Eds.) Natural Products: Drug Discovery and Therapeutics Medicines (pp 275-294). Totowa, New Jersey: Humana Press.

Newman, D. J., Cragg, G. M., Snader, K. M. (2003). Natural products as sources of new drugs over the period 1981-2002. *Journal of Natural Products. 66*, 1022–1037.

Newman, D. J., Cragg, G.M. and Snader, K.M. (2000). The influence of natural products upon drug discovery. *Natural Product Reports. 17*, 215-234.

Okeke, I. N., Laxmaninarayan, R., Bhutta, Z. A., Duse, A. G., Jenkins, P., O'Brien, T. F., Pablos-Mendez, A. and Klugman, K.P. (2005). Antimicrobial resistance in developing countries. Part 1: recent trends and current status. *Lancet Infectious Diseases. 5*, 481–493.

Olga, L. and Will, W. (2001). Inhibition of Efflux Pumps as a Novel Approach to Combat Drug Resistance in Bacteria. *Journal of Molecular Microbiology and Biotechnology. 3*, 225-236.

Palaniappan, K., Holley, R.A. Use of natural antimicrobials to increase antibiotic susceptibility of drug resistant bacteria. *International Journal of Food Microbiology. 140*, 164–168.

Park, K. M., You, J. S., Lee, H. Y., Baek, N. I., Hwang, J. K. and Kuwanon, G. (2003). An antibacterial agent from the root bark of Morus alba against oral pathogens. *Journal of Ethnopharmacology. 84*, 181–185.

Pettit, G. R., Knight, J. C., Collins, J. C., Herald, D. L., Young, V. G. (2000). Antineoplastic agents 430 Isolation and structure of cribrostatins 3 4 and 5 from the Republic of Maldives *Cribrochalina* sp. *Journal of Natural Products. 63*, 793–98.

Pitout, J. and Laupland, K. (2008). Extended-spectrum b-lactamase-producing Enterobacteriaceae: an emerging public-health concern. *The Lancet Infectious Diseases. 8*, 159-166.

Pohjala, L., Alakurtti, S., Ahola, T., Yli-Kauhaluoma, J. and Tammela, P. (2009). Betulin-derived compounds as inhibitors of alphavirus replication. *Journal of Natural Products. 72,*1917–1926.

Pomponi, S. A. (1999). The bioprocess-technological potential of the sea. *Journal of Biotechnology.* 70, 5–13.

Rao, M. N., Shinnar, A. E., Noecker, L. A., Chao, T. L., Feibush, B., Snyder, B., Sharkansky, I., Sarkahian, A., Zhang, X., Jones, S.R., Kinney, W.A. and Zasloff, M. (2000). Aminosterols from the dogfish shark *Squalus acanthias*. *Journal of Natural Products. 63*, 631–635.

Raskin, I. and Ripoll, C. (2004). Can an apple a day keep the doctor away?. *Current Pharmaceutical Design. 10*, 3419-29.

Salin, O., Alakurtti, S., Pohjala, L., Siiskonen, A., Maass, V., Maass, M., Kauhaluoma, J.Y. and Vuorela, P. (2010). Inhibitory effect of the natural product betulin and its derivatives against the intracellular bacterium *Chlamydia pnuemoniae*. *Biochemical Pharmacology. 80*, 1141-1151.

Sato, M., Tanaka, H., Fujiwara, S., Hirata, M., Yamaguchi, R., Etoh, H. and Tokuda, C. (2002). Antibacterial property of isoflavonoids isolated from *Erythrina variegata* against cariogenic oral bacteria. *Phytomedicine. 9,* 427–33.

Schimana, J., Gebhardt, K., Holtzel, A., Schmid, D. G., Sussmuth, R., Muller, J., *Pukall, R. and Fiedler, H. P.* (2002). Arylomycins A and B, new biaryl-bridged lipopeptide antibiotics produced by *Streptomyces* sp. Tu6075. Part I. Taxonomy, fermentation, isolation and biological activities. *The Journal of Antibiotics. 55,* 565–570.

Schlunzen, F., Pyetan, E., Fucini, P., Yonath, A. and Harms, J. M. (2004). Inhibition of peptide bond formation by pleuromutilins: the structure of the 50S ribosomal subunit from *Deinococcus radiodurans* in complex with tiamulin. *Molecular Microbiology.* 54,1287–1294.

Shan, B., Cai, Yi-Zhong, Brooks, J.D., Corke, H., 2007. The in vitro antibacterial activity of dietary spice and medicinal herb extracts. *International Journal of Food Microbiology. 117*, 112–119.

Shin, S. Y., Bajpai, V. K., Kim, H. R., Kang, S. C. (2007). Antibacterial activity of bioconverted eicosapentaenoic (EPA) and docosahexaenoic acid (DHA) against foodborne pathogenic bacteria. *International Journal of Food Microbiology. 113*, 233–236.

Shiota, S., Shimizu, M., Mizushima, M., Ito, H., Hatano, T., Yoshida, T. and Tsuchiya, T. (2000). Restoration of effectiveness of b-lactams on methicillin resistant *Staphylococcus aureus* by tellimagrandin I from rose red. *FEMS Microbiology Letters. 185*, 135–138.

Silva, M. L, Coimbra, H. S, Pereira, A. C, Almeida, V. A, Lima, T. C, Costa, E. S, Silva, R., Filho, Singh, S. B. and Barrett, J. F. (2006). Empirical antibacterial drug discovery-Foundation in natural products. *Biochemical Pharmacology. 71,* 1006-1015.

Singh, S.B. and Barrett, J.F. (2006). Empirical antibacterial drug discovery-Foundation in natural products. *Biochemical Pharmacology. 71*, 1006-1015.

Sivakumar, N., Kishore Reddy, T. V. and Sivakumar, T. (2005). Extraction and purification of antifungal compounds from garlic bulbs against *Aspergillus* sp. *Indian Journal of Applied Microbiology.* 111-114.

Stone, M. J. and Williams, D. H. (1992). On the evolution of functional secondary metabolites (natural products). *Molecular Microbiology. 6*, 29-34.

Suleiman, M.M., McGaw, L.J., Naidoo, V. and Eloff, J.N. (2010). Evaluation of several tree species for activity against the animal fungal pathogen *Aspergillus fumigatus. 76*, 64-71.

Theriault, R. J., Karwowski, J. P., Jackson, M., Girolami, R. L., Sunga, G. N., Vojtko, C. M., Coen, L. J. (1987). Tiacumicins, a novel complex of 18-membered macrolide antibiotics. Part I. Taxonomy, fermentation and antibacterial activity. *The Journal of Antibiotics. 40*, 567–574.

Tsui, V. W., Wong, R. W., Rabie, A. B. (2008). The inhibitory effects of naringin on the growth of periodontal pathogens in vitro. *Phytotherapy Research. 22,* 401–406.

Verpoorte, R., Choi, Y.H. and Kim, H.K. (2005). Ethnopharmacology and systems biology: A perfect holistic match. *Journal of Ethnopharmacology. 100*, 53–56.

Wani, M. C., Taylor, H. L., Wall, M. E., Coggon, P. and McPhail, A. T. (1971). Plant antitumor agents. VI. The isolation and structure of taxol, a novel antileukemic and antitumor agent from Taxus brevifolia. *Journal of the American Chemical Society. 93*, 2325–2327.

Wink, M., and Schimmer, O. (1999). Modes of action of defensive secondary metabolites. In: Wink, M. (Ed.), *Functions of Plant Secondary Metabolites and TheirExploitation in Biotechnology.* (pp. 17–112) Boca Raton, Florida, CRC Press.

World Health Organization, 2002. Traditional and Alternative Medicine, Fact Sheet # 271. World Health Organization, Geneva.

Zhang, L., An, A. R., Wang, J., Sun, N., Zhang, S., Hu, J. and Kuai, J. (2005). Exploring novel bioactive compounds from marine microbes. *Current Opinion in Microbiology. 8*, 276–281.

In: Natural Products and Their Active Compounds ... ISBN: 978-1-62100-153-9
Editors: M. Essa, A. Manickavasagan, and E. Sukumar © 2012 Nova Science Publishers, Inc.

Chapter 12

CELLULAR AND MOLECULAR ASPECTS OF POLYPHENOL ACTIONS ON GASTRIC CANCER

Thiyagarajan Ramesh[1] and Cinghu Senthilkumar[2]
[1]Department of Applied Biochemistry, Division of Life Science,
College of Biomedical and Health Science, Konkuk University, Chungju, Korea
[2]Department of Cell Biology,
University of Massachusetts Medical School, Worcester, US

ABSTRACT

Gastric cancer is the fourth most common cancer and the second leading cause of cancer death worldwide. Therefore it is necessary to determine an effective therapy for gastric cancer. Natural products are very important sources for the development of novel gastric cancer therapeutics. Plant polyphenols and many flavonoids have several beneficial actions on human gastric cancer. However, the actual molecular interactions of polyphenols with biological systems remain mostly speculative. This chapter deals with the potential cellular and molecular mechanisms of some selected polyphenols and its actions on gastric cancer cells. Those mechanisms include regulation of signal transduction pathways, transcription factors and related activities; modulation of cell-cycle regulation or induction of apoptosis, affecting cell differentiation, proliferation, metastasis, immune response, antioxidant, suppression of angiogenesis and chemical metabolism. A better understanding about the nature and biological consequences of polyphenol interactions with gastric cancer cell components will certainly contribute to develop nutritional and pharmacological strategies oriented to prevent the onset and/or the consequences of gastric cancer.

1. INTRODUCTION

Numerous studies have identified cellular targets which could be involved in the health promoting actions of dietary plant polyphenols. However the actual molecular mechanisms of polyphenols with those cellular targets remain mostly speculative. This chapter concentrates the most important molecular mechanisms proposed for the actions of polyphenols. While the particular focus on the effects of polyphenols on gastric cancer cells, the same potential mechanisms possibly will occur in other animal tissues and systems. Prevention of gastric cancer through dietary interventions has recently become the focus and increasing interest, dietary polyphenols have become important chemopreventive agent with potential therapeutic effects. Polyphenols are a vital portion of human diet with flavonoids and phenolic acids representing the majority of polyphenols present in food. In addition to fruits and vegetables, leaves, nuts, seeds, barks and flowers are also rich sources of polyphenols [1]. The growing awareness on cancer showed that lower incidence of gastric cancer in certain populations may possibly be due to consumption of certain nutrients, and especially polyphenol rich diets. As a result, a systematic analysis of the chemopreventive effects of polyphenolic compounds in recent years has clearly supported their health benefits, including anti-cancer properties. Given the challenges of cancer therapy, 'chemoprevention' which uses pharmacological or natural agents to impede, inhibit or reverse carcinogenesis at its earliest stages' is the most practical and promising approach for the management of gastric cancer patients. In the last decade, numerous laboratory studies, using both cancer cell lines and several animal tumor models, clearly showed that polyphenols have gastric cancer preventing activities and have been considered as promising chemopreventive agents [2]. These compounds can influence significant cellular and molecular mechanisms related to carcinogenesis such as inhibition of key proteins in signal transduction pathways, inhibition of the transcription factors and related activities, modulation of cell-cycle regulation or induction of apoptosis, affecting cell differentiation, proliferation and apoptosis, imune response, antioxidant and anti-inflammatory effects and suppression of angiogenesis and chemical metabolism.

2. GASTRIC CANCER

Gastric cancer (GC) is one of the most common malignancies affecting humans, and accounts for almost 10% of annual newly diagnosed cancers worldwide. It is the second leading cause of cancer related death worldwide. The majority of them are gastric adenocarcinomas (GA), divided into two histological types according to Laure□n's classification: intestinal and diffuse types. The intestinal type is related to corpus-dominant gastritis with gastric atrophy and intestinal metaplasia, more frequent in elderly men, whereas the diffuse type usually originates in pangastritis without atrophy and occurs in patients under the age of 50 predominantly women. The natural history of GC is complex and not completely understood. It is a multifactorial disease with significant geographical variations where environmental and host genetic factors are involved. Studies have shown that change of dietary habits and the excessive process of food are responsible for the increased tendency of gastrointestinal disease. A high intake of smoked, salted, nitrated foods, carbohydrates and low intake of vegetables, fruits, and milk, are linked to gastric cancer incidence. These diets have been shown to enhance the risk for stomach cancer [3]. In addition *Helicobacter pylori* infection is one of the most important etiological factors established as a risk factor for the development of gastric adenocarcinoma. However, the majority of infected individuals do not develop malignancy and the outcome of the infection is dependent on host and other environmental factors. There are growing evidences supporting the role of psychological stress in the gastric cancer onset and development. Several studies have demonstrated that psychological or behavioral stress factors may also accelerate the progression of gastric cancer. Previous studies proved that the accumulation of genetic alterations, such as inactivation of the tumor suppressor gene p53 plays an important role in gastric carcinogenesis, additionally several lines of evidences implicate the Wnt signaling pathway as a contributor to gastric carcinogenesis. Persons with germ-line mutation of the APC tumor suppressor gene have a 10-fold increased risk of developing gastric cancer compared with normal persons. β-catenin mutations have also been detected in intestinal-type gastric carcinoma tissues and gastric cancer cell lines [4].

Current advances in early diagnosis and treatment have resulted in significant improvement in long-term survival for gastric cancer patients. However, the prognosis for advanced gastric cancer remains poor. A majority of patients with advanced gastric cancer die due to complications caused by metastasis. Therefore, invasion and metastasis are critical determinants of gastric cancer morbidity. Incorporated research in molecular pathology over the past decade has disclosed the molecular mechanism of invasion and metastasis in gastric cancer [5]. The metastatic process consists of various steps including tumor cell detachment, local invasion, motility, angiogenesis, vessel invasion, survival in the circulation, adhesion to endothelial cells, extravasation, and regrowth in different organs. Cell motility is a fundamental process required during normal embryonic development, wound repair, inflamematory response, and tumor metastasis. When malignant tumors invade normal tissue, three independent processes are involved: the degradation of the extracellular matrix (ECM), cell metastasis, and proliferation. Metastasis has been found to be accompanied by various physiological alterations involved in the degradation of the ECM, such as the overexpression of proteolytic enzyme activity, including matrix metalloproteinases, as well as migration and invasion of tumor cells into the bloodstream or lymphatic system to spread to other tissues or organs. The ability of cancer cells to invade and metastasize represents the final and most difficult-to-treat stage of gastric cancer. Therefore, understanding the complicated mechanisms that underlie cellular motility holds great promise of identifying new anti-cancer drug targets.

3. Polyphenol

Polyphenols comprise one of the largest and ubiquitous groups of phytochemicals. There are several thousand compounds of biological interest with one or more aromatic rings and at least two hydroxyl groups, thus qualifying as polyphenols. These are the secondary metabolic products in plants. One of the primary functions of these plant- derived polyphenols is to protect plants from photosynthetic stress, reactive oxygen species, radiations, pathogens and give plants their colors. Polyphenols are also an essential part of the human diet, with flavonoids and phenolic acids being the most common ones in food. Up to date, a substantial number of studies in cultured cells, animal models and human clinical trials have illustrated a protective role of dietary polyphenols against different types of cancers. Some analyses suggest that more than 8000 different dietary polyphenols exist and can be divided into ten different general classes based on their chemical structure [6]. Phenolic acids, flavonoids, stilbenes and lignans are the most abundantly occurring polyphenols that are also an integral part of everyday nutrition in populations worldwide. Some of the common examples of the most studied and promising cancer chemopreventive polyphenols include epigallocatechin-3-gallate (from green tea), curcumin (from curry), genistein (from soy), gallic acid (from vegetables), etc. Considerable gains have been made in understanding the molecular mechanisms supporting the chemopreventive effects of polyphenols, and consequently, a broad range of molecular mechanisms and gene targets have been identified for individual compounds. Several mechanistic explanations for their chemopreventive effect include their capability to disrupt or reverse the carcinogenesis process by acting on intracellular signaling network molecules involved in the initiation and/or promotion of cancer, or their potential to inhibit or reverse the progression stage of cancer. Polyphenolic compounds may also activate apoptosis in cancer cells through the modulation of a number of key elements in cellular signal transduction pathways linked to apoptosis. In addition, the biological activities of polyphenols are mainly attributed to their antioxidant properties, which is strictly related to their chemical structure. Polyphenols prevent reactive oxygen species (ROS)-induced DNA damage by scavenging free radicals (reactive oxygen, nitrogen, and chlorine species) and by inactivating metal catalysts by chelation thereby decreasing their oxidative activity. Their capability to interact with other reducing compounds and to inhibit redox-active transcription factors may also contribute to the antioxidant properties of these molecules as well as to their ability to regulate gene expression. Paradoxically, in addition to their antioxidant effects, polyphenols have also been shown to exert pro-oxidant effects that could also be responsible for their anticancer properties [7].

4. Molecular Mechanism of Polyphenol Actions on Gastric Cancer

Polyphenols have been demonstrated to inhibit gastric cancer *in vitro* and considerable evidences indicated that they can also do so *in vivo*. Numerous experiment with different cellular models suggested that certain polyphenol could inhibit gastric cancer initiation as well as gastric cancer progression. This chapter is mainly focused on cellular and molecular mechanism of only few polyphenol (apigenin, anthocyanins, genistein 2',4'-dihydroxy-chalcone, luteolin, epigallocatechin-3-gallate, flavanone, naringenin, acacetin, curcumin and gallic acid) and its actions on gastric cancer cells.

4.1. Apigenin

Apigenin (4', 5, 7-trihydroxyflavone), (Figure 4.1) a polyphenol, is commonly found in many fruits and vegetables such as apple, orange, celery, onion, etc. Apigenin has anti-proliferative effects on numerous forms of cancer cells such as prostate cancer, breast cancer, leukemia and colon cancer cells.

Apigenin is a potential cancer suppressor which may give a new approach for the treatment of human cancers. MTT and clone-forming assay are used to perceive the growth inhibitory effect of apigenin on human gastric SGC-7901 cells. Apigenin showed considerable inhibition on the growth and clone formation of the human gastric SGC-7901 cells in a dose and time dependent manner.

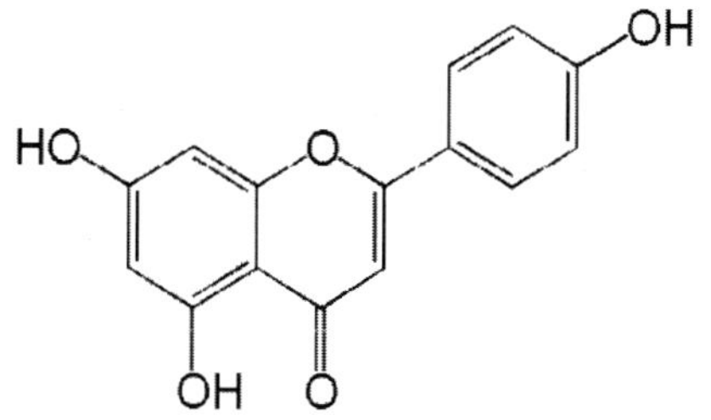

Figure 4.1. Structure of Apigenin.

After treatment of SGC-7901 cells with apigenin, clone formation revealed the proliferative capacity of tumor stem cells, which is the essential target of anticancer treatment. Suppression of stem cells are more efficient than that of common carcinoma cells during the treatment of cancer. Typical morphological changes such as the disintegrity of nuclear membrane, condensation of chromatin and broken nuclei were observed by fluorescence microscopy. FACS analysis identified special apoptosis peak which further supported the results in fluorescence morphological examination. These observations suggested that apigenin can suppress the growth of human gastric cancer SGC-7901 cells, which is coupled with its apoptosis-inducing effect [8].

4.2. Anthocyanins

Anthocyanins are naturally occurring reddish pigments that are rich in vegetables and fruits. Anthocyanins have been identified to be an effective chemo preventive agent for mammary carcinogenesis in rats. Anthocyanins illustrated inhibitory effects on the growth of some cancer cells [9]. Among the nine kinds of anthocyanins and anthocyanidins (five aglycone such as cyanidin, delphinidin, malvidin, pelargonidin, and peonidin and four glycosylated such as cyaniding-3-glucoside, malvidin-3-glucoside, pelargonidin-3-glucoside and peonidin-3-glucoside), malvidin (Figure 4.2) has more efficacy in anti-proliferation and cytotoxicity of gastric adenocarcinoma (AGS) cells with a time- and dose-dependent manner.

The capacity to induce programmed cell death in gastric carcinoma cells by anthocyanis are still questionable despite, the morphological damages of classical apoptotic cells such as cell shrinkage and membrane blabbing to the formation of apoptotic bodies were observed in malvidin.

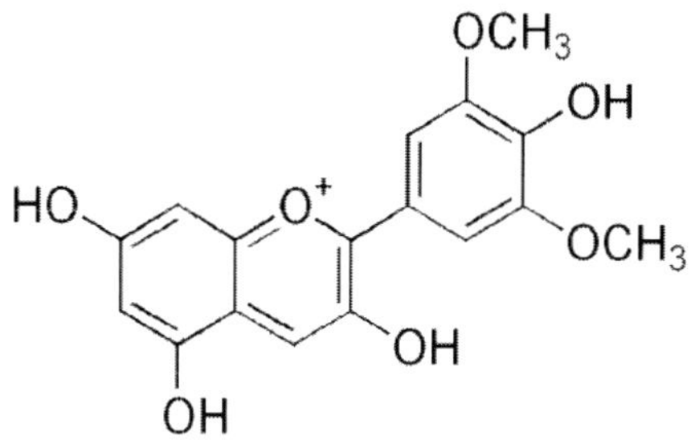

Figure 4.2. Structure of Malvidin.

This treatment significantly accumulated the percentage of cells at G0/G1 phase; and further elevated the number of sub-G1 cells, accompanied with caspase-3 expression which resulted in the proteolytic cleavage of poly (ADP-ribose) polymerase (PARP). Malvidin further affect cellular proliferation by arresting the cell cycle progression and sequentially induces apoptosis. The mitochondrial dysfunction is linked with the intrinsic pathway of apoptosis. Bcl-2 family members were classified into two major roles, the pro-apoptotic and the anti-apoptotic function. When the death signal has been taken over, the C-terminal signal anchor sequences will target them to the outer membrane of organelles included mitochondria and then shift the permeability of mitochondria. The malvidin mediate the continued diminution of mitochondrial membrane potential with increased incubation time, and enhances the ratio of Bax/Bcl-2 expression in a dose dependent manner, which represented the pro/anti-apoptotic functions [10].

Mitogen-activated protein kinases (MAPKs) are well known in the regulation of survival, proliferation and death of the cell. Malvidin treatment has an intense effect in the expression of MAPK subfamilies which consist of ERK and p38 kinase. Malvidin could steadily decrease the activities of ERK, which is responsible for the regulation of meiosis, mitosis and post-mitotic functions in cells that promote anti-proliferative effect. Malvidin arbitrated the activation of p38 enzyme expression, which is involved in apoptosis. Malvidin treatment in AGS cells upregulated the expression of p38 kinase.

There are a number of specific inhibitors that are used for assessing the precise molecular mechanism. PD98059 specifically binds to MKK1/2, the ERK upstream kinases, resulted in the inhibition of ERK phosphorylation and activation. The p38 inhibitor SB203580 abrogates malvidin-induced expression of caspase-3, while ERK inhibitor PD98059 inhibits the stimulatory effect of malvidin on caspase-3 activity. These phenomenons indicated that malvidin mediated the apoptosis of AGS cells is through the effect of pro and anti-apoptotic molecules of the MAPK family. These observations show malvidin has profound effect on the induction of the programmed cell death in human gastric adenocarcinoma AGS cells. Therefore, anthocyanidin should be good naturally occurring neutraceuticals for gastric cancer preventions [10].

4.3. Genistein

Genistein (Figure 4.3) is one of the richest isoflavones in soy. Isoflavones belong to the group of flavonoids. It is a planar molecule with an aromatic A-ring that has a second oxygen atom, and also has a molecular mass similar to those of the human estrogen, so it is also called a phytoestrogen.

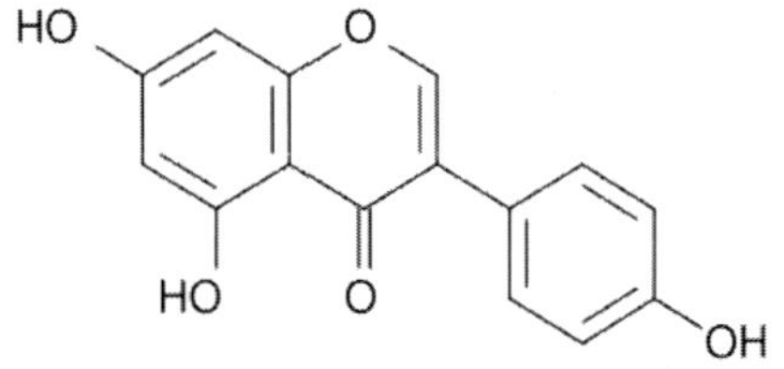

Figure 4.3. Structure of genistein.

Genistein has been shown to inhibit tumor cell proliferation and induce tumor cell differentiation, and trigger cell cycle arrest and apoptosis in some cell types. Many studies have reported that genistein is a potent cancer chemo preventive agent [11, 12]. MTT assay showed that genistein could significantly inhibit the proliferation of SGC-7901 cells in a dose and time dependent manner. When genistein treated with SGC-7901 cells illustrated significant inhibition on the expression of cyclin D1, suggested that genistein might inhibit cell proliferation of gastric carcinoma by diminishing the over-expression of cyclin D1. Cyclins are a group of proteins that control the progression of cells through the cell cycle by activating cyclin-dependent kinase (Cdk) enzymes. There are several different cyclins that are active in different parts of the cell cycle that cause the Cdk to phosphorylate different substrates. Cyclin D1 is synthesized in pre-DNA-synthetic gap (early G1 phase), that play a significant role in G1 to S phase and stimulates cells into S phase. In common, cyclin D1 is the key regulator of cell cycle progression and the key protein of the signal transduction in G1 phase cell proliferation. If cyclin D1 is overexpressed, the checkpoint of G1/S will be out of control and lose its role in the signaling of proliferation. This further promotes cell cycle progression and cell proliferation, and causes carcinomatous change of cells. Thus cyclin D1 is called the shirking protein of G1/S checkpoint. It has been proved that cyclin D1 is overexpressed in several neoplasms. Suppressed expression of cyclin D1 in cancer cells would help to recover normal cell cycle and control proliferation speed of tumor cells. Cyclin B1 and cyclin-dependent kinase 1 (Cdk1) are two proteins required for cells to traverse from G (2) into M phase. G (2) arrest occurs in response to DNA damage caused by a variety of agents and treatments. Cyclin B1 is synthesized in late S and G2 phase. It binds to Cdk1 and triggered to form maturation promoting factor. Cyclin B1 is degraded in M phase. However when different concentrations of genistein treated with SGC-7901 cells, the expression of cyclin B1 did not decrease with increased concentrations of genistein as cyclin D1, instead it increased. Some studies indicate that sustained increase of cyclin B1 causes cell cycle blockage in cell cleavage phase. Paradoxically various studies show that when cell cycle blockage occurs in G2/M phase, cyclin B1 is not degraded, but accumulated in cells. Thus confirm that G2/M blockage does not always follow the decrease of cyclin B1 expression. Genistein blocked SGC-7901 cell proliferation and increased the number of cells in G2/M phase more than three times, as well as the expression of cyclin B1. The increased cyclin B1 expression did not make cancer cells escape the regulation of checkpoint from G2 to M phase because cyclin B1 protein accumulates during interphase, while cell cycle progression is arrested at G2/M phase [13].

To evaluate the effect of genistein on cell proliferation cycle, the expression of Cyclin-Dependent Kinases Inhibitors (CKI) $p21^{waf1/cip1}$ protein was analyzed. The expression of $p21^{waf1/cip1}$ protein relates with G2/M phase arrest. $p21^{waf1/cip1}$ binds to a variety of Cdks and cyclins, and exerts inhibitory activity on cyclin/Cdk complexes, including cyclinA-Cdk1 and cyclinB1-Cdk1. Hence $p21^{waf1/cip1}$ protein has a close relationship with G2 and M phases of cell cycle. When SGC-7901 cells are treated with genistein, the expression of $p21^{waf1/cip1}$ is reduced in a dose dependent manner. All these exhibit that the inhibitory effect of genistein on human gastric carcinoma cells relates with genistein-induced expression of $p21^{waf1/cip1}$ and genistein arrests tumor cells in G2/M phase. The possible mechanism is that genistein promotes the expression of $p21^{waf1/cip1}$ and reduces the degradation of cyclin B1 protein in tumor cells. Therefore tumor cells are unable to pass the checkpoint pathway of G2/M and can not proceed to mitosis. Genistein inhibits tumor cell growth and proliferation by increasing the expression of cyclin B1 and $p21^{waf1/cip1}$ and decreasing the expression of cyclin D1 in SGC-7901 cells [13].

Genistein was able to induce apoptosis in SG7901 cells. The apoptosis might have been mediated by down-regulating the expression of the antiapoptotic gene Bcl-2 and up-regulating the expression of proapoptotic gene Bax. The Bcl-2 family plays a vital role in the control of apoptosis. It includes a number of proteins that have homologous amino acid sequences, including anti-apoptotic members such as Bcl-2 and Bcl-xL, as well as pro-apoptotic members including Bax and Bad. Overexpression of Bax has the effect of promoting cell death. On the other hand, overexpression of anti-apoptotic proteins such as Bcl-2 represses the function of Bax. Thus, the ratio of Bcl-2/Bax appears to be a significant determinant of a cell's threshold for undergoing apoptosis. Expression of Bcl-2 in the presence of genistein was diminished simultaneously the expression of bax was elevated. The density of Bcl-2 mRNA decreased gradually, and the density of bax mRNA increased gradually with genistein. The ratio of Bcl-2/Bax was diminished and activated apoptosis of transplanted tumor cells. Genistein was able to persuade apoptosis of transplanted tumor cells in nude mice. The apoptosis may have been mediated by down-regulating expression of apoptosis-regulated gene Bcl-2 and up-regulating expression of apoptosis-regulated gene Bax [14]. Hence, genistein may act as a potential chemotherapeutic drug in the anti-gastric carcinoma chemotherapy.

4.4. 2', 4'-Dihydroxychalcone (TFC)

2',4'-Dihydroxychalcone (Figure 4.4), is one of the most important components in *Herba oxytropis*, belongs to the flavonoid group, which is known to have anti-tumor activity *in vitro*. When TFC treated with human gastric cancer MGC-803 cells, the growth was inhibited in a dose and time dependent manner. The decrease in cell growth was attributable to apoptosis; using fluorescence microscopy, transmission electron microscope and flow cytometry in MGC-803 cancer cells showed characteristics of apoptosis after TFC treatment. These observations suggested that TFC is capable of suppressing the growth of MGC-803 cells and affect the continuous cell population [15].

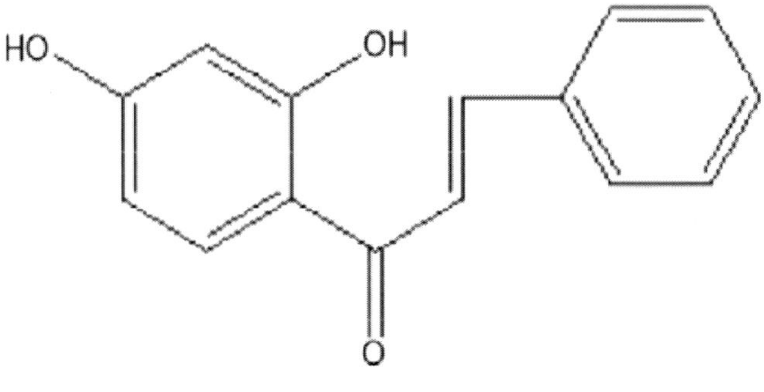

Figure 4.4. Structure of 2',4'-Dihydroxychalcone.

To reveal the molecular mechanism of apoptosis TFC treatment in MGC-803 cells, survivin expression was analyzed by RT-PCR. Survivin, is a member of the inhibitor of apoptosis family, inhibits the activation of downstream effectors of apoptosis (i.e., caspase-3 and -7) in cells exposed to apoptotic stimuli. Overexpression of survivin is present in human cancers of various origins. TFC appears to stimulate apoptosis in MGC-803 cells and to regulate survivin expression. TFC induces the reduction of surviving mRNA. Survivin is expressed in a cell-regulated mode in the G2/M phase of the cell cycle. The interaction of survivin with the mitotic spindle is important for anti-apoptotic function. Overexpression of survivin has oncogenic potential because it may overcome the G2/M checkpoint to impose development of cells through mitosis. Severe suppression of survivin by TFC prevents cell cycle development through the M phase, resulting in apoptosis. Consistantly, TFC treatment blocked the cell cycle in the G2/M phase. The decreased effects at high dose propose that G2/M phase arrest is followed by apoptotic cell death, which results in the removal of cells in the G2/M phase, thereby decreasing the cell number. The down-regulation of survivin mRNA is consequently an essential mechanism, and contributes mostly in TFC-induced apoptosis of MGC-803 cells. In addition, the increase in S-phase cell population when using low dose of TFC exhibit that TFC can induce S-phase cell cycle arrest at this concentration. TFC treatment at the low concentration possibly blocked the cell cycle in both the S and G2/M phases.

Cysteine aspartases (caspases), are proteins belong to protease family, that are necessary for apoptosis induced by various stimuli. Among mammalian caspases, caspase-3 is considered as the main effector and has been known to be triggered in response to cytotoxic drugs. Caspase-3 activation is a vital step in the execution phase of apoptosis and its inhibition blocks cell apoptosis.

Caspase-3 activity is elevated in a dose dependent manner when MGC-803 cells are treated with different TFC concentrations. Survivin can inhibit the activation of downstream effectors of caspase-3 activity. The increasing caspase-3 activity may be due to the down-regulation of survivin mRNA. These observations show that TFC can be promoted as an agent for gastric cancer prevention and/or therapy [16].

4.5. Luteolin

Luteolin (Figure 4.5) is one of the most common flavonoids that have been found to have anti-carcinogenic properties [17]. Luteolin inhibits the growth of gastric cancer AGS cells through the activation of G2/M phase arrest and regulate the expression of different enzymes related to G2/M transition.

In addition luteolin induce apoptosis and affect the levels of both pro-apoptotic and anti-apoptotic proteins, especially those in the intrinsic apoptotic pathway. Luteolin inhibit the growth of AGS cells through arresting the cell cycle progression at G2/M phases and stimulating intrinsic apoptotic pathway. In terms of the cell cycle regulation, Cyclin-dependent kinases (Cdk) and Cyclin-dependent kinases inhibitors (CKIs) play vital roles. In G2 and M phases, Cdc2 kinase is triggered by binding to Cyclin B. Cdc25C play a significant role in the dephosphorylation of Cdc2 on Thr14/Tyr15 and the blockage of its activity inhibit the consequent stimulation of Cyclin B1/Cdc2. Luteolin was found to down-regulate Cdc2, Cyclin B1, and Cdc25C protein levels, which are necessary for the development of G2/M checkpoints. $p21^{cip1}$ and $p27^{kip1}$ are two important members of CDKIs that inhibit the activity of Cdk-Cyclin complexes and negatively regulate cell cycle progression.

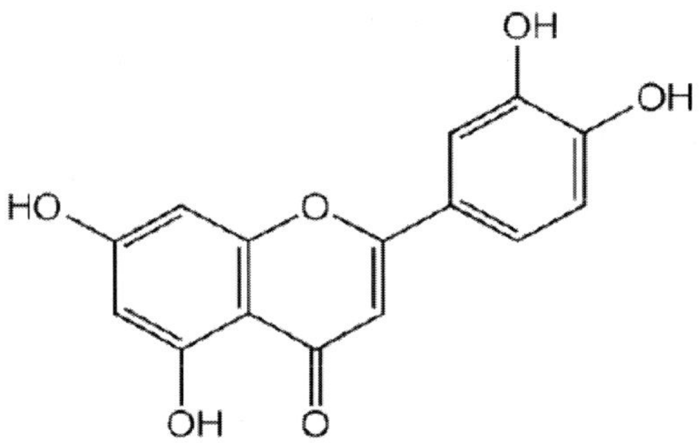

Figure 4.5. Structure of Luteolin.

It has been demonstrated that $p21^{cip1}$ regulated G1 and S progression by inhibiting cdk4 and cdk2 activities, and $p21^{cip1}$ was also involved in the suppression of G2/M transition. Luteolin also demonstrated increased $p21^{cip1}$ protein levels. All these results suggest that the regulation of G2/M transition may be an effective target to control the growth and proliferation of AGS cells by luteolin [18].

Impaired apoptosis is another critical mechanism for cancer progress. The Bcl-2 family proteins are vital in the transmission of apoptosis and can be divided into pro and anti-apoptotic members. The anti-apoptotic Bcl-2 family proteins form a heterodimer with Bax/Bak and thereby suppress the pro-apoptotic effects. Besides, Bcl-2 family proteins regulate the release of cytochrome C from the mitochondria into cytosol, which will stimulate initiator caspase-9 and induce a subsequent caspase cascade (the intrinsic apoptotic pathway). Luteolin treatment significantly enhance the levels of pro-apoptotic proteins, including Caspase-3, 6, 9, Bax, and p53, and decreased the levels of anti-apoptotic Bcl-2 protein, thus shifting the Bax/Bcl ratio in favor of apoptosis.

The levels of cytochrome C does not change when whole-cell extracts were examined. However, subcellular translocation of Cytochrome C was observed, which further support the luteolin treatment induced apoptosis in AGS cells through an intrinsic mitochondria apoptotic pathway. These observations prove the clinical application potential for luteolin in gastric cancer chemotherapy [18].

4.6. Epigallocatechin-3-Gallate (EGCG)

Epigallocatechin-3-gallate (Figure 4.6) is a main monomer in green tea polyphenol; the others are Epigallocatechin (EGC), Epicatechin-3-gallate (ECG) and Epicatechin (EC). Numerous studies have shown green tea polyphenolic compounds could inhibit tumor growth and metastasis both *in vivo* and *in vitro* [19, 20]. EGCG inhibit proliferation and induce apoptosis of several tumor cell lines *in vitro* and decrease the size of tumors in mice and rats.

In addition, EGCG can arrest mutations induced by carcinogens, inhibit transformation, tumor invasion, angiogenesis, induce cell growth arrest in G1 phase and cell apoptosis. EGCG can also arrest the growth of different digestive tumor cells, and MKN45 cells were the most sensitive one among the gastric cancer cell lines tested. EGCG can induce cancer cell apoptosis by direct binding to the Fas death receptor to initiate the caspase-8 activation and apoptosis. EGCG can also induce apoptosis by the mitochondrial pathway.

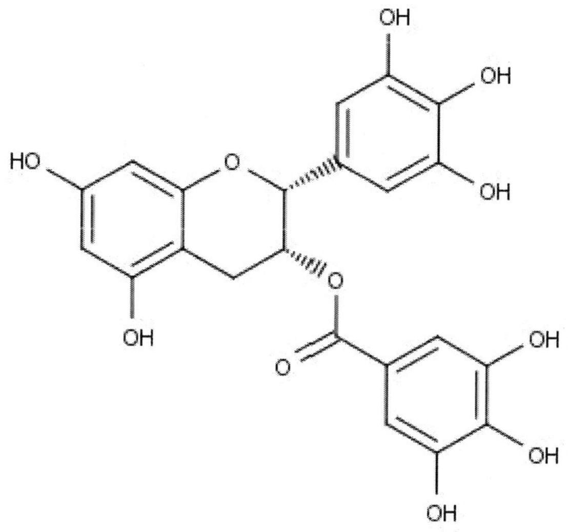

Figure 4.6. Structure of epigallocatechin-3-gallate (EGCG).

The apoptosis was induced by the effect of EGCG on human colon adenocarcinoma cells HT-29 depend on the mitochondrion to activate caspase-9 and caspase-3. EGCG could induce apoptosis of MKN45 cells, and the apoptotic rates were in a time and dose dependent manner. The growth inhibitory effect of EGCG on MKN45 cells was mediated by the apoptosis pathway and this apoptotic process was coupled with the activation of caspase-3. Caspases play an important role in the initiation and execution of apoptosis induced by various stimuli. Among them, caspase-3 is a key caspase involved in apoptosis, which is a joint effector of the endogenous and exogenous pathways. Activated caspase-3 can hydrolyze PARP in the nucleus, resulting in DNA repair inhibition and apoptosis. Specific caspase-3 substrate interacted with caspase-3 which was activated in the apoptotic pathway, and identified caspase-3 activity at different time and dose of EGCG treatment. In order to examine the upstream events of caspase-3 activation in this apoptotic pathway, a measure of change in the mitochondrial membrane potential is essential. The change of mitochondrial permeability is considered to be the earliest event in the cell apoptosis cascade. The mitochondrial membrane potential changes result in the cytochrome C release to initiate caspase-9 and -3 activation and consequent apoptosis. The change of mitochondrial membrane potential can investigate by Rhodamine 123 staining. In normal cells, mitochondria can absorb Rhodamine 123, and its absorptivity will decline while the membrane potential is decreased. When EGCG treated with MKN45 cells, the mitochondrial membrane potential has declined and the absorptivity of Rhodamine 123 reached to the lowest level. While, the activity of caspase-3 in MKN45 cells was increased after EGCG treatment. Upon addition of the caspase-3 inhibitor the activity was dramatically decreased. The change of mitochondrial membrane potential was prior than the increase of caspase-3 activity; therefore, it is considered that the apoptotic pathway activated by EGCG in MKN45 cells was dependent on the mitochondrion to activate caspase-3 [21].

The Bcl-2 family of proteins, including Bad, Bax, Bid, Bcl-2, and Bcl-xL, are key regulators of mitochondria mediated apoptosis. Pro-apoptotic Bad and Bax induce apoptotic cell death by promoting mitochondrial release of cytochrome-c and Smac, Anti-apoptotic Bcl-2 and Bcl-xL inhibit mitochondria-mediated apoptosis and promote cell survival by preventing cytochrome-c or Smac release from the mitochondria. A shift in the balance between anti- and pro-apoptotic bcl-2 family proteins towards the expression and activation of Bad and Bax proteins could lead to mitochondria-dependent caspase activation and apoptotic cell death. Thus, it is possible that EGCG induces mitochondrial release of cytochrome-C and Smac that affect the expression of the bcl-2 family of proteins. When EGCG treated with MKN45 cell the anti-apoptotic genes were down-regulated and pro-apoptotic genes were up-regulated, the level of transcription confirm the protein level. The ratio of Bcl/Bax gradually decreased as the treated time is increased. This investigation provides further evidence for its clinical application in the therapy of gastric cancer [21].

Moreover, EGCG has anti-angiogenic property and reduce the tumor growth by inhibiting angiogenesis. Angiogenesis, the growth of new blood vessels from preexisting capillaries, is essential for solid tumor growth and metastasis. Angiogenesis is activated by the release of certain angiogenic factors from tumor cells. Vascular endothelial growth factor (VEGF), has been shown to be the most potent angiogenic factor, which is coupled with tumor-induced angiogenesis. Angiogenesis is closely associated with progression and prognosis of gastric cancer, and VEGF expression is a predictive and prognostic factor of gastric cancer. Anti-angiogenic therapy targeting VEGF can inhibit growth and metastasis of gastric cancer [22]. Anti-angiogenic therapy is one of the most promising novel strategies and many challenges have been made to prevent or delay tumor growth by anti-angiogenesis. As the most potent angiogenic factor, VEGF has become one of the most common targets in anti-angiogenic therapy. Green tea has been shown to have anti-angiogenic activity and to drink tea can significantly prevent VEGF induced corneal neovascularization. As the most abundant and active component of green tea, EGCG has been shown to have anti-angiogenic property. Treatment with EGCG inhibit the secretion of VEGF both in endothelial cells and tumor cells dose-dependently. To assess the efficacy of EGCG on expression of VEGF in gastric cancer, the expression of VEGF determined in cultured cells and tumor tissues. EGCG treatment dose-dependently reduced the expression of VEGF in cultured tumor cells. In tumor tissues treated with EGCG, the expression of VEGF was markedly decreased. Treatment with EGCG also diminished VEGF secretion in tumor cells in a dose-dependent manner. EGCG dose dependently inhibited VEGF mRNA expression. These observations suggest that EGCG inhibits angiogenesis in gastric cancer by decreasing production of VEGF at transcriptional level [23].

VEGF expression is linked with a variety of transcription factors, genes and modulators. Numerous mechanisms have been proposed for the inhibitory effect of EGCG on VEGF expression. EGCG strongly inhibit the transcriptional activity of transcription factors, such as NF-kB and activator protein-1. Treatment with EGCG also reduce the constitutive activation of EGFR and the expression of protein kinase C, transcription modulators of VEGF, suggested that these alterations are associated with inhibition of VEGF promoter activity and cellular production of VEGF. EGCG strongly inhibit the constitutive activation of Stat3 in tumor cells. Stat3 is one of the key transcription factors in regulation of VEGF expression. Stat3 activation directly promotes VEGF expression and activate tumor angiogenesis. The higher activation level of Stat3 was found in gastric cancer; the abnormally activated Stat3 expression significantly correlates with VEGF expression and microvessel density, and is an independently prognostic factor of poor survival. EGCG dose-dependently inhibited activation of Stat3. In tumor cells and tissues treated with EGCG, the protein level of phosphorylated Stat3 was markedly reduced, but the total Stat3 level remain unchanged, that suggest EGCG down-regulates VEGF expression at least in part by inhibiting Stat3 activation in gastric cancer [23].

Upon treatment with EGCG, VEGF is suppressed and induced HUVEC proliferation in a time- and dose dependent manner and endothelial cell migration and tuber formation were also abrogated by EGCG, which suggested that EGCG suppressed angiogenesis in gastric cancer by inhibiting VEGF-induced angiogenesis and decreasing production of VEGF. VEGF stimulated angiogenesis by binding to its receptors on endothelial cell surface. It showed that EGCG dose-dependently inhibit VEGF binding to its receptors on endothelial cell surface and EGCG is the only catechin of green tea that could interrupt VEGF receptor binding. Prominently, EGCG at low doses markedly inhibit formation of VEGFR-2 complex. The exact molecular mechanism in the suppression of VEGF receptor binding is not well known, despite the interruption of VEGF receptor binding would block receptor tyrosine phosphorylation and VEGF induced growth and survival signaling. Treatment with EGCG suppresses VEGF-induced VEGFR-1 and VEGFR-2 phosphorylation on endothelial cells in a time- and dose-dependent manner. The interruption diminish PI3-kinase activity in a dose-dependent manner and inhibit VE-cadherin tyrosine phosphorylation and Akt activation, and also reduce the PI3 kinase-dependent activation and DNA-binding ability of NF-kB, that suggest EGCG inhibit VEGF-induced angiogenesis by disrupting VEGF signaling pathway. Many studies reported that EGCG has anti-cancer effect without any severe side effects. Taken together, as a natural and non-toxic product, EGCG might be a promising candidate for anti-angiogenic treatment of gastric cancer [23].

4.7. Flavanone and Naringenin

Flavanones (Figure 4.7a) are a type of flavonoids. Flavonoids are polyphenolic compounds found in various foods of plant origin. It is well known that a high dietary intake of flavonoids is associated with low cancer prevalence in humans. Various molecular mechanisms by which the flavonoids may affect tumorigenesis were discussed, including anti-oxidant activities, the scavenging effect on activated mutagens and carcinogens, interaction with proteins that control cell cycle progression, and altered gene expression. Naringenin (4',5,7-trihydroxyflavanone), (Figure 4.7b) aglycone of naringin, is a major flavanone in grape fruits and tomatoes.

It has been shown to have a number of biological effects, such as an anticancer, antimutagenic, and anti-inflammatory activity. β-catenin/Tcf signaling plays an important role in gastric cancer cells. The antitumor effects of flavanone and naringenin are mediated by its ability to down regulate the β-catenin/Tcf signaling. Previous studies reported the potential role of flavanone and naringenin in the repression of the β-catenin/Tcf signaling and its inhibitory mechanism. The luciferase activity show that β-catenin/Tcf-driven transcription was suppressed effectively by flavanone and naringenin in AGS gastric cancer cells in a dose-dependent manner. Flavanone and naringenin also inhibit a constitutively active mutant β-catenin/Tcf signaling in HEK293 cells and the β-catenin proteins in AGS cells were not degraded by flavanone and naringenin treatment. These observations strongly suggest that the inhibitory mechanism of flavanone and naringenin is not related to the upstream regulators of the β-catenin/Tcf pathway but to β-catenin itself or to the downstream components.

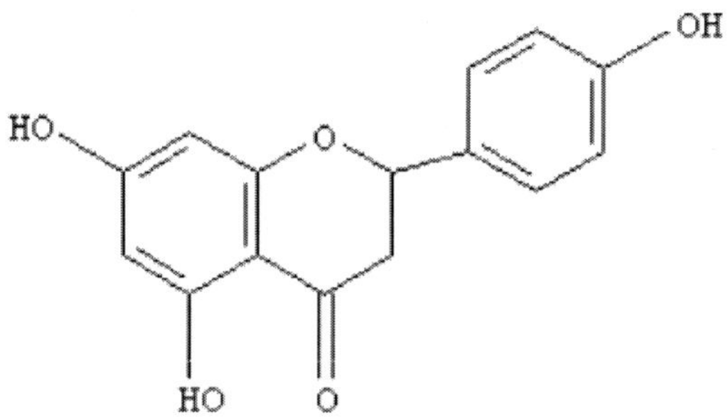

Figure 4.7a. Structure of flavanone.

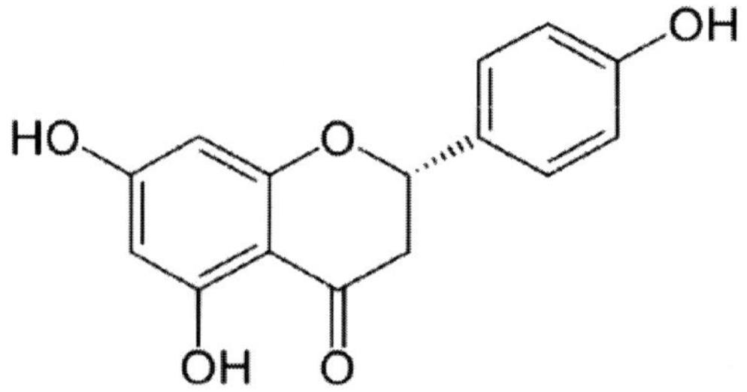

Figure 4.7b. Structure of Naringenin.

The β-catenin distribution change caused by flavanone and naringenin affected β-catenin/Tcf signaling. β-catenin is ubiquitous and move freely in a cell. It contribute to the cell–cell adhesion in the membrane and functions as a transcriptional activator in the nucleus. These observations help to reveal the molecular mechanism underlying the anti-tumor effect of flavanone and naringenin in gastric cancer cells [24, 25].

4.8. Acacetin

Acacetin (5,7-dihydroxy-4'-methoxy flavone), (Figure 4.8) is a flavonoid compound. Acacetin has been reported to exhibit many biological effects including anticancer activity. Studies clarified the molecular mechanism by which acacetin activated human gastric carcinoma AGS cells undergo apoptosis. Acacetin can suppress the proliferation of human gastric carcinoma cells through stimulating apoptosis. Acacetin was the potent suppressor of cell viability and cause the potent and rapid induction of apoptosis, concurrent with DNA ladders, sub-G1 peak appearance, chromatin condensation, and apoptotic appearance in AGS cells.

This induction of apoptosis occurred within hours, consistent with the view that acacetin induced apoptosis by activating preexisting apoptosis machinery. Indeed, treatment with acacetin caused an induction of caspase-3, but not caspase-1, associated with the degradation of DFF-45 and PARP, which preceded the inception of apoptosis. Pretreatment with the caspase-3 inhibitor Z-VAD-FMK arrested acacetin-induced apoptosis, suggest that apoptosis induced by acacetin involve a caspase-3-mediated mechanism.

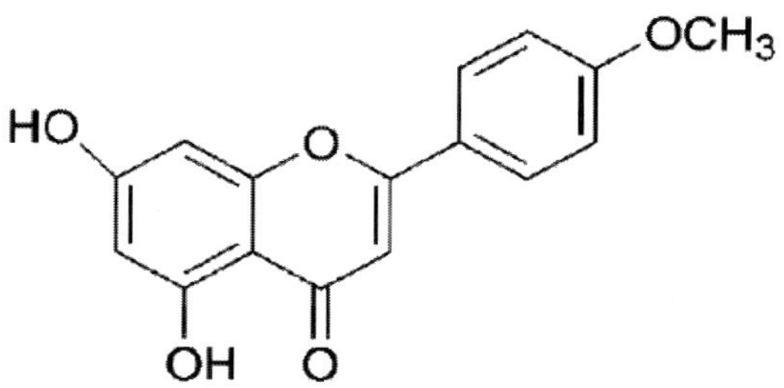

Figure 4.8. Structure of Acacetin.

Mitochondrial transmembrane potential is often employed as a marker of cellular viability, and its interruption has been implicated in a variety of apoptosis phenomena. Mitochondria have also been implicated as a source of reactive oxygen species (ROS) during apoptosis. Reduced mitochondria membrane potential has been shown to lead to increased generation of ROS and apoptosis. Acacetin could interrupt the functions of mitochondria at the early stages of apoptosis and subsequently coordinate caspase-9 activation, but not caspase-1, through the release of cytochrome C. AGS cells show increased ROS production after acacetin treatment. The raise in ROS was probably due to the affected mitochondria cycle the dioxygen through the electron transport assembly, and generate ROS by one-electron-transfer mitochondria could be a main target of nonspecific damage through oxidative stress at the level of the outer and inner membranes. As a consequence of oxidative membrane damage, membrane potential and permeability-barrier function are impaired, leading to further mitochondrial damage. Oxidative damage to the mitochondrial membrane due to increased generation of ROS has been shown to play a role in apoptosis. Therefore, the intracellular generation of ROS could be an important factor in acacetin-induced apoptosis.

The receptor-mediated signaling transduction pathway of apoptosis is another major pathway in activating caspase cascades. When acacetin treated with AGS cells the enhancement expression of Fas and FasL were observed. Fas and FasL binding with its receptor induce apoptosis. Acacetin stimulat a marked increase in caspase-8 activity. Caspase-3 can stimulate caspase-8. Caspase-8 is implicated in cytochrome *C*-mediated apoptosis and participates in a feedback amplification loop involving caspase-3 in acacetin treated AGS cells. Bcl-2 family proteins might participate in the event that controlled the change in mitochondrial membrane potential and activate cytochrome C release during apoptosis induced by acacetin. The down-regulation of the Bcl-2 expression and the cleavage of the Bad were observed in acacetin-treated cells. The up-regulation of Bax expression and the cleavage of Bcl-xL during acacetin-induced apoptosis were observed in AGS cells. However, the ratio between Bcl-2 and Bax and Bcl-xL cleavage determines cell survival or death. The p53 tumor suppressor is predominantly a nuclear transcription factor, stimulated by various stresses including chemopreventive agents. Normal p53 function acts as tumor suppressor inducing both growth arrest and apoptosis. p53 activate the Fas gene in response to DNA damage by anticancer drugs. Treatment of the AGS cells with acacetin result in an increase in the level of p53 protein. Acacetin induced apoptosis might enhance the expression of Bax and Fas protein dependent on the p53 protein that affects mitochondrial function, raising the possibility that the expression of Bax or Fas could be transcriptionally regulated in response to acacetin treatment. These observations clearly demonstrate that the gastric cancer preventive action of acacetin in AGS cells [26].

4.9. Curcumin

Curcumin (Figure 4.9) is a natural phenolic coloring compound that is found in the rhizomes of *Curcuma longa* L, commonly called turmeric. Curcumin has a wide range of biological and pharmacological activities, including antioxidant properties, anti-inflammatory properties, anti-mutagenic activity, and anti-carcinogenic properties. Curcumin inhibited the proliferation, invasion and metastasis of different cancers through interaction with multiple cell signaling proteins. It was found that curcumin could inhibit recepteur d'origine nantais (RON) tyrosine kinase, Akt, NF-kB and p38 MAPK to affect the biological function of cancer cells, such as apoptosis and invasion. In addition, curcumin inhibit the proliferation of gastric cancer cells and the inhibition correlated with the down regulation of the expression of cyclin D1. Cyclin D1 is required to mediate the G1 to S transition, in turn it lead to DNA synthesis and cell cycle progression.

Curcumin makes G1 cells increase with S-cells decrease, it is reasonable that the antiproliferative effects of curcumin are related to the down regulation of cyclin D1 expression. However, a time lag between the inhibition of transition of the cells from G1–S and the down regulation of Cyclin D1 expression. Many reports suggest that cyclin D1 transcription could be up regulated through multiple signaling pathways, including NF-κB. p21-activated kinase 1 (PAK1) signaling has been shown to modulate NF-κB activation, when human gastric cancer cell treated with curcumin the activity of PAK1 is inhibited, it is concluded that curcumin downregulate the expression of cyclin D1 through PAK1 signaling pathway. Considering that PAK1 is implicated in biological process ranging from cytoskeletal dynamics and motility to tumorigenesis the activation of PAK1 kinase activity contribute to the actin reorganization and cell migration, thus the suppression of PAK1 activity lead to the decrease of proliferation and invasion in gastric cancer cells.

Human Epidermal growth factor Receptor 2 (HER2), a hypotype of_*epidermal growth factor receptor* (EGFR) family, encodes a 185-kDa transmembrane glycoprotein with ntrinsic tyrosine kinase activity that has been revealed to be overexpressed, amplified or both, in a number of human malignant cancer. Overexpression of the HER2 receptor is linked with increased progression and metastasis. There are lot of evidence to explain the role of HER2 overexpression in patients with gastric cancer. Both homo and hetero dimeric receptor complexes are formed upon cell activation by growth factor, but the signal is generated by the activated Erb1/Erb2 heterodimers that led to greater tumorigenic cell growth than homodimer induced signal. The physical and functional interaction of HER2 and EGFR leads to the formation of a highly active, heterodimeric tyrosine kinase complex which synergistically stimulate cellular transformation.

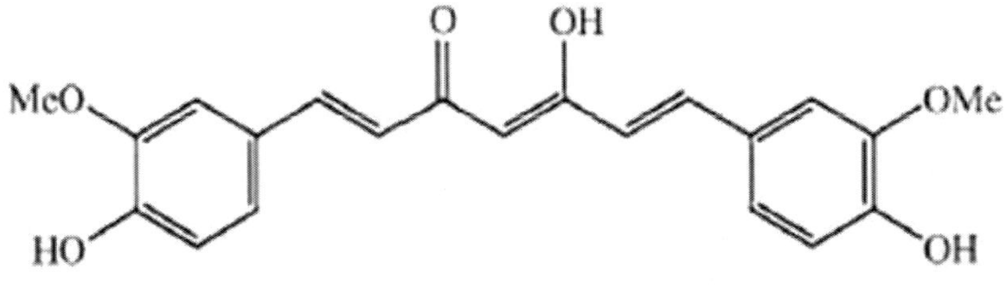

Figure 4.9. Structure of Curcumin.

Besides, it is known that the EGFR-initiated signaling pathway is a potent inducer of c-Src and PAK1 pathways, as well as reorganization of the cytoskeleton, which allow increased tumor invasiveness. Hence, suppressive expression of the upstream protein HER2 by curcumin is accompanied by inhibition of PAK1 activation and invasiveness of cancer cells. Curcumin strongly inhibit the activity of PAK1 when HER2 was silenced. Curcumin is capable to inhibit the expression and the tyrosine phosphorylation of EGFR, these reports indicate that curcumin is an effective inhibitor of EGFR a direct and potent receptor of EGF.

PAK1 play a vital role in motility, invasion and cell survival in human cancer, and the curcumin does efficiently inhibit PAK1 activity induced by EGF. Interestingly, curcumin suppress the kinase activity of PAK1 not only when it is added in cells, but also when incubated directly together with PAK1, ATP, ^{32}P and substrates. The predicted binding modes of curcumin in the PAK ATP binding domain also indicate that curcumin has the capacity of acting as an ATP-competitive inhibitor. Since curcumin exist in two tautomeric forms: ketone and enol, both were used in molecular docking studies.

Enol and ketone curcumin bind in same position of ATP binding site, however, their binding orientation and affinity were different, as ketone curcumin showed higher putative affinity towards PAK1 than enol curcumin. Hence, the role of the ATP-competitive inhibitor of curcumin and its suppressive effect on HER1/2 both contribute its suppression of PAK1 activity. Collectively, these observations provided novel insights into the mechanisms of curcumin inhibition of gastric cancer cell growth and potential therapeutic strategies for gastric cancer [27].

4.10. Gallic Acid (GA)

Gallic acid (Figure 4.10) is found abundantly in vegetables. GA has strong anticancer properties, including cytotoxic effects and inhibition of cell migration through the suppression of a number of signaling pathways. And these effects might lead to the suppression of metastatic capacity of AGS cells. Various studies of tumor invasion and metastasis have supported the degradation of the extracellular matrix metallo proteinases (MMPs) play a central role. Two of these enzymes, MMP-2 and MMP-9, associate with the progressions of tumor cell invasion and metastasis in human cancers. GA inhibited MMP-2 and MMP-9 expression. Besides, MMP gene expression is chiefly regulated by transcriptional factors like nuclear factor kappa-light-chain-enhancer of activated B cells (NF-kB) and activator protein-1 (AP-1) *via* the phosphatidylinositol 3-kinase (PI3K)/Akt pathway.

NF-kB is a multi-subunit transcription factor involved in cellular responses to viral infection and inflammation. NF-kB is retained in the cytoplasm through interactions with an inhibitor of NF-kB (IkB). Upon dissociation, NF-kB move into the nucleus and promote cancer cell proliferation, angiogenesis, and metastasis. The quantity of IkB could control NF-kB nuclear translocation and consequently influence the expression of MMP-2 in several types of human cells. There was a dose-dependent increase in IkB when AGS cells were treated with GA. However, the protein expression of NF-kB was not changed under the same condition.

This observation suggested that increase of cytoplasmic IkB could bind with NF-kB and then inhibit of NF-kB activity. PI3K signal transduction pathways regulate cell metastasis of melanoma that is closely associated with the development and progress of various tumors.

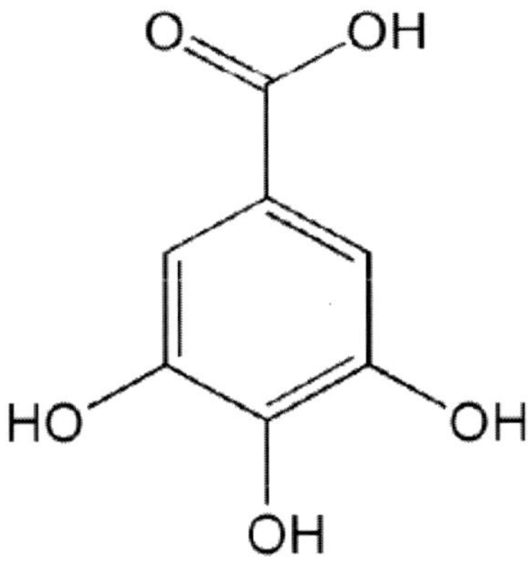

Figure 4.10. Structure of gallic acid.

Therefore, PI3K/Akt pathway is constitutively active in most tumors. In addition to its role in tumor cell invasion, this pathway also regulate many cellular processes implicated in tumorigenesis, cell size/growth, proliferation, survival, glucose metabolism, genome stability, metastasis, and angiogenesis. Activated Akt can direct bind with the IkB kinase (IKK) then active it, finally let IkB degrade. These protein levels of PI3K, Akt-1 and P-Akt also decreased in a dose-dependent manner when treated with GA. This observation indicated that GA might enhance IkB binding to NF-k then suppress the PI3K/Akt pathway subsequently lead to anti-metastasis of AGS cell.

Ras-homologous (Rho) GTPases play a key role in the regulation of several cellular functions linked with malignant transformation and metastasis. Members of the Rho family of small GTPases are pivotal regulators of actin reorganization, cell motility, cell–cell and extracellular matrix (ECM) adhesion as well as of cell cycle progression, gene expression, and apoptosis. RhoB shares 86% amino acid sequence identity with RhoA, yet the roles of the low-molecular-weight GDP/GTP binding GTPases in oncogenesis are quite different. While RhoA, like other GTPase family members such as Ras, Rac1, and cdc42, promote oncogenesis invasion, and metastasis emerging evidence point to a tumor-suppressive role for RhoB. Besides, PI3K is an effector of Ras function and has been shown to be necessary for both the development and maintenance of tumors driven by mutant H-Ras. When GA treated with AGS cells were showed diminished protein levels of Ras, cdc42, Rac1 and RhoA. However, the protein levels of RhoB were increased. All these findings are showing that GA is able to suppress gastric cancer metastasis and thus it can be used as a potential chemotherapeutic agent for gastric cancers [28].

Conclusion

Polyphenols provide a broad spectrum of biological actions that are potentially beneficial for gastric cancer prevention. Experimental evidence gathered in the recent years from various pre-clinical and clinical studies clearly support the idea that dietary polyphenols have potentially prevented gastric cancer. This chapter provides a novel perspective on the potential chemoprevention by polyphenols (apigenin, anthocyanins, genistein 2',4'-Dihydroxychalcone, luteolin, epigallocatechin-3-gallate, flavanone, naringenin, acacetin, curcumin and gallic acid), as the extensive studies summarized here recommend that beneficial effects of different dietary polyphenols may in part be attributable to their gastric cancer preventive properties, including regulation of signal transduction pathways, transcription factors and related activities; modulation of cell-cycle regulation or induction of apoptosis, affecting cell differentiation, proliferation, metastases, angiogenesis and suppression of imune response. The efficiency of polyphenols against gastric cancer progression represent one of the most important future applications of this knowledge. A complete understanding of their valuable effects may not yet be fully realized, the established observations of numerous studies points to a need for further investigations that can expand our understanding of the dynamic role these dietary polyphenols play in the reduction of certain risk factors associated with gastric cancer. A better knowledge on polyphenol interactions with various cell components will positively contribute to develop nutritional and pharmacological strategies oriented to prevent the onset and/or the consequences of gastric cancer.

REFERENCES

[1] Ramos S. Cancer chemoprevention and chemotherapy: dietary polyphenols and signalling pathways. *Mol. Nutr. Food Res.* 2008, 52, 507-526.

[2] Lambert, JD; Hong, J; Yang, G; Liao, J; Yang CS. Inhibition of carcinogenesis by polyphenols: evidence from laboratory investigations. *Am. J. Clin. Nutr.* 2005, 81, 284S-291S.

[3] Serafini, M; Bellocco, R; Wolk, A; Ekstrom, AM. Total antioxidant potential of fruit and vegetables and risk of gastric cancer. *Gastroenterology.* 2002, 123, 985–991.

[4] Park, WS; Oh, RR; Park, JY; Lee, SH; Shin, MS; Kim, YS; Kim, SY, Lee, HK; Kim, PJ; Oh, ST; Yoo, NJ;. Lee, JY. Frequent somatic mutations of the b-catenin gene in intestinal type gastric cancer, *Cancer Res.* 1999, 59 4257–4260.

[5] Werner, M; Becker, KF; Hofler, H. Gastric adenocarcinoma: pathomorphology and molecular pathology. *J. Cancer Res. Clin. Oncol.* 2001,127, 207–216.

[6] Bravo L. Polyphenols: chemistry, dietary sources, metabolism, and nutrition- al significance. *Nutr. Rev.* 1998, 56, 317–333.

[7] Murzakhmetova, M; Moldakarimov, S; Tancheva, L; Abarova, S; Serkedjieva, J. Antioxidant and prooxidant properties of a polyphenol-rich extract from Geranium sanguineum L. in vitro and in vivo. *Phytother. Res.* 2008, 22, 746–751.

[8] Wu, K; Yuan, LH; Xia, W. Inhibitory effects of apigenin on the growth of gastric carcinoma SGC-7901 cells *World J. Gastroenterol.* 2005, 11, 4461-4464.

[9] Kamei, H; Kojima, T; Hasegawa, W; Koide, T; Umeda, T; Yukawa, T; Terebe, K. Suppression of tumor cell growth by anthocyanins in vitro. *Cancer Invest.* 1995, 13, 590–594.

[10] Shih, PH; Yeh, CT; Yen, GC. Effects of anthocyanidin on the inhibition of proliferation and induction of apoptosis in human gastric adenocarcinoma cells. *Food Chem. Toxicol.* 2005, 43, 1557–1566.

[11] Davis, JN; Kucuk, O; Sarkar, FH. Genistein inhibits NF-kappa B activation in prostate cancer cells. *Nutr. Cancer.* 1999, 35, 167-174.

[12] Davis, JN; Singh, B; Bhuiyan, M; Sarkar, FH. Genistein-induced upregulation of p21WAF1, downregulation of cyclin B, and induction of apoptosis in prostate cancer cells. *Nutr. Cancer.* 1998, 32, 123-131.

[13] Cui, HB; Na, XL; Song, DF; Liu Y. Blocking effects of genistein on cell proliferation and possible mechanism in human gastric carcinoma. *World J. Gastroenterol.* 2005, 11, 69-72.

[14] Zhou, HB; Chen, JM; Cai, JT; Du, Q; Wu, CN. Anticancer activity of genistein on implanted tumor of human SG7901 cells in nude mice. *World J. Gastroenterol.* 2008, 14, 627-631.

[15] Lou, C; Wang, M; Yang, G; Cai, H; Li, Y; Zhao, F; Yang, H; Tong, L; Cai, B. Preliminary studies on anti-tumor activity of 20,40-dihydroxychalcone isolated from Herba Oxytropis in human gastric cancer MGC-803 cells. *Toxicol. In Vitro.* 2009, 23, 906–910.

[16] Lou, C; Yang, G; Cai, H; Zou, M; Xu, Z; Li, Y; Zhao, F; Li, W; Tong, L; Wang, M; Cai, B. 2',4'-Dihydroxychalcone-induced apoptosis of human gastric cancer MGC-803 cells via down-regulation of survivin mRNA. *Toxicol. In Vitro.* 2010, 24, 1333–1337.

[17] Zhang, FF; Shen, HM; Zhu, XQ. Research progress on antitumor effects of luteolin. *Zhejiang Da Xue Xue Bao Yi Xue Ban.* 2006, 35, 573–578.

[18] Wu, B; Zhang, Q; Shen, W; Zhu, J. Anti-proliferative and chemosensitizing effects of luteolin on human gastric cancer AGS cell line. *Mol .Cell Biochem.* 2008, 313, 125–132.

[19] Leone, M; Zhai, D; Sareth, S; Kitada, S; Reed, JC; Pellecchia, M. Cancer prevention by tea polyphenols is linked to their direct inhibition of antiapoptotic Bcl-2-family proteins. *Cancer Res.* 2003, 63, 8118-8121.

[20] Yang, CS; Chung, JY; Yang, G; Chhabra, SK; Lee, MJ. Tea and tea polyphenols in cancer prevention. *J. Nutr.* 2000, 130, 472S-478S.

[21] Ran, ZH; Xu, Q; Tong, JL; Xiao SD. Apoptotic effect of Epigallocatechin-3-gallate on the human gastric cancer cell line MKN45 via activation of the mitochondrial pathway. *World J. Gastroenterol.* 2007, 13, 4255-4259.

[22] Shishido, T; Yasoshima, T; Denno, R; Sato, N; Hirata, K. Inhibition of liver metastasis of human gastric carcinoma by angiogenesis inhibitor TNP-470. *Jpn. J. Cancer Res.* 1996, 87, 958-962.

[23] Zhu, BH; Zhan, WH; Li, ZR; Wang, Z; He, YL; Peng, JS; Cai, SR; Ma, JP; Zhang CH. (-)-Epigallocatechin-3-gallate inhibits growth of gastric cancer by reducing VEGF production and angiogenesis. *World J. Gastroenterol.* 2007, 13:1162-1169.

[24] Park, CH; Hahm, ER; Lee, JH; Jung, KC; Yang, CH. Inhibition of *b*-catenin-mediated transactivation by flavanone in AGS gastric cancer cells. *Biochem. Biophys. Res. Commun.* 2005, 331, 1222–1228.

[25] Lee, JH; Park, CH; Jung, KC; Rhee, HS; Yang, CH. Negative regulation of *b*-catenin/Tcf signaling by naringenin in AGS gastric cancer cell. *Biochem. Biophys. Res. Commun.* 2005, 335, 771–776.

[26] Pan, MH; Lai, CS; Hsu, PC; Wang, YJ. Acacetin Induces Apoptosis in Human Gastric Carcinoma Cells Accompanied by Activation of Caspase Cascades and Production of Reactive Oxygen Species. *J. Agric. Food Chem.* 2005, 53, 620-630.

[27] Cai, XZ; Wang, J; Li, XD; Wang, GL; Liu, FN; Cheng, MS; Li, F. Curcumin *Cancer Biol. Ther.* 2009, 8, 1360-1368.

[28] Ho, HH; Chang, CS; Ho, WC; Liao, SY; Wu, CH; Wang, CJ. Anti-metastasis effects of gallic acid on gastric cancer cells involves inhibition of NF-kappaB activity and downregulation of PI3K/AKT/small GTPase signals. *Food Chem. Toxicol.* 2010, 48, 2508-2516.

In: Natural Products and Their Active Compounds … ISBN: 978-1-62100-153-9
Editors: M. Essa, A. Manickavasagan, and E. Sukumar

Chapter 13

BIOACTIVE COMPOUNDS FROM *TRIBULUS TERRESTRIS* L. (ZYGOPHYLLACEAE)

Sardar A. Farooq, Talat T. Farook and Salim H. Al-Rawahy
Department of Biology,
Sultan Qaboos University,
Al Khod, Oman

ABSTRACT

This chapter discusses the role of bioactive compounds from *Tribulus terrestris* L. (Zygophyllaceae) (TT) in disease prevention and treatment. *Tribulus terrestris* is gaining popularity in the media because it is projected as libido enhancer, treats male sexual dysfunction, gives stamina and confidence to perform. A large number of research papers have been published on this plant and there is a need to critically review, analyze and summarize the literature to draw a line between myth and reality. Since ancient times *Tribulus terrestris* has been used in the Indian and Chinese traditional medicine to treat hypertension, premature ejaculation, erectile dysfunction, vitiligo, and kidney and eye problems. It has anti-urolithiatic, diuretic, antiacetylcholine, aphrodisiac properties and can stimulate spermatogenesis and libido.

The phytochemistry of the extract reveals the presence of alkaloids, steroidal saponins, furostanol saponins, flavonoid glycosides which impart the medicinal properties to this plant.

The research papers published until 2010 on *Tribulus terrestris* are thoroughly reviewed and summerized under the following captions: Botany, Phytochemisty, the effects on hypertension, hormones, oxalate metabolism, endocrine sensitive organs, androgen receptors, diuretic and contractile effects, its use as an aphrodisiac, protective agent in diabetes, use as a nutritional supplement, its anti cancer activity, anti microbial activity, analgesic, anthelmintic properties and its side effects. In each section emphasis is laid on identifying the bioactive compound, its specific use and the mode of action.

Introduction

Plants have a significant importance in human health since time immemorial. Human reliance on plants for medicine continues even today. The world Health Organization reports that 80% of the world population relies primarily on plant based crude drugs and traditional systems of medicine. The most commonly used treatments for many diseases are derived from ethanobotanically important plants. The medicinal value of the plants lies in certain chemical substances produced by them, which were thought to be waste products in their metabolism. These are called secondary metabolites (natural products) which have a definite physiological action in the disease prevention and treatment. The most important of these bioactive constituents of plants are alkaloids, tannins, flavonoids, and phenolic compounds.

From ancient times, man has been in search of plants to boost up his sexual powers and performance. *Tribulus terrestris* L. (TT) is one of the earliest potent aphrodisiac used by man. It is a drought-tolerant, annual creeping herb with prostrate hairy branches belonging to plant family Zygophyllaceae. It is distributed worldwide in the tropical and subtropical regions. In the traditional folk medicine the leaves, fruits and roots have been extensively used to treat urinary infections, inflammations, abdominal distention, emission, morbid leucorrhea, sexual dysfunction, veiling, oedema and ascites (Chopra et al., 1958; CHEM-EXCIL 1992; Adaikan et al., 2001).

Indian ayurveda system of medicine has recognized the importance of this herb long ago and used in tonics, under the Sanskrit name, "gokshura" to energize, vitalize and improve sexual function and physical performance in males (Anonymous 2010). Chinese use this plant for centuries in the treatment of dizziness, liver problems, premature ejaculation and headaches and eye problems. In different regions the common names of this plant are Puncture vine, Caltrop, Yellow vine, Goathead, Burra Gokharu and bindii.

Botanical Description

Habitat: Tribulus terrestris L. is an annual plant found in Asia, Europe, Africa, Australia and some parts of USA mostly as a weed. It belongs to family Zygophyllaceae. It is an annual herb about 30–70 cm tall with pinnate leaves and bright yellow flowers. The fruits are stellate with spikes at angles, giving the plant its common name puncture vine.

Plant: Stem is profusely branched, semi erect or prostrate reaching up to 2m long. The root is slender, cylindrical, some what fibrous, Leaves are opposite, paripinnate, each consists of 4-8 pairs of linear or oblong leaflets with hairy margins. Stems are round and hairy. Flowers are yellow, solitary with 5 petals. The fruit is globose with 5-12 woody cocci with sharp spines of unequal length, giving it a star shape. The seeds are 1.5-3 mm long, yellowish, enclosed with in 5-7 mm long carpels; up to 5 in each chamber (Ross, 2001).

Medicinal Uses

The plant *Tribulus terrestris* (TT) or its products have been extensively used both in the Indian traditional medicine Ayurveda and Chinese medicine for the treatment of various ailments such as urinary, cardiovascular, and gastrointestinal disorders. The usable parts from plant are the fruits, seed and the leaves, besides these the flowers and the root is also used. In Ayurveda TT is known for its anti-urolithiatic, diuretic and aphrodisiac properties (Sivarajan and Balachandran, 1994). Pharmacological studies reported in the literature (Anand et al., 1994, Ross, 2001) have confirmed these properties. The seed is abortifacient, alterative, anthelmintic, aphrodisiac, astringent, carminative, demulcent, diuretic, emmenagogue, galactogogue, pectoral and tonic and is reported to have antiseptic and anti-inflammatory properties (Warrier, 1997, Williamson, 2002).

It stimulates blood circulation. A decoction is used in treating impotency in males, nocturnal emissions, gonorrhea and incontinence of urine. It has also proved effective in treating painful urination, gout and kidney diseases (Sangeeta et al., 1994). The plant has shown anticancer activity (Neychev et. al., 2007). The flowers are used in the treatment of leprosy. The stems are used in the treatment of scabious skin diseases and psoriasis and also for relieving rheumatic pain and as an analgesic. The dried and concocted fruits are used in the treatment of congestion, gas, headache, liver, ophthalmia and stomatitis (Warrier, 1997). it is described as a highly valuable drug used to restore the depressed liver for the treatment of fullness in the chest and mastitis and also used to dispel the wind and clear the eyes for the treatment of acute conjunctivitis, with anti-inflammatory and immunosuppressive activities, headache and vertigo (Xie, Z. F., and Huang, X. K. (1988). *Dictionary of Traditional Chinese Medicine*, Commercial Press, LtdHong Kong, p. 205.Xie and Huang, 1988).

TT is also reported to have antimicrobial, antihypertension, diuretic, antiacetylcholine and haemolytic activity and stimulate spermatogenesis and libido and premature ejaculation. It is also helpful in the areas of building muscles Milasius (2010). TT has been used as an herbal agent for years to treat hypertension with no serious side effects (Chui et al., 1992; Lu et al., 1994). Monson and Schoenstadt (2008) have warned that TT could theoretically increase the risk of prostate cancer or might worsen prostate cancer. It may also worsen an enlarged prostate. It might increase the risk of low blood sugar (hypoglycemia) in people with diabetes. However they did not substantiate these statements with experimental data. In sheep limb paresis (staggers) occurs due to Tribulus trestles alkaloids beta-carboline alkaloid harman (harmane) and norharman (norharmane). The dried leaves contain 44 mg/kg alkaloid (Bourke, 1992).

Different organs of *Tribulus terrestris* plant are used to treat various ailments. It is hypotensive and has cardiac depressant effects and contractile activities on smooth muscles (Mossa et al., 1983). It has been found to be effective in treating angina pectoris by dilating the coronary arteries and improving the cardiac circulation (Wag et al., 1990). TT has been commonly used in folk medicine in Turkey as diuretic and against colicky pains, hypertension and hypercholesterolemia (Arcasoy et al., 1998).

It has been shown to increase the free serum testosterone (Brown et al., 2001) and to be effective in the treatment of sexual and erectile dysfunction by conversion of its phytochemical derivative, protodioscine to De Hydro Epi Androsterone (DHEA) (Adiomoelja, 2000).

It has protective effect on genetic damage (Liu et al., 1995) and stimulates melanocyte proliferation in the treatment of vitiligo (Lin et al., 1999) and it has an antibacterial and cytotoxic activity (Ali et al., 2001). A nematocidal activity has also been reported (Nandal and Bhatti, 1983). With regard to urinary stones, it has been shown to reduce the amount of urinary oxalate in rats (Sangeeta et al., 1994), and it has an antiurolithic activity in experimentally induced urolithiasis in rats too (Anand et al., 1994).

This plant is given by Iraqi herbalists in the form of beverages (bush teas) prepared by soaking the leaves and fruits in water or boiling them.

Tribulus has several steroidal saponins that have been used in many pharmaceutical preparations to treat cardiovascular diseases, the mode of its action and efficacy remains uncertain and controversial. (Wu et. al., 1999 and Xu et al., 2000) The plant extracts have different effects on animals (Gauthaman et al., 2002; Gauthaman et al., 2003; Arcasoy et al., 1998) and men (Brownet al., 2000; Brown et al., 2001; Kohut et al., 2003; Antonio et al., 2000). Several active compounds have been isolated from this plant (Huang et al., 2003; De Combarieu et al., 2003; Cai et al., 2001; Conrad et al., 2004).

Tribulus has been shown to enhance sexual behaviour in an animal model by stimulating androgen receptors in the brain (Gauthaman et al., 2002). It affects strongly the androgen metabolism by increasing significantly testosterone or testosterone precursor levels dihydrotestosterone and dehydroepiandrosterone (Gauthaman and Ganesan, 2008) and produces effects suggestive of aphrodisiac activity. However, there are conflicting reports on its efficacy. Neychev and Mitev (2005) found that the chronic ingestion of either 10 or 20 mg/kg body weight of TT extract did not influence androgen production either directly or indirectly in young males. It did not influence the androgen production or LH in young men in young men. (Brown et al., 2001 and Neychev and Mitev, 2005). The active chemical in T. terrestris is protodioscin (PTN), a cousin to DHEA (Gauthaman et al., 2003). In a study with mice, *Tribulus* was shown to enhance mounting activity and erection better than testosterone cypionate. But it is unconvincing and needs further probe. The proerectile aphrodisiac properties are most likely due to the release of nitric oxide from the nerve endings innervating the corpus cavernosum penis.

The extract is known to increase the level of testosterone in the body and helps in enhancing sexual performance to maintain a healthy sex drive in men. The extract is claimed to prevent osteoporosis, lowers the risks of type 2 diabetes, obesity, depression, anxiety and early mortality. Furthermore it is also helpful in building muscles Milasius (2010).

The steroidal saponins of are responsible for the biological activity of TT. The amount of active ingredient varies in plants from different geographical regions. Dinchev et al., (2008) analysed samples of TT collected from many parts of Europe and Asia such as Bulgaria, Greece, Serbia, Macedonia, Turkey, Georgia, Iran, Vietnam and India and found distinct differences in the concentration of protodioscin, prototribestin, pseudoprotodioscin, dioscin, tribestin and tribulosin depending on location, plant part and its developmental stage. There was a similarity in the chemical profile of the samples from Bulgaria, Turkey, Greece, Serbia, Macedonia, Georgia and Iran with little quantitative differences.

The Vietnamese and Indian samples lack prototribestin and tribestin, but has high amount of tribulosin. Depending on this, Dinchev et al., (2008) suggested the presence of two different chemotypes in TT. The fruit morphology and the burrs do not have any correlation with the chemical composition.

Protodioscin was also detected from the samples from Indonesia (Adimoelja, 2000) and Singapore (Adaikan et al., 2000) that is claimed to improve sexual desire and enhance erection by converting into dehydroepiandrosterone (DHEA). Protodioscin is a complex molecule (C51H84O22) and has at its core the furost- 5-ene-3, 22, 26-triol nucleus. There are sugar moieties attached at C-3 and the C-26 hydroxyl group (Shao et al., 1997).

PHYTOCHEMISTRY

Tribulus terrestris is reported to have antimicrobial, antihypertension, diuretic, antiacetylcholine and haemolytic activity and can stimulate spermatogenesis and libido (Jit and Nag, 1986; Bose et al., 1964; Tomova, 1988; Sharma et al., 1977 and Wu et al., 1999). Alkaloids, steroids, flavonoids and carbohydrates Table-1) have been isolated from this plant (Duan and Zhou, 1993; Chiang, 1977; Mahato et al., 1981b; Saleh and El-Hadidi, 1982; Saleh et al., 1982; Mahato et al., 1978; Perepelitsa and Kintya, 1975; Tomova et al., 1981; Zafar et al., 1989; Chiu and Chang, 1986; Nag et al., 1979; Mahato et al., 1981a; Mathur et al., 1977; Vasi and Kalintha, 1982 and Wu et al., 1999).

Table 1. Phytochemicals isolated from *Tribulus terrestris*

S. No.	Compound	Reference	
1	Alkaloids, steroids, flavonoids and carbohydrates	Duan and Zhou, 1993; Chiang, 1977; Mahato et al., 1981b; Saleh and El-Hadidi, 1982; Saleh et al., 1982; Mahato et al., 1978; Perepelitsa and Kintya, 1975; Tomova et al., 1981; Zafar et al., 1989; Chiu and Chang, 1986; Nag et al., 1979; Mahato et al., 1981a; Mathur et al., 1977; Vasi and Kalintha, 1982 and Wu et al., 1999	Whole plant
2	Steroidal saponins: terrestrosins A, B, C, D and E, desgalactotigonin, F-gitonin, desglucolanatigonin, gitonin,	Yan et al., 1996	Fruits
3.	Diosgenin, hecogenin, neotigogenin.	Wu et al., 1996. Joshi et al., 2005.	Arial parts
4.	Alkaloids, Phytosterols: b-sitosterol, stigmasterol terrestriamide and 7-methyl-hydroindanone	Joshi et al., 2005	
5.	Steroidal glycosides: Hecogenin.	Wu et al., 1996.	Arial parts
6.	Terrestribisamide, 25*R*-spirost-4-en-3, 12-dione tribulusterine, *N*-*p*-coumaroyl-tyramine, terrestriamide, hecogenin, aurantiamide acetate, xanthosine, ferulic acid, vanillin, *p*-hydroxybenzoic acid and β-sitosterol	Wu et al., 1999.	Dried fruits
7.	Furostanol saponins	Wang et al., 1997.	Dried fruits
8.	Flavonoid glycosides kaempferol, quercetin and isorhamnetin, with the 3-gentiobiosides as the major glycosides.	Saleh et al., 1982.	Tribulus pentandrus and T.terrestris.
9.	Flavonoids Kaempferol, kaempferol-3-glucoside, kaempferol-3-rutinoside and tribuloside (kaempferol-3- β-d-(6″-p-coumaroyl) glucoside).	Bhutani et al., 1969	Leaves and Fruits
10.	Steroidal saponins	Su et al., 2009.	Fruits
11.	Furostanol saponins: terrestroside.	Yuan et al., 2008.	Fruits
12.	Furostanol saponin 1, Tribol, spirostanol saponins 2 and 3 and sitosterol glucoside.	Conrad et al., 2004.	Aerial parts
13.	Methyl-protodioscin, protodioscin, methylprototribestin prototribestin, dioscin and sitosterol glucoside	Conrad et al., 2004.	Aerial parts
14.	Furostanol glycosides, tribufurosides 1, 2.	Xu et al., 2009.	Fruits

The constituents of *Tribulus terrestris* also include steroidal saponins (Yan et al., 1996) for example, terrestrosins A, B, C, D and E, desgalactotigonin, F-gitonin, desglucolanatigonin, gitonin, etc. The hydrolyzed products include diosgenin, hecogenin, neotigogenin, etc. Joshi et al., (2005). The minor constituents of *Tribulus terrestris* also include alkaloids, common phytosterols, viz, b-sitosterol, stigmasterol a cinnamic amide derivative such as terrestriamide and 7-methylhydroindanone.

Wu et al., (1996) have reported the following steroidal glycosides from *Tribulus terrestris* hecogenin, 3-O-β-D-glucopyranosyl(1→4)- β-D-galactopyranoside; 26-O- β- D-glucopyranosyl- 3-O- [{ β-D-xylopyranosyl(1→3)}{β-O-galactopyranosyl(1→2)}- β - D-glucopyranosyl (1→4)-β-D-glucopyranosyl]-5α-furost-20(22)en-12-one-3β, 26-diol and 26-O- β -D-glucopyranosyl-3-O-[{β - D- xylopyranosyl(1 → 3)}{ β-O- galactopyranosyl(1 → 2)}- β- D- glucopyranosyl(1→4)- β- D- glucopyranosyl] - 5 α- furostan- 12- one - 3 β,22,26 - triol. Wu et al., (1999) also isolated terrestribisamide, 25*R*-spirost-4-en-3, 12-dione and tribulusterine, *N-p*-coumaroyltyramine, terrestriamide, hecogenin, aurantiamide acetate, xanthosine, fatty acid ester, ferulic acid, vanillin, *p*-hydroxybenzoic acid and *β*-sitosterol from the dried fruits of *Tribulus terrestris.*

Wang et al., (1997) isolated six furostanol saponins from the fruits of *Tribulus terrestris* 26-*O*-β- D-glucopyranosyl (25*R*)-furostane-2α,3β,22α,26-tetrol-3-*O*-βD -glucopyranosyl (1–4)-β-D -galactopyranoside, 26- *O*-βD -glucopyranosyl (25*R*,*S*)-5α-furostane-2α,3β,22α,26-tetrol-3-*O*-βD -galactopyranosyl(1–2)-βD - glucopyranosyl(1–4)-βD -galactopyranoside, 26-*O*-β-D -glucopyranosyl (25*R*,*S*)-5α-furostane-3β,22α,26-triol-3- *O*-βD -galactopyranosyl(1–2)-βD -glucopyranosyl(1–4)-βD -galactopyranoside, 26-*O*-βD -glucopyranosyl (25 *R*,*S*)-5α-furostan-12-one-3β,22α,26-triol-3-*O*-β-D -galactopyranosyl(1–2)-βD -glucopyranosyl (1–4)-βD - galactopyranoside, 26-*O*-βD -glucopyranosyl (25*R*,*S*)-furost-5-ene-3β,22α,26-triol-3-*O*-β-D -galactopyranosyl(1–2)- βD -glucopyranosyl(1–4)-βD -galactopyranoside, 26-*O*-β-D -glucopyranosyl (25*R*)-5α-furost-20(22)-en-12-one-3β,26- diol-3-*O*-βD -galactopyranosyl(1–2)-β-D -glucopyranosyl(1–4)-βD -galactopyranoside, named terrestrosin F-K, respectively.

Saleh et al., (1982) detected 25 flavonoid glycosides from *Tribulus pentandrus* and *T. terrestris.* The glycosides belong to the common flavonols, kaempferol, quercetin and isorhamnetin, with the 3-gentiobiosides as the major glycosides. Flavonoids were also isolated from the leaves and fruits (Bhutani et al., 1969) which include Kaempferol, kaempferol-3-glucoside, kaempferol-3-rutinoside and tribuloside (kaempferol-3- β-d-(6″-p-coumaroyl) glucoside). Steroidal saponins were isolated by Su et al., (2009) from the fruits of TT and their structures were established by spectroscopic and chemical analysis as (23S,25S)-5alpha-spirostane-24-one-3beta,23-diol-3-O-{alpha-l-rhamnopyranosyl-(1→2)-O-[beta-d-glucopyranosyl-(1→4)]-beta-d-galactopyranoside} (1), (24S,25S)-5alpha-spirostane-3beta, 24-diol-3-O-{alpha-l-rhamnopyranosyl-(1→2)-O-[beta-d-glucopyranosyl-(1→4)]-beta-d-galactopyranoside} (2), 26-O-beta-d-glucopyranosyl-(25R)-5alpha-furostan-2alpha,3beta, 22alpha, 26-tetraol-3-O-{beta-d-glucopyranosyl-(1→2)-O-beta-d-glucopyranosyl-(1→4)-beta-d-galactopyranoside} (3), 26-O-beta-d-glucopyranosyl-(25R)-5alpha-furostan-20(22)-en-2alpha,3beta,26-triol-3-O-{beta-d-glucopyranosyl-(1→2)-O-beta-d-glucopyranosyl-(1→4)-beta-d-galactopyranoside} (4), and 26-O-beta-d-glucopyranosyl-(25S)-5alpha-furostan-12-one-22-methoxy-3beta,26-diol-3-O-{alpha-l-rhamnopyranosyl-(1→2)-O-[beta-d-glucopyranosyl-(1→4)]-beta-d-galactopyranoside} (5). Yuan et al., (2008) identified two furostanol saponins from the fruits namely 3-*O*-{β-*D*-xylopyranosyl(1→3)-[β-*D*-xylopyranosyl(1→2)]-β-*D*-glucopyranosyl (1→4)-[α-*L*-rhamnopyranosyl (1→2)]-β-*D*-

galactopyranosy}-26-*O*-β-*D*-glucopyranosyl-5α-furost-12-one-22-methoxyl-3β,26-diol(named terrestroside A) and 3-*O*-{β-*D*-xylopyranosyl(1→3)-[β-*D*-xylopyranosyl(1→2)]-β-*D*-glucopyranosyl(1→4)-[α-*L*-rhamnopyranosyl(1→2)]-β-*D*-galactopyranosy}-26-*O*-β-*D*-glucopyranosyl-5a-furost-22-methoxyl-3β,26-diol and concluded terrestroside as a new furostanol saponin.

Tribulus terrestris is a rich source of furostanol and spirostanol saponins and flavonoids. The phytochemical studies on the aerial parts of Bulgarian TT have resulted in the isolation of the furostanol saponin 1, named tribol, along with spirostanol saponins 2 and 3 and sitosterol glucoside. The structure of tribol was determined as (25R)-furost-5(6)-ene-3β,16,26-triol-3-*O*-α-rhamnopyranosyl-(1→2)-[α-rhamnopyranosyl-(1→4)]-β-glucopyranoside (**1**) by spectral analysis, including extensive 1D and 2D-NMR experiments.. Further, methyl-protodioscin, protodioscin, methylprototribestin and prototribestin, dioscin and sitosterol glucoside have also been isolated (Conrad et al., 2004).

Xu et al., (2009) isolated two new furostanol glycosides, (tribufurosides 1, 2), were isolated from the fruits of *Tribulus terrestris*. Their structures were established as 26-O-β-D-glucopyranosyl-(25S)-5α-furost-12-one-2α, 3β, 22α,26-tetraol-3-O-β-D-glucopyranosyl (1→2)-β-D-glucopyranosyl (1→4)-β-D-galactopyranoside (1) and 26-O-β-D-glucopyranosyl-(25R)-5α-furost-20(22)-en-12-one-2α,3β,26-triol-3-O-β-Dglucopyranosyl(1→4)-β-D-galacto-pyranoside.

EFFECT OF *TRIBULUS TERRESTRIS* ON HYPERTENSION

The crude saponin fraction of *Tribulus terrestris* has significant effect in the treatment of various cardiac diseases including hypertension, coronary heart disease, myocardial infarction, cerebral arteriosclerosis and thrombosis (Yang et al., 1991; Chui et al., 1992; Lu et al., 1994). These beneficial effects have partly been attributed to its ability to increase nitric oxide (NO) release from the endothelium and nitrergic nerve endings (Adaikan et al., 2000) and direct smooth muscle relaxant effects (Arcasoy et al., 1998). It also has been shown that the aqueous extract of TT fruits had significantly inhibited angiotensin converting enzyme (ACE) activity (Somanandhan et al., 1999). However the mechanism responsible for this antihypertensive activity is still not clear. Sharifi et al., (2003) reported a significant antihypertensive effect of aqueous extract in renin-dependent (2K-1C) model of hypertension. Phillips et al., (2006) are of the opinion that the reduction in blood pressure involved arterial vasodilation through nitric oxide release and membrane hyperpolarization. The BP lowering effect of the extracts was not attenuated by blocking the autonomic ganglia or by blocking β -adrenoceptors or histamine H1-receptors with propranolol and mepyramine, respectively, indicating that these extracts did not act via stimulating the ganglion, β-adrenoceptors or histamine H1-receptors. There was considerable debate over aqueous and methanolic extracts. Somanadhan et al., (1999) and Sharifi et al., (2003) have reported that the inhibition of ACE activity varied with the solvent used in the extraction. Philips et al., (2006) tested various doses of the extract on the perfused rat mesenteric vascular bed and observed a dose dependent increase in basal perfusion pressure, hinting that the methanolic extract contained a vasoconstrictor substance. The vasodilator effect of aqueous and methanolic extracts of *Tribulus terrestris* involved hyperpolarization of the membrane in addition to releasing NO from the vascular endothelium. The major chemical constituents of TT from aqueous or methanolic extracts are the steroidal saponins. Yan et al., (1996) reported the isolation and characterization of steroidal saponins including terrestrosin A, B, C, D and E, desgalacto-tigonin, gitonin, desglucolanatigonin and F-gitonin. Recently, other steroidal saponins including terrestrinins A and B, (Huang et al., 2003); protodioscin and their respective sulfates, (De Combarieu et al., 2003); and spirostanol type saponin, tribulosin and beta-sitosterol-d-glucoside (Deepak et al., 2002) and Conrad et al., 2004) have been isolated and characterized. Tuncer et al., (2009) indicated that dietary intake of TT significantly lowered serum lipid profiles. There is also increasing evidence for anti-cancer properties of saponins from this plant (Yang et al., 2005). Despite the various steroidal saponins that have been isolated (Deepak et al., 2002; Huang et al., 2003; Conrad et al., 2004), their mechanisms of action remains obscure. It has also been suggested that TT could lead to a relaxation in smooth muscle cells in sheep ureter and rabbit jejunum (Arcasoy et al., 1998).

In several studies, it has been reported that saponins derived from *Tribulus terrestris* can exhibit cytotoxic effects on various cancer cell lines from breast, liver, malignant melanoma and ovary (Sun et al., 2003 and 2004). Saponins, one of the active components of TT, have been proposed to affect smooth muscle cells (Arcasoy et al., 1998) and to regulate lipid metabolism (Yang et al., 1999), hyperglycemia (Li et al., 2002) and also to have anti-fungal properties (Zhang et al., 2006).

Tuncer et al., (2009) revealed that *Tribulus terrestris* is very effective in lowering high levels of cholesterol in serum and triglyceride levels (Chu et al., 2003). In addition, it has also been suggested that saponins derived from TT also improved cardiac function in early stages of myocardial infarction (Guo et al., 2007).

Tribulus terrestris has been used as an herbal agent for years to treat hypertension with no serious side effects (Chui et al., 1992; Lu et al., 1994). TT is extremely rich in substances having potential biological significance, including: saponins, phytosteroids, flavonoids, alkaloids, glycosides (Wu et al., 1996) and other nutrients (Wang et al., 1997). The quantities and presence of these important metabolites depend on the various parts of the plant used. The fruit and root of *Tribulus terrestris* contains pharmacologically important metabolites. Shafifi et al (2003) investigated the antihypertensive mechanism of TT in 2K1C renovascular hypertensive model measuring circulatory and local ACE activity in aorta, heart, kidney and lung. The research indicated that TT had a significant effect in reducing blood pressure in renovascular hypertensive rats. In spite of antihypertensive effect of this herb, its precise mechanism of lowering blood pressure is unknown. It may be partly be due to inhibition of ACE activity in serum and various tissues of 2K1C hypertensive rats. Surprisingly, the effect of TT on ACE activity in various tissues was even lower than control level which was corresponding to lower blood pressure even below control at the end of the fourth week of treatment.

EFFECT OF *TRIBULUS TERRESTRIS* ON OXALATE METABOLISM

Sangeeta et al., (1993) studied the effect of TT extract on oxalate metabolism in male rats fed on sodium glycolate that resulted in hyperoxaluria as well as increased activities of oxalate synthesizing enzymes in the liver i.e. glycolate oxidase (GAO), glycolate dehydrogenase (GAD) and lactate dehydrogenase (LDH), and decreased kidney LDH activity. TT administration to sodium glycolate fed rats produced a significant decrease in urinary oxalate excretion, and a significant increase in urinary glyoxylate excretion. Besides, it also caused a reduction in liver GAO and GAD activities, whereas liver LDH activity remained unaltered. Extract of TT was effective in reversing hyperoxaluria in 24 h to normal, from 1.97±0.314 to 0.144±0.004 mg/mg creatinine.

HORMONAL EFFECTS OF TRIBULUS TERRESTRIS

Gauthaman and Ganesan (2008) evaluated the hormonal effects of the extract of TT in a comparative study with primates, rabbits and rats in the management of erectile dysfunction. TT extract was administered intravenously, as a bolus dose of 7.5, 15 and 30 mg/kg, in primates for acute study while Rabbits and normal rats were treated with 2.5, 5 and 10 mg/kg of TT extract orally for 8 weeks. In addition, castrated rats were treated either with testosterone cypionate (10 mg/kg, subcutaneously; biweekly for 8 weeks) or TT orally (5 mg/kg daily for 8 weeks). Blood samples were analyzed for testosterone (T), dihydro-testosterone (DHT) and dehydroepiandrosterone sulphate (DHEAS) levels. In primates, significant increases were found in all the hormone levels. In castrated rats, significant increase in T levels was observed with TT extract. The testosterone and DHT increased significantly, which could possibly be due to the presence of steroidal glycosides, among them PTN, is the major active principle in TT. The steroidal nature of this compound may facilitate its role as an intermediary in the steroidal pathway of androgen production. It may act either by binding to hormone receptors or to enzymes that metabolize hormones. Most biological actions of plant-derived compounds are brought about by these mechanisms (Baker, 1995). The presence of protodioscin in the extract indicates that TT may be useful in mild to moderate cases of erectile dysfunction. In the extract protodioscin is about 5% (Dikova and Ognyanova, 1993). The levels of testosterone and lutinizing hormone are increased following treatment with PTN for a period of 30–90 days in patients with hypogonadism (Koumanov et al., 1982). Improvement in sperm count and motility has been reported in patients with low seminological indices following treatment with TT for 3 months (Balanathan et al., 2001). TT also increased the sexual behavior parameters in castrated rats compared to the normal rats (Gauthaman et al., 2002). The administration of Tribestans a commercial product containing 250 mg of TT to humans and animals for a period of 60–90 days was found to improve testosterone levels, libido and promote spermatogenesis (Tomova et al., 1981; Koumanov et al., 1982). Apart from the hormones, the blood pressure in primates had a transient fall immediately following the intravenous administration of the extract that returned to normal in less than a minute. A similar response was found in mongrel dogs, when the aqueous extract was given in the dose of 80 mg/kg (Bose et al., 1963). In another study on dogs, a dose of 20 mg/kg produced a sharp fall in blood pressure (20–50mmHg) that lasted for about 3 min before returning to baseline values (Chakraborty and Neogi, 1978).

EFFECTS OF TRIBULUS TERRESTRIS (TT) EXTRACT ON ENDOCRINE SENSITIVE ORGANS

Although experimental and clinical studies have partially confirmed the effects of TT on libido and sperm production, there is still much debate regarding possible mechanisms of action as well as therapeutic applications (Andrade et al., 2010). Gauthaman et al., (2002, 2003) indicated that TT can improve some aspects of male sexual behavior and enhance spermatogenesis in rats. In addition, there are clinical data indicating stimulatory effects of TT on sperm quantity and quality and improved sexual response in men (Arsyad, 1996). Reports of increased androgen levels have also been reported following TT administration to nonhuman primates, rats and rabbits (El- Tantawy et al., 2007; Gauthaman and Ganesan, 2008). But most of these effects were short-lived and showed no clear dose–response relationships. In addition, there is no consensus on the exact mechanisms underlying TT effects on sexual performance and spermatogenesis. It is believed that the steroidal saponins present in TT extracts, particularly protodioscin, can increase endogenous androgen production by increasing luteinizing hormone (LH) release from the pituitary gland. Alternatively, it has also been proposed that TT active components might be enzymatically converted into weak androgens like dehydroepiandrosterone (DHEA), which could in turn be converted into more potent androgens like testosterone in the gonads and peripheral tissues (Adaikan et al., 2001). However, changes in endogenous hormone levels following TT administration is still controversial (Neychev and Mitev, 2005).

The presence of steroidal saponins also raises the question whether TT extracts might have intrinsic hormonal activity that could directly stimulate male and/or female endocrine sensitive tissues such as the prostate, seminal vesicle, uterus and vagina (Nian et al., 2006). This is considered as an important issue for both efficacy and safety assessment of TT. Gauthaman et al., (2002) reported that administration of TT to castrated rats was able to improve sexual behavior and increase prostate weight and intracavernous pressure in relation to castrated controls. TT has been indicated as an alternative treatment to hormone replacement therapy in aging men and women (Andrade et al., 2010).

Tribulus terrestris has been popularly used as an aphrodisiac and enhancer of sperm production and more recently as an alternative to hormone replacement therapy in aging men and women (Gauthaman and Ganesan, 2008). In a study conducted by Andrade et al., (2010) demonstrated that TT has no intrinsic hormonal activity, being unable to stimulate endocrine sensitive organs in both male and female rats. Further, administration of TT to intact male rats for 28 days did not change serum testosterone levels and did not produce any quantitative change in the fecal excretion of androgenic metabolites in Wistar rats, indicating lack of androgenic and estrogenic activity in vivo. Andrade et al., (2010) also showed a positive effect of TT administration on rat sperm production, associated with unchanged levels of circulating androgens. However, it is important to note that conflicting results following TT administration in experimental animals may arise as a consequence of differences in species or strains used, duration of treatment, and content of active components in the extract. Additional clinical and experimental studies using extended treatment periods and a larger number of individuals are necessary in order to clearly establish the effects of TT on sperm production and sex steroid levels

EFFECTS OF TRIBULUS TERRESTRIS (TT) EXTRACT ON ANDROGEN RECEPTORS

Neychev and Mitev (2005) evaluated the effect of TT on nicotinamide adenine dinucleotide phosphate-diaphorase (NADPH-d) activity and androgen receptor (AR) immunoreactivity in rat brain. Adult male Sprague-Dawley rats were treated with TT (5mg/kg/day) orally for 8 weeks, the brain tissue was removed and sections of the para-ventricular (PVN) area of hypothalamus were studied for NADPH-d and AR immuno-staining. There was a significant increase in both NADPH-d (67%) and AR immunoreactivity (58%) in TT treated group. Androgens are known to increase both AR and NADPH-d positive neurons either directly or by its conversion to oestrogen. The mechanism for the observed increase is probably due to the androgen increasing property of TT. Androgens are sex hormones that play an essential role in male reproductive function. They are known to act centrally and peripherally in the initiation and maintenance of sexual functions. The interaction of these hormones with the androgen receptors (AR) and their downstream signaling cascade influence the final outcome of their effects in the system. Though androgen increase has a positive correlation with the increase in AR immunoreactivity, the NOS activity has been reported to decrease in response to increased androgens in the brain. This contradicts with the findings of Vincent and Kimura (1992) where NADPH-d positive neurons were increased, which are same as those containing NOS. However, it should be noted that many of the actions of testosterone in the brain are mediated by its conversion to estrogen by the enzyme aromatase (Roselli et al., 1997). Estrogen is found to have opposite effect compared to testosterone; it increases the NOS activity in vari us regions of the brain (Ceccatelli et al., 1996). It is possible that the increase in NADPH-d positive neurons following TT is brought about by a similar mechanism. These findings add further support to the aphrodisiac claims of TT and its androgen releasing property. The aphrodisiac effects of TT are probably mediated by increase in both AR and NOS containing neurons (Gauthaman and Adaikan, 2005).

The extract obtained from the air-dried aerial parts of the plant contains mainly steroidal glycosides (saponins) of furastanol type, Protodioscin (PTN) that forms 45% (dry weight) of the TT extract (Tomova et al., 1981; Dikova and Ognyanova, 1983). PTN is found to increase the levels of testosterone, lutinizing hormone (Koumanov et al., 1982), dehydroepiandros-terone (Adimoelja and Adaikan, 1997), dihydrotestosterone and dehydroepiandrosterone sulphate (Gauthaman et al., 2000). Administration of TT to humans and animals improved libido and spermatogenesis (Tomova et al., 1981); it increased the sexual behaviour and intracavernous pressure in male rats (Gauthaman et al., 2002; Gauthaman et al., 2003), and also had a proerectile effect on isolated rabbit corpus cavernosal tissues in vitro (Adaikan et al., 2000); most of these responses of TT are attributed to the presence of protodioscin and its phytoandrogenic properties.

TRIBULUS TERRESTRIS AS AN APHRODISIAC

Tribulus terrestris has long been used in the traditional Chinese and Indian systems of medicine for the treatment of various ailments and is popularly claimed to improve sexual functions in man. Gauthaman et al., (2002) have studied the sexual behavior and intracavernous pressure (ICP) in both normal and castrated Sprague-Dawley rats to understand the role of TT containing protodioscin (PTN) as an aphrodisiac. Significant increase in the sexual behavior parameters was noticed after TT treatment in the castrated groups of rats as evidenced by an increase in mount and intromission frequencies and increase in mount, intromission and ejaculation latencies (ML, IL, EL) as well as post-ejaculatory interval (PEI) and prostate weight. They concluded that TT extract possed aphrodisiac activity probably due to androgen increasing property of TT because it contained protodioscin (PTN), a steroidal saponin that forms 45% (dry weight) of the extract. PTN, has been shown to increase the DHEA level in man (Gauthaman et al., 2002). DHEA is considered as a neurosteroid that acts centrally as a gamma amino butyric acid antagonist to facilitate sexual function (Majewska, 1995). Possible increase in DHEA and its subsequent conversion to testosterone and its metabolites may account for these effects. However it is not clear how such increase in DHEA is brought about by TT extract. Yet, TT may contribute as a conditioner for the treatment of mild to mid-level erectile dysfunction. Gauthaman et al., (2002) discussed androgens and the regulation of penile erection and observed that the TT extract has similar properties to androgens and confirmed the role of TT as an aphrodisiac in the traditional medicine.

DIURETIC AND CONTRACTILE EFFECTS OF *TRIBULUS TERRESTRIS*

Calcium oxalate monohydrate (COM) and calcium oxalate dihydrate (COD) containing stones (calculi) are commonly found as urinary stones. TT is used as herbal medicine for urinary calculi in India.

Inhibition of COM crystal growth was observed by TT extract. In another study, the administration of *Tribulus terrestris* to sodium glycolate fed rats produced a significant decrease in urine oxalate excretion and a significant increase in urinary glycoxylate excretion (Sangeeta et al., 1994). However, the occurrence of calcium oxalate calculi in the body is a much more complex phenomenon and the extracts contain many complex macro-biomolecules and the roles of these molecules are very important in the growth inhibition (Joshi et al., 2004).

The aqueous extract of TT was tested in Wistar male rats at an oral dose of 5 g/kg that elicited a positive diuresis and an increase in Na+, K+ and Cl+ concentrations in the urine. In addition to its diuretic activity TT had evoked a contractile activity on Guinea pig ileum. This indicates its potential in propelling urinary stones. Al Ali et al., (2003) evaluated the diuretic activities and pharmacological effects of TT and compared with the diuretic activity of maize extract.

The patients passed renal and/or ureteric stones on taking TT either alone or with maize. Arcasoy et al., (1998) have shown that T. terrestris L. saponin mixture caused a significant decrease in the peristaltic movements of the isolated sheep ureter and rabbit jejunum preparations in a dose-dependent manner, and there was no effect on the isolated rabbit aorta and its contractile response to KCl or noradrenaline.

TT has been found safe without any side effects in a large cohort of patients treated for ischaemic heart disease (Wag et al., 1990). However, Bourke (1983) has reported that it caused some hepatopathy in sheep. The diuretic action of this plant makes it useful as an anti-hypertensive agent, and its hypotensive activity was reported by Mossa et al., (1983). However, Al Ali et al., (2003) have used high furosemide doses in their experiments and demonstrated that TT has significant diuretic and contractile effects which makes it useful in the propulsion of urinary stones. However, Tatasnaz et al., (2010) reported a case of TT induced hepatotoxicity, nephrotoxicity and neurotoxicity in an Iranian male patient who used the plant's extract to prevent kidney stone formation.

It resulted in high serum aminotransferases and creatinine after consuming TT for 2 days. But discontinuation of TT resulted in normalization of liver enzymes. Kavita and Jagadeesan (2006) reported that the oral administration of methanolic fraction of TT at dose 6 mg/kg body weight provided protection against the mercuric chloride induced toxicity in the mice.

THE PROTECTIVE EFFECT OF TRIBULUS TERRESTRIS (TT) IN DIABETES

Amin et al., (2007) investigated the protective effects of TT in diabetes mellitus (DM). Diabetes is known to increase reactive oxygen species level that subsequently contributes to the pathogenesis of diabetes. Rats with streptozotocin induced diabetes were treated with TT for 30 days. Levels of serum alanine aminotransferase (ALT) and creatinine were estimated. In addition, levels of malondialdehyde (MDA) and reduced glutathione (GSH) were assayed in the liver. The rats treated with TT extract significantly decreased the levels of ALT and creatinine in the serum in diabetic rats and lowered the MDA level in liver.

On the other hand, level of reduced GSH in liver was significantly increased in diabetic rats treated with TT. Histopathological examination revealed recovery of liver in TT treated rats suggesting its protective effect possibly mediated by inhibiting oxidative stress. In another study (El-Tantawy and Hassanin, 2007) TT extract significantly decreased fasting glucose level in diabetic rats.

TT extract showed significant reduction in glucose level. The extracts also decreased the levels of glycosylated hemoglobin, total cholesterol, triglycerides and LDL-cholesterol. This indicates the hypoglycemic activity in type-1 model of diabetes.

TRIBULUS TERRESTRIS AS A NUTRITION SUPPLEMENT

The role of *Tribulus terrestris* as a nutritional supplement for the athletes is highly debated regarding its physiological and actual effects. The main claimed effect is an increase of testosterone anabolic and androgenic action through the activation of endogenous testosterone production is not entirely proven. The short-term treatment with T. terrestris showed no impact on the endogenous testosterone metabolism (Saudan et al., 2008). According to this report, TT may not be a direct precursor of testosterone or as a stimulating agent of endogenous testosterone production. However, it may be possible that daily intake of the plant extract over a longer period of time could lead to new equilibrations of the $^{13}C/^{12}C$ ratio for endogenous steroids.

Tribulus is a popular food supplement used by a large number of athletes. TT increases protein synthesis and muscular mass and facilitates recuperation after physical loads.

However, data on the efficacy is not sufficient. Milasius (2010) found that the Tribulus food supplement increased mixed anaerobic alactic glycolytic muscular power at 30-s work and reduced lactate concentration in the blood and concluded that the Tribulus food supplement has a positive influence on athletes'anaerobic alactic glycolytic power and aerobic capacity when energy is produced in the aerobic way. Rogerson (2007) reported that that the aromatisation pathway is stimulated to increase blood estradiol when the TT food supplement was given to healthy young males. There was no significant effect on plasma total testosterone, luteinizing hormone or the urinary testosterone/epitestosterone ratio. TT did not produce the large gains in strength or lean muscle mass and did not alter the urinary T/E ratio.

Singh et al., (2009) have shown that TT extract affects the body metabolism and reproductive functions. In albino rats there was a highly significant increase in body weight, organ weight, sperm count, total cholesterol, while a highly significant decrease in serum total protein, albumin, globulin, A/G ratio, creatinine and LDL was noticed which is attributed to the active constituent like protodioscin and saponin.

Wang et al., 2009 observed the effect of TT gross saponins on protein kinase Cepsilon (PKCepsilon) and apoptosis-associated protein in the apoptosis of cultured cardiocyte apoptosis induced by hydrogen peroxide (H_2O_2). TT decreased the apoptotic percentage in cardiocytes but also reduced protein contents of Bax and cleaved caspase-3, and increased protein content of phospho-PKCepsilon and Bcl-2 significantly.

Ethanolic extract of TT was found to contain adaptogenic (anti-stress) property. There was a marked increase in anoxia stress tolerance time and swimming endurance time. Similarly, a marked decrease in blood glucose, cholesterol, triglycerides and BUN level as compared to stress control in both immobilization stress and cold stress models. (Sivakumar et al., 2006).

ANTI MICROBIAL ACTIVITY OF *TRIBULUS TERRESTRIS*

Antifungal and antibacterial activity of natural products has been attributed to saponins from TT extract. Zhang et al., (2006) have isolated eight steroid saponins from Tribulus which were identified as hecogenin- 3-O-β-d-glucopyranosyl (1→4)- β -d-galactopyranoside (TTS-8), tigogenin-3-O- β -d-glucopyranosyl (1→4)- β -d-galactopyranoside (TTS-9), hecogenin-3-O- β -d-glucopyranosyl (1→2)- β -d-glucopyranosyl (1→4)- β -d-galactopyranoside (TTS-10), hecogenin- 3-O- β -d-xylopyranosyl (1→3)- β -d-glucopyranosyl (1→4)- β -d-galactopyranoside (TTS-11), tigogenin-3-O- β -d-xylopyranosyl (1→2)-[_-d-xylopyranosyl (1→3)]- β -d-glucopyranosyl (1→4)-[β-l-rhamnopyranosyl (1→2)]-β-d-galactopyranoside (TTS-12), 3-O-{ β -d-xylopyranosyl (1→2)-[_-d-xylopyranosyl (1→3)]- β -d-glucopyranosyl (1→4)-[β -l-rhamnopyranosyl (1→2)]- β -dgalactopyranosyl}- 26-O-- β d-glucopyranosyl-22-methoxy-(3 β ,5 β,25R)-furostan-3,26-diol (TTS-13), hecogenin-3-O- β -d-glucopyranosyl (1→2)-[_-d-xylopyranosyl (1→3)]- β -d-glucopyranosyl (1→4)- β -d-galactopyranoside (TTS-14), tigogenin-3-O- β -d-glucopyranosyl (1→2)-[β -d-xylopyranosyl (1→3)]- β -d-glucopyranosyl (1→4)- β -d-galactopyranoside (TTS-15) (Dinchev et al., 2005). The in vitro antifungal activities of the saponins against five species of yeast were studied using microbroth dilution assay. The results showed that TTS-12 and TTS-15 were very effective against several pathogenic species. Xu et al., (2000) also isolated and identified 10 compounds from TT and tested their antifungal properties. In traditional Chinese medicine, TT has long been used for the treatment of cutaneous pruritus, edema and inflammation. (Chu et al., 2003). Earlier studies showed that *Tribulus terrestris* L. contained flavanoids, steroid saponins, alkaloids and polysaccharides (Bourke et al., 1992; Wu et al., 1996; Yan et al., 1996; Li et al., 1998; Liu et al., 2003; Conrad et al., 2004).

A series of experiments were conducted to investigate antifungal activities of the above steroid saponins, the results showed that only TTS-12 and TTS-15 had significant antifungal activities against the five yeasts tested. Saponins exert antifungal activity by inhibiting fungal hyphae and destroying the ultra structure of fungi. Bedir et al., (2002) also studied the antifungal activities of steroid saponins from TT and they were inactive against fungi.

Al-Bayati et al., (2008) reported antimicrobial activity of TT extracts from fruits, leaves and roots against 11 species of pathogenic and non-pathogenic microorganisms: *Staphylococcus aureus, Bacillus subtilis, Bacillus cereus, Corynebacterium diphtheriae, Escherichia coli, Proteus vulgaris, Serratia marcescens, Salmonella typhimurium, Klebsiella pneumoniae, Pseudomonas aeruginosa and Candida albicans.*

The ethanol extract showed antimicrobial activity against both Gram-negative and Gram-positive bacteria and antifungal activity. According to Verma et al., (2009) methanolic extract of TT showed growth inhibitory effects and antimicrobial activity against Salmonella and E. coli and no microbial activity was observed for *Bacillus badius*, Lacto bacillus *Plantarum* and *Lactococcus lactus*.

ANTHELMINTIC PROPERTIES OF *TRIBULUS TERRESTRIS*

Helminthiasis affects over 2 billion people world wide. Worm infestation is also a major health problem among the livestock. Deepak et al., (2002) obtained TT extracts using petroleum ether, chloroform, 50% methanol, water and tested for anthelmintic activity in-vitro using the nematode *Caenorhabditis elegans*. The spirostanol type saponin, tribulosin and b-sitosterol-D-glucoside from the extract exhibited anthelmintic activity.

Singh et al., (2008) showed that TT extract showed strong larvicidal, properties with100 per cent mortality in the 3(rd)-instar larvae in the bioassys with mosquitoes Anopheles and Culex, while a strong repellent activity was recorded against adult mosquitoes. Pandey and Singh (2009) found TT extract can be used effectively as molluscicide.

TRIBULUS TERRESTRIS CAUSES POISONING IN GOATS

Aslani et al., (2004) have reported clinical signs of toxicity in the goats fed on dried TT for 8 weeks that included weight loss, depression, ruminal stasis, icterus and elevation of body temperature. Haematological studies revealed a declining of packed cell volume (PCV) and plasma total protein and elevation of total and direct bilirubin, blood urea nitrogen (BUN), creatinin and potassium concentrations and serum aspartate amino transferase (AST) activity. The affected goats showed gross pathological changes and marked microscopic lesions in liver and kidneys including generalized icterus, hepatocellular degeneration and necrosis, biliary fibrosis and proliferation, renal tubular necrosis and crystalloid materials in bile ducts and renal tubules.

The outbreaks of photosensitivity disease occurred in weaner sheep in Australia due to *Tribulus terrestris* (Glastonbury, 1984). Histopathological studies revealed Ochre and khaki discolouration in the liver and kidneys, apart from the presence of acicular, cholesterol-like clefts in the lumens of bile ducts and in the cytoplasm of hepatocytes and Kupffer cells. Bourke, (1984) reported staggers in sheep associated with the ingestion of TT. The unusual locomotory disturbance of sheep was detected when sheep grazed on TT for long time. The disease was characterised by a slowly developing, irreversible, asymmetrical, weakness of the hindlimbs. The clinical signs suggested that a lesion of the thoraco-lumbar spinal cord region was present.

Grazing on TT induces a hepatogenous photosensitization in sheep and goats known as Geeldikkop, Tribulosis ovis or yellow big head (Hennings, 1932; Van Tonder et al., 1972; Jacob and Peet, 1987). In this disease phylloerythrin, a photodynamic porphyrin derived from the degradation of chlorophyll by micro-organisms in the rumen, is believed to be retained as a result of occlusion of bile ducts by birefringent crystalloid materials (Kellerman et al., 1991).

Biliary crystalloid materials are the calcium salts of the β-D-glucuronide of epismilagenin and episarsasapogenin (Miles et al., 1994a) which are metabolites of diosgenin and yamogenin, steroidal saponins of T. terrestris, respectively (Miles et al., 1994b). The disease has also been reported in Australia (Bourke, 1983; Glastonbury et al., 1984; Glastonbury and Boal, 1985; Jacob and Peet, 1987), USA (McDonough et al., 1994), Argentina (Tapala et al., 1994) and Iran (Amjadi et al., 1979; Aslani et al., 2002 and 2004).

Xu et al., (2010) have isolated some steroidal saponins that are used in many pharmaceutical preparations and food supplements. e.g. Tribestane and Vitanone, are used to treat impotency, tribusaponins and Xin-nao-shutong, are used for the treatment of cardiovascular diseases (Kostava et al., 2005 and Li and Yang, 2006). The inhibition of cell growth and cytotoxicity of these compounds has been reported. Crude saponin fraction of TT showed significant effects in the treatment of coronary heart disease, myocardial infarction, cerebral arteriosclerosis and the sequelae of cerebral thrombosis (Xu et al., 2010).

These biomacromolecules seem to play an important role in the inhibition of COM crystals. However, the occurrence of calcium oxalate calculi in the body is a much more complex phenomenon and the extracts contain many complex macro-biomolecules and the roles of these molecules are very important in the growth inhibition (Joshi et al., 2004).

ANTI CANCER ACTIVITY OF *TRIBULUS TERRESTRIS*

Neychev et al., (2007) studied the effect of TT saponins on normal human skin fibroblasts and compared it with their anticancer properties. They found that TT is less toxic for normal human skin fibroblasts in comparison to many cancer lines investigated in previous studies. The molecular mechanism of this cytotoxicity involves up- and downregulation of polyamines' homeostasis, suppression of proliferation, and induction of apoptosis. Tribulus contains Sterols like betasitosterols or stigma. These substances protect the prostate from swelling and in combination with the X steroidal saponins, protect the prostate from cancer.

THE ANALGESIC EFFECT OF *TRIBULUS TERRESTRIS* EXTRACT

The analgesic effect of methanolic extract of TT on male albino mice was evaluated by formalin and tail flick test (Heidari et al., 2007). The results showed that a dose of 100 mg/kg of percolated extract had the highest significant analgesic effect, nevertheless lower than morphine, 2.5 mg/kg. The gastric ulcerogenecity of TT extract was lower than the indomethacin in the rat's stomach.

SIDE EFFECTS OF TRIBULUS TERRESTRIS

Tribulus terrestris may cause stomach problems, enhance heart beat, restlessness and induce warmer feeling. Adverse effects from supplementation with Tribulus terrestris are rare and tend to be insignificant. However, some users reported stomach upset, gynaecomastia(abnormal enlargement of breasts in men), which supports the androgenic-anabolic effects of this plant (Jameel et. al., 2004). Uncommon reports of increase in prostate weight, insomnia, menorrhagia (heavy menstrual bleeding), pneumothorax (air in the chest cavity from injury to the lung) and reduced levels of glucose and total cholesterol after taking TT extract (Qazi, 2008).

REFERENCES

Adaikan PG, Gauthaman K, Prasad RNV, Ng SC. 2000. Proerectile pharmacological effects of *Tribulus terrestris* extract on the rabbit corpus cavernosal smooth muscle in vitro. Annals Academy of Medicine, Singapore. 29: 22-26.

Adaikan, P.G. K. Gauthaman and R.N. Prasad, 2001 History of herbal medicines with insight on the pharmacological properties of *Tribulus terrestris*, The Aging Male 4:163–169.

Adimoelja, A., 2000. Phytochemicals and the breakthrough of traditional herbs in the management of sexual dysfunction. *Intl. J. Androl.* 23: 82-84.

Adimoelja, A., Adaikan, P.G., 1997. Protodioscin from herbal plant *Tribulus terrestris* L. improves male sexual functions possibly via DHEA. *International Journal of Impotence Research*. 9: 64.

Al-Ali, M., S. Wahbi, H. Twaij, A. Al-Badr. 2003. *Tribulus terrestris*: preliminary study of its diuretic and contractile effects and comparison with Zea mays. *Journal of Ethnopharmacology*. 85: 257–260.

Al-Bayati, F. A., Al-Mola, H. F. 2008. Antibacterial and antifungal activities of different parts of *Tribulus terrestris* L. growing in Iraq. *Journal of Zhejiang University-Science B.* Vol. 9: 154-159.

Ali, N.A., Julich, W.D., Kusnick, C., 2001. Screening of Yemeni medicinal plants for antibacterial and cytotoxic activities. *Journal of Ethnopharmacology.* 74: 173-179.

Amin, A. M. Lotfy, M. Shafiullah,. and E. Adeghatec. 2006. The Protective Effect of *Tribulus terrestris* in Diabetes. *Ann. N.Y. Acad. Sci.* 1084: 391–401.

Anand, R., Patnek, G.K., Kulshreshtha, D.K., Dawan, B.N., 1994. Activity of certain fractions of *Tribulus terrestris* fruits against experimentally induced urolithiasis in rats. *Indian Journal of Experimental Biology.* 32: 548–552.

Andrade,M., Anderson J., Morais, Rosana N., Spercoski, Katherinne M., Rossi, Stefani C., Vechi, Marina F., Golin, Munisa , Lombardi, Natalia F., Greca, Claudio S., Dalsenter, Paulo R. 2010. Effects of *Tribulus terrestris* on endocrine sensitive organs in male and female Wistar rats. *Journal of Ethnopharmacology.* 127: 165-170.

Anonymous. 2010 http://www.toddcaldecott.com/index.php/herbs/learning-herbs/366-gokshura.

Antonio, J., J. Uelmen, R. Rodriguez and C. Earnest, 2000. The effects of *Tribulus terrestris* on body composition and exercise performance in resistance-trained males. *International Journal of Sport Nutrition and Exercise Metabolism.* 10: 208-215.

Arcasoy, H.B, Erenmemisoglu A, Tekol Y, Kurucu S, Kartal M. Effect of *Tribulus terrestris* L. 1998. Saponin mixture on some smooth muscle preparations: a preliminary study. *Boll. Chim. Farm.* 137: 473-5.

Arsyad K.M. 1996. Effect of protodioscin on the quantity and quality of sperms from males with moderate idiopathic oligozoospermia. *Medika.* 22: 614-618.

Aslani, M.R., A.R. Movassaghi, M. Mohri, V. Ebrahim-pour, A.N. Mohebi. 2004. Experimental *Tribulus terrestris* poisoning in goats. *Small Ruminant Research.* 51: 261-267.

Aslani, M.R., Movassaghi, A.R., Mohri, M., Pedram, M., Abavisani, A., 2002. Experimental *Tribulus terrestris* poisoning in sheep: clinical, laboratory and pathological findings. *Vet. Res. Commun.* 27, 53–62.

Balanathan, K., Omar, M.H., Zainul Rashid, M.R., Ong, F.B., Nurshaireen, A., Jamil, M.A., 2001. A clinical study on the effect of *Tribulus terrestris* (Tribestan) on the semen profile in males with low sperm count and low motility. *Malay. J. Obstet. Gynaecol.* 7: 69-78.

Bedir, E., Khan, I.A., Walker, L.A., 2002. Biologically active steroidal glycosides from *Tribulus terrestris*. *Pharmazie.* 57, 491–493.

Bhutani, S.P., S.S. Chibber, T.R. Seshadri 1969. *Phytochemistry.* Volume 8: 299-303.

Bose, B.C., Saifi, A.Q., Vijayvargiya, R. and Bhatnager, J. N. 1963. *Indian Journal of Medical Sciences.* 17: 291.

Bourke C.A, Stevens GR, Carrigan MJ 1992. Locomotor effects in sheep of alkaloids identified in Australian *Tribulus terrestris*. *Australian Veterinary Journal*. 69: 163-165.

Bourke, C. A. 1983. Hepatopathy in sheep associated with *Tribulus terrestris*. *Australian Veterinary Journal.* 60: 189.

Bourke, C. A. 1984. Staggers in sheep associated with the ingestion of Tributes terrestris. *Australian Veterinary Journal.* 61: 360–363.

Bourke, C.A., 1983. Hepatopathy in sheep associated with *Tribulus terrestris*. *Australian Veterinary Journal.* 60: 189.

Brown G.A, Vukovich MD, Martini ER, Kohut ML, Franke W.D, Jackson D.A, King D.S. 2001. Endocrine and lipid responses to chronic androstenediol-herbal supplementation in 30 to 58 year old men. *J. Am. Coll. Nutr.* 20: 520-8.

Brown G.A, Vukovich MD, Reifenrath TA, Uhl NL, Parsons KA, Sharp RL, King DS. 2000. Effects of anabolic precursors on serum testosterone concentrations and adaptations to resistance training in young men. *International Journal of Sport Nutrition and Exercise Metabolism.* 10: 340-59.

Cai L. Cai, Y. Wu, J. Zhang, F. Pei, Y. Xu, S. Xie and D. Xu 2001. Steroidal saponins from *Tribulus terrestris, Planta Medica.* 67: 196-198.

Ceccatelli, S., L. Grandison, R.E. Scott, D.W. Pfaff and L.M. Know, 1996. Estradiol regulation of nitric oxide synthase mRNAs in rat hypothalamus, *Neuroendocrinology.* 64 (1996), pp. 357-363.

Chakraborty, B., Neogi, N.C., 1978. Pharmacological properties of *Tribulus terrestris* Linn. *Indian J. Pharm. Sci.* 40: 50-52.

Chemexcil, 1992. *Tribulus terrestris* Linn. (N.O.-Zygophyllaceae). In: Selected Medicinal Plants of India (A Monograph of Identity, Safety and Clinical Usage). Compiled by Bharatiya Vidya Bhavan's Swamy Prakashananda Ayurveda Research Centre for CHEMEXCIL. Tata Press, Bombay. pp. 323-326 (Chapter 10).

Chiang Su New Medicinal College (ed.) 1977. Dictionary of Chinese Crude Drugs, Shanghai Scientific Technologic Publisher, Shanghai, p. 1274.

Chiu, N. Y., and Chang, K. H. 1986. The Illustrated Medicinal Plants of Taiwan, 3, P118, SMC Publishing INC. Taipei, Taiwan.

Chopra, R.N., Chopra, I.C., Handa, K.L., and Kapoor, L.D., 1958. Chopra's Indigenous drugs of India, Second edition, U.N. Dhur and Sons Private Ltd., Calcutta, pp. 430-431.

Chu, S., W. Qu, X. Pang, B. Sun and X. Huang, 2003. Effect of saponin from *Tribulus terrestris* on hyperlipidemia, *Zhong Yao Cai.* 26: 341-344.

Chui, S.Z., Liao, C.X., Jiao, Q.P., Zhu, H.M., Chen, S.Y., Chou, Z.J., 1992. Xinnao shutong for coronary heart disease in 41 patients. *New Drugs and Clinical Remedies.* 11, 202-204.

Conrad J. Conrad, D. Dinchev, I. Klaiber, S. Mika, I. Kostova and W. Kraus 2004. A novel furostanol saponin from *Tribulus terrestris* of Bulgarian origin, Fitoterapia 75: 117-122

De Combarieu, E., Fuzzati, N., Lovati, M., Mercalli, E., 2003. Furostanol saponins from *Tribulus terrestris. Fitoterapia.* 74: 583-591.

Deepak, M., Dipankar, G., Prashanth, D., Asha, M.K., Amit, A., Venkataraman BV, 2002. Tribulosin and beta-sitosterol-d-glucoside, the anthelmintic principles of *Tribulus terrestris. Phytomedicine.* 9: 753–756.

Dikova, N., Ognyanova, 1993. Pharmacokinetic studies of Tribestan. Anniversary Scientific Session'35 Chemica.

Dikova, N., Ognyanova, V. 1983. Pharmacokinetic studies of Tribestan, Anniversary Scientific Session-35, Chemical Pharmaceutical Research Institute, Sofia. pp. 1-7.

Dinchev D, Ivanova A, Kostova I, Evstatieva L (2005) Comparative GC-MS study of the volatiles of *Tribulus terrestris* L. from different geographical regions. Dokladi na Bolgarskata Akademiya na Naukite 58, 1173-1178. Contact: Dinchev, D. ; Bulgarian Acad. Sci. Inst. Organ. Chem. Acad. G. Bonchev. Str. Bl 9, BU-1113 Sofia, Bulgaria.

Dinchev D, Janda B, Evstatieva L, Oleszek W, Aslani MR, Kostova I. 2008. Distribution of steroidal saponins in *Tribulus terrestris* from different geographical regions. *Phytochemistry.* 69: 176-86.

Duan, J. A. and Zhou, R. H. 1993. *Zhongcaoyao.* 24: 320.

El-Tantawy W.H, Temraz A, El-Gindi O.D 2007. Free serum testosterone level in male rats treated with Tribulus Alatus extracts. *Int. Br. J. Urol.* 33: 554-559.

Gauthaman K and Ganesan A.P 2008. The hormonal effects of *Tribulus terrestris* and its role in the management of male erectile dysfunction – an evaluation using primates, rabbit and rat. *Phytomedicine.* 15: 44-54.

Gauthaman K, Adaikan P.G, Prasad R.N. 2002. Aphrodisiac properties of *Tribulus terrestris* extract (Protodioscin) in normal and castrated rats. *Life Sciences.* 71: 1385-96.

Gauthaman K, Ganesan AP and Prasad RN 2003. Sexual effects of puncturevine (*Tribulus terrestris*) extract (protodioscin): An evaluation using a rat model. *Journal of Alternative and Complementary Medicine.* 9: 257-265.

Gauthaman, K. and P.G. Adaikan, 2005. Effect of *Tribulus terrestris* on nicotinamide adenine dinucleotide phosphate-diaphorase activity and androgen receptors in rat brain. *J. Ethnopharmacol.* 96: 127-132.

Gauthaman, K., Adaikan, P.G., Prasad, R.N.V., Goh, V.H.H., Ng, S.C., 2000. Changes in hormonal parameters secondary to intravenous administration of *Tribulus terrestris* extract in primates. *International Journal of Impotence Research.* 12: 6.

Glastonbury JR, Doughty FR, Whitaker SJ, Sergeant E. 1984. A syndrome of hepatogenous photosensitisation, resembling geeldikkop, in sheep grazing *Tribulus terrestris. Aust. Vet. J.* 61: 314-6.

Guo Y., D.Z. Shi, H.J. Yin and K.J. Chen, 2007. Effects of Tribuli saponins on ventricular remodeling after myocardial infarction in hyperlipidemic rats. *Am. J. Chin. Med.* 35: 309-316.

Halvorson W.L. and P. Guertin. 2003. Research SpecialistFactsheet for *Tribulus terrestris* L. U.S. Geological Survey, National Park Service. Pp 1-29.

Heidari, M.R., M. Mehrabani, A. Pardakhty, P. Khazaeli. 2007. The analgesic effect of *Tribulus terrestris* extract and comparison of gastric ulcerogenicity of the extract with Indomethacine in animal experiments. *Ann. N.Y. Acad. Sci.* 1095: 418–427.

Hennings, M.W., 1932. Animal diseases in South Africa. In: Virus and Deficiency Diseases, Plant Poisonings, vol. II. Gengra New Agency, South Africa, pp. 855–867.

Huang J.W., Huang, C.H. Tan, S.H. Jiang and D.Y. Zhu, 2003. Terrestrinins A and B, two new steroid saponins from *Tribulus terrestris*, *Journal of Asian Natural Products Research.* 5: 285-290.

Jacob, R.H., Peet, R.L., 1987. Poisonings of sheep and goats by *Tribulus terrestris* (caltrop). *Aust. Vet. J.* 64: 288.

Jameel, J.K, Kneeshaw PJ, Rao VS, Drew. PJ 2004. Gynaecomastia and the plant product Tribulis terrestris. *Breast.* 13: 428-430.

Jit, S., and Nag, T. N. 1985. *Indian Journal of Pharmaceutical Sciences,* 47: 101

Joshi, V.S. B.B. Parekha, M.J. Joshia,_, A.B. Vaidyab. 2005. Herbal extracts of *Tribulus terrestris* and Bergenia ligulata inhibit growth of calcium oxalate monohydrate crystals in vitro. *Journal of Crystal Growth.* 275: e1403–e1408.

Kavitha, A. V. and Jagadeesan, G. 2006. Role of *Tribulus terrestris* (Linn.) (Zygophyllaceae) against mercuric chloride induced nephrotoxicity in mice, Mus musculus (Linn.) *Journal of Environmental Biology.* Vol. 27: 397-400.

Kohut J.R. Thompson, J. Campbell, G.A. Brown, M.D. Vukovich, D.A. Jackson and D.S. King, 2003. Ingestion of a dietary supplement containing dehydroepiandrosterone (DHEA) and androstenedione has minimal effect on immune function in middle-aged men, *Journal of the American College of Nutrition.* 22: 363-371.

Koumanov, F., Bozadjieva, E., Andreeva, M., Platonova, E., Ankova, V., 1982. *Clinical trial of Tribestan. Exp. Med.* 4: 211-215.

Li, J.X., Shi, Q., Xiong, Q.B., Prasain, J.K., Tezuka, Y., Hareyama, T., Wang, Z.T., Tanaka, K., Namba, T., Kadota, S., 1998. Tribulusamide A and B, new hepatoprotective lignanamides fromthe fruits of *Tribulus terrestris*: indications of cytoprotective activity in murine hepatocyte culture. *Planta Medica.* 64, 628–631.

Li, M., W. Qu, Y. Wang, H. Wan and C. Tian, 2002. Hypoglycemic effect of saponin from *Tribulus terrestris*, *Zhong Yao Cai.* 25: 420-422.

Lin, Z.X., Hoult, J.R., Raman, A., 1999. Sulphorhodamine B assay for measuring proliferation of a pigmented melanocyte cell line and its application to the evaluation of crude drugs used in the treatment of vitiligo. *Journal of Ethnopharmacology.* 66 (2), 142-150.

Liu, J., Chen, H.S., Xu, Y.X., Zhang, W.D., Liu, W.Y., 2003. Studies on chemical constituents of *Tribulus terrestris* L. *Di Er Jun Yi Da Xue Xue Bao.* 24, 221–222.

Liu, Q., Chen, Y., Wang, J., Chen, X., Han, Y., 1995. Preventive effect of *Tribulus terrestris* L. on genetic damage. *Zhonggno Zhong Yao Za Zhi.* 20: 477-479.

Lu, S.B., B.J. Lu, M.Z. Shen and Y.Z. Rong, 1994. The clinic report of Xinnao shutong on myocardial infarction. Acta Universitatis Medicinalis Secondae Shanghai 14: 78-79.

Mahato, B., Shau, N. P., Pal, B. C., Chakravarti, R. N., Chakravaryi, D., and Ghosh, A. 1978. *Journal of Institution and Chemists.* 50: 49.

Mahato, S. B., Niranjan, P., Ganguly, A. N., Miyaharo, K., and Kawasaki, T. 1981. *Journal of Chemical Society Perkin Transactions.* 1: 2405.

Majewska MD. 1995. Neuronal actions dehydroepiandrosterone. In: Bellino FL, Daynes RA, Hornsby PJ, Lavrin DH, Nestler JE, editors. Dehydroepiandrosterone (DHEA) and Ageing. The New York Academy of Sciences, 774: 111-20.

Mathur, G. S., Nag, T. N., and Goyal, S. C. 1977. Food Farming Agric, 9: 11.

McDonough, S.P., Woodbury, A.H., Galey, F.D., Wilson, D.W., East, N., Bracken, E., 1994. Hepatogenous photosensitization of sheep in California associated with ingestion of *Tribulus terrestris* (puncture vine). *J. Vet. Diagn. Invest.* 6, 392–395.

Milašius, K., M. Pečiukonienė, R. Dadelienė, J. Skernevičius. 2010. Efficacy of the *Tribulus* food supplement used by athletes. *Acta Medica Lituanica.* 17: 65–70.

Miles, C.O., Wilkins, A.L., Erasmus, G.L., Kellerman, T.S., 1994a. Photosensitivity in South Africa. VIII. Ovine metabolism of *Tribulus terrestris* saponons during experimentally induced geeldikkop. *Onderstepoort J. Vet. Res.* 61, 351–359.

Miles, C.O., Wilkins, A.L., Erasmus, G.L., Kellerman, T.S., Coetzer, J.W.A., 1994b. hotosensitivity in South Africa. VII. Chemical composition of biliary crystals from a sheep with experimentally induced geeldikkop. *Onderstepoort J. Vet. Res.* 61, 215–222.

Monson, K. and A. Schoenstadt, 2008 Is *Tribulus terrestris* Safe? http://men.emedtv.com/tribulus-terrestris/is-tribulus-terrestris-safe.html

Mossa, J.S., Al-Yahya, M.A., Al-Meshal, I.A., Tariq, M., 1983. Phytochemical and biological screening of Saudi medical plants. *Fitoterapia.* 54: 147–152.

Nag, T. N., Mathur, G. S., and Goyal, S. C. 1979. *Comparative Physiology and Ecology.* 4: 157.

Nandal, S.N., Bhatti, D.S., 1983. Preliminary screening of some weed shrubs for their nematocidal activity against Meloidogyne Javanica. *Indian Journal of Nematology.* 13: 123-127.

Neychev V.K, Mitev V.I. 2005. The aphrodisiac herb *Tribulus terrestris* does not influence the androgen production in young men". *Journal of Ethnopharmacology*. 101: 319-23.

Neychev VK, Nikolova E, Zhelev N, Mitev VI. 2007. Saponins from *Tribulus terrestris* L are less toxic for normal human fibroblasts than for many cancer lines: influence on apoptosis and proliferation. *Exp. Biol. Med.* (Maywood). 232:126-33.

Nian, H., L.P. Qin, W.S. Chen, Q.Y. Zhang, H.C. Zheng and Y. Wang, 2006. Protective effect of steroidal saponins from rhizome of Anemarrhena asphodeloides on ovariectomy-induced bone loss in rats, *Acta Pharmacologica Sinica*. 27: 728-734.

Pandey, J. K. and Singh, D. K. 2009. Molluscicidal activity of Piper cubeba Linn., Piper longum Linn. and *Tribulus terrestris* Linn. and their combinations against snail Indoplanorbis exustus Desh. *Indian Journal of Experimental Biology*. 47: 643-648.

Perepelitsa, E. D., and Kintya, P. K. 1975. Chemical study of steroid glycosides of *Tribulus terrestris* part 4 steroid saponins. *Khimiya Prirodnykh Soedinenii.* 11: 260-261.

Phillips, Oludotun A. Mathew, Koyippalli I, Oriowo, Mabayoje A. 2006. Antihypertensive and vasodilator effects of methanolic and aqueous extracts of *Tribulus terrestris* in rats. *Journal of Ethnopharmacology*. 104: 351-355.

Qazi 2008. http://www.theherbs.org/articles/article-6.html.

Rogerson S, Riches Ch, Jennings C, Weatherby R, Meir R, Marshall-Gradisnik S. 2007. The effect of five weeks of *Tribulus terrestris* supplementation on muscle strength and body composition during preseason training in elite rugby league players. *J. Strength Condit. Res.* 21: 348-353.

Roselli, C.E., S.E. Abdelgadir and J.A. Resko, 1997. Regulation of aromatase gene expression in the adult rat brain. *Brain Research Bulletin.* 44: 351-357.

Ross, I.A., 2001. Medicinal plants of the world Vol. 2. Humana Press. Pp 412.

Şahin, A. 2009. Effects of dietary *Tribulus terrestris* L. Powder on Growth Performance, Body Components and Digestive System of Broiler Chicks. *Journal of Applied Animal Research.* 35: 193-195.

Saleh, N. A. M., Ahmed, A. A., and Abdalla, M. F. 1982. Flavonoid glycosides of Tribulus pentandrus and T. Terrestris. *Phytochemistry.* 21: 1995-2000.

Saleh, N. A. M., and El-Hadidi, M. N. and Ahmed, A.A, 1982. The chemosystematics of Tribulaceae. *Biochemical Systematics and Ecology.* 4: 313-317.

Sangeeta D, Sidhu H, Thind SK. 1994. Effect of *Tribulus terrestris* on oxalate metabolism in rats. *J. Ethnopharmacol.* 44: 61-66.

Sangeeta, D, Sidhu, H., Thind, S.K, Nath, R and Vaidyanathan, S. (1993). Therapeutic response of *Tribulus terrestris* (Gokhru) aqueous extract on hyperoxaluria in male adult rats. *Phytothe. Res.* 7: 116-119.

Saudan C, Baume N, Emery C, Strahm E, Saugy M. 2008. Short term impact of *Tribulus terrestris* intake on doping control analysis of endogenous steroids. *Forensic Sci. Int.* 178:e7-10.

Shao, Y., Poobrasert, O., Kernelly, E.J., Chin, C.K., Ho, C.T., Huang, M.T., Garrison, S.A., Cordell, G.A., 1997. Steroidal saponins from Aspargus officinalis and their cytotoxic activity. *Planta Med.* 63: 258-262.

Sharifi, A.M., Darabi, R., Akbarloo, N., 2003. Study of antihypertensive mechanism of *Tribulus terrestris* in 2K1C hypertensive rats: role of tissue ACE activity. *Life Science.* 73: 2963-2971.

Sharma, H. C., Norula, J. L., and Varadarajan, R. 1977. *Chemical Era.* 13: 261.

Shivakumar, H. J. Talha, Prakash, T., Rao, R. N., Swamy, B. H. M. Jayakumar, G. A. Veerana. 2006. Adaptogenic activity of ethanolic extract of *Tribulus terrestris* L. *Journal of Natural Remedies.* 6: 87-95.

Singh R., S. Pervin, J. Shryne, R. Gorski and G. Chaudhuri, 2000 Castration increases and androgens decrease nitric oxide synthase activity in the brain: Physiologic implications, Proceedings of the National Academy of Sciences of the United States of America 97: 3672-3677.

Singh, P. K., A. P. Gupta, A. K. Chaudhary and Seema. 2009. Beneficial effects of aqueous fruit extract of *Tribulus terrestris* on testicular and serum biochemistry of albino rats. *Journal of Ecophysiology and Occupational Health.* Vol. 9: 217-223.

Singh, S. P., Raghavendra, K., Singh, R. K., Mohanty, S. S. and Dash, A. P. 2008. Evaluation of *Tribulus terrestris* Linn (Zygophyllaceae) acetone extract for larvicidal and repellence activity against mosquito. *Journal of communicable diseases.* 40: 255-261.

Sivarajan, V.V. and Balachandran, I. 1994. Ayurvedic Drugs And Their Plant Sources, Oxford and IBH Publishing Co.

Somanadhan, B., Varughese, G., Palpu, P., Sreedharan, R., Gudiksen, L., Smitt, U.W., Nyman, U., 1999. An ethnopharmacological survey for potential angiotensin converting enzyme inhibitors from Indian medicinal plants. *Journal of Ethnopharmacology.* 65: 103-112.

Su L, Chen G, Feng SG, Wang W, Li ZF, Chen H, Liu YX, Pei YH. 2009. Steroidal saponins from *Tribulus terrestris. Steroids.* 74: 399-403.

Sun B., W. Qu and Z. Bai, 2003. The inhibitory effect of saponins from *Tribulus terrestris* on Bcap-37 breast cancer cell line in vitro, *Zhong Yao Cai.* 26: 104-106.

Sun, B., W.J. Qu, X.L. Zhang, H.J. Yang, X.Y. Zhuang and P. Zhang, 2004. Investigation on inhibitory and apoptosis-inducing effects of saponins from *Tribulus terrestris* on hepatoma cell line BEL-7402, *Zhongguo Zhong Yao Za Zhi.* 29: 681-684.

Talasaz, A.H. M. Reza Abbasi, S. Abkhiz and S. Dashti-Khavidaki. 2010. *Tribulus terrestris*-induced severe nephrotoxicity in a young healthy male *Nephrol. Dial Transplant.* 25: 3792–3793.

Tomova, M. 1987. Tibestan a preparation from Tribulus terrestris. *Farmastiya.* 37: 40-42.

Tomova, M., Gjulemetova, R., Zarkova, S., Peeva, S., Pangarova, T., Simova, M., 1981. Steroidal saponins from *Tribulus terrestris* L. with a stimulating action on the sexual functions. In: International Conference of Chemistry and Biotechnology of Biologically Active Natural Products, Varna, Bulgaria, 3: 298–302.

Tuncer, M. A, B. Yaymacib, L. Satic, S. Caylic, G. Acard, T. Altuge and R. Demirc. 2009. Influence of *Tribulus terrestris* extract on lipid profile and endothelial structure in developing atherosclerotic lesions in the aorta of rabbits on a high-cholesterol diet. *Acta Histochemica.* 111: 488-500.

Van Tonder, E.M., Basson, P.A., Van Renburg, I.B.G., 1972. Geeldikkop: experimental induction by feeding the plant *Tribulus terrestris* L. (Zygophyllacae). *J. S. Afr. Vet. Assoc.* 43: 363–373.

Vasi, I.G., and Kalintha, V. P. 1982. Chemical examination of the fruits of Tribulus. *Comparative Physiology and Ecology.* 7: 68.

Verma, N., Kumar, A., Saggoo, M. I. S., Kaur, G., Singh, M., Kumar, V., Kumar and U. 2009. Phytochemical analysis and antimicrobial activity of crude extract of *Tribulus terrestris* fruits; a known medicinal weed. *Advances in Plant Sciences.* 22: 155-158.

Vincent, S.R. and H. Kimura, 1992. Histochemical mapping of nitric oxide synthase in the rat brain. *Neuroscience.* 46: 755-784.

Wag, B., Ma, L., Liu, T., 1990. 406 cases of angina pectoris in coronary heart disease treated with saponin of *Tribulus terrestris. Zhong Xi Yi Jie He Za Zhi.* 10: 85–87.

Wang, Y., Ohtani, K., Kasai, R., Yamasaki, K., 1997. Steroidal saponins from fruits of *Tribulus terrestris. Phytochemistry.* 45: 811-817.

Warrier, P. K., V. P. K. Nambiar, C. Ramankutty, R. Vasudevan Nair. 1997. Indian medicinal plants: a compendium of 500 species, Orient Longman Publications, Hyderabad, India. Volume 5: 311-312.

Williamson E.M. (ed.), 2002. Major herbs of ayurveda compiled by Dabur Research Foundation and Dabur Ayurvet Ltd., Churchill Livingstone.

Wu G, Jiang S, Jiang F, Zhu D, Wu H, Jiang S. 1996. Steroidal glycosides from *Tribulus terrestris. Phytochemistry.* 42:1677-81.

Wu T. S., Shi L. S. and Kuo S. C. 1999. Alkaloids and other constituents from *Tribulus terrestris. Phytochemistry.* 50: 1411-15.

Xie, Z. F. and Huang, X. 1988. Dictionary of Traditional Chinese Medicine. Commercial Press Ltd. Hong Kong. p 194.

Xu Y. X., Chen H. S., Liang H. Q., Gu Z. B., Lui W. Y., Leung W. N. and Li T. J., 2000. Three new saponins from *Tribulus terrestris. Planta Medica.* 66: 545-50.

Xu Y.J., T. H. Xu, J. Y. Yang, S. X. Xie, Y. Liu, Y. S. Si and D. M. Xu. 2010. Two new furostanol saponins from *Tribulus terrestris* L. *Chinese Chemical Letters.* 21: 580-583.

Xu, T. Y. Xu, Y. Liu, S. Xie, Y. Si and D. Xu. 2009 Two new furostanol saponins from *Tribulus terrestris* L. *Fitoterapia.* 80: 354-357.

Yan W, Ohtani K, Kasai R, Yamasaki K.1996. Steroidal saponins from fruits of Tribulus terrestris. *Phytochemistry.* 42: 1417.

Yan W., Ohtani K., Kasai R., Yamasaki K. 1996. Steroidal saponins from fruits of *Tribulus terrestris. Phytochemistry.* 42: 1417-1422.

Yan W., Ohtani K., Kasai R., Yamasaki K. Steroidal saponins from fruits of *Tribulus terrestris* 1996. *Phytochemistry.* 42: 1417-1422.

Yang H.J., W.J. Qu and B. Sun, Experimental study of saponins from *Tribulus terrestris* on renal carcinoma cell line, 2005. *Zhongguo Zhong Yao Za Zhi.* 30: 1271-1274.

Yang, X.Y., B.F. Hang, R. Jiang. 1991. Shutong therapy in 50 patients with cerebral arteriosclerosis and the sequelae of cerebral thrombosis. *New Drugs and Clinical Remedies.* 10: 92–95.

Yang, Y., T. Wu, K. He and Z.G. Fu, 1999. Effect of aerobic exercise and ginsenosides on lipid metabolism in diet-induced hyperlipidemia mice, *Zhongguo Yao Li Xue Bao.* 20: 563-565.

Yuan Wei-Hua, Wang Nai-Li, Yi Yang-Hua, Yao Xin-Sheng. 2008. *Chinese Journal of Natural Medicines.* Vol. 6: 172-175.

Zafar R and Lalwani M. 1989. *Tribulus terrestris* Linn-a review of the current knowledge. *Indian Drugs.* 1989; 27: 148-153.

Zafar, R., Lalwani, M., and Siddiqui, A. 1989. Hecogenin and neotigogenin from the root of *Tribulus terrestris Indian Drugs.* 26: 460.

Zhang, J.D., Z. Xu, Y.B. Cao, H.S. Chen, L. Yan and M.M. An et al., 2006. Antifungal activities and action mechanisms of compounds from Tribulus terrestris L, *J. Ethnopharmacol.* 103: 76-84.

In: Natural Products and Their Active Compounds … ISBN: 978-1-62100-153-9
Editors: M. Essa, A. Manickavasagan, and E. Sukumar © 2012 Nova Science Publishers, Inc.

Chapter 14

POLYPHENOLIC TEA CATECHINS: HEALTH PROMOTION, DISEASE PREVENTION AND MOLECULAR MECHANISMS OF ACTION

K. M. Umar, S. M. Abdulkarim*[*], A. Abdul Hamid, Son Radu and N. B. Saari

Department of Food Science,
Faculty of Food Science and Technology,
University Putra Malaysia, Serdang,
Selangor, Malaysia

ABSTRACT

Catechins belong to the flavan-3-ol class of flavonoids, the most abundant polyphenolic compounds found in green tea (*Camellia sinensis*) that have been shown to bioactively affect the pathogenesis of several diseases. Catechins have drawn a lot of attention due to the variety of properties they possess. Among the most studied properties are induction of apoptosis, antioxidative, anti-inflammatory, antiviral, antimicrobial, antiobesity and anti-diabetic properties.

Several *in vitro* and *in vivo* studies have attributed the molecular mechanisms by which polypenolic tea catechins (PTCs) exhibit these properties with their potential to increase the expression and phosphorylation of certain proteins, cytokines, upregulation of endogenous free radical scavengers and regulation of signal transduction pathways. This chapter will give a narrative review of the various health promotion, disease prevention properties and molecular mechanisms of action of polyphenolic tea catechins present in a selected database.

Keywords: polyphenolic tea catechins, apoptosis, antioxidative, anti-inflammatory, antiviral, antimicrobial, antiobesity

[*] ak@food.upm.edu.my; abdulk3@yahoo.co.uk

INTRODUCTION

Catechins belong to the flavan-3-ol class of flavonoids, the most abundant polyphenolic compounds found in green tea (*Camellia sinensis)* that have been shown to bioactively affect the pathogenesis of several diseases. All catechins contain a benzopyran skeleton with a phenyl and an ester groups substituted at position 2 and 3 respectively. The major green tea catechins include (-) epicatechin (EC), (-) epigallocatechin (EGC), (-) epigallocatechin-gallate (EGCG) and (-) epicatechin-gallate (ECG) of which EGCG is the most abundant in *Camellia sinensis*.

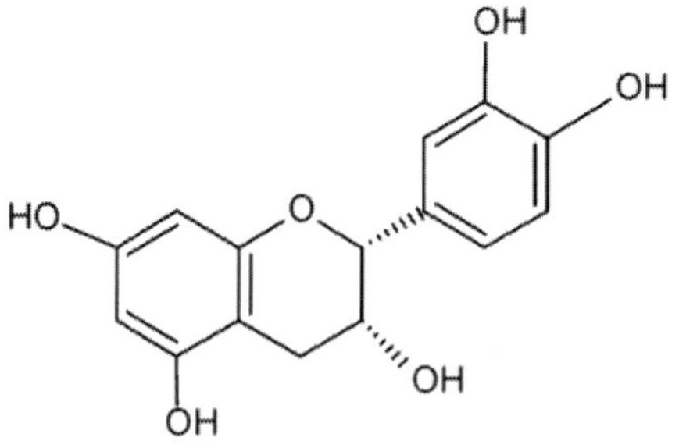

(-) Epicatechin

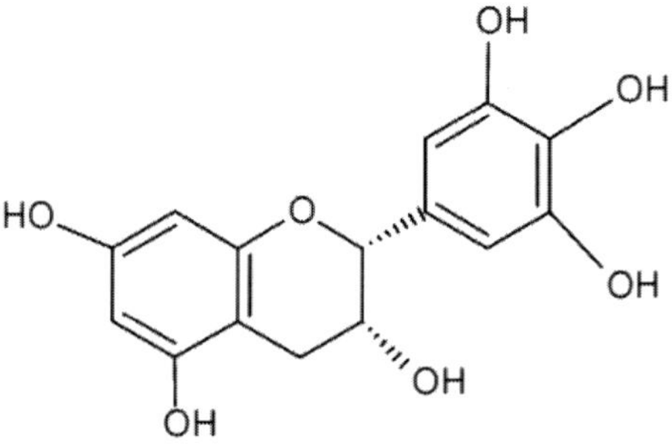

(-) Epigallocatechin

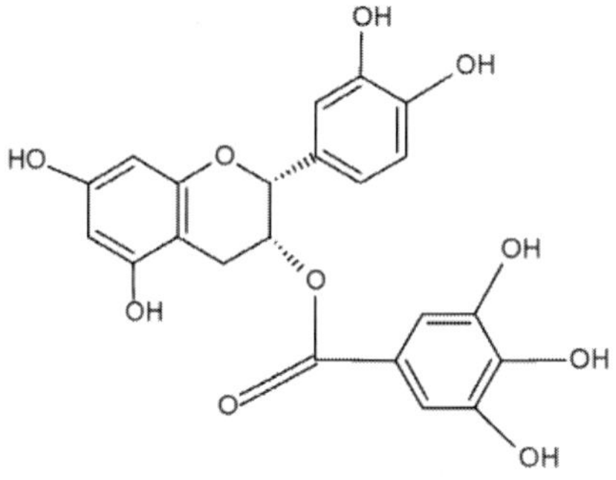

(-) Epicatechin-gallate

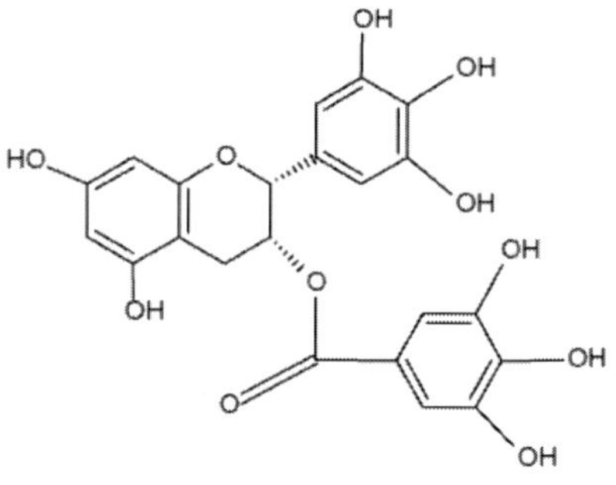

(-) Epigallocatechin-gallate

In recent years, there has been an increasing interest in polyphenolic tea catechins (PTC) due to the variety of health promotion and disease prevention properties they possess. These characteristics made them potential candidates as functional foods in food stuffs and feed stuffs that improve quality through creation of value added food products with increased shelf life by inhibiting lipid oxidation (reviewed by Yilmaz, 2006). These potentials have led to the development of food products like tea beverages, cereal bars, ice creams, confectionaries and pet foods that contain tea as active ingredient (Ferruzzi and Green, 2006). In addition, a number of methods have also been developed to maximise the content of PTCs in food drinks (Copland et al., 1998; Labbe *et al*., 2005).

Among the most studied properties are anti-inflammatory (Yang, 1993; Katiyar and Mukhtar, 1996; Yang *et al.,* 2002; Adhami *et. al.,* 2003; Mittal *et al.,* 2004; Thangapazham *et al.,* 2006; Nichols and Katiyar, 2010), induction of apoptosis (Ahmad *et al.,* 1997; Ahmad *et al.,* 2000), antioxidative (Sinha *et al.,* 2010 ; Lee *et al.,* 2003; Erba *et al.,* 2005; Nichols and Katiyar, 2010), antiobesity (Klaus *et al.,* 2005; Wolfram *et. al.,* 2005; Ikeda *et al.,* 2005; Bose *et al.,* 2008), anti-diabetic (Wolfram *et al.,* 2006) and many more.

More recently, literature has emerged from numerous research explaining the molecular mechanism of actions of PTCs. The mechanisms have been associated with PTCs potential to increase the expression and phosphorylation of certain protein, cytokines and regulation of signal transduction pathways. Moreover, studies have also demonstrated the up-regulation of endogenous free radical scavengers like glutathione. This chapter will give a narrative review of the various health promotion, disease prevention properties and molecular mechanisms of action of polyphenolic tea catechins present PubMed. However, literature that did not specifically deal with polyphenolic catechins or in combination with other polyphenols were not used for this write up.

INDUCTION OF APOPTOSIS

Apoptosis is a coordinated energy dependent process which involves activation of a group of cysteine proteases (called caspases) and complex cascade of events that link initiating stimuli and cell death leading to the elimination of mutated preneoplastic and hyperproliferating cells from the system (reviewed by Elmore, 2007). Antiapoptotic and proapoptotic proteins of Bcl-2 family, an oncoprotein of which its down-regulation has been implicated in tumor regression (Kluck *et al.*, 1997; Sedlak *et al*., 1995), modulate the process of apoptosis.

Studies sought to provide evidence for the molecular mechanisms whereby EGCG resulted in apoptosis of proliferating vascular smooth muscle cells show the collective activation of tumor suppressor p53 and transcription factor nuclear factor kappaB (NF-*k*B). The induction of p53 is attributed to increased transcription of its target, cyclin-dependent kinase inhibitor ($p21^{CIP1}$), which is one of the key mediators of growth arrest. While that of NF-κB activity, which consisted predominantly of classical p50/p65 complexes, was likely due to its release upon degradation of its cytoplasmic inhibitor IκB, allowing for its translocation to the nucleus where it can induce expression of κB element containing target genes (Hofmann and Sonenshein, 2003).

With a view to investigate the influence of EGCG on signalling molecules directly involved in apoptotic pathway, Chen *et al.,* (2003) compared the anti-proliferation effect of EGCG on normal colon epithelial cells and colon carcinoma cells, the morphological changes after EGCG treatments, and the influence of EGCG on caspases, cytochrome c, mitochondria, as well as mitogen-activated protein kinases (MAPKs). EGCG was found to attenuate the promotion and progression of adenoma and carcinoma cells, induce oxidative stress via activation of stress signals that damage mitochondria releasing cytochrome c. This infers that pro-apoptotic effect of EGCG may in part contribute to its overall chemo-preventive function against colonic carcinogenesis (2003).

Baliga and co-workers (2005) have suggested the induction of apoptosis and inhibition of cell proliferation and metastasis of highly metastatic mouse breast cancer cells through disruption of mitochondrial pathway in both *in vitro* and *in vivo* models. This study further elucidated the molecular mechanisms by which apoptosis is initiated. The results revealed the down-regulation of Bcl-2 protein expression coupled with increased expression of Bax, the ratio of which has been found to increase during apoptosis in previous studies (Oltvai *et al.*, 1993; Sedlak *et al.*, 1995). Additionally, a dose- and time-dependent increase in levels of cytochrome c, the adaptor Apaf-1 and activated cleaved caspase 3 were observed in *in vitro* models. The activated caspase 3 is the key executioner of cell apoptosis that cleaves intracellular proteins vital to cell survival and growth, such as PARP, an important marker of apoptosis (Darmon *et al.*, 1995), hence strengthening the conviction that EGCG mediates apoptosis via mitochondrial disruption pathway.

For the *in vivo* models, emphasis on different surrogate markers of apoptosis confirmed the increase in Bax/Bcl-2 ratio upon PTCs administration in addition to the inhibition of PCNA expression in tumors. PCNA, a subunit of DNA polymerase, plays a crucial role in DNA synthesis and serves as a biomarker of proliferation.

ANTI-INFLAMMATORY PROPERTY

Inflammation is a primary response through secretion of a variety of cytokines, chemokines, lipid mediators and bioactive amines upon exposure to insult or injury that normally resolves with no detriment to the host. However, chronic inflammation can result to deleterious consequences if pro-inflammatory signaling pathways and removal of inflammatory cells failure to restore normal tissue conditions (reviewed by Lawrence and Fong, 2010).

Consistent with the findings of other *in vitro* studies, EGCG inhibits cell proliferation, survival and phenotype transformation of mouse mammary tumor virus (MMTV)-Her-2/neu NF639 cells via down-regulation of phosphatidylinositol 3- kinase, Akt kinase to NF-*k*B pathway. This reduced signaling has been attributed to of inhibition of basal Her-2/neu receptor tyrosine phosphorylation in addition to basal receptor phosphorylation in SMF and Ba/F3 2 + 4 cells (Pianetti *et. al.,* 2002).

Findings have emerged to explain the possible mechanisms by which PTCs attenuate the degree of inflammation in animal models of carrageenan-induced pleurisy. These studies revealed the down regulation of intercellular adhesion molecule (ICAM-1), tumor necrosis factor alpha (TNF-alpha) and signal transducers and activators of transcription (STAT) factors activation. ICAM-1 is an endothelial- and leukocyte-associated trans-membrane protein that facilitates leukocyte endothelial transmigration to inflammation sites enabling the firm adhesion and diapedesis of leukocytes. Therefore, the down regulation of surface expression of ICAM-1 on endothelial cells will prevent the infiltration of neutrophils at inflamed sites (Di Paola *et al.*, 2005).

Studies have shown that PTCs regulate TNF-alpha gene expression by modulating NF-kB activation through their antioxidant properties (Yang *et al.*, 1998). TNF-alpha is a cytokine involved in systemic inflammation produced predominantly by activated macrophages and lymphocytes that that play a pivotal role in the pathogenesis of inflame-matory autoimmune diseases (Brennan and Feldman 1996). Following stimulation, the increase in cytokines occurs as a result of gene expression and *de novo* synthesis. An oxidative stress - sensitive nuclear transcription factor (NF-kB), controls the expression of many genes including the TNF-alpha gene (Yang et. al., 1998).

Signal transducers and activators of transcription (STAT) factors are a family of cytoplasmic transcription factors that mediate intracellular signaling initiated at cytokine cell surface receptors and transmitted to the nucleus. Recent reports have indicated EGCG as a potent inhibitor of STAT-1 phosphorylation and activation (Menegazzi *et al.*, 2001). The JAK/STAT pathway has been shown to be essential for human and murine iNOS expression (Gao *et al.*, 1997) and in ICAM-1 induction (Roy *et al.*, 2001). In addition, pretreatment with PTCs reduced STAT-1 activation in carrageenan-treated mice. The fall in STAT-1 activity may further account for the repression of ICAM-1 and iNOS lung expression with decrease in nitrite/nitrate concentration in the pleural exudates (Di Paola *et al.*, 2005).

Antioxidative Property

The most important cell targets of reactive oxygen species (ROS) are proteins, carbohydrates, lipids and DNA (Sies, 1986). Several defence mechanisms present in living species eliminate, neutralize, repair lesions and compensate metabolic pathways caused by these ROS (Vasseur and Cossu-Leguile, 2003) by mounting a number of free radical scavengers known as antioxidants from dietary sources (Food and Nutrition Board, 1998), and endogenous sources which include glutathione (Meister and Anderson, 1983, Miyamoto *et al.,* 2003), catalase (Scott *et al.,* 1991) and superoxide dismutase (Noguchi and Niki, 1998). Under physiologic conditions, these mechanisms maintain a stable state called redox homeostasis (Dröge, 2002). An imbalance between the unregulated production of free radicals and the deficiency of antioxidants results in oxidative stress (Bowen *et al.*, 2001).

Efforts to investigate the anti-oxidative properties of green tea (containing approximately 250 mg of PTCs) determined through a variety of variables in a dietary intervention study have indicated a significant increase and decrease in plasma total antioxidant activity and plasma peroxides level respectively. In addition, significant decreases in induced DNA oxidative damage in lymphocytes and LDL cholesterol were observed with respect to control upon administration of green tea (Erba *et al.*, 2005).

Product analysis of oxidation products formed by reactions of PTCs with with peroxyl radicals generated by thermolysis of the initiator 2,2′-azobis(2,4-dimethylvaleronitrile) (AMVN) in oxygenated acetonitrile have provided proofs that the principal site of antioxidant reactions on the EGCG molecule is the trihydroxyphenyl B ring, rather than the 3-galloyl moiety. These products include a seven-membered B-ring anhydride ($C_{22}H_{16}O_{12}$) and a novel dimmer ($C_{44}H_{34}O_{23}$) (Valcic *et al.*, 1999).

Three major mechanisms have been identified by *in vitro* studies to be the molecular mechanisms of PTCs anti-oxidative property. The first being the inhibition of the redox-sensitive transcription factors, nuclear factor-*k*B and activator protein-1 via their potentials to act as kinase inhibitors in complex signaling pathways. The second is inhibition of "prooxidant" enzymes, such as inducible nitric oxide synthase, lipoxygenases, cyclooxygenases and xanthine oxidase. Lastly is the induction of phase II detoxification enzymes (e.g. glutathione S-transferases and superoxide dismutases) (reviewed by Frei and Higdon, 2003).

Antiviral Property

Systematic research has revealed the virucidal activity of PTCc, with the catechin gallates showing more potency over the non gallates. Studies on MDCK cell culture have correlated the inhibition of influenza viral subtypes A/H1N1, A/H3N2 and B virus replication with inhibition of hemagglutination, neuraminidase activity and affects viral RNA synthesis at high concentration. This suggests that, besides the known inhibitory activity on viral attachment of host cells, the antiviral activities of polyphenols are associated with various steps in the influenza virus life cycle (Song *et al.*, 2005).

In another study, Weber *et al.* (2002) determined the anti-viral activity of PTCs on adenovirus infection and viral protease adenine in cell culture. Findings indicated the suppression of different stages of viral life cycle and adenain activity with EGCG being more effective.

Concerning HIV infection, Yamaguchi *et al.*, (2002) have provided facts relating to the anti- human immunodeficiency virus type-1 (HIV-1) properties of PTCs in two different host cell systems, T-lymphoid (H9) and monocytoid (THP-1) cell lines. ECGC was found to inhibit post-adsorption entry and reverse transcription in addition to destruction of virions in a dose- dependent manner through the deformation the phospholipids. In another report, EGCG obstructed gp120 binding to CD4 molecules by regulating CD4 expression (Kawai *et al.*, 2003).

Anti Parasitic Property

Evidence is emerging on the anti-parasitic property of PTCs. Studies by Paveto *et al.*, (2004) established a trypanocidal effect against both the infective, non-replicative trypomastigote form and the replicative amastigote stage.

At low concentrations, aqueous extract of *C. sinensis* strongly inhibited the activity of *Trypanosoma cruzi* arginine kinase (AK) that plays a fundamental role in the intracellular energy flow, acting as a reservoir for ATP through the reversible formation of a phosphagen guanidino derivative. However, parasite lysis was not a consequence of AK inhibition because of the non specific binding of catechins to proteins and peptides which hardly occurs at low concentrations.

EGCG and ECG have been shown to demonstrate an anti-malarial activity by inhibition of *Plasmodium falciparum* growth *in vitro.* Both of these catechins were found to increase the antimalarial effects of artemisinin when administered at sub-lethal doses without interfering with the folate pathway (Sanella *et al.*, 2007).

ANTIMICROBIAL PROPERTY

Studies on the antimicrobial activity of PTCs have demonstrated the suppression of virulence in a number of bacterial species (*Staphylococcus aureus* including the methicilin resistant, *Helicobacter pylori*, *Prevotella spp.*, *Porphyromonas gingivalis* and *α-haemolytic streptococci*) with modest antibacterial activity while favouring the growth of beneficial species (lactobacilli and bifidobacteria).

Research has revealed that the effect of catechins against pathogenic microbes are due to bacterial phenotype modifications like denaturation or deconformation of protein ligands or inhibition of virulence factors associated with aetiology of the diseases caused by the organism. Examples of these virulence factors are glucosyl transferase, toxic end metabolites, protein tyrosine phosphatase and gingipains. Furthermore, extracts of green tea have the capacity to reverse methicillin resistance in MRSA isolates at concentrations much lower than those needed to produce inhibition of bacterial growth (Taylor *et al.*, 2005).

Despite their known biological activity against a variety of diseases, the instability due to enzymatic or non-enzymatic cleavage of the 3-galloyl group has hampered the clinical use of EGCG and ECG. However, Park and Cho (2010) synthesized a series of catechin analogues to enhance not only antimicrobial activities but also stability.

The synthesized compounds showed increased levels of antimicrobial activities due to their accelerated adsorption on the surface of bacterial membranes which leads to disruption of the membrane by increasing the lipophilicity. Recent investigations have also shown the inhibition of established biofilms and maintenance of *Candida albicans* via impairment of proteasomal activity upon treatment with EGCG (Evensen and Braun, 2009).

ANTI OBESITY PROPERTY

Previous studies have attributed the anti-obesity properties of PTCs (especially EGCG) with reduction in body weight, percent body fat, visceral fat accumulation (Ikeda *et. al.*, 2005), metabolic syndrome, fatty liver (Bose *et al.*, 2008) as well as decreasing energy absorption and increasing oxidation of fats (Klaus *et al.*, 2005).

Recently, animal studies have shown that PTCs (especially EGCG) have the potentials of preventing and reversing diet induced obesity through reduced expression of fatty acid synthase and acetyl-CoA carboxylase-1 mRNA levels thereby directly affecting adipose tissues (Wolfram *et al.*, 2005).

Other studies reported the down regulation of peroxisome proliferator-activated receptor-gamma (PPAR-gamma), CCAAT enhancer-binding protein-alpha (C/EBP-alpha), regulatory element-binding protein-1c (SREBP-1c), adipocyte fatty acid-binding protein (aP2), lipoprotein lipase (LPL) and fatty acid synthase (FAS) as well as upregulation of carnitine palmitoyl transferase-1 (CPT-1), uncoupling protein 2 (UCP2), hormone sensitive lipase (HSL) and adipose triglyceride lipase (ATGL).

Taken all together, regulation of these multiple genes by PTCs (EGCG) effectively reduces adipose tissues in addition to improvement of plasma lipid profiles (Lee *et al.*, 2009).

REFERENCES

Baliga, M. S., Meleth, S., and Katiyar, S. K. (2005). Growth Inhibitory and Antimetastatic Effect of Green Tea Polyphenols on Metastasis-Specific Mouse Mammary Carcinoma 4T1 Cells In vitro and In vivo Systems. *Clinical Cancer Research,* 11(5), 1918-1927.

Bowen, R. S., Moodley, J., Dutton, M. F., and Theron, A. J. (2001). Oxidative stress in pre-eclampsia. *Acta Obstetricia et Gynecologica Scandinavica, 80*(8), 719-725.

Brennan F. M., Feldmann M. (1996) Cytokines in autoimmunity. *Current Opinion in Immunology.* 8(6): 872 – 877.

Chen, C., Shen, G., Hebbar, V., Hu, R., Owuor, E. D., and Kong, A.-N. T. (2003). Epigallocatechin-3-gallate-induced stress signals in HT-29 human colon adenocarcinoma cells. *Carcinogenesis, 24*(8), 1369-1378.

Darmon, A. J., Nicholson, D. W., and Bleackley, R. C. (1995). Activation of the apoptotic protease CPP32 by cytotoxic T-cell-derived granzyme B. *Nature, 377*(6548), 446-448.

Dröge, W. (2002). Free Radicals in the Physiological Control of Cell Function. *Physiological Reviews, 82*(1), 47-95.

Erba, D., Riso, P., Bordoni, A., Foti, P., Biagi, P. L., and Testolin, G. (2005). Effectiveness of moderate green tea consumption on antioxidative status and plasma lipid profile in humans. *The Journal of Nutritional Biochemistry, 16*(3), 144-149.

Evensen N.A., Braun P.C. (2009). The effects of tea polyphenols on Candida albicans: inhibition of biofilm formation and proteasome inactivation *Canadian Journal of Microbiology*, 55(9):1033-9.

F Frei, B., and Higdon, J. V. (2003). Antioxidant Activity of Tea Polyphenols In Vivo: Evidence from Animal Studies. *The Journal of Nutrition, 133*(10), 3275S-3284S.

Gao, J., Morrison, D. C., Parmely, T. J., Russell, S. W., and Murphy, W. J. (1997). An Interferon-γ-activated Site (GAS) Is Necessary for Full Expression of the Mouse iNOS Gene in Response to Interferon-γ and Lipopolysaccharide. *Journal of Biological Chemistry, 272*(2), 1226-1230.

Food and Nutrition Board, Institute of Medicine (1998) Dietary Reference Intakes. National Academic Press, Washington, DC.

Hofmann C. S., and Sonenshein G. E. (2003) Green tea polyphenol epigallocatechin-3 gallate induces apoptosis of proliferating vascular smooth muscle cells via activation of p53. *FASEB Journal.* 17:702 – 704.

Kawai K., Tsuno N. H., Kitayama J., Okaji Y., Yazawa K., Asakage M., Hori N., Watanabe T., Takahashi K., Nagawa H (2003) Epigallocatechin gallate, the main component of tea polyphenol, binds to CD4 and interferes with gp120 binding. *The Journal of Allergy and Clinical Immunology,* 112, 951–957.

Kluck R. M., Bossy-Wetzel E., Green D. R., Newmeyer D. D. (1997). The release of cytochrome c from mitochondria: a primary site for Bcl-2 regulation of apoptosis. *Science.* 275(5303):1132–6.

Lawrence, T., and Fong, C. (2010). The resolution of inflammation: Anti-inflammatory roles for NF-[kappa]B. *The International Journal of Biochemistry and Cell Biology, 42*(4), 519-523.

Lee M.S., Kim C.T., and Kim Y. (2009). Green tea (-)-epigallocatechin-3-gallate reduces body weight with regulation of multiple genes expression in adipose tissue of diet-induced obese mice. *Annals of Nutrition and Metaboism,* 54(2):151-7.

Meister, A., and Anderson, M. E. (1983). Glutathione. *Annual Review of Biochemistry, 52*(1), 711-760.

Menegazzi, M., Tedeschi, E., Dussin, D., Carcereri de Prati, A., Cavalieri, E., Mariotto, S., et al. (2001). Anti-interferon-g action of epigallocatechin-3-gallate mediated by specific inhibition of STAT1 activation. *The FASEB Journal.*

Miyamoto, Y., Koh, Y. H., Park, Y. S., Fujiwara, N., Sakiyama, H., Misonou, Y., et al. (2003). Oxidative Stress Caused by Inactivation of Glutathione Peroxidase and Adaptive Responses. [doi: 10.1515/BC.2003.064]. *Biological Chemistry, 384*(4), 567-574.

Nichols J. A., Katiyar S. K. (2010). Skin photoprotection by natural polyphenols: anti-inflammatory, antioxidant and DNA repair mechanisms *Archives of Dermatological Research,* 302(2):71-83.

Noguchi, N., and Niki, E. (Eds.). (1999). *Chemistry of active oxygen species and antioxidants.* London, UK. . CRC Press LLC :In Meister A, Anderson M. E (1983) Glutathione. *Ann. Rev. of Biochem.* 52:711–760.

Oltvai Z. N., Milliman C. L., and Korsmeyer S. J. (1993). Bcl-2 heterodimerizes in vivo with a conserved homolog, Bax, that accelerates programmed cell death. *Cell.* 74:609–19.

Park K. D., and Cho S. J. (2010). Synthesis and antimicrobial activities of 3-O-alkyl analogues of (+)-catechin: improvement of stability and proposed action mechanism. *European Journal of Medicinal Chemistry,* 45(3):1028-33.

Paveto C., María C. G., Mo□nica I. E., Virginia M., Jorge C., Mirtha M. F., and Hector N. T. (2004). Anti-Trypanosoma cruzi Activity of Green Tea (*Camellia sinensis*) Catechins *Antimicrobial Agents and Chemotherapy*, 48(1) 69–74.

Pianetti S., Shangqin G., Kathryn T. K., and Gail E. S. (2002). Green Tea Polyphenol Epigallocatechin-3 Gallate Inhibits Her-2/Neu Signaling, Proliferation, and Transformed Phenotype of Breast Cancer Cells. *Cancer Res.* 62: 652–655.

Rizvi S. I., and Zaid M. A. (2001) Intracellular reduced glutathione content in normal and type 2 diabetic erythrocytes: effects of insulin and (-) epicatechin. *Journal of physiology and Pharmacology,* 52(3): 483 - 488.

Rosanna D. P., Emanuela M., Carmelo M., Tiziana G., Marta M., Raffaela Z., Hisanory S., and Salvatore Cuzzocrea (2005). Green tea polyphenol extract attenuates lung injury in experimental model of carrageenan-induced pleurisy in mice *Respiratory Research,* 6:66 doi:10.1186/1465-9921-6-66.

Roy, J., Audette, M., and Tremblay, M. J. (2001). Intercellular Adhesion Molecule-1 (ICAM-1) Gene Expression in Human T Cells Is Regulated by Phosphotyrosyl Phosphatase Activity. *Journal of Biological Chemistry, 276*(18), 14553-14561.

Sannella A. R., Luigi M., Angela C., Franco F. V., Anna R. B., Giancarlo M., Carlo S. (2007). Antimalarial properties of green tea. *Biochemical and Biophysical Research Communications,* 353 177–181.

Scott M. D., Lubin B. H., Lin Zuo., and Kuypers, FA. (1991) Erythrocyte defence against hydrogen peroxide pre-eminent importance of catalase. *Journal of Laboratory Clinical Medicine,* 118. 7-16.

Sedlak, T. W., Oltvai, Z. N., Yang, E., Wang, K., Boise, L. H., Thompson, C. B., et al. (1995). Multiple Bcl-2 family members demonstrate selective dimerizations with Bax. *Proceedings of the National Academy of Sciences, 92*(17), 7834-7838.

Sies, H. (1986). Biochemistry of Oxidative Stress. *Angewandte Chemie International Edition in English, 25*(12), 1058-1071.

Sinha D., Roy S., and Roy M. (2010). Antioxidant potential of tea reduces arsenite induced oxidative stress in Swiss albino mice *Food and Chemical Toxicology,* 48(4):1032-1039.

Song Jae-Min, Kwang-Hee Lee, Baik-Lin Seong (2005) Antiviral effect of catechins in green tea on influenza virus. *Antiviral Research.* 68 66–74.

Susan E. (2007) Apoptosis: A Review of Programmed Cell Death. *Toxicology and Pathology,* 35(4): 495–516.

Taylor P. W., Jeremy M. T., Hamilton M., and Paul D. S. (2005) Antimicrobial properties of green tea catechins. *Food Science and Technology Bulletin,* 2: 71–81.

Valcic S., Muders A., Jacobsen N. E., Liebler D. C., Timmermann B. N. (1999). Antioxidant chemistry of green tea catechins. Identification of products of the reaction of (–)-epigallocatechin gallate with peroxyl radicals. *Chemical Research and Toxicology,* 12:382–6.

Vasseur, P., and Cossu-Leguille, C. (2003). Biomarkers and community indices as complementary tools for environmental safety. *Environment International, 28*(8), 711-717.

Weber J. M., Angelique R. U., Lise I., and Sucheta S. (2003). Inhibition of adenovirus infection and adenain by green tea catechins. *Antiviral Research,* 58(2): 167 – 173.

Wolfram, S., Raederstorff, D., Wang, Y., Teixeira, S., Elste, V., and Weber, P. (2005). TEAVIGO (epigallocatechin gallate) supplementation prevents obesity in rodents by reducing adipose tissue mass. *Annals of Nutrition and Metabolism, 49* (1), 54-63.

Yamaguchi K., Honda M., Ikigai H., Hara Y., Shimamura T. (2002). Inhibitory effects of (−)-epigallocatechin gallate on the life cycle of human immunodeficiency virus type 1 (HIV-1). *Antiviral Research,* 53: 19–34.

Yang F., Willem J. S., Villiers D. E, Craig J. M., and Gary W. V. (1998). Green Tea Polyphenols Block Endotoxin-Induced Tumor Necrosis Factor- Production and Lethality in a Murine Model. *Journal of Nutrition,* 128: 2334–2340.

In: Natural Products and Their Active Compounds ... ISBN: 978-1-62100-153-9
Editors: M. Essa, A. Manickavasagan, and E. Sukumar © 2012 Nova Science Publishers, Inc.

Chapter 15

FENUGREEK (*TRIGONELLA FOENUM GRAECUM*) SEEDS: IN HEALTH AND DISEASE

Subramanian Kaviarasan*
Department of Occupational and Environmental Health,
Nagoya University Graduate School of Medicine, Nagoya, Japan

ABSTRACT

Plant based medicinal research is well recognized world over as a viable healthcare component. An overwhelming body of evidence has collected in recent years to show the immense potential of the medicinal plants used in various traditional systems. Fenugreek, an annual herb of the *Leguminosae* family has been quoted in Indian, Arabic and Chinese medicine as a treatment for diabetes and as a general tonic to improve metabolism and health. This plant has received attention as an antidiabetic agent and has undergone extensive research in clinical and animal models of diabetes which have clearly documented its blood glucose lowering property. However the plant, especially the seeds, has many benefits beyond that. The present chapter aims to compile the data on wide range functional benefits of this plant demonstrated through the research activities using modern scientific approaches and innovative tools.

INTRODUCTION

Fenugreek (*Trigonella foenum graecum*) is an annual herb that belongs to the family Leguminosae. The plant, native to Southern Europe and Western Asia, is now cultivated in Argentina, France, India, North Africa, United states and in the Mediterranean countries. The name fenugreek comes from *foenum-graecum*, meaning Greek hay, as the plant was traditionally used to scent inferior hay. Fenugreek seeds are brown in color and are frequently used as a flavoring agent for artificial maple syrup [1]. Fenugreek leaves are used as

* Corresponding author: Dr. S. Kaviarasan; Department of Occupational and Environmental Health; Nagoya University Graduate School of Medicine; Nagoya, Aichi; Tel.:+81-52-744-2124; fax: +81-52-744-2126; E-mail: kavi_sing@yahoo.com

vegetable while the seeds which mature from white flowers in long pods are commonly used for flavoring and seasoning due to their strong flavor and aroma.

Fenugreek has a long history of traditional use as a medicinal herb. For centuries fenugreek has been used in folk medicine to heal aliments ranging from indigestion to baldness [2]. The seeds are assumed to possess nutritive and restorative properties and to stimulate digestive process [3, 4]. It's been used as a tonic, carminative, aperient and diuretio, useful in dropsy, chronic cough, in external and internal swellings and in hair decay [5]. The seeds are described in the Greek and Latin Pharmacopoeias to possess antidiabetic activity [6-8]. In traditional Chinese medicine, fenugreek seeds are used as a tonic, as well as a treatment for weakness and edema of the legs [9]. In India, fenugreek is commonly consumed as a condiment and is used medicinally as a lactation stimulant [10]. They are known to be important constituents of the traditional food (*Methipak*) consumed during lactation. In Egypt, fenugreek is used as a supplement in wheat and maize flour for bread-making [11] and its use is still continued. In Arabic countries the seeds are used for preparation of hot beverages after adding sugar [12, 13]. In Sudan, it is used in porridge and dessert and consumed mostly by lactating mothers [14] (Figure 1).

Figure 1. Fenugreek (*Trigonella foenum graecum*) plant and seeds.

COMPOSITION OF SEEDS

Fenugreek seeds are not cotyledenous but endospermic in nature and are rich in protein, fibre and gum. The nutrient composition of fenugreek seeds (g/100g) is moisture 2.4; protein 25.4; fat 7.9; saponins 4.8; total dietary fibre 53.9 (consisting of gum 19.0, hemicellulose 23.6, cellulose 8.9 and lignin 2.4) and ash 3.9 [15].

The protein fraction contains essential amino acids and a unique non-protein amino acid 4-hydroxyisoleucine (4-OH-Ile). The content of essential amino acids is comparable to soy protein. The amino acid content of protein fraction in fenugreek seeds [16] is given in the Table 1.

Table 1. Protein content in Fenugreek seeds

Content in g amino acid /16 g N			
Asp	11.1	Ileu	4.6
Thr	3.5	Leu	6.9
Ser	5.0	Try	2.8
Glu	17.1	Phe	4.0
Pro	4.5	His	2.4
Gly	4.6	Lys	6.2
Ala	3.6	Arg	9.4
Val	3.6	Try	0.8
Cys	1.4	OHIleu	2.7
Met	1.1		

Fenugreek seed oil has a golden yellow color, the fatty acid composition of which has been analysed. The oil contains linoleic (33.7%) oleic (35.1%) and linolenic (13.8%) acids. The percentage differs according to place and conditions of cultivation of plant [17].

Fenugreek seeds are reported to be rich in flavonoids (100mg/g) [18]. Five different flavonoid compounds namely vitexin, tricin, naringenin, quercetin and tricin-7-*O*-beta-D-glucopyranoside are reported to be present in fenugreek seeds [19]. Later seven compounds N,N' – dicarbazyl, glycerol monopalmitate, stearic acid, beta-sitosteryl glucopyranoside, ethyl-alpha- D- glucopyranoside, D-3-O-methyl-chiroinsitol and sucrose were identified by the same group [20]. HPLC analysis of aqueous extract showed the presence of gallic acid, *O*-coumaric acid, *p*-coumaric acid, rutin and caffeic acid [21].

The presence of choline, essential fatty acids, lecithin, trimethyl aminophytosterols, tannic acid, fixed and volatile essential oils and bitter extractive, diosgenin, alkaloids (trigonelline and gentianine), trigocoumarin and trigomethyl coumarin in fenugreek seeds has been reported [22, 23]. Fenugreek leaf is rich in vitamins A, D, B_1, B_2, B_3, B_6, B_{12}, biotin, pantothenic acid, folic acid, para aminobenzoic acid, choline and minerals such as iron, selenium, phosphorus and potassium [24].

APPLICATIONS TO HEALTH PROMOTION AND DISEASE PREVENTION

Fenugreek is used to treat diabetes, migraines, allergies, elevated cholesterol, epilepsy, paralysis, gout, dropsy, chronic cough, piles and constipation [25]. Researchers have convincingly shown several beneficial properties of fenugreek in experimental and human diseases. In this chapter, we discuss most, if not all, the popular available reports on fenugreek hit her to published.

HYPOGLYCAEMIC ACTIVITY

The hypoglycemic properties of fenugreek seeds have been known for many years and are well documented in experimental diabetic rats [26], dogs [27] and mice [28] and in healthy volunteers [24], in type 1[29] and 2 diabetics [30]. Both the seeds and leaves are found to possess antidiabetic property [24]. However the seeds have been subjected to extensive investigation.

In a clinical trial involving type 2 diabetic patients, administration of fenugreek seeds improved the symptoms such as polydipsia and polyuria in majority of the patients inspite of reducing the antidiabetic drug dose [31]. The improvement in clinical symptoms followed the alterations in biochemical parameters such as reduction in blood glucose level and urinary excretion of sugar.

Gupta *et al* [32] conducted a double blind placebo controlled study involving twenty-five type 2 diabetic subjects. Fenugreek considerably improved insulin sensitivity as measured by Homeostasis Model Assessment (HOMA), insulin metabolism and lipid profile. The patients received either 1 g daily of a hydroalcoholic extract of fenugreek seeds the mean fasting blood glucose levels were reduced. Patients are advised to eat less fat, but overweight or obese patients, consume a diet high in fat and low in fruits and vegetables and engage in very little physical exercise. It was suggested that fenugreek seed extract and diet/exercise may be equally effective strategies for attaining glycemic control in type 2 diabetes. Raghuram *et al* [33] reported that administration of 25 g powdered fenugreek seeds to type 2 diabetic patients improved the glucose tolerance test scores and serum-clearance rates of glucose. Neeraja and Rajyalakshmi [34] reported that fenugreek seeds lower post-prandial hyperglycemia in human subjects with diabetes.

Ground seeds of fenugreek administered to diabetic rats decreased the postprandial glucose levels [35]. Supplementation of fenugreek seeds in diets of alloxan-diabetic dogs have also confirmed this property [27]. In these studies the seeds have been shown to lower blood glucose level and partially restore the activities of key enzymes of carbohydrate and lipid metabolism close to normal values.

In human studies, fenugreek reduced the area under the plasma glucose curve and increased the number of insulin receptors [29]. Fenugreek is reported to exert its hypo-glycemic effect, by increasing the metabolic clearance rate of glucose and by delaying the gastric emptying with a direct interference with glucose absorption [24, 31, 33].

Fenugreek significantly reduced the postprandial glucose and insulin levels in non-insulin dependent diabetics [30]. In humans, fenugreek seeds exert hypoglycemic effects by stimulating glucose-dependent insulin secretion from pancreatic beta cells as well as by inhibiting the activities of alpha-amylase and sucrase, two intestinal enzymes involved in carbohydrate metabolism [38].

The identification of components responsible and the mechanism by which fenugreek exerts these effects has been of interest to researchers.

It was originally believed that major alkaloid trigonelline, found in these seeds are responsible for the hypoglycemic effect [39]. However studies by Shani [8] in diabetic rats and humans led to the supposition that the active principle of trigonella might not be trigonelline. It was suggested that nicotinic acid and coumarins present in the seeds have mild hypoglycemic action in diabetic rats.

The high level of fibre contributes to a secondary mechanism for the hypoglycemic effect of fenugreek seeds. The defatted powder of fenugreek seeds contains approximately 50 percent fiber (30 percent soluble fiber and 20 percent insoluble fiber). It is well known that gel forming dietary fibre reduces the release of insulinotropic hormones and gastric inhibitory polypeptide (GIP) and slows down the rate of post-prandial glucose absorption. Addition of fibre and gum to the diet was found to reduce the postprandial glycaemia and urinary glucose excertion in diabetic patients [40, 41]. In addition to these properties, recent research suggests that fenugreek gum may also be surface active. Garti *et al* [42] found that stable emulsions with a relatively small droplet size (3 μm) could be formed using purified fenugreek gum.

Much work on 4-hydroxyisoleucine (4-OH-Ile), a unique amino acid found in fenugreek seeds and its hypoglycemic properties have been carried out. 4-OH-Ile, extracted and purified from fenugreek seeds, displayed an insulinotropic property *in vitro*, stimulated insulin secretion *in vivo* and improved glucose tolerance in normal rats and dogs and in rat model of type 2 diabetes mellitus [43, 44]. The effect of 4-OH-Ile was both dose and glucose-dependent, and was shown to stimulate insulin secretion as a result of direct β-cell stimulation only under the hyperglycemic condition [44]. Besides 4-OH-Ile, arginine and tryptophan are the other amino acids in fenugreek seeds that possess antidiabetic and hypoglycemic effect. Many trace elements, which are the components of fenugreek, have antidiabetic effects [45].

Recently the seeds have been subjected to detailed scientific investigation by mechanism based *in vitro* and *in vivo* assays to elucidate the action at the cellular and molecular level [46]. An aqueous extract of fenugreek seeds was dialysed for 24 h to eliminate small molecules and the hypoglycemic potential of this dialysed extract (FSE) was investigated *in vivo* in alloxan-induced diabetic mice. FSE significantly improved glucose homeostasis in these diabetic mice and in normal glucose-loaded mice by effectively lowering blood glucose levels. This effect of FSE on glucose levels was found to be comparable to that of insulin.

FSE stimulated insulin signaling pathways in adipocytes and liver cells. FSE induced a rapid, dose-dependent stimulatory effect on cellular glucose uptake by activating cellular responses that lead to glucose transporter type 4 (GLUT4) translocation to the cell surface. The investigators suggested that fenugreek might act independently of insulin to enhance glucose transporter-mediated glucose uptake. FSE activated the tyrosine phosphorylation of insulin receptor beta (IR-β), subsequently enhancing tyrosine phosphorylation of insulin receptor substrate 1 (IRS-1) and the p85 subunit of Phosphatidylinositol 3-kinases (PI-3 kinase), but had no effect on basal insulin receptor alpha (IR-α) phosphorylation. This suggests that adipocytes and liver cells could be target sites for FSE and that it exerts its effects by activating insulin signaling pathways. The authors also suggested that the seeds are capable of specifically activating the insulin receptor (IR) and its downstream signaling molecules in adipocytes and liver cells and do not act as general sensitizer of receptor tyrosine kinase domains [46].

Fenugreek seeds might have an effect on the glyoxylase system and a link between antidiabetic action and glyoxylase system has been suggested [47]. Fenugreek (at 1% and 2% in diet for mice) enhanced glyoxylase I system, prevented accumulation of methyl glyoxal, a key player in advanced glycation end products (AGE) formation, oxidative stress and diabetic complications [48].

Hypolipidemic Activity

Fenugreek seeds also possess hypocholesterolemic effect. The active component is found to be associated with the defatted part. Fenugreek seeds influence plasma lipid profile and tissue lipid levels both in human diabetics and experimental animals. For instance, Sharma [29] has shown a decrease in total LDL and VLDL cholesterol and triglycerides in both insulin dependent and non-insulin dependent diabetics by the administration of fenugreek seeds.

In another study Sharma *et al* [49] investigated the hypolipidemic effect in 15 nonobese, asymptomatic, hyperlipidemic adults. After supplementation of 100 g defatted fenugreek powder per day for three weeks, triglyceride (TG) and LDL-C levels were lower than baseline values. Slight decreases in HDL-C levels were also noted. In another study, A decrease in total cholesterol levels in five diabetic patients treated with fenugreek seed powder (25 g per day oral) for 21 days was also observed [50].

Bordia *et al* [51] reported that fenugreek seed powder (2.5 g twice daily for three months) showed significant decrease in the TC and TG levels, with no change in HDL-C level in a subgroup of 40 subjects who had coronary artery disease and type 2 diabetes.

The active components of fenugreek seeds has been found to be associated with the defatted part, rich in fibre containing steroid saponins and protein [16]. The gum isolate constituting 19.2% of defatted part has strong hypocholesterolemic activity. The galacto-mannans present in the gum could increase the viscosity of digesta which may inhibit the absorption of cholesterol from the small intestine and also the reabsorption of bile acids from terminal ileum resulting in a decrease in serum cholesterol. The absorbed bile acids would be lost by faecal excretion and this loss would be offset by conversion of cholesterol into bile acids by liver.

Saponins present in the defatted part (4.8%) reduced hypercholesterolemia and hypertriglyceridemia of alloxan-diabetic dogs [52]. Saponins present in fenugreek could get transformed in the gastrointestinal tract into sapogenins. It is further suggested that saponins and cholesterol form insoluble complexes and retard cholesterol absorption in the intestine [53]. The hypocholesterolemic effect was not observed with the saponin-free subfractions.

The lipid-lowering effect of fenugreek might also be attributed to its estrogenic constituent, indirectly increasing thyroid hormone T4. The flavonoids present in the seeds may also be responsible for these activities.

PHARMACOLOGICAL ACTIVITIES

Immunomodulatory and Anti-Inflammatory Effects

Total aqueous extract of fenugreek plant has immune potentiating functions when administered to Swiss albino mice at different-doses [54]. The extract exhibited stimulating effect on macrophages and showed a positive effect on specific and non-specific immune functions of the lymphoid organs such as thymus, bone marrow, spleen and liver. High quantity of mucilage and iron in the organic form in fenugreek seeds can facilitate stimulation of macrophages and the hemopioetic system.

Ahamadini *et al* (2001) [55] evidenced the anti-inflammatory activity of fenugreek leaves that could inhibit the rat paw inflammation induced by formalin. They also reported the antipyretic effect of leaves extract as demonstrated by the reduction of yeast-induced hyperthermia. In their study the effect of leaves was more potent than that of sodium salicylate and was observed on single and chronic administration.

Antinociceptive Effect

Reports stating that the antinociceptive effects of the plant in Iranian folk medicine which suggest the use of fenugreek leaves as pain killers [56, 57]. Javan *et al.*(1997) [58] reported that the leaf extract produces antinociceptive effect in a dose-dependent manner in both tail flick and formalin test models and proposed that its activity results from both central and peripheral actions of the active constituents.

Inotropic and Cardiotonic Effect

Ion pump Na^+/K^+- ATPase is an ubiquitous membrane protein and is an active transporter of Na^+ and K^+ ions across the cell. Polyphenol fractions from fenugreek seeds dose-dependently inhibited the erythrocyte membrane Na^+/K^+- ATPase activity [59]. While going through the mechanism of action of polyphenols it was suggested that decreased activity of ATPase might be due to the conformational changes in the structure of the enzyme brought about by the extract. Flavones and flavonols containing hydroxyl groups inside the phenyl radical in ortho- and vicinal positions exhibit high inhibitory effects on the membrane enzymes. The presence of dimethyl allyl groups in flavonones leads to the inhibitory activity [60]. Further the polyphenolic structures of flavonoids similar to cholesterol partition into hydrophobic core of the membrane and cause modulation in lipid fluidity [61]. This could sterically hinder diffusion of ions and other transport processes.

Quercetin, a flavonol, richly present in fenugreek seeds produces inotropism in the frog and rabbit heart and pig kidney medulla [62]. Quercetin inhibits Na^+/K^+- ATPase pump and stimulates β-adreno receptors involving adenylate cyclase- cAMP system ultimately increasing the availability of calcium from intracellular sites. These findings suggest that fenugreek has a positive inotropic effect due to the presence of quercetin and may be potentially useful as cardiotonic agent.

Antitumour Effects

Most of the anti-inflammatory agents of plant origin have antitumour activity and it could be expected that fenugreek seeds exhibit antineoplastic effects. Sur *et al.* [63] evaluated the antineoplastic activity of the seeds by studying the inhibition of tumor cell growth in Ehrlich ascites carcinoma (EAC) model *in vivo* in mice by fenugreek seed extract. The alcoholic extract showed 70% inhibition of tumor cell growth and enhanced peritoneal exudation and macrophage cell count indicating activation of macrophages and anti-inflammatory action.

Epidemiological evidences suggest that dietary spices and fibre prevent colon carciniogenesis. Flavonoids and fibre can potentially act as anticarcinogenic agents by binding to free carcinogens and/or carcinogenic metabolites. This prevents their access to colonic mucosa and enhances fecal excretion [64, 65]. Hydrolyses of the protective mucin coat in the intestinal wall by Microfloral enzyme, Mucinase results in exposing mucosa to carcinogens that are released by β -glucuronidase activity. Thus, gut microfloral β -glucuronidase and mucinase activity were increased are considered precancer events associated with increased risk of colon cancer. The high content of indigestible polysaccharides in the seeds can act as substrates to mucinase and this also helps to prevent the hydrolysis/degradation of mucin, thus contributing to the anticarcinogenic potential of the seeds [66]. Futhermore, the modulation of cholesterol and phospholipids metabolism in target organs by the seeds could in turn prevent 1,2-dimethyl hydrazine-induced colon cancer [67].

Antioxidant Activity

Fenugreek appears to have antiradical property and ability to prevent lipid peroxidation [68]. Supplementation studies have shown that dietary fenugreek seed can lower blood and tissue lipid peroxidation and augument the antioxidant potential in alloxan diabetic rats [69] and experimental colon cancer [70].

An aqueous extract of the seeds inhibited the lipid peroxidation stimulated by Fe^{2+}-ascorbate system and by glucose *in vitro* in rat liver homogenate [71]. The aqueous extract inhibits lipid peroxidation *in vitro* in presence of promoters of lipid peroxidation and the effect was comparable with α-tocopherol and reduced glutathione [72]. A polyphenol-rich extract of fenugreek seed protect erythrocytes from peroxide-induced oxidative hemolysis and lipid peroxidation [73].

Dixit *et al* [21] have pointed out that germinated fenugreek seeds have several beneficial properties over ungerminated. Germination improves *in vitro* protein digestibility, as well as fat absorption capacity [74] and decreases the levels of total unsaturated fatty acids, total lipid, triglycerides, phospholipids and unsaponifiable matter while saturated fatty acids are increased. Dixit *et al* [21] examined the antioxidant activities of different fractions from the powder of germinated seeds of fenugreek as well as that of two of its active chemical constituents, namely trigonelline and diosgenin. The aqueous extract of fenugreek had a high phenolic and flavonoid content. These substances exhibit high antioxidant capacity in terms of scavenging H_2O_2 and increased 2,2-diphenyl-1-picrylhydrazyl (DPPH) and 2,2'-azino-bis(3-ethylbenzthiazoline-6-sulphonic acid) (ABTS) radicals, inhibition of lipid peroxidation and inhibition of oxidation of β-carotene by hydroxyl free radical. Germinated fenugreek seeds showed a good antioxidant potential in terms of Oxygen Radical Absorbance Capacity (ORAC) values, ABTS and Fluorescence recovery after photobleaching (FRAP).

Dietary administration of fenugreek seeds resulted in an increase of glutathione (GSH) levels glutathione-S-transferase (GST) activity in liver of mice fed 1%, 2% and 5% fenugreek seeds [48].

The mode and magnitude of effect seems to depend on the dose of fenugreek and the tissue studied. Fenugreek seeds for instance appear to have a dual effect on the tissues. The antioxidant status is enhanced at lower levels (1% and 2% of fenugreek) while its pro-oxidant action is observed at high dose (5% and 10% of fenugreek). Enhancement of antioxidant status (both enzymic and non-enzymic) in blood of normal and alloxan diabetic animals have been reported by administration of 2% dose of fenugreek [48].

Therefore fenugreek especially the seeds can be considered as health food and can be used as supplement in the possible prevention of many human diseases in which free radicals are implicated.

Alcohol Toxicity

Administration of an aqueous extract of fenugreek seeds augmented the status of ethanol-induced damage in experimental rats [72]. Treatment with extract reduced fatty changes and portal inflammation in liver of ethanol-treated rats and there was a reduction in spongiosis in brain. This study point out that the seeds can confer protection against ethanol toxicity by enhancing antioxidant potential. It has been reported that fenugreek seed polyphenolic extract protected Chang liver cells (normal hepatic cell line) against ethanol-induced cytotoxicity and apoptosis [75]. The presence of flavonoids in the seeds could be responsible for cytoprotection. The polyphenolic extract significantly increased cell viability, prevented oxidative damage, redox changes and LDH enzyme leakage.

Drug Metabolism

The seeds of fenugreek are found to modulate the detoxication process. Fenugreek seeds when fed through diet generally stimulated the hepatic mixed function oxygenase system (MFOS), the cytochrome P_{450} –dependent aryl hydroxylase, cytochrome P_{450} and b_5. The stimulation of the hepatic xenobiotic metabolism under normal conditions by fenugreek may be implicated in its potential as a hepatoprotective /detoxifying agent [76].

Gastrointestinal Function

In the Indian system of medicine the fenugreek seeds have been used to treat a number of gastrointestinal disorders and have been well recognized to stimulate digestion. They also possess carminative, tonic and galactogogue properties. It has been used to check dysentery, diarrhea and dyspepsia with loss of appetite. A study by Platel and Srinivasan [77] revealed that diet containing 2% fenugreek seed improved the intestinal function by enhancing the activities of terminal enzymes of digestive process such as lipases, sucrase and maltase. They are supposed to stimulate appetite and improve feeding behaviour through an endocrine response [78].

Safety/Adverse Effects and Drug Interactions

Traditionally fenugreek seeds have been considered safe and well tolerated. The dose used in several diabetic clinical studies varied from 2.5 to 15gm daily in powdered or defatted form. About one gram to three gram mixed with food and taken at meal time appears to be the suggested dose for diabetics. Acute, subchronic and long term studies in animals and normal subjects have not indicated hepatic or renal toxicity or hematological abnormalities and is well tolerated upto 10% dietary level [79]. In an animal study, the acute oral LD50 was found to be >5 g/kg in rats, and the acute dermal LD50 was found to be >2 g/ kg in rabbits [80].

Some side effects such as GI tract upset, transient diarrhea and flatulence and dizziness [81] have been associated with its use. Use of fenugreek is warranted in patients who are known to be allergic to it or who are allergic to chickpeas because of possible cross-reactivity [82]. Fenugreek contained in curry powder was found to be an allergen in a patient who reported severe bronchospasm, wheezing, and diarrhea [83]. It should not be used during pregnancy because of its uterine stimulation properties [84]. Reduction in the serum levels of potassium, T_3, T_3/T_4 ratio, and increase in the levels of T_4 has been observed in fenugreek-fed mice and rat models possibly through its estrogenic activity [85]. The high fiber content of the seeds can interfere with the absorption of oral medication and should be taken 2 hours after other medication [86].

It has been concluded here that from the above chapter it's clear that fenugreek possesses numerous potential for improving health and preventing disease progress. It's strongly supported to be a right supplement in diet to help with variety of health conditions.

References

[1] Natarajan, C.P. and Shankaracharg, N.B. Chemical composition of raw and roasted fenugreek seeds. *J. Food Sci. Tech.*1973, 10, 179-181.

[2] Fazli, F.R.Y. and Hardman, R. The spice fenugreek (Trigonella foenum-graecum L.). Its commercial varieties of seed as a source of diosgenin. *Trop. Sci.* 1968, 10, 66-78.

[3] Moissides, M. Le fenugreec autrifois et aujourd'hui. *Janus.* 1939, 43, 123-130.

[4] Sorengarten, FJr. The book of spices. Wynnewood, PA: Living ston Publishing Company. 1969, 250-3.

[5] Watt, GA. Dictionary of the Economic Products of India (Periodical Experts, Delhi), 1972, p.86.

[6] Loeper,M. and Lemaire, A. Sun quelques points delaction generale des amers. *Presse. Med.* 1931, 24, 433-35.

[7] Bever,B.O. and Zahnd, G.R. Plants with oral hypoglycemic action.*Quart. J. Crude. Drug Research.* 1979, 17, 139-196.

[8] Shani,J. Goldschmied, A. Joseph,B. Ahronson, Z. Sulman, F.G. Hypoglycaemic effect of trigonella foenumgraecum and lupinus termis (Leguminosae) seeds and their major alkaloid in alloxan diabetic and normal rats. *Arch. Int. Pharmacodyn. Ther.* 1974, 210, 27-36.

[9] Yoshikawa, M. Murakami, T. Komatsu, H. Murakami, N. Yamahara, J. Matsuda, H. Medicinal foodstuffs. IV. Fenugreek seed. (1): Structures of trigoneosides Ia, Ib, IIa, IIb and IIIb, new furostanol saponins from the seeds of Indian Trigonella foenum-graecum L. *Chem. Phar. Bull.* 1997, 45, 81-7.

[10] Mital, N. and Gopaldas, T. Effect of fenugreek (Trigonella foenum granum) seed based diet on the birth outcome in albino rats. *Nutr. Rep. Int.* 1986, 33, 363-69.

[11] Morcos, S.R. Elhawary, Z. Gabrial, G.N. Protein-rich food mixtures for feeding the young in Egypt. 1. formulation. *Z.Ernahrungswiss.* 1981, 20, 275-82.

[12] Elmadfa, I. On the Fenugreek Protein Trigonella-Foenum-Graecum. *Nahrung.* 1975, 19, 683-6.

[13] Elmadfa, I and Kuhl, B.E. The quality of fenugreek seeds protein tested alone and in a mixture with corn flour. *Nutr. Rep. Int.* 1976, 14, 165-72.

[14] Abuzied, A.N. Al-hilba (fenugreek) in: Al-nabatat Wa-alaashab al-tibeeva Das Albihar Beirut.1986, 223-224. Leabanon (in Arabic).

[15] Shankaracharya, N.B. and Natarajan, C.P. Fenugreek-chemcial composition and use. *Indian Spices.* 1972, 9, 2-12.

[16] Valette, G. Sauvaire, Y. Baccou, J.C. Ribes, G. Hypocholesterolaemic effect of fenugreek seeds in dogs. *Atherosclerosis.* 1984, 50, 105-111.

[17] Badami, R.C. and Kalburgi, G.S.. Component acids of Trigonella foenum graecum (fenugreek) sed oil. The Karnatak University Journal Science XIV.1969, 16-19.

[18] Gupta, R. and Nair, S. Antioxidant flavonoids in common Indian diet. *South Asian J. Prev. Cardiol.* 1999, 3, 83-94.

[19] Shang, M. Cai, S. Han, J. Li, J. Zhao, Y. Zheng, J. Namba, T. Kadota, S. Tezuka, Y. Fan, W. Studies on flavonoids from fenugreek (*Trigonella foenum graecum* L). *Zhongguo Zhong Yao Za Zhi.* 1998, 23, 614-616.

[20] Shang, M. Cai, S. Wang, X. Analysis of amino acids in Trigonella foenumgraecum seeds. *Zhong Yao Cai.* 1998, 21,188-90.

[21] Dixit, P. Ghaskadbi, S. Mohan, H. Devasagayam, T.P.A. Antioxidant properties of germinated fenugreek seeds. *Phy. Res.* 2005, 19, 977-983.

[22] Petit, P.R. Sauviaire, Y.D. Hillaire-Buys, D.M. Leconte, O.M. Baissac, Y.G. Ponsin, G.R. Ribes, G.R.. Steroid saponins from fenugreek seeds: extraction, purification and pharmacological investigation on feeding behaviour and plasma cholesterol. *Steroids.* 1995, 10, 674-680.

[23] Jayaweera, D.M.A. Medicinal plant: Part III. Peradeniya, Srilanka: Royal Botanic Garden.1981, p.225.

[24] Sharma, R.D. Effect of fenugreek seeds and leaves on blood glucose and serum insulin responses in human subjects. *Nutr. Res.* 1986, 6, 1353-1364.

[25] CCRUM Standardisation of single drugs of Unani medicine: Part I. New Delhi: Central Council for Research in Unani medicine, Ministry of Health and Family Welfare, Government of India.

[26] Madar, Z. Fenugreek (*Trigonella foenum graceum*) as a means of reducing post prandial glucose level in diabetic rats. *Nutr. Rep. Int.* 1984, 29, 1267-1273.

[27] Ribes, G. Sauvaire, Y. Costa, C.D. Baccou, J.C. Mariani L.M.M. Antidiabetic effects of sub fractions of fenugreek seeds in diabetic dogs. *Proc. Soc. Exp. Biol. Med.* 1986,182, 159-166.

[28] Ajabnoor, M.A.and Tilmisany, A.K. Effect of *Trigonella foenum graceum* on blood glucose levels in normal and alloxan-diabetic mice. *J. Ethnopharmcol.* 1988, 22, 45-49.

[29] Sharma, R.D. Raghuram, T.C. and Rao, N.S. Effect of fenugreek seeds on blood glucose and serum lipids in type 1 diabetes. *Eur. J. Clin. Nutr.* 1990, 44, 301-306.

[30] Madar, Z. Abel, R. Samish, S. Arad, J. Glucose-lowering effect of fenugreek in non-insulin dependent diabetics. *Eur. J. Clin. Nutr.* 1988, 42, 51-54.

[31] Sharma, R.D. and Raghuram T.C. Hypoglycemic effect of fenugreek seeds in non-insulin depenedent diabetic subjects. *Nutr. Res.* 1990, 10, 731-739.

[32] Gupta, A. Gupta, R. Lal, B. Effect of *Trigonella foenum-graecum* (fenugreek) seeds on glycaemic control and insulin resistance in type 2 diabetes mellitus: a double blind placebo controlled study. *J. Assoc. Phy. Ind.* 2001, 49, 1057-1061.

[33] Raghuram,T.C. Sharma, R.D. Sivakumar, B. Sahay, B.K. Effect of fenugreek seeds on intravenous glucose disposition in non-insulin dependent diabetic patients. *Phytother. Res.* 1994, 8, 83-86.

[34] Neeraja, A. and Rajalakshmi, P. Hypoglycemic effect of processed fenugreek seeds in humans. *J. Food Sci. Techol.* 1996, 33, 427-430.

[35] Vats, V. Grover, J.K. Rathi, S.S.. Evaluation of anti-hyperglycemic and hypoglycemic effect of *Trigonella foenum graceum* Linn. *Ocimum sanctum* Linn and *Pterocarpus mersupium* Linn in normal and alloxan diabetic rats. *J. Ethanopharmacol.* 2002, 79, 95-100.

[36] Moorthy, R. Prabhu, K.M. Murthy, P.S. Studies on the isolation and effect of an orally active hypoglycemic principle from the seeds of fenugreek (*Trigonella foenum graecum*). *Diab. Bull.* 1989,.9, 69-72.

[37] Sharma, R.D. Sarkar, A. Hazra, D.K. Mishra, B. Singh, J.B. Sharma, S.K. Maheshwari, B.B. Maheshwari, P.K. Use of fenugreek seed powder in the management of non-insulin dependent diabetes mellitus. *Nutr. Res.* 1996, 16, 1331-1339.

[38] Amin, R. Abdul-Ghani, A.S. Suleiman, M.S.. Effect of *Trigonella foenum graecum* on intestinal absorption. Proc. of the 47th Annual Meeting of the American Diabetes Association (Indianapolis U.S.A). *Diabetes.* 1987, 36, 211a.

[39] Mishkinsky, S.J. Joseph, B. Sulman, F.G. Hypoglycemic effect of trigonelline. *Lancet.* 1967, 2, 1311-1312.

[40] Jenkins, D.J.A. Wolever, T.M.S. Taylor, R.H. Reynolds, D. Nineham, R. Hockaday, T.D.R.. Diabetic glucose control, lipids and trace elements on long term guar. *Br. Med. J.* 1980, 280, 1353-1354.

[41] Anderson, J.W.. High-fibre diets for obese diabetic men on insulin therapy: Short-term and long term effects: In: Bjoerntorp P, Vahouny GV, Kritchevsky D, eds. *Current topics in nutrition and disease.* New York: Alan R Riss Inc, 1985, 14, 49-68.

[42] Garti, N. Madar, Z. Aserin, A. and Sternheim, B.. Fenugreek galactomannans as food emulsifiers. *Lebensmittel-Wissenschaft und Technologie,* 1997, 30, 305–311.

[43] Sauvaire, Y. Petit, P. Broca, C. Manteghetti, M. Baissac, Y. Fernandez-Alvarez, J. Gross, R. Roye, M. Leconte, A. Gomis, R. Ribes, G.. 4-Hydroxyisoleucine: a novel amino acid potentiator of insulin secretion. *Diabetes.*1998, 47, 206-210.

[44] Broca, C. Gross, R. Petit, P. Sauvaire, Y. Manteghetti, M. Tournier, M. Masiello, P. Gomis, R. Ribes, G.. 4-Hydroxyisoleucine: experimental evidence of its insulinotropic and antidiabetic properties. *Am. J. Physiol.*1999, E617-E623.

[45] Broca, C. Manteghetti, M. Gross, R. Baissac, Y. Jacob, M. Petit, Y. Sauvaire, Y. Ribes, G.. 4-Hydroxyisoleucine: effects of synthetic and natural analogues on insulin secretion. *Eur. J. Pharmacol.* 2000, 390 (3), 339–345.

[46] Vijayakumar, M.V. Sandeep, S. Chhipa, R.R. Manoj, K.B. The hypoglycemic activity of fenugreek seed extract is mediated through the stimulation of an insulin signaling pathway. *Br. J. Pharmacol.* 2005, 146: 41-48.

[47] Raju, J. Gupta, D. Rao, A.R. Baquer, N.Z. Effect of antidiabetic compounds on glyoxylase I activity in experimental diabetic rat liver. *Ind. J. Exp.Biol.* 1999, 37, 193-195.

[48] Choudhary, D. Chandra, D. Choudhary, S. Kale, R.K. Modulation of glyoxylase, glutathione-S-transferase and antioxidant enzymes in the liver, spleen and erythrocytes

of mice by dietary administration of fenugreek seeds. *Food Chem. Toxicol.*2001, 39, 989-997.

[49] Sharma, R.D. Raghuram, T.C. Dayasagar Rao, V. Hypolipidaemic effect of fenugreek seeds. A clinical study. *Phytother. Res.* 1991, 3, 145-147.

[50] Sharma, R.D. Sarkar, A. Hazra, D.K.. Hypolipidaemic effects of fenugreek seeds: A chronic study in non-insulin dependent diabetic patient. *Phytother. Res.* 1996, 10, 332-334.

[51] Bordia, A. Verma, S.K. Srivastava, K.C. Effect of ginger (*Zingiber officinale Rosc.*) and fenugreek (*Trigonella foenum graecum L.*) on blood lipids, blood sugar and platelet aggregation in patients with coronary artery disease. *Prost. Leu. Ess. Fatty Acids.* 1997, 56,379-384.

[52] Sauvaire, Y. Ribes, G. Baccou, J. Loubafieres, M. Implication of steroid-saponins and sapogenins in the hypocholesterolemic effect of fenugreek. *Lipids.* 1991, 26, 191-197.

[53] Stark, A.and Madar, Z.. The effect of an ethanol extract derived from fenugreek (*Trigonella foenum-graecum*) on bile acid absorption and cholesterol levels in rats. *Br. J. Nutr.* 1993, 69, 277-287.

[54] Hafeez, B.B. Haque, R. Parvez, S. Pandey, S. Sayeed, I. Raisuddin, S. Immunomodulatory effects of fenugreek (*Trigonella foenum graecum* L) extract in mice. *J. Ethanopharmacol.* 2003, 3, 257-265.

[55] Ahmadiani, A. Javan, M. Semnanian, S. Barat, E. Kamalinejad, M. Anti-inflammatory and antipyretic effects of *Trigonella foenum greaecum* leaves extract in the rat. *J. Ethanopharmacol.* 2001, 75, 283-286.

[56] Zargari, A. Medicinal plant. Vol. I. Tehran University Press. Iran.1989, pp.637-639.

[57] Mirhaydar, H. Plant information. Vol I. Nashre Farhang Eslami Press, Iran, 1993, p.145.

[58] Javan, M. Ahmadiani, A. Semnanian, S. Kamalinejad, M. Antinociceptive effects of *Trigonella foenum greaecum* leaves extract. *J. Ethanopharmcol.* 1997, 58, 125-129.

[59] Anuradha, C.V. Kaviarasan, S. Vijayalakshmi, K.. Fenugreek seed polyphenols inhibit RBC membrane Na^+/K^+-ATPase activity. *Orient. Phar. Exp. Med.* 2003, 3, 129-133.

[60] Umarova, F.T. Khushbactova, Z.A. Batirov, E.H. Mekler, V.M. Inhibition of Na+/K+-ATPase by flavonoids and their inotropic effect. Investigation of the structure-activity relationship. *Cell. Biol.* 1998, 12, 27-40.

[61] Arti, A. Byren, T.M. Nair, M.G. Strasburg, G.M. Modulation of liposomal membrane fluidity by flavonoids and isoflavonoids. *Arch. Biochem. Biophys.* 2000, 373, 102-109.

[62] Bhansali, B.B. Vyes, S. Goyal, R.K. Cardiac effects of quercetin on isolated rabbit and frog heart preparations. *Ind. J. Pharmacol.* 1989, 19, 100-107.

[63] Sur, P. Das, M. Gomes, A. Vedasiromoni, J.R. Sahu, N.P. Banerjee, S. Sharma, R.N. Ganguly, D.K. Trigonella foenum greaecum (Fenugreek) seed extract as an antineoplastic agent. *Phy. Res.* 2001, 15, 257-259.

[64] Bobek, P. and Galbavy, S.. Influence of insulin on dimethylhydrazine induced carcinogenesis and antioxidant enzymatic system in rat. *Biol. Bratislava.* 2001, 56, 287-291.

[65] Fujiki, H. Horiuchi, T. Yamashita, K.. Inhibition of tumor promotion by flavonoids. *Prog. Clin.Biol. Res.*1986, 213, 429-440.

[66] Devasena, T. and Menon, V.P.. Fenugreek affects the activity of b-glucuronidase and mucinase in the colon. *Phy. Res.* 2003, 17, 1088-1091.

[67] Devasena, T. Gunasekaran, G. Viswanathan, P. Menon, V.P.. Chemoprevention of 1,2-dimethylhydrazine-induced colon carcinigenesis by seeds of *Trigonella foenum greaecum* L. *Biol. Bratislava.* 2003, 58, 357-364.

[68] Kaviarasan, S. Naik, G.H. Gangabhagirathi, R. Anuradha, C.V. Priyadarsini, K.I.. In vitro studies on antiradical and antioxidant activities of fenugreek (*Trigonella foenum graecum*) seeds. *Food Chemistry.* 2007, 103, 31-7.

[69] Anuradha, C.V. and Ravikumar, P. Effect of fenugreek seeds on blood lipid peroxidation and antioxidants in diabetic rats. *Phyto. Res.* 1999, 13, 197-201.

[70] Devasena, T. and Menon, V.P. Enhancement of circulatory antioxidants by fenugreek during 1,2-dimethylhydrazine-induced rat colon carcinogenesis. *J. Biochem. Mol. Biol. Biophys.* 2002, 6, 289-292.

[71] Anuradha, C.V. and Ravikumar, P. Anti-lipid peroxidative activity of seeds of fenugreek (*Trigonella foenum graecum*). *Med. Sci. Res.* 1998, 26, 317-321.

[72] Thirunavukkarasu, V. Viswanathan, P. Anuradha, C.V.. Protective effect of fenugreek (*Trigonella foenum graecum*) seeds in experimental ethanol toxicity. *Phy. Res.* 2003, 17, 737-743.

[73] Kaviarasan, S. Vijayalakshmi, K. and Anuradha, C.V. A polyphenol-rich extract of fenugreek seeds protect erythrocytes from oxidative damage. *Plant Food Hum. Nutr.* 2004, 59, 143-147.

[74] Mansour, E.H. El-Adawy, T.A. Nutritional potential and functional properties of heat treated and germinated fenugreek seeds. *Lebensm Wiss Technol.* 1994, 27, 568-572.

[75] Kaviarasan, S. Nalini. R. Gunasekaran, P. Varalakshmi, E. Anuradha, C.V.. Fenugreek (*Trigonella foenum graecum*) seed extract prevents chang liver cells against ethanol-induced toxicity and apoptosis. *Alcohol Alcohol.* 2006, 41(3), 267-73.

[76] Kaviarasan, S. Nalini, R. Gunasekaran, P. Varalakshmi, E. Anuradha, C.V.. Induction of alcohol-metabolizing enzymes and heat shock protein expression by ethanol and modulation by fenugreek seed polyphenols in Chang liver cells. *Toxicol. Mech. Methods.* 2009, 19,116-22.

[77] Platel, K. and Srinivasan, K.. Influence of dietary spices or their active principles on digestive enzymes of small intestinal mucosa in rats. *Int. J. Food. Sci. Nutr.* 1996, 47, 55–59.

[78] Petit, P. Sauvaire, Y. Ponsin, G. Manteghetti, M. Fave, A. Ribes, G.. Effects of fenugreek seed extract on feeding behaviour in the rat: metabolic-endocrine correlates. *Pharm. Biochem. Behav.* 1993, 45, 369–374.

[79] Muralidhara, K. Narasimhamurthy, S. Viswanatha, B.S. Ramesh. Acute and subchronic toxicity assessment of debitterized fenugreek powder in the mouse and rat. *Food Chem. Toxicol.* 1999, 37, 831-838.

[80] Opdyke, D.L.. Fenugreek absolute. *Food Cosmet. Toxicol.* 1978, 16, S755-S756.

[81] Abdel-Barry, J.A. Abdel-Hassan, I.A. Al-Hakiem, M.H. Hypoglycemic and antihyperglycemic effects of *Trigonella foenum graceum* leaf in normal and alloxan induced diabetic rats. *J. Ethanopharmacol.* 1997, 58, 149-155.

[82] Patil, S.P. Niphadkar, P.V. Bapat, M.M.. Allergy to fenugreek (*Trigonella foenum graecum*). *Ann. Aller. Asth. Immunol.* 1997, 78, 297-300.

[83] Ohnuma, N. Yamaguchi, E. Kawakami, Y.. Anaphylaxis to curry powder. *Allergy.* 1998, 53, 452-454.

[84] Abdo, M.S. al-Kafawi, A.A.. Experimental studies on the effect of *Trigonella foemun graecum*. *Planta. Med.* 1969, 17, 14-18.
[85] Panda, S. Tahiliani, P. Kar, A.. Inhibition of triiodothyronine production by fenugreek seed extract in mice and rats. *Pharmacol. Res.* 1999, 40, 405-9.
[86] Blumenthal M, Busse WR, Goldberg A, Gruenwald J, Hall T, Riggins CW, Rister RS (eds). The complete German commission E Monographs: Therapeutic Guide to Herbal Medicine. Klein S, trans. Austin, Tex. American Botanical Council: 1998.

In: Natural Products and Their Active Compounds ... ISBN: 978-1-62100-153-9
Editors: M. Essa, A. Manickavasagan, and E. Sukumar © 2012 Nova Science Publishers, Inc.

Chapter 16

NATURAL COMPOUNDS AS STEM CELL STIMULATORS IN REGENERATIVE AND CELL-BASED THERAPIES

***Senthil Kumar Pazhanisamy*[1,*], *Annamalai Prakasam*[2] *and Vinu Jyothi*[3]**
[1]Baylor College of Medicine, Houston, TX, US
[2]Medical University of South Carolina, Charleston SC, US
[3]School of Public Health, The University of Texas Houston, TX, US

ABSTRACT

Adult stem cells hold great promise for the treatment of a spectrum of disorders including chronic, degenerative and malignant diseases. A growing body of evidence indicates that natural compounds with potent stem cell stimulatory mechanisms are invaluable for therapeutic strategies as an alternative or adjuvant to tissue transplantation therapy. Although many herbal stem cell stimulators have been identified, comprehensive knowledge on the development of natural therapeutics is fairly limited. Moreover, in spite of recent developments, underlying cellular and molecular mechanisms by which certain natural products enhance the self-renewal and proliferation of tissue stem cells are still poorly understood. Unraveling the mechanisms of herbal stem cell stimulators provide promising therapeutic approaches for spectrum of disorders. This chapter illustrates emerging trends in the identification and characterization of medicines derived from natural components in regenerative medicine and stem cell-based therapies. Our main intention in this chapter is to highlight the enormous potential of herbal stem cell stimulators that remain unexplored despite their substantial medicinal values.

[*] Corresponding author: E-mail:scendhil@gmail.com or senthilp@bcm.edu

Keywords: Stem cell stimulators; Natural compounds; Herbal medicine; regenerative medicine

INTRODUCTION

The chemotherapeutic approaches currently available for most debilitating disorders including cancer, diabetes, stroke, amyotrophic lateral sclerosis (ALS), Parkinson's disease and premature aging disorders either have significant long-term side effects or fail to provide holistic treatment. Stem cells, with their unique self-renewal and proliferation characteristics, not only revolutionized the way treatment strategies are designed but also paved the way for innovative cell based therapies in regenerative medicine [1-4]. The exquisite feature of pluripotent stem cells lie in their ability to differentiate and give rise to any cell type under specific circumstances [5]. Interestingly, the regenerative and tissue repair potential of stem cells are not restricted to their local milieu but also to tissues of distal organs via pro-inflammatory cytokines and growth factors [6]. And yet, translating the stem cell potential into clinical benefits encounters enormous challenges. These include poor engraftment of stem cells, transplantation complications, defective differentiation and high therapeutic cost. Regardless of their current clinical usage, challenging issues still remain: transplantation of stem cells remains a risky treatment as it is accompanied by high treatment-related mortality caused by veno-occlusive disease, mucositis, infections (sepsis), Graft-versus-tumor effect (GVT) and graft-versus-host disease [7]. Besides, transplantation techniques pose major immunoreactivity problems to the host body system. Stimulation or activation of endogenous quiescent stem cells is a viable alternative option to transplantation issues. An array of factors including haematopoietic growth factors and cytokines such as interleukins, granulocyte-colony stimulating factor (G-CSF), granulocyte, macrophage-colony stimulating factor (GM-CSF), stem cell factor (SCF) were identified to stimulate the hematopoietic stem cell system [8]. However, studies indicate potential disadvantages including clinical risks and high costs associated with these factors. Use of chemotherapy to stimulate stem cells, on the other hand, not only fail to treat certain disorders but also result in treatment-induced secondary disorders as seen in chemotherapy-induced secondary tumors [9-11]. Arguably, chemical therapy lacks holistic approach as they often restricted to alleviating the symptoms rather than eradicating disorder *per se*. Further, they also lead to untoward side-effects. Such therapeutic limitations are applicable even to clinically successful chemotherapeutic drugs in practice. For instance, Amifostine, the only FDA approved radioprotector, not only has short-term efficacy against radiation injury but also leads to severe side effects (e.g. nausea, vomiting, diarrhea, hypotension, hypocalcaemia, nephro- and neurotoxicity) at the doses used in clinical practice [12]. This roadblock drives an enormous attention to natural compound-stimulators to activate quiescent stem cells as a systematic approach to develop novel therapeutic agents or synergistic adjuvant formula to currently available therapies.

In regenerative medicine, one underexplored strategy is stimulation of endogenous stem cells by using natural compounds to gain enhanced therapeutic benefits. Such herbal compounds possess pharmacological activities that typically impact self-renewal, differentiation and proliferation pathways in stem cells. As a consequence, stem cells are triggered to generate of multiple progenitor population to replenish or rejuvenate the damaged organ or diseased tissue [4]. In this perspective, herbal medicines/natural compounds have significant advantage over chemical or other relevant hematopoietic stimulators due to their profound pharmacological actions coupled with low clinical side-effects. One of the most fascinating insights in recent investigations is that natural compounds can elicit remarkable responses with more potency compared to chemical compounds. In cancer treatment, stem cell therapy has an advantage over conventional radiotherapy and chemotherapy which inadvertently compromise normal hematopoietic cells while targeting cancer cells. There is strong evidence, however, that the co-administration of herbal medicine with chemotherapy and/or radiation therapy can reduce the side effects of these treatments and promote healing in patients [13]. Given that tissue injuries induced by radiation or chemotherapy could be efficiently mitigated or even protected by replenishing the tissue progenitor cells through stem cell stimulation, it is conceivable that radioprotectants developed from herbal or natural compounds with potent stem cell stimulatory mechanisms will show impressive results. Similarly, this scenario fits well with curative strategies directed at relevant degenerative disorders as they often depend on replenished cells generated by the existing stem cell population to promote tissue repair. This suggests that medicaments derived from natural/herbal products provide holistic therapy that unleashes the body's own regenerative potential by facilitating tissue repair and replenishing the organ without compromising its functionalities.

In perspective, herbal medicine based stem cell stimulators provide promising therapy for the treatment of radiation injury or radiotherapy induced secondary tumors. It is worth mentioning that the benefits derived from herbal medicine are many-fold. Unlike chemical drugs, active ingredients in medicinal plants barely produce any side-effects or clinical complications. Certain plant extracts display multiple pharmacological actions such as antioxidant, immune-stimulatory, proliferative, anti-inflammatory and antimicrobial behaviors and holistically cure relevant complications. Interestingly, some natural components may also synergize the potency of medications isolated from other plant or natural ingredients. In traditional Ayurveda, Chinese, Japanese, Korean, Siddha, European, Tibetan and Unani systems of medicine it is common practice to use multi-plant formulations for treating diseases.

Identification and Implication of Natural Compounds as Potent Stem Cell Stimulators

In this section, several plant and natural compounds which elicit potent stimulatory actions on stem or progenitor cells are enlisted. Recent study delineates that NT-020, a proprietary nutraceutical formulation comprised of blueberry, green tea, Vitamin D_3 and carnosine, promotes the proliferation of human hematopoietic stem cells *in vitro* and protects stem cells from oxidative stress when given chronically to mice *in vivo* [14]. Interestingly, when an ethanol extract of blue green algae (Aphanizomenon flos-aquae; AFA) is added to NT-020, a significant synergistic effect is observed in the proliferation of human hematopoietic stem cells in culture. When used alone, some components from this NT-020 can have potential stimulatory effect on the progenitor cell population. For instance, Vitamin D_3 has a dramatic effect on the proliferation of various multipotent stem cell populations [15,16]. Similarly, Carnosine also has an impressive ability to rejuvenate the senescent cells and extend their cellular lifespan [17,18]. Diet supplemented with 2% blueberry extract has not only shown remarkable neuro-protective and neuro-restorative functions but also exhibits enhanced neurogenesis in aged rat brain [19]. Studies also report that polysaccharide preparations isolated from food-grade microalgae are more than one thousand times potent in activating human monocytes/macrophages. Active ingredient ginsenoside Rg1 isolated from the herbal medicine *Panax ginseng (C.A. Meyer)* has shown to have a stimulatory effect on stem cells. Similarly, ginsenosides also demonstrated pro-angiogenic activity [20,21]. Investigations on the Japanese traditional medicine *Kampo* (*Juzen-Taiho-Tohor*; TJ-48) reveals its exposure could dramatically enhance the peripheral blood cell population in cancer patients already treated with chemotherapy and or radiotherapy. Besides, not only does TJ-48 have the capacity to accelerate recovery from hematopoietic injury in irradiated mice but it also prolongs the survival of mice treated with anti-cancer drug mitomycin C (MMC). In another study, TJ-48 displayed a stimulatory immune-modulatory effect in mice [22]. In this, TJ-48 stimulated HSCs (hematopoietic stem cells) to generate multilineage colonies selectively on bone marrow stromal feeder layer but failed to reconstitute in methylcellulose medium supplemented with various combinations of early- to late-stage-acting cytokines. It seems, oleic acid and linolenic acid, elaidic acid and behenic acid are essential for the proliferation and/or differentiation of HSCs [23].

Intriguingly, certain herbal compounds not only display repair activities on damaged cells, but also demonstrate a protective effect. For instance, the extract of *Acanthopanax senticosus* (*Shigoka*) exhibits a radioprotective effect on hemopoiesis of irradiated mice (CBA/olac) exposed to ^{60}Co irradiation. The radioprotective effect was attributed to the stimulation of both proliferation and self-renewal of HSCs [24]. The active ingredients in numerous herbal plants demonstrated this radioprotective capability via their stimulatory functions on the haemopoietic sytem. For instance, *Panax ginseng*, *Acanthopanax senticosus* [24], *Ginkgo biloba* [25], *Hippophae rhamnoides* [26], *Podophyllum hexandrum* [27], *Tinospora cordifolia* [28], *Boerhaavia diffusa* [29] and Spirullina [30,31] can naturally activate the hematopoietic progenitor cells. α-tocopherol succinate has significant radioprotective characteristic as it protects mice against lethal gamma irradiation when administered subcutaneously 24 hr before irradiation. Some herbal extracts possess sufficient potency to exert radioprotection when used alone (*e.g.* ginsan, a purified polysaccharide isolated from *Panax ginseng* and glycyrrhizic acid [32]). However, some herbal extracts demonstrate their potency additively with other herbal constituents as seen in many traditional medicines. For instance, medicines such as Liv52 [33], Abana [34] and Chayawanprash [35] in Ayurveda, Si-Jun-Zi-Tang [36], Si-Wu-Tang [37], Jeng-Sheng-Yang-Yung-Tang [38], Lifukang [39] in Chinese medicine and TJ-48 [22,40] in Japanese medicine have shown to facilitate the hematopoietic recovery or mitigate the injury from radiation damage. A single dose of an aqueous extract of *Centella asiatica* was sufficient to protect Sprague Dawley rats from radiation-induced body weight loss and conditioned taste aversion initiated by 2 Gy low-dose ionizing radiation. In another study, an ethanol extract of *Ginkgo biloba* was effective against radiation induced damage [41]. Oral administration was shown to be a therapeutically viable option to mitigate the recovery in patients victimized by the Chernobyl accident site [42]. Administration of an aqueous-alcohol extract of *Hippophae rhamnoides* prior to a lethal dose (10 Gy) of whole-body ^{60}Co γ-irradiation enhanced the survival rate of mice (82%). The potency of *Hippophae rhamnoides* is exemplified by the fact that this drug rescued 100% mortality at 10 Gy radiation exposure [26].

In addition to radioprotection, stem cell stimulators can also be exploited for the treatment of injury caused by chemotherapy or exposure to harmful chemicals. Certain natural compounds such as anti-cancer polysaccharides isolated from Lentinan have successfully rescued the damage induced by chemotherapy by stimulating the proliferation of immature erythroid progenitors. Lentinan also accelerated the recovery of colony-forming unit erythroid (CFU-E) dramatically reduced by 5-fluorouracil (5-FU) exposure in mice [26,43]. Extract of herbal medicine Ninjin'yoeito (NYT) has shown to accelerate hematopoietic recovery in a murine model of syngeneic bone marrow transplantation (BMT). Study by Fujii *et al.* [44] indicate that NYT pretreatment was profoundly effective in enhancing the total number of CFU-E and colony-forming unit granulocyte-macrophage (CFU-GM) per marrow and spleen over a prolonged period following total body irradiation and BMT. NYT pretreatment significantly accelerated recovery of not only erythrocyte and leukocyte counts but also platelet counts after transplantation with a limited number of BM cells [44]. Study by Hatano *et al.* reports that the polysaccharides in the root of *Angelica acutiloba Kitagawa* (AR) significantly improve recovery from 5-FU-induced anemia by actively promoting hematopoiesis [45]. This component is reported to be clinically active against erythropoietin (EPO)-resistant anemia in chronic renal failure.

UNDERLYING MECHANISMS OF NATURAL COMPOUND STIMULATORS

In recent years, enormous attention has been focused on understanding the mechanisms of natural compounds to enhance therapeutic strategies against disorders. Although, underpinning mechanisms of the NT-020 were not clearly established, recent examination delineates that NT-020 could possibly rejuvenate stem cells by reducing the oxidative stress-induced apoptosis in cultured murine bone marrow stem cells, neurons and microglial cells *in vitro*. The robust generation of differentiated neural cells even after an ischemic stroke signifies active neurogenesis induced by NT-020, thereby promoting repair. In other studies, AFA has shown to modulate the CXCR4 expression on CD34+ bone marrow cells *in vitro* which as a consequence, triggers the mobilization of CD34+ CD133+ and CD34+ CD133- cells *in vivo* [46,47]. Polysaccharides that are currently being clinically used for cancer immunotherapy, including AFA, have substantially increased mRNA levels of interleukin and tumor necrosis factor-a (TNF-a) [31]. Ginsenosides elicit the pro-angiogenic activity via the phosphatidylinositol-3ˉkinase-*Akt* pathway [20]. Overexpression of *Akt* in rat mesenchymal stem cells transplanted into the ischemic rat myocardium by genetic manipulation inhibited cardiac remodeling by reducing intramyocardial inflammation, collagen deposition and cardiac myocyte hypertrophy [48;49]. Ginsenosides *Rh1* and *Rh2* isolated from *Panax ginseng* were found to induce differentiation of F9 (embryonal carcinoma cells) in endoderm-like cells [50] and the rodent hippocampal progenitor cells both *in vitro* and *in vivo* by stimulating the nuclear translocation of glucocorticoid receptor.

Mechanistic investigation reveals that the stimulatory mechanism of TJ-48 depends on the microenvironment of bone marrow stromal feeder layer wherein it can successfully proliferate HSCs. Intriguingly, failure of TJ-48 to reconstitute the hematopoietic system in methylcellulose supplemented with cytokines supports the notion that the regulatory mechanism from stromal cells impact proliferation of HSCs. Possibly, TJ-48 actively modulates the regulatory mechanisms of the stromal layer and or HSCs to enhance their self-renewal and proliferation. Experiments on MS-5 cells indicate that treatment with oleic acid and linolenic acid enhances the expression of adhesion molecules (CD54 and CD106e) and MHC class I as well as II antigens on the stromal cells [23]. Moreover, linolenic acid has also been proven to stimulate the production of early-acting cytokines in MS-5 cells. Antioxidants and diets supplemented with rich oxygen radical absorbance capacity (ORAC) have significantly reduced ageing in neuronal cells. The underlying mechanism is that these antioxidants could efficiently rescue the age-associated decline in cellular β-adrenergic receptor function and minimize the levels of proinflammatory cytokines in the cerebellum.

The radioprotective feature of the *Shigoka* extract is attributed to the enhanced proliferation and self-renewal of stem cells. Similarly, the radioprotective effect of Tocols has coupled with peripheral blood cells recovery by hematopoietic cytokine induction. *Ginkgo biloba* extract (Egb761) has been shown to prevent mitochondrial aging by rescuing oxidative damage [51]. Studies demonstrated that active constituents of *H. rhamnoides* extracts impact the compaction of chromatin organization and block the cell cycle at G_2-M phase by interfering with topoisomerase-I activity [26], to render their radioprotective mechanisms. Thus, *H. rhamnoides* appears to be a promising herb in the prophylactic treatment of radiation-induced damage but further research is necessary to identify appropriate dosing regimens and to characterize the active constituents. Administration of *Lentinan* reduces the stem cell inhibitory factor (SCIF) and enhances the CFU-S formation. Furthermore, *Lentinan* also augments the effects of erythropoietin (Epo) to increase the femoral marrow and splenic CFU-E formation significantly better than Epo administered alone in 5-FU-treated mice[43].

The drug NYT has the potential to protect against hematotoxicity induced by 5-fluorouracil (5-FU) and also hastens recovery of anemia through the stimulation of immature erythroid progenitor cell differentiation [52]. These results suggest that NYT stimulates early differentiation of erythroid lineage cells in bone marrow after 5-FU toxicity, thereby promoting hematopoiesis. Altogether, such herbal stimulators activate endogenous stem cell populations to either replace the damaged cells with replenished progenitor cells or to mitigate the genotoxic stress effects in tissue by anti-oxidant/anti-inflammatory mechanisms. Stimulation and/or restoration of the hematopoietic system allow the differentiated hematopoietic cells to better defend their cellular functions and rejuvenate the tissue, which in turn enhance their survival after genotoxic stress induced by injuries.

THERAPEUTIC USES

The dissemination of precursor cells from ESCs or adult SCs is invaluable for the development of disease models of various tissues and can also be utilized for tissue rejuvenation *in situ*. Thus, hES cells could be directed by natural compound stimulators to establish a large number of progenitor cells for drug discovery, development of novel therapeutics for diseases and toxicity studies. Transplantation of mesenchymal stem cells derived from adult hematopoietic tissue has been proposed as a strategy for cardiac repair following myocyte hypertrophy [53]. However, the major limitation lies in the poor cell viability associated with transplantation which diminishes the therapeutic potential of tissue regeneration *in vivo*. Ginsenosides from ginseng may circumvent this potential issue by promoting the differentiating signals with enhanced cell viability to facilitate the repair of tissues such as cardiac myocytes. The ginsenoside Rg1 might provide a therapeutic way to cure age-related neurodegenerative diseases and ageing disorders [54]. Gingenosides can promote the differentiation of mesenchymal stem cells to repair infarcted myocardium, prevent remodeling and promote cardiac performance after injury. NT-020 could be exploited for promoting tissue healing or regeneration via stimulation of various adult stem cell populations. Diet supplemented with nutritive extracts can serve as an efficient ischemic stroke adjunct therapy.

Ohnishi Y *et al.* observed that oral administration of TJ-48 prior to syngenic bone marrow transplantation (BMT) enhances the reconstitution of stem cells [40]; TJ-48 could possibly be exploited as an adjuvant to current therapeutics available for BMT. This is evident in cancer patients treated with chemotherapy and radiotherapy as J-48 showed to have therapeutic potential to promote healing. TJ-48 is proven to facilitate hematopoiesis even after tissue injury which in turn prolongs the survival of mice. Agents that inhibit free radical-mediated apoptosis are known to provide radioprotection, which explains the radioprotective effect of *Ginkgo biloba*. The extract of *G. biloba* is also useful in the treatment of aging and hypoxia-induced cerebral disorders [55]. An intravenous infusion of an ethanol extract of *G. biloba* leaves protected patients from vasogenic edema observed after irradiation of the brain [56] *G. biloba* extract also provided protection to brain neurons against oxidative stresss [57]. The radioprotective efficiency of the Shigoka extract is intriguing as enhanced proliferation and self-renewal of stem cells guarantee the rapid recovery of tissues damaged by radiation exposure [24]. One of the attractive features of the *Shigoka* extract is that administration of this plant extract can be done either prior to or after radiation exposure to control damage.

Study reports that Ninjin-Youei-To dramatically accelerated the proliferation of the oligodendrocyte precursor cells in aged-rat brains. This potent mitotic effect on oligo-dendrocyte precursor cells in aged-rat brains could be exploited for anti-ageing therapeutics [58]. *Centella asiatica* could be useful in preventing radiation-induced behavioral changes during clinical radiotherapy [59]. These studies suggest that α-tocopherol succinate can be potentially used as an adjunct in cancer chemotherapy [60]. The synergistic effect of Lentinan with Epo can be used to maintain normal erythropoiesis in the course of anemia. This Lentinan used as an adjuvant with other compounds also paves the way for enhanced recovery in cancer or other relevant disorders. Extract of *Actinidia arguta* stems promoted proliferation and differentiation of cultured bone marrow cells. Besides, (+)-Catechin isolated from *Actinidia arguta* not only stimulated the formation of myeloid colonies but also augmented the effect of interleukin-3 (IL-3) to increase the number of colony forming-units in culture (CFU-c) [61]. This stem cell stimulatory effect has been exploited to treat the hematotoxicity of 5-fluorouracil (5-FU) in mice [13](Takano *et al.*, 2004).

PERSPECTIVE AND FUTURE DIRECTIONS

Our understanding that bioactive principles from natural resources are successful against oxidative stress, inflammation, tissue damage and cancer growth tempt us to speculate that such compounds can be exploited for therapeutic benefits in conditions like ageing, cell death, tissue inflammation and radiation damage. Seemingly, these approaches stand for how we can identify and utilize plant or other natural sources for drug development and discovery. What is the trend in natural resource based drug discovery in the last few years? It is undeniable that emerging studies report a spectrum of pharmacological activity from various natural resources including plants. However, there are numerous natural compounds or resources yet to be recognized for their potent medicinal values. For instance, although lichens have been used for their antitumor, immunomodulatory, antiviral, cardiovascular, hypercholesterolemic, antibacterial, antiparasitic and memory enhancing capabilities, merely 100 out of the 13,500 species have been tested for their pharmacological activity [62]. Similarly, a vast majority (~90%) of mushrooms are largely unknown despite being potentially useful [63]. This suggests that we have a monumental task to characterize such unexplored species. Unfortunately, even for the relatively few known herbal or natural sources, we still have poor knowledge in regard to their cellular, molecular and pharmacological mechanisms. This is attributed to insufficient identification strategies, poor clinical trials, inadequate bioassays and lack of appropriate tissue or cell model systems selectively for natural compounds. How to improve this scenario? In this perspective, potential identification assays such as high-throughput library screening alone or in combination with other assays will rapidly accelerate drug discovery from a plethora of natural compounds. Development of efficient and reliable bioassays will help us to systematically assess the bioactive compound-evoked responses. More importantly, suitable *in vivo* and *in vitro* models should be utilized to gain more insight about the molecular and cellular mechanisms. It is likely that the lack of large-scale production of bioactive compounds will dampen our efforts to treat a large number of patients. To circumvent this issue, biotechnological interventions for large-scale production can be used to reduce the reliance on field-grown plants and to generate novel biomolecules with substantially improved pharmacological activities. More importantly, it is necessary to create public awareness regarding the benefits we could derive from these natural resources. Undoubtedly, these systematic efforts will provide us numerous therapeutic gains, help us fight many debilitating disorders and enhance our healthy lifestyle in harmony with nature.

REFERENCES

[1] Kondo M, Wagers AJ, Manz MG, Prohaska SS, Scherer DC, Beilhack GF, et al. Biology of hematopoietic stem cells and progenitors: implications for clinical application. *Annu. Rev. Immunol.* 2003;21:759-806.

[2] Passier R, van Laake LW, Mummery CL. Stem-cell-based therapy and lessons from the heart. *Nature.* 2008 May 15;453(7193):322-9.

[3] Mimeault M, Batra SK. Concise review: recent advances on the significance of stem cells in tissue regeneration and cancer therapies. *Stem Cells.* 2006 Nov;24(11):2319-45.

[4] Pazhanisamy SK. Stem cells, DNA damage, ageing and cancer. *Hematol. Oncol. Stem Cell Ther.* 2009;2(3):375-84.

[5] Wagers AJ, Weissman IL. Plasticity of adult stem cells. *Cell.* 2004 Mar 5;116(5): 639-48.

[6] Sauvageau G, Iscove NN, Humphries RK. In vitro and in vivo expansion of hematopoietic stem cells. *Oncogene.* 2004 Sep 20;23(43):7223-32.

[7] Davies SM, Kollman C, Anasetti C, Antin JH, Gajewski J, Casper JT, et al. Engraftment and survival after unrelated-donor bone marrow transplantation: a report from the national marrow donor program. *Blood.* 2000 Dec 15;96(13):4096-102.

[8] Levesque JP, Winkler IG, Larsen SR, Rasko JE. Mobilization of bone marrow-derived progenitors. *Handb. Exp. Pharmacol.* 2007;(180):3-36.

[9] O'Brien MM, Donaldson SS, Balise RR, Whittemore AS, Link MP. Second malignant neoplasms in survivors of pediatric Hodgkin's lymphoma treated with low-dose radiation and chemotherapy. *J. Clin. Oncol.* 2010 Mar 1;28(7):1232-9.

[10] Hudson MM, Poquette CA, Lee J, Greenwald CA, Shah A, Luo X, et al. Increased mortality after successful treatment for Hodgkin's disease. *J. Clin. Oncol.* 1998 Nov;16(11):3592-600.

[11] van Leeuwen FE, Klokman WJ, Veer MB, Hagenbeek A, Krol AD, Vetter UA, et al. Long-term risk of second malignancy in survivors of Hodgkin's disease treated during adolescence or young adulthood. *J. Clin. Oncol.* 2000 Feb;18(3):487-97.

[12] Cairnie AB. Adverse effects of radioprotector WR2721. *Radiat. Res.* 1983 Apr;94(1):221-6.

[13] Takano F, Tanaka T, Aoi J, Yahagi N, Fushiya S. Protective effect of (+)-catechin against 5-fluorouracil-induced myelosuppression in mice. *Toxicology.* 2004 Sep 1;201(1-3):133-42.

[14] Yasuhara T, Hara K, Maki M, Masuda T, Sanberg CD, Sanberg PR, et al. Dietary supplementation exerts neuroprotective effects in ischemic stroke model. *Rejuvenation Res.* 2008 Feb;11(1):201-14.

[15] Meyer C. Scientists probe role of vitamin D: deficiency a significant problem, experts say. *JAMA.* 2004 Sep 22;292(12):1416-8.

[16] Mathieu C, van EE, Decallonne B, Guilietti A, Gysemans C, Bouillon R, et al. Vitamin D and 1,25-dihydroxyvitamin D3 as modulators in the immune system. *J. Steroid Biochem. Mol. Biol.* 2004 May;89-90(1-5):449-52.

[17] Hipkiss AR, Preston JE, Himsworth DT, Worthington VC, Keown M, Michaelis J, et al. Pluripotent protective effects of carnosine, a naturally occurring dipeptide. *Ann. N. Y. Acad. Sci.* 1998 Nov 20;854:37-53.

[18] Holliday R, McFarland GA. A role for carnosine in cellular maintenance. *Biochemistry.* (Mosc) 2000 Jul;65(7):843-8.

[19] Joseph JA, Shukitt-Hale B, Denisova NA, Bielinski D, Martin A, McEwen JJ, et al. Reversals of age-related declines in neuronal signal transduction, cognitive, and motor behavioral deficits with blueberry, spinach, or strawberry dietary supplementation. *J. Neurosci.* 1999 Sep 15;19(18):8114-21.

[20] Sengupta S, Toh SA, Sellers LA, Skepper JN, Koolwijk P, Leung HW, et al. Modulating angiogenesis: the yin and the yang in ginseng. *Circulation.* 2004 Sep 7;110(10):1219-25.

[21] Shytle DR, Tan J, Ehrhart J, Smith AJ, Sanberg CD, Sanberg PR, et al. Effects of blue-green algae extracts on the proliferation of human adult stem cells in vitro: a preliminary study. *Med. Sci. Monit.* 2010 Jan;16(1):BR1-BR5.

[22] Komatsu Y, Takemoto N, Maruyama H, Tsuchiya H, Aburada M, Hosoya E, et al. Effect of Juzentaihoto on the anti-SRBC response in mice. *Jpn. J. Inflamm.* 6, 405-408. 1986.

[23] Hisha H, Yamada H, Sakurai MH, Kiyohara H, Li Y, Yu C, et al. Isolation and identification of hematopoietic stem cell-stimulating substances from Kampo (Japanese herbal) medicine, Juzen-taiho-to. *Blood.* 1997 Aug 1;90(3):1022-30.
[24] Miyanomae T, Frindel E. Radioprotection of hemopoiesis conferred by Acanthopanax senticosus Harms (Shigoka) administered before or after irradiation. *Exp. Hematol.* 1988 Oct;16(9):801-6.
[25] Gohil K, Moy RK, Farzin S, Maguire JJ, Packer L. mRNA expression profile of a human cancer cell line in response to Ginkgo biloba extract: induction of antioxidant response and the Golgi system. *Free Radic. Res.* 2000 Dec;33(6):831-49.
[26] Goel HC, Prasad J, Singh S, Sagar RK, Kumar IP, Sinha AK. Radioprotection by a herbal preparation of Hippophae rhamnoides, RH-3, against whole body lethal irradiation in mice. *Phytomedicine.* 2002 Jan;9(1):15-25.
[27] Goel HC, Sajikumar S, Sharma A. Effects of Podophyllum hexandrum on radiation induced delay of postnatal appearance of reflexes and physiological markers in rats irradiated in utero. *Phytomedicine.* 2002 Jul;9(5):447-54.
[28] Goel HC, Prasad J, Singh S, Sagar RK, Agrawala PK, Bala M, et al. Radioprotective potential of an herbal extract of Tinospora cordifolia. *J. Radiat. Res.* (Tokyo) 2004 Mar;45(1):61-8.
[29] Thali S, Thatte U, Dhakanukar S. The potential of Boerhaavia diffusa in radiation induced haemopoietic injury. *Amala Res. Bull.* 1998;18:20-2.
[30] Qishen P, Guo BJ, Kolman A. Radioprotective effect of extract from Spirulina platensis in mouse bone marrow cells studied by using the micronucleus test. *Toxicol. Lett.* 1989 Aug;48(2):165-9.
[31] Pugh N, Ross SA, ElSohly HN, ElSohly MA, Pasco DS. Isolation of three high molecular weight polysaccharide preparations with potent immunostimulatory activity from Spirulina platensis, aphanizomenon flos-aquae and Chlorella pyrenoidosa. *Planta Med.* 2001 Nov;67(8):737-42.
[32] Lin IH, Hau DM, Chen WC, Chen KT, Lin JG. Effects of glycyrrhizae and glycyrrhizic acid on cellular immunocompetence of gamma-ray-irradiated mice. *Chin. Med. J.* (Engl.) 1996 Feb;109(2):138-42.
[33] Saini MR, Kumar S, Jagetia GC, Saini N. Whole body radiation-induced damage to the peripheral blood and protection by Liv 52. *Radiobiol. Radiother.* 26, 487-493. 1985.
[34] Chandra JG, Rajanikant GK, Rao SK, Shrinath BM. Alteration in the glutathione, glutathione peroxidase, superoxide dismutase and lipid peroxidation by ascorbic acid in the skin of mice exposed to fractionated gamma radiation. *Clin. Chim.* Acta. 2003 Jun;332(1-2):111-21.
[35] Agarwal GN, Katiyar GK, Arora D, Kachroo P. Usefulness of Dabur Chyawanprash special Ayurvedic medicine in prevention of early reactions during radiotherapy. *Antiseptic.* 100, 189-192. 2003.
[36] Hsu HY, Yang JJ, Lian SL, Ho YH, Lin CC. Recovery of the hematopoietic system by Si-Jun-Zi-Tang in whole body irradiated mice. *J. Ethnopharmacol.* 1996 Nov;54(2-3):69-75.
[37] Hsu HY, Ho YH, Lin CC. Protection of mouse bone marrow by Si-WU-Tang against whole body irradiation. *J. Ethnopharmacol.* 1996 Jun;52(2):113-7.

[38] Hsu HY, Lin CC, Hau DM. Restoration of radiation injury in mice by two Chinese medicinal prescriptions Kuei-Pi-Tang and Jeng-Sheng-Yang-Yung-Tang. *Phytother. Res.* 6, 294-299. 1992.
[39] Kim JH, Kim SH, Lee EJ. Radioprotective effect of Lifukang, a Chinese medicinal plant prescription. *Nat. Prod. Sci.* 4, 26-31. 1998.
[40] Ohnishi Y, Yasumizu R, Fan HX, Liu J, Takao-Liu F, Komatsu Y, et al. Effects of juzen-taiho-toh (TJ-48), a traditional Oriental medicine, on hematopoietic recovery from radiation injury in mice. *Exp. Hematol.* 1990 Jan;18(1):18-22.
[41] Emerit I, Arutyunyan R, Oganesian N, Levy A, Cernjavsky L, Sarkisian T, et al. Radiation-induced clastogenic factors: anticlastogenic effect of Ginkgo biloba extract. *Free Radic. Biol. Med.* 1995 Jun;18(6):985-91.
[42] Emerit I, Oganesian N, Sarkisian T, Arutyunyan R, Pogosian A, Asrian K, et al. Clastogenic factors in the plasma of Chernobyl accident recovery workers: anticlastogenic effect of Ginkgo biloba extract. *Radiat. Res.* 1995 Nov;144(2):198-205.
[43] Takatsuki F, Miyasaka Y, Kikuchi T, Suzuki M, Hamuro J. Improvement of erythroid toxicity by lentinan and erythropoietin in mice treated with chemotherapeutic agents. *Exp. Hematol.* 1996 Feb;24(3):416-22.
[44] Fujii Y, Imamura M, Han M, Hashino S, Zhu X, Kobayashi H, et al. Recipient-mediated effect of a traditional Chinese herbal medicine, ren-shen-yang-rong-tang (Japanese name: ninjin-youei-to), on hematopoietic recovery following lethal irradiation and syngeneic bone marrow transplantation. *Int. J. Immunopharmacol.* 1994 Aug;16(8):615-22.
[45] Hatano R, Takano F, Fushiya S, Michimata M, Tanaka T, Kazama I, et al. Water-soluble extracts from Angelica acutiloba Kitagawa enhance hematopoiesis by activating immature erythroid cells in mice with 5-fluorouracil-induced anemia. *Exp. Hematol.* 2004 Oct;32(10):918-24.
[46] Jensen GS, Hart AN, Zaske LA, Drapeau C, Gupta N, Schaeffer DJ, et al. Mobilization of human CD34+ CD133+ and CD34+ CD133(-) stem cells in vivo by consumption of an extract from Aphanizomenon flos-aquae--related to modulation of CXCR4 expression by an L-selectin ligand? *Cardiovasc. Revasc. Med.* 2007 Jul;8(3):189-202.
[47] Bickford PC, Tan J, Shytle RD, Sanberg CD, El-Badri N, Sanberg PR. Nutraceuticals synergistically promote proliferation of human stem cells. *Stem Cells Dev.* 2006 Feb;15(1):118-23.
[48] Mangi AA, Noiseux N, Kong D, He H, Rezvani M, Ingwall JS, et al. Mesenchymal stem cells modified with Akt prevent remodeling and restore performance of infarcted hearts. *Nat. Med.* 2003 Sep;9(9):1195-201.
[49] Shytle RD, Ehrhart J, Tan J, Vila J, Cole M, Sanberg CD, et al. Oxidative stress of neural, hematopoietic, and stem cells: protection by natural compounds. *Rejuvenation Res.* 2007 Jun;10(2):173-8.
[50] Lee YN, Lee HY, Chung HY, Kim SI, Lee SK, Park BC, et al. In vitro induction of differentiation by ginsenoides in F9 teratocarcinoma cells. *Eur. J. Cancer.* 1996 Jul;32A(8):1420-8.
[51] Sastre J, Millan A, Garcia dlA, Pla R, Juan G, Pallardo, et al. A Ginkgo biloba extract (EGb 761) prevents mitochondrial aging by protecting against oxidative stress. *Free Radic Biol. Med.* 1998 Jan 15;24(2):298-304.

[52] Takano F, Ohta Y, Tanaka T, Sasaki K, Kobayashi K, Takahashi T, et al. Oral Administration of Ren-Shen-Yang-Rong-Tang 'Ninjin'yoeito' Protects Against Hematotoxicity and Induces Immature Erythroid Progenitor Cells in 5-Fluorouracil-induced Anemia. *Evid. Based Complement Alternat. Med.* 2009 Jun;6(2):247-56.
[53] Pittenger MF, Martin BJ. Mesenchymal stem cells and their potential as cardiac therapeutics. *Circ. Res.* 2004 Jul 9;95(1):9-20.
[54] Shen LH, Zhang JT. Ginsenoside Rg1 promotes proliferation of hippocampal progenitor cells. *Neurol. Res.* 2004 Jun;26(4):422-8.
[55] Duche JC, Barre J, Guinot P, Duchier J, Cournot A, Tillement JP. Effect of Ginkgo biloba extract on microsomal enzyme induction. *Int. J. Clin. Pharmacol. Res.* 1989;9(3):165-8.
[56] Hannequin D, Thibert A, Vaschalde Y. [Development of a model to study the anti-edema properties of Ginkgo biloba extract]. *Presse Med.* 1986 Sep 25;15(31):1575-6.
[57] Oyama Y, Chikahisa L, Ueha T, Kanemaru K, Noda K. Ginkgo biloba extract protects brain neurons against oxidative stress induced by hydrogen peroxide. *Brain Res.* 1996 Mar 18;712(2):349-52.
[58] Kobayashi J, Seiwa C, Sakai T, Gotoh M, Komatsu Y, Yamamoto M, et al. Effect of a traditional Chinese herbal medicine, Ren-Shen-Yang-Rong-Tang (Japanese name: Ninjin-Youei-To), on oligodendrocyte precursor cells from aged-rat brain. *Int. Immunopharmacol.* 2003 Jul;3(7):1027-39.
[59] Shobi V, Goel HC. Protection against radiation-induced conditioned taste aversion by Centella asiatica. *Physiol. Behav.* 2001 May;73(1-2):19-23.
[60] Singh VK, Shafran RL, Jackson WE, III, Seed TM, Kumar KS. Induction of cytokines by radioprotective tocopherol analogs. *Exp. Mol. Pathol.* 2006 Aug;81(1):55-61.
[61] Takano F, Tanaka T, Tsukamoto E, Yahagi N, Fushiya S. Isolation of (+)-catechin and (-)-epicatechin from Actinidia arguta as bone marrow cell proliferation promoting compounds. *Planta Med.* 2003 Apr;69(4):321-6.
[62] Olafsdottir ES, Ingolfsdottir K. Polysaccharides from lichens: structural characteristics and biological activity. *Planta Med.* 2001 Apr;67(3):199-208.
[63] Wasser SP. Medicinal mushrooms as a source of antitumor and immunomodulating polysaccharides. *Appl. Microbiol. Biotechnol.* 2002 Nov;60(3):258-74.

In: Natural Products and Their Active Compounds ... ISBN: 978-1-62100-153-9
Editors: M. Essa, A. Manickavasagan, and E. Sukumar © 2012 Nova Science Publishers, Inc.

Chapter 17

NUTRITIONAL AND MEDICINAL VALUES OF PAPAYA (CARICA PAPAYA L.)

Amanat Ali[1*], Sankar Devarajan[2], Mostafa I. Waly[1], Mohamed M. Essa[1,3,4] and M. S. Rahman[1]

[1]Department of Food Science and Nutrition, College of Agricultural and Marine Sciences, Sultan Qaboos University, Al-Khoud, Muscat, Oman
[2]Department of Cardiovascular Diseases, Fukuoka University Chikushi Hospital, Fukuoka, Japan and WHO Regional Office, Japan
[3]Neuropharmacology group, Department of Pharmacology, College of Medicine, University of New South Wales, Sydney, Australia
[4]Developmental Neuroscience Lab, NYSIBR, Staten Island, NY, US

ABSTRACT

Papaya (*Carica papaya* L.) is a deliciously sweet tropical fruit with musky undertones and a distinctive pleasant aroma. It was first cultivated in Mexico several centuries ago but is currently being cultivated in most of the tropical countries. Everything in papaya plant such as roots, leaves, peel, latex, flower, fruit and seeds have their nutritional and medicinal significance. Papaya can be used as a food, a cooking aid, and in medicine. Papaya is considered as a low calorie nutrient dense fruit. The fresh fruit is commonly used as a carminative, stomachic, diuretic and antiseptic in many parts of the world. The nutrients and phytochemicals contained in papaya help in digestion, reduce inflammation, support the functioning of cardiovascular, immune and digestive systems and may also help in prevention of colon, lung and prostate cancers. Overall, the papaya can act as a detoxifier, activator of metabolism, rejuvenating the body and in the maintenance of body's homeostasis because it is rich in antioxidants, B vitamins, folate and pantothenic acid, and potassium and magnesium as well as fiber. Because of its high vitamin A and carotenoids contents, it can help in preventing the cataract and age-related macular degeneration. Papaya pastes can be used externally as a treatment for skin wounds and burns. This paper discusses the nutritional and medicinal value of papaya (*Carica papaya* L.) and its relationship to human health.

* Corresponding author: Dr. Amanat Ali, email: amanat@squ.edu.om

INTRODUCTION

The papaya (*Carica papaya* L.) is a tropical fruit that is native to the tropics of South America. According to the historical reports, it was first cultivated in Mexico several centuries ago but is currently being cultivated in most of the tropical countries. The current largest commercial producers of papaya include the United States, Mexico and Puerto Rico. Currently many genetically modified hybrid varieties are commercially available for cultivation, which are more resistant to diseases (Jiao et al., 2010). Over the past 40 years the production of papaya has increased drastically. The estimated production in the year 2009 was 10.21 million tons (FAO, 2010).

Papaya is normally a single stem plant that can reach up to 10 meters, with spirally arranged leaves confined to the top of the trunk (Rice et al., 1987). It is a highly frost sensitive plant. The plant grows rapidly and starts fruiting within one and half to 3 years. The productive life of the plant is about three and a half year. Although there is a slight seasonal peak in its production in early summer and fall, yet the papaya tree can produce the fruit year round. The flowers appear on the axils of the leaves, maturing into large spherical, pear-shaped fruit whose length can vary from 7 to 20 inches and can reach up to 2.5kg in weight. Papaya fruit has normally greenish yellow, yellow or orange color.

The fruit is climacteric and exhibits an increase in respiration and ethylene production during ripening (Koslanund, 2003). The fruit ripens rapidly at room temperature. It is ripe when it feels soft and its colour changes to amber or orange hue. The shelf life of the ripened fruit is short only 2 to 3 days (Archbold et al., 2003). The two flesh colours (red and yellow) of papaya fruit are controlled by the same single gene, however the yellow colour is dominant (Yamamoto, 1964). The red colour of papaya fruit is due to the accumulation of lycopene, whereas the yellow colour is the result of conversion of lycopene to β-carotene and β-cryptoxanthin (Hirschberg, 2001). As the fruit ripens, its colour changes, which is caused by the breakdown and disappearance of chlorophyll. The flesh colour of papaya fruit is considered a quality trait that correlates with its nutritional value and is linked to shelf life of the fruit. The full genomic sequences of both yellow and red fleshed papayas have been reported to be identical (Skelton et al., 2006). No significant compositional differences have been reported between the transgenic and non-transgenic papaya (Jiao et al., 2010).

GENERAL CHARACTERISTICS AND USES OF PAPAYA

Papaya is a deliciously sweet fruit with musky undertones and a distinctive pleasant aroma. It has a soft texture with butter-like consistency (Bari et al., 2006). Its taste and sweetness increases with the ripening process of fruit. However, the overripe fruit quickly starts deteriorating in quality. This is a greatly loved tropical fruit that was sensibly called "The Fruit of Angels" by Christopher Columbus. Papaya plant is also called a "tree of health" and its fruit is termed as a "fruit of long life". Ripe papaya flesh has a rich orange color with either yellow or pink hues. The inner cavity contains a wealth of black round seeds, encased in a gelatinous-like substance. The leaves, stem, and unripe fruit of papaya release a whitish milky fluid (latex) that consists of proteins, alkaloids (mainly carpaine), starches, sugars, oils, tannins, resins, pectins and gums, which coagulate on exposure to air. Green papaya is a rich source of papain and chymopapain. Papain, the proteolytic enzyme, is regarded as vegetable pepsin that helps in the digestion of proteins in acid, alkaline or neutral medium. Papain is used like bromelain, a similar enzyme found in pineapple, to treat sports injuries, other causes of trauma and allergies. Papain is also used in tenderizing meat and other proteins as it has the ability to break down the tough meat fibres and is used since thousands of years. Papain is included as a component in powdered meat tenderizers. Papain is used in chewing gum, in brewing/wine and beer making, textile and tanning industries (Bruneton, 1999, Bhattacharjee, 2001, Oloyede, 2005). Papain and chymopapain have been shown to help lower inflammation and to improve healing from burns in addition to helping in digestion of proteins.

The ripe fruit is usually eaten raw, without skin or seeds. The unripe green fruit can be eaten both as raw and cooked but is usually eaten as cooked in curries, salads and stews. Unripe fruit has a relatively high amount of pectin, which can be used to make jellies. In some parts of Asia, the young leaves of papaya are steamed and eaten like spinach. The leaves are also made into tea to be used as a preventive agent for malaria, although there is no real scientific evidence for the effectiveness of this treatment. The black seeds are edible and have a sharp, spicy taste. The ground seeds are sometimes used as a substitute for black pepper. The stem and bark are also used in rope production. Papaya is also frequently used as a hair conditioner. Papaya is used in making soft drinks, jam, ice cream, and flavouring of crystallized fruit and canned in syrup. Papaya pulp nectar prepared using irradiation and mild heat treatment retained its fresh flavour and nutritional qualities closest to untreated controls and was found to be microbiologically safe with acceptable enzyme levels (Parker et al., 2010). Everything in papaya plant such as roots, leaves, peel, latex, flower, fruit and seeds have their nutritional and medicinal significance. Papaya can be used as a food, a cooking aid, and in medicine. In general, the papaya promotes proper functioning of pancreas, alleviates indigestion, protects against infection, aids in diabetics and hepatitis patients. The consumption of ripe papaya is thought to help in the prevention of cancer in organs and glands with epithelial tissue. Papaya has rejuvenating properties that especially assist in controlling the early ageing process.

Overall the papaya acts as a detoxifier, activates the metabolism, rejuvenates the body and maintains the body's homeostasis because it is rich in antioxidants, B vitamins, folate and pantothenic acid, the minerals potassium and magnesium, and fiber. Papaya juice is a popular beverage and can assist in mitigating infections of the colon and breaking down the pus and mucus. It is believed that it can act as a useful tonic for the stomach and intestines, if consumed alone for at least 3 days. In the folklore medicine, sometimes it is suggested to go for "Papaya Therapy" once a year i.e., to eat one or two papayas daily for 2 to 4 week to benefit from its healing properties.

Chemical Composition and Nutritional Quality of Papaya

The nutritional qualities and medicinal value of papaya are closely related. The papaya can be considered as a nutrient dense food, as it provides many more nutrients on per calorie basis as compared to other foods. The chemical composition as well as the mineral and vitamin composition of fresh papaya is given in Tables 1 and 2. It contains only small amounts of protein and is almost free from cholesterol and fat. The carbohydrate content of ripe papaya mainly consists of invert sugars, which are easily digestible and absorbed. Ripe fruit can therefore easily boost body's energy.

The whole papaya fruit is an excellent source of dietary fiber and therefore can also help in preventing the constipation. The fiber content of papaya can help in lowering the high blood cholesterol levels. Papaya is rich in vitamins C and A. One serving of papaya can provide about 100% daily requirement for vitamin C and 30% of vitamin A. It also contributes to small quantities of vitamin E, K, thiamine, riboflavin, niacin, pyridoxine and folate. Folic acid is needed for the conversion of homocysteine to cysteine and methionine. The increased level of homocysteine in blood is considered a significant risk factor for a heart attack or stroke as it can directly damage the wall of blood vessels (Antoniades et al., 2009, Seo et al., 2010). Papaya could be a candid source to reverse the homocysteine mediated cardiovascular diseases since it has profuse amount of folic acid. The nutrients contained in papaya can also help to prevent the oxidation of cholesterol. The oxidized cholesterol sticks to the internal lining of blood vessels, forming dangerous plaques that can eventually cause heart attacks or strokes. Data from various studies indicate that dietary vitamin E and C may exert some effect in preventing the oxidation of cholesterol because of their suggested association with paraoxonase, an enzyme that inhibits the oxidation of LDL and HDL cholesterol (Jarvik et al., 2002, Schürks, et al., 2010, Gaby, 2010).

Table 1. The chemical composition of fresh papaya fruit

Parameters	Range
Energy	39.0 - 41.4 (kcal/100g)
Moisture	86.9 – 89.8 %
Crude protein	0.5 - 0.6 (g/100g)
Total fat	0.1 - 0.14 (g/100g)
Ash	0.5 - 0.7 (g/100g)
Crude fibre	0.4 - 0.8 (g/100g)
Dietary fibre	0.5 - 2.2 (g/100g)
Carbohydrates	7.5 - 10.98 (g/100g)
Total Sugars	7.2 - 9.8 (g/100g)
Sucrose	1.9 - 6.1 (g/100g)
Glucose	2.6 -3.4 (g/100g)
Fructose	2.1 - 2.6 (g/100g)

The values are calculated based on the data reported by Adetuyi et al. (2008), Gouado et al. (2007), Nakamura et al. (2007), Nguyen and Schwartz (1999), Sirichakwal et al. (2005), Wall (2006), Wall et al. (2010), Veda et al. (2007).

Papaya also contains small amounts of calcium, magnesium, potassium, iron, manganese, zinc, copper, boron and selenium. It has low levels of sodium and high levels of potassium and can therefore be helpful for the hypertensive people to balance their overall daily dietary intake of sodium.

Table 2. Vitamin and Mineral Composition of fresh papaya fruit

Parameters	Range
Vitamin A (RAE)*	23 – 55 (μg/100g)
Vitamin E	3.13 – 5.3 (mg/100g)
Vitamin K	2.3 – 2.9 (μg/100g)
Vitamin C	57 – 108 (mg/100g)
Thiamine (vitamin B_1)	0.04 – 0.05 (mg/100g)
Riboflavin (vitamin B_2)	0.05 - 0.07 (mg/100g)
Niacin	0.34 – 44 (mg/100g)
Pyridoxine	0.1 – 0.15 (mg/100g)
Folate	39 – 55 (μg/100g)
Calcium	17 – 24 (mg/100g)
Phosphorous	5 – 9 (mg/100g)
Magnesium	10 – 33 (mg/100g)
Sodium	3 – 24 (mg/100g)
Potassium	90 – 257 (mg/100g)
Iron	0.23 - 0.66 (mg/100g)
Manganese	0.01 – 0.03 (mg/100g)
Zinc	0.06 – 0.09 (mg/100g)
Copper	0.06 – 0.14 (mg/100g)
Boron	0.01 – 0.21 (mg/100g)
Selenium	1.2 – 1.5 (μg/100g)

The values are calculated based on the data reported by Adetuyi et al. (2008), Gouado et al. (2007), Nakamura et al. (2007), Nguyen and Schwartz (1999), Sirichakwal et al., (2005), Wall (2006), Wall et al. (2010), Veda et al. (2007).

PHYTOCHEMICAL COMPOSITION OF PAPAYA

In addition to its nutritional contents, papaya also contains many bioactive phytochemicals with diverse structure and functional properties which have not yet been fully exploited for their potential health benefits. The phytochemical composition of fresh papaya is given in Table 3. It contains substantial amounts of carotenoids, flavonoids and polyphenols. It contains relatively high levels of beta-carotene, which the body converts to vitamin A. Papaya contains about 6% of the level of beta carotene found in carrots (USDA, National Nutrient Database for Standard Reference, 2006). Red flesh papaya has been reported to contain significant quantities (4.1 mg/100g flesh) of lycopene (Nguyen and Schwartz, 1999).

Because of its high phytochemical contents, it shows significant antioxidant activities. During the growing process, papaya produces some specific compounds (such as benzyl-isothiocyanate; BITC and carpaine) to protect itself against the attacks of insects, pests and herbivores.

Table 3. Phytochemicals Composition of fresh papaya fruit

Parameters	Range
α- carotene	16 – 31 (μg/100g)
β- carotene	130 – 730 (μg/100g)
Lycopene	113 – 4138 (μg/100g)
β-Cryptoxanthin	124 – 3799 (μg/100g)
Zeaxanthin	19 – 27 (μg/100g)
Total provitamin A	256 – 890 (μg/100g)
Lutein	93 – 318 (μg/100g)
Total carotenoids	321.2 – 7210 (μg/100g)
Total Polyphenols	51 – 59 (mg GAE/100g)[1]
Total Antioxidant Activity - ORAC	250 – 350 (μmol TE/100g)[2]
Total Antioxidant Activity - FRAP	350 – 430 (μmol TE/100g)[3]
Phytate	1.22- 1.45 (g/100g)
Oxalate (g/100g)	0.45- 57 (g/100g)
Condensed tannins (g/100g)	0.062 -0.087 (g/100g)
Hydrolysable tannins (g/100g)	0.021- 033 (g/100g)

[1] GAE = Gallic acid equivalent.

[2] Oxygen radical absorbance capacity (ORAC) expressed as μmol of Trolox Equivalent (TE) per 100 g of fresh weight.

[3] Ferric reducing antioxidant power (FRAP), expressed as μmol of Trolox Equivalent (TE) per 100 g of fresh weight.

The values are calculated based on the data reported by Adetuyi et al. (2008), Gouado et al (2007), Nakamura et al. (2007), Nguyen and Schwartz (1999), Sirichakwal et al., (2005), Wall (2006), Wall et al. (2010), Veda et al. (2007).

These natural toxins may exert some adverse effects on human health when consumed in large quantities. However, their levels in mature papaya fruit are low and therefore are considered as safe for humans (Roberts et al., 2008).

Benzyl isothiocyanate (BITC) isolated from the extracts of papaya whole fruit has shown potent diverse biological activities in inducing the phase 2 detoxifying enzymes and apoptosis (Nakamura et al., 2000 and 2002, Miyoshi et al 2004 and 2007). Nakamura et al. (2007) reported that papaya seeds represent a rich source of biologically active isothiocyanates and the n-hexane extract of papaya seeds homogenate was highly effective in inhibiting the superoxide generation and apoptosis induction in HL-60 cells, the activities of which are comparable to those of authentic benzyl isothiocyanate. In contrast, the papaya pulp contained an undetectable amount of bezyl-glucosinolate. They showed that papaya seeds and not the papaya pulp is a rich source of biologically active isothiocyanate, especially the BITC and its precursor glucosinolate, which are as high as those in *Brassica* vegetables. BITC is formed from benzyl glucosinolate in papaya seeds (Bennett et al., 1997).

Papaya is also a good source of carpaine. It is one of the major alkaloid components of papaya leaves that have been studied for its cardiovascular effects in male Wistar rats. Increasing dosages of carpaine from 0.5 mg/kg to 2.0 mg/kg resulted in progressive decrease in systolic, diastolic, and mean arterial blood pressure. It was concluded that carpaine directly affects the myocardium. The effects of carpaine may be related to its macrocyclic dilactone structure, a possible cation chelating structure (Burdick, 1971, Hornick et al., 1978). The extracts of unripe papaya have been reported to contain terpenoids, alkaloids, flavonoids, carbohydrates, glycosides and steroids (Ezike et al., 2009). The papaya lipase is currently considered as a "naturally immobilized" biocatalyst (Dominguez de Maria et al., 2006).

The papaya also contains some other anti-nutrient compounds such as phytate, oxalate and tannins. The levels of these anti-nutrients (phytate, oxalate, hydrolysable tannins and condensed tannins) and antioxidants (vitamin C, tocopherols, total phenols, and carotenoids) contents of papaya (*Carica papaya*) can decrease significantly ($P < 0.05$) with increased storage period and temperature (Adetuyi et al., 2008). Simirgiotis et al (2009) reported the presence of 10 low molecular weight quercetin glycoside derivatives in the fruit of mountain papaya (*Vasconcellea pubescens* A DC) grown in Chile. It is also called "cold papaya" as it grows in relatively cooler climates as compared to the popular and widely cultivated *Carica papaya* L. The fruits of mountain papaya are mostly consumed after processing and are also used in the production of jams, beverages, cold drinks and cocktails (Idstein et al., 1985, Moya-Leon et al., 2004). Oliveira et al (2010) observed that the nutritional quality of papaya (*Carica papaya* L.) in terms of its vitamin C and carotenoids contents showed excellent stability under the usual handling conditions employed in commercial restaurants.

MEDICINAL AND HEALTH EFFECTS OF PAPAYA

The overall nutritional and health benefits of papaya are because of the interactions of its nutrients and phytochemicals present in whole fruit, rather than due to a single "active" component. Because of its high antioxidant contents, papaya can prevent cholesterol oxidation and can be used as a preventive treatment against atherosclerosis, strokes, heart attacks and diabetic heart disease. The fresh fruit is commonly used as a carminative, stomachic, diuretic and antiseptic in many parts of the world (Iwu, 1993, Bhattacharjee, 2001). A number of studies have reported the various pharmacological properties of *C. papaya* (fruit and seeds) such as histaminergic action, inhibition of rabbit jejunal contractions, and tocolytic and antihelminthic activities and anti-ulcer properties (Adebiyi et al., 2003, 2004, Adebiyi and Adeikan, 2005, Okeniyi et al., 2007, Ezike et al., 2009). Papaya can strengthen the immune system therefore can help in preventing the recurrent colds and flu. After treatment with antibiotics eating papaya or drinking its juice can help to replenish the intestinal microflora. Papaya has also been reported to have significantly high hydroxyl radical and hydrogen peroxide scavenging activity (Murcia et al., 2001). The fermented papaya products showed free radical scavenging activity and were effective in providing protection against various pathological disorders including tumors and immunodeficiency (Osanto et al., 1995). The fermented papaya products improved the antioxidant defense in elderly patients without any overt antioxidant deficiency state at the dose of 9g/day orally (Marotta et al., 2006a, b).

Many active components from papaya have been isolated and studied. Papain, the main proteolytic enzyme in papaya, is also being studied for relief of cancer therapy side effects, especially in relieving the side effects such as difficulty in swallowing and mouth sores after radiation and chemotherapy as well as boosting up the immune system and helping body to fight against the cancer cells. The supplements produced by dehydrating and concentrating the whole fruit may be helpful for people who do not get enough of these components in their daily diet. Dried papaya is marketed in tablet form to remedy digestive problems. Papaya pills, juices, and whole food supplements containing papaya are currently being promoted as weight loss aids, digestive aids, and natural pain relievers, as well as for many other health benefits. Whether the papaya pills and supplements can prevent or counteract these ailments is still to be validated through well-designed controlled studies. Papaya leaves are poultice on to nervous pains and elephantoid growths. They are also dried and infused to make a tea, which is used as a vermifuge, an amoebicide, a purgative, to prevent and treat malaria, and as a treatment for gastritis and genitor-urinary ailments.

Promotes the Functioning of Digestive System

The consumption of papaya after a meal can improve digestion, prevent bloating and chronic indigestion as well as may help in preventing nausea, vomiting and morning sickness in pregnant women. Papaya has also been reported to increase the absorption of iron from rice based meals in Indian women (Ballot et al., 1987). Papaya is used in the preparation of antacids, ulcer treatment and to prevent constipation. The unripe papaya extracts have shown cytoprotective and antimotility properties, suggesting the effectiveness of unripe papaya as an anti-ulcer fruit (Ezike et al., 2009).

Anthelmintic and Anti-Amoebic Properties

Human intestinal parasitosis constitutes a significant global health problem with enormous financial implications, in particular in developing countries. In folk medicine, *C. papaya* seeds have been used to treat the antheminthiasis (Bhattacharjee, 2001, Okeniyi et al., 2007). The papaya fruit, seeds, latex, and leaves contain carpaine, an anthelmintic alkaloid that can remove the parasitic worms from the body. The carpaine can however be dangerous in high doses. In folklore medicine, the papaya seeds when taken with honey are known to be anthelmintic for expelling the worms. The latex of papaya as well as the aqueous extracts of papaya seeds have shown potent anthelmintic and anti-amoebic properties (Satrija et al., 1994 and 1995). Among others, the active chemical agents contained in *Carica papaya* seeds and fruit are benzyl isothiocyanate and papain that have proven anthelmintic properties (Kumar et al., 1991, Tona et al., 1998, Ghosh et al., 1998). The air dried papaya seeds offer a cheap, harmless and preventive strategy against intestinal parasitosis, especially in tropical communities (Okeniyi et al., 2007).

Anti-Inflammatory and Wound Healing Properties

Papaya is thought to contain some natural pain relieving abilities. The unique protein digesting enzymes (papain and chymopapain) have been reported to help in lowering the inflammation and healing of burns. Papaya can lower inflammation in the body, alleviate pain and edema caused by sport injuries. The antioxidant nutrients found in papaya, including vitamin C, vitamin E, and beta-carotene can also help in reducing inflammation. Because of its anti-inflammatory properties papaya can relieve the severity of rheumatoid arthritis, polyarthritis and osteoarthritis. It has been reported that people who consumed the lowest amounts of vitamin C-rich foods were more than three times more likely to develop arthritis than those who consumed the highest amounts (Pattison et al., 2004). This may explain why the people with diseases that are worsened by inflammation, such as asthma, osteoarthritis and rheumatoid arthritis, find a relief in their sufferings when they get more of these nutrients. Collard and Roy (2010) observed that oral supplementation of fermented papaya preparation (FPP) in adult obese diabetic (db/db) mice specifically improved the response of wound-site macrophages and subsequent angiogenic response. They suggested that the beneficial effects of FPP should be tested in clinical trials for its beneficial effects on diabetic wound-related outcomes as it has a long track record of safe human consumption.

Papaya is also a rich source of fibrin, an important factor in the blood clotting process and can therefore help in quick healing of wounds. This factor is not common in the plant kingdom and is mainly formed in the body of animals, papaya being an exception. Papaya paste/ointment, made from fermented papaya flesh, is used traditionally for the relief of burns, cuts, rashes and stings. The latex can however, cause irritation and may provoke some allergic reaction in hypersensitive people. Data from the preliminary studies suggest that the treatment with papaya preparations and aqueous extracts may help to facilitate the wound-healing response (Nayak et al., 2007, Pieper and Caliri 2003). Papaya derived enzyme papain has been shown to facilitate the enzymatic wound debridement when applied topically (Telgenhoff et al., 2007). The aqueous extracts have also shown significant wound healing properties in diabetic rats (Dawkins et al., 2005, Nayak et al., 2007). The papaya fruit is used in topical ulcer dressing and for burn dressings to treat the wounds. The possible mechanism of action may be due to proteolytic enzymes chymopapain and papain as well as the antimicrobial activity (Hewitt et al., 2002, Starley et al., 1999). The fruit can also be directly applied to skin sores and seeds could also be effectively used for treating chronic skin ulcers.

PROTECTION AGAINST AGE-RELATED MACULAR DEGENERATION (AMD)

Papaya is considered as the top ranking fruit in terms of its carotenoids, flavonoids, fibre, vitamin A, ascorbic acid, folate, niacin, thiamin, riboflavin, iron, calcium and fibre contents per serving (USDA, National Nutrient Database for Standard Reference, 2006, Luximon-Ramma et al., 2003, Lim et al., 2007). The consumption of papaya is therefore recommended to prevent the vitamin A deficiency, a cause of childhood blindness in many tropical and subtropical developing countries (Chandrika et al., 2003, Gouado et al., 2007).

Our bodies need vitamin A for the maintenance of epithelial surfaces, for immune competence, for normal functioning of retina, as well as for growth, development and reproduction. Dark green as well as the coloured fruits and vegetables provide provitamin A carotenoids. Reducing the vitamin A deficiency can save the eyesight and lives of countless children and adults.

Higher dietary intake of lutein/zeaxanthin and vitamin E from foods and supplements have been reported to be associated with significantly decreased risk of neovascular age-related macular degeneration (AMD), geographic atrophy, large or extensive intermediate drusen and cataract (AREDS Research Group, 2007, Christen et al., 2008).

Arscott et al. (2010) observed that diets supplemented with papaya, oranges, mangoes and tangerines were able to prevent vitamin A deficiency in Mongolian gerbils. They suggested that these fruits could be an effective part of food-based interventions to support vitamin A nutrition in developing countries and worldwide.

Eating 3 or more servings of fruit per day may lower the risk of age-related macular degeneration (AMD), the primary cause of vision loss in older adults, by 36%, compared to persons who consume less than 1.5 servings of fruit daily. Although the intakes of vegetables, antioxidant vitamins and carotenoids were not strongly related to the incidence of either form of AMD, fruit intake was definitely protective against the severe form of this vision-destroying disease.

The bioavailability of carotenoids from fresh papaya fruit and its juice were found to be better as compared to dry slices. It was suggested that fresh papaya fruit and its juice can efficiently contribute to improve vitamin A status of the population. The high carotenoid contents of papaya can provide a better vitamin A value and anti-oxidative capacity to the people living in vitamin A deficient areas. Veda et al. (2007) reported that the bioaccessibility of β-carotene from two different varieties of papaya was similar (31.4 to 34.3%). Addition of milk increased the bioaccessibility by 19 and 38% in these two varieties. They suggested that the addition of milk to mango and papaya pulp is advantageous to derive provitamin A activity.

Protection Against Lung and Prostate Cancers

Data from the *in-vivo* animal studies suggests that there is a relationship between vitamin A, lung inflammation, and emphysema. Laboratory animals fed on vitamin A-deficient diets developed emphysema (Li et al., 2003). Cigarette smoke contains a common carcinogen (benzo-(a)-pyrene) that can induce vitamin A deficiency.

Eating vitamin A rich foods such as papaya can help to counteract the effects of vitamin A deficiency caused by the benzo(a)pyrene in cigarette smoke, and can thus greatly reduce the emphysema.

A number of studies have shown the protective effects of dietary intake of supplementary vitamin A as well as provitamin A containing vegetables and fruits in lung cancer. Jin et al. (2007) reported an inverse association between the consumption of vitamin A and pro-vitamin A rich vegetable and lung cancer in Taiwan. In a case control study, Jian et al (2007) observed that regular intake of lycopene-rich fruits (tomatoes, apricots, pink grapefruit, watermelon, papaya, and guava) and drinking green tea may greatly reduce the risk of developing prostate cancer in men.

They suggested that the synergistic protective effect of regular consumption of both the green tea and lycopene rich fruits was stronger than the protection provided by either of them individually.

ANTIFUNGAL AND ANTIBACTERIAL PROPERTIES

The ripe and unripe fruits have been reported to have significant antibacterial activity against *Staphylococcus aureus*, *Bacillus cereus*, *Bacillus subtilus Eschericia coli*, *Pseudomonas aeruginosa*, *Proteus vulgaris*, *Salmonella typhi*, and *Schigella flexneri* as well as antimicrobial activity against *Trichomonas vaginalis* trophozoites (Emeruwa, 1982, Osato et al., 1993, Dawkins et al., 2005). The latex proteins are considered to have the antifungal action (Giordani et al., 1996). The papaya chitinase has been reported to have antifungal and antibacterial activities (Chen et al., 2007). The bacteriostatic activity of papaya may also be attributed to its free radical scavenging potential. Papaya latex was effective in inhibiting the growth of *Candida albicans* and showed synergistic action when used mixed with Fluconazole. Papaya seed can be used as an antibacterial agent for *Eschericia coli*, *Staphylococcus aureus* or *Salmonella typhi*. However, further research is needed before advocating large-scale therapy. Being a good bleaching agent, papaya forms a vital ingredient in liquid and bar bath soaps, hand washes, astringents and even detergent bars.

ANTIFERTILITY AND CONTRACEPTIVE USE OF PAPAYA

In India, Bangladesh, Pakistan, Sri Lanka, and other countries, green papaya is traditionally used as a folk remedy for contraception and abortion, as its phytochemicals can negate the effects of progesterone (Oderinde, 2002). Ripe fruit however did not show any contraceptive properties. Unripe papaya is especially effective in large amounts or high doses. Ripe papaya is not teratogenic and will not cause miscarriage in small amounts. It is speculated that unripe papayas may cause miscarriage due to latex content, which may cause uterine contractions leading to a miscarriage.

Papaya seed extracts in large doses have a contraceptive effect on rats and monkeys, but in small doses have no effect on the unborn animals. In animal model studies on rats, langur monkeys and rabbits, the methanol sub-fraction of the extracts from seeds of *Carica papaya* L. has been shown to possess 100% contraceptive efficacy by inhibiting the sperm motility without any adverse systemic side effects on libido and toxicity and has been identified as a putative candidate for male contraception (Lohiya et al., 1999, Udoh et al., 2005, Lohiya et al., 2005 and 2008). Goyal et al. (2010) reported that the long term oral daily administration of methanol-sub-fraction (MSF) of the extract of seeds of *Carica papaya* affected the sperm parameters without any adverse systemic side effects and is clinically safe to use as male contraceptive.

These results verify the traditional use of papaya as a potential male contraceptive in some parts of Assam, India (Tiwari, 1982). The aqueous seed extracts have shown abortifacient properties in female Sprague Dawley rats, whereas the ether, alcoholic and aqueous extracts inhibited the ovulation in rabbits (Oderinde et al., 2002, Kapoor et al., 1974). The normal consumption of ripe papaya during pregnancy does not pose any significant danger. Only small quantities of ripe papaya fruit should be consumed during pregnancy, as green papaya and papaya seeds can cause miscarriage, particularly in large amounts, due to their contraceptive and abortifacient competence. However, the use of unripe or semi-ripe papaya could be considered as unsafe in pregnancy (Adebiyi et al., 2002).

Support for the Immune System and Anti-Mutagenic Properties

Vitamin C and vitamin A are both needed for the proper functioning of a healthy immune system. Papaya contains significant quantities of vitamin C and provitamin A (beta-carotene) and therefore may be a healthy fruit choice for preventing such illnesses as recurrent ear infections, colds and flu. The nutrients in papaya have also been shown to be helpful in the prevention of colon cancer. Rahmat et al. (2002) reported that both pure and extracted lycopene as well as papaya juice showed antiproliferative and anticancer properties on liver cell line (Hep G2) and the juice appeared to be more effective than the extracted lycopene in inhibiting the cancer cell growth. Papaya's fiber is able to bind to cancer-causing toxins in the colon and keep them away from the healthy colon cells. The other nutrients provide synergistic protection for colon cells from free radical damage to their DNA.

The fermented papaya products (FFP) have shown the ability to modulate the oxidative DNA damage due to H_2O_2 in rat pheochromocytoma tumor cells and protection of brain against oxidative damage in hypertensive rats (Aruoma et al., 1998). The FFP also exhibit supportive role in reducing the oxidative inflammatory damage in cirrhosis caused by hepatitis C virus (Marotta et al., 2007). The papaya juice when compared with standard antioxidant (vitamin E) showed comparable efficacy and safety in reducing the oxidative stress (Mehdipour et al., 2006). It has therefore been suggested that FPP (because their free radical scavenging potential) can be used as a prophylactic food against age-related and neurological diseases and may improve the lipid profile by inhibiting the lipid peroxidation (Imao et al., 1998, Rahmat et al., 2004).

The fruit juice of papaya may contain some anti-hypertensive agents, which may modulate the α-adenoreceptor activity and may help in lowering the blood pressure (Eno et al., 2000). The ethanol extracts of papaya seeds showed some diuretic properties equivalent to that of hydrochlorothiazide (Sripanidkulchai et al., 2001). The nutrients and phytochemicals contents of papaya may be effective in the prevention of diabetes mellitus complications (Savickiene et al 2002). The ethanol and aqueous extracts of papaya have also reported to have hepatoprotective activities against carbon tetrachloride (CCl_4) induced hepatotoxicity (Rajkapoor et al., 2002). Papaya seed extract may be nephroprotective in toxicity-induced kidney failure. Papaya seed extracts are currently being marketed as nutritional supplements with the claims to improve immunity and body functioning as they have been reported to have imunomodulatory and anti-inflammatory actions (Mojica-Henshaw et al., 2003).

Conclusion

Papaya can help in the digestion of food proteins, renewal of muscle tissues, revitalization of human body and slowing of ageing process as well as in the maintenance of body's homeostasis. The nutrients and phytochemicals contained in papaya can reduce inflammation, support the functioning of cardiovascular, immune and digestive systems and may also help in prevention of colon, lung and prostate cancers. Papaya pastes can be used externally as a treatment for skin wounds and burns. Because of its high vitamin A and carotenoids contents, it can reduce the cataract and age-related macular degeneration. More extensive studies and well-designed randomized clinical trials are however, required to explore the role of papaya in the prevention of different forms of cancers, cardiovascular diseases, age-related macular degeneration and the ailments of gastrointestinal system.

REFERENCES

Adebiyi, A, Adaikan, PG, (2005). Modulation of jejjunal contraction by extract of *Carica papaya* Linn. seeds. *Phytother. Res.,* 19: 628-632.

Adebiyi, A, Adaikan, PG, Prasad, RNV. (2003). Tocolytic and toxic activity of Papaya seed extracts on isolated rat uterus. *Life Sci.,* 74: 581-592.

Adebiyi, A, Adaikan, PG, Prasad, RNV. (2004). Histaminergic effect of a crude papaya latex on isolated guinea pig ilea strips. *Phytomed.* 11: 65-70.

Adetuyi, FO, Akinadewo, LT, Omosuli, SV, Lola, A.(2008). Antinutrient and antioxidant quality of waxed and unwaxed *Carica papaya* pawpaw fruit stored at different temperatures. *Afric. J. Biotech.* 7: 2920-2924.

Antoniades, C, Antonopoulos, AS, Tousoulis, D, Marinou, K, Stefandis, C. (2009). Homocysteine and coronary atherosclerosis: from folate fortification to the recent clinical trials. *Europe. Heart. J.* 30: 6-15.

Archbold, DD, Koslanund, R, Pomper, KW. (2003). Ripening and post-harvest storage of pawpaw. *Horticlculture Technology.* 13: 439-441.

ARDES Report No 22. (2007). The relationship of dietary carotenoids and vitamin A, E and C intake with age related macular degeneration in case-control study. *Arch. Opthalmology.* 125: 125-1232.

Arscott, SA, Howe, JA, Davis, CR, Tanumihardijo, SA. (2010). Carotenoid profiles in provitamin A- containing fruits and vegetables affect the bioefficacy in Mongolian gerbils. *Exp. Bio. Med.* 235: 839-848.

Aruoma, OI, Colognato, R, Fontana, I, Gartlon, J, Miglor, I, Koike, K, Coecke, S, Lame, E, Mersch-Sundermann, V, Laurenza, I, Benzi, I, Yoshino, F, Kobayashi, K, Lee, MC (1998). Molecular effects of fermented papaya preparation on oxidative damage, MAP kinase activation and modulation of the benzo(a)pyrene mediated genotoxicity. *Biofactor.* 26: 147-159.

Ballot, D, baynes, RD, Bothwell, TH, Gillooby, M, Macfarlane, BJ, MacPhail, AP, Lyons, G, Derman, DP, Bezwoda, WR, Torrance, JD. (1987). The effect of fruit juices and fruits on the absorption of iron from rice meal. *Br. J. Nutr.* 57; 331-343.

Bari, JM, Hasa, P, Absar, N, Haque, ME, Khuda, MIIF, Pervin, MM, Khatun, S, and Hussain, MI. (2006). Nutritional analysis of two local varieties of papaya (*Carica papaya* L.) at different maturity stages. *Pak. J. Bio. Sci.* 9: 137-140.

Bennett, RN, Kiddle, G, Wallsgrove, RM. (1997).Biosynthesis of benzylglucosinilates, cynogenic glucocides and phenyle prepanoids in *Carica papaya* Phytochemistry.45: 59-66.

Bhattacharjee, SK. (*Carica papaya.* In: hand book of medicinal plants, 3rd revised edition, 2001 by Shashi Jain (Ed), Pointer Publisher, Jaipur. Pp1-71.

Bruneton, J. (1999). *Carica papaya.* In: Pharmacognosy, Phytochemistry of medicinal plants, 2nd ed, Techniques and Documentation, France. Pp 221-223.

Burdick, Everette M. (1971). "Carpaine. An alkaloid of *Carica papaya*. Chemistry and Pharmacology. *Economic Botany*. 25: 363-365.

Chandrika, UG, Jansz, ER, Wickramasinghe, SMDN, Warnsurriya, ND (2003). Carotenoids in yellow and red fleshed papaya (*Carica papaya* L.). *J. Sci. Food Agric.* 83: 1279-1282.

Chen, YT, Hsu, LH, Huang, IP, Tsai, TC, Lee, GC, Shaw, JF. (2007). Gene cloning and characterization of a novel recombinant antifungal chitanase from papaya (*Carica papaya*). *J. agric. Food Chem.* 55(3): 714-722.

Christen, W, Liu, S, G, RJ, Gaziano, JM, Buring, JE. (2008). Dietary carotenoids, Vitamin C and E, and risk of cataract in woman. *Arch. Opthalmology*. 1261: 102-109.

Collard, E, Roy, S. (2010). Improved function of diabetic wound –site macrophages and accelareted wound closure in response to oral supplementation of a fermented papaya preparation. *Antioxidant Redox Signaling*. 13: 599-606.

Dawkins, G, Hewitt, H, Wint, Y, Obiefuna, PC, Wint, B.(2003). Antibacterial effect of *Carica papaya* fruit on common wound organisms. *West Indian Med. J.* 52(4): 290-292.

Dominguez, de Maria, P, Sinisterra, JV, Tsai, SW, Alcantara, AR. (2006). *Carica papaya* lipase (CPL) , an emerging and versatile biocatalyst. Biotechno. Adv. 24(5): 493-499.

Emeruwa, AC. (1982). Antibacterial substances from fruit extracts. *J. Nat. Products.* 45: 123-127.

Eno, AF, Owo, OI, Itam, EH, Konya, RS. (2000). Blood pressure depression by the fruit juice of *Carica papaya* L.nn. in renal and DOCA-induced hypertension in the rat. *Phytother. Res.* 14: 235-239.

Ezike, AC, Akah, PA, Okoli, CO, Ezeuchenne, NA, Ezeugwu, S. 2009. *Carica papaya* (Pawpaw) unripe fruit may be benifevcial in ulcer. *J. Medicinal Food.* 12(6): 1268-1273.

FAO, (2010). http://faostat.fao.org/site/567/desktopdefault.aspx?pageID=567#ancor (last accessed on 24.11.2010).

Gaby, AR. (2010). Nutritional treatment for acute myocardial infraction. *Alternative Med.* Rev.15: 113-123.

Ghosh, NK, Babu, SPS, Sukul, NC (1998). Antifilarial effect of a plant *Carica papaya*. *Jpn. J. Trop. Med. Hyg.* 26: 117-119.

Giordani, R, Cardenas, ML, Moulin-Tiraffort, J, Regli, P. (1996). Antifungal action of *Carica papaya* latex, isolation of fungal cell wall hydrolyzing enzymes. *Mycoses*. 34: 467-477.

Gouado, I, Schweigert, FJ, Ejoh, RA, Tchouanguep, MF, Camp, JV. (2007). Systematic levels of carotenoids from mangoes and papaya consumed in three forms (juice, fresh and dry slice). *Eurp. J. Clin. Nutr.* 61: 1180-1188.

Goyal, S, Manivannan, B, Ansari, AA, Jain, SC, Lohiya, NK. (2010). Safety evaluation of long term oral treatment of methanol sub-fraction of the seeds of *Carica papaya* as male contraceptives in male albino rats. *J. Ethnopharmacology.* 127: 286-291.

Hewitt, H, Whittle, S, Lopez, S, Bailey, E, Weaver, S. (2002). Tropical use of papaya in chronic ulcer therapy in Jamaica. *West Indian Med. J.* 25(7): 636-639.

Hirschberg, J. (2001). Carotenoid biosynthesis in flowering plants. *Curr. Opin. Plants Biol*. 4: 210-218.

Hornick, CA, Sanders, LI., Lin, YC. (1978). Effect of carpaine, a papaya alkaloid, on the Circulatory function in the rat. Research Communications in Chemical Pathology and *Pharmacology*. 22: 277-289.

Idstein, H, Keller, T, Schrejer, P. (1985). Volatile constituents of mountain papaya (*Carica candamarcensis* synonym *Carica pubescens*) fruit. *J. Agric. Food Chem*. 33(4): 663-66.

Imao, K, Wang, H, Komatsu, M, Hiramatsu, M. (1998). Free radical scavenging activity of fermented papaya preparation and its effect on lipid per oxidation level and superoxide dismutase activity in iron-induced epileptic foci of rats. *Biochem. Mol. Bio. Int.*, 45: 11-23.

Iwu, MM. (1993). Handbook of African Medicinal Plants. CRC press, BOCA Raton, FL, pp 141-143.

Jarvik, GP, Tsai, NT, Mckinstry, LA. (2002). Vitamin C and E intake is associated with increased paraoxonase activity. *Arteriscler. Thomb. Vasc. Biol.* 22: 1329-1333.

Jian L, Lee AH, Binns CW.2007. Tea and lycopene protect against prostate cancer. *Asia Pac. J. Clin. Nutr.* 16 (Suppl 1): 453-457.

Jiao, Z, Deng, J, Li, Gongke, Zhang, Z, Cai, Z. (2010). Study on the compositional differences between transgenic and non-transgenic papaya (*Carica papaya* L.). *J. Food Comp. Anal.* 23: 640-647.

Jin, YR, Lee, MS, Lee, JH, Hsu, HK, Lu, JY, Chao, SS, Chen, KT, Liou, SH, Ger, LP. (2007). Intake of vitamin A-rich foods and lung cancer risk in Taiwan: with special reference to garland chrysanthemum and sweet potato leaf consumption. *Asia Pacific J. Clin. Nutr.,* 16: 477-488.

Kapoor, M, Garg, sk, Mathur, VS. (1974). Antiovulatory activity of five indigenous plants in rabbit. Ind. *J. Med. Res.* 62(8): 1225-1225.

Koslanund, R. (2003). Ethylene production, fruit softening and their manipulation during pawpaw ripening. Ph.D diss. Uni. Ky., Lexington.

Kumar, D, Misra, SK, Tripathi, HC. (1991). Mechanism of anthelimintic action of benzylisothiocynate. *Fitoterapia.* 62: 403-410.

Li T, Molteni, A, Latkovich P, Castellani W, Baybut RC. (2003). Vitamin A depletion induced by ciggarate smoke is associated with the development of emphysema in rats. *J. Nutr.* 133: 2629-2634.

Lim, YY, Lim, TT, Tee, JJ. (2007). Antioxidant properties of several tropical fruits: A comparative study. *Food Chem.* 103: 1003-1008.

Lohiya, NK, Manivannan, B, Goyal, S, Ansari, AS. (2008). Sperm motility inhibitory effect of the benzene chromatographic fraction of the chloroform extract of seed of *Carica papaya* in langur monkey. *Presbytisentellu s entellus. Asian J. Androl.* 10: 298-306.

Lohiya, NK, Misra, RK, Pathak, N , Maniyanann, B. (1999). Reversible contraception with chloroform rextract of *Carica papaya* Linn. Seeds in male rabbits. *Reprod. Toxico.* 13: 59-66.

Lohiya, NK, Misra, RK, Pathak, N , Maniyanann, B, Bhande, SS, Panneerdoss, S, (2005). Efficacy trial on the purified compounds of the seeds of *Carica papaya* for male contraception in albino rats. *Reproductive Toxicology.* 20: 135-148.

Luximonn-Ramma, A, Bahourn, T, Crozier, A. 2003. Antioxidant action and phenolic and vitamin C content of common Mauritian exotic fruits. *J. Sci. Food Agric.* 83: 496-502.

Marotta, F, Yoshida, C, Barreto, R, Naito, Y, Packer, I. (2007). Oxidative-Inflammatory damage in cirrhosis: effect of vitamin E and A fermented papaya preparation. *J. Gastroentero. Hepatol.* 22: 697-703.

Marotta, F. Pavasuthipaisit, K, Yoshida, C, Alberjaci,F, M, Marandola, P. (2006a) Relationship between aging and susceptibility of erythrocytes to oxidative damage, in review of nutraceutical intervention. *Rejuven. Res.* 9: 227-230.

Marotta, F. Weksler, M, Naito, Y, Yoshida, C, Yoshioka, M, Marandola, P. (2006b). Nutraceutical supplementation, effect of fermented papaya preparation on redox status and DNA damage in healthy elderly individuals and relationship with GSTM 1 genotype, a randomized, placebocontrolled cross-over study. *Ann. N. Y. Acad. Sci.* 1067: 400-407.

Mehdipour, S, yasa, N, Dehgan, G, Khorasani, R, Mohammadirad, A, Rahimi, R, Abdollahi, M. (2006). Antioxidant potential of Iranian *Carica papaya* juice in-vitro and in-vivo are comparable to alpha-tocopherol. *Phytother. Res.* 20: 591-594.

Miyoshi, N, Takabayashi, S, Osawa, T, Nakamura, Y. (2004). Benzyl isothiocynate inhibits excessive superoxide generation in inflammatory leucocytes: implication of prevention against inflammation-related carcinogenesis. *Carcinogenesis.* 25: 567-575.

Miyoshi, N, Uchida, K, Osawa, T, Nakamura, Y. (2007). Selective cytotoxicity of benzyl isothiocynate in proliferating fibroblastoid cells. 120: 482-492.

Moijca-Henshaw, MP, Fransisco, Ad, De Guzman, F , Tingo, XT. (2003). Possible immunumodulatory action of *Carica papaya* seed extract. *Clin. Hemorheol. Microcirc.* 29: 219-229.

Moya-Leon, MA, Moya, M, Herrera, R. (2004). Ripening of mountain papaya (*Vasconcellea pubscens*) and ethylene dependence of some ripening events. *Postharvest Bio. Techno.* 34: 211-218.

Murcia, MA, Jimnez, AM, Martinez-Tome, M. 2001. Evaluation of the antioxidant properties of Mediterranean and tropical fruits compared with common food additives. *J. Food Protection.* 64: 2037-2046.

Nakamura, Y, Kawakami, M, Yoshihiro, A, Miyoshi, N, Ohigashi, H, Kawai, K, Osawa, T, Uchida, K. 2000. Involvement of the mitochondrial death pathway in chemo preventive benzyl isothiocynate-induced apoptosis. *J. Bio. Chem.* 277: 8492-8499.

Nakamura, Y, Ohigashi, H, Masuda, S, Murakami, A, Morimitsu, Y, Kawamoto, Y, Osawa, T, Imagawa, M, Uchida, K. (2002).Redox regulation of glutathione s transferase induction by benzyl isothiocynate : correlation enzyme induction with the formation of reactive oxygen intermediates. *Cancer Res.* 60: 219-225.

Nakamura, Y, yoshimoto, M, Murrata, Y, Shimoishi, Y, Asai, Y, Park, EU, Sato, K, Nakamura, Y. (2007).Papaya seed represents a rich source of biologically active isothiocynate. *J. Ag Food Chem.* 55: 4407-4413.

Nayak, SB, Pinto Pereira, L, maharaj, D. 2007. Wound healing activity of (*Carica papaya* L.) in experimentally induced diabetic rats. *Ind. J. Exp. Bio.* 45: 739-743.

Nguyen, ML, Schwartz, SJ. 1999. Lycopene: Chemical and biological properties. *Food Tech.* 53: 38-45.

Oderinde, O, Noronha, C, Kusemiju, T, Okanlawon, OA. (2002). Abortifacient properties of aqueous extracts of *Carica papaya* Linn. seeds on female Sprague-Dawley rats. *Niger. Postgrad. Med.* J. 9: 95-98.

Okeniyi, JOA, Ogunleshi, TA, Oyelami, OA, Adeyemi, LA. 2007. Effectiveness of dried *Carica papaya* seeds against human intestine parasitosis: A pilot study. *J. Medicinal Food.* 10: 194-196.

Oliveira, DDA, Lobato, AL, Ribeiro, SMR, Santana, AMC, Chaves, JSP, Pinheiro-Santana, M. (2010). Carotenoids and vitamin C during handling and distribution of guava(Psidium gujava L) mango (mangifera indica L.) and papaya (Carica papaya L.) at commercial restaurant. *J. Agric. Food Chem.* 58: 6166-6172.

Oloyede, OI. 2005. Chemical profile of unripe pulp of *Carica papaya* . *Pak. J. Nutr.* 4(6): 379-381.

Osanto, JA, Korkina, LG, Santigo, LA, Afanas'ev, IB. 1995. Effects of bio-normalizer (a food supplementation) on free radical production by human blood neutrophils, erythrocytes and rat peritoneal macrophages. *Nutrition.* 11: 568-572.

Osato, JA, santigo, LA, Remo, GM, Cuadra, MS, Mori, A. (1993). Antimicrobial and antioxidant activities of unripe papaya. *Life Sci.* 53: 1383-1389.

Parker, TL, Esgro, ST, Miller, SA, Myers, LE, Meister, RA, Toshkov, SA, Engeseth, NJ. 2010. Development of an optimized papaya pulp nectar using a acombination of irradiation and mild heat. *Food Chem.* 118: 861-869.

Pattison DJ, Silman AJ, Goodson NJ, Lunt M, Bunn D, Luben R, Welch A, Bingham R. (2004). Inflammatory polynuritis: Prospected nested case-control study. *Ann. Rheum. Dis.* 63: 843-847.

Pieper, B, Caliri, MH. (2003). Non-traditional wound care: a review of evidence for the use of sugar, papaya/ papain and fatty acids. *J. Wound Ostomy Continence Nurs.* 30: 175-183.

Rahmat, A, Abu Bakar, MF, Faezah, N, Hambali, Z. (2004). the effects of consumption of guava (Psidium guajava) or p[apaya (*Carica papaya*) on total antioxidant and lipid profile in normal amle youth. *Asia Pac. J. Clin. Nutr.* 13 suppl. S106.

Rahmat, A, Rosli, R, Zain, WNWM, Endrini, S, Sani, A. (2002). Antiproleferative activity of pure lycopene compared to both extracted lycopene and juices from watermelon (Citrullus vulgarize) and papaya (Carica papaya) on human breast and liver cancer cell lines. *J. Med. Sci.* 2: 55-58.

Rajkapoor, B, Jaykar, B, Kavimani, S, Murgesh, N. (2002). Effect of dried fruits of *Carica papaya* Linn. on hepatotoxicity. *Pharma Bull.* 25: 1345-1346.

Roberts, M, Minott, DA, Tannant, PF, Jackson, JC. (2008). Assessment of compositional changes during repining of transgenic papaya modified for protection against papaya ring spot virus. *J. Sci. Food Che.* 88: 1911-1920.

Satrija, F, Nansen, P, Bjorn, H, Murtini, S, He, S. (1994). Effect of papaya latex against *Ascaris suum* in naturally infected pigs. *J. Helminthol.* 64: 343-346.

Satrija, F, Nansen, P, Murtini, S, He, S. (1995). Anthelmic activity of papaya latex against *Heligmosomoides polygyrus*.infections in mice. *J. Ethnopharmocol.* 48: 161-164.

Savickiene, N, Dagilyte, A, LukoSius, A, Zitkevicius, V. 2002. Importance of biologically ative components and plants in the prevention of complications of diabetes mellitus. Medicine (Kaunas). 38: 970-975.

Schürks, M, Glynn, R, Rist, PM, Tzourio, C, Kurth, T. (2010). Effects of vitamin E on stroke subtypes: meta-analysis of randomized controlled trials. *BMJ.* 341: c5702.

Seo, H, Oh, H, Park, H, park, M, Jang, Y, Lee, M. (2010). Contribution of dietary intake s of antioxidants to homocystein-induced low density lipoprotein (LDL) oxidation in atheroscelerotic patients. *Yonsei Med. J.* 51: 526-533.

Simirgiotis, MJ, caligari, PDS, Schmeda-Hirschmann, G. 2009. Identification of phenolic compounds from the fruit of mountain papaya *Vasconcellea pubescens* A. DC grown in Chile by liquid chromatography-UV. *Food Chem.* 115: 775-784.

Sirichakwal, PP, puwastien, P, polngam, J, Kongkachuichai, R. (2005). Selenium content of Thai foods. *J. Food Comp. Ana.* 18: 47-59.

Skelton, RL, Yu, Q, Srinivasan, R, Manshardt, R, Moore, PH, Ming, R. (2006). Tissue differential expression of lycopene ß-cyclase gene in papaya. *Cell Res.* 16: 731-739.

Sripanidkulchai, B, Wongpanich, V, Laupattarakasem, P, Suwansaksari, J, and Jirakulsomchok, D, (2001). Diuretic effects of selected Thai indigenous medicinal plants in rats. *J. Ethnophramacology.* 75: 185-190.

Starley, IF, Mahammed, P, Schneider, G, Bickler, SW. (1999). The treatment of pediatrics burns using tropical papaya. *Burns.* 25: 636-639.

Telgenhoff, D, Lam, K, Ramsy, S, Vasquez, V, Villareal, K, Slusarewicz, P, Attar, P, Shroot, B. (2007). Influence of papain urea copper chlorophyll in on wound matrix remodeling. *Wound Repair Regen.* 15: 727-735.

Tiwari, KC, Majumdar, R, Bhattarayajee, S. (1982). Folklore information from Assam for family planning and birth control. *Int. J. Crude Drug Res.* 20: 133-137.

Tona, L, Kambu, K, Ngimbi, N, Kimanga, K, Vlietinck, AJ. (1998). Antiamoebic and phytochemical screening of some Congolese medicinal plants. *J. Ethnpharmacology.* 61: 57-65.

Udoh, P, Essien, I, Udoh, F. (2005). Effect of *Carica papaya* (pawpaw) seeds extract on the morphology of pituitary –gonadal axis of male Wistar rats. *Phytother. Res.* 19: 1065-1068.

USDA National Nutrient Database for Standard Reference, (2006). http://www.nal.usda.gov/fnic/foodcomp/search

Veda, S, Patel, K, Sriniwasan, K. (2007). Varietal differences in the bioaccessibility of ß-carotene from mango (*Mangifera indica*) and papaya (*Carica papaya*) fruits. *Agri. Food Chem.* 55: 7931-35.

Wall, MM. (2006). Ascorbic acid, vitamin A and mineral composition of banana (*Musa* sp.) and papaya (*Carica papaya*) cultivars grown in Hawaii. *J. Food Comp. and Analysis.* 19: 434-445.

Wall, MM, Nishijima, KA, Fitch, MM, and Nishijima, WT (2008). Physicochemical, nutritional and microbial quality of fresh-cut and frozen papaya prepared from cultivars with varying resistance to internal yellowing disease. *J. Food Chem.* 33: 131-149.

Yamamoto, H. (1964). Differences in carotenoid composition between red and yellow fleshed papaya. *Nature.* 201: 1049-50.

In: Natural Products and Their Active Compounds ... ISBN: 978-1-62100-153-9
Editors: M. Essa, A. Manickavasagan, and E. Sukumar

Chapter 18

HYPOLIPIDAEMIC AND ANTIHYPERLIPIDAEMIC EFFECT OF *HIBISCUS SABDARIFFA* LEAVES IN EXPERIMENTAL HYPERAMMONEMIC RATS

Mohamed M. Essa[1,2,3,*], Gilles J. Guillemin[2], F. Lukmanul Hakkim[1] and P. Subramanian[4]

[1]Dept of Food Science and Nutrition, College of Agriculture and Marine Sciences, Sultan Qaboos University, Oman
[2]Neuropharmacology group, Dept of Pharmacology, College of Medicine, University of New South Wales, Sydney, Australia
[3]Developmental Neuroscience Lab, NYSIBR, Staten Island, NY, US
[4]Department of Biochemistry and Biotechnology, Faculty of Science, Annamalai University, Annamalainagar, Tamil Nadu, India

ABSTRACT

Effect of *Hibiscus sabdariffa* (is an edible medicinal plant, indigenous to India, China and Thailand and is used in Ayurveda and traditional medicine), leaf extract (HSEt) on the levels blood ammonia, and serum lipid profiles (cholesterol, triglycerides, phosphor lipids, free fatty acids) were studied for its protective effect during ammonium chloride induced hyperammonemia in Wistar rats. Ammonium chloride (AC) treated rats showed a significant increase in the levels of circulatory ammonia and lipid profiles. These changes were significantly decreased in HSEt and AC treated rats. Our results indicate that HSEt offers protection by influencing the levels of ammonia and lipid profiles in experimental hyperammonemia and this could be due to its (i) ability to detoxify excess ammonia, urea and creatinine, (ii) free radical scavenging property both in vitro and in vivo by means of reducing lipid peroxidation and the presence of natural antioxidants. Hence, it may be concluded that the hypolipidaemic and antihyperlipidaemic effects produced by the HSEt may be due to the presence of flavonoids and other polyphenolic compounds. But the exact underlying mechanism is remains to be elucidated.

[*] Corresponding author : drmdessa@gmail.com

Keywords: Hyperammonemia, *Hibiscus sabdariffa*, Cholesterol, ammonia, triglycerides, phospholipids, free fatty acids

INTRODUCTION

Hyperammonemia is a major contributing factor to neurological abnormalities observed in hepatic encephalopathy and in congenital defects of ammonia detoxication [1]. Ammonia is a neurotoxin that has been strongly implicated in the pathogenesis of hepatic encephalopathy [2]. Ammonia has also been a major pathogenetic factor associated with inborn errors of urea cycle, Reye's syndrome, organic acidurias and disorders of fatty acid oxidation [3]. It was reported that elevated levels of ammonia causing irritability, somnolence, vomiting, seizures, derangement of cerebral function, coma and death [1, 4]. Ammonia toxicity results in lipid peroxidation and free radical generation, which cause hepatic dysfunction and failure and significantly increase number of brain peripheral benzodiazepine receptors and could increase the affinity of ligands for these receptors that might enhance GABA (gamma amino butyric acid) adrenegergic neurotransmission. These changes probably contribute to deterioration of intellectual function, decreased consciousness, coma and death [5, 6, 7].

In recent years of scientific investigations, attention has been drawn to the health promoting activity of plant foods and its active components. *Hibiscus sabdariffa* (Linn) (family Malvaceae), is an annual dicotyledonous herbaceous shrub popularly known as 'Gongura' in Hindi or 'Pulicha keerai' in Tamil. This plant is well known in Asia and Africa and is commonly used to make jellies, jams and beverages. In the Ayurvedic literature of India, different parts of this plant have been recommended as a remedy for various ailments such as hypertension, pyrexia, liver disorders and antidotes to poisoning chemicals (acids, alkali, pesticides) and venomous mushrooms [8]. Anthocyanins, flavonols, protocatechuic acid (PCA), along with others, have been identified as contributors to the observed medicinal effect of *Hibiscus sabdariffa* [9]. Anthocyanin and PCA have been shown to have antioxidant activity and to offer protection against atherosclerosis and cancer [10]. Compared to common antioxidants such as ascorbate, anthocyanins were found to be much more potent antioxidant [11]. It is well-documented that most medicinal plants are enriched with phenolic compounds and bioflavonoids that represent potent antioxidants [12]. There is currently a growing body of evidence that supplementing the human diet with antioxidants is of major benefit for human health and well-being.

Nowadays, the use of complementary/alternative medicine and especially the consumption of botanicals have been increasing rapidly worldwide, mostly because of the supposedly less frequent side effects when compared to modern western medicine. Both in conventional and traditional medicines, plants continue to provide valuable therapeutic agents [13]. Doubts about the efficacy and safety of currently available anti - hyperammonemic agents have prompted the search for safer and more effective alternatives [14]. To our best knowledge no scientific data regarding the hypolipidaemic and antihyperlipidaemic effect of *H. sabdariffa* leaves during hyperammonemia are available except in the treatise of Ayurvedic medicine. Thus, the present study was undertaken to evaluate the *protective effect of* ethanolic extract of *H. sabdariffa* leaves on lipid profile in ammonium chloride induced experimental hyperammonemic Wistar rats.

Materials and Methods

Animals

Adult male albino Wistar rats, weighing 180-200 g bred in the Central Animal House, Rajah Muthiah Medical College, Annamalai University, were used. The animals were housed in polycarbonate cages in a room with a 12 h day-night cycle, temperature of 22 ± 2°C and humidity of 45-64%. Animals were fed with a standard pellet diet (Hindustan Lever Ltd., Mumbai, India) and water *ad libitum.* All animal experiments were approved by the ethical committee, Annamalai University and were in accordance with the guidelines of the National Institute of Nutrition (NIN), Indian Council of Medical Research (ICMR), Hyderabad, India.

Plant Material and Preparation of Extract (HSEt)

The mature green leaves of *Hibiscus sabdarifa* were collected from Chidambaram, Cuddalore District, Tamil Nadu, India. The plant was identified and authenticated at the Herbarium of Botany Directorate in Annamalai University. A voucher specimen (No.3648) was deposited in the Botany Department of Annamalai University. The shade-dried and powdered leaves of *Hibiscus subdariffa* were subjected to extraction with 70% ethanol under reflux for 8 h and concentrated to a semi solid mass under reduced pressure (Rotavapor apparatus, Buchi Labortechnik AG, Switzerland). The yield was about 24% (w/ w) of the starting crude material. In the preliminary phytochemical screening, the ethanolic extract of HSEt gave positive tests for glycosides, anthocyanins, polyphenols and flavones [15]. The residual extract was dissolved in sterile water and used in the investigation.

Experimental Design

Hyperammonemia was induced in Wistar rats by daily intraperitoneal injections of ammonium chloride at a dose of 100mg/kg body weight for 8 consecutive weeks [16]. In this experiment, a total of 24 rats were used. Rats were divided into 4 groups, 8 animals each. Group 1: Control rats. Group 2: Rats orally administered with HSEt (250 mg/kg body weight) [17]. Group 3: Rats intraperitoneally treated with ammonium chloride (100 mg/kg body weight) [16]. Group 4: Rats treated with ammonium chloride (100 mg/kg) + HSEt (250 mg/kg). At the end of 8 weeks, all the rats were killed by decapitation after giving (Pento-barbitone sodium) anesthesia (60 mg/kg). Blood samples were collected for various biochemical estimations.

Blood ammonia levels were estimated by the method of Wolheim [18]. Circulatory Cholesterol levels were analyzed by the method of Zlatkis [19]. Serum triglycerides were analyzed by method of Foster and Dunn [20]. Phospholipids were analysed by the method of Zilversmit and Davis [21] and the serum free fatty acids were estimated by the method of Falholt [22].

Statistical Analysis

Statistical analysis was carried out by analysis of variance (ANOVA) and the groups were compared using Duncan's Multiple Range Test (DMRT).

RESULTS

Table 1 shows the levels of blood ammonia of control and experimental animals. Circulatory ammonia levels increased significantly and the levels reduced significantly in ammonium chloride and HSEt treated rats. Normal rats treated with HSE*t* showed no significant differences in levels of ammonia when compared with control rats (Table 1). The effect of HSEt on the levels of serum cholesterol, triglycerides, free fatty acids and phospholipids in normal and experimental groups illustrated in Table 1. The levels of all these lipids were significantly increased in AC induced hyperammonemic rats whereas the administration of HSEt to AC treated rats significantly restored all these changes to almost normal levels (Table 1).

Table 1. Effect of HSEt on changes in the blood ammonia and serum lipid profiles of normal and experimental rats

Group	Cholesterol (mg/dl)	Triglycerides (mg/dl)	Free fatty acids (mg/dl)	Phospholipids (mg/dl)	Blood ammonia (μmol/L)
1. Normal	84.61 ± 4.03^{a}	60.17 ± 3.60^{a}	73.2 ± 3.95^{a}	101.6 ± 5.2^{a}	88.28 ± 6.72^{a}
2. Normal + HSEt	79.30 ± 3.61^{a}	58.89 ± 3.33^{a}	70.3 ± 3.81^{a}	97.3 ± 4.92^{a}	82.30 ± 6.27^{a}
3. AC treated	171.25 ± 12.5^{b}	91.85 ± 5.45^{b}	149.2 ± 7.92^{b}	171.1 ± 11.7^{b}	331.21 ± 25.22^{b}
4. AC + HSEt	113.25 ± 9.86^{c}	71.24 ± 4.54^{c}	89.4 ± 4.99^{c}	113.25± 9.86^{c}	136.70 ± 12.46^{c}

Values are given as mean ± S.D from six rats in each group.
ANOVA followed by Duncan's multiple range test.
Values not sharing a common superscript (a, b, c) differ significantly at $p \leq 0.05$.

DISCUSSION

Ammonia is removed either in the form of urea in periportal hepatocytes and/or as glutamine in perivenous hepatocytes in liver [23]. An increased level of circulatory ammonia might indicate a hyperammonemic condition in the rats treated with ammonium chloride [6, 7, 16, 24]. Decreased levels of blood ammonia, HSEt and ammonium chloride treated rats show the significant anti-hyperammonemic activity of this plant and it was reported that *H. sabdariffa* normalized the levels of ammonia, urea and creatinine during hyperammonemic conditions [17]. Our present findings have an in agreement with these reports and the exact mechanism remains to be explored.

Oxidative stress mediated lipid peroxidation was also shown as one of the characteristic features of hyperammonemia [6, 7, 25]. Free radical damage to cellular components and decomposition of hydroperoxide formed from oxidative breakdown of poly unsaturated fatty acids (PUFAs) are important factors in the development of cellular toxicity and pathology caused by lipid peroxidation. In a large number of tissues, it is now generally accepted that lipid peroxides play an important role in liver, kidney and brain toxicity [26].

In the present study, elevated levels of ammonia caused a significant rise in serum lipids (cholesterol, triglycerides, phospholipids and free fatty acids) were observed. These findings indicate that hyperammonemia may be accompanied by complications of atherosclerosis. Previous studies from our lab reported that the levels of serum and tissue lipid were elevated during hyperammonemic conditions [6, 24, 27, 28]. Previous studies have indicated that viral infections, aspirin treatment and hyperammonemia are associated with Reye's syndrome. It has also been reported that free fatty acids in serum and total lipids in the liver of Reye's syndrome patients are elevated during illness [29].

It was reported that ammonium (chloride/acetate) salts may deplete levels of α-KG and other Krebs cycle intermediates [6, 7, 27, 30, 31] and thus elevate the levels of acetyl coenzyme A. The elevated levels of acetyl coenzyme A may increase levels of lipid profile (free fatty acids, triacylglycerols, phospholipids, and cholesterol), as observed in our study. Another important function of α-KG occurs in the formation of carnitine [6, 7, 21, 23, 27, 30, 31]. Carnitine acts as a carrier of fatty acids into cell mitochondria so that proper catabolism of fats can proceed [23, 27, 30, 31].

The decreased α-KG and other Krebs cycle intermediates levels in rats treated with AC might have led to accumulation of fatty acids [27].

Lowering of serum lipid levels through dietary or drug therapy seems to be associated with a decrease in the risk of various vascular diseases [32]. In the present study, HSEt treatment to hyperammonemic rats caused a significant decrease in serum lipids (cholesterol, triglycerides, phospholipids and free fatty acids).

The effect of HSEt on controlled mobilization of serum cholesterol, triglycerides, phospholipids and free fatty acids presumably mediated possibly by controlling the hydrolysis of certain lipoproteins and their selective uptake and metabolism by different tissues. It was reported that the aqueous extracts from the dried calyx of *H. sabdariffa* possess both antioxidant effects and hypolipidemic effects against LDL oxidation in vivo [33].

Previous reports suggest that *H. sabdariffa* could exhibit antihypertensive and cardio-protective effects in vivo and support the public belief that *H. sabdariffa* may be a useful antihypertensive agent [34, 35, 36].

In conclusion, the alterations in serum lipids during experimental hyperammonemia were restored to near normal levels by HSEt treatment. Phytochemical studies of *H. sabdariffa* plant extract revealed the presence of flavonoids and other polyphenolics in various concentrations. Several authors reported that, flavonoids and phenolic compounds have hypolipidaemic and antihyperlipidaemic effects [37, 38]. Hence, it may be concluded that the hypolipidaemic and antihyperlipidaemic effects produced by the HSEt may be due to the presence of flavonoids and other polyphenolic compounds.

References

[1] Rodrigo R, Montoliu C, Chatauret N, Butterworth R, Behrends S, Olmo JAD, Serra MA, Rodrigo JM, Erceg S, Felipo V. Alterations in soluble guanylate cyclase content and modulation by nitric oxide in liver disease. *Neurochem. Intl.* 2004; 45: 947-953.

[2] Norenberg MD, Rama Rao KV, Jayakumar AR. Ammonia Neurotoxicity and the Mitochondrial Permeability Transition. *J. Bioenerget. Biomemb.* 2004; 36: 303-307.

[3] Qureshi IA,. Rama Rao KV. Decreased brain cytochrome C oxidase activity in congenitally hyperammonemic spf mice: Effects of acetyl-L-carnitine. In (R.L. Mardini, ed.) Advances in Hepatic Encephalopathy and Metabolism in Liver disease, Ipswich Book Company, UK, 1997; 385–393.

[4] Murthy CR, Rama Rao KV, Bai G,. Norenburg MD. Ammonia induced production of free radicals in primary cultures of rat astrocytes. *J. Neurosci. Res.* 2001; 66: 282-288.

[5] Kosenko E, Kaminsky Y, Lopata O, Muravyov N, Kaminsky A, Hermenegildo C, Stavroskaya IG, Felipo V. Nitroarginine, an Inhibitor of Nitric Oxide Synthase, Prevents Changes in Superoxide Radical and Antioxidant Enzymes Induced by Ammonia Intoxication. *Metabolic Brain Dis.* 1997; 13: 29-41.

[6] Lena PJ, Subramanian P. Effects of melatonin on the levels of antioxidants and lipid peroxidation products in rats treated with ammonium acetate. *Pharmazie.,* 2004; 59: 636-639.

[7] Essa MM, Subramanian P. Pongamia pinnata modulates the oxidant–antioxidant imbalance in ammonium chloride-induced hyperammonemic rats. *Fund Clin. Pharmacol.* 2006; 20:299–303.

[8] Chifundera K, Balagizi K, Kizungu B. Les empoisonnements et leurs antidotes en me□ decine traditionnelle au Bushi, Zaire. *Fitoterapia.* 1994:65:307-313.

[9] Seca AML, Silva AMS, Silvestre AJD, Cavaleiro JAS, Domingues FMJ, Neto CP. Phenolic constituents from the core of kenaf (Hibiscus cannabinus). *Phytochemistry.* 2001:56:759-767.

[10] Satue-Gracia MT, Heinonen M, Frankel EN. Anthocyanins as antioxidants on human low-density lipoprotein and lecithin- liposome systems. *J. Agric. Food Chem.* 1997:45:3362-3367.

[11] Wang H, Cao G, Prior RL. Oxygen radical absorbing capacity of anthocyanins. *J. Agric. Food Chem.* 1997:45:302-309.

[12] Shirwaikar A, Malini S, Kumari SC. Protective effect of Pongamia pinnata flowers against cisplatin and gentamicin induced nephrotoxicity in rats. *Ind. J. Exp. Biol.* 2003:41(1):58-62.

[13] Hu X, Sato J, Oshida Y, Yu M, Bajotto G, Sato Y. Effect of Goshajinki-gam on insulin resistance in STZ induced diabetic rats. *Diab. Res. Clin. Pract.* 2003:59: 101-103.

[14] Essa MM, Subramanian P, Suthakar G, Manivasagam T, Dakshayani KB. Protective influence of Pongamia pinnata (Karanja) on blood ammonia and urea levels in ammonium chloride-induced hyperammonemia. *J. Appl. Biomed.* 2005:3:133-138.

[15] Trease CE, Evan VC. Pharmacopoeial and related drugs of biological origin. Part V Pharmacognosy. Saunders, London 1959, 161:466.

[16] Essa MM, Subramanian P, Suthakar G, Manivasagam T, Dakshayani KB. Protective influence of Pongamia pinnata (Karanja) on blood ammonia and urea levels in ammonium chloride-induced hyperammonemia. *J. Appl. Biomed.* 2005; 3:133-138.

[17] Essa MM, Subramanian P. *Hibiscus sabdariffa* Affects Ammonium Chloride-Induced Hyperammonemic Rats. *eCAM.* 2007 4(3):321-325.

[18] Wolheim DF. Preanalytical increase of ammonia in blood specimens from healthy subjects. *Clin. Chem.* 1984; 30:906-908.

[19] Zlatkis A, Zak B, Bogle GJ. A method for the determination of serum cholesterol. *J. Clin. Med.* 1953, 41:486-492.

[20] Foster LB, Dunn RT. Stable reagents for determination of serum triglycerides by colorimetric hantzsch condensation method. *Clin. Chem.* 1973 ; 19:338-340.

[21] Zilversmit DB, Davis AK. Micro determination of phospholipids by TCA precipitation. *J. Lab. Clin. Med.* 1950; 35:155-159.

[22] Falholt K, Falholt W, Lund B. An easy colorimetric method for routine determination of free fatty acids in plasma. *Clin. Chim. Acta.* 1973 ; 46:105-111.

[23] Nelson DL, Cox MM. Lehninger Principles of Biochemistry, Macmillan, London, 2000.

[24] Lena PJ, Subramanian P. Evaluation of the antiperoxidative effects of melatonin in ammonium acetate-treated Wistar rats. *Pol. J. Pharmacol.* 2003; 55:1031-36.

[25] Vidya M, Subramanian P. Effects of α-ketoglutarate on antioxidants and lipid peroxidation products in rats treated with sodium valproate. *J. Appl. Biomed.* 2006; 4:141-146.

[26] Shafiq-ur-Rehman, Mahdi AA, Hasan M. Trace metal-induced lipid peroxidation in biological system. *SFRR-India Bull.* 2003, 2:12-18.

[27] Velvizhi S, Dakshayani KB, Subramanian P. Protective influences of alpha-keto glutarate on lipid peroxidation and antioxidant status ammonium acetate rats. *Ind. J. Exp. Biol.* 2002: 40:1183-1186..

[28] Dakshayani KB, Velvizhi S, Subramanian P. Effects of ornithine alpha ketoglutarate on circulatory antioxidants and lipid peroxidation products in ammonium acetate treated rats. *Ann. Nutr. Metab.* 2002; 46:93-96.

[29] Deshmukh DR, Deshmukh GD, Shope TC, Radin NS. Free fatty acids in an animal model of Reye's syndrome. *Biochimica et Biophysica Acta.* 1983; 753(2):153-158.

[30] Hwu W, Chiang SC, Chang MH, Wang TR. Carnitine transport defect presenting with hyperammonemia: Report of one case. *Chung Hua Min Muo Hsiao Erh Ko I Heueh Hui Tsa Chih*, 2000; 41:36-38.

[31] Yamamoto H. Hyperammonemia increased brain and aromatic amino acids in mice. *Toxicol. Appl. Pharmacol.* 1989; 99:412-420.

[32] Brown GB, Xue-Qiao Z, Sacco DE, Alberts JJ. Lipid lowering and plaque regression. New insights into prevention of plaque disruption and clinical events in coronary disease. *Circulation.* 1993; 87:1781-1791.

[33] Hirunpanich V, Utaipat A, Morales NP, Bunyapraphatsara N, Sato H, Herunsale A, Suthisisang C. Hypocholesterolemic and antioxidant effects of aqueous extracts from the dried calyx of Hibiscus sabdariffa L. in hypercholesterolemic rats. *J. Ethnopharmacol*. 2006, 103:252-260.

[34] Faraji M, Tarkhani AH.The effect of sour tea (Hibiscus sabdariffa) on essential hypertension. *J. Ethnopharmacol.* 1999, 65:231-236.

[35] Odigie IP, Ettarh RR, Adigun S. Chronic administration of aqueous extract of Hibiscus sabdariffa attenuates hypertension and reverses cardiac hypertrophy in 2K-1C hypertensive rats. *J. Ethnopharmacol.* 2003, 86:181-185.

[36] Herrera-Arellanoa A, Flores-Romerob S, Chavez-Sotoc MA, Tortorielloa J. Effectiveness and tolerability of a standardized extract from Hibiscus sabdariffa in patients with mild to moderate hypertension: a controlled and randomized clinical trials. *Phytomedicine.* 2004, 11:375-382.

[37] Rupasinghe HP, Jackson CJ, Poysa V, DiBerado C, Bewely JD, Jenkinson J. Soyasapogenol A and B distribution in Soybean (Glycine Max L.Merr) in relation to seed physiology, genetic variability and growing location. *J. Agri. Food Chem.* 2003; 51:5888–5894.

[38] Leontowicz H, Gorinstein S, Lojek A, Leontowicz M, Ciz M, Soliva-Fortuny R. Comparative content of some bioactive compounds in apples, peaches, and pears and their influence on lipids and antioxidant capacity in rats. *J. Nutr. Biochem.* 2002; 13:603-610.

In: Natural Products and Their Active Compounds … ISBN: 978-1-62100-153-9
Editors: M. Essa, A. Manickavasagan, and E. Sukumar © 2012 Nova Science Publishers, Inc.

Chapter 19

WALNUTS (JUGLANS REGIA LINN) AND ITS HEALTH BENEFITS

Mohamed M. Essa*[1,2,3*]*, G. J. Guillemin*[2]*, Amani S. Al-Rawahi*[1]*,
***V. Singh*[1]*, N. Guizani*[1]* and A. M. Memon*[4]**

[1]Department of Food Science and Nutrition,
College of Agriculture and Marine Sciences,
Sultan Qaboos University, Oman
[2]Neuropharmacology group, Department of Pharmacology,
College of Medicine, University of New South Wales, Sydney, Australia
[3]Developmental Neuroscience Lab, NYSIBR,
Staten Island, NY, US
[4]Washington State University, Pullman, WA, US

ABSTRACT

Walnuts *(Juglans regia* L.) belong to the family Juglandaceae and are a good source of fat, protein, vitamins and minerals along with phenolics which act as antioxidants. Walnuts are ranked second after blackberries for their antioxidant activities. Walnuts have high levels of melatonin (sleep hormone and antioxidant) and vitamin E. They are also a rich source of L-arginine, phospholipids, proteins, tocopherols, polysterols, squalene and unsaturated fatty acids. The whole walnut tree including the seeds, leaves, husks, nuts and kernels are rich in active phytochemicals and natural antioxidants, which have been proven to ameliorate/prevent many diseases including cardiovascular disease, diabetes, obesity, cancer, neurological diseases, etc. Walnuts are unique among nuts in that they are highest in alpha-linolenic acid (ALA - omega-3 fatty acid) levels. Hence, it they are gaining more importance in medicinal and pharmaceutical industries. In the last few decades, there has been tremendous interest in the use of walnuts as evidenced by the voluminous scientific work along with consumer's health awareness. This chapter will describe the medicinal properties and importance of walnuts.

[*] Corresponding author: drmdessa@gmail.com

Keywords: Walnuts, *Juglans regia,* health benefits, Melatonin, Vitamin E, natural products

INTRODUCTION

As modern medicines are associated with many side effects there is always a need for complementary and alternative medicines with fewer side effects. So, traditional herbal medicines are gaining more importance in the treatment of health problems. Research work related to medicinal plants and natural products has been increased and intensified worldwide as there is a need to explore the benefits from medicinal plants in order to decrease dependency on synthetic drugs (Amadou, 1998). Nowadays, nuts are gaining more importance as they are considered an important part of a healthy diet. Many studies have shown a positive link between increased nut consumption and decreased risk factors for many diseases, especially cardiovascular diseases due to high content of polyunsaturated fatty acids (Fraser *et al.* 1992, Hu *et al.* 1998, Sabate 1999).

JUGLANSREGIA LINN (WALNUTS)

Walnuts *(Juglans regia* L.) belong to the family Juglandaceae. *Juglans regia* L. is the Common walnut, Persian walnut, or English walnut and is native to a region stretching from the Balkans eastward to the Himalayas and southwest China. The largest forests are found in Kyrgyzstan (Fernandez *et al.,* 2000), where trees exist in extensive, nearly pure walnut forests at 1,000–2,000 m altitude (Hemery, 1998), specifically at Arslanbob in the Jalal-Abad Province. The walnut has spread and adapted to many regions throughout the world. The walnut tree (*Juglansregia* L.) is cultivated commercially throughout southern Europe, northern Africa, eastern Asia, the USA and western South America. Among South American cultivators, Argentina is the main producer, with about 8500 metric tons per year (Diana et al., 2008). According to 2008 Food and Agricultural Organization statistics, the production of whole walnut was around 1.5×10^6 tons globally (FAO, 2008). China is the top world producer and is followed by the USA, Iran, Turkey, Ukraine, Romania, France and India.

Taxonomy

Kingdom	Plantae
Division	Angiospermae
Class	Eudicots
Order	Fagales
Family	Juglandaceae
Genus	Juglans
Species	J. regia
Bionomical name	JuglansregiaLinn

BOTANICAL DESCRIPTION AND PHYTOCHEMICALS

Juglans regia is a large deciduous tree attaining heights of 25–35 m and a trunk up to a 2 m diameter. Commonly, the trunk is short with a broad crown. It requires full sunlight to grow well. Young bark is smooth and olive-brown in color and changes to silvery-grey when old. It has broad fissures with a rougher texture. Leaves are alternate, odd-pinnate with 5–9 leaflets. Margins of the leaflets are entire and 25–40 cm in length. The male flowers are in drooping catkins 5–10 cm long, and the female flowers are terminal and in clusters of two to five. The tree ripens in the autumn, displaying a fruit with a green, semi-fleshy husk and a brown corrugated nut. The whole fruit with the husk falls in autumn. The seed is large, with a relatively thin shell. It is edible with a rich flavor.

The scientific name *Juglans regia* is from Latin Jupiter *glans*, "Jupiter's acorn", and *regia*, meaning "royal" in Latin. Its common name, Persian walnut, indicates its origins in Persia in southwest Asia. 'Walnut' derives from the Germanic wal- for "foreign", recognizing that it is not a nut native to Western Europe. Synonyms for walnut include: *Juglans duclouxiana* Dode; *J. fallax* Dode; *J. kamaonia* (C. de Candolle) Dode; *J. orientis* Dode; *J. regia* var. *sinensis* (C. de Candolle); *J. sinensis* (C. de Candolle) Dode.

Walnuts are a rich source of unsaturated fatty acids, such as linoleic and oleic acid, which are susceptible to oxidation. Walnuts have lower content of α-tocopherol (an antioxidant) than other nuts, such as almonds, hazelnuts, peanuts, etc. (Kagawa, 2001), but they are readily preserved. This suggests that the nuts have enough antioxidants that they inhibit the lipid auto-oxidation. Serrano et al., 2005 studied the effect of addition of walnut to restructured beef steak by comparing amino acid content and fatty acid composition, including cholesterol content. They found that walnut products have significantly lower lysine/arginine ratios, significantly higher monounsaturated (MUFA) and n3 polyunsaturated (PUFA) fatty acids (mainly α-lionolenic acid), lower n6/n3 PUFA ratios, and higher ($P < 0.05$) poly-unsaturated/saturated fatty acid ratios. Further, there was significant decrease in the cholesterol content and significant increase in α-tocopherol after replacing raw meat with walnut. Moreover, as compared to the control sample, iron, calcium, magnesium and manganese were significantly increased in a sample with added walnuts. This study supports that walnuts are a good source of PUFA (Serrano et al., 2005).

Most of the polyphenolics are present in the seed coat of the walnut that is composed principally of ellagitannins (Fukuda et al., 2003; Colaric et al., 2005; Zhang et al., 2009). Fukuda et al., 2003 isolated three hydrolyzable tannins, glansrins A–C, together with adenosine, adenine, and 13 known tannins from the n-BuOH extract of walnuts (the seeds of *Juglans regia* L.).

They characterized glansrins A–C as ellagitannins with a tergalloyl group, or a related polyphenolic acyl group, on the basis of spectral and chemical evidence. Walnut polyphenols have superoxide dismutase (SOD)-like activity with EC_{50} 21.4-190 mM and also possess significant radical scavenging effect against DPPH with EC_{50} 0.34–4.72 mM.

Stampar et al., 2006 studied the phenolic composition of walnut husk of a Slovenian cultivar traditionally used for liqueurs. They identified thirteen phenolic compounds in walnut husks using HPLC and a PDA detector. Chlorogenic acid, caffeic acid, ferulic acid, sinapic acid, gallic acid, ellagic acid, protocatechuicacid, syringic acid, vanillic acid, catechin, epicatechin, myricetin, and juglone were identified in the walnut husks. Moreover, they also identified 1,4-naphthoquinone in walnut liqueur. Juglone was the major phenolic compound in the husk but its content was low in the liqueur. The phenolic content was quite low in the liqueur as compared to the source. This may be due to traditionally used methods for the preparation of liqueur (Stampar et al., 2006). Pereira et al., 2007 identified and quantified ten compounds from the leaves of walnuts (*Juglans regia* L.) of six different cultivars (Lara, Franquette, Mayette, Marbot, Mellanaise and Parisienne) from Portugal. They used reverse phase HPLC/DAD methods for the identification of the compounds: 3- and 5-caffeoylquinic acids, 3- and 4-p-coumaroyl-quinicacids, p-coumaric acid, quercetin 3-galactoside, quercetin 3-pentoside derivative, quer-cetin 3-arabinoside, quercetin 3-xyloside and quercetin 3-rhamnoside.

Global Usage of Walnuts

Especially within the food industry, walnuts (the seeds of *Juglans regia* L.) are of high economic interest. They are consumed fresh or toasted, alone or mixed with other products. Fresh seeds are consumed mainly as whole nuts or used in different confectioneries as they are good sources of fat, protein, vitamins and minerals (Martinez *et al.,* 2010). In addition to this, walnuts are considered a good source of polyphenolics including phenolic acids which are important for human health due to their antioxidant, antiatherogenic, anti-inflamatory and antimutagenic properties (Anderson *et al.,* 2001; Carvalho *et al.,* 2010). Green walnuts, shells, kernels, bark, green walnut husks (epicarp) and leaves along with dry fruit (nuts) are used in both the cosmetic and pharmaceutical industries in Portugal, (Stampar *et al.,* 2006).

Walnuts are considered a highly nutritious food. They are used in traditional medicine for cough and stomach ache treatments (Perry, 1980) and also in the treatment of cancer in Asia and Europe (Duke, 1989). Besides polyphenolics compounds, tocopherols occurring in walnut oil are important in providing protection against oxidation (Savage et al., 1999; Demirand Cetin, 1999). In England, walnuts are preserved in vinegar and used as pickles. In America, they are preserved in sugar syrup. In Italy, liquors are flavored with walnuts and walnut sauce is also available in Georgia and other places. In India, people offer dried walnuts to Mother Goddess *Vaisnav Devi* (Stone *et al.,* 2009). In Oman, the people of Jabal Akhdar eat walnuts with fresh pomegranate seeds. In Syria, Turkey and Iran, walnuts are used as a garnish on traditional sweets (Pereira *et al.,* 2008). Walnuts are eaten across the world as fresh or salted nuts or even mixed and flavored with other food ingredients such as olive oil and mayonnaise (Miele*et al.,* 2010; Jimenez-Gomez, *et al.,* 2009; Cortes *et al.,* 2006). Walnut oil has been produced around the world, but with minimal use due to its high prices. Walnut oil has also been employed in oil paint, as an effective binding medium, and is known for its clear, glossy consistency and non-toxicity. Some countries like Egypt, the US and Europe are employing walnut wood in the furniture industry.

Antioxidant Activities

Walnuts can be considered a good source of antioxidants, containing about 20mmol/100 grams most of which is found in its pellicles (Blomhoff et al., 2006). Walnuts ranked second after blackberries in antioxidant content among 1113 different foods tested in 2006 (Halvorsen et al., 2006). Reiter et al., 2005 reported that the melatonin concentration in walnuts was 3.5+/-1.0ng/g. The blood melatonin concentration increased after the consumption of walnuts in a rat model. The rats also demonstrated increased antioxidant capacity confirmed by trolox equivalent antioxidant capacity and ferric-reducing-ability of serum. The green husk of walnuts is a by-product of walnut production used very rarely. Utilization of this by-product as a source of phyto-chemicals will increase the value of walnut production. Further, other walnut products like fruits (Espín*et al.,* 2000; Li *et al.,* 2006, 2007; Pereira *et al.,* 2008), leaves (Pereira *et al.,* 2007b) and liqueurs from the green fruits (Stampar *et al.,* 2006) also showed antioxidant activities.

Pereira et al., 2007 studied the antioxidant activities of the walnut leaves (*Juglans regia* L.) of six different cultivars (Lara, Franquette, Mayette, Marbot, Mellanaise and Parisienne) from Portugal. They assessed the antioxidant activity using reducing power assay, DPPH (2,2-diphenyl-1-picrylhydrazyl) assay and β-carotene linoleate model system. They found that all the cultivars of the walnut leaves are high in antioxidant activity (EC_{50} values lower than 1 mg/mL). Out of all the cultivars, Lara was found to be the most effective in terms of antioxidant effect. Walnuts and almonds are nutritionally valuable in terms of chemical composition. Labuckas et al., 2008 studied the antioxidant capacity of the hulls and walnut flour along with the extraction of phenolic fraction from them. They also evaluated the effect of removing the hull from walnut flour on the solubility of protein fractions. They found that hull extract is rich in phenolics and possess higher antioxidant activity than walnut flour extract. Presence of phenolic compounds decreases the solubility of protein in the walnut flour. Recovery of protein improved significantly after dehulling of kernels. But, this recovery is influenced by the solvent system. Oliveira *et al.,* 2008 studied antioxidant activities in walnut (*Juglans regia* L.) green husks of five different cultivars (Franquette, Mayette, Marbot, Mellanaise and Parisienne). They analyzed the total polyphenolics and the antioxidant activities of aqueous extracts of walnut green husks. They found the total phenolic content ranged from 32.61 mg/g of GAE in Mellanaise cultiver to 74.08 mg/g of gallic acid equivalent (GAE) in Franquette cultivar. Further, they studied the antioxidant capacity of aqueous extracts using reducing power assay and also studied its scavenging effects on DPPH (2, 2-diphenyl- 1-picrylhydrazyl) radicals and β-carotene linoleate model system. They found the concentration-dependent antioxidative capacity in reducing power and DPPH assays, with EC_{50} values lower than 1 mg/ ml for tested extracts in all the cultivers. Walnut green husks may become an important source of compounds with health protective potential due to its antioxidant activity. Further, Mishra *et al.,* 2010 also supported the antioxidant nature of walnut. They studied the antioxidant activities of different dried fruits including almonds, walnuts, cashews, raisins and chironji by using different assays like reducing power, lipid peroxidation damage in bio-membranes, and determination of antioxidant enzyme activity (SOD and CAT) including Folin–Ciocalteu assay to characterize total polyphenols. Based on lipid peroxidation assay they found higher values of antioxidant activity of the methanolic extract of walnuts as compared to other dried fruits. In addition to this, walnut also showed the highest phenolic content, followed by almonds, cashews, chironji and raisins. Overall, walnuts possess the best antioxidant properties, confirmed by the lower EC_{50} values in all assays except in antioxidant enzymatic activity.

Qamar and Sultana, 2010 found that the total polyphenolic content of *J. regia* kernel extract was 96+0.81 mg GAE/g dry weight of extract which is higher than the total phenolic content of walnut green husks. Further, they found high free radical scavenging activity of the extract measured by DPPH (2,2-Diphenyl-1-Picrylhydrazyl) assay. Therefore, *J. regia* was found to possess antioxidant activity. Salcedo et al., 2010 selected walnuts and almonds as model systems in order to characterize the possible interactions among oxidisable substances, pro-oxidant species and antioxidants. Lipoxygenase activities in a direct micelle system were determined for these nuts according to the globulin contents in their soluble protein fraction. Walnut lipoxygenase activity was 1.5 times higher than that of almonds. Hydrolysable and condensed phenolic compounds were analyzed in hydrophilic as well as lipophilic fractions along with the antioxidant activities. They found that almonds have significantly lower phenolics and lower antioxidant activity as compared to brown skins and whole walnuts. Further, lipoxygenases of walnuts and almonds acted as a catalyst towards the oxidation of linoleic acid, its main lipid component.

Further, Carvalho et al., 2010 analyzed the methanolic and petroleum ether extract of walnut seeds, green husks and leaves (*Juglans regia* L) for the total polyphenolic content using antioxidant assay (DPPH), which included 2,2'-azobis(2-amidinopropane) dihydrochloride (AAPH)-induced oxidative hemolysis of human erythrocytes. Methanolic seed extract exhibited the highest total phenolic content (116 mg GAE/g of extract) and DPPH scavenging activity (EC_{50} of 0.143 mg/mL), followed by leaf and green husk extract. On the other hand, antioxidant action was lower or absent in petroleum ether extracts. Although all the methanolic extracts significantly protected the erythrocyte membrane from hemolysis, the inhibitory efficiency of the leaf extract was much stronger (IC_{50} of 0.060 mg/mL) as compared to green husks and seeds (IC_{50} of 0.127 and 0.121 mg/mL, respectively) in both a time- and concentration-dependent manner. This research indicates that walnut tree comprises a good source of natural antioxidants.

ANTIMICROBIAL ACTIVITY

Walnuts have been found to have high antimicrobial activities. Pereira et al., 2007 screened the antimicrobial capacity of the leaves of walnut (*Juglans regia* L.) of six different cultivars (Lara, Franquette, Mayette, Marbot, Mellanaise and Parisienne) from Portugal. They studied against Gram positive (*Bacillus cereus, B. subtilis, Staphylococcus aureus*), Gram negative bacteria (*Pseudomonas aeruginosa, Escherichia coli, Klebsiella pneumoniae*) and fungi (*Candida albicans, Cryptococcus neoformans*). Leaf extract from all the cultivars only inhibited the growth of Gram positive bacteria, with *B. cereus* the most susceptible with MIC 0.1 mg/mL. Gram negative bacteria and fungi were resistant to the leaf extract even at 100 mg/ml (Pereira et al., 2007).

Later, Oliveira et al., 2008 studied antimicrobial activities in walnut (*Juglansregia* L.) green husks of five different cultivars in aqueous extracts (Franquette, Mayette, Marbot, Mellanaise and Parisienne). They screened the antimicrobial capacity against Gram positive and Gram negative bacteria and fungi. All the extracts inhibited the growth of Gram positive bacteria. Out of all, *Staphylococcus aureus* was the most susceptible with MIC of 0.1 mg/ml for all the extracts. The results indicated that walnut green husks may be an important source of compounds with health protective potential due to their antimicrobial activity.

ANTI-INFLAMMATORY ACTIVITIES

Excess free radicals are especially generated under conditions of chronic inflammation. They are highly deleterious and can lead to lung injuries, arthritis, etc. Therefore, their control is important for the betterment of health. Qamar and Sultana, 2010 evaluated the protective effects of *Juglans regia* kernel extract against cigarette smoke extract (CSE)-induced lung toxicities in wistar rats.

Methanolic extract of *J. regia* kernel was given to wistar rats for 1 week prior to CSE exposure in a dose dependent manner. *J. regia* extract not only significantly decreased the levels of lung injury markers (i.e., lactate dehydrogenase (LDH), total cell count, total protein and reduced glutathione (GSH) in bronchoalveolar lavage fluid (BALF)), but it also restored the levels of glutathione reductase (GR) and catalase and significantly reduced the xanthine oxidase (XO) activity in lung tissue.

The above results suggest the protective role of *J. regia* extract against CSE-induced acute lung toxicity. Further, the investigators found that the total polyphenolic content of *J. regia* kernel extract was 96+0.81 mg GAE/g dry weight of extract. The extract also possessed high free radical scavenging activity measured by DPPH (2,2-Diphenyl-1-Picrylhydrazyl) assay. Therefore, *J. regia* was found to possess anti-inflammatory and antioxidant activity against lung injuries.

ANTI-ALLERGIC ACTIVITY

Tree nuts are among the top ranked foods causing allergic reactions; walnuts ranked second in this aspect (Teuber et al., 2003; Bock et al., 2007). Many tree nuts, including walnuts, can cause food allergies. Anderson and Teuber, 2010 investigated whether walnut kernel polyphenolics and purified ellagic acid (EA) were able to alter cytokine production and cellular proliferation from stimulated human peripheral blood mononuclear cells (PBMC). They found decreased production of IL-13 and TNF-α without any change in IL-4 production. EA and the walnut polyphenolics not only significantly inhibited the stimulated [phytohemagglutin (PHA), α-CD3, and phorbolmyristate acetate (PMA)/ionomycin] PBMC proliferation in a dose dependent manner, but also increased the production of IL-2.

The investigators concluded that walnut polyphenolics are unable to distort a cytokine response toward Th2 in an *in vitro* environment. There was no decrease in production of IL-4 or IL-2, but the immunomodulatory effects were present along with inhibition of cellular proliferation (Anderson and Teuber, 2010). Further, Comstock et al., 2010 concluded the extracted polyphenolic compounds from walnuts could promote IgE production in a mice model. They found that there were increased serum concentrations of antigen-specific IgE and IgG1 when they treated the mice with polyphenolic-enriched extract of walnut with antigen ovalbumin (OVA) and alum (AL).

Polyphenolic extract enhanced OVA-specific IgE and IgG1 insignificantly as compared to that induced by OVA alone. There were no differences between serum IgG2a/2b levels between mice receiving OVA/AL and OVA/AL with polyphenolics. This suggested that walnut polyphenolic may act as an antiallergen (Comstock et al., 2010).

Cardiac Protection

For optimum cardiac health, to the body must respond to stimuli in a healthy way, requiring the presence of antioxidant and anti-inflammatory nutrients, proper blood composition, correct balance in inflammation-regulating molecules, and proper composition and flexibility in our blood vessel walls. So far researchers have found walnuts to be an excellent source of all these aspects. It decreases LDL and total cholesterol and increases gamma-tocopherol and omega-3 fatty acids (alpha-linolenic acid) in red blood cells (Zhao et al., 2004; Rajaram et al., 2009; Banel and Hu in 2009). Nuts are beneficial to many cardio-vascular risk biomarkers like LDL oxidizability, soluble inflammatory molecules, and endothelial dysfunction. Anderson et al., 2001 analyzed walnut extract and ellagic acid for their ability to inhibit *in vitro* plasma and LDL oxidation and their effects on LDL alpha-tocopherol during oxidative stress. They also analyzed walnut extract using Trolox equivalent antioxidant activity (TEAC) and liquid chromatography electrospray detection mass spectrometry (LC-ELSD/MS). They found that ellagic acid and walnut extract significantly inhibited 2,2'-Azobis'(2-amidino propane) hydrochloride (AAPH)-induced LDL oxidation and copper-mediated LDL oxidation by 87 and 38% and 14 and 84%, respectively. Moreover, TBARS formation was significantly inhibited by walnut extracts and ellagic acid in a dose-dependent manner and TEAC values of the extracts were greater than that of alpha-tocopherol (a standard antioxidant). They identified ellagic acid monomers, polymeric ellagitannins and other phenolics, principally nonflavonoid compounds, from walnuts using LC-ELSD/MS analysis. These results showed that walnut can be considered antiatherogenic due to its polyphenolic content. Nut intake is protective against LDL oxidation, but this aspect still needs to be supported by more clinical studies. Further, omega-3s also reduce inflammation that prevents artery-clogging plaque formation. L-Arginine, an essential amino acid, is present in walnuts and is converted into nitric oxide. Nitric oxide aids in relaxation of blood vessels. In hypertensive people it is very difficult to maintain normal nitric oxide levels that can lead to diabetes and heart problems. Therefore, walnut consumption can be beneficial (Morgan et al., 2002). The presence of omega-3, an essential fatty acid of walnuts, favors the consumption of walnut for prevention of cardiovascular risk factors (Lorigeril and Salen, 2004). Nuts are complex mixtures of biotic compounds including vegetable protein, fiber, minerals, tocopherols, and phenolic compounds. After one daily serving of mixed nuts, oxidized LDL concentrations are lowered. Many cross-sectional studies have shown that nut consumption also lowers the concentrations of inflammatory molecules and increases plasma adiponectin, a potent anti-inflammatory adipokine (Zhao et al., 2004). Walnuts (not other nuts) have been studied for their effects on endothelial function, but there were inconclusive results for reduced inflammatory cytokine concentrations (Ros et al., 2004). However, researchers observed favorable vasoreactivity after not only prolonged walnut diets, but also single walnut meals. There was decreased expression of endothelin 1 (a potent endothelial activator) after walnut consumption in an atherosclerotic animal model. The positive effects of walnuts on vascular reactivity may be attributed to its components including L-arginine, the precursor of nitric oxide, α-linolenic acid, and phenolic antioxidants. Blomhoff et al., 2006 found U-shaped association with nut/peanut butter consumption and the hazard ratio for total death rates. They also found constant reduction of deaths due to cardiovascular and coronary heart diseases with increased consumption of nut/peanut butter. Studies are needed to explain the contibution of antioxidants to the clear beneficial health effect of nuts.

Walnuts are a rich source of α-linolenic acid (ALA) that is linked to the prevention of cardiovascular diseases. Marangoni et al., 2007 evaluated levels of ALA and its metabolic derivatives in blood after walnut consumption. The study included 10 volunteers who consumed 4 walnuts per day (along with their regular diet) for 3 weeks. Researchers found that consumption of a few walnuts per day for 3 weeks not only increased ALA significantly, but also increased its long chain derivative eicosapentaenoic acid (EPA). Bes et al., 2007 completed a prospective cohort study of 8865 adult men and women over 28 months. Semiquantitative food-frequency questionnaires were used to assess dietary habits. During the follow up period, 937 participants gained a weight of $\geq$ 5 kg. Subjects who consumed nuts two or more times per week had significantly lower risk of weight gain in comparison to those who did not consume or rarely consumed nuts. Those subjects with less or no consumption gained an average of 424 grams more as compared to regular eaters. The investigators concluded that consumption of nuts can reduce the risk of weight gain (5 kg or more). Therefore, regular nut consumption can be included in cardio protective diets.

Evidence from epidemiological and clinical studies suggests that for the prevention and treatment of cardiovascular diseases (CVD) dietary intake of n-3 fatty acids, including alpha-linolenic acid (ALA), eicosapentaenoic acid (EPA), and docosahexaenoic acid (DHA), should be recommended (Gebauer et al., 2006). For the prevention of deficiency symptoms, the amount of n-3 fatty acid required is 0.6-1.2% of energy for ALA, with up to 10% provided by EPA or DHA. Flaxseed and flaxseed oil, walnuts and walnut oil, and canola oil are suggested in order to achieve recommended intake of ALA. Many health agencies recommend 500 mg/day of EPA and DHA for CVD reduction and 1 g/day as a treatment for existing CVD (Gebauer et al., 2006). Walnut diet is rich in polyunsaturated fatty acids which may improve blood lipids and other cardiovasculardisease risk factors. A meta analysis done by Banel and Hu in 2009 concluded that walnut rich diets not only help in improving cardiovascular risk factors by decreasing total and LDL cholesterol significantly, but also significantly decrease inflammation and oxidative stress, probably due to its high antioxidant capacity. They reviewed thirteen studies, overall including 365 participants. When they compared the walnut rich diet with the control diet, they found that walnut supplemented diets significantly decreased the total cholesterol and LDL-cholesterol. However, walnut diet did not significantly affect the HDL cholesterol and triglycerides when compared to the control diet. Walnut intake showed important benefits for antioxidant capacity and inflammatory markers without having any adverse effect on body weight. Although these findings proved that a walnut rich diet helps in improving cardiac health long term trials are still required to clearly justify the effect of walnut consumption on cardiovascular diseases and obesity (Banel and Hu, 2009).

Increase in the consumption of n-3 (omega-3) fatty acids can help in improving coronary heart disease (CHD). Rajaram et al., 2009 accomplished a randomized crossover trial with 25 normal to mildly hyperlipidemic adults. The subjects were given a control diet (without nuts or fish), a walnut diet (42.5 g walnuts/10.1 mJ) or a fish diet (113 g salmon, twice/week). Each diet provided the same amount of calories for four weeks. Subjects with the walnut diet showed lower total cholesterol and LDL cholesterol concentrations as compared to the control. There was no change in triglyceride and HDL-cholesterol concentrations. On the other hand, subjects on the fish diet showed decreased serum triglyceride and increased HDL-cholesterol concentrations as compared to the control. The walnut diet group had significantly lowered ratios of total cholesterol: HDL cholesterol, LDL cholesterol: HDL cholesterol, and apolipoprotein B: apolipoprotein A-I as compared with those on the control and fish diets. Therefore, walnut serves an important role in improving cardiac health.

Metabolic syndrome (MetS) is a group of metabolic irregularities, including central obesity, dyslipidemia, elevated blood pressure, and hyperglycemia. It is also an important risk factor for type 2 diabetes and cardiovascular disease (Grundy et al., 2005). Lifestyle education programs can be helpful in preventing or controlling the epidemic trend of MetS. However, it is not clear whether nuts and seeds may have some benefit on MetS.

More studies are needed to justify whether nut consumption is beneficial for maintaining optimum cardiac health apart from its cholesterol lowering properties (Ros Emilio, 2010).

Dietary intake of nuts significantly contributes to antioxidant level. Further, obesity and diabetes are the important risk factors for cardiovascular diseases. There is a greater chance of diabetes or obesity with high consumption of highly processed foods and decreased consumption of whole grains and nuts. Many observational studies suggest that nut consumption may reduce calorie consumption along with providing superior satiation, decreasing the risk of type 2 diabetes. But, there is lack of evidence from randomized, interventional studies. Brennan et al., 2010 designed a randomized, double-blind, crossover study of walnut consumption with 20 subjects, including both men and women. Subjects were given isocaloric diets in a clinical research center for 4 days. For breakfast, they were provided with either walnuts or placebo (shakes were standardized for calories, carbohydrate, and fat content). The investigators measured appetite, insulin resistance, and metabolic parameters. There was no change in resting expenditure, hormones known to mediate satiety, or insulin resistance with the walnut diet as compared to the placebo diet. But, they found that there was significant increase in the level of satiety and sense of fullness in prelunch questionnaires of those having the walnut breakfast as compared to the placebo breakfast. This effect was significantly higher on days 3 and 4. Long term studies are needed to support the physiologic role of walnuts and to explain the underlying mechanisms (Brennan et al., 2010).

Wu et al., 2010 did a 3-arm, randomized, controlled trial among 283 participants. Subjects were screened for Metabolic syndrome (MetS) using the updated National Cholesterol Education Program Adult Treatment Panel III criteria for Asian Americans. Subjects were consigned to Lifestyle Counselling (LC) on the AHA guidelines. They were either given LC + flaxseed (30 g/day) (LCF) or LC + walnuts (30 g/day) (LCW) for 12-weeks. The investigators found that MetS decreased significantly in all groups, but the number of subjects who no longer met the MetS criteria at 12 wk was not significantly different among groups. There were significantly higher reversion rates of central obesity in the LCF and LCW groups than in the LC group. Metabolic variables like weight, waist circumference, serum glucose, total cholesterol, LDL cholesterol, apolipoprotein (Apo) B, ApoE, and blood pressure were significantly reduced in all 3 groups. However, MetS severity was significantly reduced in the LCW group as compared to the LC group. Present research suggests that lifestyle education programs are efficient in MetS management, but diet supplementation with flaxseed and walnut may improve central obesity. Further long term studies with larger samples are needed to examine the role of these foods in the management of MetS (Wu et al., 2010).

Din et al., 2011 studied moderate walnut consumption on the improvement of lipid profile, arterial stiffness and platelet activation. They did a single-blind randomized controlled crossover trial with thirty healthy male volunteers. Subjects were given 4 weeks of dietary walnut supplementation (15 g/day). They found no difference in lipid profile, augmentation index or augmented pressure during or after 4 weeks of walnut supplementation as compared with control (no walnuts). Moreover, platelet-monocyte aggregation was not affected by the walnut supplementation. The investigators suggested that low level consumption of walnuts may not be helpful for maintaining cardiac health but large intake has beneficial effects.

NEUROLOGICAL DISEASES

Fruits and vegetables rich in antioxidants like blueberries, strawberries, walnuts, and Concord grape juice help in reducing the oxidative stress involved in aging (Joseph et al., 2009). Many neurochemical changes including neural cell populations and metal concentration are associated with aging. These may lead to many diseases including Alzheimer's disease (AD). Decrease in the level of melatonin occurs with both aging and AD. Decreased malatonin, degenerated cholinergic neurons of basal forebrain and deposition of amyloid beta peptides (a-beta) can lead to the development of cognitive symptoms of dementia. In patients with Alzheimer's disease (AD), amyloid beta-protein (Ab) is the major component of amyloid fibrils that compose neuritic plaques and cerebrovascular amyloid in the brain. Chauhan et al., 2004 studied the effect of walnut extract on Aß fibrillization by using Thioflavin T fluorescence spectroscopy and electron microscopy. They reported that extract of walnut not only repressed Aß fibril formation, but also was able to defibrillize Ab preformed fibrils in a concentration and time- dependent manner. They reported maximum defibrillization at 91.6% when preformed Aß fibrils were incubated for 2 days with 10 ml of methanolic extract of walnut (MEOW). This research suggested that the onset of Alzheimer's disease may be prolonged by walnuts as they maintain Aß in the soluble form. The present research further concluded that an amyloidogenic compound found in walnuts is an organic compound, is neither a lipid nor a protein, and has a molecular weight less than 10 kDa. While Aßb fibrillization is inhibited by MEOW and its 10 kDa filtrate, walnut chloroform extract showed no effect. This anti-amyloidogenic activity of walnuts may be due to the presence of polyphenolic compounds such as flavonoids (Chauhan et al., 2004). Lahiri et al., 2005 studied the neuroprotective function of melatonin, which is important for delaying aging, increasing life span, and maintaining the overall health of aged persons. Melatonin can neutralize the toxic effect of heavy metals. Its dietary supplementation restores age-related neural loss, including effects from AD. Walnuts are a rich source of melatonin. Increased brain melatonin significantly reduced the levels of Abeta peptides in mice. Therefore, dietary intake of melatonin reduced the metal toxicity, lipid peroxidation, and losses in cholinergic signaling. Further, authors suggested that if FDA-approved drugs for AD were combined with melatonin or other neuroprotectants they can greater boost the functioning of the central nervous system (Lahiri et al., 2005). Later, Chauhan et al., 2010 continued their research of Alzheimer's disease in a mice model. They investigated the improvements in memory deficits and learning skills by feeding a walnut rich diet to mice. Animals were fed with 6% or 9%

walnut rich diet. Diets for the control and treatment groups were balanced in terms of total calories, protein, carbohydrate and fat contents. The investigators found that walnut rich diets (6% or 9%) significantly improved the memory, learning ability, anxiety and motor development in treatment mice as compared to control. The above research suggests that the dietary intake of walnuts delays the onset or slows the progression of Alzheimer's disease (Chauhan, et al., 2010). Recently, Muthaiyah and Essa MM, et al., 2011 reported that walnut extract offers protection to Aß induced cytotoxicity in PC12 cells in a dose dependant manner. Walnuts are a rich source of the polyunsaturated fatty acids alpha-linolenic acid and linoleic acid. This can alter many processes in the brain, including microglial activation. This can result in the generation of many age-related and neurodegenerative conditions. Lipopolysaccharide (LPS) can induce microglial activation. Willis et al., 2010 studied the effect of walnut extract exposure on LPS-induced activation in BV-2 microglial cells. Production of nitric oxide and expression of nitric oxide synthase were attenuated when cells were treated with walnut extract prior to stimulation. The investigators found that the extract stimulated the internalization of the LPS receptor, toll-like receptor 4. This study suggests walnuts have anti-inflammatory effects on microglia, a fact that can be helpful in the treatment and prevention of neurodegenerative diseases (Willis et al., 2010).

Yang et al., 2010 isolated a new neuroprotective diaryl l heptanoid, juglanin C (1), along with three known diaryl heptanoids, juglanin A (2), juglanin B (3), and (5R)-5-hydroxy-7-(4-hydroxy-3-methoxy-phenyl)-1(4-hydroxyphenyl)-3-heptanone (4), from the 80% methanolic leaves and twigs extract of *Juglans sinensis* by using bioactivity-guided fractionation and chromatographic techniques. They found compounds 1 and 2 had neuroprotective activities against glutamate-induced toxicity that reduced the production of cellular peroxide in HT22 cells. Moreover, under glutamate-induced oxidative stress, these two compounds maintained antioxidative defense systems, including glutathione, glutathione reductase, and glutathione peroxidase in HT22 cells.

Anticancer Activity

Many studies suggest positive associations between intake of nuts—mainly walnuts—and reduction of many diseases, including cardiovascular disease and cancer. Carvalho et al., 2010 assayed methanolic extracts of walnut seed, green husk and leaf (Juglansregia L.) using methanol for antiproliferative effectivity using human renal cancer cell lines A-498 and 769-P and the colon cancer cell line Caco-2. They found that all extracts showed growth inhibition toward human kidney and colon cancer cells in a concentration-dependent manner.

All extracts exhibited similar growth inhibition activity (IC_{50} values between 0.226 and 0.291 mg/mL) for the A-498 renal cancer cells while walnut leaf extract showed a higher antiproliferative efficiency (IC_{50} values of 0.352 and 0.229 mg/mL, respectively) for 769-P renal and Caco-2 colon cancer cells as compared to green husk or seed extracts. This research indicated that walnut tree comprises a good source of chemopreventive agents.

Juglans mandshurica Maxim is known as the Manchurian walnut. It has been found to have several bioactivities, including anti-tumor, due to the presence of Juglone, its major chemical constituent. Xu et al., 2010 studied the molecular mechanism of Juglone-induced apoptosis in human leukemia HL-60 cells. They incubated HL-60 cells with various concentrations of Juglone. Hoechst 33342 staining and flow cytometry were used to detect the occurrence of apoptosis.

Quantitative polymerase chain reaction (qPCR) was used for the determination of expression of Bcl-2 and Bax mRNA. They found that Juglone inhibited the growth of human leukemia HL-60 cells in dose- and time-dependent manners. Further, they studied the topical morphological changes of apoptotic body formation after the treatment with Juglone by Hoechst 33342 staining. They applied the concentrations of Juglone (0, 0.5, 1.0 and 1.5 μg/ml) and found that the percentages of Annexin V-FITC-positive/PI negative cells were 7.81%, 35.46%, 49.11% and 66.02%, respectively.

They reported that Juglone could induce mitochondrial membrane potential ($\Delta\Psi m$) loss that is followed by release of cytochrome-c (Cyt c), Smac and apoptosis inducing factor (AIF) to the cell cytoplasm. After the Juglone treatment, they also found that there was a marked increase of Bax mRNA and appearance of that protein with a simulateneous decrease of Bcl-2 mRNA and protein.

The above events were paralleled with activation of caspase-9, -3 and PARP cleavage. Further, Juglone induced apoptosis was blocked by z-LEHD-fmk (a caspase-9 inhibitor). The investigators reported that both mitochondrial dysfunction and the elevated ratio of Bax/Bcl-2 induced by Juglone in HL-60 cells caused events responsible for mitochondrial dependent apoptosis pathways (Xu *et al.,* 2010).

Yang et al., 2011 studied bioassay-guided fractionation of MeOH extract of leaves and twigs of *Juglans inensis*. They isolated four new triterpenes (1-4) and 17 known triterpenes. New triterpenes were 1-oxo-3β,23-dihydroxyolean-12-en-28-oic acid 28-O-β-d-glucopyranoside (1), 1-oxo-3β-hydroxyolean-18-ene (2), 3β,23-dihydroxyurs-12-en-28-oic acid 28-O-β-d-glucopyranoside (3), and 3β,22α-dihydroxyurs-12-en-28-oic acid 28-O-β-d-glucopyranoside (4). Further, they found that triterpenes had antiproliferative activities in HSC-T6 cells through apoptosis.

CONCLUSION

Pharmacological studies conducted on walnuts indicate the medicinal potential of this plant and its parts especially seed, shells, kernels, bark, leaves and green husk. Walnuts have been found to have potential in the treatment of many disease conditions, such as cardiovascular diseases, inflammatory ailments including allergies, liver and kidney disorders, neuronal disorders, diabetes, cancer, and microbial infections including fungal and bacterial infections.

Use of walnut seeds may be increasing in cosmetic, nutraceutical and pharmaceutical industries. Epidemiological studies of fruits and vegetables show that it contains many phytonutrients which may be beneficial in protecting the human body against damage by reactive oxygen and nitrogen species (Diplock et al., 1998; Halliwell, 1997) during various disease conditions.

All of these studies are only on the experimental level and there is a need for complete and comprehensive clinical studies to prove the aforementioned effects. The modern world is trying to return to nature's way, possibly due to increased prevalence in the number of communicable and non-communicable diseases. We believe this review may be useful in this regard.

REFERENCES

Anderson, K., J., Teuber, S., S., 2010.Ellagic acid and polyphenolics present in walnut kernels inhibit in vitro human peripheral blood mononuclear cell proliferation and alter cytokine production. *Ann. N.Y. Acad. Sci.* 1190: 86–96.

Anderson, K., J., Teuber, S., S., Gobeille, A., Cremin, P., Waterhouse, A.,L., and Steinberg, F.,M., 2001. Walnut polyphenolics inhibit *in vitro* human plasma and LDL oxidation. *J. Nutr.* 131: 2837–2842.

Banel, D., K., and Hu, F., B. 2009. Effects of walnut consumption on blood lipids and other cardiovascular risk factors: a meta-analysis and systematic review. *Am. J. Clin. Nutr.* 90: 56–63.

Brennan, A., M., Sweeney, L., L., Liu, X., and Mantzoros, C., S. 2010. Walnut consumption increases satiation but has no effect on insulin resistance or the metabolic profile over a 4-day period. *Obesity.* 18: 1176–1182.

Bes, R., M., Sabate, J., Gomez, G., E., Alonso, A., Martinez, J., A., Martinez, G., M.,A., 2007. Nut consumption and weight gain in a Mediterranean cohort: The SUN study. *Obesity (Silver Spring.)*15: 107-16.

Blomhoff, R., Carlsen, M., H., Andersen, L., F., Jacobs, D., R., 2006. Health benefits of nuts: potential role of antioxidants. *Br. J. Nutr.* 96 Suppl 2: S52-60.

Carvalho, M., Ferreira, P., J., Mendes, V., S., Silva, R., Pereira, J.,A., Jeronimo, C., *et al.*, 2010. Human cancer cell antiproliferative and antioxidant activities of *Juglansregia*L. *Food Chem. Toxicol.*48: 441–447.

Chauhan, A., Essa, M., M., Muthaiyah, B., Chauhan, V., Kuar, K., Lee, M., 2010. Walnuts-rich diet improves memory deficits and learning skills in transgenic mouse model of Alzheimer's disease. International Conference on Alzheimer's Disease, Honolulu.

Chauhan, N., Wang, K., C., Wegiel, J., and Malik, M., N. 2004. Walnut Extract Inhibits the Fibrillization of Amyloid Beta-Protein, and also Defibrillizes its Preformed Fibrils. *Current Alzheimer Research.*1: 183-188.

Comstock, S., S., Gershwin, L., J., and Teuber, S., S. 2010. Effect of walnut (Juglansregia) polyphenolic compounds on ovalbumin-specific IgE induction in female BALB/c mice. *Ann. N.Y. Acad. Sci.* 1190: 58–69.

Colaric, M., Veberic, R., Solar, A., Hudina, M., and Stampar, F., 2005. Phenolic acids, syringaldehyde, and juglone in fruits of different cultivars of *Juglansregia*L. *J. Agric. Food Chem.* 53:6390–6396.

Cortes, B., Nunez, I., Cofan, M., Gilabert, R., Perez,H., A., Casals, E., Deulofeu, R., Ros, E., 2006. Acute effects of high-fat meals enriched with walnuts or olive oil on postprandial endothelial function. *J. Am. Coll. Cardiol.* 48:1666-71.

Demir, C., and Cetin, M. 1999. Determination of tocopherols, fatty acids and oxidative stability of pecan, walnut and sunflower oils. *Dtsch. Lebensm. Rundsch.*95: 278–282.

Diana, O., L., Damian, M., M., Milton, P., Marcela, L., Martınez, Alicia, L., Lamarque Duke, J., A., 1989. Handbook of Nuts. CRC Press, London.

Diplock, A., Charleux, J., Grozier-Willi, G., Kok, K., Rice-Evans, C., Roberfroid, M., Stahl, W., andVina-Ribes, J. 1998. Functional food sciences and defence against reactive oxidative species. *British Journal of Nutrition.* 80: 77–82.

Fernandez, L., J., Aleta, N., and Alias, R., 2000. *Forest Genetic Resources Conservation of Juglansregia L.* IPGRI, Rome.

Fraser, G., E., et al. 1992. A possible protective effect of nut consumption on risk of coronary heart disease. The Adventist Health Study. *Arch. Intern. Med.*152: 1416-1424.

Fukuda, T., Ito, H., Yoshida, T, 2003. Antioxidative polyphenols from walnuts (Juglansregia L.). *Phytochemistry.* 63: 795–801.

Gebauer, S., K., Psota, T.,L., Harris, W.,S., Kris, E., P., M. 2006. n-3 fatty acid dietary recommendations and food sources to achieve essentiality and cardiovascular benefits. *Am. J. Clin. Nutr.*83(6 Suppl): 1526S-1535S.

Grundy, S., M., Cleeman, J., I., Daniels, S.,R., Donato, K.,A., Eckel, R.,H., Franklin, B.,A., Gordon, D., J., Krauss, R., M., Savage, P., J., et al. 2005. Diagnosis and management of the metabolic syndrome. An American Heart Association/ National Heart, Lung, and Blood Institute Scientific Statement. Executive summary. *Cardiol. Rev.*13: 322–7.

Halvorsen, B., L., Carlsen, M., H., Phillips, K., M., Bøhn, S.,K., Holte, K., Jacobs, D., R., and Blomhoff, R., 2006. Content of redox-active compounds (ie, antioxidants) in foodsconsumed in the United States. *Am. J. Clin. Nutr.*84: 95–135.

Halliwell, B.1997. Antioxidants and human disease: A general introduction. *Nutrition Reviews.* 55, 44–52.

Hemery, G., E. 1998.Walnut seed-collecting expedition to Kyrgyzstan in Central Asia. *Quarterly Journal of Forestry.* 92: 153-157.

Hu, F., B. *et al.* 1998. Frequent nut consumption and risk of coronary heart disease in women: prospective cohort study. *Brit. Med. J.* 317: 1341–1345.

Jimenez, G., Y., Lopez, M., J., Blanco, C., L.,M., Marin, C., Perez, M., P., Ruano, J., Paniagua Rodriguez, J., A., Egido, F., J., Perez, J., F., 2009. Olive oil and walnut breakfasts reduce the postprandial inflammatory responce in mononuclear cells compared with a butter breakfast in healthy men. *Atherosclerosis.*204: e70-e76.

Joseph J., A., Shukitt, H., B., Willis, L.,M., 2009. Grape Juice, Berries, and Walnuts Affect Brain Aging and Behavior. *The Journal of Nutrition.*1813S-17S.

Kagawa, Y., 2001. Standard Tables of Food Composition in Japan. Kagawa Nutrition University Press, Tokyo.

Lahiri, D.,K., Chen, D.,M., Lahiri, P., Bondy, S., Greig, N.,H. 2005. Amyloid, cholinesterase, melatonin, and metals and their roles in aging and neurodegenerative diseases. *Ann. N.Y. Acad. Sci.* 1056: 430-49.

Morgan, J., M., Horton, K., Reese, D., 2002. Effects of walnut consumption as part of a low-fat, low-cholesterol diet on serum cardiovascular risk factors. *Int. J. Vitam. Nutr. Res.* 72:341-7.

Martinez, M., L., Labuckas, D., O., Lamarque, A., L., Maestri, D., M. 2010. Walnut (*Juglansregia*L.): genetic resources, chemistry, by-products. *J. Sci. Food Agric.* 90: 1959–1967.

Mishra, N., Dubey, A., Mishra, R., Barik, N., 2010. Study on antioxidant activity of common dry fruits. *Food and Chemical Toxicology.* 48: 3316–3320.

Marangoni, F., Colombo, C., Martiello, A., Poli, A., Paoletti, R., Galli, C., 2007.Levels of the n-3 fatty acid eicosapentaenoic acid in addition to those of alpha linolenic acid are significantly raised in blood lipids by the intake of four walnuts a day in humans. *Nutr. Metab. Cardiovasc. Dis.*17: 457-61.

Miele, N., A., Monaco, R., Cavella, S., Masi, P., 2010. Effect of meal accompaniments on the acceptablity of a walnut oil-inriched mayonnaise with and without a heath claim. *Food Quality and Preference*. 21: 470-77.

Muthaiyah, B., Essa, M., M., Chauhan, V., Chauhan A., 2011. Protective effects of walnut extract against amyloid beta peptide-induced cell death and oxidative stress in PC12 cells. *Neurochem Res*. 36(11): 2096-103.

Oliveira, I., Sousa, A., Ferreira, I., C., F., R., Bento, A., Estevinho, L., Pereira, J., A, 2008. Total phenols, antioxidant potential and antimicrobial activity of walnut (*Juglansregia* L.) green husks. *Food and Chemical Toxicology*. 46: 2326–2331.

Perry, L., M., 1980.Medicinal Plants of East and Southeast Asia. MIT Press, Cambridge.

Pereira, J., A., Oliveira, I., Sousa, A., Ferreira, I., Bento, A., Estevinho, L., 2008. Bioactive properties and chemical composition of six walnut *(Juglansregia L.)* cultivars. *Food and Chemical Toxicology*. 46: 2103-11.

Pereira, J., A., Oliveira, I., Sousa, A., Valentao, P., Andrade, P., B., Ferreira, I.,C.,F.,R., Ferreres, F., Bento, A., Seabra, R., Estevinho, L., 2007. Walnut (Juglansregia L.) leaves: Phenolic compounds, antibacterial activity and antioxidant potential of different cultivars. *Food and Chemical Toxicology*. 45: 2287–2295.

Qamar, W., and Sultana, S., 2010. Polyphenols from Juglansregia L. (walnut) kernel modulate cigarette smoke extract induced acute inflammation, oxidative stress and lung injury in Wistar rats. *Human and Experimental Toxicology*. 000: 1–8.

Rajaram, S., Haddad, E., H., Mejia, A., and Sabate, J. 2009. Walnuts and fatty fish influence different serum lipid fractions in normal to mildly hyperlipidemic individuals: a randomized controlled study. *Am. J. Clin. Nutr.* 89 (suppl): 1657S–63S.

Reiter, R., J., Manchester, L., C., Tan, D., X., 2005. Melatonin in walnuts: Influence on levels of melatonin and total antioxidant capacity of blood. *Nutrition*. 21: 920–924.

Ros, E., Nunez, I., Perez, H., A., Serra, M., Gilabert, R., Casals, E., Deulofeu, R., A., 2004. Walnut diet improves endothelial function in hypercholesterolemic subjects: a randomized crossover trial. *Circulation*. 109:1609-14.

Ros, E., 2010. Nuts and novel biomarkers of cardiovascular disease. *The American Journal of Clinical Nutrition*. 89: 1649S-56S.

Sabate, J. 1999. Nut consumption, vegetarian diets, ischemic heart disease risk, and all-cause mortality: evidence from epidemiologic studies. *Am. J. Clin. Nutr.*70: 500S–503S.

Serrano, A., Cofrades, S., Ruiz-Capillas, C., Olmedilla-Alonso, B., Herrero-Barbudo, C., Jiménez-Colmenero, F. 2005. Nutritional profile of restructured beef steak with added walnuts. *Meat Science*. 70:647–654.

Stampar, F., Solar, A., Hudina, M., Veberic, R., Colaric, M., 2006.Traditional walnut liqueur – cocktail of phenolics. *Food Chem.* 95: 627–631.

Stone, D.,O.,S., Tripp, E., Luis, Gios, Manos, P., 2009. Natural history,distribution, phylogenetic relationships, and conservation of Central American black walnuts (*Juglans sect. Rhysocaryon*). *Journal of the Torrey Botanical Society*. 136: 1-25.

Savage, G., P., Dutta, P., C., and McNeil, D.,L., 1999. Fatty acid and tocopherol contents and oxidative stability of walnut oils. *J. Am. Oil Chem. Soc.* 76:1059–1063.

Wu, H., Pan, A., Yu, Z., Qi, Q., Lu, L., Zhang, G., Yu, D., Zong, G., Zhou, Y., Chen, X., Tang, L., Feng, Y., Zhou, H., Chen, X., Li, H., Wahnefried, D., Hu, F., B., Lin, X., 2010.

Lifestyle Counseling and Supplementation with Flaxseed or Walnuts Influence the Management of Metabolic Syndrome. *The Journal of Nutrition.*1937-42.

Willis, L., M., Bielinski, D.,F., Fisher, D., R., Matthan, N., R., Joseph. 2010. Walnut Extract Inhibits LPS-induced Activation of Bv-2 Microglia via Internalization of TLR4: Possible Involvement of Phospholipase D2. *Inflammation.* 33: 325-333.

Xu, H., L., Yu, X., F., Qu, S., C., Zhang, R., Qu, X., R., Chen, Y., P., Ma, X.,Y., Sui, D.,Y., 2010. Anti-proliferative effect of Juglone from *Juglansmandshurica* Maxim on human leukemia cell HL-60 by inducing apoptosis through the mitochondria-dependent pathway. *European Journal of Pharmacology.* 645: 14–22.

Yang, H., Sung, S., H., Kim, J., Kim, Y., C. 2010.NeuroprotectiveDiarylheptanoids from the Leaves and Twigs of Juglanssinensis against Glutamate-Induced Toxicity in HT22 Cells. *Planta Med.* [Epub ahead of print]

Yang, H., Jeong, E., J., Kim, J., Sung, S.,H., Kim, Y.,C. 2011. AntiproliferativeTriterpenes from the Leaves and Twigs of Juglanssinensis on HSC-T6 Cells. *J. Nat. Prod.* [Epub ahead of print]

Zhang, Z., Liao, L., Moore, J., Wu, T. and Wang, Z., 2009. Antioxidant phenolic compounds from walnut kernels (*Juglansregia*L). *Food Chem.*113:160–165.

Zhao, G., Etherton, T., D., Martin, K.,R., West, S.,G., Gillies, P.,J., Kris, E., P.,M., 2004. Dietary α-Linolenic Acid Reduces Inflammatory and Lipid Cardiovascular Risk Factors in Hypercholesterolemic Men and Women. *J. Nutr.* 134: 2991-2997.

In: Natural Products and Their Active Compounds ... ISBN: 978-1-62100-153-9
Editors: M. Essa, A. Manickavasagan, and E. Sukumar © 2012 Nova Science Publishers, Inc.

Chapter 20

COFFEE, CAFFEINE AND HUMAN HEALTH

Mostafa Waly[1,2], Amanat Ali[1], Mohamed M. Essa[1,5,6], Mohamed Farhat [2,3] and Yahya Al-Farsi[4]

[1]Department of Food Science and Nutrition, College of Agricultural and Marine Sciences, Sultan Qaboos University, Muscat, Oman
[2]Department of Community Nutrition, College of Applied Medical Sciences, King Saud University, Riyadh, Saudi Arabia.
[3]Nutrition Department, High Institute of Public Health, Alexandria University, Egypt
[4]Department of Family Medicine and Public health, College of Medicine and Health Sciences, Sultan Qaboos University, Muscat, Oman
[5]Neuropharmacology group, Department of Pharmacology, College of Medicine, University of New South Wales, Sydney, Australia.
[6]Developmental Neuroscience Lab, NYSIBR, Staten Island, NY, US

INTRODUCTION

The high prevalence of coffee drinking and coronary heart diseases (CHD) in many developed countries has led to studies on coffee drinking as an etiological factor for CHD. The first major epidemiological study suggests a coffee-coronary disease link was the work of Paul *et al* [1].

Two large case control studies were conducted in the early 1970s, suggested a positive association between coffee drinking and myocardial infarction [2, 3]. In the ensuing decade, additional reports [4, 5] were essentially negative, resulting in abatement of concern about the role of coffee in CHD. Interest was reawakened in the 1980s by reports of an association between coffee drinking and higher serum cholesterol level [6-11]; additional studies attributed the relation to high low-density lipoprotein cholesterol [10,11].

Several reports since 1987 [12-15] have supported that coffee was related to CHD. In addition controlled clinical trials showed that abstaining from boiled coffee resulted in great decrease in serum cholesterol of volunteers and in hypercholesterolemic men [16-19]. A possible explanation for the coffee-cholesterol association is that heavy coffee drinkers consume a more atherogenic diet and that it is this diet rather than the coffee per se that accounts for the increased total and LDL- cholesterol levels. This finding may explain why coffee- cholesterol association is a point of controversy where, in some populations coffee drinking may be associated with ingestion of a more atherogenic diet and hence, indirectly with hypercholestrolemia, whereas in other populations it may not be [20-23].

Differences in coffee-brewing methods may complicate the puzzle, where investigations in which coffee was prepared by boiling revealed increased serum total cholesterol, LDL-cholesterol, whereas filtered caffeinated coffee did not increased these values [24], where it was found that the consumption of Scandinavian-style "boiled coffee" i.e. coffee prepared by boiling ground coffee beans with water and decanting the fluid into a cup without filtration is associated with elevated levels of serum cholesterol [25].

In Norway, where coffee is often consumed as boiled coffee, coffee consumption is associated with an increased risk of coronary heart disease [26]. In Finland, 40% of the decline in serum cholesterol over the last 25 years has been attributed to the switch from boiled to filtered coffee, leading to 7% reduction in cardiovascular disease [27]. Also, a daily intake of six or more cups of boiled coffee increases the serum cholesterol concentration in healthy subjects [28]. It was found that both caffeinated and decaffeinated coffee caused an increase in serum lipids mainly cholesterol and low density lipoprotein [29,30], thus it does strongly suggest that caffeine is not the component of coffee responsible for its serum lipids elevating effect and in support of this finding is the fact that although both of tea and cola contain caffeine, but none of it have any significant raising effect on serum lipids [31,32].

Accordingly the simplest answer for the question of why serum lipids (mainly serum cholesterol) were significantly associated with coffee consumption but not tea or cola consumption, is that either coffee contains many substances other than caffeine that might influence serum lipid levels[33], tea and cola contain other substances that balance the hypercholesterolemic effect of caffeine [34-36] or coffee stimulates lipolysis with elevation of plasma unesterified fatty acids which is a substrate for hepatic synthesis of lipoproteins, that may accelerate rate of synthesis and release of lipoproteins and raise serum lipid levels [37,38].

COFFEE

A. History

Throughout the centuries humans have searched for stimulants in their food and beverages. The coffee bean was discovered in Arabia, the tea leaf in china, the Kola nut in West Africa, the Cocoa bean in Mexico, the ilex plant which provides mate in Brazil, and the cassina or Christmas berry tree in North America. All contain caffeine, which accounts for their stimulant properties and, at least to some extent, their consumer acceptance. The popular coffee beverage spread from Arabia to Ethiopia, Turkey, then to Near Eastern and other North African regions, and finally to Europe, North and South America. When initially introduced, coffee was considered potentially dangerous to health, but, despite early medical warnings and efforts to suppress it, coffee became a welcomed and socially acceptable drink. However, the controversy over the health aspects of coffee continues [39].

There are numerous conjectural statements and mythical stories about the earliest history of these beverages. Legend credits the head of an Arabian monastery for the discovery of coffee, when Shepherds reported nightly friskiness and frolic in goats who had eaten berries of the coffee plant, the Abbott instructed his Shepherds to pick the berries so that he might make a beverage from them and this drink proved very useful in helping him endure his long nights of prayer, but stirred fierce opposition among the more orthodox priests who were opposed to this use of antisoporifics. Despite their severe criticism, his experiment was an obvious success, as evidenced by the popularity of coffee. Coffee beans are the seeds of coffee trees that are widely cultivated in the tropics, they have been grown in Ethiopia and Arabia for hundreds of generations and there are large plantations in India, Indonesia, and Africa and especially in Brazil. A variety of trees are grown, most are 10 to 15 ft high with ever green leaves. Shade trees are necessary to protect them from excessive sun and they bear clusters of white flowers which develop into the beans. A tree provides the best crop when it between 6 and 14 years old and it requires regular and skillful pruning [40]. Furthermore, the word coffee comes from the Arabic qahwah, coffee was first a food, then a wine, a medicine and finally a beverage and in the beginning the dried coffee berries were crushed and mixed with fat to form food balls, then a wine was made from the raw beans and dried skins. The roasting of the beans began in the 13th century [41]. The beverage was introduced from Arabia into Turkey in 1554, Venice in 1615, France in 1644, England and Vienna in 1650, and North America in 1668.

B. Composition of Green Coffee

Coffee is derived from a small evergreen tree or shrub of which there are two main species, *Coffee Arabia and C. robusta*, a third species *C. Liberica* is of little commercial importance. The fruits (Crimson in colour) are often called "cherries" which they resemble in appearance and each contains two hemispherical seeds, the adjacent sides of which are flattened and embedded in a pulpy mass. All the outer layers (pulp, skin, mucilage, parchment, etc.) are discarded during preparation for market leaving the cleaned greenish-yellowish seeds as the coffee beans of commerce [41].

Table 1 reveals that raw green coffee contains water, protein, caffeine, oil, various carbohydrates and acids (mainly soluble and non volatile), trigonelline and mineral matter, meanwhile roast coffee contains reducing sugars, caramelised sugars, hemi-cellulose, fiber, protein, non volatile acids, caffeine, oil and the ash [42].

C. Coffee Roasting

The raw green beans are roasted by heating to 180-230°C for 15-20 minute and the seed increases in size, due to the production of carbon dioxide within the bean, which acts as a preservative until released by grinding [41]. In fact the roasting process is actually a mild pyrolysis during which some of the coffee components are decomposed at rates in accordance with their relative stabilities towards the heat applied, during roasting the bean moisture content is reduced to about 2 to 3% and there is a total loss in weight of about 16% (this loss is made up of original moisture and of volatile decomposition products). The colour of the beans changes to brown, and their volume increases from 50 to 100% [42].

Table 1. Composite analysis of green coffee

Component	(%)	Mg per 100 gm
Water	8-12	
Oil (ether extract)	4-18	
Unsaponifiable	0-2	
Nitrogen	1.8-2.5	
Protein*	9-16	
Caffeine	0-2	
Chlorogenic acid	2-8	
Trigonelline	1-3	
Ash:	2.5-4.5	
Calcium		85-100
Phosphorus		130-165
Iron		3-10
Sodium		4
Manganese		1-45
Rudidium		traces
Copper		traces
Fluorine		traces
Tannin	2	
Caffetannic acid	8-9	
Caffeic acid	1	
Pentosans	5	
Starch	5-23	
Dextrin	0.85	
Sucrose	5-10	
Reducing sugars	0-5	
Cellulose	10-20	
Hemicellulose	20	
Lignin	4	
Vitamins (present in small amounts):		
Carotene, thiamin, riboflavin, folic acid, niacin, pantothenic acid, citrovorum factor, B-6, and B-12. Choline: 60 mg per 100 gm.		

* Amino acids from protein: alanine, aspartic acid, glumatic acid, glycine, leucine, phenylalanine, serine, theronine, valine, crystine, methionine, and proline.

D. Manufacture

Most commercial coffee is derived from the prepared seeds of coffee arabica and C. robusta.

(I). Instant Coffee

This is the modern form of coffee prepared in the factory; it is prepared in a battery of column extractors in which the finely ground roasted beans are extracted with water under pressure at 175°C for a period of about 4 hr. The extract is then either spray dried to give the familiar free-flowing powder or freeze dried and agglomerated to give a product looking more like ground coffee [40]. At home it is conveniently made into a drink by simply adding warm water and it contains 20-40 mg caffeine / gm of coffee powder [40].

(II). Decaffeinated Coffee

Many consumers have tried to eliminate caffeine from their diets by selecting decaffeinated coffee. Just when they though it was safe, news surfaced of the potential dangers of the chemical solvents used to remove caffeine from coffee, where the standard substance used to remove caffeine from coffee beans is methylene chloride, used in either two ways, directly or indirectly. Both methods leave traces in the final product, the FDA estimates that the average cup of coffee treated by this way contains a methylene chloride concentration of about 0.1 ppm, and a person drinking 2 1/2 cup of decaffeinated coffee containing 0.25 ppm methylene chloride every day for a lifetime has a one-in-a-million chance of developing cancer from it [41] .Legally decaffeinated coffee must not contain more than 0.1% caffeine (approximately 2 mg/cup).

(III). Coffee Substitutes

In 1977, the world price of coffee beans rose very sharply and trigged off request for substitute materials, having the same flavoring characteristics for the partial complete replacement of genuine coffee. Numerous alternatives are now available based on roasted cereals (e.g. corn, soy-beans or other edible plants and seeds) together with caramel color, sugar and small percentage of malic acid. Caffeine may be incorporated if desired and imitation coffee flavoring is added to give a full coffee flavored product, unlike other canned products, a bulging can does not indicate spoilage, but rather that there is a release of carbon dioxide [40,42].

E. Adulteration

Coffee may be adulterated and especially in times of scarcity, a great variety of substances are used for this purpose like: chicory, dandelion and other roots, acorns, figs and cereals, the most important is chicory which is frequently added to French coffee. Chicory is the root of a wild endive; it is dried, partly caramelized and then added to coffee in proportions varying from 10 to 80 percent. Chicory's great attraction is its cheapness, furthermore, there is no reason to think that it is any way injurious to health [40,42].Table 2 illustrates the components of chicory and dandelion roots [41]. Table 2 illustrates the components of chicory and dandelion roots.

CAFFEINE

Even if all contaminants, pesticides, residues, and additives were removed, foods would still contain thousands of different chemicals, because they come from the tissues of plants and animals.

Table 2. Analytical figures (as percentages) for raw and roasted coffee and roasted chicory and retail samples of dandelion root and dandelion and chicory

	Raw coffee			Roasted coffee			Roasted chicory			Roasted Dandelion	Roasted Dandelion and chicory
	min	max	aver	min	max	aver	min	max	aver		
Moisture	8.2	13.8	10.3	0.3	5.6	2.2	2.5	12.0	5.5	6.9	7.3
Total ash	3.0	4.5	4.0	3.4	4.9	4.3	4.0	6.7	5.0	15.7	9.6
Water soluble ash	-	-	2.9	3.0	3.6	3.2	1.6	3.3	2.8	1.5	0.9
Ditto (as % of total ash)	-	-	-	65	85	75	-	-	55	8.8	9.4
Alky of sol ash (as K_2CO_3)	-	-	-	1.9	3.2	2.4	-	-	-	0.4	0.1
Acid insoluble ash (as % of total ash)	-	-	-	-	-	-	10	35	20	-	-
Crude fibre	-	-	-	10.5	15.3	13.0	-	-	6.9	12.5	15.1
Tannia	-	-	9.0	-	-	4.6	-	-	-	1.5	2.7
Total nitrogen	-	-	2.7	2.1	3.3	2.6	-	-	1.4	0.87	1.17
Caffeine	1.1	1.8	1.3	0.9	1.8	1.2	-	-	Nil	Nil	Nil
Starch	-	-	-	0.9	3.5	2.3	-	-	2.1	-	-
Ether extract	11.4	13.7	12.2	8.0	14.2	13.5	0.9	3.9	2.1	1.3	2.3
Extractives of dry sample*	25	34	-	23	33	*	70	78	*	50.5	61.6

* Depends to some extent on extraction method.

Only about 40 of the chemicals found in foods serve as nutrients, most of the others have no known nutritional value. Some of the no nutrients, however, have other effects of interest. Caffeine is one such non nutrient.

Water extracts of widely distributed plants have served as beverages for man since ancient times. The most popular ones contain caffeine, theophylline and theobromine, three closely related alkaloids, which possess important pharmacological properties.

These alkaloids are valuable therapeutic agents, coffee made from the seeds of *coffee arabica* and the related species, contain caffeine, the leaves of *Thea sinesis*, used for making tea, contain caffeine and theophyllinem and Cocoa, obtained from the seeds of *theobroma cocao*, contains caffeine and theobromine [40,42].

A. Chemical Structure of Caffeine

Caffeine is a white odorless powder with better taste, it is an organic chemical belonging to a class of chemicals called purines some of which are constituents of the nucleic acids RNA and DNA. The proper chemical name for caffeine is 1,3,7-trimethylxanthine . Theophylline and theobromine are chemical cousins of caffeine, differing only in the number and placement of the methylgroups (CH_3), where theophylline is 1,3-dimethylxanthine and theobromine is 3,7- dimethylxanthine [43].

Ingested caffeine is rapidly absorbed from the intestine and within a few minutes caffeine enters all organs and tissues. Within one hour after ingestion, it is distributed in the body tissues in proportion to their water content. The metabolic half-life of, caffeine is about 3 hours, hence, there is no day-to-day accumulations as it almost completely disappears from the body overnight.

Most of the ingested caffeine is excreted in the urine, and the major excretory product in man is 1-methyl uric acid formed by the demethylation (removal of CH_3 groups) of caffeine [43]. Clearance of caffeine may be either slowed by liver diseases, pregnancy and oral contraceptive use or accelerated by smoking. The blood concentration of caffeine is rarely greater than 10 mg/liter [44], such levels was not considered deleterious to health.

C. Caffeine Sources

Today, for many persons, the consumption of caffeine in one form or another begins at an early age and continues for much of their life times. Most people are aware that a cup of coffee or tea contains caffeine.

Perhaps not as many persons realize that this drug is also found in many soft drinks, and even fewer may realize that they are ingesting caffeine when they drink a cup of cocoa, eat a chocolate bar, or take pills for a headache or cold. Coffee was by far the most important source, after considering losses during preparation and serving, coffee accounted for about 75-80% of total caffeine consumption in the world. Tea was next in importance while cola beverages ranked third [40]. Table 3 represents the caffeine content of beverages, foods and over-the-counter drugs [41].

Table 3. Caffeine content in food and drugs

Item	Measure	Caffeine (mg)
Percolated roasted and ground coffee[1]	Cup[3]	76-155
Instant coffee	Cup[3]	66
Decaffeinated coffee	Cup[3]	2-5
Tea[1]	Cup[3]	20-100
Instant tea	Cup[3]	24-131
Hot cocoa	Cup[3]	5
Soft drinks[2]	12 oz can	26-34
Milk chocolate candy	2 oz	12
Sweet or dark chocolate	2 oz	40
Baking chocolate	2 oz	70
Alertness tablets (stay-awake tablets)	tablet	100-200
Pain relievers	tablet	32-65
Cold allergy relief remedies	tablet	15-32

[1] The longer coffee or tea is brewed, the greater its caffeine content.
[2] Include primarily cola and pepper drinks, but other flavors of soft drinks may also contain caffeine.
[3] One cup = 240 ml; one oz = 30 ml.

D. Caffeine Effects on Human Health

In the amounts normally consumed, caffeine acts as a drug, indeed much of its popularity is owed to its stimulant effect on the central nervous system. Like many drugs, users develop a dependence on caffeine when consumed in amounts equivalent to more than 4 cups of coffee per day. Furthermore, abstinence from caffeine - containing substances for a day or two may cause the development of withdrawal symptoms - headaches, irritability, restlessness, or fatigue [45].

The pharmacologically active dose of caffeine is about 200 mg and while the effects of caffeine vary from person to person, a dose of 1000 mg or more will generally produce adverse effects such as insomnia, restlessness, excitement, trembling, rapid heart beats, in creased breathing, desire to urinate, ringing in the ears and heart burn. Furthermore a fatal dose of caffeine appears to be more than 10 gm or 170 mg/kg of body weight - about 80 to 100 cups of coffee in one sitting [46]. Over the years, caffeine has been implicated in birth defects [47], heart diseases [48], hypertension [49], cancer [50], peptic ulcer [50], breast disease [51] and behavioral effects [50].

(1). Birth Defects

Numerous animal studies since 1946 have demonstrated in a variety of animals that large doses of caffeine can cause birth defects [52]. Human epidemiological studies of caffeine consumption and birth defects, were support the findings of the animal studies, because of this concern the Food and Drug Administration (FDA) counsels pregnant women to avoid caffeine-containing foods and drugs, indeed it remove caffeine from the list of substances generally recognized as safe [53].

Breast feeding mothers are discouraged from drinking large quantities of coffee or other caffeine containing beverages, where larger doses of coffee may interfere with the iron availability from the milk and impair the infant's iron status [54].

(2). Cardiovascular Disease

For forty years, studies on caffeine's relationships with heart disease have produced various findings, but no study indicates that moderate caffeine intake causes cardiovascular disease. Some studies suggest that coffee raises heart rate, blood pressure, and blood lipid levels [55-57]. Some do not [58-60].

Some comparing decaffeinated brews with coffee, suggest that a component other than caffeine is responsible for the effects seen, also during the first several days or weeks after a myocardial infraction, caffeine-containing beverages are often routinely restricted as a precautionary measure because of their suspected arrhythmogenic potential effect [61]. The relation of coffee consumption to cardiovascular disease has been examined in the Framingham study [20]. Coffee intake was studied in relation to total coronary heart disease, angina pectoris, myocardial infarction, sudden death, and death from all causes and a statistically significant increase was observed with increasing coffee consumption only in the category "death from all causes" [21].

Although caffeine may not cause heart disease directly but it may mask the warning signs, where caffeine oppose the effect of adenosine compounds (which is responsible for the occurrence of chest pain "the prime symptom of heart disease") thus caffeine blunt the sensation of pain and override adenosine's warning [46].

(3). Hypertension

In sensitive individuals, caffeine consumption can temporarily elevate blood pressure. However, in regular users, 2 to 3 cups of coffee at a time does not increase blood pressure [49]. Others found that caffeine had no effect on blood pressure [62].

(4). Cancer

Coffee drinking has been implicated in the development of pancreatic cancer [63], while caffeine per se seems to have anticancer properties since in experimental settings it seems to increase the effectiveness of radiation and dry therapy by preventing cancer from repairing themselves [50]. But further testing of this indication is needed Furthermore, a study has cited evidence that men with pancreatic cancer consumed more decaffeinated coffee than control subjects, suggesting that an agent other than caffeine present in coffee may exhibit a carcinogenic effect [63].

(5). Peptic Ulcer

Caffeine - containing beverages may stimulate gastric secretion for both acid and pepsin. For this reason, it has been postulated that caffeine is an etiologic factor in ulcerogenesis, and elimination of coffee and other caffeine - containing beverages is often recommended in the management of gastric or duodenal ulcers [50].

(6). Fibrocystic Breast Disease

The elimination of methylxanthines from the diet has been reported to reduce the severity of fibrocystic breast disease in women [51]. However, the results of another study suggest that caffeine - free diets are not associated with a substantial improvement in this disease [64]. Moreover, there is currently no scientific basis for associating methylxanthine consumption with fibrocystic disease of the breast and it has been suggested that lumpy, fibrous breast tissue in women is normal and represents a response to physiological hormonal variation [65].

(7). Behavioral Effects

Caffeine is a central nervous system (CNS) stimulant, but the molecular mechanisms of the behavioral actions of caffeine and other methylxanthines have not been fully elucidated [66]. It is known that methylxanthines can inhibit adenosine monophosphate (AMP) phosphodiesterase, thus preventing the formation of cyclic AMP (cAMP) which is the essential signal in (CNS) stimulant effect [67]. But, the concentrations of caffeine required for significant phosphodiesterase inhibition, are much greater than those found in brain tissue at behaviourally effective caffeine doses [68]. Thus, it has been though, that the behavioural action of caffeine may not involve the inhibition of brain phosphodiesterase, but it involves a blockade of central extracellular adenosine receptors [69], this is supported by findings that adenosine receptor activity is blocked by methylxanthines, including caffeine, in concentrations similar to those produced by pharmacologic doses [69,70].

It has been shown that the consumption of substantial amounts of caffeine or caffeine - containing foods can produce significant neurochemical changes in the rats by enhancing the synthesis and metabolism of serotonin in the brain which is generally associated with the wakefulness caused by pharmacologic doses of caffeine [70].

(8). Caffeine and Obesity

Caffeine stimulates thermogenesis, where it was found that the dose in 1 cup of coffee speeds metabolism slightly for 1-2 hours; this fact leads researchers to speculate that caffeine may reduce body fat stores [71].

Indeed, a study states that caffeine promotes weight loss by reducing lipid stores because of increased energy expenditure but without decreasing energy intake. An increase in plasma triglyceride was observed due to the lipolysis in adipose tissue and subsequent hepatic rectification, i.e. formation of triglyceride (under the effect of caffeine), this may explain why the increase in triglyceride plasma level correlates with the thermic effect of caffeine [72].

(9). Caffeine and Nutrition

Some studies suggest that caffeine may hinder the availability of certain nutrients such as calcium and iron, evidence is scarce and inconsistent but worthly of brief mention here. Concerning calcium, caffeine has been considered a possible risk factor for the development of osteoprosis [73], other study indicate that the effect of caffeine on calcium balance may be deleterious only when calcium intake is low, where caffeine promotes the release of calcium from the sacroplasmic reticulum and inhibits its reuptake into the cell [74]. Regarding iron absorption, the polyphenol and tannic acid content of coffee not the caffeine content are responsible for impairment of iron absorption [50].

REFEERNCES

[1] Paul, O; Lepper M.H. and Phelan W.H. A longitudinal study of coronary heart diseases. *Circulation.* 1963; 28: 20-31.

[2] Boston Collaborative Drug Surveillance Program. Coffee drinking and acute myocardial infarction; report from the Boston collaborative drug surveillance program. *Lancet.* 1972; 2: 1278-1281

[3] Jick, H; Miettinen, O.S. and Neff RK. Coffee and myocardial infarction. *N. Engl. J. Med.* 1973; 289: 63-67.

[4] Yano, K; Rhoads, G.G.and Kagan, A. Coffee, alcohol and risk of coronary heart disease among Japanese men living in Hawaii. *N. Eng. J. Med.* 1977; 201: 547-552.

[5] Hennekens, C.H.; Drolette, M.E.and Jesse, M.J. Coffee drinking and death due to coronary heart disease. *N. Eng. J. Med.* 1977; 297: 405-409.

[6] Klatsky, A.L.; Pettiti D.B.and Armstrong M.A. Coffee, tea and cholesterol. *Am. J. Cardiol.* 1985; 55: 577-588.

[7] Mathias, S; Garland, C and Barrett-Connor E. Coffee, plasma cholesterol and lipoproteins: A population study in adult community. *Am. J. Epidemiol.* 1985; 121: 869-905.

[8] Haffner, S.M.; Knappj, A and Stern M.P. Coffee consumption, diet and lipids. *Am. J. Epidemiol.* 1985; 122: 1-12.

[9] Williams, P.T; Wood, P.D.and Vranizan, K.M. Coffee intake and elevated cholesterol and apolipoprotein B levels in men. *JAMA*. 1985; 253: 1407-1411.

[10] Kark, J.D; Friedlander, Yand Kauffman, N.A. Coffee, tea and cholesterol: The jerusalem lipid research clinical prevalence study. *Br. Med. J.* 1985; 291: 699-704.

[11] Curb, J.D; Reed, D.Mand Kautz, J.A. Coffee, caffeine and serum cholesterol in Japanese men in Hawaii. *Am. J. Epidemiol.* 1986; 123: 648-655.

[12] Rosenberg, L; Werler, M.M and Kaufman, D.W. Coffee drinking and myocardial infarction in young women. *Am. J. Epidemiol.* 1987; 126: 147-149.

[13] LeGrady, D; Dyer, A.R.and Shekelle RB. Coffee consumption and mortality in the Chicago Western electron company study. *Am. J. Epidemiol.* 1987; 126: 803-812.

[14] Rosenberg, L; Palmer, J.Rand Kelly, J.P. Coffee drinking and non fatal myocardial infarction in men under 55 years of age. *Am. J. Epidemiol*. 1988; 128: 570-578.

[15] La Vecchia, C; Gentili, Aand Negri E. Coffee consumption and myocardial infarction in women. *Am. J. Epidemiol.* 1989; 130: 481- 485.

[16] Aro, A; Tuomilehto, Jand Kostianien E. Boiled coffee raises serum low-density lipoprotein concentration. *Metabolism.* 1987; 36: 1027-1030.

[17] Bonaa, K; Arneson, E and Thelle, D.S. Coffee and cholesterol: Is it all in the brewing? The Tromso study. *Br. Med. J.* 1988; 297: 1103-1104.

[18] Bak, A.Aand Grobbee, D.E. The effect on serum cholesterol of coffee brewed by filtering or boiling. *N. Engl. J. Med.* 1989; 321: 1432-1437.

[19] Klatsky, A.L.; Friedman, G.D.and, Armstrony M.Y. Coffee use prior to myocardilal infarction restudied: heavier intake may increase the risk. *Am. J. Epidemiol.* 1990; 132: 479-487.

[20] Dawber, T.R; Kannel, W.Band Gordon, T. Coffee and cardiovascular disease: observation from the Framingham study. *N. Engl. J. Med.* 1974; 291: 871-74.

[21] Murray, S.S; Bielke, Eand Gibson, R.W. Coffee consumption and mortality from ischemic heart disease and other causes: Results from the Lutheran Brotherhood study 1966-1978. *Am. J. Epidemiol.* 1981; 113: 661-667.

[22] Forde, O.H.; Krutsen, S.F and Arnesen, E. Coffee consumption and serum lipid concentrations in men with hypercholesterolemia: A randomized intervention study. *Br. Med. J.* 1985; 290: 893-895.

[23] La Croix, A.Z.; Mead, L.A and Liang, K,Y. Coffee consumption and the incidence of coronary artery disease. *N. Engl. J. Med.* 1986; 315: 977-982.

[24] Fried, R.E; Pearson, T.Aand Levine, D.M. The effect of filtered coffee consumption on plasma lipids: Results of a randomized clinical trial. *JAMA.* 1992; 267; 811-815.

[25] Zock, P.L; Katan, M.Band Merkjeas, M.P. Effect of a Lipid-rich fraction from boiled coffee on serum cholesterol. *Lancet.* 1990; 335: 1235-1237.

[26] Tverdal, A; Stensvold, Iand Solvoll K. Coffee consumption and death from coronary heart diseases in middle aged Norwegian men and women. *Br. Med. J.* 1990; 30: 566-69.

[27] Salomen, J.T; Happonen, Pand Salon R. Interdependence of association of physical activity, smoking and alcohol and coffee consumption with scrum high density lipoprotein and-non-high density lipoprotein cholesterol: A population study in eastern Finland. *Prev. Med. J.* 1987; 290: 893-895.

[28] Thelle, D.S; Arnesen, Eand Forde, O.H. The Tromso heart study: Does coffee raise serum cholesterol?. *N. Engl. J. Medicine.* 1983; 300: 1454-1457.

[29] Dusseldorp, M.V; Katan, M.Band Demacker, P.N. Effects of decaffeinated versus regular coffee on serum lipoprotein. *Am. J. Epidemiol.* 1990;132: 33-40.

[30] Superko, H.R; Bortz, W and Williams, P.T. Caffeinated and decaffeinated coffee effects on plasma lipoprotein, cholesterol, apolipoprotein: A controlled randomized trail. *Am. J. Clin. Nutr.* 1991; 54: 599-605.

[31] Aro, A; Kostiainen, E and Huttunen, J.K. Effects of coffee and tea on serum Lipoproteins. *Atherosclerosis.* 1985; 57: 123-128.

[32] Kark, J.D; Friedlander, Yand Kaufman, N.A. Coffee, tea and plasma cholesterol: The Jerusalem lipid research clinic prevalence study. *Br. Med. J.* 1985; 291: 699-704.

[33] Superko, H.R; Bortz,Wand Williams, P.T. Caffienated and decaffeinated coffee effects on plasma lipoprotein, cholesterol, apolipoportein: A controlled randomized trila. *Am. J. Clin. Nutr.* 1991;54:599-605.

[34] Hrubec, Z. Coffee drinking ischemic heart disease. (letter). *Lancet.* 1973; 1: 548.

[35] Hofman, A; Vanlaar, Aand Klein F. Coffee and cholesterol (Letter). *N. Engl. J. Med.* 1983; 309: 1245.

[36] Shekelle, R.G; Gale, Mand Paul O. Coffee and cholesterol (Letter). *N. Engl. J. Med.* 1983; 309: 1249-1250.

[37] Arnesen, E; Forde, O.H and Thelle, D.S. Coffee and serum cholesterol. *Br. Med. J.* 1984; 288: 1960.

[38] Little, J.Aand Shanoff, H.M. Serum lipid in coronary heart disease. *Lancet.* 1966; 1: 732-734.

[39] Weininger, J and Briggs, M.G. Nutrition update. Volume 1. John Wiley and Sons. Inc. 1983; 4-8.

[40] Sheilla Bingham. Dictionary of nutrition (consumer's guides to the facts of foods) 1977, by Barrie and Jenkins Ltd., pp 68-70.

[41] Hareld Egan, Ronald S.Kirk, Ronald Sanyer. Pearson's chemical analysis of foods. Longman group limited. 1981; pp. 292-299.

[42] Henny BH, and Pharm B. Source book of flavors. Avi Publishing Company, INC. 1987: 163-160.

[43] Delvin, TM. Text book of Biochemistry with clinical correlation. 1982, pp.464.

[44] Smith, J.M; Pearson, S and Marks, V. Plasma caffeine concentration in out patients. *Lancet.* 1982; (Letter) 985-986.

[45] Bergman, Jand Dews, P.B. Dietary caffeine and its toxicity in nutritional toxicology. Volume II. JN Hathcock, Editor, Academic Press, New York. 1987; 199-221.
[46] Council on Scientific Affairs Report. Caffeine labeling. *JAMA*. 1984; 252: 803-6.
[47] Rosenberg, L; Mitchell, A.Aand Shapine S. Selected birth defects in relation to caffeine containing beverages. *JAMA*. 1982; 247: 1429-1432.
[48] Nguyen, P.Vand Myers, M.G. Cardiovascular effects of caffeine and nifedipine. *Clin. Pharm. Ther. J.* 1988; 44: 315-319.
[49] Bak, A.A and Grobbee, D.E. Caffeine, blood pressure and serum lipids. *Am. J. Clin. Nutr.* 1991; 53: 971-975.
[50] Curatolie, P.Wand Robertson, D. Health consequences of caffeine. *Ann. Inter. Med.* 1983; 98: 641- 653.
[51] Minton, J.P; Foecking, M.Kand Webster, D.J. Caffeine, cyclic nucleotides and breast diseases. *Surgery*. 1979; 86: 105-109.
[52] Naismith, D.J; Akinyanjn, P.A and Yudkin, J. Influence of caffeine containing beverages on the growth, food utilization and plasma lipids of the rats. *J. Nutr.* 1969; 97: 375-381.
[53] Food and nutrition Board. Nutrition during pregnancy. National Academic Press, Washington DC. 1990. pp. 399.
[54] Food and Nutrition Board Nutrition during lactation. National Academic Press, Washington DC. 1991; 176.
[55] Grobbe, D.E; Rimm, E.B and Giovannuccin E. Coffee, caffeine and cardiovascular diseases in men. *N. Engl. J. Med.* 1990; 323; 102-32.
[56] Rosenberg, L; Slone, D and Shapine S. Coffee drinking and myocardial infarction in young women. *Am. J. of Epidemiol.* 1980;111: 675-81.
[57] Klatsky, A.L; Friedman, G.D and Siegelaub, A.B. Coffee drinking prior to acute myocardial infarction: results from the Kaiser-permanent epidemiologic study of myocardial infarction. *JAMA*. 1973; 5: 540-43.
[58] Myron, Wand Vorisol L. Effects of caffeine in man. *Nutr. Rev.* 1970; 28: 38-40.
[59] Hennekers, C.H; Drolette, M.Eand Jesse, M.J. Coffee drinking and death due to coronary heart disease. *N. Engl. J. Med.* 1976; 294: 633-636.
[60] Kannel, W.B and Dawber TR. Coffee and coronary heart disease. *N. Engl. J. Med.* 1973; 289: 100-101.
[61] Dobmeyer, D.J; Stine, R.Aand Leier, C.V. The arrhythmogenic effect of caffeine in human beings. *N. Engl. J. Med.* 1983; 308: 814-816.
[62] Sung, B.H. Effect of caffeine on blood pressure response during exercise in normotensive health young men. *Am. J. Cardiol.* 1990; 65: 909-913.
[63] Lin, Kessler .H. A multifactorial model for pancreatic cancer in man. *JAMA*. 1981; 245: 147-152.
[64] Ernster, V.L; Mason, L and Goodson W. Effects of caffeine-free diet on benign breast disease: A randomized trial. *Surgery*. 1982; 91: 263-267.
[65] Love, S.M; Gelman, R.Sand Silen W. Fibrocystic disease of the breast: A non disease?. *N. Eng. J. Med.* 1982; 307: 1010-14.
[66] Rall, T.W. Central nervous system stimulants. 6th Ed. Macmillan Publishing Co. Inc, New York 1980; pp594.

[67] Smellie, F.W; Davis, C.Wand Daly JW. Inhibition of adenosine-elicited accumulation of cyclic AMP in brain slices and of brain phosphodiesterase activity. *Life Science.* 1979; 24: 2475-2482.

[68] Butcher, R.Wand Sutherland, E.W. Adenosine 3',5'phosphate in biologic materials. *J. Biol. Chem.* 1962; 237: 1244-1250.

[69] Sattin, Aand Rall, T.W. The effect of Adenosine and Adenine nucleotides on the cyclic adenosine 3',5' phosphate content of Guinea Pig cerebral cortex slices. *Molecular Pharmacology.* 1970; 6: 13- 23.

[70] Yokogoshi, H; Tani, Sand Amano N. The effects of caffeine and caffeine containing beverages on the disposition of brain sertonin in rats. *Agri. Biol. Chem.* 1983; 51: 3281 -86.

[71] Dulloo, A.G. Normal caffeine consumption: Influence on thermogenic and daily energy expenditure in lean and post obese human volunteers. *Am. J. Clin. Nutr.* 1989; 49: 44 -50.

[72] Astrup, Aand Toubro, S. Caffeine: A double-blind placebo-controlled study of its thermogenic, metabolic and cardiovascular effects in healthy volunteers. *Am. J. Clin. Nutr.* 1990; 51: 759-767.

[73] Arnaud, C.Dand Sanchez, S.D. The role of calcium in osteoperosis. *Ann. Rev. Nutr.* 1990; 10: 397-414.

[74] Barger-Lux, M.J; Heaney, R.Pand Stegma, M.R. Effects of moderate caffeine intake on the calcium economy of premanopausal women. *Am. J. Clin. Nutr.* 1992; 52: 722-5.

In: Natural Products and Their Active Compounds ... ISBN: 978-1-62100-153-9
Editors: M. Essa, A. Manickavasagan, and E. Sukumar

Chapter 21

HERBS, SUPPLEMENTS AND NUTRIENTS FOR THE TREATMENT OF CARDIOVASCULAR DISEASES

A. R. Mullaicharam[*]
Oman Medical College, Sultanate of Oman

ABSTRACT

Herbal medicines are regulated in many countries and accepted to be integrated in healthcare system. Herbal medicine is an affordable health care resource for many countries. Among the World Health Organization (WHO) efforts for promoting the use of alternative medicines is the creation of awareness about safe and effective alternative medicine therapies among the public and consumers. For cardiovascular diseases, herbal treatments have been used in patients with atherosclerosis, (which occurs when fatty deposits clog and harden arteries), coronary heart disease, (caused by the reduced blood supply to the heart muscle), stroke, (caused by inadequate blood flow to the brain leading to the death of brain cells), hypertension, (occurs when blood pressure is higher than the normal range), cardiac arrhythmias, (which are irregular or abnormal heartbeats).

INTRODUCTION

Significant lifestyle changes in the second half of the 20th century have greatly contributed to the emerging epidemic of chronic diseases such as cardiovascular diseases (CVD). Currently, 15.3 million people are estimated to die from cardiovascular diseases every year; that represents one-third of all global deaths from all causes. In the next two decades, the increasing burden of cardiovascular diseases will be borne mostly by developing countries [1].

Herbs have been used as medical treatments since the beginning of civilization and some derivatives have become mainstays of human pharmacotherapy.[2] Antioxidants are believed to help prevent heart disease by fighting free radicals, substances that harm the body when

[*] For Correspondence: mullaicharam@yahoo.com

left unchecked. These nutrients and dietary supplements are on a constant search and destroy mission, fighting the continuous onslaught of free radicals.

HERBAL REMEDIES FOR HEART CARE

Garlic (Allium Sativum)

Garlic has been used longer for more purposes in more places than any other plant. (Cultivated in the Middle East over 5,000 years ago and found in the Egyptian tomb of King Tutankhamen). In the circulatory system, Garlic lowers cholesterol, low density lipoproteins, blood lipids and blood pressure and also raises high density lipoproteins. The sulphides present in garlic help in bringing down high blood pressure and thus, assist in blood pressure management. In addition, the herb inhibits certain compounds, like natural ACE inhibitors (gamma-glutamyl peptides and flavonolic compounds), magnesium (natural calcium channel blocker), phosphorus, adenosine and allicin. These compounds are known for widening arteries and making blood flow easier. Garlic is available as a food, as a spice in powder form, and as a supplement.

Eating garlic has helped to lower cholesterol in some research,[3] though several double-blind trials have not found garlic supplements to be thusly effective.[4-6] Garlic prevent the oxidation of LDL cholesterol, may prevent the liver from producing excess fat and cholesterol. In one study, adding as little as two ounces of garlic juice to a fatty, cholesterol-laden meal was found to actually lower the cholesterol by up to 7 percent. Another study found that 600 mg of garlic powder a day could push the total cholesterol down by some 10 percent. Other research has corroborated these findings reporting that garlic can lower both total and LDL cholesterol while raising the HDL ("good") cholesterol. A 10-month study found that eating three cloves of garlic a day keeps the cholesterol down for extended periods. And because it contains ajoene and other substances, garlic also helps to keep the blood "thin" and free of potentially deadly blood clots. Garlic reduces cholesterol, strengthens the circulation, and acts as a decongestant. Reports on many double-blind garlic trials performed through 1998 suggested that cholesterol was lowered by an average of 9 to 12% and triglycerides by 8 to 27% over a one-to-four month period.[7-9] Most of these trials used 600 to 900 mg per day of garlic supplements. More recently, however, several double-blind trials have found garlic to have minimal success in lowering cholesterol and triglycerides.[10-14] One negative trial has been criticized for using a steam-distilled garlic “oil” that has no track record for this purpose, [15] while the others used the same standardized garlic products as the previous positive trials. Based on these findings, the use of garlic should not be considered a primary approach to lowering high cholesterol and triglycerides. [16]

Part of the confusion may result from differing effects from dissimilar garlic products. In most but not all trials, aged garlic extracts and garlic oil (both containing no illicit) have not lowered cholesterol levels in humans.[17,18] Therefore, neither of these supplements can be recommended at this time for cholesterol lowering. Odor-controlled, enteric-coated tablets standardized for illicit content are available and, in some trials, appear more promising.[19] Doctors typically recommend 900 mg per day (providing 5,000 to 6,000 mcg of illicit), divided into two or three administrations.

Green Tea

It is popular in Asia for centuries. Green tea helps to keep blood pressure under control. It also may help keep cholesterol from clogging arteries. The tea contains Epigallocatechin Gallate (EGCG) and other substances that protect the body against the dangers of oxidation, while helping to keep the harmful LDL cholesterol down and the helpful HDL cholesterol up. They also assist in keeping blood pressure under control. Green tea has been shown to lower total cholesterol levels and improve people's cholesterol profile, decreasing LDL cholesterol and increasing HDL cholesterol according to preliminary studies.20-23 However, not all trials have found that green tea intake lowers lipid levels.24 Much of the research documenting the health benefits of green tea is based on the amount of green tea typically drunk in Asian countries—about three cups per day, providing 240 to 320 mg of polyphenols.

An extract of green tea, enriched with a compound present in black tea (theaflavins), has been found to lower serum cholesterol in a double-blind study of people with moderately high cholesterol levels.25 The average reduction in total serum cholesterol during the 12-week study was 11.3%, and the average reduction in LDL cholesterol was 16.4%. The extract used in this study provided daily 75 mg of theaflavins, 150 mg of green tea catechins, and 150 mg of other tea polyphenols.

Tea Helps the Heart

A research team at Brigham and Women's Hospital in Boston has found that heart-attack risk in people drinking one or more cups of tea per day was about half that of those who drank no tea.

The team's findings suggest that "tea may be beneficial because it contains flavonoids which reduce platelet aggregation and inhibit LDL-cholesterol oxidation," according to Peter Wehrwein, in an article "More evidence that tea is good for the heart," in the January 30, 1999 issue of The Lancet, the journal of the British Medical Association. The research team found that "all teas are not equal." Gary Beecher of the US Department of Agriculture's Agricultural Research Service reported on catechin (flavan-3-ol) concentrations in different forms of tea. Several brands of black tea had substantial amounts of catechins. However, there were reduced amounts in decaffeinated black tea, and catechins were undetectable in herbal teas.

Hawthorn (Crataegus Monogyna)

The herb is known to expand arteries and hence, improve coronary blood flow, reducing blood pressure. Hawthorn is also used as cardio tonic that helps in strengthening the heart muscle and promotes forceful contractions. In all, it helps in improving heart function.

It contains a combination of flavonoids that can protect the heart against oxygen deprivation and the development of abnormal rhythms. It dilates coronary blood vessels, improving the flow of blood to the heart. It strengthens the heart muscle and works to help the body rid itself of excess salt and water. It reduces blood levels of cholesterol and triglycerides, and brings down high blood pressure. Choose a standardized extract containing 1.8 percent vitexin-2 rhamnosides.

Hawthorn Berries (Crataegus Oxycantha)

Hawthorn regulates high and low blood pressure, arrhythmic heartbeat, irregular pulse and prevents the hardening of the arteries. It can also treat arteriosclerosis and cools inflammation in heart muscle in inflammatory conditions. If used regularly Hawthorn can strengthen the heart muscles and nerves of the heart.

Arjuna (Terminalia Arjuna)

Arjuna, an important Ayurvedic herb, is a coronary vasodilator. It protects the heart, strengthens circulation, and helps to maintain the tone and health of the heart muscle. It is also useful in stopping bleeding and to promote healing after a heart attack. Current scientific research has proved that T.Arjuna contains specific medically active constituents namely triterpine glycosides like arjunetosides I,II,III, IV, arjunine and arjunetein.Bark of Arjuna tree has been found to be rich in Co-enzyme Q-10 which is highly prescribed in cardiology departments now a days to prevent heart problems.

Ginger (Zingiber Officinale)

Ginger is an important herb for a healthy heart. Ayurvedic physicians suggest that eating a little bit of ginger every day will help to prevent heart attack. It reduces cholesterol. It also reduces blood pressure and prevents blood clots. Ginger is known to help improve blood circulation and thus, lower blood pressure in a safe and effective manner. In addition, it also relaxes the muscles around the blood vessel, purifies the blood and lowers cholesterol. Make fresh ginger tea by boiling a little grated or sliced ginger in a cup or two of water. We can also grate a little ginger and add it to rice and/or soup. The volatile oils in ginger stimulate both the circulatory and respiratory systems. Ginger's heart-helping attributes are similar to that of garlic. Ginger interferes with the long sequence of events necessary for blood clots to form. This helps to prevent clots that can lodge in narrowed coronary arteries and set off a heart attack.

Ashwagandha (Withania Somnifera)

Ashwagandha is a popular herb that helps in counteracting high blood pressure and thus aids in lowering it. However, a doctor should be consulted before taking this herb, as it can react or interfere with the medication that an individual is already taking.It is a unique herb with anti-stress adaptogenic action that leads to better physical fitness and helps cope with life's daily stress. It is especially beneficial in stress related disorders such as arthritis, hypertension, diabetes and general debility.

Kelp (Fucus Visiculosis)

Kelp is seaweed that helps in reducing the raised levels of cholesterol and blood pressure, by thinning the blood. It is available in the temperate and coastal oceans.

Cinnamon (Cinnamomum Zeylanicum)

Cinnamon is known to reduce the levels of LDL cholesterol in the body, due to its antioxidant properties. It is truly a healing herb that helps use the hormone insulin in the body more efficiently.

Turmeric

It lowers blood cholesterol levels by stimulating the production of bile. It also prevents the formation of dangerous blood clots that can lead to heart attack.

Onions (Allium Cepa)

Onions contain adenosine and other "blood thinners" that help to prevent the formation of blood clots. In addition to thinning the blood, onions can help keep the coronary arteries open and clear by increasing the HDL. Eating half a raw onion every day can increase HDL by 20 to 30 percent. Throughout recorded history onion has been used as a heart tonic, blood purifier, antiseptic, digestive aid, sedative and aphrodisiac.

An onion a day will boost their high density lipoprotein levels which will help remove cholesterol and low density lipoproteins.

Adenosine has been proven to lower blood pressure by inhibiting platelets to stick together and form clots. Onions will also aid the body in dissolving already formed clots by stimulating the bodies' fibrinolytic system.

Ginkgo Biloba

It improves the flow of blood throughout the body. It is also an antioxidant. Ginkgo biloba can benefit the cardiovascular system by preventing the formation of free radicals. The ginkgo extract containing 24-percent of ginkgo flavone glycosides.

Alfalfa

Alfalfa leaves and sprouts help reduce the blood cholesterol levels and plaque deposits on artery walls.

Citrin (Garcinia Cambogia)

It is an extract from the plant Garcinia cambogia, inhibits the synthesis of fatty acids in the liver. It helps to prevent the accumulation of potentially dangerous fats in the body.

Grape Seed Extract

Grape seed extract with oligomeric proanthocyanidins (OPCS) may lower high blood pressure, which can cause heart disease.

Soy

Soy had been long popular in Asia. It has been proven to be heart protectors also.

When people with high cholesterol are put on a low-fat, low-cholesterol diet, their cholesterol levels usually drop. But if we replace the animal protein in their diet with soy protein, their cholesterol levels are found to drop significantly lower. One study has showed that soy protein could cancel out the effect of 500 mg of cholesterol deliberately added to the daily diet.

Although soy can lower cholesterol levels in those with normal levels, it works best in people with elevated cholesterol.

Brewer's Yeast

Brewer's yeast can lower the total cholesterol and LDL while raising the helpful HDL. In one study with normal- and high-cholesterol patients, 11 healthy volunteers were given brewer's yeast. Eight weeks later, 10 of the 11 people with normal cholesterol levels had even lower total cholesterol levels and increased HDL levels. Among the 15 volunteers with high cholesterol, eight enjoyed the same beneficial results.

Cordyceps

Cordyceps is a Chinese herb. It can slow the heart rate, increase blood supply to the arteries and heart, and lower blood pressure.

Artichoke Leaf Extract

It reduces blood cholesterol and protects the liver. This herb has antioxidant activity and may inhibit the oxidation of cholesterol, a factor in atherosclerosis.

Cat's claw

It contains a variety of valuable phytochemicals that inhibit the processes involved in the formation of blood clots. It increases circulation and inhibits inappropriate clotting. Thus, it may help to prevent stroke and reduce the risk of heart attack.

Oat straw and kava kava are tonics for the nervous system.

White Willow Bark

It contains salicin, an aspirinlike compound. It has been used for centuries much as aspirin is today. Aspirin is often recommended for cardiovascular condition. This herb may provide the same protection without stomach upsets associated with aspirin.

Note: Do not take this herb if you are allergic to aspirin.

Cayenne – (Capsicum Annum)

It is the one of the most effective stimulants for both the digestive and circulatory systems. Cayenne regulates blood pressure, strengthens the pulse, feeds the heart, lowers cholesterol, thins the blood, and cleanses the circulatory system. Cayenne helps cell structure in the arteries aiding the rebuilding process of damaged arteries due to heart disease.

Gotu Kola (Centella Asiatica)

The volatile oils in Gotu Kola have a diuretic and blood purifying property which helps lower serum and cholesterol levels because they contain beta-sitoerol.

Valerian (Valeriana Officinalis)

Hundreds of experiments have been done on Valerian in Germany and Russia for its effectivemeness in the treatment of nervous, circulatory, digestive and sleep disorders. Studies have shown Valerian to reduce hypertension, slow the heart rate and increase the power of each beat. Valerian is great natural treatment for palpitations and nervousness.

Rudraksha

Rudraksha are the dried seeds of the rudraksha tree. They are good for the heart both physically and spiritually. They are often used in meditation (like the rosary by Catholics). It is believed to "open the heart chakra." Rudraksha can be used in a number of forms. It can be worn in a necklace of the beads externally, in front of the heart. We can also soak a rudraksha bead overnight in water and drink the water in the morning. Drinking rudraksha water is believed to reduce blood pressure and strengthen the heart.

Psyllium

Use of psyllium has been extensively studied as a way to reduce cholesterol levels. An analysis of all double-blind trials in 1997 concluded that a daily amount of 10 grams psyllium lowered cholesterol levels by 5% and LDL cholesterol by 9%. Since then, a large controlled trial found that use of 5.1 grams of psyllium two times per day significantly reduced serum cholesterol as well as LDL-cholesterol. Generally, 5 to 10 grams of psyllium are added to the diet per day to lower cholesterol levels. The combination of psyllium and oat bran may also be effective at lowering LDL cholesterol.

Guggul

This ayurvedic herb is derived from a type of myrrh tree. It has been shown to lower blood-fat levels while raising levels of HDL, the so called "good cholesterol."

Note: Do not use this herb if you have a thyroid disorder. It is a mixture of substances taken from a plant, is an approved treatment for elevated cholesterol in India and has been a mainstay of the Ayurvedic approach to preventing atherosclerosis. One double-blind trial studying the effects of guggul reported that serum cholesterol dropped by 17.5%. In another double-blind trial comparing guggul to the drug clofibrate, the average fall in serum cholesterol was slightly greater in the guggul group; moreover, HDL cholesterol rose in 60% of people responding to guggul, while clofibrate did not elevate HDL. A third double-blind trial found significant changes in total and LDL cholesterol levels, but not in HDL. However, in another double-blind trial, supplementation with guggul for eight weeks had no effect on total serum cholesterol, but significantly increased LDL-cholesterol levels, compared with a placebo. Daily intakes of guggul are based on the amount of guggulsterones in the extract. The recommended amount of guggulsterones is 25 mg taken three times per day. Most extracts contain 5 to 10% guggulsterones, and doctors familiar with their use usually recommend taking guggul for at least 12 weeks before evaluating its effect.

Achillea Wilhelmsii

In a double-blind trial, people with moderately high cholesterol took a tincture of Achillea wilhelmsii, an herb used in traditional Persian medicine. Participants in the trial used

15 to 20 drops of the tincture twice daily for six months. At the end of the trial, participants experienced significant reductions in total cholesterol, LDL cholesterol and triglycerides, as well as an increase in HDL cholesterol compared to those who took placebo. No adverse effects were reported.

Artichoke

Artichoke has moderately lowered cholesterol and triglycerides in some, but not all, human trials. One double-blind trial found that 900 mg of artichoke extract per day significantly lowered serum cholesterol and LDL cholesterol but did not decrease triglycerides or raise HDL cholesterol. Cholesterol-lowering effects occurred when using 320 mg of standardized leaf extract taken two to three times per day for at least six weeks.

Berberine

Berberine, a compound found in certain herbs such as goldenseal, barberry, and Oregon grape, has been found to lower serum cholesterol levels. In a study of people with high cholesterol levels, 500 mg of berberine taken twice a day for three months lowered the average cholesterol level by 29%. No significant side effects were reported, except for mild constipation.

Fenugreek Seeds

Fenugreek seeds contain compounds known as steroidal saponins that inhibit both cholesterol absorption in the intestines and cholesterol production by the liver. Dietary fiber may also contribute to fenugreek's activity. Multiple human trials (some double-blind) have found that fenugreek may help lower total cholesterol in people with moderate atherosclerosis or those having insulin-dependent or non-insulin-dependent diabetes. One human double-blind trial has also shown that defatted fenugreek seeds may raise levels of beneficial HDL cholesterol. One small preliminary trial found that either 25 or 50 grams per day of defatted fenugreek seed powder significantly lowered serum cholesterol after 20 days. Germination of the fenugreek seeds may improve the soluble fiber content of the seeds, thus improving their effect on cholesterol. Fenugreek powder is generally taken in amounts of 10 to 30 grams three times per day with meals.

Fo-Ti (Ho Shou Wu, Polygonum Multiflorum)

It combats the symptoms of heart disease, helping to reduce blood pressure and blood-cholesterol levels.Preliminary Chinese research has found that high doses (12 grams per day) of the herb fo-ti may lower cholesterol levels. Double-blind or other controlled trials are needed to determine fo-ti's use in lowering cholesterol. A tea may be made from processed roots by boiling 3 to 5 grams in a cup of water for 10 to 15 minutes. Three or more cups should be drunk each day.

Wild yam has been reported to raise HDL cholesterol in preliminary research. Doctors sometimes recommend 2 to 3 ml of tincture taken three to four times per day, or 1 to 2 capsules or tablets of dried root taken three times per day.

Other Herbs which are used for Heart disease are Cardamom (Elettaria cardamomum) , Gotu Kola (Hydrocotyle asiatica), Skullcap (Scutellaria), Valerian (Valeriana), Manjishta (Rubia tinctoria), Jatamamsi (Nardostachys jatamamsi), Shankapushpi (Clitoria ternatea) and Burdock (Arctum lappa) barberry, black cohosh, butcher's broom, cayenne (capsicum), dandelion, and ginseng.

Caution: Do not use barberry or black cohosh during pregnancy. Do not use ginseng if you have high blood pressure. Also avoid the herbs ephedra (ma huang) and licorice, as they cause a rise in blood pressure.

Nutritional Supplements

Lecithin (Natural Occurring Fat)

Lecithin breaks down fat and cholesterol enabling the body to use what it needs and discarding the rest. It therefore cleanses the circulatory system of cholesterol deposits while providing essential nutrients for proper brain and nervous system function.

Red Yeast Rice

Researchers have determined that one of the ingredients in red yeast rice, called monacolin K, inhibits the production of cholesterol by stopping the action of the key enzyme in the liver (in other words, HMG-CoA reductase) that is responsible for manufacturing cholesterol.

Monacolin K is the came compound as lovastatin (Mevacor), a prescription drug used to treat high cholesterol. However, the amount per volume of monacolin K in red yeast rice is small (5 mg per 2.4 grams of red yeast rice) when compared to the 20 to 40 mg of lovastatin typically used to lower cholesterol levels.[26] It appears that other monacolin compounds present in red yeast rice work together with monacolin K to produce a greater cholesterol-lowering effect than would be expected from the small amount of monacolin K alone.The red yeast rice used in various studies was a proprietary product called Cholestin®, which contains ten different monacolins.

Note: Cholestin has been banned in the United States, as a result of a lawsuit alleging patent infringement.

Other red yeast rice products currently on the market differ from Cholestin in their chemical makeup. None contain the full complement of ten monacolin compounds that are present in Cholestin, and some contain a potentially toxic fermentation product called citrinin. Despite these concerns, other red yeast rice products are being widely used and both anecdotal reports and clinical research suggest that they have a similar safety and efficacy profile as that of Cholestin.

Animal studies suggest that the mushroom maitake may lower fat levels in the blood. This research is still preliminary and requires confirmation with controlled human trials.

Animal studies indicate that saponins in alfalfa seeds may block absorption of cholesterol and prevent the formation of atherosclerotic plaques. However, consuming the large amounts of alfalfa seeds (80 to 120 grams per day) needed to supply high doses of these saponins may potentially cause damage to red blood cells in the body.

Dietary fibre is also a major factor in reducing total cholesterol in the blood and LDL cholesterol in particular. Eating a diet high in fibre and wholegrain cereals can reduce the risk of coronary heat disease.

An intake of 0.8 mg of folic acid could possibly reduce the risk of coronary heart disease (reduced blood supply to the heart muscle) by 16% and the risk of stroke by 24%. Flavonoids, compounds that occur in a variety of foods such as tea, onions and apples, could also possibly reduce the risk of coronary heart disease. There is insufficient evidence to support the theory that antioxidants such as Vitamin E, Vitamin C or b-carotene might reduce the risk of cardiovascular diseases (CVD).

A high intake of salt (sodium) has been linked to high blood pressure, a major risk factor for stroke and coronary heart disease.

Consumption of fruits and vegetables has been widely associated with good health. Recent studies show a protective effect against coronary heart disease, stroke and high blood pressure.

Fish consumption also reduces the risk of coronary heart disease. The benefits are most evident in high risk groups. For these groups, consuming 40-60g of fish per day would lead to a 50% reduction in the number of deaths form coronary heart disease. Other dietary factors may also contribute to reducing the risk.

Nuts are high in unsaturated fatty acids and low in saturated fats, which contribute to lowering cholesterol levels. Several animal experiments have suggested that isoflavones, present in soy products, may provide protection against coronary heart disease.

Alcohol can have both a damaging and protective role in the development of cardiovascular disease. Despite convincing evidence that low to moderate alcohol consumption reduces the risk of coronary heart disease, consumption should be limited because of the risk of other cardiovascular diseases and health problems.

Coffee beans contain a substance called cafestol, which can raise the level of cholesterol in the blood and may increase the risk of coronary heart disease. The amount of cafestol in the cup depends on the brewing method: zero for paper-filtered drip coffee and high for unfiltered coffee which is widely drunk in Greece, the Middle East and Turkey.

Dash

The National Institute of Health (U.S.A) recommends the Dietary Approaches to Stop Hypertension (DASH) eating plan as a key strategy for reducing high blood pressure. This diet calls for a reduction in salt or sodium as an essential first step, followed by lowering fat and sugar intake and increasing fiber through whole grains, fruits and vegetables. The slogan "five a day" was once thought sufficient to cover the recommended intake of fruits and veggies for all people, but as a result of the NIH's DASH study, scientists now know that five may not be the right number for every person. The right amount will depend on the age, physical activity level and sex of the individual.

Potassium

1. Research has shown that potassium may not only help prevent high blood pressure, but may, in fact, contribute to lowering blood pressure. Bananas, apricots, prunes, dates, cantaloupe, watermelon, strawberries and tomatoes are rich in potassium. If we get three servings of these per day we are probably getting enough potassium. But not to supplement our diet with potassium, as too much dietary potassium can have deleterious effects, particularly on those who are elderly or have kidney disorders.

Calcium

2. One fruit stands out as a good source of calcium. The venerable orange provides a good calcium charge served whole or in juice.

Magnesium

3. Magnesium is harder to come by in fruit, but two stands out as good sources. Bananas are an excellent source of magnesium as are avocados. Avocados are, like tomatoes, a crossover fruit/veggie that serves admirably.

Research has shown that potassium may actually help lower blood pressure, but potassium should be considered as only part of our total dietary pattern. In fact, the dietary pattern may be more important than the individual elements of the diet. Factors such as salt intake, amount and type of dietary fat, cholesterol, protein and fiber, as well as minerals such as potassium, calcium and magnesium appear to work together to affect blood pressure. Researchers attribute changes in blood pressure to certain patterns of food consumption rather than to individual foods.

VITAMINS

Glucomannan is a water-soluble dietary fiber that is derived from konjac root. Controlled[27,28] and double-blind[29,30] trials have shown that supplementation with glucomannan significantly reduced total blood cholesterol, LDL cholesterol, and triglycerides, and in some cases raised HDL cholesterol. Effective amounts of glucomannan for lowering blood cholesterol have been 4 to 13 grams per day.

Test tube and animal studies indicate that policosanol is capable of inhibiting cholesterol production by the liver. [31,32]

Vitamin C appears to protect LDL cholesterol from damage.[33] In some clinical trials, cholesterol levels have fallen when people with elevated cholesterol supplement with vitamin C.[34] Some studies report that decreases in total cholesterol occur specifically in LDL cholesterol.[35] Doctors sometimes recommend 1 gram per day of vitamin C. A review of the disparate research concerning vitamin C and heart disease, however, has suggested that most protection against heart disease from vitamin C, is likely to occur with as little as 100 mg per day.[36]

Pantethine, a byproduct of vitamin B5 (pantothenic acid), may help reduce the amount of cholesterol made by the body. Several preliminary[37-41] and two controlled[42, 43] trials have found that pantethine (300 mg taken two to four times per day) significantly lowers serum cholesterol levels and may also increase HDL. However, one double-blind trial in people whose high blood cholesterol did not change with diet and drug therapy, found that pantethine was also not effective.[44] Common pantothenic acid has not been reported to have any effect on high blood cholesterol.

Chromium supplementation has reduced total cholesterol,[45,46] LDL cholesterol[47,48] and increased HDL cholesterol[49,50] in double-blind and other controlled trials, although other trials have not found these effects.[51,52] One double-blind trial found that high amounts of chromium (500 mcg per day) in combination with daily exercise was highly effective, producing nearly a 20% decrease in total cholesterol levels in just 13 weeks.[53]

Brewer's yeast, which contains readily absorbable and biologically active chromium, has also lowered serum cholesterol.[54] People with higher blood levels of chromium appear to be at lower risk for heart disease.[55] A reasonable and safe intake of supplemental chromium is 200 mcg per day. People wishing to use brewer's yeast as a source of chromium should look for products specifically labeled "from the brewing process" or "brewer's yeast," since most yeast found in health food stores is not brewer's yeast, and does not contain chromium. Optimally, true brewer's yeast contains up to 60 mcg of chromium per tablespoon, and a reasonable intake is 2 tablespoons per day.

High amounts (several grams per day) of niacin, a form of vitamin B3, lower cholesterol, an effect recognized in the approval of niacin as a prescription medication for high cholesterol.[56] The other common form of vitamin B3—niacinamide—does not affect cholesterol levels. Some niacin preparations have raised HDL cholesterol better than certain prescription drugs.[57] Some cardiologists prescribe 3 grams of niacin per day or even higher amounts for people with high cholesterol levels. At such intakes, acute symptoms (flushing, headache, stomachache) and chronic symptoms (liver damage, diabetes, gastritis, eye damage, possibly gout) of toxicity may be severe. Many people are not able to continue taking these levels of niacin due to discomfort or danger to their health. Therefore, high intakes of niacin must only be taken under the supervision of a doctor.

Symptoms caused by niacin supplements, such as flushing, have been reduced with sustained-release (also called "time-release") niacin products. However, sustained-release forms of niacin have caused significant liver toxicity and, though rarely, liver failure.[58-62] One partial time-release (intermediate-release) niacin product has lowered LDL cholesterol and raised HDL cholesterol without flushing, and it also has acted without the liver function abnormalities typically associated with sustained-release niacin formulations.[63] However, this form of niacin is available by prescription only.

In an attempt to avoid the side effects of niacin, alternative health practitioners increasingly use inositol hexaniacinate, recommending 500 to 1,000 mg, taken three times per day, instead of niacin.[64,65] This special form of niacin has been reported to lower serum cholesterol but so far has not been found to cause significant toxicity.[66] Unfortunately, compared with niacin, far fewer investigations have studied the possible positive or negative effects of inositol hexaniacinate. As a result, people using inositol hexaniacinate should not take it without the supervision of a doctor, who will evaluate whether it is helpful (by measuring cholesterol levels) and will make sure that toxicity is not occurring (by measuring liver enzymes, uric acid and glucose levels, and by taking medical history and doing physical examinations).

Soy supplementation has been shown to lower cholesterol in humans.[67] Soy is available in foods such as tofu, miso, and tempeh and as a supplemental protein powder. Soy contains isoflavones, naturally occurring plant components that are believed to be soy's main cholesterol-lowering ingredients. A controlled trial showed that soy preparations containing high amounts of isoflavones effectively lowered total cholesterol and LDL ("bad") cholesterol, whereas low-isoflavone preparations (less than 27 mg per day) did not.[68] However, supplementation with either soy[69] or non-soy isoflavones (from red clover) [70] in pill form failed to reduce cholesterol levels in a group of healthy volunteers, suggesting that isoflavone may not be responsible for the cholesterol-lowering effects of soy. Further trials of isoflavone supplements in people with elevated cholesterol, are needed to resolve these conflicting results. In a study of people with high cholesterol levels, a soy preparation that contained soy protein, soy fiber, and soy phospholipids lowered cholesterol levels more effectively than isolated soy protein.[71]

Soy contains phytosterols. One such molecule, beta-sitosterol, is available as a supplement. Beta-sitosterol alone, and in combination with similar plant sterols, has been shown to reduce blood levels of cholesterol in preliminary[72] and controlled[73,74] trials. This effect may occur because beta-sitosterol blocks absorption of cholesterol.[75] In studying the effects of 0.8, 1.6, and 3.2 grams of plant sterols per day, one double-blind trial found that higher intake of sterols tended to result in greater reduction in cholesterol, though the differences between the effects of these three amounts were not statistically significant.[76]

A synthetic molecule related to beta-sitosterol, sitostanol, is available in a special margarine and has also been shown to lower cholesterol levels. In one controlled trial, supplementation with 1.7 grams per day of a plant-sterol product containing mostly sitostanol, combined with dietary changes, led to a dramatic 24% drop in LDL ("bad") cholesterol compared with only a 9% decrease in the diet-only part of the trial.[77] Other controlled and double-blind trials have confirmed these results.[78-83] A review of double-blind trials on sitostanol found that a reduction in the risk of heart disease of about 25% may be expected from use of sitostanol-containing spreads, a larger clinical effect than that produced by people reducing their saturated fat intake.[84] Supplementation with sitostanol in the amount of 1.8 grams per day for six weeks has also been shown to enhance the cholesterol-lowering effect of statin drugs.[85]

Tocotrienols, a group of food-derived compounds that resemble vitamin E, may lower blood levels of cholesterol, but evidence is conflicting. Although tocotrienols inhibited cholesterol synthesis in test-tube studies,[86,87] human trials have produced contradictory results. Two double-blind trials found that 200 mg per day of either gamma-tocotrienol[88] or total tocotrienols[89] were more effective than placebo, reducing cholesterol levels by 13–15%. However, in another double-blind trial, 200 mg of tocotrienols per day failed to lower cholesterol levels,[90] and a fourth double-blind trial found 140 mg of tocotrienols and 80 mg of vitamin E (d-alpha-tocopherol) daily resulted in no changes in total cholesterol, LDL cholesterol, or HDL cholesterol levels [91].

In a double-blind study of people with elevated blood levels of cholesterol or triglycerides, supplementation with 1 to 3 grams krill oil from Antarctic krill (a zooplankton crustacean) for three months decreased levels of total cholesterol, LDL cholesterol, and triglycerides, and increased HDL-cholesterol levels. Krill oil was significantly more effective than either a placebo or small amounts of regular fish oil containing 900 mg per day of omega-3 fatty acids [92].

Activated charcoal has the ability to adsorb (attach to) cholesterol and bile acids present in the intestine, preventing their absorption.[93,94] Reducing the absorption of bile acids results in increased cholesterol breakdown by the liver. In controlled studies of people with high cholesterol, activated charcoal reduced total- and LDL-cholesterol levels, when given in amounts from 4 to 32 grams per day. Larger amounts were more effective: reductions in total and LDL cholesterol were 23% and 29%, respectively, with 16 grams daily, and 29% and 41% with 32 grams daily.[95] Similar results were reported in other controlled[96] and preliminary[97] studies using 16 to 24 grams per day, but one small double-blind trial found no effect of either 15 or 30 grams per day in patients with high cholesterol [98].

Deficiency of the trace mineral, copper, has been linked to high blood cholesterol.[99,100] In a controlled trial, daily supplementation with 3 to 4 mg of copper for eight weeks decreased blood levels of total cholesterol and LDL cholesterol, in a group of people over 50 years of age [101].

Beta-glucan is a type of soluble fiber molecule derived from the cell wall of baker's yeast, oats and barley, and many medicinal mushrooms, such as maitake. Beta-glucan is the key factor for the cholesterol-lowering effect of oat bran.[102-105] As with other soluble-fiber components, the binding of cholesterol (and bile acids) by beta-glucan and the resulting elimination of these substances in the feces is very helpful for reducing blood cholesterol.[106-108] Results from a number of double-blind trials with either oat- or yeast-derived beta-glucan indicate typical reductions, after at least four weeks of use, of approx-imately 10% for total cholesterol and 8% for LDL ("bad") cholesterol, with elevations in HDL ("good") cholesterol ranging from zero to 16%.[109-113] For lowering cholesterol levels, the amount of beta-glucan used has ranged from 2,900 to 15,000 mg per day.

Some preliminary [114] and double-blind[115,116] trials have shown that supplemental calcium reduces cholesterol levels. Possibly the calcium is binding with and preventing the absorption of dietary fat.[117] However, other research has found no substantial or statistically significant effects of calcium supplementation on total cholesterol or HDL ("good") cholesterol.[118] Reasonable supplemental levels are 800 to 1,000 mg per day.

In one double-blind trial, [119] vitamin E increased protective HDL cholesterol, but several other trials,[120-122] found no effect of vitamin E. However, vitamin E is known to protect LDL cholesterol from damage. [123] Many cardiologists believe that only damaged LDL increases the risk of heart disease. Studies of the ability of vitamin E supplements to prevent heart disease have produced conflicting results, [124] but many doctors continue to recommend that everyone supplement 400 IU of vitamin E per day to lessen the risk of having a heart attack.

L-carnitine is needed by heart muscle to utilize fat for energy. Some,[125,126] but not all, preliminary trials report that carnitine reduces serum cholesterol.[127] HDL cholesterol has also increased in response to carnitine supplementation.[128,129] People have been reported in controlled research to stand a greater chance of surviving a heart attack if they are given L-carnitine supplements. [130] Most trials have used 1 to 4 grams of carnitine per day.

Magnesium is needed by the heart to function properly. Although the mechanism is unclear, magnesium supplements (430 mg per day) lowered cholesterol in a preliminary trial. [131] Another preliminary study reported that magnesium deficiency is associated with a low HDL cholesterol level. [132] Intravenous magnesium has reduced death following heart attacks in some, but not all, clinical trials. [133] Though these outcomes would suggest that people with high cholesterol levels should take magnesium supplements, an isolated double-blind trial reported that people with a history of heart disease assigned to magnesium supplementation experienced an increased number of heart attacks. [134] More information is necessary before the scientific community can clearly evaluate the role magnesium should play for people with elevated cholesterol.

Chondroitin sulfate has lowered serum cholesterol levels in preliminary trials.[135,136] Years ago, this supplement dramatically reduced the risk of heart attacks in a controlled, six-year follow-up of people with heart disease. [137] The few doctors aware of these older clinical trials sometimes tell people with a history of heart disease or elevated cholesterol levels, to take approximately 500 mg of chondroitin sulfate three times per day.

Although lecithin has been reported to increase HDL cholesterol and lower LDL cholesterol, [138] a review of the research found that the positive effect of lecithin was likely due to the polyunsaturated fat content of the lecithin. [139] If this is so, it would make more sense to use inexpensive vegetable oil, rather than take lecithin supplements. However, an animal study found a cholesterol-lowering effect of lecithin independent of its polyunsaturate content. [140] A double-blind trial found that 20 grams of soy lecithin per day for four weeks had no significant effect on total cholesterol, LDL cholesterol, HDL cholesterol, or triglycerides. [141] Whether taking lecithin supplements is a useful way to lower cholesterol in people with elevated cholesterol levels remains unclear.

The fiber-like supplement chitosan appears to reduce the absorption of bile acids or cholesterol; either of these effects may cause a lowering of blood cholesterol. [142].This effect has been repeatedly demonstrated in animals, and a preliminary human study showed that 3 to 6 grams per day of chitosan taken for two weeks resulted in a 6% drop in cholesterol and a 10% increase in HDL ("good") cholesterol. [143] Another preliminary trial showed a 43% lowering of total cholesterol in people being treated for kidney failure with dialysis who took 4 grams per day of chitosan for 12 weeks. These people also appeared to have improved kidney function and less severe anemia after chitosan treatment. [144] In a double-blind trial, however, administration of 2.4 grams of chitosan per day for three months to people with high cholesterol had no effect on their cholesterol levels. [145] Another study also found no cholesterol-lowering effect of chitosan when taken in amounts up to 6.75 g per day for 8 weeks. [146].

Chitosan in large amounts, given with vitamin C, has been shown to reduce dietary fat absorption in animals fed a high-fat diet. [147-149] However, the absorption of minerals and fat-soluble vitamins was also reduced by feeding animals large amounts of chitosan. [150] In studies in humans, chitosan did not reduce the absorption of dietary fat. [151,152]. Royal jelly has prevented the cholesterol-elevating effect of nicotine153 and has lowered serum cholesterol in animal studies.[154] Preliminary human trials have also found that royal jelly may lower cholesterol levels.[155] An analysis of cholesterol-lowering trials shows that 50 to 100 mg per day is the typical amount used in such research.[156]

A double-blind trial found that 20 grams per day of creatine taken for five days, followed by ten grams per day for 51 days, significantly lowered serum total cholesterol and triglycerides, but did not change either LDL or HDL cholesterol, in both men and women. However, another double-blind trial found no change in any of these blood levels in trained athletes using creatine during a 12-week strength training program.

Creatine supplementation in this negative trial was lower—only 5 grams per day were taken for the last 11 weeks of the study. Homocysteine, a substance linked to heart disease risk, may increase the rate at which LDL cholesterol is damaged. While vitamin B6, vitamin B12, and folic acid lower homocysteine, a recent trial found no effect of supplements of these vitamins on protecting LDL cholesterol, even though homocysteine was lowered.

Omega-3 Fatty Acids - More than 4,500 studies over the last 25 years have shown how vital Omega-3 Fatty Acids are to both preventing and treating cardiovascular-related diseases. Few of us eat enough fish to provide necessary levels of these essential fatty acids. Taking them in supplement form reduces both heart disease and sudden cardiac death.

Supplements that Open Blood Vessels

Our research indicates that when it comes to heart disease and dietary supplements used for opening blood vessels, the following are some of the best and, as such, may be a part of your preventive strategies against heart disease. Ginkgo biloba is well renowned for improving blood flow throughout the body, including the heart muscle. Ginkgo is also a powerhouse antioxidant and it appears to reduce blood stickiness, which lowers the risk of blood clots. Fish oil is a rich source of omega-3 fatty acids (DHA and EPA) that benefits heart health. Fish oil helps prevent platelets in the blood from clumping together, reducing the risk that blood clots will form. It has also been shown to reduce blood pressure, lower triglycerides (blood fats) levels, and improve blood flow. Indeed, fish oil omega 3's are praised by many experts as being one of the best heart disease and dietary supplements, meaning it should be a part of your preventive strategies against heart disease.

Policosanol -- Some studies have shown that policosanol can lower one's bad cholesterol (LDL) by up to 20% and raise beneficial cholesterol (HDL) by 10%.

Guggulipid is prized for its ability to lower bad cholesterol (LDL) levels as well as high blood triglyceride levels. It has also shown to boost the levels of good cholesterol (HDL).

Vitamin B Complex, particularly vitamins B6, B12, and folic acid reduce levels of homocysteine.

Chromium is a mineral that plays a role in helping to manage cholesterol levels. In addition, it can help improve blood sugar control for diabetes sufferers.

Antioxidants

Antioxidants are believed to help prevent heart disease by fighting free radicals, substances that harm the body when left unchecked. These nutrients are on a constant search and destroy mission, fighting the continuous onslaught of free radicals. The following dietary supplements help fight free radicals and, as such, should be a part of your preventive strategies against heart disease.

Other possible heart muscle strengtheners include: L- Carnitine and Potassium.

Preventing and treating heart disease in some patients could be as simple as supplementing their diet with extra vitamin D, according to two new studies at the Intermountain Medical Center Heart Institute in Murray, Utah.

Researchers at the Intermountain Medical Center Heart Institute last fall demonstrated the link between vitamin D deficiency and increased risk for coronary artery disease. These new studies show that treating vitamin D deficiency with supplements may help to prevent or reduce a person's risk for cardiovascular disease and a host of other chronic conditions. They also establish what level of vitamin D further enhances that risk reduction.

Dietary Suggestions to Reduce CVD Risk

- Eat more fruits and vegetables.
- Eat more whole grains.
- Use low-fat snacks.
- Reduce fat intake to 25-30 percent of the diet.
- Reduce cholesterol intake to less than 300 mg. per day.
- Reduce consumption of egg yolks to three to five per week.
- Minimize use of whole milk and its products; use low-fat or nonfat milk products.
- Avoid red meats; eliminate all cured meats and lunchmeats.
- Limit the use of nuts and seeds, not more than a handful daily.
- Avoid excess intake of avocados, olives, crab, and shrimp.
- Eat more coldwater fish, such as sardines and salmon.
- Use fresh, monounsaturated, mechanically pressed oils, such as olive or flaxseed oils, to provide the essential fatty acids.

Dangerous Dietary Supplements and Heart Disease

(LifeWire) - The US government's bold action in 2003 banning all dietary supplements containing ephedra -- often used in weight loss or "athletic performance" products -- targeted Americans' cardiac health following the implication that the use of ephedra triggers heart attacks and strokes. But those supplements are not the only ones that can affect people with heart disease. Many products still on the market have potentially dangerous consequences.

Dietary supplements other than vitamins are hugely popular among Americans hoping to either enhance their health or prevent disease. An estimated 10 to 19% of all adults take these supplements, which are not regulated by the US FDA. They typically contain minerals, herbs or other botanicals, fibers, proteins, amino acids, organ tissues or metabolites.

According to a 2006 report that examined more than 31,000 people's health habits, more than 1 in 5 who take supplements also use prescription drugs, but don't inform their doctors that they're combining both of these together. Also, 1 out of every 6 is being treated for congestive heart failure or heart disease or has a history of heart attacks.

Supplements that are potentially dangerous to cardiac health often interact negatively with prescription medications for heart disease patients. They include:

- "Sexual enhancement" supplements, such as Zimaxx, Libidus, Vigor-25 and 4EVERON. These products contain ingredients similar to the prescription erectile dysfunction drug Viagra (Sildenafil). These supplements may dangerously interact with medications taken by those with hypertension or high cholesterol; one particular danger is that they may create hypotension -- a medically unacceptable low level of blood pressure. Those who take nitrates for heart conditions may be particularly at risk.
- St. John's wort, used for mild depression, lowers the effectiveness of a wide range of medications, including drugs to treat heart failure, strokes or heart attacks.
- More than 180 supplements -- including anise, dong quai, ginger and gingko -- can prevent blood thinners from working properly or can enhance their effectiveness to what may be dangerous levels. Blood thinners, which help prevent heart attack and stroke, include aspirin, warfarin, clopidogrel and ticlopidine.[156]

CONCLUSION

To promote cardiovascular heath, the following recommendations are important.

1) Drink skim milk and buy low-fat cheese, yogurt and margarine.
2) Eat less fat (especially butter, coconut and palm oil, saturated or hydrogenated vegetable fats such as Crisco, animal fats in meats, fats in dairy products).
3) Limit cholesterol consumption
4) Eat a bare minimum of saturated fats and trans-fatty fats.
5) Reduce salt intake.
6) Water is vital to life. Staying hydrated makes feel energetic and eat less. Drink 32 to 64 ounces of water daily.
7) Reduce the intake of salt.

REFERENCES

[1] WHO/FAO Diet, Nutrition and the prevention of chronic diseases; Available at http// www.who.int/entity/nutrition/topics/5-popular nutrition.

[2] Nick H.Mashour, MD; George I.Lin, MD; William H.Frishman, MD, (1998).Herbal Medicine for the treatment of Cardiovascular Disease, *Arch. Intern. Med.* 158, 9,.

[3] Warshafsky S, Kamer RS, Sivak SL. (1993).Effect of garlic on total serum cholesterol—a meta-analysis. *Ann. Intern. Med.* 119:599–605.

[4] McCrindle BW, Helden E, Conner WT.(1998). Garlic extract therapy in children with hypercholesterolemia. *Arch. Pediatr. Adolesc. Med.*152:1089–94.

[5] Isaacsohn JL, Moser M, Stein EA, et al.(1998). Garlic powder and plasma lipids and lipoproteins. *Arch. Intern. Med.*158:1189–94.

[6] Berthold HK, Sudhop T, von Bergmann K. (1998).Effect of a garlic oil preparation on serum lipoproteins and cholesterol metabolism. *JAMA*. 279:1900–2.

[7] Warshafsky S, Kamer R, Sivak S.(1993). Effect of garlic on total serum cholesterol: a meta-analysis. *Ann. Int. Med.* 119(7)599–605.

[8] Silagy C, Neil A. (1994).Garlic as a lipid-lowering agent—a meta-analysis. *J. R. Coll. Phys* London; 28(1):39–45.

[9] Neil HA, Silagy CA, Lancaster T, et al.(1996). Garlic powder in the treatment of moderate hyperlipidaemia: a controlled trial and a meta-analysis. *J. R. Coll. Phys.* 30:329–34.

[10] Gardner CD, Lawson LD, Block E, et al.(2007). Effect of raw garlic vs commercial garlic supplements on plasma lipid concentrations in adults with moderate hypercholesterolemia: a randomized clinical trial. *Arch. Intern. Med.* 167:346–53.

[11] McCrindle BW, Helden E, Conner WT.(1998). Garlic extract therapy in children with hypercholesterolemia. *Arch. Pediatr. Adolesc. Med.* 152:1089–94.

[12] Isaacsohn JL, Moser M, Stein EA, et al. (1998).Garlic powder and plasma lipids and lipoproteins. *Arch. Intern. Med.* 158:1189–94.

[13] Berthold HK, Sudhop T, von Bergmann K.(1998).Effect of a garlic oil preparation on serum lipoproteins and cholesterol metabolism. *JAMA*. 279:1900–2.

[14] Superko HR, Krauss RM. (2000).Garlic powder, effect on plasma lipids, postprandial lipemia, low-density lipoprotein particle size, high-density lipoprotein subclass distribution and lipoprotein(a*). J. Am. Coll. Cardiol.* 35:321–6.

[15] Lawson L.(1998). Garlic oil for hypercholesterolemia—negative results. *Quart. Rev. Natural Med. Fall.* 185–6.

[16] Lawson LD. (1998).Garlic powder for hyperlipidemia—analysis of recent negative results. *Quart. Rev. Natural Med. Fall.* 187–9.

[17] Berthold HK, Sudhop T, von Bergmann K. (1998).Effect of a garlic oil preparation on serum lipoproteins and cholesterol metabolism. *JAMA*. 279:1900–2.

[18] Silagy C, Neil A. (1994).Garlic as a lipid-lowering agent—a meta-analysis*. J. R. Coll. Physicians.* London. 28:39–45.

[19] Silagy C, Neil A. (1994).Garlic as a lipid-lowering agent—a meta-analysis. *J. R. College Phys.* London. 28:39–45.

[20] Kono S, Shinchi K, Ikeda N, et al.(1992). Green tea consumption and serum lipid profiles: a cross-sectional study in Northern Kyushu, Japan. *Prev. Med.* 21:526–31.

[21] Yamaguchi Y, Hayashi M, Yamazoe H, et al.(1991). Preventive effects of green tea extract on lipid abnormalities in serum, liver and aorta of mice fed an atherogenic diet. Nip Yak Zas;97(6):329–37.

[22] Sagesaka-Mitane Y, Milwa M, Okada S. (1990).Platelet aggregation inhibitors in hot water extract of green tea. *Chem. Pharm. Bull.* 38(3):790–3.

[23] Stensvold I, Tverdal A, Solvoll K, et al. (1992). Tea consumption. Relationship to cholesterol, blood pressure, and coronary and total mortality. *Prev. Med.* 21:546–53.

[24] Tsubono Y, Tsugane S. (1997).Green tea intake in relation to serum lipid levels in middle-aged Japanese men and women. *Ann. Epidemiol.* 7:280–4.

[25] Maron DJ, Lu GP, Cai NS, et al. (2003).Cholesterol-lowering effect of a theaflavin-enriched green tea extract: a randomized controlled trial. *Arch. Intern. Med.* 163:1448–53.

[26] Heber D, Yip I, Ashley JM, et al.(1999). Cholesterol-lowering effects of a proprietary Chinese red-yeast-rice dietary supplement. *Am. J. Clin. Nutr.* 69:231–6.

[27] Vuksan V, Jenkins DJ, Spadafora P, et al. (1999).Konjac-mannan (glucomannan) improves glycemia and other associated risk factors for coronary heart disease in type 2 diabetes. A randomized controlled metabolic trial. *Diabetes Care.* 22: 913–9.

[28] Zhang MY, Huang CY, Wang X, et al.(1990). The effect of foods containing refined Konjac meal on human lipid metabolism. *Biomed. Environ. Sci.* 3:99–105.

[29] Arvill A, Bodin L.(1995). Effect of short-term ingestion of konjac glucomannan on serum cholesterol in healthy men. *Am. J. Clin. Nutr.* 61:585–9.

[30] Walsh DE, Yaghoubian V, Behforooz A.(1984). Effect of glucomannan on obese patients: a clinical study. *Int. J. Obes.* 8:289–93.

[31] Menendez R, Arruzazabala L, Más R, et al. (1997). Cholesterol-lowering effect of policosanol on rabbits with hypercholesterolaemia induced by a wheat starch-casein diet. *Br. J. Nutr.* 77:923–32.

[32] Gouni-Berthold I, Berthold HK. (2002).Policosanol: clinical pharmacology and therapeutic significance of a new lipid-lowering agent. *Am. Heart J.* 143:356–65 [review].

[33] Frei B.(1991). Ascorbic acid protects lipids in human plasma and low-density lipoprotein against oxidative damage. *Am. J. Clin. Nutr.* 54:1113–8S.

[34] Simon JA.(1992). Vitamin C and cardiovascular disease: a review. *J. Am. Coll. Nutr.* 11:107– 27.

[35] Gatto LM, Hallen GK, Brown AJ, Samman S. (1996). Ascorbic acid induces a favorable lipoprotein profile in women. *J. Am. Coll. Nutr.* 15;154–8.

[36] Balz F. (1999).Antioxidant Vitamins and Heart Disease. Presented at the 60th Annual Biology Colloquium, Oregon State University, February 25,

[37] Galeone F, Scalabrino A, Giuntoli F, et al. (1983).The lipid-lowering effect of pantethine in hyperlipidemic patients: a clinical investigation. *Curr. Ther. Res.* 34: 383–90.

[38] Miccoli R, Marchetti P, Sampietro T, et al.(1984). Effects of pantethine on lipids and apolipoproteins in hypercholesterolemic diabetic and non diabetic patients. *Curr. Ther. Res.* 36:545–9.

[39] Avogaro P, Bon B, Fusello M.(1983). Effect of pantethine on lipids, lipoproteins and apolipoproteins in man. *Curr. Ther. Res.* 33;488–93.

[40] Coronel F, Tornero F, Torrente J, et al. (1991). Treatment of hyperlipemia in diabetic patients on dialysis with a physiological substance. *Am. J. Nephrol.* 11:32–6.

[41] Arsenio L, Bodria P, Magnati G, et al.(1986). Effectiveness of long-term treatment with pantethine in patients with dyslipidemia. *Clin. Ther.* 8:537–45.

[42] Prisco D, Rogasi PG, Matucci M, et al.(1987). Effect of oral treatment with pantethine on platelet and plasma phospholipids in IIa hyperlipoproteinemia. *Angiology.* 38: 241–7.

[43] Gaddi A, Descovich GC, Noseda G, et al.(1984). Controlled evaluation of pantethine, a natural hypolipidemic compound, in patients with different forms of hyperlipoproteinemia. *Atherosclerosis.* 50:73–83.

[44] Da Col PG, et al. (1984). Pantethine in the treatment of hyper-cholesterolemia: a randomized double-blind trial versus tiadenol. *Curr. Ther. Res.* 36:314.

[45] Anderson RA, Cheng N, Bryden NA, et al.(1997). Elevated intakes of supplemental chromium improve glucose and insulin variables in individuals with type 2 diabetes. *Diabetes.* 46:1786–91.

[46] Offenbacher EG, Pi-Sunyer FX.(1980). Beneficial effect of chromium-rich yeast on glucose tolerance and blood lipids in elderly subjects. *Diabetes.* 29:919–25.

[47] Press RI, Geller J, Evans GW.(1990). The effect of chromium picolinate on serum cholesterol and apolipoprotein fractions in human subjects. *West J. Med.* 152:41–5.

[48] Hermann J, Chung H, Arquitt A, et al.(1998). Effects of chromium or copper supplementation on plasma lipids, plasma glucose and serum insulin in adults over age fifty. *J. Nutr. Elderly.* 18:27–45.

[49] Riales R, Albrink MJ. (1981).Effect of chromium chloride supplementation on glucose tolerance and serum lipids including high-density lipoprotein of adult men. *Am. J. Clin. Nutr.* 34:2670–8.

[50] Roeback JR, Hla KM, Chambless LE, Fletcher RH.(1991). Effects of chromium supplementation on serum high-density lipoprotein cholesterol levels in men taking beta- blockers. *Ann. Intern. Med.* 115:917–24.

[51] Uusitupa MI, Kumpulainen JT, Voutilainen E, et al.(1983). Effect of inorganic chromium supplementation on glucose tolerance, insulin response, and serum lipids in noninsulin- dependent diabetics. *Am. J. Clin. Nutr.* 38:404–10.

[52] Uusitupa MI, Mykkanen L, Siitonen O, et al.(1992). Chromium supplementation in impaired glucose tolerance of elderly: effects on blood glucose, plasma insulin, C-peptide and lipid levels. *Br. J. Nutr.* 68:209–16.

[53] Boyd SG, Boone BE, Smith AR, et al.(1998). Combined dietary chromium picolinate supplementation and an exercise program leads to a reduction of serum cholesterol and insulin in college-aged subjects. *J. Nutr. Biochem.* 9:471–5.

[54] Wang MM, Fox EA, Stoecker BJ, et al. (1989).Serum cholesterol of adults supplemented with brewer's yeast or chromium chloride. *Nutr. Res.* 9:989–98.

[55] Newman HA, Leighton RF, Lanese RR, Freedland NA. (1978).Serum chromium and angiographically determined coronary artery disease. *Clin. Chem.* 541–4.

[56] Brown WV. (1995).Niacin for lipid disorders. *Postgrad. Med.* 98:185–93 [review].

[57] Guyton JR, Blazing MA, Hagar J, et al. (2000). Extended-release niacin vs gemfibrozil for the treatment of low levels of high-density lipoprotein cholesterol. Niaspan-Gemfibrozil Study Group. *Arch. Intern. Med.* 160:1177–84.

[58] McKenney JM, Proctor JD, Harris S, Chinchili VM. (1994). A comparison of the efficacy and toxic effects of sustained- vs immediate-release niacin in hypercholesterolemic patients. *JAMA.* 271:672–7.

[59] Knopp RH, Ginsberg J, Albers JJ, et al. (1985). Contrasting effects of unmodified and time- release forms of niacin on lipoproteins in hyperlipidemic subjects: clues to mechanism of action of niacin. *Metabolism.* 34:642–50.

[60] Gray DR, Morgan T, Chretien SD, Kashyap ML. (1994).Efficacy and safety of controlled- release niacin in dyslipoproteinemic veterans. *Ann. Intern. Med.* 121:252–8.

[61] Rader JI, Calvert RJ, Hathcock JN. (1992). Hepatic toxicity of unmodified and time-release preparations of niacin. *Am. J. Med.* 92:77–81 [review].

[62] Knopp RH. (1989). Niacin and hepatic failure. *Ann. Intern. Med.* 111:769 [letter].

[63] Goldberg A, Alagona P Jr, Capuzzi DM, et al.(2000). Multiple-dose efficacy and safety of an extended-release form of niacin in the management of hyperlipidemia. *Am. J. Cardiol.* 85:1100–5.

[64] Head KA. (1996). Inositol hexaniacinate: a safer alternative to niacin. *Alt. Med. Rev.* 1:176–84 [review].

[65] Murray M. (1995).Lipid-lowering drugs vs. Inositol hexaniacinate. *Am. J. Natural Med.* 2:9– 12 [review].

[66] Dorner Von G, Fisher FW.(1961). Zur Beinflussung der Serumlipide und-lipoproteine durch den Hexanicotinsaureester des m-Inositol. *Arzneimittel Forschung.* 11:110–3.

[67] Carrol KK, Kurowska EM. (1995).Soy consumption and cholesterol reduction: review of animal and human studies. *J. Nutr.* 125:594–7S.

[68] Crouse JR 3rd, Morgan T, Terry JG, et al.(1999). A randomized trial comparing the effect of casein with that of soy protein containing varying amounts of isoflavones on plasma concentrations of lipids and lipoproteins. *Arch. Intern. Med.* 159:2070–6.

[69] Nestel PJ, Yamashita T, Sasahara T, et al.(1997). Soy isoflavones improve systemic arterial compliance but not plasma lipids in menopausal and perimenopausal women. *Arterioscler. Thromb. Vasc. Biol.* 17:3392–8.

[70] Samman S, Lyons, Wall PM, et al. (1999). The effect of supplementation with isoflavones on plasma lipids and oxidisability of low density lipoprotein in premenopausal women. *Atherosclerosis.* 147:277–83.

[71] Hoie LH, Morgenstern EC, Gruenwald J, et al.(2005). A double-blind placebo-controlled clinical trial compares the cholesterol-lowering effects of two different soy protein preparations in hypercholesterolemic subjects. *Eur. J. Nutr.* 44:65–71.

[72] Lees AM, Mok HY, Lees RS, et al.(1977). Plant sterols as cholesterol-lowering agents: clinical trials in patients with hypercholesterolemia and studies of sterol balance. *Atherosclerosis*. 28:325–38.

[73] Pelletier X, Belbraouet S, Mirabel D, et al.(1995). A diet moderately enriched in phytosterols lowers plasma cholesterol concentrations in normocholesterolemic humans. *Ann. Nutr. Metab.* 39:291–5.

[74] Korpela R, Tuomilehto J, Hogstrom P, et al. (2006).Safety aspects and cholesterol-lowering efficacy of low fat dairy products containing plant sterols. *Eur. J. Clin. Nutr.* 60:633–42.

[75] Grundy SM, Ahrens EH Jr, Davignon J.(1969). The interaction of cholesterol absorption and cholesterol synthesis in man. *J. Lipid Res.* 10:304–15 [review].

[76] Hendriks HF, Weststrate JA, van Vliet T, Meijer GW.(1999). Spreads enriched with three different levels of vegetable oil sterols and the degree of cholesterol lowering in normocholesterolaemic and mildly hypercholesterolaemic subjects. *Eur. J. Clin. Nutr.* 53:319– 27.

[77] Jones PJ, Ntanios FY, Raeini-Sarjaz M, Vanstone CA.(1999). Cholesterol-lowering efficacy of a sitostanol-containing phytosterol mixture with a prudent diet in hyperlipidemic men. *Am. J. Clin. Nutr.* 69:1144–50.

[78] Blair SN, Capuzzi DM, Gottlieb SO, et al. (2000).Incremental reduction of serum total cholesterol and low-density lipoprotein cholesterol with the addition of plant stanol ester- containing spread to statin therapy. *Am. J. Cardiol.* 86:46–52.

[79] Jones PJ, Raeini-Sarjaz M, Ntanios FY, et al. (2000). Modulation of plasma lipid levels and cholesterol kinetics by phytosterol versus phytostanol esters. *J. Lipid Res.* 41: 697–705.

[80] Hallikainen MA, Sarkkinen ES, Uusitupa MI.(2000). Plant stanol esters affect serum cholesterol concentrations of hypercholesterolemic men and women in a dose-dependent manner. *J. Nutr.* 130:767–76.

[81] Vuorio AF, Gylling H, Turtola H, et al. (2000). Stanol ester margarine alone and with simvastatin lowers serum cholesterol in families with familial hypercholesterolemia caused by the FH-North Karelia mutation. *Arterioscler. Thromb. Vasc. Biol.* 20:500–6.

[82] Nguyen TT, Dale LC, von Bergmann K, Croghan IT. (1999).Cholesterol-lowering effect of stanol ester in a US population of mildly hypercholesterolemic men and women: a randomized controlled trial. *Mayo Clin. Proc.* 74:1198–206.

[83] Hyun YJ, Kim OY, Kang JB, et al. (2005).Plant stanol esters in low-fat yogurt reduces total and low-density lipoprotein cholesterol and low-density lipoprotein oxidation in normocholesterolemic and mildly hypercholesterolemic subjects. *Nutr. Res.* 25:743–55.

[84] Law M.(2000). Plant sterol and stanol margarines and health. *BMJ.* 320:861–4.

[85] Goldberg AC, Ostlund RE Jr, Bateman JH, et al. (2006).Effect of plant stanol tablets on low- density lipoprotein cholesterol lowering in patients on statin drugs. *Am. J. Cardiol.* 97:376–9.

[86] Parker RA, Pearce BC, Clark RW, et al.(1993). Tocotrienols regulate cholesterol production in mammalian cells by post-transcriptional suppression of 3-hydroxy-3-methylglutaryl- coenzyme A reductase. *J. Biol. Chem.* 268(15):11230–8.

[87] Pearce BC, Parker RA, Deason ME, et al. (1992). Hypocholesterolemic activity of synthetic and natural tocotrienols. *J. Med. Chem.* 35:3595–606.

[88] Qureshi AA, Bradlow BA, Brace L, et al. (1995). Response of hypercholesterolemic subjects to administration of tocotrienols. *Lipids.* 30:1171–7.

[89] Qureshi AA, Qureshi N, Wright JJ, et al. (1991).Lowering serum cholesterol in hypercholesterolemic humans by tocotrienols (palmvitee). *Am. J. Clin. Nutr.* 53: 1021–6S.

[90] Wahlqvist ML, Krivokuca-Bogetic A, Lo CS, et al. (1992). Differential serum response of tocopherols and tocotrienols during vitamin supplementation in hypercholesterolemic individuals without change in coronary risk factors. *Nutr. Res.* 12:S181–201.

[91] Mensink RP, van Houwelingen AC, Kromhout D, Hornstra G. (1999). A vitamin E concentrate rich in tocotrienols had no effect on serum lipids, lipoproteins, or platelet function in men with mildly elevated serum lipid concentrations. *Am. J. Clin. Nutr.* 69:213–9.

[92] Bunea R, El Farrah K, Deutsch L.(2004). Evaluation of the effects of Neptune Krill Oil on the clinical course of hyperlipidemia. *Altern. Med. Rev.* 9:420–8.

[93] Krasopoulos JC, De Bari VA, Needle MA. (1980).The adsorption of bile salts on activated carbon. *Lipids.* 15:365–70.

[94] Tishler PV, Winston SH, Bell SM.(1987). Correlative studies of the hypocholesterolemic effect of a highly activated charcoal. *Methods Find Exp. Clin. Pharmacol.* 9:799–806.

[95] Neuvonen PJ, Kuusisto P, Vapaatalo H, Manninen V. (1989).Activated charcoal in the treatment of hypercholesterolaemia: dose-response relationships and comparison with cholestyramine. *Eur. J. Clin. Pharmacol.* 37:225–30.

[96] Park GD, Spector R, Kitt TM.(1988). Superactivated charcoal versus cholestyramine for cholesterol lowering: a randomized cross-over trial. *J. Clin. Pharmacol.* 28:416–9.

[97] Neuvonen PJ, Kuusisto P, Manninen V, et al. (1989). The mechanism of the hypocholesterolaemic effect of activated charcoal. *Eur. J. Clin. Invest.* 19:251–4.

[98] Hoekstra JB, Erkelens DW. (1988). No effect of activated charcoal on hyperlipidaemia. A double-blind prospective trial. *Neth. J. Med.* 33:209–16.

[99] Davis GK, Mertz W. Copper. (1987). In: Mertz W, ed. Trace elements in human and animal nutrition, vol. 1. 5th ed. San Diego: Academic Press, , 301–64 [review].

[100] . Klevay LM. (1987).Dietary copper: a powerful determinant of cholesterolemia. *Med. Hypotheses.* 24:111–9 [review].

[101] Hermann J, Chung H, Arquitt A, et al.(1998). Effects of chromium or copper supplementation on plasma lipids, plasma glucose and serum insulin in adults over age fifty. *J. Nutr. Elderly.* 18:27–45.

[102] Bell S, Goldman VM, Bistrian BR, et al.(1999). Effect of beta-glucan from oats and yeast on serum lipids. *Crit. Rev. Food Sci. Nutr.* 39:189–202 [review].

[103] Behall KM, Scholfield DJ, Hallfrisch J. (1997).Effect of beta-glucan level in oat fiber extracts on blood lipids in men and women. *J. Am. Coll. Nutr.*16:46–51.

[104] Braaten JT, Wood PJ, Scott FW, et al. (1994).Oat beta-glucan reduces blood cholesterol concentration in hypercholesterolemic subjects. *Eur. J. Clin. Nutr.* 48:465–74.

[105] Davidson MH, Dugan LD, Burns JH, et al. (1991). The hypocholesterolemic effects of beta- glucan in oatmeal and oat bran. A dose-controlled study. *JAMA.* 265:1833–9.

[106] Wood PJ.(1990). Physicochemical properties and physiological effects of the (1----3) (1---- 4)-beta-D-glucan from oats. *Adv. Exp. Med. Biol.* 270:119–27.

[107] Uusitupa MI, Miettinen TA, Sarkkinen ES, et al. (1997). Lathosterol and other non-cholesterol sterols during treatment of hypercholesterolaemia with beta-glucan-rich oat bran. *Eur. J. Clin. Nutr.* 51:607–11.

[108] Lia A, Hallmans G, Sandberg AS, et al. (1995).Oat beta-glucan increases bile acid excretion and a fiber-rich barley fraction increases cholesterol excretion in ileostomy subjects. *Am. J. Clin. Nutr.* 62:1245–51.

[109] Bell S, Goldman VM, Bistrian BR, et al.(1999). Effect of beta-glucan from oats and yeast on serum lipids. *Crit. Rev. Food Sci. Nutr.* 39:189–202 [review].

[110] Nicolosi R, Bell SJ, Bistrian BR, et al. (1999).Plasma lipid changes after supplementation with beta-glucan fiber from yeast. *Am. J. Clin. Nutr.* 70:208–12.

[111] Behall KM, Scholfield DJ, Hallfrisch J. (1997).Effect of beta-glucan level in oat fiber extracts on blood lipids in men and women. *J. Am. Coll. Nutr.* 16:46–51.

[112] Braaten JT, Wood PJ, Scott FW, et al.(1994). Oat beta-glucan reduces blood cholesterol concentration in hypercholesterolemic subjects. *Eur. J. Clin. Nutr.* 48:465–74.

[113] Uusitupa MI, Ruuskanen E, Makinen E, et al. (1992)A controlled study on the effect of beta-glucan-rich oat bran on serum lipids in hypercholesterolemic subjects: relation to apolipoprotein E phenotype. *J. Am. Coll. Nutr.* 11:651–9.

[114] Yacowitz H, Fleischman AI, Bierenbaum ML. (1965). Effects of oral calcium upon serum lipids in man. *Br. Med. J.* 1:1352–4.

[115] Bell L, Halstenson CE, Halstenson CJ, et al. (1992). Cholesterol-lowering effects of calcium carbonate in patients with mild to moderate hypercholesterolemia. *Arch. Intern. Med.* 152:2441–4.
[116] Karanja N, Morris CD, Illingworth DR, (1987). Plasma lipids and hypertension: response to calcium supplementation. *Am. J. Clin. Nutr.* 45:60–5.
[117] Denke MA, Fox MM, Schulte MC. (1993). Short-term dietary calcium fortification increases fecal saturated fat content and reduces serum lipids in men. *J. Nutr.* 123:1047–53.
[118] Bostick RM, Fosdick L, Grandits GA, et al. (2000). Effect of calcium supplementation on serum cholesterol and blood pressure. *Arch. Fam. Med.* 9:31–9.
[119] Cloarec MJ, Perdriset GM, Lamberdiere FA, et al., (1987). Alpha-tocopherol: effect on plasma lipoproteins in hypercholesterolemic patients. *Isr. J. Med. Sci.* 23:869–72.
[120] Kesaniemi YA, Grundy SM. (1982). Lack of effect of tocopherol on plasma lipids and lipoproteins in man. *Am. J. Clin. Nutr.* 36:224–8.
[121] Kalbfleisch JH, Barboriak JJ, Else BA, et al. (1986).alpha-Tocopherol supplements and high-density-lipoprotein-cholesterol levels. *Br. J. Nutr.* 55:71–7.
[122] Stampfer MJ, Willett W, Castelli WP, et al. (1983). Effect of vitamin E on lipids. *Am. J. Clin. Pathol.* 79:714–6.
[123] Belcher JD, Balla J, Balla G, et al. (1993).Vitamin E, LDL, and endothelium: brief oral vitamin supplementation prevents oxidized LDL-mediated vascular injury in vitro. *Arterioscler. Thromb.* 13:1779–89.
[124] Traber MG. (2001).Does vitamin E decrease heart attack risk? summary and implications with respect to dietary recommendations. *J. Nutr.* 131:395S–7S. [review].
[125] Pola P, Savi L, Grilli M, et al. (1980). Carnitine in the therapy of dyslipidemic patients. *Curr. Ther. Res.* 27:208–16.
[126] Stefanutti C, Vivenzio A, Lucani G, et al. (1998). Effect of L-carnitine on plasma lipoprotein fatty acids pattern in patients with primary hyperlipoproteinemia. *Clin. Ter.* 149:115–9.
[127] Maebashi M, Kawamura N, Sato M, et al. (1978) Lipid-lowering effect of carnitine in patients with type-IV hyperlipoproteinaemia. *Lancet.* ii: 805–7.
[128] Rossi CS, Siliprandi N. (1982). Effect of carnitine on serum HDL-cholesterol: report of two cases. *Johns Hopkins Med. J.* 150:51–4.
[129] Pola P, Savi L, Grilli M, et al. (1980).Carnitine in the therapy of dyslipidemic patients. *Curr. Ther. Res.* 27:208–16.
[130] Davini P, Bigalli A, Lamanna F, Boehm A.(1992). Controlled study on L-carnitine therapeutic efficacy in post-infarction. *Drugs Exptl. Clin. Res.* 18:355–65.
[131] Davis WH, Leary WP, Reyes AJ, Olhaberry JV.(1984). Monotherapy with magnesium increases abnormally low high density lipoprotein cholesterol: a clinical assay. *Curr. Ther. Res.* 36:341–6.
[132] Nozue T, Kobayashi A, Uemasu F, et al. (1995).Magnesium status, serum HDL cholesterol, and apolipoprotein A-1 levels. *J. Pediatr. Gastroenterol. Nutr.* 20:316–8.
[133] . Baxter GF, Sumeray MS, Walker JM. (1996).Infarct size and magnesium: insights into LIMIT-2 and ISIS-4 from experimental studies. *Lancet.* 348:1424–6.
[134] Galloe A, Rasmussen HS, Jorgensen LN, et al. (1993). Influence of oral magnesium supplementation on cardiac events among survivors of an acute myocardial infarction. *BMJ.* 307:585–7.

[135] Izuka K, Murata K, Nakazawa K, et al.(1968). Effects of chondroitin sulfates on serum lipids and hexosamines in atherosclerotic patients: With special reference to thrombus formation time. *Jpn Heart J.* 9:453–60.

[136] Nakazawa K, Murata K.(1979). Comparative study of the effects of chondroitin sulfate isomers on atherosclerotic subjects. *ZFA*.34:153–9.

[137] Morrison LM, Enrick NL.(1973). Coronary heart disease: reduction of death rate by chondroitin sulfate A. *Angiology*. 24:269–87.

[138] Childs MT, Bowlin JA, Ogilvie JT, et al.(1981). The contrasting effects of a dietary soya lecithin product and corn oil on lipoprotein lipids in normolipidemic and familial hypercholesterolemic subjects. *Atherosclerosis*. 38:217–28.

[139] Knuiman JT, Beynen AC, Katan MB. (1989).Lecithin intake and serum cholesterol. *Am. J. Clin. Nutr.* 49:266–8.

[140] Wilson TA, Meservey CM, Nicolosi RJ. (1998). Soy lecithin reduces plasma lipoprotein cholesterol and early atherogenesis in hypercholesterolemic monkeys and hamsters: beyond linoleate. *Atherosclerosis.* 140:147–53.

[141] Oosthuizen W, Vorster HH, Vermaak WJ, et al.(1998). Lecithin has no effect on serum lipoprotein, plasma fibrinogen and macro molecular protein complex levels in hyperlipidaemic men in a double-blind controlled study. *Eur. J. Clin. Nutr.* 52:419–24.

[142] Koide SS.(1998). Chitin-chitosan: properties, benefits and risks. *Nutr. Res.* 18:1091-101 [review].

[143] Maezaki Y, Tsuji K, Nakagawa Y, et al.(1993). Hypocholesterolemic effect of chitosan in adult males. *Biosci. Biotech .Biochem*. 57:1439-44.

[144] Jing SB, Li L, Ji D, et al.(1997). Effect of chitosan on renal function in patients with chronic renal failure. *J. Pharm. Pharmacol.* 49:721-3.

[145] Metso S, Ylitalo R, Nikkila M, et al. (2003).The effect of long-term microcrystalline chitosan therapy on plasma lipids and glucose concentrations in subjects with increased plasma total cholesterol: a randomised placebo-controlled double-blind crossover trial in healthy men and women. *Eur. J. Clin. Pharmacol.* 59:741–6.

[146] Tapola NS, Lyyra ML, Kolehmainen RM, et al.(2008). Safety aspects and cholesterol-lowering efficacy of chitosan tablets. *J. Am. Coll. Nutr.* 27:22–30.

[147] . Deuchi K, Kanauchi O, Imasato Y, et al.(1995). Effect of the viscosity or deacetylation degree of chitosan on fecal fat excreted from rats fed on a high-fat diet. *Biosci. Biotech. Biochem.* 59:781-5.

[148] Deuchi K, Kanauchi O, Imasato Y, et al.(1994). Decreasing effect of chitosan on the apparent fat digestibility by rats fed on a high-fat diet. *Biosci. Biotech. Biochem.* 58:1613-6.

[149] Kanauchi O, Deuchi K, Imasato Y, et al.(1994). Increasing effect of a chitosan and ascorbic acid mixture on fecal dietary fat excretion. *Biosci. Biotech. Biochem.* 58:1617-20.

[150] Deuchi K, Kanauchi O, Shizukuishi M, et al. (1995).Continuous and massive intake of chitosan affects mineral and fat-soluble vitamin status in rats fed on a high-fat diet. *Biosci. Biotech. Biochem.* 59:1211-6.

[151] . Gades MD, Stern JS.(2003). Chitosan supplementation and fecal fat excretion in men. *Obes. Res.* 11:683–8.

[152] Gades MD, Stern JS.(2002). Chitosan supplementation does not affect fat absorption in healthy males fed a high-fat diet, a pilot study. *Int. J. Obes. Relat. Metab. Disord.* 26:119–22.

[153] Abou-Hozaifa BM, Badr El-Din NK. (1995). Royal jelly, a possible agent to reduce the nicotine-induced atherogenic lipoprotein profile. *Saudi Med. J.* 16:337–42.

[154] Abou-Hozaifa BM, Roston AAH, El-Nokaly FA. (1993). Effects of royal jelly and honey on serum lipids and lipoprotein cholesterol in rats fed cholesterol-enriched diet. *J. Biomed. Sci. Ther.* 9:35–44.

[155] Cho YT. (1977). Studies on royal jelly and abnormal cholesterol and triglycerides. *Am. Bee J.* 117:36–9.

[156] Carola R, et al. (1995).Prescription for Nutritional Healing. 2nd ed. Garden City Park, NY: Avery Publishing Group: 23,27.

In: Natural Products and Their Active Compounds ... ISBN: 978-1-62100-153-9
Editors: M. Essa, A. Manickavasagan, and E. Sukumar

Chapter 22

BIOACTIVE MOLECULES FROM THE SEA: SUCCESS, CHALLENGES AND FUTURE PERSPECTIVES

Sergey Dobretsov[1] and Bassam Soussi[2,3,*]
[1]Department of Marine Science and Fisheries, Sultan Qaboos University, Oman
[2]UNESCO Chair in Marine Biotechnology, Sultan Qaboos University, Oman
[3]Institute of Clinical Sciences, University of Gothenburg, Sweden

ABSTRACT

Bioactive compounds from marine organisms have attracted attention of scientists for only a few decades. Nowadays, more than 10,000 new bioactive molecules that exhibited anti-microbial, anti-viral, anti-fungal, anti-cancer, anti-inflammation, anti-fouling and other properties have been isolated from the marine organisms.

Only a few of these compounds have been transformed into drugs that have appeared on the market or undergone clinical trials. Some of the marine derived compounds, like omega-3 fatty acids, are important nutraceuticals that provide health and medical benefits including treatment of diseases. In this chapter we reviewed some of the marine derived pharmaceuticals, highlighted challenges of marine drug discovery and outlined important future directions.

Keywords: marine environment, marine organisms, secondary metabolites, drug discovery, pharmaceuticals

INTRODUCTION

Terrestrial plants and animals have long been used as a source of food, fragrances, pigments, insecticides, and medicines due to their high accessibility (Thoms and Schupp, 2005). Nowadays, prescribed drugs are mostly come from terrestrial plants or micro-organsims. The successful examples include the anti-cancer drug paclitaxel (Taxol®) from the

---

* Correspondent author: Bassam.soussi@gu.se

yew tree *Taxus brevifolia*, eirinotecan (Camptosar®) from the cancer tree *Camptotheca acuminate* and penicillin from the mould *Penicillium notatum* (Cragg and Newman 2005).

Compared to the terrestrial environment, the marine environment covers two-thirds of Earth's surface and has diverse environmental conditions (temperatures, pressure, nutrient and light). This variety of environmental conditions facilitates specialization and diversity of marine organisms and affects the chemistry of molecules. The halogens, such as iodine, bromine and chlorine, are highly dominant in sea water and they play an important role as a subsistence of marine-derived molecules and bioactive compounds (Fenical and Jensen 2006). Despite these facts, natural products from marine environment have not been well investigated so far. According to Scopus database, only 7.8% of natural products publications are dealing with marine derived chemical compounds (Scopus data on 12.03.08). This may be due to the difficulties involved in the collection of marine organisms.

The discovery of unusual nucleosides in the 50s with anti-vial properties is considered as the starting point for the search of "drugs from the sea" (Proksch et al. 2002; Thoms and Schupp 2005). The number of publications dealing with marine natural products are increasing (Scopus data on 16.02.10) and more than 70 research articles were published in 2009 (Figure 1). Recent investigations by leading research groups in the United States, Europe and Japan have resulted in the discovery of more than 10,000 new compounds from different marine organisms (Sennett et al. 2002), which have been published in 6,500 scientific publications and resulted in 42,037 patents (Scopus data on 12.03.08). These publications clearly show that marine organisms possess a rich chemistry that has never been seen before, and the appearance of new drugs and bioactive compounds from the sea is expected in the future (Fenical and Jensen 2006).

The aims of this publication are to review the current status of pharmaceuticals from the marine environment and to discuss future directions of drug discovery.

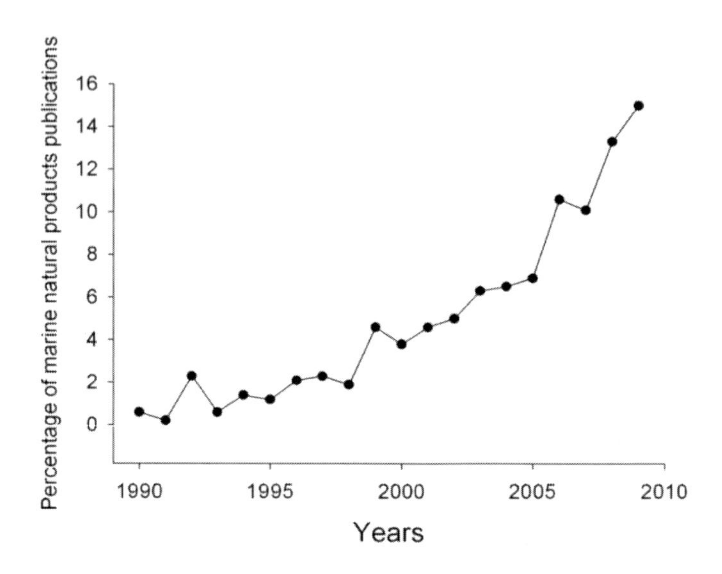

Figure 1. Marine natural products –related publication trends in the scientific literature. To access the frequency of publications, we run a search on the Scopus for the period from 1990 to 2009. Our search terms include “marine natural products”. Data on 16.02.10.

BIOACTIVE COMPOUNDS FROM MARINE ORGANISMS

Bioactive compounds from terrestrial environment are mostly isolated from plants and microorganisms, while the majority of bioactive compounds from marine organisms have been isolated from sponges (Figure 2). The amount of drugs produced from marine algae and fungi remains quite low. In contrast, significant amount of marine natural products were isolated from marine bacteria (Figure 2). These facts can be explained by a difference in chemistry of marine and terrestrial plants and lack of correspondent investigations.

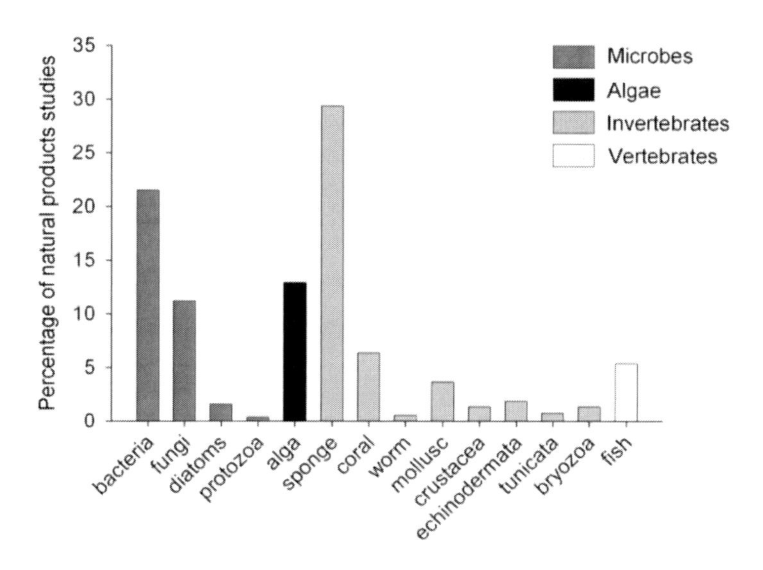

Figure 2. The main topics of marine natural products publications. Scopus data on 16.02.10 for the period of 20 years. Our search terms were “marine natural products” plus specific names of marine organisms. Because some articles considered multiple variables the bars sum to more than 100%.

Sessile organisms, such as sponges and tunicates, remain an important source of bioactive compounds (Figure 2; Table 1) because these organisms lack mechanical defense and protective structures (Proksch et al. 2002). It has been repeatedly shown that chemical defense through accumulation of toxic or repelling chemical compounds in tropical organisms is an effective strategy to compete for space and to protect organisms from predators and pathogens (Paul et al. 2007). Screening of marine organisms for bioactive compounds has resulted in finding new pharmaceuticals (see Table 1). Secondary metabolites from the tropical sponge *Cryptotethia (Tethya) crypta* have been used as commercial anti-viral and anticancer drugs available on the market now. Another compound available on the market is Ziconotide® (or Prialt®; Table 1). It is a short peptide found in the venom of the Indo-Pacific molluscs *Conus margus* and *C.geographicus*. These molluscs use their venom in order to paralyse their preys and to protect themselves from predators. Results from clinical trials have suggested that Ziconotide® is 1,000 times more active than morphine and it works by blockage of calcium channels that are necessary for signal transmission (Terlau and Olivera 2004). In contrast to morphine, Ziconotide® is non-addictive and thus may be suitable for long term use. This makes this drug one of the best in the row of pain killers for cancer patients and people suffering from chronic neuropathic pain.

Table 1. Status of marine-derived natural products

Source of drug	Compound's name	Properties	Status of the product	References	Comments
Sponge *Cryptotethia (Tethya) crypta*	Ara-A (Acyclovir)	Anti-viral	Market	Newman and Cragg 2007	Pharmaceutical
Soft coral *Pseudoterigorgia elisabethae*	Pseudopterosins	Anti-inflammatory, analgesic	Market	Look et al. 1986	Used for Estee Lauder® skin care and cosmetics products
Snail *Conus magnus*	Ziconotide (Prialt)	Pain killer	Market	Terlau and Olivera 2004	Pharmaceutical
Seaweeds *Gelidium, Pterocladia, Gelidiella* and *Gracilaria*	Agar	Thickening and gelling agent	Market	FAO 1987, 1990	Used in food industry
Brown macro-algae (Phaeophyceae)	Alginates	Thickening agent, water absorbent, drug delivery	Market	Berndt 2009	Used in food industry, medicine
Red seaweeds (*Rhodophyceae*)	Carrageenans	Thickening agent Anti-viral	Market	Luescher-Mattli 2003; Alawi et al 2011	Food industry, medicine
Crustaceans, green alga *Haematococcus pluvialis*	Astaxanthin	Antioxidant	Market	Kobayashi et al. 1997	
Fishes	Gelatin	Thickening agent, antioxidant	Market	Cole 2000	Food industry, medicine and cosmetics
Fish oil	Eicosapentaenoic and docosahexaenoic acids, omega 3-fatty acids	Antioxidant, prevent atherosclerosis and thrombosis	Market	Dyerberg et al. 1978; Gunnarsson et al. 2006	Food industry and medicine
Shark	Neovastat AE941	Anti-cancer	Phase III	Cho and Kim 2002	Defined mixture of <500 kDa from cartilage
Tunicate *Ecteinascidia turbinata*	Ecteinascidin 743	Anti-cancer	Phase II-III	Fenical and Jensen 2006	Produced by partial synthesis from microbial metabolite
Tunicate *Aplidium albicans*	Aplidine, Dihydrodidemnin B	Anti-cancer	Phase II-III	Proksch et al. 2002	Made by total synthesis
Mollusc sea hare *Dolabella auricularia*	Dolastation 10	Anti-cancer	Phase II	Gademann and Portmann 2008	Marine microbe derived. Made synthetically. No further trials known

Source of drug	Compound's name	Properties	Status of the product	References	Comments
Bryozoan *Bugula neritina*	Bryostatin	Anti-cancer	Phase II	Kraft et al. 1986	Produced by bacterial symbionts
Tunicate *Trididemnum solidum*	Didemnin B	Anti-cancer	Phase II	Newman and Cragg 2007	
Mollusk *Eylisia rufescens,* green alga *Bryopsis* sp.	Kahalalide F	Anti-cancer	Phase II	Hamann and Scheuer 1993	
Shark *Squalus acanthias*	Squalamine	Anti-cancer	Phase II	Cho and Kim 2002	
Sponge *Lissodendoryx* sp.	E7389	Anti-cancer	Phase II	Litaudon et al. 1997	Synthetic derivate halichondrin B
Sponge *Agelas mauritianus*	KRN7000	Anti-cancer	Phase I	Natori et al. 2000	
Sponge *Petrosia contignata*	IPL576092; HMR-4011A; IPL-550,260; IPL-512,602	Anti-inflammation/ asthmatic	Phase I	Abad et al. 2008	Derivative of contignasterol
Soft coral *Pseudopterogorgia elisabethae*	Methopterosin	Anti-inflammation	Phase I	Abad et al. 2008	
Sponge *Luffariella variabilis*	Manoalide	Anti-inflammation	Phase I	Abad et al. 2008	Discontinued due to formulation problem
Worm *Amphiporus lactifloreus*	GTS-21	Against Alzheimer disease	Phase I	Kem 1997	
Sponge *Discodermia dissoluta*	Discodermolide	Anti-cancer	Phase I	Gunasekera et al. 1990	
Bacterium *Salinispora tropica*	Salinosporamide A	Anti-cancer	Phase I	Williams et al. 2005	
Cyanobacteria *Lyngbya majuscula*	Curacin A	Anti-cancer	Pre clinical	Nagle et al. 1995	Synthesized
Tunicate *Didemnum cucliferum*	Vitilevuamide	Anti-cancer	Pre clinical	Ireland and Fernandez 1998	
Sponge *Cacospongia mycofijiensis*	Laulimalide	Anti-cancer	Pre clinical	Mooberry et al. 1999	Synthesized

Most of the bioactive compounds isolated from marine organisms are undergoing medical trials or have been selected for extended medical evaluation (Table 1). ET-743 (Ecteinascidin or Yondelis®) is one of the most promising anti-cancer drugs (Fenical and Jensen 2006). This compound was isolated from the tunicate *Ecteinascidia turbinata* that lives in the Caribbean and Mediterranean Seas. Nowadays, the drug is synthesized by PharmaMar company (Spain). Early trial results indicated that this compound can help patients with advanced-stage breast, colon, ovarian and lung cancer, melanoma, and several types of sarcoma, including soft tissue sarcoma (Delaloge et al. 2001).

Although marine natural products from tunicates have not been investigated intensively so far (Figure 2); these organisms have been shown to produce different bioactive compounds including alkaloids and peptides, such as didemnins and aplidine (Table 1). Didemnin B was isolated from the Caribbean tunicate *Trididemnum solidum* and it has a pronounced anti-tumor activity but, unfortunately, was found to be toxic to humans and was dropped after clinical trial II (Newman and Cragg 2007). On the other hand, aplidine from the Mediterranean sponge *Aplidium albicans*, which provides higher activity and has lower side effects, was selected for Phase III clinical trials (Proksch et al. 2002).

Most of the drugs from marine organisms that entered clinical trials exhibit anti-cancer activity (Table 1). Does this mean that most of marine animals are responsible mainly for the production of anti-cancer compounds? High amount of anti-cancer compounds isolated from marine organisms is due to an intensive screening rather than specific activity of organisms. This is supported by the facts that in addition to anti-cancer drugs marine organisms can produce anti-inflammatory, anti-malaria and anti-viral compounds, antibiotics, analgetics and antifoulants (Table 1) (Newman and Cragg 2007).

Marine Bioactive Compounds in Cosmetic and Food Industry

Marine derived compounds can be used not only in medicine but also in the cosmetics industry. Estee Lauder® is successfully using pseudopterosin from the Carribbean sea whip (gorgonian coral) *Pseudopterogorgia elisabethae* as a skin-care additive to its products (Kohl et al. 2003). These compounds have anti-inflammatory and analgesic properties (Look et al. 1986). Another attractive area for application of marine micro-organisms is production of polyhydroxyalkanoates (bioplastics) (Luengo et al. 2003). It is highly possible that in the nearest future these biodegradable, biocompatible bioplastics made of carbon dioxide and water using sunlight energy will replace traditional non-biodegradable petrochemical-based plastics in many industrial areas.

Application of bioactive compounds from marine organisms in food industry is another promising area. Macro-algae (seaweeds) can be used in a wide range of products. Agar and agarose are used in the human food industry as thickening and gelling agent. Usually, agar is produced from seaweeds belonging to the different species of *Gelidium, Pterocladia, Gelidiella* and *Gracilaria* (FAO 1987, 1990). Other useful products from the brown macro-algae (Phaeophyceae) are alginates (Berndt 2009). They are chain-forming hetero-polys-accharides made up of blocks of mannuronic acid and guluronic acid. Alginates are used in industry as water absorbing agents, manufacture of paper and textile and for thickening drinks and yogurts. Additionally, they are used in medicine for drug delivery.

All red seaweeds (*Rhodophyceae*) contain carrageenans, which are polysaccharides prepared by alkalinization extraction. There are several types of carrageenans, differing in their chemical structure and properties, and therefore in their uses (Alawi et al 2011). The carrageenans of commercial interest are called iota, kappa and lambda (Table 1). All these carrageenans have the ability to form thick solution or gels in the presence of water. The carrageenan composition in red seaweeds differs from one species to another. For example, in the Irish moss (*Chondrus crispus*) mixture of kappa and lambda carrageenan can be found, in

Kappaphycus alvarezii only kappa carrageenan is obtained, in *Eucheuma denticulatum* mainly iota carrageenan is dominated, while the red macro-alga *Gigartina skottsbergii* contains mainly kappa carrageenan. In opposite, the alga *Sarcothalia crispate* is contained mixture of kappa and lambda carrageenans. Carrageenans are used mainly in food industry, such as cream dairy products, jellies and toothpastes, while there are reports that carrageenans have antiviral properties and may offer some protection against HIV, HSV and HPV (Luescher-Mattli 2003).

Most crustaceans, including shrimps, crawfishes, crabs and lobsters, are tinted red by accumulated pigment astaxanthin. This naturally occurring carotenoid is a powerful biological antioxidant. It is known to protect against oxidative damage of LDL-cholesterol (Kobayashi et al. 1997). Most of astaxanthin are obtained synthetically or from the green micro-alga *Haematococcus pluvialis* (Olaizola 2000). Another compound that synthesized by crustaceans is chitin. It is a polysaccharide composed from N-acetyl-D-glucosamine units (Table 1). Chitin can be readily extracted, and is widely used industrially, e.g. in foods and pharmaceuticals.

Marine species have been shown to be a rich source of fatty acids, such as eicosa-pentaenoic acid (EPA) and docosahexaenoic acids (DHA). EPA and DHA are essential unsaturated omega 3-fatty acids that are used by our body. EPA from marine organisms has been shown to reduce cholesterol, triglyceride, and low-density lipoprotein levels in blood thus preventing atherosclerosis and thrombosis (Dyerberg et al. 1978). Fat fish species, such as salmon, herring and anchovies, are known to be rich in omega-3 fatty acids. The beneficial health effects of fish have been usually attributed to the action of omega -3 fatty acids. True that many studies have shown the positive effect of omega 3-fatty acids in a wide range of diseases, mainly cardiovascular as well as its importance for the foetus brain. However, in recent studies, Gunnarsson et al (2006) have shown that the aqueous phase of herring muscle press juice processes powerful antioxidative properties as evidenced from experiments in a human monocyte cell model in vitro (Gunnarsson et al 2006) and in an *in vivo* animal model (Omerovic et al 2008). These experiments explain the beneficial effect of fish rich diet, probably through a synergistic mechanism of omega 3-fatty acids and intrinsic antioxidants in the water soluble phase

PROBLEMS OF MARINE DRUG DISCOVERY

The Question of Supply

Bioactive compounds in marine organisms are often present in small quantities and in order to obtain sufficient amounts of compounds, investigators need to harvest and extract many marine organisms. For example, in order to obtain 300 mg of Halichondrin B the New Zealand government gave a permission to collect 1 metric tonne of the deep sea sponge *Lissodendoryx* sp. (Thoms and Schupp 2005). If Halichondrin B will come to the market as a new anti-cancer drug the annual need for this compound will be in the range of 3,000 – 6,000 metric tonnes of sponges per year. Such high biomass of sponges and other marine organisms cannot be harvested from the ocean without a risk of species extinction. Therefore, new sources of drug supplies from marine organisms are urgently needed.

Some of the pharmaceuticals from marine organisms can easily be synthesized in a laboratory. For example, small peptide ziconotide® from the mollusc *C. magnus* can be industrially synthesised from available amino acids. Most of the other marine derived compounds, (i.e. halichondrin, bryostatin, etc.) are too complex in order to be synthesized. However, these complex bioactive molecules can be synthesized by using some molecules obtained from microrganisms. For example, the anti-cancer compound ET-734 from the tunicate *E.turninata* can be synthesized by using safracin B produced by the bacterium *Pseudomonas fluorescens* (Sainz-Diaz et al. 2003).

Aquaculture of marine organisms is another promising way of getting large biomasses of marine organisms for isolation of bioactive compounds. Experiments with the deep-sea sponge *Lissodendoryx* sp. in New Zealand demonstrated that this sponge could be successfully cultivated at 10 m depth and produce anti-cancer compound Halichondrin B with similar concentrations as those of sponges grown under the original conditions (Thoms and Schupp 2005). The secondary compounds from the bryozoan *Bugula neritina* have received a great attention and bryostatins appeared to be useful in fighting some types of cancers, like leukaemia (Table 1; Figure 3). The quantity of bryostatin 1 in the bryozoan is quite low and successful cultivation of this animal in aquaculture helps to obtain sufficient amount of bryostatins necessary for clinical trials. It should be pointed out that, some marine organisms cannot be cultivated in aquaculture conditions. Marine organisms that can be cultivated are largely affected by diseases and dramatic changes in environmental conditions, like storms and cyclones, reduce yields of aquaculture.

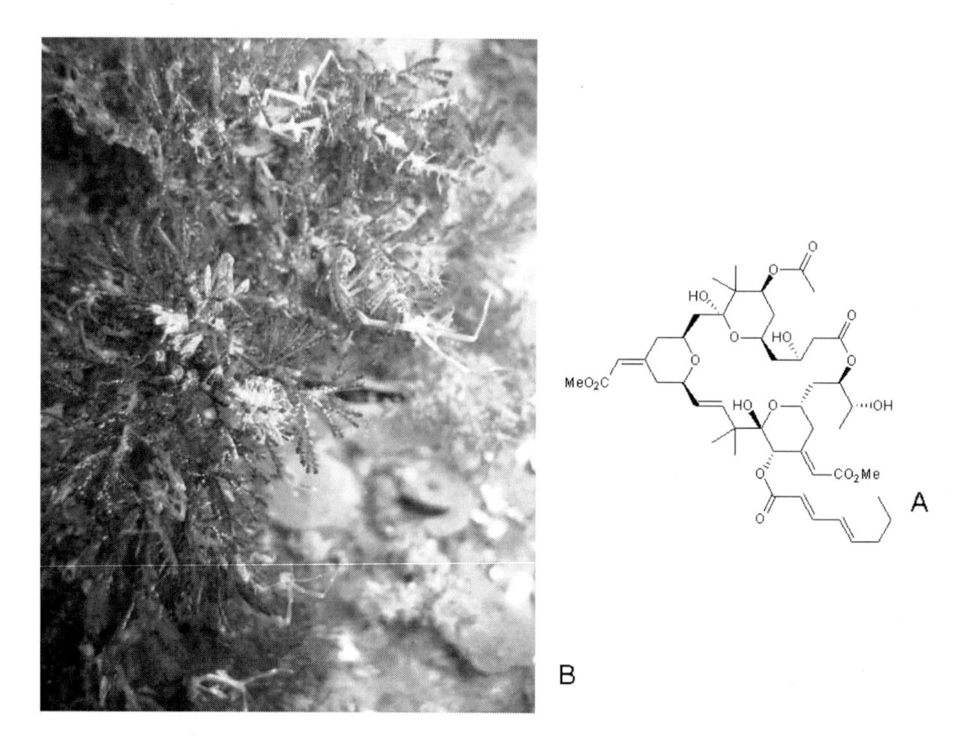

Figure 3. Structure of the anticancer drug bryostatin (A) produced by the symbiont *Endobugula sertula* of the bryozoan *Bugula neritina* (B).

The Real Source of Bioactive Compounds

The surfaces and internal spaces of marine organisms often containing numerous species of microorganisms, such as bacteria, microalgae and fungi (Hentschel et al. 2006; Newman and Hill 2006; Dobretsov and Qian 2006; Dobretsov et al. 2006). In some organisms, such as sponges, the amount of symbiotic bacteria is quite high and can reach up to 40 % of total host body weight (Wehrl et al. 2007).

As more evidence is being obtained, it is becoming clearer that microorganisms associated with macroorganisms may have highly specific, symbiotic relationships with their "hosts" (Hentschel et al. 2002). In many cases, these microorgansims may be considered as a true source of bioactive compound that can be isolated from the host organism. For example, an uncultivable symbiotic bacterium *Endobugula sertula* has been shown to be the real source of the anti-cancer compound bryosatin from *B.neretina* (Hildebrand et al. 2004) (Figure 3).

The cyanobacterium *Symplotica hydroides* is a true source of anti-cancer drug Dolastatin 10 (Table 1), while this compound accumulates in the body of the mollusc *Dolabella auricularia* (Gademann and Portmann 2008). Similar examples have been shown for different species of sponges (Fusetani 1996). By separation of cells of the tropical sponge *Dysidea herbacea* from its cyanobacterial symbiont *Oscillatoria spongeliae* it has been shown that the major products of this sponge, such as sesquiterpenes spirodysin and herbadysidolide, are actually produced by the symbiont (Unson et al. 1994). Okadaic acid – a protein phosphatase inhibitor – is well known for the sponges of the genes *Halichondria* (Sugiyama et al. 2007). Later, it was shown that this compound is produced not by the sponges but by dinoflagellates of the genus *Prorocentrum* (Proksch et al. 2002).

Finding the true source of bioactive compounds is necessary in order to scale up the production of marine drugs. In a case of microbial origin, bioactive compounds can be obtained using biotechnological methods. Unfortunately, only less than 1% of marine microorganisms can be cultivated under "standard" laboratory conditions (Okami 1986). In order to solve the problem of compound production from "uncultivable" microorganisms new methods should be developed either to allow the cultivation of these microbes (Stevenson et al. 2004), or to transfer responsible for production genes into cultivable hosts.

In a case of the uncultivable symbiotic bacterium *Endobugula sertula* it was possible to isolate a gene bryA that is responsible for the production of bryostatin (Hildebrand et al. 2004; Sudek et al. 2007). This example clearly shows that functional genes involved in the biosynthesis of bioactive compounds indeed can be isolated and expressed in cultivable hosts once the process of biochemical synthesis is known.

FUTURE PERSPECTIVES

It is here revealed that bioactive compounds from marine organisms are diverse and proven to be an important source of pharmacological compounds. Mostly, marine compounds have been isolated from marine sponges and to a less extent from marine microorganisms (Blunt et al. 2007). Increasing evidence that micro-organisms but not macro-organisms are responsible for the production of bioactive compounds or their predecessors (see: the question of supply) suggests that marine micro-organisms may hide a huge potential for bioactive compound isolation in the future. Marine microbes have certain advantages as a source of bioactive compounds compared to other marine organisms. This includes the high diversity of secondary metabolites due to high microbial diversity and the production of large quantities of compounds thus facilitating their isolation and identification. Another advantage of using microbes as a source of bioactive compounds is that microorganisms can produce compounds much more rapidly and in larger amounts compared to other marine organisms. Additionally, microorganisms can easily be genetically and chemically modified in order to increase compounds yield and bioactivity (König et al. 2006).

In a case of uncultivable microbial strains it is possible to transfer the target gene clusters responsible for biosynthesis of bioactive compounds into cultivable strains for large scale fermentation. All this makes microorganisms highly important in the production of bioactive compounds.

Nowadays, a large majority of industrial companies are looking for drugs against chronic diseases, such as cancer, that could give them high profit (Fenical and Jensen 2006). This review clearly shows that marine organisms have the ability to provide new pharmaceuticals against cancer (Table 1). Increasing problems of modern medicine highlight the importance of wide screening of marine organisms for pharmaceuticals applicable in diverse therapeutic areas, such as pain relief, Alzheimer's disease, infections diseases and malaria (Fenical and Jensen 2006; Blunt et al. 2007).

Many of the marine derived bioactive compounds have multiple biological properties. For example, diketopiperazines from marine microorganisms have antitumor, antibacterial, antifungal and antifouling properties (Li et al. 2006). Antifouling properties and ability to disrupt attachment and metamorphose of invertebrate larvae have been found for several commercially available pharmaceuticals (Rittschof et al. 2003). In this investigation, antihistamine drug Zyrtec® had very low toxicity to barnacle larvae and inhibited their attachment at nano gram concentrations which are comparable with existing toxic antifouling agents. Investigation of additional bioactive properties of already discovered drugs should be an important future direction (Dobretsov et al. 2006).

At present, most marketed marine drugs are derived from shallow waters. Recently, studies of marine organisms that live in extreme conditions (extremely hot or cold, high-pressure, anoxic, hypersaline) have led to significant discoveries of unusual natural products with novel biotechnological potential (Munro et al. 1999).

This is due to the fact that organisms that live in extreme environment have developed unusual strategies to enable themselves to adapt and thrive in such a harsh environment, and their metabolism may vary greatly from that of other organisms (Shilo 1980). More knowledge about bioactive compounds of marine extremophyles is needed and is going to facilitate discovery of novel bioactive compounds.

In conclusion, marine organisms, especially microorganisms, will be an important source of biologically active metabolites in the future. Development of new drugs and optimisation of production, yield and activity of already discovered ones require a deep knowledge of biochemical mechanisms underlying the expression of genes and synthesis of bioactive compounds in marine organisms. This process requires a join effort of biochemists, microbiologists, marine biologists, industrial and biotechnology companies. Over the next decade, we will see a significant growth of marine-targeted biotechnological companies which will result in an increase of marine derived drugs available on a market.

ACKNOWLEDGMENTS

Supported by SQU internal grant IG/AGR/FISH/09/03 (SD) and the HM Sultan Qaboos Fund for Strategic Research SR/AGR/FOOD/05/01 (BS) and The Sahlgrens's Academy, University of Gothenburg (BS).

REFERENCES

Abad, M.J., Bedoya, L.M. and Bermejo, P. (2008). Natural marine anti-inflammatory products. *Mini Reviews in Medicinal Chemistry, 8*, 740-754.

Alawi, A., Marhubi, I., Belushi, M. and Soussi, B. (2011). Characterization of carrageenan extracted from *Hypnea bryoides* in Oman. *Marine Biotechnology,* doi:10.1007/s10126-010-9350-7.

Berndt, R.H.A. (2009). Alginates: Biology and Applications. *Microbiology Monographs, Vol. 13*, Springer, Berlin.

Blunt, J.W., Copp, B.R., Hu, W-P, Munro, M.H.G., Northcote, P.T. and Prinsep, M.R. (2007). Marine natural products. *Natural Product Reports, 24* (*1*), 31-60.

Cho, J.J. and Kim, Y.T. (2002). Sharks: a potential source of antiangiogenic factors and tumor treatments. *Mar Biotech*, *4*, 521-525.

Cole, CGB. (2000). Gelatin. In: Francis F. J., editor. *Encyclopedia of Food Science and Technology, 2nd edition, 4 Vols*. New York: John Wiley and Sons; 1183-1188.

Cragg, G.M. and Newman, D.J. (2005). Plants as a source of anti-cancer agents. *Journal of Ethnopharmacology*, *100* (*1-2*), 72-79.

Delaloge, S., Yovine, A., Taamma, A., Riofrio, M., Brain, E., Raymond, E., Cottu, P., et al. (2001). Ecteinascidin-743: A Marine-Derived Compound in Advanced, Pretreated Sarcoma Patients--Preliminary Evidence of Activity. *J Clin Oncol*, *19* (*5*), 1248-1255.

Dobretsov, S., Dahms, H-U. and Qian, P-Y. (2006). Inhibition of biofouling by marine microorganisms and their metabolites. *Biofouling*, *22* (*1*), 43-54.

Dobretsov, S. and Qian, P-Y. (2006). Facilitation and inhibition of larval attachment of the bryozoan Bugula neritina in association with mono-species and multi-species biofilms. *Journal of Experimental Marine Biology and Ecology*, *333* (*2*), 263-274.

Dyerberg, J., Bang, H. O., Stoffersen E., Moncada, S., Vane, J. R. (1978). Eicosapentaenoic acid and prevention of thrombosis and atherosclerosis. *Lancet*, *312*, 117-119.

FAO. (1987). A guide to the Seaweed Industry", Fisheries Technical Paper. 441.

FAO. (1990). Technical Resource papers Regional Workshop on the Culture and Utilization of Seaweeds, Volume II.

Fenical, W. and Jensen, P.R. (2006). Developing a new resource for drug discovery: marine actinomycete bacteria. *Nat Chem Biol 2* (*12*), 666-673.

Fusetani, N. (1996). Bioactive substances from marine sponges. *Journal of Toxicology-Toxin Reviews*, *15* (*2*), 157-170.

Gademann, K. and Portmann, C. (2008). Secondary Metabolites from Cyanobacteria: Complex Structures and Powerful Bioactivities. *Current Organic Chemistry*, *12*, 326-341.

Gunasekera, S.P., Gunasekera, M., Longley, R.E., and Schulte, G.K. (1990). Discodermolide:

a new bioactive polyhydroxylated lactone from the marine sponge *Discodermia dissolute. Journal Organic Chemistry*, *55* (*16*), 4912–4915.

Gunnarsson, G., Undeland, I., Sannaveerappa, T., Sandberg, A.S., Lindgård, A., Hultén, L.M. and Soussi, B. (2006). Inhibitory effect of known antioxidants and of press juice from herring (Clupea harengus) light muscle on the generation of free radicals in human monocytes. *Journal Agricultural Food Chemistry*, *54*(*21*), 8212-8221.

Hamann, M.T. and Scheuer, P.J. (1993). Kahalalide F: a bioactive depsipeptide from the sacoglossan mollusk *Elysia rufescens* and the green alga *Bryopsis* sp. *Journal American Chemical Society*, *115*, 5825–5826.

Hentschel, U., Hopke, J., Horn, M., Friedrich, A. B., Wagner, M., Hacker, J., and Moore, B. S. (2002). Molecular evidence for a uniform microbial community in sponges from different oceans. *Applied and Environmental Microbiology*, *68* (*9*), 4431-4440.

Hentschel, U., Usher, K.M. and Taylor, M.W. (2006). Marine sponges as microbial fermenters. *FEMS Microbiology Ecology*, *55* (*2*), 167-177.

Hildebrand, M., Waggoner, L.E., Liu, H., Sudek, S., Allen, S., Anderson, C., Sherman, D.H. and Haygood, M. (2004). bryA: An unusual modular polyketide synthase gene from the uncultivated bacterial symbiont of the marine bryozoan *Bugula neritina. Chemistry and Biology*, *11* (*11*), 1543-1552.

Ireland, C.M. and Fernandez, A. (1998). Cyclic peptide antitumor agent from an ascidian. US patent 5830996, 3 November 1998.

Kem, W.R. (1997). Alzheimer's drug design based upon an invertebrate toxin (anabaseine) which is a potent nicotinic receptor agonist. *Invertebrate Neuroscience*, *3*, 251-259.

Kobayashi, M., Kakizono, T., Nishio, N., Nagai, S., Kurimura, Y. and Tsuji, Y. (1997). Antioxidant role of astaxanthin in the green alga *Haematococcus pluvialis. Applied Microbiology Biotechnology*, *48*, 351-356.

Kohl, A.C., Ata, A. and Kerr, R.G. (2003). Pseudopterosin biosynthesis—pathway elucidation, enzymology, and a proposed production method for anti-inflammatory metabolites from Pseudopterogorgia elisabethae. *Journal of Industrial Microbiology and Biotechnology*, *30* (*8*), 495-499.

König, G.M., Kehraus, S., Seibert, S.F., Abdel-Lateff, A. and Müller, D. (2006). Natural Products from Marine Organisms and Their Associated Microbes. *ChemBioChem*, *7* (*2*), 229-238.

Kraft, A.S., Smith, J.B. and Berkow, R.L. (1986). Bryostatin, an activator of the calcium phospholipid-dependent protein kinase, blocks phorbol ester-induced differentiation of human promyelocytic leukemia cells HL-60. *PNAS*, *83*, 1334-1338.

Li, X., Dobretsov, S., Xu, Y., Xiao, X., Hung, O.S. and Qian, P.Y. (2006). Antifouling diketopiperazines produced by a deep-sea bacterium, *Streptomyces fungicidicus. Biofouling*, *22* (*3*), 201-208.

Litaudon, M., Hickford, S.J.H., Lill, R.E., Lake, R.J., Blunt, J.W. and Munro, M.H.G. (1997). Antitumor polyether macrolides: new and hemisynthetic halichondrins from the New Zealand deep-water sponge *Lissodendoryx* sp. *Journal Organic Chemistry*, *62* (*6*), 1868–1871.

Look S.A., Fenical W., Jacobs R.S. and Clardy J. (1986). The pseudopterosins: anti-inflammatory and analgesic natural products from the sea whip *Pseudopterogorgia elisabethae. PNAS*, *83*, 6238-6240.

Luengo, J.M., García, B., Sandoval, A., Naharro, G. and Olivera, E.R. (2003). Bioplastics from microorganisms. *Current Opinion in Microbiology*, *6* (*3*), 251-260.

Luescher-Mattli. (2003). Algae, a possible source for new drugs in the treatment of HIV and other viral diseases. *Current Medicinal Chemistry - Anti-Infective Agents*, *2*, 219-225.

Mooberry, S.L., Tien, G., Hernandez, A.H., Plubrukarn, A. and Davidson, B.S. (1999). Laulimalide and isolaulimalide, new paclitaxel-like microtubule-stabilizing agents. *Cancer Research*, *59*, 653-657.

Munro, M.H.G., Blunt, J.W., Dumdei, E.J., Hickford, S.J.H., Lill, R.E., Li, S., Battershill, C.N. and Duckworth, A.R. (1999). The discovery and development of marine compounds with pharmaceutical potential. *Journal of Biotechnology*, *70* (*1-3*), 15-25.

Nagle, D.G., Geralds, R.S., Yoo, H-D, Gerwick, W.H., Kim, T-S, Nambu, N. and White, J.D. (1995). Absolute configuration of curacin A, a novel antimitotic agent from the tropical marine cyanobacterium *Lyngbya majuscula. Tetrahedron Letters*, *36*, 1189-1192.

Natori, T., Motoki, K., Higa, T. and Koezuka, Y. (2000). KRN7000 as a new type antitumor and immunostimulatiory drug. In: Fusetani, editor. *Drugs from the sea*. Karger, Basel; 2000; 86-97.

Newman, D. J. and Cragg, G.M. (2007). Natural Products as Sources of New Drugs over the Last 25 Years. *Journal of Natural Products*, *70* (*3*), 461-477.

Newman, D. and Hill, R. (2006). New drugs from marine microbes: the tide is turning. *Journal of Industrial Microbiology and Biotechnology*, *33* (*7*), 539-544.

Olaizola, M. (2000). Commercial production of astaxanthin from Haematococcus pluvialis using 25,000-liter outdoor photobioreactors. *Journal of Applied Phycology*, *12*, 499–506.

Okami, Y. (1986). Marine microorganisms as a source of bioactive agents. *Microbial Ecology*, *12* (*1*), 65-78.

Omerovic, E., Lindbom, M., Ramunddal, T., Lindgard, A., Undeland, I., Sandberg, A.S., and Soussi B. (2008). Aqueous fish extract increases survival in the mouse model of acute cytostatic toxicity. *Journal of Experimental and Clinical Cancer Research*, *27*, 81, doi:10.1186/1756-9966-27-81.

Paul, V. J, Arthur, K.E., Ritson-Williams, R., Ross, C. and Sharp, K. (2007). Chemical defenses: From compounds to communities. *Biological Bulletin*, *213* (*3*), 226-251.

Proksch, P., Edrada, R. and Ebel, R. (2002). Drugs from the seas - current status and microbiological implications. *Applied Microbiology and Biotechnology*, *59* (*2*), 125-134.

Rittschof, D., Lai, C-H., Kok, L-M. and Teo, S.L.M. (2003). Pharmaceuticals as antifoulants: concept and principles. *Biofouling*, *19* (*1 supp 1*), 207-212.

Sainz-Diaz, C. I., Manzanares, I., Francesch, A. and Garcia-Ruiz, J. (2003). The potent anticancer compound ecteinascidin-743 (ET-743) as its 2-propanol disolvate. *Acta Crystallographica Section C Crystal Structure Communications*, *59* (*4*), 197-198.

Scopus database (www.scopus.com/scopus/home.url) accessed data on 12/03/08.

Sennett, S., McCarthy, P., Wright, A. and Pomponi, S. (2002). Natural Products from Marine Invertebrates: The Harbor Branch Oceanographic Institution Experience. *Pharmaceutical News 9* (*6*), 483.

Shilo, M. (1980). Strategies of adaptation to extreme conditions in aquatic microorganisms. *Naturwissenschaften*, *67* (*8*), 384-389.

Stevenson, B. S., Eichorst, S.A., Wertz, J.T., Schmidt, T.M. and Breznak, J.A. (2004). New strategies for cultivation and detection of previously uncultured microbes. *Applied Environmental Microbiology*, *70* (*8*), 4748-4755.

Sudek, S., Lopanik, N.B., Waggoner, L.E., Hilderband, M., Abderson, C., Liu, H., Patel Am

Sherman, D.H. and Haygood, M.G. (2007). Identification of the putative bryostatin polyketide synthase gene cluster from "*Candidatus Endobugula sertula*", the uncultivated microbial symbiont of the marine bryozoan *Bugula neritina*. *Journal Natural Products*, *70*, 67-74.

Sugiyama, N., Konoki, K. and Tachibana, K. (2007). Isolation and characterization of okadaic acid binding proteins from the marine sponge *Halichondria okadai*. *Biochemistry*, *46* (*40*), 11410-11420.

Terlau, H. and Olivera, B.M. (2004). Conus venoms: a rich source of novel ion channel-targeted peptides. *Physiological Reviews*, *84* (*1*), 41-68.

Thoms, C. and Schupp, P. (2005). Biotechnological potential of marine sponges and their associated bacteria as producers of new pharmaceuticals (Part I). *Journal of International Biotechnology Law*, *2* (*5*), 217-220.

Unson, M. D., Holland, N. D. and Faulkner, D. J. (1994). A brominated secondary metabolite synthesized by the cyanobacterial symbiont of a marine sponge and accumulation of the crystalline metabolite in the sponge tissue. *Marine Biology*, *119* (*1*), 1-11.

Wehrl, M., Steinert, M. and Hentschel, U. (2007). Bacterial uptake by the marine sponge *Aplysina aerophoba*. *Microbial Ecology*, *53* (*2*), 355-365.

Williams, P.G., Buchanan, G.O., Feling, R.H., Kauffman, C.A., Jensen, P.R. and Fenical, W. (2005). New cytotoxic salinosporamides from the marine actinomycete *Salinispora tropica*. *Journal Organic Chemistry*, 70 (*16*), 6196–6203.

In: Natural Products and Their Active Compounds … ISBN: 978-1-62100-153-9
Editors: M. Essa, A. Manickavasagan, and E. Sukumar

Chapter 23

FUNCTIONALITY OF POMEGRANATE (*PUNICA GRANTUM L.*) IN HUMAN HEALTH

Amani S. Al-Rawahi, Mohammad Shafiur Rahman [*] ***and Mohamed M. Essa***
Department of Food Science and Nutrition,
Sultan Qaboos University, Oman

ABSTRACT

The fruit of *Punica granatum L. Punicaccae* has been widely used since ancient times. In the past decade; scientists have been researching on pomegranate fruit by analyzing its nutrition, chemistry, pharmaceuticals, medical properties, or even cosmetic properties. These studies are the results of increasing awareness of the health benefits of functional fruits. Pomegranate is a nutritious fruit containing carbohydrates, minerals, vitamins and most importantly, antioxidants.

Evidence from literature showed medicinal effects of different parts of pomegranate to prevent varied chronic or common illnesses. In this chapter, a focus on the potential of pomegranate in medicine, food industry, and pharmaceuticals is presented, with emphasis on its nutrition and active components, and possible toxicity of the pomegranate skin. Pomegranate showed protective effects, such as breast cancer, menopausal syndrome, thyroid dysfunctions, injured cells, diabetes, inflammations, influenza, prostate cancer, brain damage, Alzheimer's disease, hyperlipidemia, hypertension, artery stenosis, dental problems, male infertility, erectile dysfunction, obesity and other health issues.

INTRODUCTION

Pomegranate, attractive delicious and somewhat thorny large shrub or small tree (*Punica granatum L),* belonging to the family *Punicaceae*, native to semitropical Asia (Iran) and naturalized in the Mediterranean region. It has long been cultivated as an ornamental edible

[*] Corresponding author: E-mail: shafiur@squ.edu.om

fruit. The fruit, about the size of an apple, bears many seeds, each within a fleshy crimson seed coating, enclosed in a tough yellowish to deep red rind. Pomegranates are either eaten fresh or used for making syrup. The astringent properties of the rind and bark have been valued medicinally for several thousand years. The pomegranate is now cultivated in most warm climates [Anonymous, 2009].

The fruit has been used since ancient times. It has been mentioned in many holy books and was therefore; well received by many people. Scientists from all over the world have been studying pomegranate's diversified properties and functionalities.

Pomegranates are classified in the division *Magnoliophyta*, class *Magnoliopsida*, and order Mortals, family *Punicaceae*. The leaves are opposite or sub-opposite, glossy, narrow oblong, entire, 3-7 cm long and 2 cm broad. The flowers are bright red, 3 cm in diameter, with five petals (often more on cultivated plants).

Flowers are hermaphrodite (intersexual) with 4-8 leathery sepals and equal number of red petals, numerous stamens and variable number of carpals, which together make lower ovary [Tous and Ferguson, 1996].

The fruit is a berry, between an orange and a grapefruit in size, 7-12 cm in diameter with a rounded hexagonal shape, and has thick reddish skin and many seeds [Anonymous, 2009]. Pomegranate has a leathery-skinned berry containing many seeds, each surrounded by a juicy, fleshy aril.

Eighty per cent of pomegranate fruit are comprised of 80% juice and 20% seed. Pomegranate fruit is a rich source of two types of polyphenolic compounds: anthocyanins (such as delphinidin, cyanidin, and pelargonidin), which give the fruit and juice its red color; and hydrolyzable tannins (such as punicalin, pedunculagin, punicalagin, gallagic, and ellagic acid esters of glucose), which account for 92% of the antioxidant activity of the whole fruit [Singh et al. 2002; Shanshan et al. 2007; Aslam et al. 2006].

Chemical Composition and Its Nutritional Value

Pomegranate is considered as highly nutritious fruit. Seeds are rich source of carbohydrates, vitamins and minerals; yet very low in saturated fat, cholesterol and sodium [Anonymous, 2009]. Pomegranate fruit contains important amounts of phenolics including flavonoids (anthocyanins, catechins and other complex flavanoids) and hydrolyzable tannins (punicalin, pedunculagin, punicalagin, gallagic and ellagic acid esters of glucose), which account for 92% of its antioxidant activity [Syed et al. 2007; Afaq et al. 2005].

There is high amount of nutrients both in arils and peel, especially micronutrients (Zn, Cu, Mn, Fe and B) and macronutrients (P, K, N, Mg, Ca and Na) in edible parts that play a valuable role in daily requirement of minerals [Mirdehghan and Rahemi, 2007].

The peel of pomegranate consists of considerable amounts of total phenolics, flavonoids, proanthocyanidins and ascorbic acid. Pomegranate seeds are rich in sugars, polyunsaturated (n−3) fatty acids, vitamins, polysaccharides, polyphenols and minerals and showed high antioxidant activity. The oil in seeds contains 80% punicic acid, 18-carbon fatty acid, along with the isoflavone genistein, phytoestrogen coumestrol and sex steroid estrone [Syed et al. 2007].

Table 1 illustrates the nutritional value of pomegranate per 100 g edible portion.

Table 1. Nutritional value of Pomegranate (per 100g edible portion)

Composition	Unit	Quantity
Calories	kcal	63-78
Moisture	g	72.6-86.4
Protein	g	0.05-1.6
Fat	g	0.9
Carbohydrates	g	15.4-19.6
Fiber	g	3.4-5.0
Ash	g	0.36-0.73
Calcium	mg	3-12
Phosphorus	mg	8-37
Iron	mg	0.3-1.2
Sodium	mg	3
Potassium	mg	259
Manganese	mg	3.0
Zinc	mg	0.12
Magnesium	mg	0.15
Copper	mg	0.07
Selenium	mg	0.6
Carotene	mg	None to trace
Thiamin	mg	0.003
Riboflavin	mg	0.012-0.03
Niacin	mg	0.180-0.3
Panthothenic acid	mg	0.596
Ascorbic acid	mg	4.0-4.2
Citric acid	mg	0.46-3.6
Boric acid	mg	0.005

Source: Anonymous, 2009.

HEALTH FUNCTIONAL COMPONENTS OF POMEGRANATE

Active chemical constituents are identified in all parts of pomegranate including leaf, seed, juice, husk and peel [Lansky and Newman, 2007; Singh et al. 2002; Gil et al. 2000]. Diverse phenolic compounds are present in pomegranate juice, including punicalagin isomers, ellagic acid derivatives and anthocyanins (delphinidin 3-glucoside, cyanidin and pelargonidin 3-glucosides and 3, 5-diglucosides). These compounds are known for their properties in scavenging free radicals and inhibiting lipid oxidation in vitro [Gil et al. 2000; Hernandez et al. 1999; Noda et al. 2002]. Anthocyanins potent antioxidant flavonoids produce pomegranate juice with its brilliant color, and its intensity increases during ripening [Hernandez et al. 1999], and declines after pressing [Perez-Vicente et al. 2002; Miguel et al. 2004].

Table 2. Main antioxidants present in different parts of pomegranate

Antioxidants	Part of extraction	References
Total Phenolics (mg/g)	Pulp, peel, juice	Yunfeng et al., (2006); Cam and Hisil, (2010); Tezcan et al., (2009); Tehranifar, et al., (2010)
Total Flavonoids (mg/g)	Pulp, peel, juice	Yunfeng et al., (2006); Cam and Hisil, (2010), Tezcan et al., (2009); Tehranifar, et al., (2010)
Proanthocyanidins (mg/g)	Pulp, peel	Yunfeng et al., (2006)
Ascorbic acid (mg/g)	Pulp, peel	Yunfeng et al., (2006); Tehranifar, et al., (2010)
Polyphenols (mg/g)	Peel, juice	shanshan et al., (2007); Tehranifar, et al., (2010)
Total tannins (w/w, %)	Peel, juice	shanshan et al., (2007)
Hydrolyzable tannins (mg TAE/g)	Peel	Cam and Hisil, (2010)
Condensed tannins (mg CE/g)	Peel	Cam and Hisil, (2010)
B-Carotene	Fermented juice (W)	Schubert et al., (1999)
ABTS, DMPD, FRAP	Commercial PJ (whole fruit squeezed), flower, seed, pulp, peel	Gil et al., (2000); Kaur et al., (2006); Tezcan et al., (2009); Guo et al., (2003); Madrigal-Carballo, et al., (2009)
DPPH	Commercial PJ (whole fruit squeezed), flower, peel, seed	Gil et al., (2000); Kaur et al., (2006); Okonogi et al., (2007); He et al., (2010); Tezcan et al., (2009); Madrigal-Carballo, et al., (2009)
Low density lipoprotein (LDL)	Peel, seed, flower	Singh et al. (2002), Kaur et al., (2006)
TEAC, inhibition of peroxidation of phosphatidylcholine liposomes	Peel, seed	Plumb et al., (2002); Okonogi et al., (2007); He et al., (2010)
H2O2-induced LDL oxidation	Whole pomegranate fruit	Noda et al., (2002)
Ellagic acid	Peel	Panichayupakaranant et al., (2010); Seeram, et al., (2005)
Caffeic acid glycoside dimmer	Seed	He et al., (2010)
Pentagalloylglucopyranose	Seed	He et al., (2010)
Feuric acid derivate	Seed	He et al., (2010)
Tetragalloylglucopyranose	Seed	He et al., (2010)
Coumaric acid derivative	Seed	He et al., (2010)
Trigalloylglucopyranose	Seed	He et al., (2010)
Caffeic acid dimglu dimer	Seed	He et al., (2010)
2,3-O-hexahydroxydiphenoyl-glu	Seed	He et al., (2010)
Pedunculagin	Seed	He et al., (2010); Madrigal-Carballo, et al., (2009)
Procyanidin timer type C	Seed	He et al., (2010)
Procyanidin drimer type B	Seed	He et al., (2010)
Kaempferol 3-O-rutinoside	Seed	He et al., (2010)
Kaempferol 3-O-glucoside	Seed	He et al., (2010)
Quercitrin 3-O-rhamnoside	Seed	He et al., (2010)

Table 2 illustrates the antioxidant components in different parts of pomegranate fruit identified by different researchers.

The husk of the fruit is a very rich in ellagic acid derivatives, such as ellagitannins punicalagin, as well as less amounts of punicalin, gallic acid, ellagic acid and EA-glycosides (hexoside, pentoside, rhamnoside etc.) [Lu et al. 2007].

These ellagitannins are extracted in significant levels into juice from the husk during industrial hydrostatic processing methods. Punicalagin levels are widely variable in pomegranate juice and range from 0.017 to 2 g/L pomegranate juice depending on the fruit cultivar as well as processing and storage conditions.

When comparing the antioxidant fractions present in the peel and pulp extracts, it was found that the total phenolics content of peel extract was nearly 10-fold as high as that of pulp extract. The contents of flavonoids and proanthocyanidins were also higher in peel extract than does pulp extract. This clearly indicates that peel extract contains more antioxidants than the pulp extract. It was also found that peel tissues usually contained larger amount of phenolics, anthocyanins and flavonols than did flesh tissues in nectarines, peaches and plums. However, flavonoids or proanthocyanidins account for only a small part of total phenolics present in the peel extract. In addition, both peel and pulp extracts contained a small amount of ascorbic acid. Therefore, ascorbic acid could not be an important antioxidant, either in the peel or pulp extract [Tomas-Barberan et al. 2001; Yunfeng et al. 2006]. Pomegranate seed, a by-product of pomegranate juice processing, contains a range of nutraceutical components, such as sterols, γ-tocopherol, punicic acid and hydroxybenzoic acids [Liu et al. 2009]. Phenyl aliphatic glycosides as phenethyl rutinoside were also found in pomegranate seed [Wang et al. 2004].

MEDICINAL VALUES OF POMEGRANATE

Pomegranate husk is a traditional herbal medicine in many cultures, such as Chinese, Arabian, and Persian. It has been administrated in heeling various illnesses in humans since prehistoric ages supported by public acceptance and positive believe in the medications derived from this fruit. Some literature reveals that the tannins from the pericarp of pomegranate exhibit antiviral activity against the genital herpes virus [Zhang et al. 1995]. The pomegranate rind extract is also shown to be a potent virucidal agent [Stewart et al. 1998] and has been used as a constituent of antifungal and antiviral preparations [Jassim, 1998]. There are reports on the use of a water decoction of pomegranate husk powder as a multifunctional vaginal suppository for contraception and for the prevention and cure of venereal disease [Zhan, 1995]. Pomegranate husk is reported as a part of a preparation used for treating the infection of male or female sexual organs, mastitis, acne, folliculitis, pile, allergic dermatitis, tympanitis, and scald for curing diarrhoea and dysentery, and oral diseases [Hu, 1997; Fengchun et al. 1997]. The pomegranate husk extract, when introduced into juice, improves clarity of juice by accelerating precipitation of the haze-forming substance [Kvasenkov et al. 1999]. Commercial pomegranate juice showed one of the highest antioxidant activities compared to other fruit juices, red wine and green tea [Gil et al. 2000]. This can be attributed to its high content of polyphenols including ellagic acid (free and bound forms), gallotannins and anthocyanins, and other flavonoids [Seeram et al. 2005]. The most abundant of these polyphenols is punicalagin. Punicalagin levels are widely variable in pomegranate juice and can range from as low as 0.017 to 1.5 g/L of pomegranate juice, depending on the fruit cultivar as well as processing and storage conditions. Therefore, the regular consumption of pomegranate juice may provide significant amounts of this water-soluble hydrolyzable ellagitannin [Gil et al. 2000; Seeram, 2004]. The medicinal properties of pomegranate are presented in the following sections.

Effects on Depressive State and Bone Properties

Pomegranate is known to contain estrogens (estradiol, estrone, and estriol) and showed estrogenic activities in mice. Mori-Okamoto et al. [2004] investigated whether pomegranate's extract was effective on experimental menopausal syndrome in ovariectomized mice. Administration of pomegranate extract (juice and seed extract) for 2 weeks to ovariectomized mice prevented the loss of uterus weight and shortened the immobility time compared with 5% glucose-dosed mice (control). In addition, ovariectomy-induced decreased bone density, which was normalized when pomegranate extract was used in the meal. The bone volume and the trabecular number were significantly increased and the trabecular separation was decreased in the pomegranate-dosed group as compared with the control. Some histological bone formation/resorption parameters were significantly increased by ovariectomy but were normalized by administration of the pomegranate extract. These changes suggested that the pomegranate extract inhibited ovariectomy-stimulated bone turnover. It is thus conceivable that pomegranate is clinically effective on a depressive state and bone density loss in menopausal syndrome in women.

Cancer and Chemoprevention

The effective approaches for the prevention of cancer have become an important goal to reduce cancer burden [Syed et al. 2007]. One strategy could be through chemoprevention, preferably by the use of non-toxic dietary substances and natural products. Pomegranate is now being recognized as a potential chemopreventive and anticancer agent. The emerging data in the literature provided new insights into the molecular framework which is needed to establish novel mechanism-based chemopreventive strategies for various human cancers [Syed et al. 2007].

Effects on Skin Cancer

In recent years, research has focused on the use of naturally occurring botanicals for their potential preventive effect against UV damages [Afaq et al. 2002]. Some studies using normal human epidermal keratinocytes (NHEK) showed remarkable photochemopreventive effects of pomegranate against UV-A and UV-B radiation promoting tumor [Syed et al. 2006; Afaq et al. 2005]. They also found that pomegranate fruit may have a regulatory effect on the rate of protein synthesis and activation of tumor cell proliferation. Another study showed the positive effects of pomegranate on protecting human skin against UV-B mediated DNA damage [Zaid et al. 2007].

In mouse model, oral feeding of pomegranate extract to mice exposed to UV-B showed to inhibit skin derma, hyperplasia, infiltration of leukocytes, lipid peroxidation and hydrogen peroxide generation in the skin [Afaq et al. 2006]. Further studies showed the inhibition effect of pomegranate extract on UV-induced skin pigmentation in brown guinea pigs [yoshimura et al. 2005]. Oral administration of pomegranate to humans showed the protective effect of sunscreens and photoprotection from UV-B [Murad and Shellow, 2001].

Effects on Prostate Cancer

In the United States and other Western countries, prostate cancer is the second-leading cause of cancer-related death [Jurenka, 2008]. *In vitro* studies showed that several pomegranate flower extracts inhibited prostate cancer cell growth, induce apoptosis of several prostate cancer cell lines (including highly aggressive PC-3 prostate carcinoma cells), suppressed invasive potential of PC-3 cells, and decreased proliferation of DU-145 prostate cancer cells [Lansky et al., 2005; Albrecht et al. 2004; Malik et al. 2006]. Equal amounts of fermented pomegranate juice, pomegranate peel extract, and pomegranate seed oil extracts resulted in a 99% suppression of DU-145 prostate cancer cell invasion across a Matrigel matrix [Lansky et al. 2005].

Pomegranate seed oil extract or fermented pomegranate juice extract alone resulted in 60% suppression of invasion, and when combining any two extracts induced 90% suppression. Studies in mice also demonstrated that pomegranate flower extract inhibited prostate tumor growth and decreased PSA levels [Malik, et al. 2006, 2005]. These promising results led some of the same researchers to conduct a two-stage phase II clinical trial in men with recurrent prostate cancer and rising PSA levels. All eligible patients had previous surgery or radiation therapy for prostate cancer, Gleason scores (a grading system for predicting the behavior of prostate cancer) ≤ 7, rising PSA value of 0.2-5.0 mg/mL, no prior hormonal therapy, and no evidence of metastases. Twenty two participants were consumed eight ounces pomegranate juice (570 mg total polyphenol gallic acid equivalents) daily until they meet disease progression endpoints.

Endpoints measured showed an effect on PSA levels, serum lipid peroxidation and nitric oxide levels, *in vitro* induction of proliferation and apoptosis of LNCaP cells in patient serum containing pomegranate constituents, and overall safety of extract administration [Pantuck, et al. 2006].

Effects on Lung Cancer

Several studies using *in vitro* and *in vivo* models were conducted to explore the chemopreventive or therapeutic potential of pomegranate against lung cancer [Khan at al. 2007]. Interestingly, pomegranate extract treatment was found to result in a dose dependent decrease in the growth and viability of A549 cancer cells in vitro, but insignificant inhibition in the progression of tumor growth in vivo [Khan et al. 2007].

Another study demonstrated that pomegranate inhibited lung tumorigenesis by targeting multiple signalling pathways in the mouse model and merits consideration for development as a potential chemopreventive agent against human lung cancer [Khan et al. 2007].

Effects on Colon Cancer

Phytochemicals from pomegranate have been shown to inhibit colon cancer proliferation and apoptosis through the modulation of cellular transcription factors and signalling proteins

[Adams et al. 2006]. This cancer proliferation was related to the significant cell interactions with bioactive polyphenols present in pomegranate juice such as anthocyanins and flavonols. In another study, dietary administration of pomegranate seed oil rich in conjugated linolenic acid significantly inhibited the development of azoxymethane-induced colonic adeno-carcinomas in male F344 rats without causing any adverse effects [Kohno et al. 2004].

Suppressive Effects on Human Breast Cancer Cells

There are evidences of suppressive effects of pomegranate extracts on human breast cancer cells [Elswijk et al. 2004]. Three estrogenic compounds, i.e. luteolin, quercetin and kaempferol, were detected and identified by comparing molecular weights and negative ion APCI MS/MS spectra. Although it is well known in literature and widely distributed in nature, the presence of these phytoestrogenic compounds in pomegranate peel extract has been reported recently. Compared to traditional screening approaches of complex mixtures, often it was characterized by a repeating cycle of HPLC fractionation and biological screening, LC-BCD–MS.

Antioxidant Functionality

Guo et al. [2008] studied the effects of pomegranate juice (high in antioxidant capacity) and apple juice (low in antioxidant activity) on improving antioxidant function in the cases of two groups of elderly subjects. Results showed that apple juice consumption presented a less antioxidant effects in elderly subjects as compared to pomegranate juice. Guo et al. [2008] then concluded that daily consumption of pomegranate juices is potentially better than apple juice in improving antioxidant function in the elderly. This was identified by the significant difference between the plasma antioxidant capacity, ascorbic acid, vitamin E, and reduced glutathione contents between the 2 groups of the study. It was concluded that the phenolics may be the functional components contained in pomegranate juice that accounted for the observations.

Pomegranate juice is being increasingly proposed as a nutritional supplement to prevent atherosclerosis in humans. Sestili et al. [2007] worked on the antioxidant capacity of pomegranate by comparing with their cytoprotective – *bona fide* antioxidant – activity in cultured human cells (U937 promonocytes and HUVEC endothelial cells) exposed to an array of oxidizing agents. Pomegranate derivatives were pomegranate juice, arils juice and aqueous rinds extract. All the preparations displayed good antioxidant capacity in the order as: rinds extract > pomegranate juice > arils juice. On the contrary, only rinds extract was capable of preventing the deleterious effects – cytotoxicity, DNA damage and depletion of no protein sulphydrils (NPSH) pool – caused by treatment of cells with H_2O_2, *tert*-butylhydroperoxide (tB-OOH) or oxidized lipoproteins (Ox-LDL) via a mechanism which is likely to involve both direct scavenging of radical species and iron chelation. Surprisingly, arils juice and pomegranate juice slightly sensitized cells to the cytotoxic effects of the three agents. Then it would appear that arils juice, the major and tasty part of pomegranate juice, did not contain sufficient amount of ellagic acid and punicalagin (i.e. the polyphenols highly represented in rinds extract which are reputed to be responsible for the antioxidant capacity) to exert cytoprotection in oxidatively injured, living cells. Based on these results, the development and evaluation of rinds-only based derivatives for antiatherogenic preventive purposes in humans should be encouraged.

Pomegranate peels extract showed the highest free radical scavenging capacity among the medicinal plants, which are being used traditionally for treatment of diabetes in Jordan. Althunibat et al. [2010] aimed to investigate the antioxidant effect of methanolic extract of pomegranate peel against oxidative damage in streptozotocin-induced diabetic rats. Their results revealed that intraperitoneal administration of 10 and 20 mg/kg (body weight) of pomegranate peel extract for 4 weeks significantly enhanced the activities of antioxidant enzymes in liver, kidney and red blood cells of streptozotocin induced diabetic rats. The extract also caused a significant reduction in malondialdehyde; a lipid peroxide marker, in diabetic rat tissues and elevated the total serum antioxidant capacity in dose dependent manner. Thus, pomegranate peel extract showed protective role against the oxidative damage in streptozotocin induced diabetic rats.

Most studies investigated the physiological effects of pomegranate extracts in vitro and in animals using different models such as α-carotene-linoleate and 1,1-diphenyl-2-picryl hydrazyl (DPPH) model systems to prevent lipid peroxidation [Singh et al. 2002]. Recently, pomegranate juice and even fermented pomegranate juice were demonstrated to be high in antioxidant activity [Gil et al. 2000; Schubert et al. 1999]. Pomegranate juice also displayed potent antiatherogenic action in atherosclerotic mice and humans [Aviram et al. 2000; Kaplan et al. 2001]. All these activities may be related to diverse phenolic compounds present in pomegranate juice, including punicalagin isomers, ellagic acid derivatives and anthocyanins (delphinidin, cyanidin and pelargonidin 3-glucosides and 3,5-diglucosides). These compounds are known for their properties in scavenging free radicals and inhibiting lipid oxidation in vitro [Gil et al. 2000; Noda et al. 2002]. Literature results indicated that pomegranate peel showed an enriched source of the antioxidants, thus exhibiting higher activity as compared to seeds. The difference in the antioxidant activity of the peel and seed was attributed to their different phenolic compositions. In another study, the ROS scavenging ability and protective effect of nine extracts from sour pomegranate, red pomegranate, and white pomegranate on DNA damage were evaluated by utilizing in vitro chemiluminescent method. Results showed that all nine extracts could scavenge ROS and prevented DNA damage. The data also indicated that both the varieties of pomegranate and the parts could significantly influence the antioxidant activity and prevented DNA damage [Shanshan et al. 2007]. Malonaldehyde and 4-hydroxyalkenals levels in rat brain homogenates treated with pomegranate extract, delphinidin, cyanidin, and pelargonidin were estimated as an index of lipid peroxidation. Pomegranate extract showed potent scavenging activity for superoxide radicals. A filtered sample from the extract only slightly decreased the activity when compared to that of the untreated sample. These results suggested that the scavenging activity of pomegranate extract was due to its relatively smaller molecular weight components ($MW < 10,000$ [Noda et al. 2002]. The scavenging or preventive capacity of peel and pulp extracts against several common free radicals was tested in vitro by Yunfeng et al. [2006]. The superoxide anion is a well-recognized free radical species and was generated continuously by several cellular processes, including the microsomal and mitochondrial electron transport systems. Although the superoxide anion was limited in activity, it may combine with other reactive species, such as nitric oxide, produced by macrophages, to yield a more reactive species [Fridovich, 1995]. The peel extract presented rather more superoxide radical-scavenging ability than the pulp extract based on the inhibition of superoxide radical-related formosan production. At a concentration of 50 g/l, the inhibition activities were 43.0% and 37.7% for the peel and pulp extracts, respectively. Oxidative tension is potent non-specific metabolic trigger for both inflammation and angiogenic processes [Hayden and Tyagi, 2004; Karageuzyan, 2005; Kapoor et al. 2005], which are key factors in cancer initiation and promotion [Dobrovolskaia and Kozlov, 2005; Garcea et al. 2005; Ohshima et al. 2005]. Since pomegranate's antioxidative efficacy clinically may be impaired by poor bioavailability of active compounds [Cerda et al. 2004; 2006], strengths of the extracts need to be considered. In general, comparable juice or extracts from other common fruits showed antioxidant activity inferior to that of the pomegranate [Halvorsen et al. 2002; Kelawala and Ananthanarayan, 2004; Xu et al. 2005].

In vivo and clinical findings have been suggested as stemming from antioxidant effects. Examples of *in vivo* studies of beneficial effects of pomegranate antioxidant activity include: protection of rat gastric mucosa from ethanol or aspirin toxicity [Khennouf et al. 1999; Ajaikumar et al., 2005], protection of neonatal rat brain from hypoxia [Loren et al. 2005], prevention of male rabbit erectile tissue dysfunction [Azadzoi et al. 2005], and abrogation of ferric nitrilotriacetate (Fe-NTA) induced hepatotoxicity evidenced by mitigated hepatic lipid peroxidation, actions of glutathione, catalase, glutathione peroxidase, glutathione reductase, glutathione-*S*-transferase, serum aspartate aminotransferase, alanine aminotransferase and alkaline phosphatase, bilirubin and albumin levels, hepatic ballooning degeneration, fatty changes, and necrosis [Kaur et al. 2006]. Cardiovascular effects of pomegranate juice in man, which may or may not involve redox-linked biochemical pathways, include lowering of LDL and total cholesterol [Esmaillzadeh et al. 2004], ameliorating systolic hypertension [Aviram and Dornfeld, 2001], and reducing carotid arterial stenosis [Aviram et al. 2004; Lansky and Newman, 2007].

Atherosclerosis and Thyroid Dysfunctions

Fruit peels are generally considered as waste, thus value added or health functional products could be developed from peels. Parmar and Kar [2007] conducted a study on the protective role of orange, pomegranate, and banana peel extracts in diet-induced atherosclerosis and thyroid dysfunction in rats. They suggested the protective role of these fruit peels against diet-induced atherosclerosis and thyroid dysfunction. They concluded that pomegranate peel have the potential to prevent hyperlipidemia and may be further investigated for its use as potential nutraceutical supplements for the prevention of cardiovascular diseases and or thyroid abnormalities.

Effective Role on Hyperlipidaemia

Pomegranate flowers have been used in both the Unani and Ayurvedic systems of medicine as a remedy for diabetes. Pomegranate flower extract showed to activate peroxisome proliferator-activated receptor (PPAR-α), a cardiac transcription factor involved in myocardial energy production via fatty acid uptake and oxidation. PPAR-αactivation decreased cardiac uptake and circulation of lipids. Triglyceride content, plasma total cholesterol and NEFA were decreased after four weeks of treatment [Huang, et al., 2005]. A pilot study involving 22 type 2 diabetic patients (8 men and 14 women) were conducted to identify the cholesterol lowering effects of 40 g concentrated pomegranate juice for eight weeks. Statistically significant decreases were observed in total cholesterol (from 202 to 191 mg/dL), LDL cholesterol (124.4 mg/dL to 113 mg/dL), total/HDL cholesterol ratio (5.5 to 5.1), and LDL/HDL ratio (3.4 to 3.0). They also observed lower absorption and higher fecal excretion of cholesterol, as well as possible effects on HMG-CoA reductase and sterol O-acyltransferase, two key enzymes for cholesterol metabolism [Esmaillzadeh, et al. 2006].

Anti-Inflammatory Properties

Larrosa et al. [2010] evaluated the effects of pomegranate intake and its main microbiota derived metabolite urolithin-A (UROA) on colon inflammation and assessed whether UROA is the main anti-inflammatory compound. They also explored the effect of inflammation on the phenolic metabolism.

It was observed that both pomegranate extract and UROA decreased inflammation markers (iNOS, cycloxygenase-2, PTGES and PGE2 in colonic mucosa) and modulated favorably the gut microbiota. The G1 to S cell cycle pathway was regulated in both groups. UROA group showed various down regulated pathways, including that of the inflammatory response. PE, but not UROA, decreased oxidative stress in plasma and colon mucosa. Only UROA preserved colonic architecture. The normal formation of urolithins in PE-fed rats was prevented during inflammation. It was suggested that UROA could be the most active anti-inflammatory compound derived from pomegranate ingestion in healthy subjects, whereas in colon inflammation, the effects could be due to the nonmetabolized ellagitannin-related fraction.

Antimicrobial Activity

Inactivation of Influenza Virus

Pomegranates contain high levels of polyphenols, thus it could be a rich source of antiviral compounds. Sundararajan et al. [2010] evaluated the direct anti-influenza activity of three commercially available pomegranate extracts: pomegranate juice, a concentrated liquid extract (POMxl), and a 93% pomegranate peel powder extract (POMxp).

The acidity of pomegranate Juice and POMxl solutions contributed to rapid anti-influenza activity. POMxp (800 mg/ml polyphenols) at room temperature showed at least a 3 log reduction in the titers of influenza viruses PR8 (H1N1), X31 (H3N2), and a reassortant H5N1 virus derived from a human isolate. However, the antiviral activity was observed to be less against a coronavirus and reassortant H5N1 influenza viruses derived from avian isolates. The loss of influenza infectivity was frequently accompanied by loss of hemagglutinating activity.

The peel powder treatment decreased Ab binding to viral surface molecules, suggesting some coating of particles, but this did not always correlate with loss of infectivity. Electron microscopic analysis indicated that viral inactivation by polyphenols was primarily a consequence of virion structural damage. The direct anti-influenza activity of pomegranate polyphenols was substantially modulated by small changes in envelope glycoproteins.

Anti-Bacterial Properties

The antibacterial properties of pomegranate extracts have focused on human oral bacteria [Menezes, et al. 2006; Vasconcelos, et al. 2003; Sastravaha, et al. 2003; 2005]. However, several *in vitro* assays demonstrated its bacteriocidal activity against several highly pathogenic and sometimes antibiotic-resistant organisms. The synergistic effect of a pomegranate methanolic extract with five antibiotics on 30 clinical isolates of methicillin-resistant *Staphylococcus aureus* (MRSA) and methicillin-sensitive *S. aureus* were evaluated [Machado, et al. 2002]. Antibiotics tested were chloramphenicol, gentamicin, ampicillin, tetracycline, and oxacillin. Although synergistic activity between the pomegranate extract and all five antibiotics was noted in the *S. aureus* isolates, synergy with ampicillin was the most pronounced. A combination increased the lag time to bacterial growth by three hours (over that of ampicillin alone) and was also bacteriocidal as evidenced by a 72.5% reduction in methicillin-sensitive organisms and a 99.9% reduction in MRSA. The ellagitannin and punicalagin are thought to be the primary constituent responsible for the observed anti-bacterial effects [Braga, et al. 2005]. Another organism that can commonly infected by humans is enterohemorrhagic *Escherichia coli* (*E. coli* O157:H7), which can cause diarrhoea, haemorrhagic colitis, thrombocytopenic purpura, and haemolytic uremic syndrome. Pomegranate and seven other medicinal plant extracts were tested for *in vitro* activity against *E. coli* O157:H7 and it was observed that ethanolic pomegranate peel extract showed most effective extracts against *E. coli* O157: H7 [Voraavuthikunchai and Limsuwan, 2006].

Neurological Effects

Effects on Learning and Memory in Rats

Adiga et al. [2010] evaluated potential memory enhancing effect of pomegranate peel extract on rats. There was a definite trend of memory improvement by peel extracts with more marked improvement on spatial learning tendency and long term memory than on retention capacity.

Effects on Brain Injury

Neonatal hypoxic-ischemic (HI) brain injury in severely preterm, very low birth-weight infants has been a major cause of infant illness and death [Huang and Castillo, 2008], and it has been associated with an increase in reactive oxygen species [Gulcan et al. 2005]. Two studies in which pregnant mice were given pomegranate juice in drinking water revealed the neonatal offspring, when subjected to experimentally-induced HI brain injury, had significantly less brain tissue loss (64% decrease) and significantly decreased hippocampal caspase-3 activity (84% decrease) as compared to neonates [Loren et al. 2005; West et al. 2007]. These results suggested pomegranate juice has an antioxidant-driven neuroprotective effect conferred from mother to neonate.

Effects on Alzheimer's Disease

The neuroprotective properties of pomegranate polyphenols were evaluated in an animal model of Alzheimer's disease. Transgenic mice with Alzheimer's like pathology treated with pomegranate juice showed 50% less accumulation of soluble amyloid-beta and less hippocampal amyloid deposition than mice consuming sugar water. This evidence suggested that pomegranate juice may be neuroprotective. Animals also exhibited improved learning of water maze tasks and swam faster than control animals [Hartman, et al. 2006].

Effects on Hypertension

Hypertension (HTN) is the most common disease found in patients in primary care. It eventually requires medication if lifestyle modifications are not initiated or do not control the blood pressure well enough. The majority of patients would prefer not to be medicated to manage their disease, and hypertension can be found to be comorbidity along with diabetes and cardiovascular diseases. Side effects, forgetfulness and patient ignorance are multiple reasons for the hesitancy to begin drug management. Pomegranate juice is rich in tannins, possesses anti-atherosclerotic properties, has anti-aging effects, and potent anti-oxidative characteristics. Pomegranate juice consumption was found to have a potential effect in reducing systolic blood pressure, inhibits serum ACE activity, and is convincingly a heart-healthy fruit. Pomegranate juice consumption inhibited serum angiotensin converting enzyme activity and reduced systolic blood pressure.

A clinical trial demonstrated that pomegranate juice inhibited serum ACE and reduced systolic blood pressure in hypertensive patients. It was observed that 70% of the patients tested experienced a 36% decrease in serum ACE activity and a small, but significant, five-percent decrease in systolic blood pressure [Aviram et al. 2001]. Chronic administration of pomegranate juice extract (100 mg/kg and 300 mg/kg; for 4 weeks) reduced the mean arterial blood pressure and vascular reactivity changes to various catecholamine and also reversed the biochemical changes induced by diabetes and Ang II. Pomegranate juice treatment also caused a significant decrease in levels of thiobarbituric acid reactive substances (TBARS) in kidney and pancreas while activities of enzymes superoxide dismutase (SOD), catalase (CAT), and glutathione reductase (GSH) were significantly increased. The cumulative concentration response curve (CCRC) of Ang II was shifted towards right in rats treated with pomegranate juice using isolated strip of ascending colon. In histopathological examination, pomegranate juice treatment prevented the tubular degenerative changes induced by diabetes. The results suggested that the pomegranate juice extract could prevent the development of high blood pressure induced by Ang II in diabetic rats probably by combating the oxidative stress induced by diabetes and Ang II and by inhibiting serum ACE activity [Mohan et al. 2009].

Effects on Carotid Artery Stenosis

In a small, long-term three year study, the effect of pomegranate juice on Artery Stenosis was investigated. Results showed that the control subjects demonstrated a mean nine-percent increase in intima-media thickness (IMT) of left and right carotid arteries during the first year. Conversely, those consumed pomegranate juice had reduced IMT at 3, 6, 9, and 12 months ranging from 13 percent at three months to 35 percent at one year compared to baseline values [Aviram et al. 2004]. Most serum biochemistry parameters remained unchanged by pomegranate juice consumption over the first year, with the exception of triglyceride concentrations, which increased 16 percent but remained in the normal range.

Serum lipid peroxidation in subjects consuming pomegranate juice was significantly reduced by 59 percent after one year, and levels of LDL-associated lipid peroxides were also decreased by as much as 90 percent after six months of supplementation. Body mass index did not change in treated subjects but systolic blood pressure was reduced an average of 16 percent during the three-year study [Aviram et al. 2004]. In addition to previous reports of reduced systolic blood pressure [Aviram, et al., 2001], and inhibition of lipid peroxidation [Aviram et al. 2000], it was demonstrated that pomegranate juice consumption (via antioxidative mechanisms) significantly reduced various aspects of IMT in patients with severe carotid artery stenosis.

Dental Functions

Topical applications of pomegranate preparations have been found to be particularly effective for controlling oral inflammation, as well as bacterial and fungal counts in periodontal disease and Candida-associated denture stomatitis.

Dental Plaque

A hydroalcoholic extract of *Punica granatum* fruit (HAEP) was investigated for antibacterial effect on dental plaque microorganisms. Results showed that HAEP decreased the number of colony forming units (CFU) of dental plaque bacteria by 84% in comparison to chlorhexidine (79% inhibition), but significantly better than the control rinse (11% inhibition). Both HAEP and chlorhexidine were effective against *Staphylococcus, Streptococcus, Klebsiella*, and *Proteus* species, as well as *E. coli*. The ellagitannin, puni-calagin, is thought to be the fraction responsible for pomegranate's antibacterial activity [Menezes et al. 2006].

Periodontal Disease

A preliminary and follow-up study by a group of Thai researchers investigated the effect of biodegradable chips impregnated with Asiatic Pennywort and pomegranate pericarp on periodontal disease in 20 patients with gum pocket depths of 5-8 mm. All treatment sites demonstrated a trend toward decreasing plaque and significant improvements were noted in pocket depth and attachment level at three months compared to placebo [Sastravaha et al., 2003]. In the follow-up study, 15 patients who had completed standard periodontal therapy but still had pocket depths of 5-8 mm were implanted with the same medicated chips. Significant improvement was noted in all re-measured parameters and confirmed by significant decreases in IL-1β and IL-6 at three and six months as compared to baseline [Sastravaha et al. 2005].

Denture Stomatitis

The primary etiologic factors for denture stomatitis are poor oral hygiene, inflammation from ill-fitting dentures, and Candida infection, [Bergendal and Lsacsson, 1983; Lacopino and Wathen, 1992] which manifest as swelling, pain, burning in the mouth, and aphthous ulcers [Allen, 1992]. In a randomized, double-blind study of 60 subjects (ages 19-62) with candidiasis confirmed via mycologic examination, the effect of a gel-based pomegranate bark extract (GPBE) was evaluated for its effect on healing of oral lesions and direct fungicidal effect. Interestingly, despite randomized subject placement, there were three times more subjects with good oral hygiene scores in the miconazole group compared to the GPBE group, possibly accounting for the superior results observed by miconazole therapy. The initial step in the development of Candida denture stomatitis is adherence of organisms to dentures and the miconazole gel was stickier than GPBE, contact duration of miconazole was longer. A stickier GPBE might result in improved clinical response [Vasconcelos et al. 2003].

Effects on Male Fertility

A study using a rabbit model of arteriogenic erectile dysfunction (ED) measured the effect of pomegranate juice concentrate on intracavernous blood flow and penile erection. Azadzoi et al. (2005) found eight weeks administration of 3.87 mL pomegranate juice concentrate (112 μmol polyphenols per day has significantly increased intracavernous blood flow and smooth muscle relaxation, probably due to its antioxidant effect on enhanced NO preservation and bioavailability. Another research on rats demonstrated that pomegranate juice consumption improved epididymal sperm concentration, spermatogenic cell density, diameter of seminiferous tubules, and sperm motility, and decreased the number of abnormal sperm compared to control animals. An improvement in antioxidant enzyme activity in both rat plasma and sperm was also noted [Turk et al. 2008].

Effect on Obesity

Administrating pomegranate flower extract on obese hyperlipidemic mice for five weeks caused significant decreases in body weight, percentage of adipose pad weights, energy intake, and serum cholesterol, triglyceride, glucose, and total cholesterol/HDL ratios. Decreased appetite and intestinal fat absorption were also observed, improvements mediated in part by inhibition of pancreatic lipase activity also observed [Lei et al. 2007].

Ingredients for Food Stability

Pomegranate has been used in different trial studies as a stabilizing agent. Most studies found pomegranate to be effective in controlling oxidative changes during processing and storage of some foods, such as meat and chicken. The antioxidant potential of pomegranate juice, rind powder extract and butylated hydroxyl toluene (BHT) in cooked chicken patties during refrigerated storage were evaluated.

Results showed that pomegranate or rind powder at a level of 10 mg equivalent phenolics/100 g meat would be sufficient to protect chicken patties against oxidative rancidity for periods longer than the most commonly used synthetic antioxidant like BHT [Naveena et al. 2008]. In another study; effects of salt, kinnow and pomegranate fruit by-product powders on color and oxidative stability of raw ground goat meat stored at 4 ± 1 °C was evaluated. Results suggested that these powders showed potential to be used as natural antioxidants to minimize the auto-oxidation and salt induced lipid oxidation in raw ground goat meat [Devatkal and Naveena, 2010].

To overcome the disadvantages of using synthetic anti-oxidants in meat products, an investigation was carried out to evaluate the anti-oxidant effect of extracts of fruit by-products viz., kinnow rind powder, pomegranate rind powder and pomegranate seed powder in goat meat patties. Total phenolics content, DPPH radical scavenging activity and effect of these extracts on instrumental color, sensory attributes and TBARS values during storage (4 ± 1°C) of goat meat patties were evaluated.

The extracts of fruits by product powders exhibited potential to be used as natural anti-oxidants in meat products [Devatkal and Borah, 2010]. Furthermore, antioxidant efficacy of pomegranate peel extracts has been utilized in stabilization of sunflower oil [Iqbal et al. 2008].

TOXICITY

Pomegranate has been widely consumed by people in many different cultures for thousands of years, largely without unpleasant incident, and thus is considered generally safe. However, some toxicity is known, and undoubtedly, more remains to be discovered. Consumption of decoction of the tree bark, and to a lesser extent, husk of the fruit, may cause severe acute gastric inflammation due to the presence of both tannins and alkaloids [Squillaci and Di Maggio, 1946].

Whole fruit extracts have showed to cause congestion of internal organs and elevated creatinine *in vivo* [Vidal et al. 2003]. Pomegranate seed oil was found nontoxic to brine shrimp larvae [Fatope et al. 2002]; however showed allergic reactions [Lgea et al. 1991; Gaig et al. 1999; Hegde et al. 2002; Ghadirian, 1987; Ghadirian et al. 1992, Lansky and Newman, 2007].

Some studies suggested that the application of pomegranate peel extract, as a natural antioxidant in a food or drug for humans, should be treated with caution [Okonogi et al. 2007]. There have been conflicting reports regarding the toxicity of punicalagin [Scalbert et al. 2002; Doig et al. 1990; Filippich et al. 1991; Lu et al. 2007]. Some of these studies have recommended that there is a need of in-depth in vitro and in vivo studies to determine the biological properties of punicalagin.

CONCLUSION

Significant research has been conducted mostly during the past two decades on pomegranate. Studies aimed to investigate the health functional properties of pomegranate on animals and humans.

Researchers have concluded that pomegranate may be used as a prevention of certain types of diseases and illnesses. Although positive results have come up in favor for the pomegranate; further long-term research is needed to be accomplished. Clinical trials on the health functionality and toxicity limits of different parts of pomegranate fruits are scarce, thus; further investigation is needed to be adopted seeking comprehensive results.

REFERENCES

Adams S., Seeram P., Aggarwal B., Takada Y., sand D., Heber D., 2006. Pomegranate juice, total pomegranate ellagitannins, and punicalagin suppress inflammatory cell signalling in colon cancer cells. *J. Agric. Food Chem.* 54: 980-985.

Albrecht M., Jiang W., Kumi-Diaka J., 2004. Pomegranate extracts potently suppress proliferation, xenograft growth, and invasion of human prostate cancer cells. *J. Med. Food.* 7: 274-283.

Adiga S., Trivedi P., Ravichandra V., Debashree D., Mehta F., 2010. Effect of *Punica granatum* peel extract on learning and memory in rats. *Asian Pacific Journal of Tropical Medicine.* 687-690.

Afaq F., Adhami M., Ahmad N., Mukhtar H., 2002. Botanical antioxidants for chemoprvention of photocarcinogenesis. *Front Biosci.* 7: d784-792.

Afaq F., Malik A., Syed D., Maes D., Matsui S., Mukhtar H., 2005. Pomegrante fruit extract modulates UV-B-mediated phosphorylation of mitogen-activated protein kinases and activation of nuclear factor factor kappa B in normal human epidermal keratinocytes. *Photochem. Photobiol.* 81:38-45.

Afaq F., Saleem M., Krueger C., Reed J., Mukhtar H., 2005. Anthocyanin- and hydrolyzable tannin-rich pomegranate fruit extract modulates MAPK and NF-kappaB pathways and inhibits skin tumorigenesis in CD-1 mice. *Int. J. Cancer.* 113: 423–433.

Afaq F., Hafeez B., Syed N., Kweon H., Mukhtar H., 2006. Oral feeding of pomegranate fruit extract inhibits early biomarkers of UVB radiation-induced carcinogenesis in SKH-1 hairless mouse epidermis. *J. Invest. Dermatol.* 126:141.

Ajaikumar K., Asheef M., Babu B., Padikkala J., 2005. The inhibition of gastric mucosal injury by *Punica granatum L.* (pomegranate) methanolic extract. *Journal of Ethnopharmacology.* 96: 171–176.

Allen C., 1992. Diagnosing and managing oral candidiasis. *J. Am. Dent. Assoc.* 123:77-78, 81-82.

Althunibat O., Al-Mustafa A., Tarawneh K., Khleifat K., Ridzwan B., Qaralleh H., 2010. Protective role of *Punica granatum L.* peel extrat against oxidative damage in experimental diabetic rats. *Process Chemistry.* 45: 581-585.

Anonymous, 2009. Project document for a regional standard for Pomegranate. Prepared by the Islamic Republic of Iran. Joint FAO/WHO Food Standards programme Coordinating Committee for the Near East. Fifth Session, Tunisia.

Aslam M., Lansky E., Varani J., 2006. Pomegranate as a cosmeceutical source: pomegranate fractions promote proliferation and procollagen synthesis and inhibit matrix metalloproteinase 1-production in human skin cells. *J. Ethnopharmacology.* 103 (3): 311-318.

Aviram M., Dornfeld L., 2001. Pomegranate juice consumption inhibits serum angiotensin converting enzyme activity and reduces systolic blood pressure. *Atherosclerosis.* 158: 195–198.

Aviram M., Dornfeld L., Rosenblat M., Volkova N., Kaplan M., Coleman R., Hayek T., Presser D., Fuhrman B., 2000. Pomegranate juice consumption reduces oxidative stress, atherogenic modifications to LDL, and platelet aggregation: studies in humans and in atherosclerotic apolipoprotein E-deficient mice. *American Journal of Clinical Nutrition.* 71: 1062–1076.

Aviram M., Rosenblat M., Gaitini D., Nitecki S., Hoffman A., Dornfeld L., Volkova N., Presser D., Attias J., Liker H., Hayek T., 2004. Pomegranate juice consumption for 3 years by patients with carotid artery stenosis reduces common carotid intima-media thickness, blood pressure and LDL oxidation. *Clinical Nutrition.* 23: 423–433.

Azadzoi K., Schulman R., Aviram M., Siroky M., 2005. Oxidative stress in arteriogenic erectile dysfunction: prophylactic role of antioxidants. *Journal of Urology.* 174: 386–393.

Bergendal T., Isacsson G., 1983. A combined clinical; mycological and histological study of denture stomatitis. *Acta Odontol. Scand*.41:33-44.

Braga L., Leite A., Xavier K., 2005. Synergic interaction between pomegranate extract and antibiotics against *Staphylococcus aureus. Can. J. Microbiol.* 51:541-547.

Cam M., Hisil Y., 2010. Pressurised water extraction of polyphenols from pomegranate peels. *J. Food Chemistry.* 123: 878–885.

Cerda B., Espin J., Parra S., Martinez P., Tomas-Barberan, F., 2004. The potent in vitro antioxidant ellagitannins from pomegranate juice are metabolised into bioavailable but poor antioxidant hydroxy-6Hdibenzopyran-6-one derivatives by the colonic microflora of healthy humans. *European Journal of Nutrition.* 43: 205–220.

Cerda B., Soto C., Albaladejo M., Martinez P., Sanchez-Gascon F., Tomas- Barberan, F., Espin J., 2006. Pomegranate juice supplementation in chronic obstructive pulmonary disease: a 5-week randomized, double-blind, placebo-controlled trial. *European Journal of Clinical Nutrition.* 60: 245–253.

Devatkal S., Narsaiah K., Borah A., 2010. Anti-oxidant effect of extracts of kinnow rind, pomegranate rind and seed powders in cooked goat meat patties. *Meat Science.* 85: 155–159.

Devatkal S., Naveena M., 2010. Effect of salt, kinnow and pomegranate fruit by-product powders on color and oxidative stability of raw ground goat meat during refrigerated storage. *Meat Science.* 85: 306-311.

Dobrovolskaia M., Kozlov S., 2005. Inflammation and cancer: when NF-kappaB amalgamatesthe perilous partnership. *Current Cancer Drug Targets.* 5: 325-344.

Doig A., Williams D., Oelrichs P., Baczynskyi L., 1990. *J. Chem. Soc. Perkin Trans.* 2317.

Elswijk D., Schobel U., Lansky E., Irth H., Greef J., 2004. Rapid dereplication of estrogenic compounds in pomegranate (*Punica grantum*) using on-line biochemical detection coupled to mass spectrometry. *Phytochemistry.* 65:233-241.

Esmaillzadeh A., Tahbaz F., Gaieni I., 2006. Cholesterol-lowering effect of concentrated pomegranate juice consumption in type II diabetic patients with hyperlipidemia. *Int. J. Vitam. Nutr. Res.* 76:147-151.

Esmaillzadeh A., Tahbaz F., Gaieni I., Alavi-Majd H., Azadbakht L., 2004. Concentrated pomegranate juice improves lipid profiles in diabetic patients with hyperlipidemia. *Journal of Medicinal Food.* 7: 305–308.

Fatope O., Al Burtomani K., Takeda Y., 2002. Monoacylglycerol from *Punica grantum* seed oil. *Journal of Agricultural and Food Chemistry.* 50:357-360.

Fengchun H., Liu X., Chen H., 1997. Medicine for treatment of infectious oral diseases. *Chinese Patent.* 1145793A.

Filippich J., Zhu J., Asalami T., 1991. Hepatotoxic and nephrotoxic principles in *Terminailia oblongata. Res. Vet. Sci.* 50: 170-177.

Fridovich, 1995. Superoxide radical and superoxide dismutase, *Annual Review of Biochemistry*. 64: 97–112.

Gaig P., Bartolome B., Lieonart R., Garcia-Ortega P., Palacios R., Richart C., 1999. Allergy to pomegranate (*Punica granatum*). *Allergy*. 54: 287–288.

Garcea G., Dennison R., Steward P., Berry P., 2005. Role of inflammation in pancreatic carcinogenesis and the implications for future therapy. *Pancreatology*. 5:514-529.

Ghadirian P., 1987. Food habits of the people of the Caspian Littoral of Iran in relation to esophageal cancer. *Nutrition and Cancer*. 9: 147–157.

Ghadirian P., Ekoe M., Thouez P., 1992. Food habits and esophageal cancer: an overview. *Cancer Detection and Prevention*. 16: 163–168.

Gil I., Tomas-Barberan A., Hess-Pierce B., Holcroft M., Kader A., 2000. Antioxidant activity of pomegranate juice and its relationship with phenolic composition and processing. *Journal of Agricultural and Food Chemistry*. 48: 4581–4589.

Gulcan H., Ozturk I., Arslan S., 2005. Alterations in antioxidant enzyme activities in cerebrospinal fluid related with severity of hypoxic ischemic encephalopathy in newborns. *Biol Neonate*. 88:87-91.

Guo C, Wei J, Yang J, et al., 2008. Pomegranate juice is potentially better than apple juice in improving antioxidant function in elderly subjects. *Nutr. Res.* 28:72-77.

Guo, Yang, Wei, Li, Xu, Jiang, 2003. Antioxidant activities of peel, pulp and seed fractions of common fruits as determined by FRAP assay. *Nutrition Research*. 23: 1719-1726.

Halvorsen L., Holte K., Myhrstad C., Barikmo I., Hvattum E., Remberg F.,Wold B., Haffner K., Baugerod H., Andersen F., Moskaug O., Jacobs Jr. R., Blomhoff R., 2002. A systematic screening of total antioxidants in dietary plants. *Journal of Nutrition*. 132: 461–471.

Hartman R., Shah A., Fagan A., 2006. Pomegranate juice decreases amyloid load and improves behaviour in a mouse model of Alzheimer's disease. *Neurobiol. Dis.* 24: 506-515.

Hayden R., Tyagi C., 2004. Vasa vasorum in plaque angiogenesis, metabolic syndrome, type 2 diabetes mellitus, and atherosclerosis: a malignant transformation. *Cardiovascular Diabetology*. 3: 1.

He L., Xu H., Liu X., He W., Yuan F., Hou Z., Gao Y, 2010. Identification of phenolic compounds from pomegranate (Punica granatum L.) seed residues and investigation into their antioxidant capacities by HPLC–ABTS+ assay. *Food Research International*: In press.

Hegde L., Mahesh A., Venkatesh P., 2002. Anaphylaxis caused by mannitol in pomegranate (*Punica granatum*). *Allergy and Clinical Immunology International*. 14: 37–39.

Hernandez F., Melgarejo P., Tomas-Barberran F., Artes F., 1999. Evolution of juice anthocyanins during ripening of new selected pomegranate (*Punica granatum*) clones. *European Food Research and Technology*. 210: 39–42.

Hu W., 1997. Skin health inflammatory inucta and producing process there of Chinese Patent 1156617A.In pancreatic carcinogenesis and the implications for future therapy. *Pancreatology*. 5: 514–529.

Huang B., Castillo M., 2008. Hypoxic-ischemic brain injury: imaging findings from birth to adulthood. *Radiographics*. 28: 417-439.

Huang T., Yang Q., Harada M., 2005. Pomegranate flower extract diminishes cardiac fibrosis in Zucker diabetic fatty rats: modulation of cardiac endothelin-1 and nuclear factor-kappaB pathways. *J. Cardiovasc. Pharmacol.* 46: 856-862.

Iacopino A., Wathen W., 1992. Oral candidal infection and denture stomatitis: a comprehensive review. *J .Am. Dent. Assoc.* 123:46-51.

Igea J., Cuesta J., Cuevas M., Elias L., Marcos C., Lazaro M., Compaired J., 1991. Adverse reaction to pomegranate ingestion. *Allergy.* 46: 472–474.

Iqbal S., Haleem S., Akhtar M., Zia-ul-Haq M., and Akbar J., 2008. Efficiency of pomegranate peel extracts in stabilization of sunfloweroil under accelerated conditions. *Food Research International.* 41: 194–200.

Jassim A., 1998. Antiviral or antifungal composition comprising an extract of pomegranate rind or other plants and method of use. U.S. Patent 5840308.

Jurenka J., 2008. Therapeutic Applications of Pomegranate (*Punica granatum* L.): A Review. *Alternative Medicine Review.* 13(2): 128-144.

Kaplan M., Hayek T., Raz A., Coleman R., Dornfeld L. and Vaya M., 2001. Pomegranate juice supplementation to atherosclerotic mice reduces macrophage lipid peroxidation, cellular cholesterol accumulation and development of atherosclerosis. *Journal of Nutrition.* 131: 2082–2089.

Kapoor M., Clarkson N., Sutherland A., Appleton I., 2005. The role of antioxidants in models of inflammation: emphasis on L-arginine and arachidonic acid metabolism. *Inflammopharmacology.* 12: 505–519.

Karageuzyan G., 2005. Oxidative stress in the molecular mechanism of pathogenesis at different diseased states of organism in clinics and experiment. *Current Drug Targets, Inflammation, and Allergy.* 4:85-98.

Kaur G., Jabbar Z., Athar M., Alam S., 2006. *Punica granatum* (pomegranate) flower extract possesses potent antioxidant activity and abrogates Fe-NTA induced hepatotoxicity in mice. *Food and Chemical Toxicology.* 44:984-993.

Kelawala S., Ananthanarayan L., 2004. Antioxidant activity of selected foodstuffs. *International Journal of Food Science and Nutrition.* 55:511-516.

Khan N., Afaq F., Kweon H., kim M., mukhtar H., 2007. Oral consumption of pomegranate fruit extract inhibits growth and progression of primary lung tumors in mice. *Cancer Res.* 67: 3475-3482.

Khan N., Hadi N., Afaq F., Syed N., Kweon H., mukhtar H., 2007. Pomegranate fruit extract inhibits prosurvival pathways in human A549 lung carcinoma cells and tumor growth in athymic nude mice. *Carcinogenesis.* 28: 163-173.

Khennouf, S., Gharzouli, K., Amira, S., Gharzouli, A., 1999. Effects of Quercusilex and *Punica granatum* polyphenols against ethanol-induced gastric damage in rats. *Pharmazie.* 54: 75–76.

Kohno H., Suzuki R., Yasui Y., Hosokawa M., Miyashita K., Tanaka, T., 2004. Pomegranate seed oil rich in conjugated linolenic acid suppresses chemically induced colon carcinogenesis in rats. *Cancer Sci.* 95: 481-486.

Kvasenkov I., Lomachinski A., Goren'Kov S. 1999. Method of producing beverages on juice base. *Russian Patent.* 2129396C1.

Lansky P., Newman A., 2007. Punica granatum (pomegranate) and its potential for prevention and treatment of inflammation and cancer. *Journal of Ethnopharmacology.* 109: 177-206.

Lansky E., Jiang W., Mo H., 2005. Possible synergistic prostate cancer suppression by anatomically discrete pomegranate fractions. *Invest. New Drugs.* 23:11-20.

Larrosaa M., González-Sarríasb A., Yáñez-Gascónb M., Selmab M., Azorín-Ortuñob M, Totia S., Tomás-Barberánb F., Dolaraa P., Espína J., 2010. Anti-inflammatory properties of a pomegranate extract and its metabolite urolithin-A in a colitis rat model and the effect of colon inflammation on phenolic metabolism. *Journal of Nutritional Biochemistry.* 21: 717–725.

Lei F., Zhang X., Wang W., 2007. Evidence of antiobesity effects of the pomegranate leaf extract in high-fat diet induced obese mice. *Int. J. Obes. (Lond)* 31:1023-1029.

Liu G., Xu X., Hao Q., Gao Y., 2009. Supercritical CO2 extraction optimization of pomegranate (pumice granatum L) seed oil using response surface methodology. *LWT-Food Science and Technology.* 42: 1491-1495.

Loren D., Seeram N., Schulman R., Holtzman D., 2005. Maternal dietary supplementation with pomegranate juice is neuroprotective in an animal model of neonatal hypoxic-ischemic brain injury. *Pediatr. Res.* 57:858-864.

Lu J., Wei Y., Yuan Q., 2007. Preparative separation of punicalagin from pomegranate husk by high-speed countercurrent chromatography. *Journal of Chromatography. B.* 857: 175-179.

Machado T., Leal I., Amaral A., 2002. Antimicrobial ellagitannin of *Punica granatum* fruits. *J. Braz. Chem. Soc.* 13: 606-610.

Madrigal-Carballob, Rodriguezb, Kruegera, Dreherc M., Reeda D., 2009. Pomegranate (Punica granatum) supplements: Authenticity, antioxidant and polyphenol composition. *Journal of Functional Foods.* 1: 3 2 4 –3 2 9.

Malik A., Mukhtar H., 2006. Prostate cancer prevention through pomegranate fruit. *Cell Cycle.* 5:371-373.

Malik A., Afaq F., Sarfaraz S., 2005. Pomegranate fruit juice for chemoprevention and chemotherapy of prostate cancer. *Proc. Natl. Acad. Sci. U. S. A.* 102:14813-14818.

Maskan M., 2006. Production of pomegranate (Punica granatum L.) juice concentrate by various heating methods: colour degradation and kinetics. *Journal of Food Engineering.* 72: 218–224.

Menezes S., Cordeiro L., Viana G., 2006. *Punica granatum* (pomegranate) extract is active against dental plaque. *J. Herb. Pharmacother.* 6:79-92.

Mertens-Talcott, U., Jilma-Stohlawetz, P., Rios, J., Hingorani, L., Derendorf, H., 2006. *J. Agric. Food Chem.* 54: 8956.

Miguel G., Fontes C., Antunes D., Neves A., Martins D., 2004. Anthocyanin concentration of "Assaria" pomegranate fruits during different cold storage conditions. *Journal of Biomedicine and Biotechnology.* 338–342.

Mirdehghan S., Rahemi M., 2007. Seasonal changes of mineral nutrients and phenolics in pomegranate *(punica grantum L.)* fruit. *Scientia Horticulturae*, 111: 120-127.

Mohan M., Waghulde H., Kasture S., 2009. Effect of pomegranate juice on Angiotensin II-induced hypertension in diabetic Wister rats. *Phytother. Res.* In press.

Mori-Okamoto J., Otawara-Hamamoto Y., Yomato H., Yoshimwa H., 2004. Pomegranate extract improves a depressive state and bone properties in menopausal syndrome model ovariectomized mice. *Journal of Ethnopharmacology.* 92:93-101.

Murad H., Shellow W., 2001. Pomegranate extract both orally ingested and topically applied to augment the SPF of sunscreens. *Cosmet. Dermatol.* 14: 43.

Naveena M., Sen R., Vaithiyanathan S., Babji, Y., Kondaiah, N., 2008. Comparative efficacy of pomegranate juice, pomegranate rind powder and BHT in cooked chicken patties. *Meat Science.* 80: 1304-1308.

Noda Y., Kaneyuka T., Mori A., Packer L., 2002. Antioxidant activities of pomegranate fruit extract and its anthocyanidins: delphinidin, cyanidin, and pelargonidin. *Journal of Agricultural and Food Chemistry*. 50: 166–171.

Ohshima H., Tazawa H., Sylla B., Sawa T., 2005. Prevention of human cancer by modulation of chronic inflammatory processes. *Mutatation Research.* 591: 110–122.

Okonogi S., Duangrat Anuchpreeda S., Tachakittirungrod S., Chowwanapoonpohn S., 2007. Comparison of antioxidant capacities and cytotoxicities of certain fruit peels. *Food Chemistry.* 103: 839–846.

Panichayupakaranant P., Tewtrakul S., Yuenyongsawad S., 2010. Antibacterial, anti-inflammatory and anti-allergic activities of standardised pomegranate rind extract. *Food Chemistry.* 123: 400–403.

Pantuck A., Leppert J., Zomorodian N., 2006. Phase II study of pomegranate juice for men with rising prostate-specific antigen following surgery or radiation for prostate cancer. *Clinical Cancer Research.* 12: 4018-4026.

Parmar H., Kar A., 2007. Protective role of *Citrus sinensism*, *Musa paradisiacam* and *Punica granatum* peels against diet-induced atherosclerosis and thyroid dysfunction in rats. *Nutrition Research.* 27:710-718.

Perez-Vicente A., Gil-Izquierdo A., Garcia-Viguera C., 2002. In vitro gastrointestinal study of pomegranate juice phenolic compounds, anthocyanins and Vitamim C. *Journal of Agricultural and Food Chemistry.* 50: 2308–2312.

Plumb W., De Pascual-Teresa, S., Santos-Buelga C., Rivas-Gonzalo, J.C., Williamson G., 2002. Antioxidant properties of gallocatechin and prodelphinidins from pomegranate peel. *Redox Report.* 7: 41–46.

Sastravaha G., Yotnuengnit P., Booncong P., Sangtherapitikul P., 2003. Adjunctive periodontal treatment with *Centella asiatica* and *Punica granatum* extracts. A preliminary study. *J. Int. Acad. Periodontol.* 5:106-115.

Sastravaha G., Gassmann G., Sangtherapitikul P., Grimm W., 2005. Adjunctive periodontal treatment with *Centella asiatica* and *Punica granatum* extracts in supportive periodontal therapy. *J. Int. Acad. Periodontol.* 7:70-79.

Scalbert A., Morand C., Manach C., Cemesy C., 2002.Absorbtion and metabolism of polyphenols in the gut and impact on health. *Biomed. Pharmacother*. 56: 276-282.

Schubert, Y., Lansky, P., Neeman, I., 1999. Antioxidant and eicosanoid enzyme inhibition properties of pomegranate seed oil and fermented juice flavonoids. *Journal of Ethnopharmacology.* 66: 11–17.

Seeram P., Adams S., Henning M., Niu Y., Zhang Y., Nair G., Heber D., 2004. *J. Nutri. Biochemistry.* 16:360.

Seeram P., Adams S., Henning M., Niu Y., Zhang Y., Nair G., Heber D., 2005. In vitro antiproliferative, apoptotic and antioxidant activities of punicalagin, ellagic acid and a total pomegranate tannin extract are enhanced in combination with other polyphenols as found in pomegranate juice. *Journal of Nutritional Biochemistry.* 16: 360–367.

Sestili P., Martinelli C, Ricci D., Fraternale D., Bucchini A., Giamperi L., 2007. Cytoprotective effect of preparations from various parts of *Punica granatum L.* fruits in oxidatively injured mammalian cells in comparison with their antioxidant capacity in cell free systems. *Pharmacol. Res.* 56:18-26.

Shanshan G., Qianchun D., Junsong X., Bijun X., Zhida S., 2007. Evaluation of Antioxidant Activity and Preventing DNA Damage Effect of Pomegranate Extracts by Chemiluminescence Method. *J. Agric. Food Chem.* 55: 3134-3140.

Singh P., Chidambara Murthy N., Jayaprakasha K., 2002. Studies on the antioxidant activity of pomegranate (Punica granatum) peel and seed extracts using in vitro models. *Journal of Agricultural and food Chemistry.* 50: 81-86.

Singh P., Chidambara Murthy N., Jayaprakasha K., 2002. Studies on the antioxidant activity of pomegranate (*Punica granatum*) peel and seed extracts using in vitro models. *Journal of Agricultural and Food Chemistry.* 50: 81–86.

Squillaci G., Di Maggio G., 1946. Acute morbidity and mortality from decoctions of the bark of *Punica Grantum.* Bolletino Societa Italiana Biologia Sperimentale. 1095–1096.

Stewart S., Jassim A., Denyer P., Newby P., Linley K., Dhir K., 1998. *J. Appl. Microbiol.* 84: 777-783.

Stowe C., 2010. The effects of pomegranate juice consumption on blood pressure and cardiovascular health. *Complementary Therapies in Clinical Practice.* In Press.

Sundararajana A., Ganapathya, R., Huana L., Dunlapb J., Webbyc R., Kotwald G., Sangster M., 2010. Influenza virus variation in susceptibility to inactivation by pomegranate polyphenols is determined by envelope glycoproteins. *Antiviral Research.* 88: 1-9.

Syed N., Malik A., Hadi N., Sarfaraz S., Afaq F., Mukhtar H., 2006. Photochemopreventive effect of pomegranate fruit extract on UVA-mediated activation of cellular pathways in normal human epidermal keratinocytes. *Photochem. Photobiol.* 82: 398-405.

Syed D., Afaq F., Mukhtar H., 2007. Pomegranate derived products for cancer chemoprevention. *Seminars in Cancer Biology.* 17: 377-385.

Tehranifara Zareia, Nematia Esfandiyaria, Reza Vazifeshenasb, 2010. Investigation of physico-chemical properties and antioxidant activity of twenty Iranian pomegranate (Punica granatum L.) cultivars. *Scientificae horticulturae.* 126: 180-185.

Tezcan , Gultekin-Ozguven , Diken, Ozcelik, Erim F., 2009. Antioxidant activity and total phenolic, organic acid and sugar content in commercial pomegranate juices. *Food Chemistry.* 115: 873–877.

Tomas-Barberan F., Gil M., Cremin P., Waterhouse A., Hess-Pierce B., Kader A., 2001. HPLC-DAD-ESI-MS analysis of phenolic compounds in nectarines, peaches, and plums. *Journal of Agricultural and Food Chemistry.* 49: 4748–4760.

Tous J., Ferguson L., 1996. Mediterranean fruits. *In: J. Janick (ed.), Progress in new crops.* 416-430.

Turk G., Sonmez M., Aydin M., 2008. Effects of pomegranate juice consumption on sperm quality, spermatogenic cell density, antioxidant activity, and testosterone level in male rats. *Clin. Nutr.* 27:289-296.

Vasconcelos L., Sampaio M., Sampaio F., Higino J., 2003. Use of *Punica granatum* as an antifungal agent against candidosis associated with denture stomatitis. *Mycoses.* 46: 192-196.

Vidal A., Fallarero A., Pena, B., Medina M., Gra B., Rivera F., Gutierrez Y., Vuorela, P., 2003. Studies on the toxicity of *Punica granatum* L. (Punicaceae) whole fruit extracts. *Journal of Ethnopharmacology.* 89: 295–300.

Voravuthikunchai S., Limsuwan S., 2006. Medicinal plant extracts as anti-*Escherichia coli* O157:H7 agents and their effects on bacterial cell aggregation. *J. Food Prot.* 69:2336-2341.

Wang R., Xie W., Zhang Z., Xing D., Ding Y., Wang W., Ma C., Du L., 2004. Bioactive compounds from the seeds of *Punica granatum* (Pomegranate). *Journal of Natural Products.* 67: 2096–2098.

West T., Atzeva M., Holtzman D., 2007. Pomegranate polyphenols and resveratrol protect the neonatal brain against hypoxic-ischemic injury. *Dev. Neurosci.* 29: 363-372.

Xu J., Guo J., Yang J., Wei Y., Li F., Pang W., Jiang G., Cheng S., 2005. Intervention of antioxidant system function of aged rats by giving fruit juices with different antioxidant capacities. *Zhonghua Yu Fang Yi Xue Za Zhi.* 39: 80–83.

Yoshimura M., Waranabe Y., Kasai K., Yamakoshi J, Koga T., 2005. Inhibitory effect on an ellagic acid-rich pomegranate extract on tyrosinase activity and ultraviolet-induced pigmentation. *Biosci. Vitaminol.* (Tokyo) 69: 2368-2373.

Yunfeng L., Changjiang G., Jijun Y., Jingyu W., Jing X., Shuang C., 2006. Evaluation of antioxidant properties of pomegranate peel extract in comparison with pomegranate pulp extract. *Food.* 96: 2, 254-260.

Zaid A., Afaq F., Khan N., Mukhtar H., 2007. Protective effects od pomegranate derived products on UVB-induced DNA damage, PCNA expression and MMPs in human reconstituted skin. *J. Invest. Dermatol.* 127: S143.

Zhan B., 1995. Multi-function vagina suppository. *Chinese Patent.* 1103789.

Zhang J., Zhan B., Yao X., Song J., 1995. Antiviral activity of tannin from the pericarp of *Punica granatum* L. against genital herpes virus in vitro. *Zhongguo Zhongyao Zazhi.* 20: 556-558.

In: Natural Products and Their Active Compounds … ISBN: 978-1-62100-153-9
Editors: M. Essa, A. Manickavasagan, and E. Sukumar

Chapter 24

OATS AND HEALTH BENEFITS

Somasundaram Mathan Kumar*
Sultan Qaboos University, Oman

ABSTRACT

The global burden of non-communicable diseases (NCDs) has been an increasing public health concern. Non-communicable diseases (NCDs) account for 60% of the global mortality. Of the 35 million deaths in attributable to NCDs annually, about 80% are in low- and middle-income countries (LMIC). From 2006 to 2015, deaths due to NCDs are expected to increase by 17%.

Dietary approaches hold promise as effective and preventive interventions for NCDs. Dietary factors represent the most potent determinants of metabolic health and have been shown to mitigate specific physiological mechanisms in various disease conditions. Recent epidemiological and experimental studies suggest that healthy dietary pattern, including increased consumption of natural products, whole grains, fruits can favorably influence the risk of NCDs.

Increase in dietary fiber (DF) intake has been recommended for a healthy life. Cereals and cereal products, particularly from whole grains forms staple diet in most countries. Moreover, in addition to being a source of carbohydrates whole grains especially wheat, rice, and oats, provides protein and essential fatty acids and possesses unique and beneficial combinations of many micronutrients, polyphenolics and DF.

Among the whole grains, Oats had gained a unique position, because of its diverse health benefits to the humans. This chapter mainly deals with the health benefits of oats in relation to the prevention of NCDs.

Keywords: Oats, NCDs, Health benefits, Whole grains, CVDs, Type II diabetes, Coeliac disease

* Correspondence: Senior Veterinarian; Department of Animal and Veterinary Sciences, College of Agricultural and Marine sciences, Sultan Qaboos University, Sultanate of Oman. Email: Mathan@squ.edu.om

INTRODUCTION

Cardiovascular diseases, Diabetes (Type-II), and Cancer, and Obesity are major health concerns in the western nations especially in United States and developing countries all around the world. [1] Cardiovascular diseases (CVDs) are the leading cause of death in the world, accounting for 30% of deaths globally, and the estimated number of deaths due to CVDs worldwide was 17.5 million in 2005 and expected to increase by 20 million in 2015. WHO estimates that between 2000 and 2030, the world population will increase by 37% and the number of people with diabetes will increase by 114%. [2, 3, 4, 5] Blood cholesterol is a major risk factor for CVDs. Similarly calorie-enriched diet intake and lack of exercise have been causing a world-wide surge of obesity and insulin resistance are also identified as the major risk factors. [2, 6].

Today we find ourselves with an epidemic of over nutrition and obesity not only in the western world but also in the emerging economies with prevalence of "super size" food portions and "all you can eat" buffets that promotes over eating. [7], causing substantial increases in morbidity and early mortality in the population with these non communicable diseases. [9] While we derive more than enough calories from food choices, either has lost many of the natural preventive substances or their inclusion is inadequate such as cereal fibers that are proven to provide health benefits against these.[2,6, 8, 9]

Since past few decades, prevention of non-communicable diseases has become the top agenda for every international and national health authorities. Real challenges are in the prevention strategies for CVDs, diabetes and cancers as these globally burdening the population. Even the etiologies of these three diseases are complex, yet preventable. Dietary intervention is the first line of approach against these major health disorders and inclusion of plant based products has successfully proven to unveil the connection between diet and disease. [1, 2, 10.] Hippocrates- also called the "Father of Medicine" - stated "Let thy food be thy medicine and thy medicine be thy food."[11] The statement remains true and need for recollection even after 2500 years. Dietary consumption of whole-grain foods such as wheat, barley, oats, etc are associated with a decreased risk of several chronic diseases, with benefits attributed to their content of both macro and micro nutrients, fiber and phytochemicals [12].

The quest for answers continues as science uncovers the mysteries between health and nutrition for the emerging lifestyle related diseases i.e. NCDs. Even though research on Oats has been happening for the four decades, as the primary report of the cholesterol reduction by oats consumption in humans [13] ever since, oats has always been in the top of the order in nutritional research, clinical trials and constantly fueling the specialty of applied human nutrition, nevertheless, significant achievements occurred in the past two decades. At presently, research is finding more and more evidence that explains the relationship between dietary inclusion of oats and the health benefits attained through physiological responses to prevent or delay the illness against these NCDs.

OATS- A BACKGROUND

Oats are a crop of Mediterranean origin, not as old as wheat and barley but their domestication dates back to ancient times. Oats rank around sixth in the world cereal production statistics following wheat, maize, rice, barley and sorghum. [14]. World oat production is generally concentrated between Latitudes 35-65°N and 20-46°S and Russia, USA, Canada, Germany and Poland account for about 75 percent of the world's supply of grain, seed and industrial grade oats.[15] In 2005, the top five oats producers according to the UN Food and Agriculture Organization (FAO) were Russia with 5.1 million metric tons, Canada with 3.3 million metric tons, United States with 1.7 million metric tons, Poland with 1.3 million metric tons, and Finland with 1.2 million metric tons. Other top producers of oats include Australia, Germany, Belarus, China and the Ukraine. [14]

In many parts of the world oats are grown for use as grain as well as for forage and fodder either as green chaff, hay and straw. Livestock grain feed is still the primary use of oat crops, accounting for an average of around 74 percent of the world's total usage.[15] Oats for grain and forage or fodder are grown on over 1.8 million hectares in Canada and 0.8 million hectares in US.

The food industry in North America uses approximately 1.7 million tons of oats while a major share goes to livestock feed industries. Oats still remain an important grain crop for people in marginal economies throughout the developing world, and in developed economies for specialized uses. Oat grains are a good source of protein, fibre, and minerals. [14, 16]

Table 1. Taxonomic information of Oats (Obtained from [17])

Botanical name	Avena sativa
Kingdom	Plantae: plants
Subkingdom	Tracheobionta: vascular plants
Super division	Spermatophyta: seed plants
Division	Magnoliophyta: flowering plants
Class	Liliopsida: monocotyledons
Subclass	Commelinidae
Order	Cyperales
Family	Poaceae: grass family
Genus	Avena: oat
Species	A. sativa: common oat, A. byzantina, A. fatua, A. diffusa, A. orientalis

Oats and Health Claim

The Food and Drug Administration (FDA) had announced its decision on Jan 1997 to authorize the health claims on the association between soluble fiber from whole oats and a reduced risk of coronary heart disease (CHD). Further, the agency had concluded that the type of soluble fiber found in whole oats, i .e. β-glucan is primarily responsible for the claimed association. Their decision was based on a review of the evidence demonstrating that consumption of whole-oat sources (including oat bran, rolled oats, and whole oat flour) decreases total cholesterol (TC) and low-density lipoprotein cholesterol (LDL-C) concentrations. However in continuation, the health claim association between oats consumption and reduced risk of CHD was also approved by The Joint Health Claims

Initiative (JHCI) in UK in the year 2004 and Ministry of Health Malaysia in the year 2006. [18, 21.]

Whole grains contain all parts of the grain, the bran, the germ, and the endosperm. Whole grains are rich in nutrients and phytochemicals with known health benefits such as dietary fibre, antioxidants including trace minerals and phenolic compounds, phytoestrogens such as lignan, vitamins and minerals. Among the whole grains, oats had gained a unique position because of its diverse health benefits attained through favorable physiological responses to combat epidemically emerging NCDs. Dietary inclusion of oats provide beneficial effects to health because of its rich macro nutrients, micro nutrients, soluble fiber (β-glucans) and the recently discovered oat poly phenolics. This chapter details out the health benefits of oats in relation with CVDs, diabetes and celiac disease.

Oats and Cardiovascular Diseases

The cardiovascular system, which is composed of the heart and blood vessels, is essential for the distribution of oxygen, nutrients, and other critical components to all organs throughout the human's body. As the heart is the sole pump for the cardio vascular system, any disruption of its function can have critical consequences for the human's life.

Cardiovascular diseases (CVDs) are the leading cause of death in the world and that includes coronary or ischemic heart disease, cerebrovascular disease or stroke, hypertension, heart failure, and rheumatic heart disease. Coronary artery disease due to advanced athero-sclerosis is the major cause of death in the United States and in most Western countries. In contrast, Asian countries have disproportionately high morbidity and mortality from stroke compared with Western countries and rise in blood cholesterol is a major risk factor for CVDs. [2, 5]. Hypercholesterolemia is caused by increased concentrations of low-density lipoprotein cholesterol (LDL-C) and very low-density lipoprotein cholesterol (VLDL-C). Elevated triglycerides (TGs) result from elevated VLDL. High TG and greater LDL-C are predictors of increased cardiovascular risk. Oxidation of low density lipoproteins (LDL, the "bad cholesterol") is a key biochemical step in the development of cardiovascular diseases. High-density lipoprotein cholesterol (HDL-C) concentrations provide the opposite relationship, with increased blood concentrations of HDL-C predicting reduced risk. [19]. Epidemiological evidence indicates that a higher intake of oats is associated with a reduced risk of coronary heart disease, and this effect of has been attributed to its cholesterol-lowering effect and improvement of vascular endothelium through its fiber and antioxidant components.[8,20]

As was seen earlier, FDA had announced its food specific health claim decision on Jan 1997 to authorize the association between soluble fiber from whole oats and a reduced risk of CHD. The intent of this benchmark is to provide a high level of confidence in the validity of the relationship. Although it does not require unanimous and incontrovertible scientific consensus, it is meant to be a strong standard based on the totality of the science, like the preliminary report by de Groot et al in the year 1963, from then until the year 1997 with several other studies and evidence based review to unveil this relation, so, there are little likelihood of chances being this relation reversed by new data. [21]

Whole grain foods are known to exhibit positive protection against CVDs through its fiber and antioxidant components and the commonly consumed ones are: dark bread, whole-grain or bran breakfast cereals, bran, popcorn, oatmeal, wheat germ and brown rice.[3] Several large population cohort studies had concluded that higher intakes of whole-grain foods were associated with lower risks of CHD and this inverse association was independent of known coronary risk factors and these are detailed out as following, In a southern Californian population-based cohort of 859 men and women aged 50-79 years with a study design as 24-hour dietary fiber intake record had concluded that dietary fiber intake of 16 gm/ 24 hours or more provided better protection against ischemic heart disease than whose intake less than 16 gm/24 hours. Further the study added that 6 gm increment in daily fiber intake was associated with a 25% reduction in ischemic heart disease. [22]. In a prospective cohort of 75521 US women aged 38 - 63 years, who completed detailed food frequency questionnaires (FFQs) and were followed up for 12 years as part of the Nurses' Health Study concluded that higher intake of whole grain foods was associated with a lower risk of ischemic stroke among women, and the inverse relation was continuous throughout the study period and accounted about a 30%-40% lower risk of ischemic stroke among women.[23]. In1986, a total of 43757 US male health professionals 40 to 75 years of age, completed a detailed 131-item dietary questionnaire to measure usual intake of total dietary fiber and specific food sources of fiber in a six year follow up study and documented 0.59% (734 cases) of myocardial infarction as a form of fatal coronary heart disease. This study concluded that within the three main food contributors to total fiber intake (vegetable, fruit, and cereal), cereal fiber was most strongly associated with a reduced risk of total myocardial infarction.[24] In the Iowa Women's Health Study (n=34492) post menopausal women followed for 6 years, concluded that a greater intake of whole grain was associated with a reduced risk of CHD death.[25] A meta-analysis of 12 studies that were conducted between 1977 and 1999, shown regular intake of whole grain foods was associated with a 26% reduction in risk for CVDs. Whole grain foods positively influence a number of other CVD risk factors such as hypertension, diabetes and obesity through reduction of LDL-C and TGs, in addition they have favorable effects on fasting and postprandial serum lipoproteins. Higher intakes of whole grains are associated with increased sensitivity to insulin, and lower plasma insulin concentrations and these effects are possibly attributed as whole grains are the rich source of magnesium, fibers and vitamin E. Nevertheless, it is highly recommended that consumption of ≥3 servings of whole grains/day, but not the refined sources to decrease the risk of CHD by ≥30%, irrespective of other lifestyle behaviors. [3, 26]

Soluble fiber fraction of oats, in particular to the (1→3, 1→4) β-*D*-glucan component reduce total and LDL cholesterol is a sum of several effects. The effect is small within the practical range of intake,for example 3 g soluble fiber from oats (3 servings of oatmeal, 28 g each) can decrease total and LDL -C by <0.13 mmol/L. Such effects include soluble fibers bind bile acids or cholesterol during the intraluminal formation of micelles and the resulting reduction in the cholesterol content of liver cells leads to an up-regulation of the LDL receptors and thus increased clearance of LDL-C. Other suggested mechanisms include inhibition of hepatic fatty acid synthesis; by products of fermentation (production of short-chain fatty acids (SCFA) such as acetate, butyrate, and propionate). Both oats and barley contains β- glucan identically in structure but vary in their quantities, however, did not evince any dissimilarity in the reduction of LDL-C in the Syrian golden F1B hamsters and the demonstrated mechanism as β-glucan inhibits absorption of cholesterol from the gut by a significant increase in the excretion of fecal cholesterol and neutral sterols.[2, 27, 28]

Physiochemical properties such as molecular weight (MW) and solubility of β- glucan favorably influence the serum cholesterol reduction; higher the MW promotes increased gut viscosity that may prevent dietary cholesterol from reaching the intestinal epithelium. In addition, this gives increased satiety and delayed return of hunger i.e. "second meal effect" through the lowered postprandial glycemic and insulinemic response. Soluble fibers may decrease absorption of dietary cholesterol by altering the composition of the bile acid pool. Oat bran increased the portion of the total bile acid pool that was deoxycholic acid (DCA) and it has been noted to decrease the absorption of exogenous cholesterol in humans. Fermentation of fibers in the large intestine may also alter cholesterol metabolism by production of SCFA, among which butyrate was found to play a major role altering the cholesterol metabolism. The role other SCFA such as propionate and acetate remain inconclusive.[8, 29, 30]

It is likely that some food constituents, such as vitamins, trace elements, phenolic compoundsand phyto estrogens found in oats also affect CVDs risk and operate via pathways other than the lipid-regulating pathway. This conclusive thought is arrived at a study which measured the Intima media thickness (IMT) of common carotid arteries ultra sonographically and fiber intake by dietary recall. The study concluded that lowered risk was associated with improved endothelial function [31]. The spectrum of health benefits that comes with the consumption of oat and oat products does not limited to the presence of β- glucans, but also due to the presence of poly phenolics in the prevention of CVDs, especially with atherosclerosis.

Oat- Avenanthramides (Avn)

Avenanthramides (Avn) are the unique group of low-molecular weight soluble phenolic compounds found exclusively in oats. Avenanthramides 2c, 2p, and 2f were the most dominant forms found in the oats. These phytochemicals have a range of biological activities, including antiatherosclerotic, anti-inflammatory, and antioxidant effects /anti scavenging properties and are concentrated in the outer layers of oat kernel. Avenanthramide-2c (Avn-c) one of three major avenanthramides in oats, comprises about one-third of the total avenanthramide concentration in oat grain, and this avenanthramide has the highest antioxidant activity in vitro. Bioavailability of avenanthramides has been demonstrated in hamsters and recently in humans [32, 33, 34,35].

The proliferation of vascular smooth muscle cells (SMC) and impaired nitric oxide (NO) production are the key patho physiological processes in the initiation and development of atherosclerosis of arterial walls. Oxidation of LDL-C is a major risk for occurrence of atherosclerosis. Avn-c, one of the major avenanthramides inhibits the serum induced proliferation of vascular smooth muscle cells (SMC) and interacts synergistically with vitamin C to protect LDL during oxidation. [19, 20, 36]

As most phenolics are located in the bran layer of grains, oats, which are normally consumed as whole-grain cereal, can be a significant dietary source of these compounds [34] clearly suggesting preferring oats in its natural form for dietary inclusion.

Taken together, these data strongly support the potential health benefits of oat consumption in the prevention of CVDs beyond the benefits from their soluble fiber content but also through the oat Avns.

OATS AND DIABETES

Of the world population between the ages of 20 and 79 years, an estimated 285 million people, or 6.6%, have diabetes [37]. World Health Organization (WHO) estimates that between 2000 and 2030, the world population will increase by 37% and the number of people with diabetes will increase by 114%[4, 10, 38].The prevalence of type II diabetes is reaching epidemic proportions, perhaps more alarming, it is estimated that more than 57 million American adults have pre diabetes, defined by impaired glucose tolerance (IGT) or impaired fasting glucose (IFG), which places them at substantially in an increased risk for developing diabetes now or in the near future giving a great need for dietary intervention.[39, 40] A strikingly conservative estimate in Asia shows that India and China will remain the two countries with the highest numbers of people with diabetes by 2030[4,10.]. Additionally, among the top ten countries, four more are in Asian continent—Indonesia, Pakistan, Bangladesh, and the Philippines, justifying Asia as the major site of a rapidly emerging diabetes epidemic in the world[10]. Type II diabetes, once virtually unrecognized in adolescence, now emerged as a major health concern and this is entirely attributable to the child obesity epidemic, of particular concern, a prediabetic state, seems to be highly prevalent among severely obese children irrespective of ethnic background worldwide. The increased emergence of type II diabetes in children represents an ominous development of multi-systemic health disorders [41, 42, 43]. Additionally, people with diabetes are at a significantly higher risk of many forms of cancer and CVDs [21,44]. Finally, the economic burden due to diabetes at personal, societal, and national levels is huge as type II diabetes and its associated complications pose major health care burden worldwide and present many challenges to patients and health-care system[4, 10, 38].

Diabetes is typically divided into 2 major subtypes, as type I and type II. Type II Diabetes is the most common form (almost 95%) and obviated through complex interaction of genetic, dietary and lifestyle factors. Diabetes often results in complications such as macro vascular e.g. heart disease, stroke, limb amputation and micro vascular e.g. kidney failure, blindness [45]. Glucose intolerance, impaired glucose tolerance and insulin resistance are associated with obesity and may be preliminary steps in the progression to type II diabetes, and improved glycemic control by diet is the essential element for optimal management and to lessen the above mentioned long term complications [46, 47, 48].

Diabetes educators build on the evidence that type II diabetes can be prevented or delayed by body weight reduction, increased physical activity and/or by the use of selected medications among the population those with pre diabetes, the improvement of lipid profile or blood glucose control is a major challenge, as type II diabetes is a major cardiovascular disease risk factor.Worldwide clinical trials and population studies are evident of effective lifestyle interventions to prevent or delay the development of type II diabetes, among which dietary intervention is the key approach. Dietary intervention principally aims at low glycemic index foods and low fat/ high fiber diet and to accomplish this, plant based diets would be the preferred choice as they are nutrient dense and calorically dilute whilst providing the larger amounts of dietary fibers and therefore blood glucose levels and improvement of lipid profile of diabetic and pre-diabetic individuals can be moderated by using dietary fiber rich food such as oats [9, 10, 39, 40, 46, 49].

Cereals and cereal products, particularly from whole grains such as oats, wheat and rice, are the most important source of dietary fiber in the Western diet; Dietary recommendations of health organizations suggest consumption of three servings a day of whole grain foods; however, Americans generally fall below this standard. In addition whole grains provide protein, essential fatty acids and may have unique and beneficial combinations of many micronutrients, antioxidants, and phytochemicals. The key content, oat soluble fiber β-glucan, a non starch polysaccharide composed of β-(1→4)–linked glucose units separated every 2–3 units by β -(1→3)–linked glucose is a boon in diabetes control owing to the potentiality of glycemic control and lipid profile modification [50, 51, 52].

Prevention of type II diabetes by dietary intervention typically aims at improved glucose metabolism such as control of postprandial dietary peak of glucose, delaying or preventing the progression of impaired glucose tolerance to insulin resistance, control of obesity. Following are the research highlights which provide the insights of dietary inclusion of oats and the benefitting mechanisms by which above mentioned prime concerns are addressed in combating the diabetes.

It is urged that consumption of wholegrain rather refined grains substantially improves glucose tolerance and reduce insulin resistance recorded in a large cohort of 75 521 women aged 38 to 63years in a ten year follow-up study,[50] similar evidence was obtained large cohort of 2286 men and 2030 women aged 40–69 years in a ten year follow-up study[53] and also in a large cohort of 42898 men in a twelve year follow-up study [54]. Three of these large cohort studies had concluded that higher intake of whole-grain foods was associated with lower risk of type II diabetes; whereas higher intake of refined grain was related to increased risk and the inverse association was independent of known risk factors. The mechanisms attributed by which grains may improve glucose metabolism and delay or prevent the progression of impaired glucose tolerance to insulin resistance are related with the physical properties such as particle size, amount and type of fiber, presence of various individual antioxidants, and phytochemicals, as well as interactions among them. Whole grains are generally digested and absorbed slowly because of their physical form and high content of viscous fiber; compared with whole-grain products, refined grains more than double the glycemic and insulinemic responses and unable to maintain the glucose homeostasis. In addition magnesium which is found in the whole grains that can act as calcium antagonist and can promote insulin sensitivity. In the milling process, the outer bran layer of whole grains is removed and the original physical form is disrupted to make the remaining starchy endosperm more easily digestible and this typically signifies the importance of dietary inclusion of whole grains rather the refined [47, 48, 50, 54, 55].

Peripheral arterial disease (PAD) is a major cause of morbidity, might arise out of complication of type II diabetes and severe disease can lead to limb amputation, for a possible prevention increasing amounts cereal fiber in the diet is suggested, as an inverse association between cereal fiber intake and PAD risk was revealed. This is possibly through food sources of cereal fiber also contains magnesium and chromium which are associated with improved insulin sensitivity [56].

Higher the recommendation for the fiber than the American Diabetes Association (ADA) suggested is proven in a clinical study that incorporated a diet containing moderate amounts of fiber (total, 24 g; 8 g of soluble fiber and 16 g of insoluble fiber), as recommended by the ADA and a high fiber diet (total, 50 g; 25 g of soluble fiber and 25 g of insoluble fiber) containing unfortified foods and concluded that high intake of dietary fiber, particularly of the soluble type, improves glycemic control, decreases hyperinsulinemia, and lowers plasma lipid concentrations in patients with type II diabetes. Therefore, dietary guidelines for patients with diabetes should emphasize an overall increase in dietary fiber through the consumption of unfortified foods such as whole oats [57].

Concerning glucose metabolism, the beneficial metabolic effects of oat β-glucan are closely linked to the β-glucan-induced increased viscosity of the meal bolus, which delays and/or reduces carbohydrates absorption [49].

Clinical studies with isolates of oat β-glucan have demonstrated that the glycemic response is regulated not by the dose of the polysaccharide but by its molecular weight and concentration in solution (extractability); this is because the glycemic response is controlled, in relation to the luminal viscosity of the fiber in the gut· Therefore, increase in intestinal viscosity due to high molecular weight β-glucan is important for achieving the positive effect of β-glucan on the peak blood glucose [29, 30, 58] which sheds a clear thought on the need for understanding the processing technologies that are keen to develop modern day fiber supplements.

Viscous fibers, including β-glucan in oat bran and its amount can favorably enhance postprandial sensations of satiety as well as decrease feelings of hunger thereby moderating both postprandial carbohydrate and lipid metabolism. This is possibly attributed through the increased nutrient-stimulated postprandial secretion of the anorexigenic hormones and greater suppression of the orexigenic hormone and also by delaying the gastric emptying rate and intestinal transit time to increase greater satiety and delay the return of hunger through viscosity of oat β- glucan [59, 60].

Dietary fiber (DF) and resistant starch (RS) from carbohydrates are fermented in the colon by the bacterial flora releases short chain fatty acids, (SCFA) mainly acetic, propionic, and butyric acids and gases (e.g. hydrogen) which typically happens after oats ingestion mainly helps in both reduction of LDL cholesterol and the glycemic response which shall be attributed through the formation of butyric acid and also at the small intestine by increased level of bile acid excretion and at large intestine increased level of bile acid concentration [61]. Any dietary inclusion however possessing the health benefits that shall agreeably palatable for it to deliver the desired benefits, in that fitting oat bran which is high in insoluble fiber and the soluble fiber β-glucan has been tested by several studies had concluded long-term acceptance of oat bran concentrate products was good and use of oat bran concentrate bread/bread products as a dietary staple was feasible,[62,63] however, an increased level of β-glucan in oat extract did produce some abdominal discomfort and increased flatulence [64].

Health Benefits of Oats on Celiac Disease

Celiac disease (CD) is a permanent intolerance to specific storage proteins in wheat (gliadin), barley (hordein) and rye (secalin), which are collectively called 'gluten'. Ingestion of gluten causes damage to the small intestinal mucosa by an autoimmune mechanism in genetically susceptible individuals and its prevalence is 1% in populations of Caucasoid descent [65,66]. In the United States, National Institute of Health (NIH) estimates more than 2 million people actually suffer from the disease, which would correspond to 1 in 133 people [66,67].Coeliac disease can occur at any age, may peak at 50's and females are more commonly affected than males [68]. Affected individuals with celiac disease have genetic markers on chromosome 6p21, called class II human leukocyte antigen (HLA), specifically HLA-DQ2 and HLA-DQ8 [69]. This gluten sensitive enteropathy (GSE) is clinically characterized by malabsorption and causing a typical histological lesion in the small intestine. The prototypical signs are oral ulcers, weight loss, diarrhea, fatigue, and abdominal bloating. Other signs such as iron deficiency anemia, folate deficiency, and osteopenic bone disease. As sequelae malignancy such as small intestinal lymphoma may occur. This disease is closely related to dermatitis herpetiformis [68]. The disease is caused by an inappropriate immune response triggered by dietary gluten proteins and the response is controlled by $CD4^+$ T cells. These T cells are specific for proline- and glutamine-rich gluten peptides. T cells may react with tissue transglutaminase (the principal component of the endomysium auto antigen), and set in motion a series of inflammatory events that result in the characteristic coeliac mucosal lesion. Histological identification such as finding villous atrophy of duodenal biopsy and serological evidence of antibodies to endomysium and gliadin are the definitive clinical diagnostic protocols [68, 69, 70, 71]. Medical nutrition therapy (MNT) is the only accepted treatment for celiac disease (CD) and strict adherence of gluten- free (GF) diet (GFD) for life is the preferred and only available "drug of choice" [65, 68, 69,70,72, 73].

Non-gluten containing cereals would be a valuable contribution to the gluten-free human diet. Among the cereals wheat, rye, and barley are harmful to persons with celiac. In contrast, the prolamins of corn (zein) and rice (orzenin) are considered harmless. In the past, oats were considered to be toxic to individuals with celiac disease and were not allowed in a gluten-free diet. However several studies were undertaken to test the oat prolamin,i.e. ave-nin, as a safe diet inclusion for coeliac patients [70, 72, 73, 74, 75]. In a study involved with ten adult coeliac disease patients as each patient consumed 50 g of oats (as porridge) daily for 12 weeks, evidence of immunological stimulation such as serology and biopsy are considered as markers of disease activation. Such evidence includes lymphocyte infiltration of the surface epithelium and the production of antibodies to endomysium and gliadin, where as in patients those who had gluten micro challenge relapsed with signs, clearly suggesting oats cereal is neither toxic nor immunogenic in coeliac disease. Seven of the patients have continued to take the same quantity of oats for more than 12 months without adverse effect [70]. Results from one study indicated that oats does not induce cellular or humoral immunological responses within 12 months in adults with CD and continued for a period of five years and the reason for non reactivity possibly due to the absence of certain amino acid sequences found in wheat gliadin, but not in oat ave-nin, and concluded even long term use of moderate amounts of oats included in a gluten free diet in adult patients with CD is safe. Additionally commented if allowed, most patients with CD preferred some oats in their diet [76]. By evidence oats appear to be safe for use by most individuals with celiac disease, but their inclusion in a gluten-free diet is limited by potential contamination with gluten during milling and processing, as commercial lots of oat flakes and flour frequently are contaminated with gluten grains, [69, 73, 77] and the contamination levels of gluten varied between 1.5 ppm and 400 ppm from a single bag and the source was suspected as barley not wheat [72,75]. The oats that are pure and uncontaminated with other gluten-containing grains is safe and a quantity for adults, up to 70 g and for children up to 25 g per day are tolerable [17, 65]. A subset of patient population had gastrointestinal (GI) discomfort as an exaggerated sensitivity to oats, but not related to CD. However patients with coeliac disease wishing to consume a diet containing oats should therefore receive regular follow-up, including small bowel biopsy, and this caution strongly aimed at the potential risk of malignancy and the risk may even be increased in patients who consume small amounts of gluten (and by extension moderate amounts of oats if contaminated with other gluten grains) [72, 74, 75, 78]. To produce pure, uncontaminated oats, the manufacturer must have a dedicated system, including fields, harvesting, production, storage, transportation, manufacturing equipment and a production plant [17,65]. It is essential that people with celiac disease must read all food labels to ensure the gluten-free status of a food item. This process has become easier since the enactment of the Food Allergen Labeling and Consumer Protection Act of 2004 in the United States requires that all food products manufactured after January 1, 2006, be clearly labeled to indicate the presence of any of the top eight food allergens such one as Wheat but the same does exclude the cereal barley and rye. Likewise in Canada, the pure oats or the products that are made from these oats must have passed a Canadian Food Inspection Agency (CFIA) field inspection both visual and chemical. In Europe, EC-Regulation 41/2009 came into force on the content and labeling of foods for individuals with CD as oat products containing less than 20 ppm gluten are now allowed to be sold as gluten-free since January 2009. Nevertheless an additional knowledge of hidden sources of gluten/prolamins that may be found in the ingredients of many processed foods in the form of additives, stabilizers, thickeners, flavorings, extracts, emulsifiers, hydrolyzed textured vegetable proteins, and certain ground spices is needed. This constitutes the most compelling reason to advocate strict adherence to the diet and possibly need for guidance from skilled dietician on a regular basis [65, 66, 79]. Despite the arising doubts about inclusion of oats in the diet of patients of CD, oats add variety, taste, satiety, dietary fiber, and other essential nutrients to the diets and may help alleviate the relative monotony of strict gluten free diet and may improve the quality of life [68, 69].

CONCLUSION

With strong totality of science, dietary inclusion oats can elicit diversified health benefits through the favorable physiological responses that keep the check points in prevention of NCDs. It is highly suggestive with the present knowledge oat nutritional research, that consumption of oats and oat-based products should be encouraged as part of an overall lifestyle medicine approach for the prevention of CVDs and diabetes [21].

As mentioned earlier in the chapter, that prevention of NCDs is the prime agenda for every national and international health authorities and the real challenges are vested in formulating the prevention strategies specific at a regional and at a national level. It is urged that, health advocacy groups work closely with national health authorities to promote the consumption of natural food products and they need to emphasize diets with increased fibre such as oats as part of a healthy lifestyle. Additionally to promote such higher consumption they should be able to propagate and publish traditional recipes with slight modifications [80] i.e. Developing newer oat recipes with (regional and national) traditional diet choices as an attempt against fast food eating habits.

The major practical translation of nutrition research to public health consists of identifying the foods that can potentially influence health and defining optimal dietary recommendations aimed to prevent disease and to promote optimal health [81]. Equating this to oats, that it is been identified with the bioactive nutrients such as β-glucans and Avns that can potentially influence the health and optimal dietary recommendations of three grams of soluble fiber advocated in FDA's approval of the health claim association between consumption of oats and reduced risk of CVDs.

However, what lies in the future is the role of biomedical agriculture is to identify the cultivars of any food crops with an added advantage of providing increased health benefits.[1] In a similar fitting for oats, such future oat cultivars can possess the increased amounts of bio active components such as β-glucans and Avns that can exponentially influence the risk factors associated with NCDs, whilst retaining the same quantities of oat consumption/dietary inclusion. Even though the large body of research evidence that is available to elucidate that dietary inclusion of oats can greatly benefit in combating type II diabetes worldwide, nutritional research on oats in Asian population is limited.

Understandably interventions that work in some societies may not work in others, because social, economic, and cultural forces influence diet and exercise [39] likewise, both population based studies and clinical trials involving oats or its bio active components are urgently called for in Asian population as NCDs are epidemically emerging in this population.

To debate, the present era research knowledge clearly suggest the variations in terms of expression and predisposing factors for NCDs among Asian population, especially the diabetes associated risk factors such as metabolic syndrome and insulin resistance and occurrence of higher rates of cerebrovascular disease (stroke) instead of coronary heart disease [5, 10, 43] compared against with the rest of the world. Such studies of future will be able to coordinate cross-cultural/ethnic involvement and will be extremely useful in defining gene-environment interactions.

It is been hypothesized that the emerging morbidity and mortality due to CVDs is that the world population has been experiencing during recent decade is due in part to the higher frequency of deleterious alleles that predispose certain ethnic groups sensitive to the influence of environmental CVD risk factors [81]. Therefore, elucidating such ethnic-specific genetic markers will be a great stride for efficacious prevention of NCDs in countries undergoing Westernization of lifestyles, specifically to Asian countries needless to say about the weight of the issue, as half of the world's population lives in Asia.

References

[1] Thompson, H.J. (2008). Biomedical Agriculture: A New Approach to Developing Human Health Optimized Staple Food Crops [abstract] In: *International Oat Conference,* June 28 - July 2, 2008, Minneapolis, Minnesota. p.12.

[2] Brown L, Rosner B, Willett W.W, Sacks F.M. (1999). Cholesterol-lowering effects of dietary fiber: a meta-analysis. *Am. J. Clin. Nutr. 69*, 30-42.

[3] Anderson, J.W. Hanna, T.J. Peng, X. and Kryscio, R.J. (2000). Whole Grain Foods and Heart Disease Risk. *J. Am. Coll. Nutr. 19*,291–99.

[4] Wild, S. Roglic, G. Green, A. Sicree, R. King, G. (2004). Global prevalence of diabetes. Estimates for the year 2000 and projections for 2030. *Diabetes Care. 27*, 1047–52.

[5] Hong.Y. (2009) Burden of Cardiovascular Disease in Asia: Big Challenges and Ample Opportunities for Action and Making a Difference. *Clinical Chemistry. 55(8)*, 1450–52.

[6] Cai. S. Huang, C. Ji, B. Zhou, F.Wise, M.L. Zhang, D.et al. (2011). In vitro antioxidant activity and inhibitory effect, on oleic acid-induced hepatic steatosis, of fractions and sub fractions from oat (*Avena sativa* L.) ethanol extract. *Food Chemistry. 124,* 900–05.

[7] Obesity Action Coalition: "Understanding Childhood Obesity" Available from: URL:http://www.obesityaction.org/educationaltools/brochures/uoseries/UCO_brochure.pdf Accessed on: 16.02.2011.

[8] Queenan, K.M. Stewart, M.L. Smith, K.N, Thomas, W. Fulcher, R.G. Slavin, J.L. (2007). Concentrated oat beta-glucan, a fermentable fiber, lowers serum cholesterol in hypercholesterolemic adults in a randomized controlled trial. *Nutr. J. 6,* 6–6.

[9] Jacobs, D.R. Jr, Haddad, E.H. Lanou, A.J. and Mark J Messina, M.J. (2009) Food, plant food, and vegetarian diets in the US dietary guidelines: conclusions of an expert panel. *Am. J. Clin. Nutr. 89(suppl)*:1549–52.

[10] Ramachandran, A. Wan Ma, R.C. Snehalatha, C. (2010) Diabetes in Asia. *Lancet. 375,* 408–418.

[11] Saunders, KK. The Vegan Diet as Chronic Disease Prevention: Evidence Supporting the New Four Food Groups. Ist edn. Brooklyn, NY11231: Lantern Books; 2003.

[12] Chen, C-Y. Milbury, P.E. Collins, F.W. and Blumberg, J.B. (2007). Avenanthramides are bio available and have antioxidant activity in humans after acute consumption of an enriched mixture from oats. *J. Nutr. 137,* 1375-1382.

[13] De Groot, A.P. Luyken, R. and Pikaar, N.A. (1963) Cholesterol-lowering effect of rolled oats. *Lancet. 2,* 303–304..

[14] Reynolds, S.G. Background to Fodder Oats Worldwide. In: Suttie JM and Reynolds SG (eds) *Fodder Oats: a world overview*. Rome: Plant Production and Protection Series, No. 33, FAO; 2004; 6-11.

[15] Stevens, E. J. Armstrong, K. W. Bezar, H.J. Griffin, W. B. and J. G. Hampton. Fodder oats: an overview. In: Suttie JM, and Reynolds SG (eds) *Fodder Oats: a world overview.* Rome: Plant Production and Protection Series, No. 33, FAO; 2004; 11-18.

[16] Fraser, J. andMcCartney, D. 2004. Fodder oats in North America. In: Suttie JM, and Reynolds SG (eds) *Fodder Oats: a world overview.* Rome: Plant Production and Protection Series, No. 33, FAO; 2004; 19-36.

[17] Butt, M. S. Nadeem, M.T. Khan, M.K. Shabir, R. Butt, M.S. (2008). Oat: unique among the cereals. *Eur. J. Nutr. 47(2),* 68-79.

[18] US Department of Health and Human Services, Food and Drug Administration. Health claims: oats and coronary heart disease—final rule. (1997) *Fed. Regist. 62, 3583*-3601.

[19] Chen, J. Huang, X.F. (2009). The effects of diets enriched in beta-glucans on blood lipoprotein concentrations. *J. Clin. Lipidol. 3,* 154–58.

[20] Nie, L. Wise, M.L. Peterson, D. M. Meydani, M. (2006). Avenanthramide, a polyphenol from oats, inhibits vascular smooth muscle cell proliferation and enhances nitric oxide production. *J. Atherosclerosis.* 186*(2),* 260-66.

[21] Andon, M. B. and Anderson, J.W. (2008). The Oatmeal-Cholesterol Connection: 10 Years Later. *Am. J. Lifestyle Med. 2,* 51-57.

[22] Khaw, K.T. and Barrett-Connor, E.B. (1987). Dietary fiber and reduced ischemic heart disease mortality rates in men and women: a 12-year prospective study. *Am. J. Epidemiol. 126(6),* 1093-1102.

[23] Liu, S. Manson, J.E. Stampfer, M .J. Rexrode, K.M. Hu, F. B. Rimm, E. B. et al. (2000). Whole Grain Consumption and Risk of Ischemic Stroke in Women: A Prospective Study. *JAMA. 284,* 1534-1540.

[24] Rimm, E.B. Ascherio, A. Giovannucci, E. Spiegelman, D. Stampfer, M.J. Willett, W.C. (1996). Vegetable, fruit, and cereal fiber intake and risk of coronary heart disease among men. *JAMA. 275,* 447-451.

[25] Jacobs, D.R. Jr, Meyer, K.A. Kushi, L.H. Folsom, A.R. (1998). Whole-grain intake may reduce the risk of ischemic heart disease death in postmenopausal women: the Iowa Women's Health Study. *Am. J. Clin. Nutr.* 68, 248–57.

[26] Anderson, J.W. (2004). Whole grains and coronary heart disease: the whole kernel of truth. *Am. J. Clin. Nutr.* 80(6), 1459-1460.

[27] Knudsen, K.B. Jensen, B.B. and Inge Hansen, I. (1993). Oat bran but not a β-glucan-enriched oat fraction enhances butyrate production in the large intestine of pigs. *J. Nutr. 123, 1235*-1247.

[28] Delaney, B. Nicolosi, R.J. Wilson, T.A. Carlson, T. Frazer, S. Zheng, G.H.et al. (2003) β-glucan fractions from barley and oats are similarly antiatherogenic in hyper cholesterolemic Syrian golden hamsters. *J. Nutr. 133,* 468–495.

[29] Wood, P.J. (2007) .Cereal β-glucans in diet and health. *J. Cereal Sci. 46,*230-238.

[30] Wood P.J. (2010). Oat and Rye β-glucan: properties and function. *Cereal Chem. 87(4),* 315-330.

[31] Wu, H. Dwyer, K.M. Fan, Z. Shircore, A. Fan, J. and James H Dwyer, J. H. (2003). Dietary fiber and progression of atherosclerosis: the Los Angeles Atherosclerosis Study. *Am. J. Clin. Nutr. 78(6),* 1085-1091.

[32] Mattila, P. Pihlava, J.M. Hellstro, J. M. (2005) Contents of phenolic acids, alkyl- and alkenyl resorcinols, and avenanthramides in commercial grain Products. *J. Agric. Food Chem. 53,* 8290-8295.

[33] Peterson, D.M.2001. Oat antioxidants. *J. Cereal Sci.,* 33(2): 115-129.

[34] Chen CY, Milbury PE, Kwak HK, Collins FW, Samuel P, Blumberg JB. Avenanthramides and phenolic acids from oats are bioavailable and act synergistically with vitamin C to enhance hamster and human LDL resistance to oxidation. *J. Nutr.* 2004; 134:1459–66.

[35] Chen C-Y, Milbury PE, Li T, O'Leary J, and Blumberg JB. (2005).Antioxidant capacity and bioavailability of oat avenanthramides. *FASEB J. 19,* A1477.

[36] Nie, L. (B), Wise, M. Peterson, D. Meydani, M. (2006). Mechanism by which avenanthramide-c, a polyphenols of oats, blocks cell cycle progression in vascular smooth muscle cells. *Free Radic. Biol. Med.* 41,702-708.

[37] International Diabetes Federation. IDF Diabetes Atlas [article online]. 4th ed. Brussels, Belgium: International Diabetes Federation; 2009. Available at: http://www. diabetesatlas.org/. Accessed: 13.02. 2011.

[38] Weickert, M.O. Möhlig, M. Schöfl, C. Arafat, A.M. Otto, B. Koebnick, C. (2006). Cereal fiber improves whole-body insulin sensitivity in overweight and obese women. *Dia. Care.* 29 , 4775-780.

[39] Knowler, W.C. Barrett-Connor, E. Fowler, S.E. et al. (2002). Reduction in the incidence of type 2 diabetes with lifestyle intervention or metformin. *N. Engl. J. Med. 346,* 393-403.

[40] Primary Prevention of Type 2 Diabetes-AADE Position Statement: American Association of Diabetes Educators (2009) *The Diabetes Educator. 35,* 57- 59.

[41] Goran, M.I. (2001). Metabolic precursors and effects of obesity in children: a decade of progress, 1990–1999. *Am. J. Clin. Nutr. 73,158*-171.

[42] Ebbeling, C.B. Pawlak, D. B. Ludwig, D.S. (2002). Childhood obesity: public-health crisis, common sense cure. *Lancet. 360,* 473–482.

[43] Ganie, M.A. (2010). Metabolic syndrome in Indian children - An alarming rise. *Indian J. Endocr. Metab. 14,* 1-2.

[44] Giovannucci, E. Harlan, D.M. Archer, M. C. Bergenstal, R.M. Gapstur, S.M. Habel, L.A. et al. Diabetes and Cancer: A Consensus Report. (2010). *CA Cancer J. Clin.* 60, 207-221.

[45] WHO Publication. Avoiding heart attacks and strokes: don't be a victim - protect yourself. [2005]. Available from: URL:http://www.who.int/cardiovascular_diseases/en/ Accessed on 13.01.2011.

[46] Wolever, T.M.S. Jenkins, D.J.A. Jenkins, A.L., Josse, R.G. (1991). The glycemic index: Methodology and clinical implications. *Am J Clin Nutr, 54,* 846-854.

[47] Hallfrisch, J. and Behall, K.M. (2000). Mechanisms of the effects of grains on insulin and glucose response. *J. Am. Coll. Nutr. 19,*320 S-325S.

[48] Skerrett, P.J. and Willett, W.C. (2010). Essentials of Healthy Eating: A Guide. *J. Midwifery Womens Health. 55,* 492-501.

[49] Cugnet-Anceau, C. Nazare, J-A. Biorklund, M. Le Coquil, E. Sassolas, A. Sothier, M. et al. (2010). A controlled study of consumption of β-glucan-enriched soups for 2 months by type 2 diabetic free-living subjects. *Br. J. Nutr. 103,* 422–428.

[50] Liu, S. Manson, J. E. Stampfer, M. J. Hu, F. B. Giovannucci, E. Colditz, G. A. et al. (2000). A Prospective Study of Whole-Grain Intake and Risk of Type 2 Diabetes Mellitus in US Women. *Am. J. Public Health. 90,* 1409-1415. 2000; 90:1409–1415)

[51] Drzikova, B. Dongowski, G. and Gebhardt, E. Dietary fibre-rich oat-based products affect serum lipids, microbiota, formation of short-chain fatty acids and steroids in rats. *Br. J. Nutr. 94,* 1012–1025.

[52] Naumann, E. B van Rees, A. Önning, G. Öste, R. Wydra, M. and Mensink, R.P. β-Glucan incorporated into a fruit drink effectively lowers serum LDL-cholesterol concentrations: *Am. J. Clin. Nutr. 83,* 3601-3605.

[53] Montonen, J. Knekt, P. Järvinen, R. Aromaa, A. and Reunanen, A. (2003). Whole-grain and fiber intake and the incidence of type 2 diabetes. *Am. J. Clin. Nutr. 77,*622–629.

[54] Fung, T.T. Hu, F.B. Pereira, M.A. Liu, S. Stampfer, M.J. Colditz, G.A. et al.(2002). Whole-grain intake and the risk of type 2 diabetes: a prospective study in men *Am. J. Clin. Nutr. 76(3),* 535-540.

[55] McCarty, M.F. (2005). Magnesium may mediate the favorable impact of whole grains on insulin sensitivity by acting as a mild calcium antagonist. *Med. Hypotheses. 64(3),* 619-627.

[56] Merchant, A.T. Hu, F.B. Spiegelman, D. Walter C. Willett, W.C. Rimm, E.B. and Ascherio, A. (2003). Dietary fiber reduces peripheral arterial disease risk in men. *J. Nutr. 133,* 3658–3663.

[57] Chandalia, M. Garg, A. Johann, D.L. Bergmann, K.V. Grundy, S. M. Brinkley, L.J. (2000). Beneficial effects of high dietary fiber intake in patients with type 2 Diabetes mellitus. *N. Engl. J. Med. 342,* 1392-1398.

[58] Tosh, S. M. Brummer, Y. Wolever, M. S. and Wood P.J. (2008). Glycemic response to oat bran muffins treated to vary molecular weight of β-Glucan. *Cereal Chem., 85(2),* 211–217.

[59] Hlebowicz, J. Wickenberg, J. Fahlström, R. Björgell, O. Almér, L.O and Darwiche G. (2007). Effect of commercial breakfast fibre cereals compared with corn flakes on postprandial blood glucose, gastric emptying and satiety in healthy subjects: a randomized blinded crossover trial. *Nutr. J. 6,* 22.

[60] Juvonen, K.R. Purhonen, A.K. Marttila, M.S. La¨ hteenma¨ ki, L. Laaksonen, D.E. Herzig, K-H. et al. (2009).Viscosity of oat bran-enriched beverages influences gastrointestinal hormonal responses in healthy humans. *J. Nutr., 139,* 461–466.

[61] Nilsson, A.C. O¨ stman, E.M. Knudsen, E.B. Holst, J.J. and Bjo¨ rck, I.M. (2010). A cereal-based evening meal rich in indigestible carbohydrates increases plasma butyrate the next morning: *J. Nutr., 140,* 1932–1936.

[62] Pick, M.E. Hawrysh, Z.J. Gee, M.I. Toth, E. Garg, M. L. and Hardin, R.T. (1996). Oat bran concentrate bread products improve long-term control of diabetes: A pilot study. *J. Am. Diet Assoc. 96,* 1254-1261.

[63] Jenkins, A. L. Jenkins, D. J. Zdravkovic, U. Würsch, P. and Vuksan, V. (2002). Depression of the glycemic index by high levels of β-glucan fiber in two functional foods tested in type 2 diabetes. *Eur. J. Clin. Nutr., 56(7),* 622-628.

[64] Behall, K. M. Scholfield, D. J. Sluijs, A. and Hallfrisch, J. (1998). Breath hydrogen and methane expiration in men and women after oat extract consumption: *J. Nutr., 128,* 79–84.

[65] Rashid, M. Butzner, D. Burrows, V. Zarkadas, M. Case, S. Molloy, M.et al. (2007). Consumption of pure oats by individuals with celiac disease: A position statement by the Canadian Celiac Association: *Can. J. Gastroenterol. 21(10),* 649-651.

[66] The struggle against Celiac Disease-Development of safe foods for Celiac patients – A multidisciplinary approach (2010). Food for thought: News letter [Issue number: 3, May 2010]: Available from URL: http://www.nfia.com/fft/201005/article7.php. Accessed on 12.02.2011.

[67] Fasano, A. Berti, I. Gerarduzzi, T. et al. (2003). Prevalence of celiac disease in at-risk and not-at-risk groups in the United States. *Archives of Internal Medicine. 163(3),* 268-292.

[68] Feighery, C. (1999). Coeliac disease. *BMJ. 319(7204),* 236–239.

[69] Niewinski, M.M. (2008). Advances in celiac disease and gluten-free diet. *J. Am. Diet Assoc., 108,* 661-672.

[70] Srinivasan, U. Leonard, N. Jones, E. et al. (1996). Absence of oats toxicity in adult coeliac disease. *BMJ. 313,* 1300–1.

[71] Sollid, L.M. and Lundin, K.E. (2009). Diagnosis and treatment of celiac disease. *Mucosal Immunol. 2,* 3–7.

[72] Kupper, C. (2005). Dietary guidelines and implementation for celiac disease. *Gastroenterology. 128,* 121–127.

[73] Mujico, J.R. Mitea, C. Gilissen, L.J. de Ru, A. van Veelen, P. Smulders, et al. (2010). Natural variation in avenin epitopes among oat varieties: implications for Celiac Disease, *J. Cereal Sci.* doi: 10.1016/j.jcs.2010.09.007.

[74] Thompson T. (1997) . Do oats belong in a gluten-free diet? *J. Am. Diet Assoc., 97.*1413–1416.

[75] Lundin, K.E. Nilsen, E.M. Scott ,H.G. Løberg, E.M. Gjøen, A. Bratlie, J. et al. (2003) Oats induced villous atrophy in coeliac disease. *Gut. 52,* 1649–1652.

[76] Janatuinen, E.K. Kemppainen, T.A. Julkunen, R.J. Kosma, V-M. Mäki, M. Heikkinen, M. et al. (2002). No harm from five year ingestion of oats in coeliac disease. *Gut. 50,* 332–335.

[77] Garsed,K. Scott, B.B. (2007). Can oats be taken in a gluten-free diet? A systematic review. *Scand. J. Gastroenterol. 42(2),* 171-178.

[78] Haboubi, N.Y. Taylor, S. Jones, S. (2006). Coeliac disease and oats: a systematic review. *Postgrad. Med. J. 82(972),* 672-678.

[79] Schwarzenberg, S. J. and Brunzell, C. (2002) Type 1 Diabetes and Celiac Disease: Overview and Medical Nutrition Therapy. *Diabetes Spectr. 15(3),* 197-201.

[80] Uusitalo, U. Pietinen, P. Puska, P. (2002). Dietary Transition in Developing Countries: Challenges for Chronic Disease Prevention: In: Globalization, Diets and Noncommunicable Diseases-Noncommunicable Diseases and Mental Health (NMH)-WHO publication Avialable from: URL: http://whqlibdoc.who.int/publications/9241590416.pdf. Accessed on 14.12.2010.

[81] Ordovas, J.M. and Corella, D. (2004). Nutritional Genomics. *Annu. Rev. Genom. Human. Genet., 5,71*-118.

In: Natural Products and Their Active Compounds ... ISBN: 978-1-62100-153-9
Editors: M. Essa, A. Manickavasagan, and E. Sukumar

Chapter 25

MANGO (MANGIFERA INDICA LINN.) AND ITS HEALTH BENEFITS

Vandita Singh[1], Mohamed M. Essa[1,2,3], Mostafa Waly[1], Amanat Ali[1], Nejib Guizani[1] and G. J. Guillemin[2]

[1]Dept of Food Science and Nutrition, College of Agriculture and Marine Sciences, Sultan Qaboos University, Oman

[2]Neuropharmacology group, Dept of Pharmacology, College of Medicine, University of New South Wales, Sydney, Australia

[3]Developmental Neuroscience Lab, NYSIBR, Staten Island, NY, US

ABSTRACT

Mangifera indica L. (Mango) belonging to family Anacardiaceae. It is an indigenous to Indian subcontinent and an important fruit crop cultivated in tropical and subtropical regions. Its each part like pulp, peel, seed, leaves, flowers and the bark are important due to their medicinal uses. Different part of mango contains many biotic compounds like polyphenolics which can control many degenerative diseases due to their antioxidant activities. Hence, it is gaining more importance in medicinal and pharmaceutical industries. There has been tremendous interest in this plant as evidenced by the voluminous work in last few decades. This chapter will cover the medicinal uses of mango.

Keywords: Mango, *Mangifera indica*, health benefits, natural products, poly phenols, mangiferin

1. INTRODUCTION

Modern medicines are associated with many side effects, so there is always a need for complementary and alternative medicines with fewer side effects. So, the traditional herbal medicines are gaining more importance in the treatment of health problems. Research work related to medicinal plants and natural products has been increased and intensified. As there is a need to explore the benefits from the medicinal plants in order to decrease the dependency on synthetic drugs (Amadou, 1998).

Mangifera Indica Linn (Mango)

Mangifera indica L. (Mango) is a fruit which is indigenous to Indian subcontinent. It is known to have originated from Asia approximately 4000 years ago. There are 69 varieties known for the species. The common Mango or the Indian Mango is the only mango tree commonly cultivated in many tropical and subtropical regions. Mangoes rank second both in quantity and value among internationally traded tropical fruits (Yaacob and Subhadrabandhu, 1995) and fifth in total production among major fruit crops worldwide. According to the Food and Agricultural Organization statistics of 2007, the production of mangoes is estimated to be over 26 million tons per annum (FAO, 2007). India accounts for 54.2% mangoes and ranked first in production among the world. Other major mango producing countries are China, Thailand, Indonesia, Philippines, Pakistan, and Mexico (Sagar et al., 1999; Masibo and He, 2010). The English word mango probably originated from the Malayalam word "maanga" [മാങ്ങ (māṅṅa)] and the Portuguese were called it as manga.

2. Dietary Uses of Mango

Mango (*Mangifera indica* L.) is very popular among consumers because of its fresh and processed products (Maneepun and Yunchakad, 2004). Mature whole fruit is the main product of mango which can be eaten as raw or processed into different products. Green unripe fruit is used to make chutney, pickles, curries, and dehydrated products (*amchoor*- raw mango powder), and *panna* (green mango beverage). Ripe fruits can be processed into canned and frozen slices, pulp, concentrate, juices, nectar, jam, leather, puree, mango cereal flakes, mango toffee and various dried products (Singh, 1990). Mango blossoms are also used in the worship of the goddess Saraswati in India. Mango leaves are used to decorate archways and doors in Indian houses and during weddings and festivals.

3. Vernacular Names of Mango

There are different local names for the Mango. Pauh, Paoh, ampilam, mangaa in Indonesia. While pauhasal (native mango) for M. Pentadra in peninsular Malasia, pahohuttan (forest Mango) for M. Altissima in Phillipines, paopong (forest Mango) for M. Minor in Flores, Lesser Sanda Islands. In Combodia it is called as Pauh. Svay srok (M. indica), svay prey-wild mango (M. caloneura), Wai (M. minor) in New Guinea. Nowadays Wai and Pau or Pauq are generic names for Mango (M. indica) and Thayet- in Myanmar. In mainland southeast Asia none of the vernacular names of common mango exhibits signs of Indian origin. The name mango is derived from the Tamil word mankay or Manga which is adapted by Portugese too.

Taxonomy

Kingdom	Plantae
Division	Angiospermae
Class	Magnoliopsida
Order	Sapindales
Family	Anacardiaceae
Genus	Mangifera
Species	M. indica
Bionomical name	M. indica Linn

3. Botanical Description and Phytochemicals

Mango trees (*Mangifera indica* L.) grow 35–40 m (115–130 ft) tall, with a crown radius of 10 m. Bark is usually dark grey-brown to black, rather smooth, superficially cracked or inconspicuously fissured, peeling off in irregular, rather thick pieces. Exudate of the live bark transparent, a dark yellowish brown, drying brown, consisting of a resin mixed with a gum. The bark contains 78% resin and 15% gum in addition to tannic acid. The leaves are evergreen, alternate, simple, 15–35 cm long and 6–16 cm broad. Leaves are variable in shapes like oval-lanceolate, lanceolate, oblong, linear-oblong, ovate, obovate-lanceolate or roundish-oblong (Singh, 1960). When the leaves are young they are orange-pink, rapidly changing to a dark glossy red, then dark green as they mature. Hermaphrodite and male flowers are produced in the same panicle. The size of both male and hermaphrodite flowers varies from 6 to 8 mm in diameter. They are subsessile, rarely pedicellate, and have a sweet smell (Barfod, 1988). The flowers are produced in terminal panicles 10–40 cm long.

Each flower is small and white with five petals 5–10 mm long, with a mild sweet odor suggestive of lily of the valley. The ripe fruit is variable in size and color and is an indeliquescent drupe and contains a single large seed surrounded by a fleshy mesocarp which is covered by a leathery skin. It carries a single flat, oblong pit that can be fibrous or hairy on the surface. The fruit takes three to six months to ripen.

Phytochemically, whole plant contains flavanoids, phenolic acids, alkaloids, vitamins and tannins (Schieber et al., 2000; Kim et al., 2007; Kozubek and Tyman, 2005; Barreto et al., 2008). Family Anacardiaceae are known to have alkyl or alkenyl derivatives of phenol, catechol, and resorcinol (Kozubek and Tyman, 2005). Particularly, these compounds are present in the latex and fruit peel in mango while it is absent in edible pulp (Bandyopadhyay et al., 1985; Ross et al., 2004). Major polyphenolics in mango pulp are gallic acids, gallo-tannins, quercetin, isoquercetin, mangiferin, ellagic acid, *b*-glucogallin, *p*-hydroxy-benzoic acid, *m*-coumaric acid, *p*- coumaric acid, and ferulic acid (Schieber et al., 2000; Kim et al., 2007). Mango peels and stones are major waste of mango processing industries which account for 35–60% of the total fruit weight (Larrauri et al., 1996). Mango peels are found to be a rich sources of gallates, gallotannins, xanthone glucosides, flavonols (Barreto et al., 2008; Ribeiro et al., 2008), ascorbic acid, carotenoids, enzymes and dietary fibre (Frenich et al., 2005; Ajila et al., 2007).

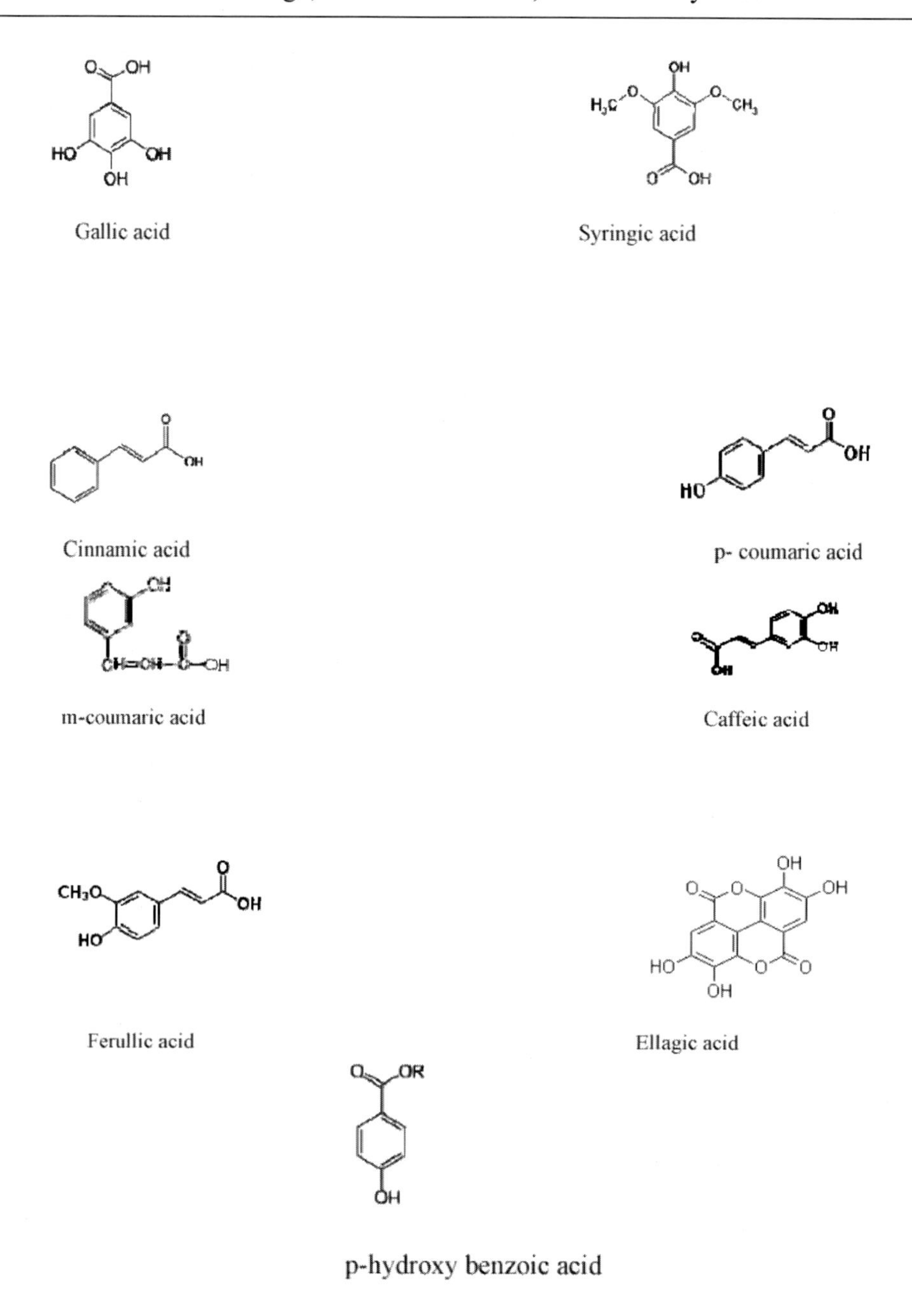

Figure 1. Structures of Phenolic acids in Mango.

Further, studies suggest that the LC-MS profile revealed major phenolic acids such as gallic acid, syringic acid, mangiferin, ellagic acid, gentisyl-protocatechuic acid, quercetin are present in the acetone extracts of raw and ripe peels of the mango and along with these the raw peel also have glycosylated iriflophenone and maclurin derivatives (Ajila et al., 2010). β-Carotene is the major carotenoid in peels which is followed by violaxanthin and lutein (Ajila et al., 2010).

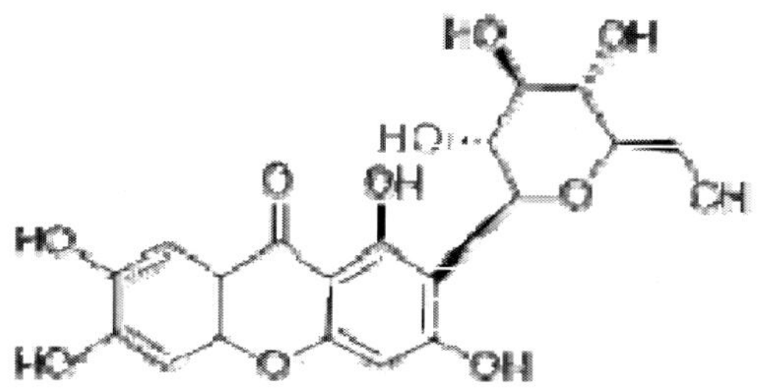

Figure 2. Structure of Mangiferin ((1,3,6,7-tetrahydroxyxanthone-C2-h-d-glucocoside).

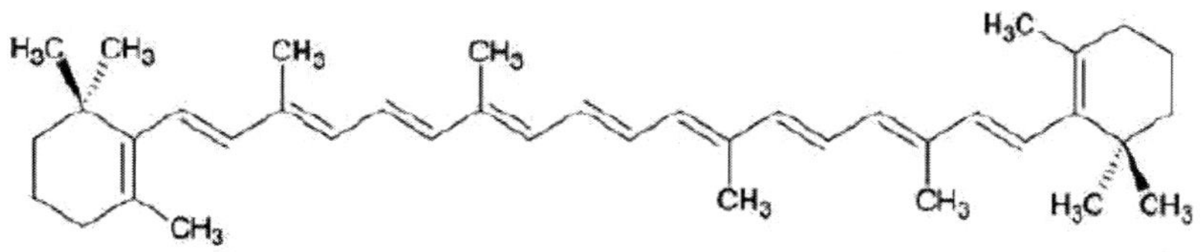

Figure 3. Structure of Carotenoid (β -carotene) present in Mango.

Mangiferin, gallic acid, 3,4-dihydroxy benzoic acid, gallic acid methyl ester, gallic acid propyl ester, (+)-catechin, (-)-epicatechin, benzoic acid and benzoic acid propyl ester are major polyphenolics in stem bark extract of mango (Alberto et al., 2002). Seed kernels of Mango contains gallic, ellagic, cinamic, caffeic, ferulic acid, coumarin, vanillin, mangiferin, gallates, tannins, gallotannins and condensed tannin-related polyphenols (Puravankara et al., 2000; Arogba, 2000; Abdalla et al, 2007). Steroids, triterpenes, phenolic compounds and flavonoids are present in the mango flowers aqueous decoction. The Caribbean population used an aqueous decoction from mango flowers for the treatment of gastritis and gastric ulcer (Robineau, 1996). The structures of important phytochemicals are also given.

Consumption and processing industries of mango fruits left annually about 3×10^5 ton of dry mango seed kernels in India. So, it could be treated as a specialized material due to high amount of bioactive compounds (Narasimha Char, Reddy, and Thirumala Roa, 1977; Narasimha Char and Azeemoddin, 1989). Hence, utilization of the seeds as a source of natural food additives and ingredients could be beneficial. Mango contains at least six vitamins including major ones are vitamin E, vitamin C and Vitamin A with small amount of vitamin B1 (thiamine), vitamin B2 (riboflavin), nicotinic acid (niacin) (Masibo and He, 2009).

4. Biological and Pharmacological Activities

4.1. Traditional Medicinal Applications

Mango and mango products are gaining experience worldwide especially in the European market (Loeillet, 1994). Fruits and other parts of mango tree have been used in traditional medicine (Khare, 2004). Mango stem bark extract (Vimang) has been traditionally used in many countries for the treatment of menorrhagia, diarrhea, syphilis, diabetes, scabies, cutaneous infections, and anemia (Scartezzini and Speroni, 2000). Stem extract which has main ingredient mangiferin is used as a nutritional supplement. It has been found to possess several pharmacological actions including antioxidant, analgesic, antidiabetic, anti-inflammatory, antitumor, immunomodulatory, and anti-HIV effects (Guha et al., 1996; Ichiki et al., 1998; Ojewole, 2005). Beverage made from mango leaves by decoction is used to manage bleeding dysentery (Quisumbing, 1978). It is scientifically proved to possess many pharmacological activities which are an indication of its usefulness in various diseases. Recently, many pharmacological studies have been conducted in *Mangifera indica* L. The major research findings can be summarized below:

4.1.1. Antioxidant Activity

Mango peel can be considered as a functional food or value added ingredient (Kim et al., 2010). Mango peel extract exhibited good antioxidant activities as compared to mango pulp extract as peel contains more polyphenols like anthocyanins, carotenoids and other flavonoids. This is supported by the fact that mango peels can effectively scavenge the free radicals like DPPH; hydroxyl radicals and alkyl radicals (Kim et al., 2010; Ajila et al., 2007). Further, Jiang et al., 2010 supported by their studies that peel extract possess significant antioxidant activities. They isolated two bioactive compounds from the peel extract namely ethyl gallate and penta-O-galloyl-glucoside which possess potent radical scavenging activities. Maisuthisakul and Gordon, 2009 reported that extracts of mango seed kernel has high total phenol content, radical scavenging, metal-chelating and tyrosinase inhibitory activities. Hence, the extract may be suitable for use in food, cosmetic, nutraceutical and pharmaceutical applications.

Stoilov et al., 2005 reported the strong antioxidant activity of Mangiferin with regard to the free radical 2, 2-diphenyl-1-picrylhydrazyl (DPPH). IC_{50} for mangiferin was 1.8 times lower than that determined for rutin (14.16 μg/ml) (Mensor et al, 2001) and 3.2 times higher than that for ascorbic acid (2.34 μg/ml) (Navarro et al., 2003). Abdalla et al., 2007 reported that the combination of both mango kernel extract and oil has potent antioxidant activities and shelf life extension effects on various edible oils like sunflower oil. Moreover, it can improve the stability of fresh and stored potato chips. The combination of both mango kernel extract and oil has optimum antioxidant activity which is much higher than each one alone. This effect is due to the polyphenolics and phospholipids as used in combination.

4.1.2. Cardiac and Renal Protection

Mango and its active ingredients offer cardiac and renal protection. DHC-1 is an herbal formulation from the plants *Mangifera indica, Bacopa monniera, Emblica officinalis, Glycyrrhiza glabra, and Syzygium aromaticum*. Pretreatment with DHC-1 formulation significantly reduce the serum markers of heart and kidney damage. This study suggests that antioxidant activity of the formulation could have a protective effect in isoproterenol-induced myocardial infarction (heart) and cisplatin-induced renal damage (kidney) (Bafna and Balaraman, 2005).

Another study suggests that mangiferin (polyphenol of mango) not only activate the mitochondrial energy metabolism by the reduction of oxidative damage but also have beneficial effect against isoproterenol-induced myocardial infarction in experimental mice model (Prabhu et al., 2006). It was reported that the rats pretreated with the stem bark extract of mango along with atherogenic diet (4% cholesterol, 1% cholic acid, 0.5% thiouracil) can prevent the elevation of lipids in the serum and heart and cause significant decrease in lipid accumulation in the liver and aorta. The stem extract significantly restore the activity of antioxidant enzymes such as superoxide dismutase, catalase, glutathione peroxidase, and glutathione and restore the antioxidant status to almost normal levels (Akila and Devaraj, 2008).

4.1.3. Antihyperglycemic and Antihyperlipidemic Activity

Potent antihyperglycemic activity of mangiferin isolated from the leaves of mango was reported recently. Chronic administration of mangiferin can significantly improve the oral glucose tolerance in glucose-loaded normal rats (Muruganandan et al., 2005). Mangeferin pretreatment offer antidiabetic activity by means of lowering the plasma glucose level experimental diabetic rats (Muruganandan et al., 2005). Further, mangiferin was found to have a significant antihyperlipidemic and antiatherogenic activity by normalizing the lipid profile and diminution of atherogenic index in diabetic rats (Muruganandan et al., 2005). Parmar and Kar (2008), reported that pretreatment of peel extracts of mango significantly decrease the levels of tissue lipid peroxidation, serum lipids, glucose, creatinine kinase-MB and increase the levels of thyroid hormones and insulin in high atherogenic diet mouse model for dyslipidemia, hypothyroidism and hyperglycemia, which may be due to the presence of polyphenols and ascorbic acid in the extract. It was reported that mangiferin not only significantly prevent progression of diabetic nephropathy, but also improves renal function (Li et al., 2010).

4.1.4. Anti-Inflammatory and Anti-Allergic Effects

Some reports suggest that mango has the anti-inflammatory and anti-allergic power. Stem bark extract of mango was found to have anti-inflammatory activities on Dextran sulfate sodium (DSS)-induced colitis in rat model. Pretreated rats with extract can improve clinical signs by reducing ulcer formation and myeloperoxidase (MPO) activity and maintained the redox balance. Inflammatory mediators such as inducible isoforms of nitric oxide synthase (iNOS) and cyclooxygenase (COX)-2, and cytokines like tumor necrosis factor (TNF)-α and TNF receptors 1 and 2 in colonic tissue were reduced along with the reduction in the serum level of Interleukin (IL)-6 and TNF-α (Márquez et al., 2010).

Vimang and the mangiferin showed anti-allergic effects. It inhibits the IgE production, anaphylaxis reaction in rats, histamine-induced vascular permeability, and the histamine release induced by compound 48/80 from rat mast cells. It also inhibits lymphocyte proliferative response which is evidenced by the reduction of the amount of B and T lymphocytes in a dose-dependent manner (Rivera et al., 2006). Present study reported that mangiferin, the major compound of Vimang, contributes to the anti-allergic effects of the extract (Rivera et al., 2006).

4.1.5. Antimicrobial and Antifungal Activity

Studies suggest that the active ingredients of mango have the antimicrobial and antifungal activities. Stoilov et al., 2005 reported that mangiferin, a polyphenol of mango was found to have antibacterial effect against both gram positive as well as gram negative bacteria like Bacillus pumilus, Bacillus cereus, Staphylococcus aureus, Staphylococcus citreus, Escherichia coli, Salmonella agona, Klebsiella pneumoniae, Saccharomyces cerevisiae.

Another study suggests that hydrolyzable tannins i.e. penta-, hexa-, and hepta-O-galloyl-glucose of mango were found to have antibacterial activity (Engels et al., 2009). Abdalla et al., 2007, reported that mango seed kernel could be useful as antimicrobial agent in foods, because it reduce the total bacterial count, inhibit the coliforms growth, showed antimicrobial activity against E.coli strain and extend the shelf-life of pasteurized cow milk. Stoilov et al., 2005 reported that mangiferin was found to have antifungal effects against *Thermoascus aurantiacus, Trichoderma reesei, Aspergillus flavus and Aspergillus fumigatus*. Kanwal et al., 2010 reported that isolated flavonoids from the leaves of mango showed potent antifungal activity against *Alternaria alternata (Fr.) Keissler, Aspergillus fumigatus Fresenius, Aspergillus niger van Tieghem, Macrophomina phaseolina (Tassi) Goid and Penicillium citrii.*

4.1.6. Anticarcenogenic Effect

The anti-carcinogenic activities of mango and its active compounds were well studied. Noratto et al., 2010 reported that the anticancer properties of polyphenolic extracts from different mango varieties (Francis, Kent, Ataulfo, Tommy Atkins, and Haden) in various cancer cell lines and they found that all the above extracts inhibit the cell growth in SW-480 colon carcinoma cells. The growth inhibition was associated with an increased mRNA expression of pro-apoptotic biomarkers and cell cycle regulators, cell cycle arrest, and a decrease in the generation of reactive oxygen species. So, they reported that the polyphenolics from several mango varieties exerted anticancer effects and among these Haden and Ataulfo mango varieties possessed superior chemo-preventive activity than others. Percival et al., 2006 did the study on the whole mango juice and juice extracts for the anticancer activity. The study was done by examining the effect of the extract on cell cycle kinetics and its ability to inhibit chemically induced neoplastic transformation of mammalian cell lines. They incubate HL-60 cells with whole mango juice and mango juice fractions which resulted in an inhibition of the cell cycle in the G(0)/G(1) phase. They found that whole mango juice was effective in reducing the number of transformed foci in the neoplastic transformation assay in a dose-dependent manner. Lupeol, a triterpene present in mango and other fruits found to possess anticancer properties in an *in-vivo* and *in-vitro* assays (Prasad et al., 2008). Another study reported that lupeol/mango pulp extract is effective in combating testosterone-induced changes in mouse prostate and it can cause apoptosis by modulating cell-growth regulators (Prasad et al., 2008). Supplementation of lupeol or the mango pulp extract can results in arrest of prostate enlargement in testosterone-treated animals. Lupeol is shown to possess apoptogenic activities in human prostate cancer cells (Prasad et al., 2008).

4.1.7. Cognitive Deficits (Memory Enhancement)

Mango and its polyphenols offer improvement in cognitive dysfunction. Kumar et al., 2009 studied the effect of ethanolic extract of mango fruit on cognitive performances in a mice model. They reported that chronic treatment with extract and vitamin C can significantly ($p < 0.05$) reversed the aging and scopolamine induced memory deficits. Also, the extract contains pharmacologically active compounds saponins, tannins, and flavonoids which can be regarded as the memory enhancer. Oral administration of mango extract can protect against neuronal cell death of the hippocampal CA1 area after ischaemia-reperfusion. As the extract can be absorbed across the blood-brain barrier and attenuates neuronal death which is due to due to its antioxidant activities (Martinez et al, 2001).

4.1.8. Hepatoprotective Effect

Mango has the liver protecting power. Pourahmad et al., 2010 studied the effect aqueous extract of mango fruit against oxidative stress toxicity induced by cumene hydroperoxide (CHP) in isolated rat hepatocytes. They found that the extracts and gallic acid (100 μM) protects the hepatocyte against all oxidative stress markers which includes cell lysis, ROS generation, lipid peroxidation, glutathione depletion, mitochondrial membrane potential decrease, lysosomal membrane oxidative damage and cellular proteolysis. Even the mango extracts were more effective than gallic acid in protecting hepatocytes against CHP induced lipid peroxidation while gallic acid proved to be more effective than mango extracts in preventing lysosomal membrane damage. Hence, it is suggested that mango extract has been found to have protective effect against liver injury associated with oxidative stress. Another study reported vimang could be a used as a natural drug for preventing oxidative damage during hepatic injury associated with free radical generation in a dose dependent manner. It can reduce transaminase levels and DNA fragmentation, can restore the cystolic Ca^{2+} levels and inhibit polymorphonuclear migration, improves the oxidation of total and non-protein sulfhydryl groups and prevent modification in catalase activity, uric acid and lipid peroxidation markers (Sánchez et al., 2003)

4.1.9. Antiulcer and Other Activities

Mango and its active polyphenols possess some antiulcer activities along with its antioxidant properties (Priya et al., 2009). Pretreatment with mango flowers aqueous decoction (AD) in mice model significantly decreased the gastric lesions and gastric index induced by ethanol and HCl/ethanol in a dose dependent manner. Further, it significantly accelerated the healing process in subacute gastric ulcer induced by acetic acid. This potential gastroprotective and ulcer-healing properties of AD strongly supports its popular use in gastrointestinal disorders in Caribbean (Limaa et al., 2006). A recent study found that the mango and derived polyphenols may affect the activity of the multidrug transporter glycoprotein (P-gp ABCB1) in HK-2 cells (Chieli et al., 2009). Javaid et al., 2010 investigated the herbicidal activity of mango leaves against parthenium weed (*Parthenium hysterophorus* L.). Aqueous mango leaf extracts significantly reduced germination, shoot length and the shoot and root biomasses of parthenium seedlings in a concentration dependent manner. Wilkinson et al., 2008 evaluated the effect of components of mango like quercetin and mangiferin and the aglycone derivative of mangiferin, norathyriol on the modulation of the transactivation of PPARs, which are transcription factors and important in many human diseases.. The study suggested that mango components may alter transcription and could contribute to positive health benefits.

Daud et al., 2010 reported that bioactive molecules from the mango extract capable of modulating endothelial cell migration which is necessary step in the formation of new blood vessels or angiogenesis. The results of this study suggest that mango extracts have a health promoting effects in diseases which are related to the impaired formation of new blood vessels like limb ischemia, coronary infarction or stroke.

Cadmium is a toxic metal which can implicates human diseases. Study suggests the protective role of mangiferin against cadmium chloride ($CdCl_2$)-induced genotoxicity. Pretreatment with mangiferin (dose of 2.5 mg/kg b.wt.) not only significantly ($p < .001$) reduce the frequency of micronucleated polychromatic (MnPCE), normochromatic erythrocytes (MnNCE), and increased PCE/NCE ratio as compared to the diseased (CDDW + $CdCl_2$)

group at all post-treatment times which indicates its antigenotoxic effect but also can decrease the lipid peroxidation in liver. Moreover, mangiferin also showed significant increase in the activity of antioxidant enzymes. Hence, mangiferin is shown to have potent antigenotoxic effect against $CdCl_2$ induced toxicity in mice model.

Kim et al., 2010 reported the significant cytoprotective effect of mango peel extracts in comparison with pulp extract against the oxidative damage, induced by H_2O_2 in cancer cell lines in a dose-dependent manner. They concluded the effect is due to presence of bioactive compounds like phenolic and flavonoid compounds in peel than in pulp. Mercury is a heavy metal and toxic for the human health. Agarwala et al., 2010 study the cytoprotective potential of mangiferin against mercuric chloride ($HgCl_2$) induced toxicity in HepG2 cell line. Pre-treatment of mangiferin significantly decreased the percentage of $HgCl_2$ induced apoptotic cells by maintaining the redox balance at the post incubation intervals. This study suggests that mangiferin has ability to quench ROS generated in the cells due to oxidative stress induced by $HgCl_2$ and can restore the mitochondrial membrane potential and normalize the cellular antioxidant levels.

Conclusion

Pharmacological studies conducted on *Mangifera indica* indicate the medicinal potential of this plant and its parts especially pulp, peels, leaves, bark, seed and seed kernel. It is found to have potential in the treatment of many diseased conditions such as cardiovascular diseases, inflammatory ailment which includes allergies, liver and kidney disorders, neuronal disorders, ulcer, diabetes, cancer, microbial infections including fungal and bacterial infections. Seed kernel extract may be suitable for use in food, cosmetic, nutraceutical and pharmaceutical applications. Epidemiological studies of fruits and vegetables show that it contains many phytonutrients which may be beneficial in protecting the human body against damage by reactive oxygen and nitrogen species (Diplock et al., 1998; Halliwell, 1997) during various disease conditions. All these studies are only in experimental level and there should be a need for complete and comprehensive clinical studies to prove the above said effects. Now the modern world is trying to go back to nature's way. This may be due to increase in the number of various communicable and non-communicable diseases. We believe that this review about mango may be useful in this regard.

References

Abdalla, A. E. M, Darwish, S.M, Ayad, E. H. E., and El-Hamahmy, R. M. 2007. Egyptian mango by-product 1. Compositional quality of mango seed kernel. *Food Chemistry.*103: 1134–1140.

Abdalla, A. E. M, Darwish, S. M, Ayad, E. H. E., and El-Hamahmy, R. M. 2007. Egyptian mango by-product 2: Antioxidant and antimicrobial activities of extract and oil from mango seed kernel. *Food Chemistry*. 103: 1141–1152.

Alberto, J. N. S., Herman, T. V. C., Juan, A., Johanes, G., Fabio, N., Francesco, D. S., and Luca, R. 2002. Isolation and quantitative analysis of phenolic antioxidants, free sugars, and polyols from mango (Mangifera indica L.) stem bark aqueous decoction used in Cuba as nutritional supplement. *Journal of Agricultural and Food Chemistry.* 50: 762–766.

Amadou, C.K., 1998. Promoting alternative medicine. *Africa health J.* 2: 20-25.

Akila, M., Devaraj, H. 2008. Synergistic effect of tincture of Crataegus and Mangifera indica L. extract on hyperlipidemic and antioxidant status in atherogenic rats. *Vascul Pharmacol.* 49: 173-7.

Ajila, C. M., Naidu, K. A., Bhat, S. G., Rao, U. J. S. P. 2007. Bioactive compounds and antioxidant potential of mango peel extract. *Food Chemistry.* 105: 982–988.

Ajila, C. M., Rao, L. J., Rao, U. J. S. P. 2010. Characterization of bioactive compounds from raw and ripe Mangifera indica L. peel extracts. *Food and Chemical Toxicology.* 48: 3406–3411.

Bandyopadhyay, C., Gholap, A.S., Mamdapur, V.R., 1985. Characterization of alkylresorcinol in mango (Mangifera indica L.) latex. J. Agric. Food Chem. 33, 377–379.

Bafna, P. A., Balaraman, R. 2005. Antioxidant activity of DHC-1, an herbal formulation, in experimentally-induced cardiac and renal damage. *Phytother Res.* 19: 216-21.

Barreto, J. C., Trevisan, M. T. S., Hull, W. E., Erben, G., Brito, E. S., Pfundstein, B., et al. 2008. Characterization and quantitation of polyphenolic compounds in bark, kernel, leaves, and peel of mango (Mangifera indica L*.). Journal of Agricultural and Food Chemistry.* 56: 5599–5610.

Chieli, E., Romiti, N., Rodeiro, I., Garrido, G. 2009. In vitro effects of Mangifera indica and polyphenols derived on ABCB1/P-glycoprotein activity. *Food Chem. Toxicol.* 47: 2703-10.

Daud, N. H., Aung, C. S., Hewavitharana, A. K., Wilkinson, A.S., Pierson, J.T., Roberts-Thomson, S.J., Shaw, P. N., Monteith, G.R., Gidley, M.J., Parat, M.O. 2010. Mango extracts and the mango component mangiferin promote endothelial cell migration. *J. Agric Food Chem.* 58: 5181-6.

Diplock, A., Charleux, J., Grozier-Willi, G., Kok, K., Rice-Evans, C., Roberfroid, M., Stahl, W., and Vina-Ribes, J. 1998. Functional food sciences and defence against reactive oxidative species. *British Journal of Nutrition.* 80: 77–82.

Engels, C., Knödler, M., Zhao, Y.Y., Carle, R., Gänzle, M.G., Schieber, A. 2009. Antimicrobial activity of gallotannins isolated from mango (Mangifera indica L.) kernels. *J. Agric. Food Chem.* 57: 7712-8.

Frenich, A. G., Torres, M. E. H., Vega, A. B., Vidal, J. L. M., and Bolaos, P. P. 2005. Determination of ascorbic acid and carotenoids in food commodities by liquid chromatography with mass spectrometry detection. *Journal of Agricultural and Food Chemistry.* 53: 7371–7376.

FAO Statistical Database—Agriculture. (Accessed 2007 August) http://www.fao.org/corp/statistics/en.

Guha, S., Ghosal, S., Chattopadhyay, U. 1996. Antitumor, Immunomodulatory and Anti-HIV *Effect of Mangiferin, a Naturally Occurring Glucosylxanthone Chemotherapy.* 42: 443–451.

Halliwell, B.1997. Antioxidants and human disease: A general introduction. Nutrition Reviews, 55, 44–52.

Ichiki, H., Miura, T., Kubo, M., Ishihara, E., Komatsu, Y., Tanigawa, K., Okada, M. 1998. New Antidiabetic Compounds, Mangiferin and Its Glucoside. *Biol. Pharm. Bull. 21*: 1389–1390.

Javaid, A., Shafique, S., Kanwal, Q., Shafique, S. 2010. Herbicidal activity of flavonoids of mango leaves against Parthenium hysterophorus L. *Nat. Prod. Res.* 24: 1865-75.

Kanwal, Q., Hussain, I., Latif Siddiqui, H., Javaid, A. 2010. Antifungal activity of flavonoids isolated from mango (Mangifera indica L.) leaves. *Nat. Prod. Res.* 24: 1907-14.

Khare, C. P., 2004. Indian Herbal Remedies. Springer Verlag, Berlin, Heidelberg. 300–303.

Kim, H., Moon, J., Y., Kim, H., Lee, D., Cho, M., Choi , H., Young Suk Kim , Y., S., Mosaddik, A., Cho, S., K., 2010. Antioxidant and antiproliferative activities of mango (Mangifera indica L.) flesh and peel. *Food Chemistry.* 121: 429–436.

Kim, Y., Brecht, K. J., Talcott, S. T. 2007. Antioxidant Phytochemical and Fruit Quality Changes in Mango (Mangifera Indica L.) Following Hot Water Immersion and Controlled Atmosphere Storage. *Food. Chem.*10:1016.

Kozubek, A., Tyman, J. H. P., 2005. Bioactive phenolic lipids. In: Atta-ur- Rahman (Ed.), *Studies in Natural Products Chemistry.* 30: 111–190.

Kumar, S., Maheshwari, K. K, and Singh, V., 2009. Effects of Mangifera indica fruit extract on cognitive deficits in mice. *J. Environ. Biol.* 30: 563-6.

Li, X., Cui, X., Sun, X., Li, X., Zhu, Q., Li, W. 2010. Mangiferin prevents diabetic nephropathy progression in streptozotocin-induced diabetic rats. *Phytother. Res.* 24: 893-9.

Limaa, Z. P., Severi, J. A., Pellizzon, C. H., Brito, A. R. M. S., Solis, P. N., Caceres, A., Giron, L. M., Vilegasb, W., Hiruma-Lima, C. A. 2006. Can the aqueous decoction of mango flowers be used as an antiulcer agent? *Journal of Ethnopharmacology.* 106: 29–37

Loeillet, D. 1994. The European mango market: a promising tropical fruit. *Fruits.* 49: 332–334.

Larrauri, J. A., Ruperez, P., and Saura-Calixto, F. 1996. Antioxidant activity from wine pomace. *American Journal of Enology and Viticulture.* 47: 369–372.

Márquez, L., Pérez-Nievas, B., J., Gárate,I., García-Bueno, B., Madrigal, J., L., M., Menchén, L., Garrido, G., Leza, J., C. 2010. Anti-inflammatory effects of Mangifera indica L. extract in a model of colitis. *World J. Gastroenterol.* 16: 4922-4931

Mensor, L., Menezes, F., Leitao, G. et al., 2001. Screening of Brazilian plant extracts for antioxidant activity by the use of DPPH free radical method. *Phytother. Res.* 15: 127-130.

Martínez, S. G., Candelario-Jalil, E., Giuliani, A., León, O. S., Sam, S., Delgado, R., Núñez, S.A.J. 2001. Mangifera indica L. extract (QF808) reduces ischaemia-induced neuronal loss and oxidative damage in the gerbil brain. *Free Radic Res.* 35: 465-73.

Masiboabc, M. and Hea, Q. 2009. Mango Bioactive Compounds and Related Nutraceutical Properties—A Review. *Food Reviews International.* 25: 346–370.

Maneepun, S., Yunchakad, M., 2004. Developing processed mango products for international markets. *Acta Hort.* 645, 93–105.

Muruganandan , S., Srinivasan, K., Gupta, S., Gupta, P. K., Lal, J. 2005. Effect of mangiferin on hyperglycemia and atherogenicity in streptozotocin diabetic rats. *J. Ethnopharmacol.* 97: 497-501.

Narasimha Char, B. L., and Azeemoddin, G. 1989. Edible fat from mango stones. *Acta Horticulture.* 231: 744–748.

Narasimha Char, B. L., Reddy, B. R., and Thirumala Roa, S. D., 1977. Processing of mango stones for fat. *Journal of the American Oil Chemists Society*. 54: 494–498.

Navarro, M., Montilla, M., Cabo, M., et al. 2003. Antibacterial, antiprotozoal and antioxidant activity of five plants used in Izabal for infectious diseases. *Phytother. Res.* 17: 325-329.

Noratto, G. D., Bertoldi, M. C., Krenek, K., Talcott, S. T., Stringheta, P. C., Mertens-Talcott, S. U., 2010. Anticarcinogenic effects of polyphenolics from mango (Mangifera indica) varieties. *J. Agric Food Chem.* 58: 4104-12.

Ojewole, J. A. 2005. Antiinflammatory, Analgesic and Hypoglycemic Effects of Mangifera Indica Linn. (Anacardiaceae) *Stem-Bark Aqueous Extract. Clin. Pharmacol.* 27: 547–554.

Parmar, H. S., Kar, A. 2008. Possible amelioration of atherogenic diet induced dyslipidemia, hypothyroidism and hyperglycemia by the peel extracts of Mangifera indica, Cucumis melo and Citrullus vulgaris fruits in rats. *Biofactors.* 33: 13-24.

Percival, S. S., Talcott, S. T., Chin, S., T., Mallak, A., C., Lounds-Singleton, A., Pettit-Moore, J. 2006. Neoplastic transformation of BALB/3T3 cells and cell cycle of HL-60 cells are inhibited by mango (Mangifera indica L.) juice and mango juice extracts. *J. Nutr.* 136: 1300-4.

Prabhu, S., Jainu, M., Sabitha, K.E., Devi, C.S. 2006. Role of mangiferin on biochemical alterations and antioxidant status in isoproterenol-induced myocardial infarction in rats. *J. Ethnopharmacol.* 107: 126-33.

Prasad, S., Kalra, N., Shukla, Y. 2008. Induction of apoptosis by lupeol and mango extract in mouse prostate and LNCaP cells. *Nutr. Cancer.* 60: 120-30.

Pourahmad, J., Eskandari, M.R., Shakibaei, R., Kamalinejad, M. 2010. A search for hepatoprotective activity of fruit extract of Mangifera indica L. against oxidative stress cytotoxicity. *Plant Foods Hum. Nutr.* 65: 83-9.

Priya, T. T., Sabu, M. C., Jolly, C. I. 2009. Role of Mangifera indica bark polyphenols on rat gastric mucosa against ethanol and cold-restraint stress. *Nat. Prod. Res.*14: 1-12.

Puravankara, D., Boghra, V., and Sharma, R. S., 2000. Effect of antioxidant principles isolated from mango (Mangifera indica L.) seed kernels on oxidative stability of buffalo ghee (butter-fat). *Journal of the Science of Food and Agriculture* .80: 522–526.

Quisumbing, E. 1978. Medicinal Plants of the Philippines; Katha Publishing Co. and JMC Press: Quezon. 538–541.

Rivera, D. G., Balmaseda, I. H., León, A. A., Hernández, B. C., Montiel, L. M., Garrido, G. G., Cuzzocrea, S., Hernández, R. D. 2006. Anti-allergic properties of Mangifera indica L. extract (Vimang) and contribution of its glucosylxanthone mangiferin. *J. Pharm. Pharmacol.* 58: 385-92.

Ribeiro, S. M. R., Barbosa, L. C. A., Queiroz, J. H., Knodler, M., and Schieber, A. 2008. Phenolic compounds and antioxidant capacity of Brazilian mango (Mangifera indica L.) varieties. *Food Chemistry.*110: 620–626.

Ross, A. B., Kamal-Eldin, A., Aman, P., 2004. Dietary alkyresorcinols: absorption, bioactivities, and possible use as biomarkers of wholegrain wheat- and rye-rich foods. *Nutr. Rev.* 62, 81–95.

Robineau, L.G.; Soejarto, D.D. Tramil. 1996. A Research Project on the Medicinal Plant Resources of the Caribbean. In *Resources of the Tropical Forest*; Balick, M.J.; Elisabetsky, E.; Laird, S.A.; Eds.; Columbia University Press: New York. 318–325.

Stoilova1, I., Gargova1, S., Stoyanova, A., Ho, L., 2005. Antimicrobial and antioxidant activity of the polyphenol mangiferin. *Herba yolonica.* 51: 37-44.

Sánchez, G. M., Rodríguez, H. M. A., Giuliani, A., Núñez, S. A. J., Rodríguez, N. P., León, F. O. S., Re, L. 2003. Protective effect of Mangifera indica L. extract (Vimang) on the injury associated with hepatic ischaemia reperfusion. *Phytother. Res.* 17: 197-201.

Scartezzini, P., Speroni, E. 2000. Review on Some Plants of Indian Traditional Medicine with Antioxidant Activity. *J. Ethnopharmacol.* 71: 23–43.

Sagar, V. R., Khurdiya, D. S., Balakrishnan, K. A. 1999. Quality of Dehydrated Ripe Mango Slices as Affected by Packaging Material and Mode of Packaging. *J. Food Sci. Technol.* 36: 67–70.

Singh, R. N. 1990. Marketing and storage in Mango; Indian Council of Agricultural Research (ICAR): New Delhi, India. 72.

Schieber, A., Ullrich, W., Carle, R. 2000. Characterization of Polyphenols in Mango Puree concentrate by HPLC with diode array and mass spectrometric detection. Innov. *Food Sc. Emerg. Techs.* 1 : 161–166.

Wilkinson, A. S., Monteith, G. R., Shaw, P. N., Lin, C. N., Gidley, M. J., Roberts-Thomson, S. J. 2008. Effects of the mango components mangiferin and quercetin and the putative mangiferin metabolite norathyriol on the transactivation of peroxisome proliferator-activated receptor isoforms. *J. Agric Food Chem.* 56: 3037-42.

Yaacob, O.,Subhadrabandhu, S. 1995. The Production of Economic Fruits in South-East Asia; Oxford University Press: Kuala Lumpur. 419.

In: Natural Products and Their Active Compounds ... ISBN: 978-1-62100-153-9
Editors: M. Essa, A. Manickavasagan, and E. Sukumar

Chapter 26

BASIL: A NATURAL SOURCE OF ANTIOXIDANTS AND NEUTRACEUTICALS

Masoud Yahya Al-Maskari[1], Muhammad Asif Hanif[2], Ahmed Yahya Al-Maskri[2] and Samir Al-Adawi[3]
[1]College of Health Sciences, University of Buraimi, Al-Buraimi, Oman
[2]College of Agricultural and Marine Sciences, Sultan Qaboos University, Muscat, Oman
[3]College of Medicine and Health Sciences, Sultan Qaboos University, Muscat, Oman

ABSTRACT

Some endemic species of medicinal and culinary herbs are of particular interest due to presence of phytochemicals with significant antioxidant capacities and health benefits. Phytochemical rich plant materials are increasingly of interest in the food and medical industry as they are helpful in oxidative retardation of lipids as well as due to their preservative action against microorganisms. Many medicinal plants are rich with large amounts of antioxidants other than vitamin C, vitamin E, and carotenoids. Basils come with loads of health benefits as it is a rich source of key nutrients like Vitamin A, Vitamin C, calcium, phosphorus, beta carotene. Basil leaves are helpful in sharpening memory. Basil is also useful in treatment of fever, common cold, stress, purifying blood, reducing blood glucose, risk of heart attacks and cholesterol level, mouth ulcer and arthritis. Anti-inflammatory properties of basil are also well known.

1. INTRODUCTION

Plants belonging to the Lamiaceae family are famous due to strong antioxidant activity [1]. The genus *Ocimum* is a member of the Lamiaceae family. This genus contains between 50 and 150 species of herbs and shrubs [2]. Basil (*Ocimum basilicum*) belongs to family Lamiaceae is a very important culinary herb. Basil has more than 60 varieties which differ in appearance and taste. Sweet basil is bright green in appearance with pungent in taste. Other varieties also offer unique tastes: lemon basil, anise basil and cinnamon basil all have flavors that subtly reflect their name.

Scientific Classification	
Kingdom:	Plantae
Phylum:	Magnoliophyta
Class:	Magnoliopsida
Order:	Lamiales
Family:	Lamiaceae
Genus:	*Ocimum*

Now Basil is cultivated al around the globe although it was first native to Asia and Africa. Basil is considered as a symbol of love in Italy and as an icon of hospitality in India. Basil is used in medical treatments for coughs, headaches, worms, diarrhea, and kidney malfunctions. Recent research studies mostly correlate the unique health benefits of basil to the presence of flavonoids and volatile oils in it. Basil essential oil has been utilized extensively in the food industry as a flavoring agent, and in perfumery and medical industries [3-5]. Oxidative deterioration of fat components in foods is responsible for the rancid odors and flavors which decrease nutritional quality. Traditionally basil is used in Mediterranean, European, American and Southeast Asian foods.

2. Etymology

The word *basil* came from the Greek βασιλεύς (basileus), meaning “king”. Basil is known as "king of herbs". Basil grows well in hot, dry conditions and is sensitive to cold. Color of leaves and size of plants is variable in different varieties.

3. Varieties

Basil varieties are usually named informally (Table 1). Most common basil varieties are *Ocimum basilicum* (Sweet Basil), *Ocimum sanctum* (Holy basil), *Ocimum canum* (Hairy Basil) and *Ocimum gratissimum* (Wild Basil). These varieties are known to have many species.

4. Basil Seeds

Basil seeds become gelatinous after soaking in water and are used in Asian drinks and desserts such as *falooda* or sherbet. Such seeds are known variously as *sabza*, *subza*, *takmaria*, *tukmaria*, *tukhamaria*, *falooda*, *selasih* (Malay/Indonesian) or *hột é* (Vietnamese).

Table 1. Widely grown Basil Varieties

Ocimum basilicum	
Common Name	Characteristics/Attributes
Basil	Height: 12"-30", 10"-12" apart, commonly used for flavoring food stuff
African Blue Basil	Height: 18"-24" , 12" apart, ornamental landscape variety
Anise Basil	Height: 15"-18" , 8"-10" apart, culinary variety
Aussie Sweet Basil	Height: ≈ 24" , 8"-10" apart, culinary variety
Baja Basil	Height: 18"-24", 12" apart, cinnamon flavor and used to flavor food.
Cinnamon Basil	Height: 18"-24", 12" apart, culinary variety with a soft cinnamon flavor.
Ocimum basilicum	
Common Name	Characteristics/Attributes
Compact Genovese Basil	Height: 12"-15", 8" apart, standard Italian basil flavor.
Genovese Basil	Height: 18"-20", 12" apart, standard basil flavor and used in food stuff.
Green Ruffles Basil	Height: 18"-24", 12" apart, culinary variety with traditional sweet basil flavor.
Lemon Basil	Height: 18"-24", 12" apart, culinary, heirloom variety, Used in chicken and fish salads.
Osmin Purple Basil	Height: 18"-24", 12" apart, culinary.
Purple Ruffles Basil	Height: 18"-24", 12" apart, culinary and ornamental.
Red Rubin Basil	Height: 18"-24", 12" apart, culinary and ornamental, leaves are coppery reddish purple.
Spicy Globe Basil	Height: 12"-15", 8"-10" apart, culinary, ornamental.
Sweet Italian Basil	Height: 18"-24", 12" apart, culinary.
Thai Basil, "Siam Queen"	Culinary, ornamental, red stems and flowers in clusters
Ocimum sanctum	
Holy Basil	Height: 18"-24", 12" apart, culinary variety with musky scent with a hint of mint. Bluish green, hairy leaf.
Holy Red and Green Basil	Height: 18"-24", 12" apart, culinary, ornamental and excellent for flavorful tea.
Italian Large Leaf Basil	Height: 18"-24", 12" apart, flavor and fragrance.
Ocimum canum	
Hairy Basil	Height: 18"-24", 12" apart, culinary and ornamental
Ocimum gratissimum	
Wild Basil	Height: 22"-26", 12" apart, culinary

Figure 1. *Ocimum basilicum* (Sweet Basil).

Figure 2. *Ocimum sanctum* (Holy basil) Synonymous: *Ocimum tenuiflorum.*

Figure 3.Ocimum canum (Hairy Basil).

Figure 4. *Ocimum gratissimum* (Wild Basil).

5. Essential Oil Contents, Chemical Compositions and Antimicrobial Activities

Basil essential oil is a natural source of antioxidants and nutraceuticals. Basil varieties significantly differ in essential oil yield and its composition. Chemotaxonomic survey of the genus *Ocimum* showed that essential oil compositions in *O. basilicum* are extremely variable. Al-Maskari et al., [6] extracted essential oil from one of the Omani basil varieties. The essential oil yield of Omani basil was found to be 0.171% on fresh shoot biomass basis. Linalool (69.9%) was identified as the major component. The other major components were geraniol (10.9%), 1, 8-cineole (6.4%), α-bergamotene (1.6%) and geranyl acetate (1.4%). French basil plant was found to contain 0.2-0.3% essential oil and methyl cinnamate (20-45%) was the major component [7]. Seasonal variation in essential oil contents of one Pakistani basil variety ranged from 0.5% to 0.8%. Maximum essential oil yield was in winter and minimum in summer. The essential oils consisted of linalool as the most abundant component (56.7-60.6%), followed by epi-α-cadinol (8.6-11.4%), α-bergamotene (7.4-9.2%) and γ-cadinene (3.2-5.4%) [8]. Essential oil yield of Indian basil on fresh shoot biomass basis was found to be 0.6-0.8%. Methyl chavicol was the major component and contributed up to 70-80% of the total basil oil [7]. The average essential oil content in 18 Turkish landraces was ranging from 0.4 to 1.5% [9]. Among 18 Turkish landraces, seven landraces contained linalool (37.7 and 60.2%) as a major component. Other important compounds were 1, 8-Cineole (0.2-14.5%), eugenole (3.1-21.1%) and δ-cadinene (7.4-8.7%). Three Turkish basil landraces have methyl cinnamate as a major compound ranging from 58.6 to 63.1%. Linalool (17.3-27.3%), α-cadinol (2.4-2.9%), γ-cadinene (1.2-1.6%), δ-cadinene (t-2.4%) and zingiberene (1.1-1.3%). Methyl eugenol (34.2%) was the major component in one landrace followed by linalool (12.3%), 1, 8-cineole (10.3%), methyl cinnamate (4.6%), eugenol (4.2%) and δ-cadinene (3.9%). Four landraces were found to contain high citral contents (56.6-65.6%). Linalool (3.2-5.3%), α-bisabolene (2.1-3.4%), geraniol (1.0-3.9%) and methyl eugenol (0.8-3.3%) were other prominent components. Methyl chavicol (60.3 and 76.3%) was dominant in two landrace. Other compounds present were linalool (1.7-8.4%), 1, 8-cineole (0.4-5.0%), δ-cadinene (t-5.4%), germacrene D (2.4-2.7%) and methyl eugenol (1.3-3.4%). One landrace contained methyl chavicol (41.8%) and citral (33.9%) as main components. Linalool (2.7%), trans-bcaryophyllene (2.5%) and α-bisabolene (2.1%) were other important components.

6. Antioxidant Activities

In particular eugenol, thymol, carvacrol and 4-allylphenol found in basil (usually present in trace quantities) exhibited varying amounts of anti-oxidative activity which may help to prevent cancer, premature aging, atherosclerosis, and diabetes [10]. Basil essential oil was found effective even at very low concentration (1%) in dropping those bacteria (number of Shigella) which may cause significant intestinal damage. Due to its anti microbial properties basil is becoming a permanent part of food items, natural food preservatives and fresh salads [11]. Anti- inflammatory effect of eugenol present in basil oil has been extensively studied as it is associated with symptomatic relief from rheumatoid arthritis or inflammatory bowel conditions. It works by causing a blockage of cyclooxygenase (COX) enzyme in the body.

Aspirin, ibuprofen and acetaminophen (non-steriodal over-the-counter anti-inflammatory medications (NSAIDS) work by inhibiting the same enzyme. This effect is somewhat controversial in case of acetaminophen and probably, its inhibition of COX occurs to a much lower degree than other drugs [12].

Javanmardi et al., [13] and Maskari et al., [6] reported that Iranian and Omani basils, respectively possess valuable antioxidant properties for culinary and possible medicinal use. It is now well know that oxidative DNA damage is responsible for aging, atherosclerosis, diabetes, neurodegenerative diseases and cancer [14]. Natural antioxidants contained in plants are useful in preventing the deleterious consequences of oxidative damage [15]. In addition to scavenging of reactive oxygen species (ROS), chelation of metal ions such as iron and copper which initiate radical reactions and inhibition of enzymes responsible for free radical generation [16], antioxidants can interfere with xenobiotic metabolizing enzymes, block activated mutagens/carcinogens, modulate DNA repair and even regulate gene expression [17]. Basil is the common medicinal and culinary herb, widely used as a potent antiseptic and preservative, digestive regulator, slight sedative, antioxidative, antiinflammatory, anti-diarrheal, antiulcer, chemopreventive, blood-sugar lowering, a nervous system stimulatory, radiation protection and diuretic and for headaches, coughs, infections of upper respiratory tract and kidney malfunction treatment [18-20]. Linalool present as one of major components in some basil species found to have protective effect against oxidative DNA damage [9, 21].

Glycosidically bounded volatile compounds are also gaining importance because of their antioxidant properties [22, 23]. Glycosidically bound volatile compounds could be interesting as hidden potential of antioxidant compounds in basil. Since volatile compounds can be released from nonvolatile glycoside precursors by enzymatic or chemical pathways during manufacturing process, these compounds can be considered as potential precursors of antioxidant substances in plant material and may contribute to the total antioxidant capacity of plants [24].

The Presence of lipophilic flavonoid aglycones on leaf surface of many species belonging to Lamiaceae presumably gives them protection against harmful UV radiation [25]. Additionally, lipophilic flavonoids may protect plants against infection by microorganisms, as several have been shown to have antibacterial [26] or antifungal activities [27]. Surface flavonoids reported from basil are nevadensin, salvigenin, hispidulin and xanthomicrol [25]. Basil species has been used for a long time as a medicinal plant, culinary herb and insect-controlling agent and is an important essential oil crop in some tropical and subtropical countries [28]. The unique array of flavonoids (especially water soluble orientin and viceninare) present in basil not only protects the cell structures but also save chromo-somes from radiation and oxygen-based damage. Basil essential oil provides protection against several unwanted microbial growths that have become resistant to commonly used antibiotic drugs [29].

The presence of basil in daily diet promotes the daily cardiovascular health. Basil is a good source of potassium, iron, magnesium, manganese, calcium, vitamin A, vitamin C and vitamin K. Carotenoids (such as beta-carotene) present in basil are not only act as "pro-vitamin A" but also act even more powerful antioxidant than vitamin A. Moreover, carotene-oids protect epithelial cells (the cells that form the lining of numerous body structures including the blood vessels) from free radical damage. Carotenoids also help in preventing free radicals from oxidizing cholesterol in the blood stream. Only build up of oxidized cholesterol in blood vessel walls could initiate atherosclerosis, whose end result can be a heart attack or stroke. Free radical damage is a contributing factor in asthma, osteoarthritis, and rheumatoid arthritis also. The beta-carotene present in basil is useful in cells protection from further damage and may help to lessen the progression of these conditions. Magnesium from basil can help in promotion of cardiovascular health by prompting muscles and blood vessels to relax [30-34].

Fresh basil is superior in flavor on dried form. Since effectiveness of basil is due to presence of volatiles in it, so it is best to add the herb near the end of the cooking process. Basil has a full array of nutrients, including carbohydrates, sugar, soluble and insoluble fiber, sodium, vitamins, minerals, fatty acids and amino acids. Presence of basil in daily diet ensures many health benefits [35].

CONCLUSION

Basil is cultivated as a culinary herb, condiment or spice; source of essential oil for use in foods, flavors, and fragrances; garden ornamental. The green aromatic leaves are used in either fresh or dried form as flavorings or spices in sauces, stews, salad dressings, vegetables, poultry, vinegar, confectionery products, and in the liqueur chartreuse. The therapeutic properties of basil oil are unlimited. In traditional medicine it is used as an antidepressant, analgesic, anti-venomous, carminative, antispasmodic, cephalic, digestive, diaphoretic, emmenagogue, febrifuge, insecticide, nervine, expectorant, stomachic, sudorific, tonic and stimulant. Future studies are needed on isolation and purification of its active components.

REFERENCES

[1] Hirasa, K., and Takemasa, M. (1998). *Spice science and technology*. Marcel Dekker: New York.

[2] Simon, J. E., Morales, M. R., Phippen, W. B., Vieira, R. F., and Hao, Z. (1999a). *Basil: a source of aroma compounds and a popular culinary and ornamental herb*. In J. Janick (Ed.), Perspectives on new crops and new uses (pp. 499–505). Alexandria, VA: ASHS Press.

[3] Simon, J.E., Quinn, J., and Murray, R.G. (1999b). *Basil: a source of essential oils*. In: Janick, J., Simon, J.E. (Eds.), Advanced in New Crops. Timber Press, Portland, OR, pp. 484-489.

[4] Exarchou, V., Nenadis, N., Tsimidou, M., Gerothanassis, I. P., Troganis, A., and Boskou, D. (2002). Antioxidant activities and phenolic composition of extracts from Greek oregano, Greek sage and summer savory. *Journal of Agricultural and Food Chemistry*. 50(19), 5294–5299.

[5] Al-Maskri, A.Y., Hanif, M.A., Al-Maskari, M.Y., Abraham, A.S., Al-Sabahi, J.N., Al-Mantheri, O., (2011). Essential oil from Omani basil (*Ocimum basilicum*): A desert crop. Natural Product Communications. 6(1-4) In press.

[6] Al-Maskari, M.Y., Hanif, M.A., Al-Maskari, A.Y., Al-Shakaily, A., Al-Maskari, A.Y. and Al-Sabahi, J.N., (2011). Essential oil composition, antimicrobial and antioxidant

activities of unexplored Omani basil. *Journal of Medicinal Plants Research.* 5(5), 751-757.

[7] Singh, S., Singh, M., Singh, A.K., Kalra, A., Yadav, A., and Patra, D.D. (2010). Enhancing productivity of Indian basil (*Ocimum basilicum* L.) through harvest management under rainfed conditions of subtropical north Indian plains. *Industrial Crops and Products.* 32, 601–606.

[8] Hussain, A.I., Anwar, F., Sherazi, S.T.H., and Przybylski, R. (2008). Chemical composition, antioxidant and antimicrobial activities of basil (*Ocimum basilicum*) essential oils depends on seasonal variations. *Food Chemistry.* 108, 986-995.

[9] Berić, T., Nikolić, B., Stanojević, J., Vuković-Gačić, B., Telci, I., Bayram, E., Yılmaz, G., and Avcı, B., (2006). Variability in essential oil composition of Turkish basils (*Ocimum basilicum* L.). *Biochemical Systematics and Ecology.* 34 (2006) 489-497.

[10] Lee, Seung-Joo., Umano, K., Shibamoto, T., and Lee, Kwang-Geun., (2005). Identification of volatile components in basil (*Ocimum basilicum* L.) and thyme leaves (*Thymus vulgaris* L.) and their antioxidant properties. *Food Chemistry,* 91, 131–137.

[11] Bagamboula, C.F., Uyttendaeleand, M. and Debevere, J. (2004). Inhibitory effect of thyme and basil essential oils, carvacrol, thymol, estragol, linalool and p-cymene towards Shigella sonnei andS. flexneri. *Food Microbiology.* 21(1):33-42.

[12] Javanmardi, J., Stushnoff, C., Locke, E., and Vivanco, J. M. (2003). Antioxidant activity and total phenolic content of Iranian *Ocimum* accessions. *Food Chemistry.* 83, 547–550.

[13] Olinski, R., Gackowski, D., Foksinski, M., Rozalski, R., Roszkowski, K. and Jaruga, P. (2002). Oxidative DNA damage: assessment of the role in carcinogenesis, atherosclerosis, and acquired immunodeficiency syndrome. *Free Radical Biology and Medicine.* 33(2), 192–200.

[14] Kris-Etherton, P.M., Hecker, K.D., Bonanome, A., Coval, S.M., Binkoski, A.E., Hilpert, K.F., Griel, A.E. and Etherton, T.D. (2002). Bioactive compounds in foods: their role in the prevention of cardiovascular disease and cancer. *American Journal of Medicine.* 113, 71–88.

[15] Edenharder, R. and Grunhage, D. (2003). Free radical scavenging abilities of flavonoids as mechanism of protection against mutagenicity induced by tert-butyl hydroperoxide or cumene hydroperoxide in Salmonella typhimurium TA102. *Mutation Research.* 540, 1–18.

[16] Nikolic□, B., Stanojevic□, J., Mitic□, D., Vukovic□-Gacic□, B., Knezevic□-Vukcevic□, J. and Simic□, D. (2004). Comparative study of the antimutagenic potential of vitamin E in different E. coli strains. *Mutation Research.* 564, 31–38.

[17] Tucakov, J., 1996. Lecˇenje Biljem, sixth ed.. In: Fitoterapija Rad, Beograd, pp. 247–248.

[18] Klem, M.A., Nair, M.G., Sraassburg, G.M. and Dewitt, D.L. (2000). Antioxidant and cyclooxygenase inhibitory phenolic compounds from *Ocimum sanctum* Linn. *Phytomedicine,* 7, 7–13.

[19] Maity, T.K., Mandal, S.C., Saha, B.P. and Pal, M. (2000). Effect of *Ocimum sanctum* roots extract on swimming performance in mice. *Phototherapy Research.* 14, 120–121.

[20] Miloshev, G., Mihaylov, I. and Anachkova, B. (2002). Application of the single cell gel electrophoresis on yeast cells. *Mutation Research.* 513, 69–74.

[21] Milos, M., Mastelic, J. and Jerkovic, I. (2000). Chemical composition and antioxidant effect of glycosidically bound volatile compounds from oregano (*Origanum vulgare* L. ssp. hirtum). *Food Chemistry.* 71, 79–83.

[22] Radonic, A., and Milos, M. (2003). Chemical composition and antioxidant test of free and glycosidically bound volatile compounds of savory (*Satureja montana* L. subsp. montana) from Croatia. *Nahrung/Food.* 47, 236–237.

[23] Politeo, O., Jukic, M., and Milos, M. (2007). Chemical composition and antioxidant capacity of free volatile aglycones from basil (*Ocimum basilicum* L.) compared with its essential oil. *Food Chemistry.* 101, 379–385.

[24] Tomds-Barberdn, F. A. and Wollenweber, E. (1990). *Plant Systematics and Evolution.* 173, 109.

[25] Harborne, J. B. and Baxter, H. (eds) (1993) *Phytochemical Dictionary.* Taylor and Francis, London and Washington, DC.

[26] Grayer, R. J. and Harborne, J. B. (1994) A survey of antifungal compounds from higher plants, 1982-1993. *Phytochemistry.* 37: 19-42.

[27] Pushpangadan, P. and Bradu, B. L. (1995) in Advances in Horticulture, Vol. 11, Medicinal and Aromatic Plants (Chadha K. L. and Gupta, R., eds), pp. 627-657. Malhotra Publishing House, New Delhi.

[28] Opalchenova, G. and Obreshkova, D. (2003) Comparative studies on the activity of basil--an essential oil from *Ocimum basilicum* L.-against multidrug resistant clinical isolates of the genera *Staphylococcus, Enterococcus* and *Pseudomonas* by usi. *Journal of Microbiology Methods.* 54(1), 105-110.

[29] Elgayyar, M., Draughon, F.A., Golden, D.A. and Mount, J.R. (2001) Antimicrobial activity of essential oils from plants against selected pathogenic and saprophytic microorganisms. *Journal of Food Protection.* 64(7):1019-24.

[30] Singh, A., Singh, S.P. and Bamezai, R. (1999). Modulatory potential of *Ocimum* oil on mouse skin papillomagenesis and the xenobiotic detoxication system. *Food Chemistry and Toxicology.* 37(6), 663-670.

[31] Orafidiya, L,O., Oyedele, A.O., Shittu, A.O. and Elujoba, A.A. (2001) The formulation of an effective topical antibacterial product containing *Ocimum gratissimum* leaf essential oil. *International Journal of Pharmacy.* 224, 177-183.

[32] Devi, P.U. (2001). Radioprotective, anticarcinogenic and antioxidant properties of the Indian holy basil, *Ocimum* sanctum (Tulasi). *Indian Journal of Experimental Biology.* 39(3):185-190.

[33] Vrinda, B. and Devi, P.U. (2001). Radiation protection of human lymphocyte chromosomes in vitro by orientin and vicenin. *Mutation Research.* 498, 39-46.

[34] Janick (Ed.), *Perspectives on new crops and new uses.* Alexandria, VA: ASHS Press. pp. 499–505.

In: Natural Products and Their Active Compounds ... ISBN: 978-1-62100-153-9
Editors: M. Essa, A. Manickavasagan, and E. Sukumar

Chapter 27

PROCESS OPTIMIZATION AND MINIMAL PROCESSING APPROACHES TO PRESERVE BIOACTIVE COMPOUNDS IN NATURAL PRODUCTS

***A. Manickavasagan*[1] *and S. Balasubramanian*[2]**

[1]Postharvest Technology and Research Laboratory,
Department of Soils, Water and Agricultural Engineering,
Sultan Qaboos University, Al-Khoud, Sultanate of Oman
[2]Central Institute of Post-harvest Engineering and Technology, Ludhiana, Punjab, India

INTRODUCTION

All over the world, health authorities and government agencies have been insisting to include more fruits and vegetables (eight to ten servings) in daily diet for the promotion and maintenance of good health. Fruits and vegetables may help in reducing the risk of high blood pressure, stroke and other cardiovascular diseases, type 2 diabetes, some cancers, developing kidney stones, bone losses and many other diseases (Harvard School of Public Health 2011; USDA 2011). Bioactive compounds and nutrients (such as potassium, dietary fiber, folate, vitamin A, vitamin E, vitamin C and so on)in fruits and vegetables are responsible for the desired health benefits(USDA 2011). In recent years, fruits and vegetables have been subjected to various treatments to develop new products, increase shelf life, blend with other products (such as dairy products) and for other reasons. During processing, the micro elements present in fruits and vegetables which are responsible for health benefits are also receiving various treatments and encounter losses and conversion into other forms. In addition to industrial processing, the storage and cooking conditions also make significant changes in these micro bioactive compounds. To gain the expected health benefits from the consumption of fruits and vegetables, the bioactive compounds should be preserved till they reach the consumer's table. Many studies are being conducted around the world about the process optimization and minimal processing approaches to prevent or minimize the losses of bioactive compounds from fruits and vegetables during processing, storage and cooking conditions. This chapter describes some of the new techniques which are becoming popular in food industries, and their role in the retention of bioactive compounds.

MINIMAL PROCESSING METHODS

Many definitions are available for the term "minimal processing "in various perspectives. It can be broadly defined as "the least possible treatment to achieve a purpose" (Manvell, 1996).Huisin't Veld (1996) defined minimal processing more specifically as "minimally influence the quality characteristics of a food whilst, giving the food sufficient shelf-life during storage and distribution". Fellows (2000) stated that minimal processing are techniques that "preserve foods but also retain to a greater extent their nutritional quality and sensory characteristics by reducing the reliance on heat as the main preservative action".

Minimal Processing by Non-Thermal Methods

In non-thermal minimal processing treatments, the food products are not heated up as in traditional heat-processing methods. Most popular non-thermal food processing methods are irradiation, high pressure and pulse electric field processing.

Irradiation Processing

During irradiation, food molecules absorb energy and form ions or free radicals which are highly reactive, and break a small percentage of chemical bonds. The cellular destruction caused by disrupting genetic material is the major effect of radiation on microorganisms in food (Ohlsson et al. 2002).Radiation processing or treatments of agricultural produces have been shown both increase and decrease of antioxidant content which is dependent on the dose delivered, exposure time, and the raw material treated. The enhancement in antioxidant capacity after irradiation is mainly due to either increased enzyme activity (such as phenyl-alanine ammonia-lyase and peroxidase activity) or to the increased extractability from tissues (Alothman et al. 2009).Furgeri et al. (2009) reported that gamma radiation processing (3, 5, 7 and 10 KGy) of mate herb (Ilex paraguariensis) did not make siginificant alterations in phenolic compounds.Alothman et al. (2009) reviewed the effect of irradiation processing on food phytochemicals. Table 1 summarizes the changes in antioxidant compounds in fruits and vegetables during irradiation processing.

High Pressure Processing (HPP)

While subjecting foods to high pressures in the range of 300 to 800 MPa, microorganisms and enzymes are inactivated without degradation in falvour and nutrients (Ohlsson et al. 2002).While evaluating the effect of high pressure treatments (400 and 600 MPa) in carrot, green bean and broccoli, it was determined that antioxidant capacity and total carotenoid content were not affected (McInerney et al. 2007). In another study, Patras et al. (2009) proved that loss of phenolic compounds was higher in thermal processing compared to HPP. They compared the effect of high pressure treatments (400, 500, 600 MPa/15 min/10-30°C) and thermal treatments (70°C /2 min) for strawberry and blackberry purees.

Table 1. Effect of irradiation on antioxidants in fruits and vegetables (Alothman et al 2009)

Product	Phytochemical	Result	Reference
Strawberries	Anthocyanins	UV-C doses at 0.25 and 1.0 kJ/m^2increased anthocyaninsconcentrations in the freshstrawberries	Baka et al. (1999)
Grape pomace	Anthocyanin	Pomace was gamma irradiated at0-9 kGy. Low doses of irradiation(below 2 kGy) prevented the loss ofanthocyanin while higher dosesdecreased the content ofanthocyanin	Ayed et al. (1999)
Strawberries	Phenolic acids	gammairradiation (1-10 kGy) led to thedegradation of cinnamic,p-coumaric, gallic, andhydroxybenzoic acids	Breitfellner et al. (2002)
Fresh-cut vegetables(Romaine, iceberg lettuce,endive)	Phenolic compounds	Gamma irradiated (0, 0.5, 1, and2 kGy) showed significant increasein the total phenolic content and antioxidant capacitycorresponding to the increasedtreatment time	Fan (2005)
Pomegranate arils (Punicagranatum cv. 'Mollar ofElche')	Anthocyanins	Exposure toUV-C(0.56-13.62 kJ/m^2)showed insignificant changes in theanthocyanins as well as theantioxidant capacity	Lopez-Rubira et al. (2005)
Tomato	p-hydroxybenz- aldehyde, p-coumaric acid, ferulic acid, rutin, naringenin	The gammairradiation treatment (2, 4,and 6 kGy) markedly reduced theconcentration of the phenoliccompounds	Schindler et al. (2005)
Fresh-cut mangoes	Phenoliccompounds, flavonoids,ascorbic acid, β-carotene	Fresh-cut mangoes UV-C irradiatedfor 0, 10, 20, and 30 min, showedincrease in phenolic compoundsand flavonoids contents with theincrease in treatment time, whileboth b-carotene and ascorbic aciddecreased	Gonzalez-Aguilaret al. (2007)
Blueberries (Vaccinium corymbosum, cvs. Collins,Bluecrop)	Anthocyanins, phenolic compounds	2 or 4 kJ/m^2 UV-C exposures didnot change the total phenolic content while it increased the totalanthocyanins content and FRAPvalues (Bluecrop cv.). For Collinscultivar, there was no significantchanges when compared with thecontrol fruits	Perkins-Veazie et al. (2008)

It was determined that antioxidant capacities of pressure treated purees were significantly higher than thermal treated purees. Therefore HPP at moderate temperatures was recommended to produce nutritious and fresh like purees. Keenan et al. (2010) compared the effect of HPP (450 MPa: 1, 3, and 5 min) and thermal treatment (end-point temperature $70^{o}C > 10$ min) on antioxidant activity in fruit smoothies. Thermally processed smoothies had a higher level of antioxidants after processing when compared to HPP. But an opposite trend was observed in total phenol content. HPP samples had higher phenolic content than thermally processed smoothies. Therefore to develop HPP processing methods, each product might require different protocols.

Pulse Electric Field Processing (PEF)

PEF concept was originally introduced in food sector to inactivate microorganism present in the food. When an electric field is applied to a food in short pulses (1 to 100 ms) form, there is a lethal effect in microorganism (Ohlsson et al. 2002). Recently, PEF technology is being investigated for various applications in food such as blanching, sterilization and partial cooking. The schematic diagram for PEF treatment of liquid foods is given in Figure 1.

Elez-Martinez and Martin-Belloso (2007) studied the effect of PEF processing on vitamin C retention and antioxidant capacity of orange juice in comparison with heat pasteurization. Vitamin C retention was higher in PEF than heat pasteurization. There were no differences in antioxidant activity between PEF processed and untreated juices whereas heat-treated foods showed lower values of antioxidant capacity. The changes in selected bioactive compounds during pulse electric treatments are summarized in Table 2.

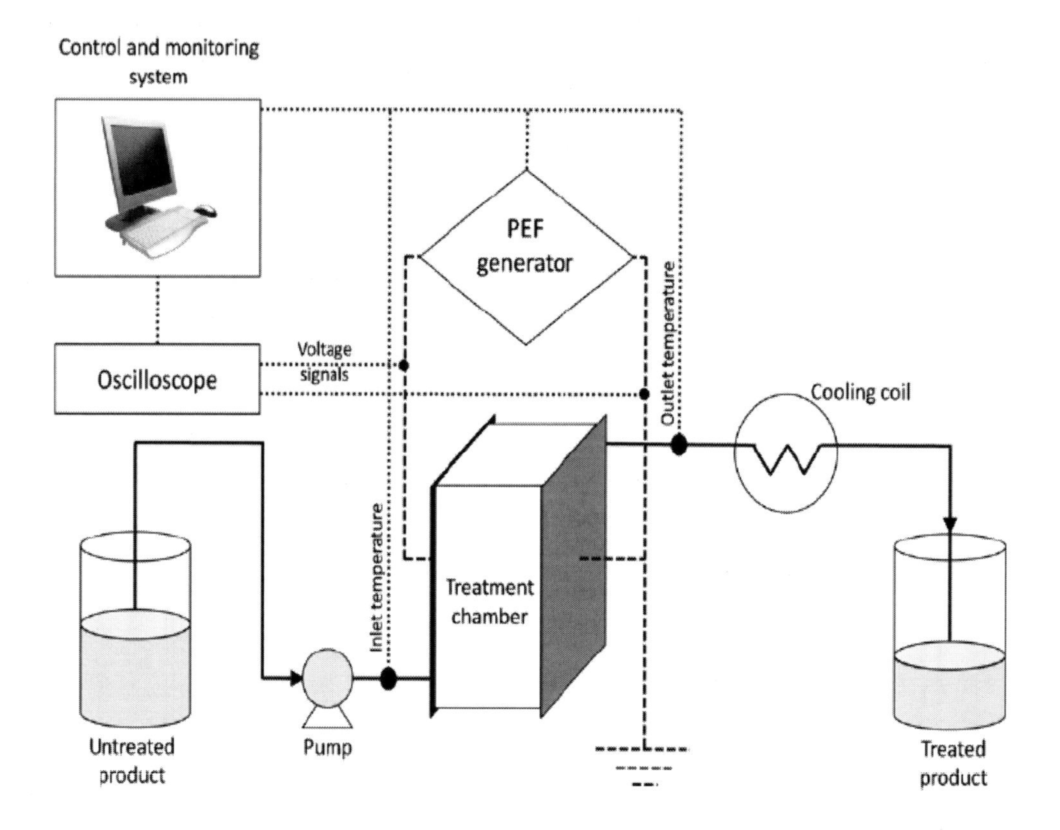

Figure 1. Schematic diagram of a pulse electric field (PEF) processing system for liquid products (Soliva-Fortuny et al. 2009).

Minimal Processing by Thermal Methods

Thermal methods have been traditionally used in various unit operations (such as blanching, drying, frying and so on) during food processing. Thermal treatments make desired changes in protein coagulation, starch swelling, textural softening, and formation of aroma compounds.

Table 2. Effect of pulse electric field processing on bioactive compounds in foods (Soliva-Fortuny et al. 2009)

Compound	Food	PEF treatment	Effects	Reference
Flavonoids	Orange juice	35 kV/cm, 750 μs	No changes in either individualflavanones nor in total content	Sanchez-Moreno et al. (2005)
Carotenoids	Orange juice	25-40 kV/cm, 30-340 ms	No significant changes in overall content.Better stability of individual compoundscompared to thermal pasteurization	Cortes et al. (2006)
	Orange-carrot juice blend	25-40 kV/cm, 30-340 μs	Rise in carotenoids content withincreasing treatment time Increase ofcompounds with provitamin A effectat 25 and 30 kV/cm compared toheat treatments	Torregrosa et al. (2005)
	Tomato juice	40 kV/cm, 57 μs	No changes in lycopene withrespect to thermal treatment	Min et al. (2003)
Vitamin B1	Milk	18-3-27.1 kV/cm, up to 400 μs	Very low or negligible reductions	Bendicho et al. (2002)
Vitamin B2	Milk	18-3-27.1 kV/cm, up to 400 μs	Very low or negligible reductions	Bendicho et al. (2002)
Vitamin C	Protein fortifiedorange juice-basedbeverage	28 kV/cm, 100-300 μs	Loss increasing from 4 to 13% as treatment time increased	Sharma et al. (1998)
	Orange juice	87 kV/cm, 40 instantcharge reversal pulses, 50°C	Very low or negligible reductions	Hodgins et al. (2002)
		15-35 kV/cm, 100-1000 μs	Loss ranging from 1.8 to 12.5%	Elez-Martınez and Martın-Belloso (2007)
	Grape juice	reversal pulses, 50°C	Very low or negligible reductions	Wu et al (2005)
	Apple juice and cider	22-35 kV/cm, 94-166 μs	Very low or negligible reductions	Evrendilek et al. (2000)
	'Gazpacho' soup	15-35 kV/cm, 100-1000 μs	Loss ranging from 2.9 to 15.7%	Elez-Martınez and Martın-Belloso (2007)
	Milk	22.6 kV/cm, 400 μs	6.6% depletion	Bendicho et al. (2002)
Vitamin D	Milk	18-3-27.1 kV/cm, up to 400 μs	Very low or negligible reductions	Bendicho et al. (2002)
Vitamin E	Milk	18-3-27.1 kV/cm, up to 400 μs	Very low or negligible reductions	Bendichoet al. (2002)

However, undesirable changes such as loss of bioactive compounds, vitamins and minerals, formation of thermal reaction components of biopolymers and loss of fresh appearance, flavour and texture also occur during thermal processing (Ohlsson and Bengtsson 2002).

Most of the traditional thermal methods involve long time-high temperature treatments especially in water medium which leads to leaching of several bioactive compound from food products. In minimal processing by thermal method approaches, the food products are heated through non-conventional methods (such as infrared heating, microwave heating, ohmic heating and so on), for a shorter period of time with or without water medium. Picouet et al. (2009) used microwaves to process apple puree in a continuous mode (625 W, 35 s). It was determined that total polyphenol content of the apple puree was not affected after microwave treatment. However there were reductions after 5 d of treatment during cold storage (5°C) and then total phenol content was stabilized after 2 weeks of storage.

PROCESS OPTIMIZATION METHODS

Plant materials are sensitive to temperature, pressure and other processing conditions. Functional properties of each food require unique set of processing conditions to be stable in their natural form.Therefore it is essential to optimize processing conditions for each product separately. Nicoli et al. (1999) explained various probable changes in the antioxidant activity of food products during processing and preservation (Table 3).

Table 3. Changes in antioxidant activity of food due to processing and preservation methods (Nicoli et al. 1999)

Change	Probable causes
No changes	No changes in naturally occurring antioxidant concentration Loss of naturally occurring antioxidants balanced by the simultaneous formation of compounds with novel or improved antioxidant properties
Increase	Improvement of antioxidant properties of naturally occurring compounds Formation of novel compounds having anti-oxidant activity (i.e. Maillard reaction products)
Decrease	Loss of naturally occurring antioxidants Formation of novel compounds having pro-oxidant activity (i.e. Maillard reaction products)

Extraction Methods

Liazid et al. (2010) studied the effect of extraction techniques on bioactivity of bioactive compounds in pine seeds. They used four extraction methods (maceration with magnetic stirring (MSAE), ultrasound-assisted extraction (UAE), microwave assisted extraction (MAE), and extraction with pressurized liquids (PLE))to obtain polyphenolic extract from pine seeds. UAE technique yielded higher recoveries (almost double) than MAE and MSAE. The greater antioxidant power of extracts was observed in UAE and PLE. Even though extracts extracted at higher temperature had greatest polyphenol content, an inverse relationship was found between degree of bioactivity and extraction temperature. Bimakra et al. (2010) compared two extraction methods (conventional soxhlet extraction (CSE) and supercritical carbon dioxide extraction (SC-CO_2)) at different temperatures (30, 40 and 50), pressures (100, 200 and 300 bar) and time (30, 60, and 90 min) for the extraction of flavonoids from spearmint (Menthaspicata L.) leaves. It was determined that CSE had a higher crude extract yield over SC-CO_2. However, the optimized SC-CO_2 was found to have more main flavonoid compounds with higher concentration than CSE. Ghafooretal. (2010) optimized the supercritical extraction parameters for grape peel using response surface methodology. The optimized parameters were 45-46°C temperature, 160-165 kg/cm^2 pressure, and 6-7% ethanol as modifier. It was recommended that CO_2 can be used as supercritical fluid (an environmental friendly solvent) which allows extracting the bioactive compounds at lower temperatures.

Cooking Conditions and Food Additives

Antioxidant properties of six varieties of pepper in 3 types of cooking conditions (microwave heating, stir-frying and boiling in water) were studied by Chuah et al. (2008). It was determined that after five minutes cooking by microwave and stir-frying, there were no significant difference in radical-scavenging activity, total polyphenol content, and ascorbic acid content. However these compounds were reduced significantly when the pepper sample was cooked by boiling in water. Robel-Sanchez et al. (2009) measured antioxidant activity of cut mango after dipping treatment with ascorbic acid + citric acid + calcium chloride and storage at 5°C. Dipped mango cubes had higher antioxidant activity than controlsafter storage.

PACKAGING AND STORAGE

Increase of consumer's awareness of health benefits of fruits and vegetables and the need for convenience due to the fast-faced life style have increased the demand for minimally processed and ready to eat fruits and vegetables (Odrizola-Serrano et al. 2008). In some cases, minimal processing of fruits and vegetables involves combination of procedures such as washing, peeling, and slicing or shredding. This may cause an increase in respiration, biochemical changes and microbial spoilages. Therefore to extend the shelf life of these products special packaging such as modified atmosphere packaging in combination with refrigerated storage is required (Plaza et al. 2011). And also to avoid cut surface browning, complementary treatments such as excluding oxygen, adding antioxidants or inhibiting the activity of responsible enzymes must be applied in addition to cold storage to facilitate distribution (Robel-Sanchez et al. 2009).

The processing industries need appropriate packaging techniques for minimally processed fruits and vegetables in order to provide high quality functional commodities (Perez-Gregorio et al. 2011).After minimal processing, the storage temperature and atmosphere and type of packaging play a vital role in the retention of bioactive compounds in food products.

In modified atmospheric packaging (MAP), the air in the package is replaced with a fixed gas mixture, then no further control of the gas composition is exercised. Oxygen, nitrogen and carbon dioxide are the major gases used in MAP of foods. In controlled atmospheric storage (CAS), the atmosphere composition is maintained and controlled throughout the storage time (sivertsvik et al. 2002). Odrizola-Serrano et al. (2008) investigated the effect of modified atmospheric storage (5% O_2 and 5% CO_2) of fresh cut tomato (7 mm thick slices) on the retention of bioactive compound during storage. There were no significant differences in lycopene, vitamin C, phenolic compounds and antioxidant capacity between whole tomato and stored cut tomato using MAP. This minimal processing approach retained the antioxidant compounds of sliced tomato for 21 days at 4°C storage.Simoes et al. (2011) investigated the effect of controlled atmosphere in phenolic compounds of baby carrot. The samples were stored under four controlled atmosphere (air, 2kPa O_2 + 15 kPa CO_2, 5 kPa O_2 + 5 kPa CO_2 and 10 kPa O_2 + 10 kPa CO_2) at 4°C. The controlled atmosphere of 5 kPa O_2 + 5 kPaCO_2 wasrecommended as an optimum storage to maintain quality and increase bioactive compounds such as chlorogenic acid.

The flavonoids changes in fresh-cut onions were analyzed during storage in different packaging systems by Perez-Gregorio et al. (2011). Fresh cut onions were packed in closed plastic cups or under vacuum conditions taking into the account of light exposure and stored at 1-2°C. Transparent polystyrene cups stored under light yielded better performance: enhanced increase of total flavonoid by 58% and an increase in total anthocyanins of 39%.

Plaza et al. (2011) studied the bioactive compounds in minimally processed orange fruits (whole fruits, hand-peeled fruits, and manually separated segments) packed under air atmosphere, and stored at 4°C for 12 days. No significant changes in total carotenoid content and vitamin A was observed for peeled or segmented samples. The antioxidant values remained stable with respect to the initial values.

In some products significant losses were recorded in MAP. The phytochemical changes in fresh-cut jack fruits during modified atmospheric storage (3kPa O_2 + 5 kPa CO_2) after pre-treatments (30 min dipping in "$CaCl_2$ + ascorbic acid + citric acid + sodium benzoate) were investigated by Saxena et al. (2009). After 35 days storage at 6°C, a loss of around 7%, 8%, 43% and 31% for total phenolics, total flavonoids, total carotenoids and ascorbic acid, respectively.

In conclusion, consumption of fruits and vegetables which were subjected to various processing with major changes in bioactive substances from their natural form would not yield expected health benefits. Research and development in minimal processing and process optimization techniques must be intensified by keeping retention of valuable bioactive compounds as prime factor. Finally implementation of those developed technologies in the industries,even if the cost of production is higher than traditional processing, is important to benefit the consumers.

REFERENCES

Alothman, M., Bhat, R., and Karim, A.A. 2009. Effects of radiation processing on phytochemicals and antioxidants in plant produce. *Trends in Food Science and Technology*. 20: 201-212.

Ayed, N., Yu, H.L. and Lacroix, M. 1999.Improvements of anthocyanin yield and self-life extension of grape pomace by gamma irradiation. *Food Research Intranational.* 32: 539-543.

Baka, M., Mercier, J., Corcuff, R., Castaigne, F. and Arul, J. 1999.Phytochemical treatment to improve storability of fresh strawberries. *Journal of Food Science*. 64: 1068-1072.

Bendicho, S., Espachs, A., Ara□ntegui, J., and Martın, O. 2002. Effect ofhigh intensity pulsed electric fields and heat treatments on vitaminsof milk. *Journal of Dairy Research.* 69: 113-123.

Bendicho, S., Estela, C., Giner, J., Barbosa-Canovas, G. V., and Martın, O. 2002. Effects of high intensity pulsed electric field andthermal treatments on a lipase from Pseudomonas fluorescens.*Journal of Dairy Science.*85: 19-27.

Bimakra, M., Rahmana, R.A., Taipa, F.S.,Ganjloo, A.,Salleh, L.M., Selamat, J., Hamid, A., and Zaidulc, I.S.M. 2010. Comparison of different extraction methods for theextraction of major bioactive flavonoid compounds fromspearmint (*Menthaspicata*L.) leaves. *Food and Bioproducts Processing.*Doi: 10.1016/j.fbp.2010.03.002.

Breitfellner, F., Solar, S. and Sontag, G. 2002.ffect of gamma radiation on phenolic acids in strawberries. *Journal of Food Science.* 67: 517-521.

Chuah, A.M., Lee, Y., Yamaguchi, T., Takamura, H., Yin, L and Motaba, T. 2008.Effect of cooking on the antioxidant properties of coloured peppers. *Food Chemistry.* 111:20-28.

Cortes, C., Esteve, M. J., Rodrigo, D., Torregrosa, F., andFrıgola, A.2006. Changes of colour and carotenoids contents during highintensity pulsed electric field treatments in orange juices. *Food andChemical Toxicology, 44*: 1932-1939.

Elez-Martınez, P., and Martın-Belloso, O. 2007. Effects of high intensitypulsed electric field processing conditions on vitamin C and antioxidantcapacity of orange juice and gazpacho, a cold vegetablesoup. *Food Chemistry.* 102: 201-209.

Evrendilek, G., Jin, Z. T., Ruhlman, K. T., and Qiu, X. 2000. Microbialsafety and shelf-life of apple juice and cider processed by benchand pilot scale PEF systems. *Innovative FoodScience and EmergingTechnologies.*1: 77-86.

Fan, X. 2005. Antioxidant capacity of fresh-cut vegetables exposed to ionizing radiation, *Journal of the Science of Food and Agriculture.* 85: 995-1000.

Fellows, P. 2000. Food processing technology: Principles and practice, Woodhead publishing Limited, Cambridge, UK.

Furgeri, C., Nunes, T.C.F., Fanaro, G.B., Souza, M.F.F., Bastos, D.H.M., and Vallavicencio, A.L.C.H. 2009. Evaluation of phenoloc compounds in mate (Ilex paraguariensis) processed by gamma radiation. *Radiation Physics and Chemistry.* 78: 639-641.

Ghafoor. K., Park, J., and Choi, Y. 2010. Optimization of supercritical fluid extraction of bioactive compounds from grape (Vitislabrusca B.) peel by using response surface methodology. *Innovative Food Science and Emerging Technologies.* 11: 485-490.

Gonzalez-Aguilar, G.A., Villegas-Ochoa, M.A., Martinez-Tellez, m.A., Gardea, A.A., and Ayala-Zavala, J.F. 2007. Improving antioxidant capacity of fresh-cut mangoes treated with UV-C. *Journal of Food Science.* 72: S197-S202.

Harvard School ofPublic Health. 2011. The Nutrition Source: Vegetables and Fruits. Available at: www.hsph.harvard.edu/nutritionsource/what-should-you-eat/ vegetables-and-fruits/index.html. Accessed on May 08, 2011.

Hodgins, A. M., Mittal, G. S., and Griffiths, M. W. 2002. Pasteurizationof fresh orange juiceusing low energy pulsed electrical field.*Journal of Food Science.* 67(6):2294-2299.

HuisIn'd Veld. 1996. Minimal processing of food: Potential, challenges and problems. EFFoST Conference on the Minimal Processing of Food, Cologne, 6-9 November.

Keenan, D.F., Brunton, N.P., Gormley, T.R., Butler, F., Tiwari, B.K., and Patras, A. 2010.Effect of thermal and high hydrostatic pressure processing on antioxidant activity and colour of fruit smoothies. *Innovative Food Science and Emerging Technologies.* 11: 551-556.

Liazid, A., Schwarz, M., Varela, R.M., Palma, M., Guillen, D.A., Brigui, J., Macias, F.A., and Barroso, C.G. 2010. Evaluation of various extraction techniques for obtaining bioactive extracts from pine seeds. *Food and Bioproducts Processing.* 88: 247-252.

Lopez-Rubira, V., Conesa, A., Allende, A., and Artes, F. 2005. Shelf life and overall quality of minimally processed pomegranate arils modified atmosphere packaged and treated with UV-C. *Postharvest Biology and Technology*. 37: 174-185.

Manvell, C. 1996. Opportunities and problems of minimal processing and minimally processed foods.*EFFoST Conference on the Minimal Processing of Food.* Cologne, 6-9 November.

McInerney, J.K., Seccafien, C.A., Stewart, C.M., and Bird, A.R. 2007.Effects of high pressure processing on antioxidant activity, and total carotenoid content and availability in vegetables. *Innovative Food Science and Emerging Technologies.* 8: 543-548.

Min, S., Jin, Z. T., and Zhang.2003. Commercial scale pulsed electricfield processing of tomato juice. Journal of Agricultural and FoodChemistry, 51(11): 3338-3344.

Nicoli, M.C., Anese, M., and Parpinel, M. 1999. Influence of processing on the antioxidant properties of fruit and vegetables. *Trends in Food Science and Technology.* 10: 94-100.

Odriozola-Serrano, I, Solvie-Fortuny, R. and Martin-Belloso, O. 2008. Effect of minimal processingon bioactive compounds and color attributes of fresh-cut tomatoes. LWT 41: 217-226.

Ohlsson, T and Bengtsson, N. 2002.Minimal processing of foods with thermal methods. In Minimal processing technologies in the food industry, edited by Ohlsson, T and Bengtsson, N.CRC Press LLC, FL, US.

Ohlsson, T, Gothenburg and Bengtsson, N. 2002.Minimal processing of foods with non-thermal methods. In Minimal processing technologies in the food industry, edited by Ohlsson, T and Bengtsson, N.CRC Press LLC, FL, US.

Patras, A., Brunton, N.P., Pieve, S.D., and Butler, F. 2009.Impact of high pressure processing on total antioxidant activity, phenolic, ascorbic acid, anthocyanin content and colour of strawberry and blackberry purees. *Innovative Food Science and Emerging Technologies.* 10: 308-313.

Perez-Gregorio, M.R., Garcia-Falcon, M.S., and Simal-Gandara, J. 2011. Flavonoids changes in fresh-cut onions during storage in different packaging systems. *Food Chemistry.* 124: 652-658.

Perkins-Veazie, P., Collins, J.K. and Howard, L. 2008.Blueberry fruit response to postharvest application of ultraviolet radiation. *Postharvest Biology and Technology.* 47: 280-285.

Picouet, P.A., Landl, A., Abadias, M., Castellari, M and Vinas, I. 2009.Minimal processing of a Granny Smith apple puree by microwave heating. *Innovative Food Science and Emerging Technologies.*10: 545-550.

Plaza, L., Crespo, I., Pascual-Teresa, S., Ancos, B., Sanchez-Moreno, C., Munoz, M., and Cano, M.P. 2011.Impact of minimal processing on orange bioactive compounds during refrigerated storage. *Food Chemistry.* 124: 646-651.

Robles-Sanchez, R.M., Rojas-Grau, M.A., Odizola-Serranao, I., Gonzalez-Anguilar, G.A., Martin-Belloso, O. 2009.Effect of minimal processing on bioactice compounds and antioxidant activity of fresh-cut 'Kent'mango (MangiferaIndica L.). *Postharvest Biology and Technology.* 51: 384-390.

Sanchez-Moreno, C., Plaza, L., Elez-Martı□nez, P., De Ancos, B.,Martı□n-Belloso, O., and Cano, M. P. 2005.Impact of high pressureand pulsed electric fields on bioactive compounds and antioxidantactivity of orange juice in comparison with traditional thermalprocessing. *Journal of Agricultural and Food Chemistry.* 53: 4403-4409.

Saxena, A., Bawa, A.S. and Raju, P.S. 2009. Phytochemical changes in fresh-cut jackfruit (Artocarpusheterophyllus L.) bulbs during modified atmospheric storage. *Food Chemistry.* 115: 1443-1449.

Schindler, M., Solar, S. and Sontag, G. 2005. Phenolic compounds in tomatoes. Natural variations and effect of gamma-irradiation. *European Food Research and Technology.* 221: 439-445.

Sharma, S. K., Zhang, Q. H., and Chism, G. W. 1998. Development ofa protein fortified fruit beverage and its quality when processedwith pulsed electric field treatment. *Journal of Food Quality.* 21(6): 459-473.

Simoes, A.D.N., Allende, A., Tudela, J.A., Puschmann, R., and Gil, M.I. 2011. Optimum controlled atmospheres minimise respiration rate and quality losses while increase phenolic compounds of baby carrot. LWT – *Food Science and Technology.* 44: 277-283.

Sivertsvik, M., Rosnes, J.T., and Bergslien, H. 2002.Modified atmosphere packaging. In Minimal processing technologies in the food industry, edited by Ohlsson, T and Bengtsson, N.CRC Press LLC, FL, USA.

Soliva-Fortuny, R., Balasa, A., Knorr, D., and Martin-Belloso, O. 2009. Effects of pulse electric fields on bioactive compounds in food: a review. *Trends in Food Science and Technology.* 20: 544-556.

Torregrosa, F., Corte□s, C., Esteve, M. J., andFrıgola, A. 2005. Effect ofhigh-intensity pulsed electric fields processing and conventionalheat treatment on orange-carrot juice carotenoids. *Journal of Agriculturaland Food Chemistry.* 53: 9519-9525.

USDA (United States Department of Agriculture). 2011. My Pyramid : Food Groups – Vegetables. Available at: http://www.mypyramid.gov/pyramid/vegetables_why.html. Accessed on May 08, 2011.

Wu, Y., Mittal, G. S., and Griffiths, M.W. 2005.Effect of pulsed electricfield on the inactivation of microorganisms in grape juices with andwithout antimicrobials.*Biosystems Engineering.* 90(1): 1-7.

In: Natural Products and Their Active Compounds ... ISBN: 978-1-62100-153-9
Editors: M. Essa, A. Manickavasagan, and E. Sukumar

Chapter 28

PROCESSING AND MEDICINAL VALUES OF SELECTED SPICES

***S. Balasubramanian*[1] *and A. Manickavasagan*[2]**

[1]Central Institute of Post-harvest Engineering and Technology,
Ludhiana, Punjab, India
[2]Postharvest Technology and Research Laboratory,
Department of Soils, Water and Agricultural Engineering,
Sultan Qaboos University, Al-Khoud, Sultanate of Oman

All over the world spices have been used in food for a long time to enhance flavor and taste. Most of the spices have potential chemo preventive properties (Lai and Roy 2004). They have been widely studied for their medicinal values such as influence in lipid metabolism, fat absorption, hypotriglyceridemic and hypocholesterolemic activity, cholesterol turnover to bile acid, anti-lithogenic activity, anti-diabetic activity, antioxidant potential, anti-inflammatory activity, anti-mutagenic and anti-carcinogenic activity, digestive stimulant action, influence in platelet aggregation, protection of erythrocyte integrity, metabolic disposition of active principles and so on (Srinivasan 2005).

Species, cultivar, agro-climatic field condition, postharvest processing, storage and cooking conditions greatly influence the availability of bioactive compounds in spices. New initiatives are being taken place in processing of spices to minimize the losses of bioactive compounds.

Grinding is a common unit operation in processing of many spices. In general, traditional milling generates heat during grinding and that leads to loss of volatile compounds. Three types of grinders such as hammer mill, attrition mill and pin mill are commonly used for spices. In modern spice milling, usually pin mills are used for better grinding performances. In addition to the cooling effect of LN_2, pin mill produces finer size particles.

Pesek and Wilson (1986) reported that cryogenically ground spices had better colour retention compared to those ground in traditional mills. Murthy *et al.* (1996) found that cryogenic grinding of black pepper in the laboratory scale grinding system resulted better product characteristics.

Singh and Goswami (1999a) studied the effect of cryogenic temperatures on volatile oil content of cumin seed and clove, and reported that cryogenic grinding improved quality of ground samples. Manohar and Sridhar (2001) explained the effect of cryogenic grinding on particle size distribution of turmeric and confirmed the usefulness of cryogrinding process for heat-sensitiveness.

Singh and Goswami (1999b) studied the effect of grinding temperature on volatile oil content of the ground powder of cumin seed. The volatile oil content decreased with increase in grinding temperature (40 to 85°C). However, increase in grinding temperature from -110 to -50°C had no significant effect in volatile oil content (13.31 to 13.16 ml/l00 g). But the volatile oil significantly decreased (11.0 to 9.3 ml / 100 g) in ambient grinding. Cryogenic grinding of cloves yielded 129.5% of volatile oil recovery compared to that of ambient grinding of clove (Singh and Goswami 1999b).

Murthy and Bhattacharya (1998) studied the effect of feed rate (7 to 60 kg/h) and product temperature (-15 to 60°C) during grinding (pin mill) of black pepper on product quality. The volatile oil yield in cryogenic grinding was higher (1.42 to 1.91 ml/100 g) than that of ambient grinding system (0.78 to 0.98 ml/100 g). The average monoterpenes content was 0.80 ml and 0.15 ml/100 g for cryogenic and for ambient ground samples, respectively. An optimum feed rate (47 to 57 kg/h) and product temperature (-15 to -20°C) were recommended for effective milling.

Table 1 explains various aspects of traditional and cryogenic grinding systems for spices.

Table 1. Comparison between traditional and cryogenic grinding (Singh and Goswami 1999a, 2000; Goswami and Manish 2003, Murthy and Bhattacharya 2008)

Parameter	Cryogenic grinding	Traditional grinding
Energy consumption	Low	High
Throughput	High	Low
Mill clogging	No Clogging	Frequent
Volatile losses	Minimum	Higher
Motor capacity	Low	High
Control on particle size	Effective	No control
Grinding of soft material	Possible	Very difficult
Fire risk	No	High
Air pollution	No	Yes
Microbial load	Does not exist	Possible

Even though several researches have been started for various aspects of grinding, literatures available for many other operations in processing of spices are very scarce. More studies have to be conducted in various unit operations in processing of spices by means of experiments and modeling approaches. This chapter explains the traditional processing, products, chemical composition and medicinal uses of five selected spices (pepper, coriander, cinnamon, fenugreek and turmeric).

PEPPER (PIPER NIGRUM)

Black pepper, known as the 'king of spices', is a flowering vine. It belongs to the family *Piperaceae*. Pepper is a tropical plant, which grows in the temperature range of 20 to 40°C and moderate winter climate. Pepper requires about 2000 mm rainfall, soil with good water holding capacity and pH of 5.5 to 6.0. It can grow well in the red dolerite soils and red andesite soils with high humus content.

Processing

Harvesting

It is an important operation in producing high quality pepper products. Generally harvesting of pepper spike is done when the berries start to turn yellow/orange colour and firm in texture. The harvesting stages vary with respect to the end product (Table 2).

Table 2. Pepper harvesting stages for different products (Govindarajan 1979)

Target products	Maturity at harvest
White pepper	Fully ripe
Black pepper	Fully mature and near ripe
Canned pepper	4 to 5 months
Dehydrated green pepper	10 to 15 days before full maturity
Oleoresin/pepper oil	15 to 20 days before maturity
Pepper powder	Fully mature with maximum starch

Brief Fermentation

Harvested spikes are kept in bags for 12 to 24 h or heaped and covered overnight for a brief fermentation which makes despiking easy.

Threshing (Decorning)

Berries are removed from the spikes by hand, beating with sticks or trampling.

Cleaning

It is done by manual or mechanical means followed by washing with water for two to three times.

Grading before Drying

The cultivars are classified into different groups based on the berry size (Table 3).

Blanching

Blanching (80°C, 2 min) is done before drying to accelerate the drying process and browning of berries to attain black shine colour (Pruthi 1992). Blanching improves colour and also removes dust and adhering microbial contamination. Volatiles and other chemical loss is minimum by this treatments (NRCS 1987). Blanching activate the phenolase enzyme (responsible for producing the black colour). It ruptures the cells and thereby accelerates the escape of moisture from inner core and simultaneously enhances the black colour with the help of resinoids inside the berry.

Drying

This is an important process to reduce the mold growth. Care should be taken to prevent over drying to avoid loss of flavor components. Generally, final moisture content of pepper should be less than 10%. The black colour that pepper acquires on drying is due to the oxidation of colourless phenolic compounds present in the skin. Polyphenolase (0-diphenol oxidase) present in the fruit wall converts these colourless phenolic substrates (3,4 dihydroxy phenyl ethanol glycoside) present in the cells to black polymeric compounds (Variyar *et al.* 1988).

Table 3. Grading of Indian pepper based on size (Gopalam *et al.* 1991)

Grade	Variety
Large (>4.25mm)	Panniyur-1 Valiakaniakkadan Vadakkan Karuvilanchi
Medium (3.25 to 4.25 mm)	Karimunda Arakulammunda Ottaplackal Kuthiravally Kaniakkadan Neelamundi Balankotta
Small (<3.25mm)	Kurialmundi Narayakodi

Sun Drying

Sun drying is widely practiced in most of the countries. The fresh and dry yield depends on temperature and humidity during flowering, soil moisture, fertilizer availability, timely rain and so on (Govindarajan 1979).

To get good quality product, it is essential to use proper drying surface. The common drying surfaces used in India are bamboo mat, cement floor and polyethylene fabric.

Grading

Generally grading is based on size, colour and relative density.

Packaging and Storage

For pepper, packaging and storage conditions are important as it is hygroscopic in nature and its starch content result in mould attack and insect infestation. Whole pepper is packed in gunny bags and polyethylene lined double burlap bags. Dried pepper (10 to11%) is stored in jute gunny bags with polyethylene lining of 0.003 inch or more thick or in laminated HP bags or similar containers (Balasubramanyam *et al.* 1978). Polypropylene is generally used for packing of ground pepper.

Products

Black Pepper

Green unripe drupes of pepper are cooked briefly in hot water, both to clean and to prepare them for drying. This thermal treatment ruptures the cell walls and catalysis browning enzymes. During drying, seed shrinks and darkens into a thin, wrinkled black layer.

Green Pepper

Dried green peppercorns are treated with sulfur dioxide or freeze-dried to retain the green colour. Pickled peppercorns are the unripe drupes preserved in brine or vinegar.

Orange Pepper/Red Pepper

Ripe red pepper drupes are preserved in brine and vinegar. It can be dried by color-preserving techniques such as sulfur dioxide, freeze-drying and so on.

Dehydrated Green Pepper

Immatured green pepper fruits are blanched (80°C, 2 min), drained, cooled and then soaked in sulphur dioxide solution to fix the green colour followed by drying at 50°C.

Canned Green Pepper

Despiked pepper fruits are soaked in water containing 20 ppm residual chlorine for 1 h and covered with 2% hot brine solution (0.2% citric acid), is exhausted at 80°C, and sealed. It is cooled immediately in a stream of running cold water and then 2% acetic acid is added to give a better colour.

Bottled Green Pepper

Despiked, clean, fresh green fruits are steeped in 20% brine solution containing citric acid for 3 to 4 weeks. The excess liquid is drained off and fresh brine of 16% concentration together with 100 ppm sulfur dioxide and 0.2% citric acid is added and the product is stored properly.

Dry Packed Green Pepper

This product is prepared just like the bottled green pepper except that the liquid at the final stage is drained off and packed in flexible pouches.

Freeze Dried Pepper

This pepper retains the original green colour and shape.

White Pepper

It is prepared by removing the outer rind of the black pepper and is sometimes used in dishes (light-colored sauces or mashed potatoes), where ground black pepper would visibly stand out. They have differing flavor due to the presence of certain compounds in the outer fruit layer of the berry that are not found in the seed.

Black and White Pepper Powder

Ground pepper is obtained by grinding pepper (hammer mill or plate mill) without adding any ingredients to it.

Cryoground Pepper

This is obtained by grinding the pepper below -100°C which will prevent the oxidation of oil.

Pepper Oil

Pepper oil is recovered by steam or water distillation.

Oleoresin

Oleoresin represents the total pungency and flavor constituents of pepper. It is obtained from ground pepper extraction using solvents like ethanol, acetone, ethylene, dichloride, ethyl acetate and so on.

Microencapsulated Flavor

It is prepared by entrapping flavor by spray drying, coacervation, polymerization and other processes, and can be release as and when required.

Heat Resistant Pepper

These are double encapsulated products in which the capsules are rendered water insoluble coating and the contained flavor will be released only at high temperatures such as in baking process.

Fat Based Pepper

Fat based pepper is a blend of pepper oil and oleoresin in an edible liquid or hydrogenated fat base formulated for use in products such as mayonnaise.

Extruded Spices

Spices can be sterilised, ground and encapsulated in a single step by this technique. The product emerges as a spice 'rope' which is cut in to pellets (Scott 1992).

Chemical Composition

The quality of pepper depends on components piperine (pungency) and volatile oil (aroma and flavor) (Traxler 1971). The most active compounds in pepper are piperine (1-[5-(1, 3-Benzodioxol-5-yl)-1-oxo-2,4-petadienyl] and piperidine $C_{17}H_{19}NO_3$) (Borges and Pino 1993).

Medicinal Values

It is used for loss of appetite, curing intestinal worms, bloating and flatulence, toothache, aches, eczema, tongue injuries, constipation, hemorrhoids, excessive thirst, pimples, reducing arthiritic pain and boosting immunity (Pruthi 1993).

It also stimulates taste buds to increase hydrochloric acid secretion, organs to produce increased flow of saliva and gastric. Piperine acts as a thermogenic compound by enhancing the thermogenesis of lipid and accelerates energy metabolism, also increases the secretion and produces β-endorphin in the brain. It reduces inflammation, has liver protective action, inhibit the enzyme, increase the bioavailability of various compound like curcumin. Piperidine is used in pharmaceutical drugs such as raloxifene and minoxidil.

Calcium and potassium present in pepper are good for health, potassium regulates blood pressure and selenium maintains bones formation, nails hair, follicles and teeth and for proper functioning of brain. Vitamin A and K of black pepper oil (β-carotene) are very good antioxidant and maintains circulatory and metabolic functions, muscles, bones and so on. Essential oil has good rubefacient and analegistic properties for skin care.

Pungency of pepper suppress all kind of infection, strengthen the nervous system, improves gastrointestinal condition, and normalizes the peristaltic system.

Antioxidant Action

Black pepper in tea can relive from arthritis, nausea, fever, migraine headaches, poor digestion, strep throat and even coma. It can reduce high-fat diet-induced oxidative stress and can treat intermittent fever, neuritis, cold, pains, diseases of throat (Dorman *et al.* 2000) and malaria (Tipsrisukond *et al.* 1998).

Antioxidant property is due to tocopherol (Saito and Asari 1976) and polyphenolic content (Revankar and Sen 1974) of pepper. Autoxidation of unsaturated fatty acids and proteins is delayed by pepper (Abdel-Fattah and El-Zeany 1979).

Analgesic and Antipyretic Actions

Pepper is used for the treatment of intermittent fever, neuritis, cold, pains and throat diseases and malaria (Lee *et al.* 1984; Nadkarni 1976).

Central Nervous System Depressant Activity

Pepper is used for the treatment of epileptic fits (Kritikar and Basu 1975).

Mutagenic and Carcinogenic Effects

Pepper prevents chemical carcinogenesis by stimulating the xenobiotic biotransformation enzymes which have chemo protective role in liver (Singh and Rao 1993).

Bio-Enhancing Action

Piperine may induce alterations in membrane dynamics and permeation characteristics along with induction in the synthesis of proteins associated with cytoskeletal function, resulting in an increase in the small intestine absorptive surface, thus assisting efficient permeation through epithelial barrier (Khajuria *et al.* 2002).

CORIANDER (CORIANDRUM SATIVUM L.)

Coriander is an annual herbaceous plant, belonging to the family *Umbelliferae*. The fresh leaves of this plant are called as cilantro. It is a soft, hairless plant grows up to 60 cm tall. The fruit is in globular shape (3 to 5 mm diameter) with two seeds.

Processing

Harvesting

a. *Leaves:* These are harvested at 60 to 75 days after sowing, bunched and stored at 90% RH, and temperature below 5°C for 24 to 36 h. These leaves can be dried or dehydrated to coarse powder.

b. *Seeds:* It is harvested early in the morning or late evening at 66% maturity. The harvesting is done by cutting the whole plant when 60% of seeds in the main umbel attain the desired size and color to minimize the breakage.

Threshing

It is done by beating the plants to remove seeds.

Drying

The plants are withered for 2 days and dried to approximately 18% moisture content. Then threshed and again dried in shade to attain 9% moisture content. Other means of mechanical drying can also be done in the temperature range of 80 to 90°C.

Essential Oil Extraction

The seed is ground immediately prior to distillation to increase oil yield and reduce the distillation time. The essential oil content is in the range of 0.1 to 1.5% and contains a range of different essential oils.

Products

Coriander powder

It is produced by grinding the seeds using hammer mills.

Coriander Oil

Coriander oil is a colourless, pale yellow liquid with a characteristic odor and taste. The seeds also contain oil (19 to 20%) which is mixture of glycerides of palmitic, oleic, linoleic and petroselinic acids. Cordiander oil is used as an ingredient in the preparation of liquors and medicine.

Chemical Composition

Coriander Leaves

It contains water, protein, fat, minerals, fiber and carbohydrates. The mineral and vitamin contents include calcium, phosphorus, iron, carotene, sodium, potassium, oxalic acid, thiamine, riboflavin, niacin and vitamin C. The taste of the fresh herb is due to an essential oil that is made up of aliphatic aldehydes.

Coriander Seed

It is composed of essential oils mainly of linalool (50 to 60%, responsible for aroma) and terpenes (20%) like pinenes, γ-tepinene, myrcene, camphene, phell andrenes, α-ter pinene, limonene, cymene, and monoterpenoid. They contain moisture (6.3%), protein (1.3%), fat (19.6%), minerals (4.4%), fiber (31.5%) and carbohydrates (24%) per 100 grams. The odor and flavor of matured seeds and fresh herbage are completely different. The main chemical components in the oil are borneol, linalool, cineole, cymene, terpineol, dipentene, phell-andrene, pinene and terpinolene. Linalool (monoterpenoid) is the primary constituent of coriander oil (Wichtl 1994; Coleman and Lawrence 1992; Leung and Foster 1996; Tashinen and Nykanen 1975; Pino *et al* 1996).

The leaf oil contain forty four compounds mostly of aromatic acids containing 2-decenoic acid (30.8%), E-11-tetradecenoic acid (13.4%), capric acid (12.7%), undecyl alcohol (6.4%), tridecanoic acid (5.5%) and undecanoic acid (7.1%). Aliphatic aldehydes (mainly C10-C16 aldehydes) are predominant in the fresh herb oil.

The seed oil contains 53 compounds mostly of linalool (37.7%), geranyl acetate (17.6%) and γ-terpinene (14.4%). The essential oil and fatty oil content varies from 0.03 to 2.6% and 9.9 to 27.7%, respectively. Other constituents are crude protein (11.5 to 21.3%), fat (17.8 to19.15%), crude fiber (28.4 to 29.1%) and ash (4.9 to 6.0%).

Medicinal Uses

It can treat fever, cold, flu, stomach disorders, digestive problem, diarrhea, dyspeptic complaints, mild gastrointestinal upsets, flatulence, colic, loss of appetite, anxiety, convulsions, dyspepsia and insomnia (Breevort 1996; De Smet 2002). It has hypolipidemic effect, reduce cholesterol and used in laxative (stimulate to empty bowels) remedies. It settles spasms in the gut and counters the effects of nervous tension.

It is used as analgesic, digestive, natural antibiotic (protection against food-borne illnesses caused by *Salmonella*), anti-rheumatic, anti-cancer (due to phthalides and poly-acetylenes, phytochemcials compounds), antioxidant, antidiabetic (Insulin-releasing and insulin-like activity), anti-inflammatory galactagogue, antispasmodic, diaphoretic, fungicidal, lipolytic. Coriander seed is a rich source of minerals like iron, magnesium and high content of dietary fiber. It is used as a drug for indigestion, against worms, rheumatism and pain in the joints (Wichtl 1994; Wangensteen *et al.* 2004).

It can control blood sugar, cholesterol and free radical production, relieve intestinal gas, refresh and uplift the mind, useful in migraine, tension, relieve wind and cramps, while revitalizing the glandular system.

Coriander oil acts as a general cleanser of the body, to get rid of toxins and fluid wastes. In vapor therapy, coriander oil can stimulate the mind and ease fatigue, while assisting with eating disorders and improving appetite. It is used as massage oil and as part of a cream or lotion, coriander oil can help with tension, mental fatigue, migraine, muscle spasms, arthritis, improves appetite and alleviates gastric insufficiency and distress (stomachic), relieves intestinal cramping and has anti-inflammatory action.

Volatile components of oil have antibacterial and antioxidative property and inhibit micro-organisms growth and lipid peroxidation (Anonymous 1950; Chopra *et al.* 1956; Ghani 2003; Yusuf *et al.* 1994).

CINNAMON (CINNAMOMUM ZEYLANICUM)

Cinnamon is an evergreen shrub or small tree of laurel family *Lauraceae*. The bark portion of this plant is being used in flavoring of foods.

Processing

Harvesting

It is harvested early in the morning during wet season (September to November) since rainfall facilitates the peeling of bark. During harvesting the stem is removed from the plant.

Preparation of Barks

For the preparation of cinnamon bark, tender stems (1.2 to 5 cm) are removed and used for mulching. Removed leaves can be used for oil distillation. Barks are produced by making longitudinal cuts (at 30 cm intervals) on either side of the stem and eased off using pointed knife (stainless steel or brass).

Stripped stem is rubbed with a brass rod to loosen the inner bark. The curled pieces of peeled bark (quills) are placed one inside another to make 1m long 'compound quills'. The best quills are placed on the outside and broken and small pieces are placed at the centre.

Drying

The 'compound quills' are placed on coir rope racks and dried in the shade to prevent warping.

Storage

Cinnamon bark can be stored in a dark, cool, dry place for 2 to 3 years.

Grading

The quality of the cinnamon depends on the bark thickness, appearance and aroma and flavor.

According to Sri Lankan grading system, cinnamon quills are classified into four groups based on diameter: a. Alba (< 6 mm) b. Continental (<16 mm) c. Mexican (<19 mm) and d. Hamburg (<32 mm).

Products

Cinnamon powder

Cinnamon powder is obtained by grinding the barks.

Cinnamon Oil

Cinnamon oil is obtained by steam or water distillation.

Chemical Composition

Cinnamon contains water (5.1%), ash (2.4%), crude protein (3.5%), crude fat (4.0%), crude fiber (33.0%), nitrogen free extract (52%) and energy (258 cal/100 g) and minerals like iron, zinc, calcium, chromium, manganese, magnesium, potassium and phosphorus.

Cinnamon oil contains many components such as cinnamaldehyde (65 to 80%), trans-cinnamic acid (5 to 10%) and eugenol (4 to 10%).

Other constituents include cinnamic alcohol, terpenes such as limonene, tannins, mucilages, oligomerprocyanidin, traces of coumarin, α-Thafone, α-pinene, benzaldehyde, heptanol, sabinene, 1-octen-3-ol, β-pinene, myrcene, p-cymene, limonene, β-phellandrene, 1,8-cineole, γ-terpinene, octanol, terpinen-4, α-terpineol, trans-carveol, nerol, geraniol, geranial, neryl acetate, cinnamyl alcohol,dihyroeugenol, ethylcis-cinnamate, t-Methyl cinnamate, iIsoeugenol, cis-Caryophyllene, t-Cinnamic acid, cinnamyl actate, α-Cary-phyllene, and E-ethyl cinnamate.

Active components in cinnamon are cinnamaldehyde, cinnamyl acetate, cinnamyl alcohol, eugenol (found mostly in the leaves), cinnamic acid, weitherhin, mucilage, diterpenes, proanthocyanidins, β-caryophyllene, linalool and methyl chavicol.

Medicinal Uses

Medicinal parts of cinnamon plant are outer bark, inner bark and leaves. It is used to treat stomach, diarrhea (Skidmore- Roth 2003), bronchitis, coughs, respiratory ailments (Martinez 1989), loss of appetite, dyspepsia, gastritis, blood circulation disturbance and inflammatory (Wang *et al.* 2009).

It is used for invigorating tonic, cold, nausea, vomiting, flatulence, asthma, paralysis, toothache, bad breath, treating diarrhea, indigestion, gas and bloating, stomach upset, gastric ulcers, nervous disorders, fatigue, hearing loss, excessive menstruation, uterus disorders, gonorrhea, uterine hemorrhage, menorrhagia, hypertension (high blood pressure), curing minor bacterial and fungal infections of skin, menopausal symptoms, rheumatic conditions, bacteria, destroying fungi, including the molds that produce carcinogenic aflatoxins, preventing the damage to cell membranes by free radicals.

It has antipyretic, antiallergenic, analgesic, antitussive (Gurdip *et al.* 2008) and chemopreventive activities (Sabulal *et al.* 2007).

Antioxidant Action

It is beneficial against free radicals (Dragland *et al.* 2003; Jayaprakasha *et al.* 2003; Lee and Shibamoto 2002).

Antimicrobial Action

Cinnamon essential oil inhibits the growth of bacteria (Bacillus cereus, *Legionella pneumophila*), fungi, yeast (*Candida)* and food borne pathogens.

Blood Sugar Control

Polyphenol in the cinnamon stimulates the insulin receptors, and also inhibiting an enzyme that inactivates them, thereby increasing cells' ability to use glucose. Cinnamon oil has the significant effect in type II diabetes (Khan *et al.* 2003).

Cinnamon's Scent Boosts Brain Function

Cinnamon odor boosts brain activity, chewing cinnamon flavored gum or cinnamon smell enhances study participants' cognitive processing, virtual recognition memory and working memory.

Calcium and Fiber Improve Colon Health and Protect against Heart Diseases

Cinnamon essential oil is a good source of manganese, iron and calcium. Calcium and fiber remove bile salt from the body thereby reducing the risk of colon cancer, prevents atherosclerosis and heart disease, and provide relief from constipation or diarrhea.

FENUGREEK (TRIGONELLA FOENUM GRAECUM)

Fenugreek is an annual herb (30 to 60 cm tall) belonging to family *Leguminosae*. It is native to southern Europe and Asia. Its flower is axillary white to yellowish, and 3 to 15 cm long thin pointed beaked pods, which contain 10 to 20 oblong greenish-brown seeds with unique hooplike groove. Both leaves and seeds of this plant are used as spices.

Processing

Harvesting

Greens are harvested at 3 or 4 leaved stage which is 120 to 150 days after sowing. The seeds are harvested 30 to 35 days after flowering or 155 to165 days after sowing.

Chemical Composition

Greens and seed contains mainly moisture (6.3%), protein (9.5%), fat (10.0%), crude fiber (18.5%), carbohydrates (42.3%), ash (13.4%), calcium (1.3%), phosphorus (0.48%), iron (0.011%), sodium (0.09%), potassium (1.7%), vitamin A 1 (040 IU/100 g), vitamin B1 (0.41 mg/100 g), vitamin B2 (0.36%), vitamin C (12.0%), niacin (6.0 mg/100 g), calorific value (370 cal/100 g), gums (23.06%), mucilage (28.0%). They are also rich in choline and contains sapogenins, trigonelline, magnesium, copper, sulphur, chlorine, magneese, zink, chromium, β-carotene, thiamine, riboflavin, nicotinic acid and folic acid. The endosperm of seed is rich in polysaccharide galactomannan. The young seeds mainly contain carbohydrates and sugar, and mature seeds contain amino acid, fatty acid, vitamins, and saponins, large quantity of folic acid (84 mg/100 g), yamogenin, disogenin, gitogenin, neogitogenin, homorientin saponaretin, neogigogenin, and trigogenin 4, 5.

Products

Fenugreek Oil

Fenugreek oil (6–8%) has a foetid odour (may resemble that of roasted coffee or maple syrup) and bitter taste.

Fenugreek Fiber

About 50% dry weight of seeds is edible dietary fiber and 30% is gel-forming soluble fiber. The remaining 20% is insoluble fiber that has bulk-forming property like wheat bran. Dietary fiber from fenugreek is very stable, more shelf life and withstands frying, baking, cooking and freezing. Dietary fiber with a high water retention capacity is made into jelly, spreads, and thickener. Flour fortified with 8 to 10% fenugreek dietary fiber has been used to prepare bakery foods like pizza, bread, muffins, and cakes.

Chemical Composition

The active components in fenugreek seed are trigonelline, galactomannan, choline, vitamine C, steroid saponins, flavonoids, trigonelline, 4-hydroxyisoleucine and sotolon.

Medicinal Uses

Immunomodulatory Effect

It has protective effect against CP-induced (Cyclophosphamide drug) urotoxicity. It also has anti-inflammatory, antipyretic, hypoglycemic, immunomodulatory, hypoglycemic and anti-diabetic properties.

Antioxidant Action

Flavonoids of fenugreek extract possess anti-oxidant and free radical scavenging activities (Bajpai *et al.* 2005). Quercetin has protective effect against CP-induced hemorrhagic cystitis, fenugreek extract prevent both lipid peroxidation (LPO) (Kaviarasan *et al.* 2004, Dixit *et al.* 2005) and hemolysis in red blood cells (RBC) protects cellular structures from oxidative damage, aqueous methanolic and polyphenolic components extract has antiradical and antioxidant activity.

Antidiabetic Action

Fenugreek seed powder has effect on enzyme changes, lowering the blood glucose and prevents the lowering of cortical thickness of the thymic lobules level in diabetic (type II diabetics or noninsulin-dependent diabetes mellitus (NIDDM)). Trigonelline, (alkaloid) present in the fenugreek reduce glycosuria in diabetes.

Anticancer Action

Flavonoids (anti-tumorigenic) induce apoptosis (death of cells) in human carcinoma cells, lung tumor cell lines, colon cancer cells, breast cancer cells, prostate cancer cells, stomach cancer cells, brain tumor cells, head and neck squamous carcinoma3 and cervical cancer cells and have stimulatory effects on macrophages. Diosgenin is useful in cancer therapy (Aggarwal and Shishodia 2004).

Chemo Preventive Action

Fenugreek seeds extract inhibit the 7, 12-Dimethylbenzanthracene (DMBA) induced mammary hyperplasia and antibreast cancer protective effect.

Complementary Cancer Therapy

Fenugreek extract prevents the cyclophosphamide-induced apoptosis caused by cyclophosphamide (CP, an anticancer drug) (Bhatia *et al.* 2006).

Gastroprotective Effect

The aqueous extract and a gel fraction isolated from seeds have significant ulcer protective effects, cytoprotective effect, anti-secretory action, effects on mucosal glycol-proteins and prevent lesion formation. Seeds prevent rise in lipid peroxidation induced there by lowering mucosal injury. Some constituent of seed stimulate the pancreas to release digestive enzymes, thereby aiding in digestion.

Hypocholesterolaemic Activity

The ethanol extract from fenugreek seeds contain hypocholesterolaemic components, which can reduce serum cholesterol (Basch *et al.* 2003; Singhal *et al.* 1982; Stark and Madar 1993 and Sharma *et al.* 1996) and other supplements can also lower tri glyceride and low-density lipoprotein (Basch *et al.* 2003).

Other than these effects fenugreek has anti-inflammatory, anti-obesity, antipyretic, antiseptic, hypocholesterolemic, aphrodisiac, astringent, urotoxic, suppurative, aperient, diuretic, emmenagogue,, demulcent emollient, expectorant, and anthelmintic action, digestive stimulant, anthelmentic and hepatoprotective action. Phyto extract of fenugreek is a good source of phytochemicals.

TURMERIC (CURCUMA LONGA)

Turmeric is a medicinal plant of family *Zingiberaceae*. It is a perennial plant having short stem with large oblong leaves and bears ovate, pyriform or oblong rhizomes, which are often branched and brownish-yellow in colour. Rhizomes may be bulbs, fingers and splits. Fingers are the secondary branches of mother rhizome (bulb).

Processing

Harvesting

Turmeric crops are generally harvested during March to July (7 to 9 months) depending upon the time of sowing and the leaves turn to dry and the color is between light brown and yellow. The land is ploughed and rhizomes are carefully lifted. Harvested rhizomes are cleaned to remove mud and other extraneous matter.

Sweating

The leaves are removed from plant and roots are washed to remove soil. Any leaf scales and long roots are trimmed off. The fingers of rhizomes are removed from main central bulb. The bulb and fingers are heaped separately, covered with leaves and left to sweat for one day.

Curing

Green rhizomes are boiled (45 to 60 min) in water to soften the roots and remove the raw colour. Boiling is stopped when froth comes out and white fumes appear, and giving out a typical odor.

Drying

Drying time can be reduced by slicing the rhizomes which improves the final product quality. They are dried to reach final moisture content of 5 to 10%. The rounds and fingers are dried separately. Rhizome pieces are kept away from the sunlight to prevent the colour loss.

Polishing

Dried rhizomes are polished to remove the rough surface by rubbing against the hard surface of drying-floor, trampled under feet or by polishing drums. Sometimes rhizomes are sprinkled with turmeric powder (mixed with little water) during the final polishing to obtain uniform colour.

Grading

Rhizomes are graded into bulb, fingers and splits.

Storage

Rhizomes are stored in cool and dry environment, away from direct sunlight.

Quality Requirement

It depends mainly on curcumin (deep, yellow colour pigment) content, low 'bitter-principle' content appearance, shape and size of rhizome and volatile oil (impart aroma and flavor). The aroma should be musky, pepper-like character and flavor should be aromatic and somewhat bitter.

Products

Dried Rhizome

Dried Turmeric is used to process powder and oleoresin. They come as bulbs fingers and splits.

Turmeric Powder

Turmeric powder is a major ingredient in curry powder and pastes. It is mostly use to colour and flavour mustard.

Oleoresins

It is obtained by solvent extraction of powdered rhizome. This process yields about 12% of an orange to red viscous liquid and contains various proportions of coloring matter (curcuminoids (40 to 55%), volatile oils (15 to 20%)) which impart flavor to the product and non-volatile fatty and resinous materials. The curcuminoids consist of curcumin that can be purified to a crystalline material.

Essential Oil

It is obtained by distillation or by supercritical fluid extraction of powdered rhizome.

Chemical Composition

Turmeric contains mainly essential oils (5%) and curcumin (5%) (polyphenol), protein (6.3%), fat (5.1%), minerals (3.5%), carbohydrates (69.4%) and moisture (13.1%). The essential oil has *a*-phellandrene (1%), sabinene (0.6%), cineol (1%), borneol (0.5%), zingiberene (25%) and sesquiterpines (53%). Curcumin (diferuloylmethane) (3 to 4%) or curcuminoid is responsible for yellow colour, and comprises curcumin I (94%), curcumin II (6%) and curcumin III (0.3%). The ketonic sesquiterpenes (*ar*-turmerone and turmerone) are responsible for aroma of turmeric (Rupe *et al.* 1934).

Medicinal Uses

Turmeric is used to cure, cuts and burns, promote healing, fever, cold, mucus in the throat, watery discharges like leucorrhea, any pus in the eyes, ears, or in wounds, dental problems (due to the presence of fluoride in turmeric), urinary disorders, diarrhea, ulcers, insanity, lactation problems, poisoning, conjunctivitis gastrointestinal upsets, colic, arthritis pain, low energy, menstrual and abdominal problems. It dispel worms, strengthen the body, dissolve gallstones, decreases congestion and inflammation from stagnant mucous membranes, purifies blood, helps in expelling gas from the intestines (as a carminative agent), protects lungs and liver from pollution, toxins (like major hepatoxins, and aflatoxin) and pathogens, rebuild the liver after being attacked by hepatoxins, protects the lungs from pollution and toxins, regulate the female reproductive system and purifies the uterus and breast milk, in men it purifies and builds semen, and helps in preventing the blockage of arteries. It is against biliary disorders, cough, anorexia, diabetic wounds, rheumatism and sinusitis (Araujo and Leon 2002). Turmeric can cure colitis, crohn's disease, post-giardia or post salmonella conditions and good effect on digestive system. Turmeric is used as a stomachic, tonic and blood purifier and in the prevention and treatment of skin diseases (Anon 2001). It has tetrahydrocurcuminoids which may have antioxidant property. Turmeric oleoresin shows inhibitory activity against different fungi. Turmerone (compound present in turmeric) is a mosquito repellent and drug for the treatment of respiratory disease and dermatophytosis. Ether, crude ethanol extract chloroform extracts and oil have a antifungal effects against *Aspergillus flavus*, *A. parasiticus*, *Fusarium moniliforme* and *Penicillium digitatum*. Turmeric oil also have antibacterial, antimutagenic, and anti-inflammatory properties.

Curcumin

It is the active compound in turmeric, and poses many medicinal properties. Anti-inflammatory activity of curcumin reduces inflammation with phenylbutazone to patients who have undergone surgery or suffered from trauma (Satoskar *et al.* 1986), reduce inflammation in arthritis, enhances wound healing. Antioxidant property of curcumin inhibit generation of superoxide, formation and generation of free radicals (Sharma 1976; Reddy and Lokesh 1992, Subramonian *et al.* 1994, Ruby *et al.* 1995), prevention of lipid peroxidation (Sreejayan and Rao 1994) and oxidative damage to the arterial wall and impose protection action against vascular dementia. About 500 mg of curcuminoids intake daily for 7 days reduces lipid peroxides by 33 % and blood cholesterol by 29 %, thereby reducing cardiovascular diseases (Soni and Kuttan 1992).

Chemopreventive and Bioprotectant Action

It has the capacity to intervene the initiation and growth of cancer cells and tumours by preventing the spread throughout the body and increases cancer cells' sensitivity to certain drugs commonly used to combat cancer, rendering chemotherapy (Stoner and Mukhtarn 1995; Khafif *et al.* 1998; Kawamori *et al.* 1999; Bush *et al.* 2001; Jung *et al.* 2005). By inhibiting the UVB radiated damage it reduces the incidence of skin cancer.

Curcumin has anti-HIV effect (Lin *et al.* 1994; Li *et al.* 1993). HIV infection is characterized by a complex command system, the structural part i.e 'long terminal repeat' (LTR), which results in virus activation or inactivation. Drugs that interfere with LTR may be of potential therapeutic value in delaying active HIV infection and the progression of AIDS. Curcumin inhibit activation of the LTR and to decrease HIV replication effectively.

Antimicrobial Action

Curcuminoids, inhibit the growth of numerous gram-positive and gram-negative bacteria, fungi and intestinal parasite, *Entamoeba histolytic* (Dhar *et al.* 1968), *Staphylococcus aureus* (Bhavani Shankar and Srinivasamurthy 1979) and inhibits *in vitro* production of aflatoxins produced by the mould Aspergillus parasiticus. Antibacterial and antiviral activities of curcumin can be enhanced significantly by illumination with visible light (Pervaiz 1990).

REFERENCES

Abdel-Fattah, L.E. and El-Zeany, B.A. 1979. Effect of spices on the autoxidation of fatty foods. *Rev. Ital. Sostanze Grasse.* 56: 441-443.

Aggarwal, B.B. and Shishodia, S. 2004. Suppression of the nuclear factor kappa B activation pathway by spice derived hytochemicals: reasoning for seasoning. *Annals of New York Academy of Sciences.* 1030: 434-441.

Anon. 2001. *Wealth of India – Raw Materials*, Volume II. Council of Scientific and Industrial Research, New Dehli, India.

Anonymous. 1950. The wealth of India: Raw materials. Vol. II, CSIR, New Delhi, India, pp: 347-50.

Araujo, C.A.C. and Leon, L.L. 2002. Biological activities of *Curcuma longa* L. *Memórias do Instituto Oswaldo Cruz.* 96: 723–728.

Bajpai, M., Mishra, A. and Prakash, D. 2005. Antioxidant and free radical scavenging activities of some leafy vegetables. *International Journal of Food Sciences and Nutrition.* 567: 473-481.

Balasubramanyam, N., Mahadevan, B. and Anandaswamy, B. 1978. Packaging and storage studies on ground black pepper (*Piper nigrum* L) in flexible consumer packages. *Indian spices.* 15 (4): 6-11.

Basch, E., Ulbricht, C., Kuo, G., Szapary, P. and Smith, M. 2003. Therapeutic applications of fenugreek. *Alternative Medicine Review.* 8(1): 20-27.

Bhatia, K., Kaur, M., Atif, F., Ali, M., Rehman, H., Rahman, S. and Raisuddin, S. 2006. Aqueous extract of *Trigonella foenum-graecum* L. ameliorates additive urotoxicity of buthionine sulfoximine and cyclophosphamide in mice. *Food and Chemical Toxicology.* 44: 1744–1750.

Bhavani Shankar, T.N. and Srinivasamurthy, V. 1979. Effect of turmeric fractions on the growth of some intestinal and pathogenic bacteria *in vitro. Indian Journal of Experimental Biology.* 17: 1363-1366.

Borges, P. and Pino, J. 1993. Preparation of black pepper oleoresin by alcohol extraction. *Nahrung.* 37(2): 127-130.

Breevort P. 1996. The U.S. botanical market: an overview. *Herbalogramm.* 36: 49- 57.

Bush, J.A., Cheung, K.J.J. and Li, G. 2001 Curcumin induces apoptosis in human melanoma cells through a Fas receptor/caspase-8 pathway independent of p53. *Experimental Cell Research.* 271(2): 305-311.

Chopra, R.N., Nayar, S.L, Chopra, I.C. 1956. Glossary of Indian medicinal plants. New Delhi, India, CSIR, , pp: 77-78.

Coleman WM, Lawrence BM. 1992. Comparative automated static and dynamic quantitative headspace analysis of coriander oil. *J. Chromatogr. Sci.*, 30: 396-98.

De Smet PA. 2002. Herbal remedies. *N. Engl. J. Med.* 19: 25-29.

Dhar, M.L., Dhar, M.M., Dhawan, B.M., Mehrotra, B.N. and Ray, C. 1968. Screening of Indian plants for biological activity. *Indian Journal of Experimental Biology.* 6: 232.

Dixit, P., Ghaskadbi, S., Mohan, H. and Devasagayam, T.P. 2005 Antioxidant properties of germinated fenugreek seeds. *Phytotherapy Research.* 19(11): 977–983.

Dorman, H.J.D., Surai, P. and Deans, S.G. 2000. *In vitro* antioxidant activity of a number of plant essential oils and phytoconstituents. *Journal of Essential Oil Research.* 12(2): 241-248.

Dragland S, Senoo H, and Wake K. 2003. Several culinary and medicinal herbs are important sources of dietary antioxidants. *Nutr.;* 133(5):1286-1290.

Ghani A. 2003. Medicinal plants of Bangladesh: Chemical constituents and uses. 2nd ed. Dhaka, Asiatic Society of Bangladesh, pp: 183.

Gopalam, A., John Zachariah, T., Nirmal Babu, K., Sadanandan, A.K. and Ramadasan, A. 1991. Chemical quality of black and white pepper. Spice India, 4: 8-10.

Goswami, T.K. and Manish, S. 2003. Role of feed rate and temperature in attrition grinding of cumin. *Journal of Food Engineering.* 59: 285-290.

Govindarajan, V.S. 1979. Pepper-chemistry, technology and quality evaluation. *CRC Crit. Rev. Food. Sci Nut.,* 9: 1-115.

Gurdip S, Kapoor IPS, Pratibha S, De Heluani CS, Marina PD, and Cesar ANC 2008. Chemistry, antioxidant and antimicrobial investigations on essential oil and oleoresins of *Zingiber officinale. Food Chemical Toxicol.*, 46(10): 3295-3302.

Jayaprakasha GK, Jagan Mohan Rao L, and Sakariah KK. 2003. Volatile Constituents from Cinnamomum zeylanicum Fruit Stalks and Their Antioxidant Activities. *J. Agric Food Chem.* 51(15): 4344-4348.

Jung, E.M., Lim, J.H., Lee, T.J., Park, J., Choi, K.S. and Kwon, T.K. 2005. Curcumin sensitizes tumor necrosis factor-related apoptosis-inducing ligand (TRAIL)-induced apoptosis through reactive oxygen speciesmediated up-regulation of death receptor 5 (DR5). *Cancer Biology.* 26(11): 1905-1913.

Kaviarasan, S., Vilayalakshmi, K. and Anuradha, C.V. 2004 Polyphenol-rich extract of fenugreek seeds protects erythrocytes from oxidative damage. *Plant Foods for Human Nutrition.* 59(4): 143-147.

Kawamori, T., Lubet, R., Vernon, E.S., Kelloff, G.J., Kaskey, R.B., Rao, V.R. and Reddy, B.S. 1999. Chemopreventive effect of curcumin, a naturally occurring anti-inflammatory agent, during the promotion/ progression stages of colon cancer. *Cancer Research.* 59: 597-601.

Khafif, A., Schantz, S.P., Chou, T.C., Edelstein, D. and Sacks, P.G. 1998. Quantitation of chemopreventive synergism between (–)-epigallocatechin-3-gallate and curcumin in normal, premalignant and malignant human oral epithelial cells. *Carcinogenesis.* 19: 419-424.

Khajuria, A., Thusu, N. and Zutshi, U. 2002. Piperine modulates permeability characteristics of intestine by inducing alterations in membrane dynamics: influence on brush border membrane fluidity, ultrastructure and enzyme kinetics. *Phytomedicine.* 9(3): 224-231.

Khan A, Safdar M, Ali Khan MM, Khattak KN, and Anderson RA. 2003. Cinnamon improves glucose and lipids of people with type 2 diabetes. *Diabetes Care.* 26(12): 3215-3218.

Kirtikar, K.R. and Basu, B.D. 1975. *Indian Medicinal Plants* III, Bishan Singh Mahendrapal Singh, Dehra Dun. India. 2133-2134.

Lai, P.K. and Roy, J. 2004. Antimicrobial and chemopreventive properties of herbs and spices. *Curr. Med. Chem.* 11: 1451-1460.

Lee, E.B., Shin, K.H. and Woo, W.S. 1984. Pharmacological study of piperine. *Archives of Pharmacaln. Research.* 7: 127-132.

Lee, K.G, and Shibamoto T. 2002. Determination of antioxidant potential of volatile extracts isolated from various herbs and spices. *J. Agric. Food Chem.*; 50(17): 4947-4952.

Leung AY, and Foster S. 1996. Encyclopedia of common natural ingredients used in food, drugs and cosmetics. 2nd ed, New York, John Wiley and Sons, pp 193-95.

Li, C.J., Zhang, L.J., Dezube, B.J. and Crumpacker, C.S. 1993. Three inhibitors of human type 1 immunodeficiency virus long terminal repeat directed gene expression and virus replication. *Proceedings of the National Academy of Science* (USA) 90: 1839-1842.

Lin, J.K., Huang, T.S., Shih, C.A. and Lin, J.Y. 1994. Molecular mechanism of action of curcumin, in food phytochemicals II: teas, spices, and herbs. *American Chemical Society.* 20: 196-203.

Manohar, B. and Sridhar, B.S. 2001. Size and shape characterization of ground turmeric (Curcuma domestica) particles. *Powder Technol.* 120, 292-297.

Martínez M. 1989. Las Plantas Medicinales de México. Mexico City: Editorial Botas.

Murthy, C. T., Krishnamurthy, N., Ramesh, T., and SrinivasaRao, P. N. 1996. Effect of grinding methods on the retention of black pepper volatiles. *Journal of Food Science and Technology*, 33(4): 299.

Murthy, C.T. and Bhattacharya, S. 1998. Moisture dependent physical and uniaxial compression properties of black pepper. *Journal of Food Eng.* 37: 193-205.

Murthy, C.T. and Bhattacharya, S. 2008. Cryogenic grinding of black pepper. *Journa of Food Eng.* 85: 18-28.

Nadkarni, K.M. 1976. *Indian Materia Medica.* Popular Prakashan, Bombay, pp: 971-972.

NRCS 1987. Annual report of National Research Centre for spices, 1986–87, NRCS, Calicut

Pervaiz, S. 1990. Antitumor and antiviral activity of curcumin and light. Proceedings of the Annual Meeting of the American Association of Cancer Research Abstract. A2325.

Pesek, C.A. and Wilson, S.A. 1986. Spice quality: Effect of cryogenic and ambient grinding on colour. *J. Food Sci.* 51: 1386.

Pino JA, Rosado A, and Fuentes V. 1996. Chemical composition of the seed oil of *Coriandrum sativum* L. from Cuba. *J Essentl Oil Res.*8: 97-98.

Pruthi, J.S. 1992. Advances in sun/solar drying and dehydratin of pepper (*Piper nigrum* L). *International Pepper News Bulletin,* 16 (2): 6-17.

Pruthi, S. 1993. Major Spices oflndia: Crop Management Post-harvest Technology. New Delhi: lCAR.

Reddy, A.C.P. and Lokesh, B.R. 1992. Studies on spice principles as antioxidants in the inhibition of lipid peroxidation of rat liver microsomes. *Molecular and Cellular Biochemistry.* III, 117-124.

Revankar, G.D. and Sen, D.P. 1974. Antioxidant effects of a spice mixture on sardine oil. *J. Food Sci. Technol.* India, 11: 31-32.

Ruby, A.J., Kuttan, G., Dinesh Babu, K., Rajasekharan, K.N. and Kuttan, R. 1995. Antitumor and antioxidant activity of natural curcuminoids (*Curcuma longa*) on iron-induced lipid peroxidation in the rat liver. *Food and Chemical Toxicology.* 32: 279-283.

Rupe, H., Clar, G., Pfau, A.St and Plattner, P. 1934. Volatile plant constituents. II. Turmerone, the aromatic principle of turmeric oils (in German). *Helvetica Chimica Acta.* 17: 372.

Sabulal B, Dan M, John JA, Kurup R, Purushothaman C, and Varughese G 2007. Phenylbutanoid-rich rhizome oil of *Zingiber neesanum* from Western Ghats, Southern India. *Flavour Fragrance J.*, 22(6): 521-524.

Saito, Y. and Asari, T. 1976. Studies on the antioxidant properties of spices. I. Total tocopherol content in spices. *J. Japanese. Soc. Food. Nutr.,* 29: 289-92.

Satoskar, R.R., Shah, S.J. and Shenoy, S.G. 1986. Evaluation of anti-inflammatory property of curcumin (diferuloyl methane) in patients with postoperative inflammation. *International Journal of ClinicalPharmacology and Theoretical Toxiclogy.* 24: 651-654.

Scott. R. 1992. Master spice: Extruded low count spices. *Proc. New Technologies Symposium,* Bristol, U.K., pp: 6-8.

Sharma, O.P. 1976. Antioxidant activity of curcumin and related compounds. *Biochemical Pharmacologym.* 25: 1811–1812.

Sharma, R.D., Sarkar, A., Hazra, D.K., Misra, B., Singh, J.B., Maheswaryi, B.B. and Sharma, S.B. 1996. Hypolipidaemic effect of fenugreek seeds. A chronic study in non-insulin dependent diabetic patients. *Phytotherapy Research.* 10(4): 332–334.

Singh, A. and Rao, A.R. 1993. Evaluation of the modulatory influence of black pepper on the hepatic detoxification system. *Cancer Lett.* 72: 5-9.

Singh, K.K. and Goswami, T.K. 1999a. Design of a cryogenic grinding system for spices. *J. Food Eng.* 39: 359-368.

Singh, K. and Goswami, T. K. 1999b. Studies on cryogenic grinding of cumin seed. *Journal of food process engineering.* 22: 175-190.

Singh K.K. and Goswami T.K. 2000. Cryogenic grinding of cloves. *Journal of Food Processing Preservation.* 24: 57-71.

Singhal, P.C., Gupta, R.K. and Joshi, L.D. 1982. Hypocholesterolmic effect of seeds. *Current Science.* 51: 136-137.

Skidmore-Roth, L. 2003. *Handbook of Herbs and Natural Supplements* 2nd Ed. St. Louis: Mosby.

Soni, K.B. and Kuttan, R. 1992. Effect of oral curcumin administration on serum peroxides and cholesterol levels in human volunteers. *Indian Journal of Physiology and Pharmacology*. 36(4): 273-275 and 239-243.

Sreejayan, N. and Rao, M.N.A. 1994. Curcuminoids as potent inhibitors of lipid peroxidation. *The Journal of Pharmacy and Pharmacology*. 46: 1013–1016.

Srinivasan, K. 2005. Spices as influencers of body metabolism: an overview of three decades of research. *Food Research International*. 38: 77-86.

Stark, A. and Madar, Z. 1993. The effect of an ethanol extract derived from fenugreeek (*Trigonella foenum graecum*) on bile acid absorption and cholesterol levels in rats. *British Journal of Nutrition*. 69: 277-287.

Stoner, G.D. and Mukhtar, H. 1995. Polyphenols as cancer chemopreventive agents. *Cell Biochemical Supplement*. 22: 169-180.

Subramonian, M., Sreejayan Rao, M.N.A., Devasagayam, T.P.A. and Singh, B.B. 1994. Diminuition of singlet oxygen induced D.N.A. damage by curcumin and related antioxidants. *Mutation Research*. 311: 249-255.

Tashinen J, and Nykanen L. 1975. Volatile constituents obtained by the extraction with alcohol-water mixture and by steam distillation of coriander fruit. *Acta Chem. Scand*. 20: 425-429.

Tipsrisukond, N., Fernando, L.N. and. Clarke, A.D. 1998. Antioxidant effects of essential oil and oleoresin of black pepper from supercritical carbon dioxide extractions in ground pork. *Journal of Agricultural and Food Chemistry*. 46(10): 4329-4333.

Traxler, J.T. 1971. Piperanine, a pungent component of black pepper. *Journal of Agricultural and Food Chemistry*. 19(6): 1135.

Variyar, P.S., Pendharkar, M.B., Banerjee, A. and Bandyopadhyay, C. 1988. Blackening in green pepper berries. *Phytochemistry*. 27: 715-717.

Wang R, Ruijiang W, and Bao Y. 2009. Extraction of essential oils from five cinnamon leaves and identification of their volatile compound compositions. *Innovative Food Sci. Emerging Technol.*, 10: 289-292.

Wangensteen, H., Samuelsen, A. B., and Malterud, K. E. 2004. Antioxidant activity in extracts from coriander. *Food Chemistry*. 88: 293-297.

Wichtl, M. W. 1994. Herbal drugs and phytopharmaceuticals. Stuttgart: Medpharm GmbH Scientific Publishers.

Yusuf M, Chowdhuty JU, Wahab MA, and Begum 1994. *J. Medicinal plants of Bangladesh*. Bangladesh, Bangladesh Council of Scientific and Industrial Research, pp: 66.

ABOUT THE AUTHORS

***Dr. Mohamed Essa* M, PhD, FLS (UK) FICS (Ind)**

Dr. Mohamed Essa obtained PhD in Biochemistry from Annamalai University, India and and did Post doctoral research in Neurosciences in NYSIBR, New York, USA. At present he is working as an Assistant Professor in Nutrition, College of Agricultural and Marine Sciences, Sultan Qaboos University, Oman. He conducts research in pharmacological approaches of natural products against neurological diseases by using various available models. Dr. Mohamed-Essa has published nearly 55 research articles in various peer-reviewed international journals and conferences and acting as an Editor in Chief for International Journal of Nutrition, Pharmacology, Neurological Diseases. He is a fellow/life member of various international scientific bodies including The Linnean Society, UK. Recently he received Best Young Researcher award from College of Agricultural and Marine Sciences, SQU and Professionals from Developing Country award given by International Society for Autism Research (INSAR), USA. His research work related with pharmacological effect of walnuts on Alzheimer's disease was press released in the International conference on Alzheimer's disease (IICAD), 2010 in Hawaii, USA. Along with these, Dr. Mohamed-Essa was appointed as a Visiting Scientist in the department of Pharmacology, University of New South Wales (until 2014), Sydney, Australia and Department of Neurochemistry, NYS Institute for Basic Research, USA. He has a strong academic and research background based on his work experiences in USA, Malaysia, India and Oman.

Dr. Manickavasagan, PhD, PEng

Obtained PhD in Biosystems Engineering from the University of Manitoba, Canada. He is a licensed professional engineer (PEng) in the province of New Brunswick, Canada. After PhD, he worked with McCain Foods Limited (Canada) as Scientist for about 3 years. At present he is working as an Assistant Professor (Postharvest Technology) at Sultan Qaboos University, Oman. He has published more than 50 scientific papers in peer reviewed journals and international conferences. He has diversified research and management experience with academic institutions and Industries.

Ethirajan Sukumar

Oobtained Ph.D. in Phytochemistry and Phytobiology from University of Madras (India) and had a brief Post Doctoral stint at University of Virginia, USA. He is Assistant Professor in the Department of Applied Sciences, Higher College of Technology, Muscat, Sultanate of Oman since October, 2005 and Hony. Visiting Faculty, Post Graduate Department of Biotechnology, The New College (University of Madras). Prior to full time teaching, he was Research Scientist in Central Research Institute of Siddha Medicine, a Government of India institution, in Chennai (India) for more than 26 years. During his research career, he worked on medicinal plants used in the indigenous medical practices and published 50 research papers in peer-reviewed journals besides presenting as many in various international and national fora. He has supervised 25 students for Doctoral, M.Phil. and Masters programs in the multi-disciplinary areas involving chemistry, biochemistry, pharmacognosy, pharma--cology and environmental science. He delivered 55 invited lectures in universities, colleges and conferences and served as Resource Person in many workshops and training programs. He is associated with leading international / national journals as Associate Editor, Editorial Board member and Peer-Reviewer. His name is included in the 'International Directory of Experts in Spices, Herbs and Medicinal Plants Research' published by University of Massachusetts, USA and 'Who is Who in Science and Engineering' brought out by Marquis Inc., USA. He was an active sports person during student days besides a health educator, Yoga teacher and social worker.

LIST OF CONTRIBUTORS

Author Name	Chapter Number
Abdul Hamid A	13
Al-Rawahi A	10,19
Ahmed Yahya masoud Almaskri	26
Mustaque Memon A	10
Anoop Austin	11
Annamalai Prakasam	7
A. R. Mullaicharam	15
B. Soussi	3,16
Cinghu Senthilkumar	21
Guillemin GJ	3, 9, 10, 24
G. Kanimozhi	5
Hemant Poudyal	1
Hiranmoy Das	12
Irfan Rahman	23
Jingwei Lu	12
K. Tamilselvam	2
K. Pushkala	17
K. M. Umar	13
Lindsay Brown	1
Masoud Yahya Al-Maskari	26
Majekodunmi O. Fatope	22
M. Amirthaveni	4
Mohamed.M.Essa	3.8, 9, 10, 14, 24
Mohamed Farhat	14
Mostafa I. Waly	8, 14, 24
M.S. Rahman	8,19
M.S. Subapriya	4
Muhammad Asif Hanif	26
N. Rajendra Prasad	5

Author Name	**Chapter Number**
N. Guizani	3,10, 24
P. Thirugnanasambantham	11
P. Subramanian	9
P. D. Gupta	17
ReevaAggarwal	12
Salim H. Al-Rawahy	18
Saleh N. Al Busafi	22
Samir Al-Adawi	26
Sankar Devarajan	8
Sardar A. Farooq	18
Saravanan Rajendrasozhan	23
S. Balasubramanian	27, 28
Senthil Kumar Pazhanisamy	7
Sergey Dobretsov	16
S. M. Abdulkarim	13
Somasundaram Mathan Kumar	20
Son Radu	13
Subramanian Kaviarasan	6
T. Manivasagam	2,27,28
Talat T. Farook	18
Thiyagarajan Ramesh	21
Vinu Jyothi	7
Vishal Diwan	1
Vandita Singh	3, 10, 24
Yahya Al-Farsi	14
Z. Al-Kharousi	3

INDEX

D

E

F

G

H

I

J

L